MW00838171

Practical
Hydrogeology

About the Author

Willis D. Weight, Ph.D, P.E., is a Professor of Engineering at Carroll College in Helena, Montana, where he leads the Environmental Engineering Track and teaches a variety of hydrology, hydrogeology, and groundwater courses. He has also been President of WDW Writing, Consulting & Planning Inc. since 1989. Dr. Weight is active in the National Ground Water Association and the American Water Resources Association and has presented at conferences around the world. He is widely recognized for his hydrogeology research and related engineering work.

Cover photos by Willis D. Weight.

Practical
Hydrogeology
Principles and Field Applications

Willis D. Weight, Ph.D., P.E.

Carroll College, Helena, Montana

Third Edition

New York Chicago San Francisco
Athens London Madrid
Mexico City Milan New Delhi
Singapore Sydney Toronto

Library of Congress Control Number: 2018964013

McGraw-Hill Education books are available at special quantity discounts to use as premiums and sales promotions or for use in corporate training programs. To contact a representative, please visit the Contact Us page at www.mhprofessional.com.

Practical Hydrogeology: Principles and Field Applications, Third Edition

1 2 3 4 5 6 7 8 9   QFR   23 22 21 20 19 18

ISBN 978-1-260-11689-2
MHID 1-260-11689-1

This book is printed on acid-free paper.

The second edition of this book was published by McGraw-Hill with the title *Hydrogeology Field Manual*. The first edition of this book was published by McGraw-Hill with the title *Manual of Applied Field Hydrogeology*.

Sponsoring Editor
Lauren Poplawski

Editorial Supervisor
Stephen M. Smith

Production Supervisor
Lynn M. Messina

Acquisitions Coordinator
Elizabeth M. Houde

Project Manager
Revathi Viswanathan,
Cenveo® Publisher Services

Copy Editor
Syed Haider,
Cenveo Publisher Services

Proofreader
Bruce Owens

Indexer
Willis D. Weight

Art Director, Cover
Jeff Weeks

Composition
Cenveo Publisher Services

To my wife Stephanie, our seven children, our sweet grandchildren,
and my mentors, clients, teachers, and students,
without whose influence this work could not have been possible.

Contents

Available online at www.mhprofessional.com/weight3e are: "Water-Quality Parameters and Their Significance," "Sources of Cave and Karst Information," and a lengthy glossary of terms.

Preface

The first edition of this book, *Manual of Applied Field Hydrogeology,* resulted from an enquiry by Bob Esposito of the McGraw-Hill Professional Book Group about Montana Tech of the University of Montana's long-running hydrogeology field camp, which was begun in 1985 by Dr. Marek H. Zaluski (Chapter 13). This course was continued under the direction of Dr. John Sonderegger and myself from 1989 to 1996, when John retired. In 1997, Chris Gammons (Chapters 8 and 9) came on board, and he assisted in directing this course until the time of the second edition printing. Bob asked whether we would consider taking the course notes and field tasks and putting them into book form. A detailed outline and proposal was submitted, which was peer-reviewed. A very distinguished reviewer was exceedingly enthusiastic about a book of this scope being produced. During his review he expressed the opinion that "every hydrogeologist with less than 10 years of experience should own this book." It was also his opinion that there was a burning need to have a reference that inexperienced field hydrogeologists could refer to that would explain things in real-world terms. This has been the perspective in putting together the chapters. Ultimately, McGraw-Hill's Senior Editor Larry Hager led the team in completing the first edition.

Feedback from readers and reviewers led to the printing format of the second edition, *Hydrogeology Field Manual.* Larry Hager again spearheaded the effort to improve the first edition. Since John Sonderegger had retired, other experts were solicited as contributors to rewrite and expand the first edition. The original chapters were revised and expanded significantly to appeal to a wider audience and broaden the scope to appeal to an international audience.

The third edition can be attributed to efforts by Wendy Fuller of McGraw-Hill. After the author ignored her enquiries to write a third edition for a few years, she managed to get his attention and encouraged him to write a proposal for changes and updating. Once again this was peer-reviewed and the leadership baton was handed over to Acquisitions Editor Lauren Poplawski. Lauren has been a joy to work with and provided valuable guidance along the way.

The third edition has been extensively revised and reorganized with significant additions and updating of all chapters with new information, a new chapter, and problems for the reader to wrestle with at the end of each chapter. This reorganization allows instructors to use the front-end chapters as an undergraduate introductory course in hydrogeology, while the other chapters are appropriate to be included in a course or two at the lower graduate level, with a strong field emphasis. The book is specifically useful for a hydrogeology field camp, with enough material to be used for other courses.

Chris Gammons, of Montana Tech's Department of Geological Engineering, added a new chapter (Chapter 8, "Water Chemistry: Theory and Application") and rewrote his chapter on water chemistry sampling (Chapter 9). Dan Stephens (Daniel B. Stephens & Associates, Inc.) from New Mexico brought on Todd Umstot, Senior Hydrogeologist at his company, to upgrade and add to the vadose zone chapter (Chapter 11). William (Bill) Woessner agreed to rewrite and refresh the groundwater/surface-water interaction chapter (Chapter 10). His strong background and knowledge brought significant changes and applications, with updated references.

Two chapters devoted to karst hydrogeology were written by David M. Bednar from Arkansas (Chapter 12) and by Thomas Aley of Ozark Underground Laboratory in Missouri (Chapter 14 on dye tracing). Thomas has performed thousands of dye-trace tests on six continents over a very distinguished career. Marek H. Zaluski, also with a very long and distinguished career, has revised his chapter on tracer-test techniques (Chapter 13). Gregory P. Wittman, a colleague with extensive field experience with up-to-date level-measuring field equipment, was brought on board to enhance the chapter on groundwater flow (Chapter 4). The result is a book intended for use by students and professionals alike as a reference that can be pulled off the shelf again and again and brought into the field. It cannot be emphasized enough how much field experience is represented by the author contributors—approximately 300 years.

Readers are encouraged to study the many examples throughout the book. They are jewels of information that provide a field context for the theory and principles presented herein. They also contain little anecdotes and solutions to problems that can help save hours of mistakes or provide an experienced perspective. Calculations in the examples are intended to show proper applications of the principles being discussed. To help illustrate field examples, the author has taken hundreds of photos and created line drawings with the hope of making the reading more understandable and interesting. Examples come from previous field camps, consulting, research, and work experience.

It is helpful for the practicing hydrogeologist to be able to "read up" on a topic in field hydrogeology without having to wade through hundreds of pages. Students and other entry-level professionals can use this reference to help them overcome the "panic" of the first few times performing a new task such as logging a drill hole, supervising the installation of a monitoring well, or analyzing slug-test data. It is felt that if this book helps someone save time in the field or reduces the panic of performing a new field task, then this effort will have been worth it.

It is the consensus of the authors that the only way to understand how to apply hydrogeologic principles correctly is to have a field perspective. Persons who use hydrogeologic data are responsible for their content, including inherent errors and mistakes that may have occurred in the field. Without knowing what difficulties there are in collecting field data and what may go wrong, an "office person" may ignorantly use poor data in a design problem. It is also our experience that there are many people performing "bad science" because the fundamentals are not well understood. When one confuses the basic principles and concepts associated with hydrogeology, all sorts of strange interpretations result. This generally leads to trouble. Like one seasoned hydrogeologist put it, "I have seen hydrogeologists over the years do all sorts of unimaginable things."

When people go fishing, they may try all sorts of bait, tackle, and fishing techniques; however, without some basic instruction on proper methodologies, the lines get tangled, the fish steal the bait or spit out the fly, or the person fishing becomes frustrated.

Even people who are considered to be fishing experts have bad days; however, they seem to catch fish most of the time. *Practical Hydrogeology: Principles and Field Applications* takes a "teaching how to fish" approach.

It is always a dilemma to decide what to include and what to leave out in writing a book like this. It is the authors' hope that the content is useful. The ideas and feedback from you, the readers, have been invaluable and led to the improvements made. A significant effort was made to acknowledge all photographs and figures contributed by others. If any errors are found, they are the responsibility of the authors. We have put our best effort into making this a useful resource.

Acknowledgments

I would like to thank the many individuals who helped review the chapters and provided some of the photos used in this book. Peter Norbeck and John Metesh (State Geologist of Montana) of the Montana Bureau of Mines and Geology in Butte, Montana, have provided several helpful comments on some of the chapters. Bill Uthman of the Montana Department of Natural Resources and Conservation (DNRC) provided some helpful suggestions for the content of Chapter 1. Mark Sholes (deceased), Diane Wolfgram, and Hugh Dresser (emeritus faculty) from the Department of Geological Engineering at Montana Tech of the University of Montana have provided helpful suggestions on Chapter 2. Patricia Heiser and Ray Breuninger provided valuable suggestions on the reorganization of Chapter 2 for the third edition. Hugh Dresser provided all of the stereo-pair photos found in the book. Peter Huntoon provided some of the photos for Chapter 2 and helpful encouragement and examples. He was also a mentor to me during my graduate-school years. John Sonderegger was gracious enough to "have others that are currently involved" replace his chapters from the first edition (first-edition Chapters 7 and 12). John was an invaluable mentor during my early years at Montana Tech. He was always good at asking the right questions.

Stephen Custer of Montana State University provided helpful comments and photos for some of the material in Chapter 9. David Nimick of the USGS in Helena, Montana, provided excellent feedback about water sampling and the geochemistry discussed in Chapter 9. Dan O'Keefe of O'Keefe Drilling in Butte, Montana, and Herb and Dave Potts (Potts Drilling in Bozeman, Montana) were very helpful in reviewing the drilling techniques section and in providing photos for Chapter 15. Other cooperating specialists, Fred Schmidt, Tom Patten, and John LaFave of the Montana Bureau of Mines and Geology in Butte, Montana, have given helpful input. Kevin Mellot was very helpful in providing information and scanning many of the slides used in several of the figures. Katie Luther from the Montana DEQ also provided some valuable editorial comments. Eric Sullivan kindly reviewed the radial flow equation in Appendix E. I would like to thank Revathi Viswanathan and her team for the excellent editing effort and book construction. Hesham Gneady was instrumental in helping with the index. Readers have provided other valuable comments and suggestions.

I would especially like to thank the chapter contributors, who have enriched this book and allowed its depth and breadth to become more useful. Curtis Link contributed all of the content for Chapter 4 in the second edition, including all of the line drawings. Chris Gammons has been a colleague and friend now for over 20 years—he contributed all of Chapters 8 and 9 and also made valuable suggestions on the introductory chapters. Dan Stephens and Todd Umstot have contributed their expertise in rewriting the vadose zone chapter (Chapter 11). Marek H. Zaluski (long-time friend) contributed Chapter 13 and some of the photos. David M. Bednar and Thomas Aley made the karst hydrogeology chapters (Chapters 12 and 14) possible. Bill Woessner, Greg Wittman, and Todd Umstot are all new contributors for the third edition. The third edition includes an additional 10-plus years' experience from the contributors over the second edition, with the added valuable experience of three new contributors. The total combined efforts of the authors represent nearly 300 years of field experience.

I wish to thank the many students over the years for being the subjects of photos and providing good questions and discussion. Kay Eccleston was instrumental in transforming all of the chapters into desktop publishing form, providing the format for all the text, and constructing the whole first edition of the book.

Finally, I wish to acknowledge the support of my wife Stephanie, our seven children (some of whom were used in the photos), and our growing number of sweet grandchildren, who make life so grand.

Willis D. Weight
Helena, Montana

CHAPTER 1

Hydrogeology and Field Work

Water is one of the natural resources unique among the substances on Earth. Water is life to us and all living things. After discounting the volumes represented by oceans and polar ice, groundwater is the next most significant source. It is approximately 60 to 70 times more plentiful than surface water (www.ngwa.org). Understanding the character, occurrence, and movement of groundwater in the subsurface and its interaction with surface water is the study of hydrogeology. Practical hydrogeology is studied with field applications; it encompasses most of the methods performed in the field to understand groundwater systems, and their connection to surface-water sources and sinks (Section 1.2; Chapter 10) and to better understand the context of the content.

A hydrogeologist must have a background in all aspects of the hydrologic cycle. They are concerned with precipitation, runoff, evaporation, water quality, surface water, and groundwater. Those who call themselves hydrogeologists may also have some area of specialization, such as the vadose zone, aquifer characterization, numerical modeling, well hydraulics, karst hydrogeology, aquifer storage and recovery (ASR), public water supply, low-temperature geochemistry, underground storage tanks, source-water protection (SWP) areas, and surface-water groundwater interaction, actually each of the chapters named in this book and beyond.

The fun and challenge of hydrogeology is that each geologic setting, each spring or hole drilled into the ground, each project, is different. Hydrogeologic principles are applied to solve problems that always have a degree of uncertainty. The reason is that no one can know exactly what is occurring in the subsurface. Hence, the challenge and fun of it. Those who are fainthearted, do not want to get their hands dirty, or cannot live with some amount of uncertainty are not cut out to be field hydrogeologists. The "buck" stops with the hydrogeologist or geologist. It always seems to be their fault if the design does not go right. If you are reading or studying about hydrogeology, you will also be heading into the field. Properly designed field work using correct principles is one key to being a successful field hydrogeologist. Another important aspect is being able to make simple commonsense adjustments in the field to allow the collection of usable data. The more levelheaded and adaptable one is, the more smoothly and cost-effectively field operations can be run.

Hydrogeology is a fairly broad topic. Entry-level professionals and even well-seasoned, practicing hydrogeologists may not have attempted one or more of the topics described in this book. No one knows it all, including the author, and that's why he solicited the help of other seasoned experts to provide a fairly comprehensive single reference. New field tasks can be stressful and having to read a large reference book on one subject can be cumbersome. The objective of this book is to provide a brief presentation on the general theory and principles associated with hydrogeology, but with field applications. Field methods, tasks, pitfalls, and examples with ideal and nonideal behavior are also presented, in the hopes of reducing stress and panic the first few times performing a new task. If this book saves you a couple of hours' time or provides you with more confidence in facing a new field task, it will have been worth the purchase.

1.1 Hydrologic Cycle

The hydrologic cycle is an open system powered by solar radiation. Water from the oceans evaporates into the atmosphere, is carried to land as precipitation, and eventually returns to the oceans. Solar radiation is more intense nearer the equator, where rising air condenses and falls back onto the world's rain forests. The movement of moisture into the atmosphere and back onto the land surface is an endless cycle. Approximately five-sixths of all water that evaporates is sourced from our oceans, while only about three-fourths of all precipitation falls directly back onto our oceans (Tarbuck and Lutgens 1993). This means approximately one-fourth of all water that falls to Earth falls on our continents. Some of this water is stored in ice caps and glaciers, some runs off from the Earth's surface in watersheds and collects in lakes and various drainage networks, some seeps into the ground, some replenishes the soil moisture, and some occurs as atmospheric moisture (USGS 2012). This is important in supplying the landmasses with freshwater. Once water reaches the land, it takes a variety of pathways back to the oceans, thus completing the hydrologic cycle (Figure 1.1). Other than ocean and saline water (97.44%) and frozen water (1.74%), groundwater (0.76%) accounts for a substantial volume of the earth's freshwater (USGS 2012).

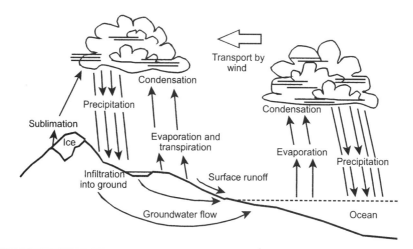

FIGURE 1.1 The hydrologic cycle. Of the five-sixths of water that evaporates from oceans, only three-fourths falls back to the oceans; therefore, one-fourth of all precipitation falls onto our landmasses. (*Tarbuck and Lutgens 1993.*)

Having a general understanding of the hydrologic cycle is important for various scale perspectives (Chapter 10), thus keeping the big picture in mind. The occurrence and behavior of groundwater in the field can be tied back to this bigger picture. For example, global climatic changes contribute to why there may be more or fewer wet years or help explain why dry years occur, which later affects water availability in storage. It is important to keep in mind that weather is on the scale of minutes to days and weeks to months, but climate is on the scale of years to decades (Whitlock et al. 2017). This difference can be visualized as a person walking their dog, where the trend of the variability in weather represents the dog tracks and the overall path of the person walking the dog represents trends in climate (Figure 1.2; skepticalscience.com).

Example 1.1   The surface sea temperature (SST) in the equatorial region is being measured in real time on a continual basis (Hayes et al. 1991; NOAA 2018). There is a pointed relationship between the atmosphere and the ocean temperatures that affect the weather around the globe. In normal years, the SST is about 8°C warmer in the west, with cooler temperatures off the coast of South America from cold-water upwelling from the deep (NOAA 2018). Cold-water upwelling brings nutrient-rich waters, important for fisheries and other marine ecosystems. These cooler waters are normally within 50 m of the surface. What generally results, then, is that the trade winds blow toward the west across the tropical Pacific, resulting in the surface-sea elevation being 0.5 m higher in Indonesia than in Ecuador (Philander 1990). During an El Niño year, the cooler waters off the coast of South America deepen to approximately 150 m, effectively cutting off the flow of nutrients to near-surface fisheries. The trade winds relax and the rainfall follows the warmer waters eastward, the jet stream shifts, resulting in changes in the distribution of higher- and lower-pressure regions in the United States (affects where wetter and drier areas occur), flooding in Peru, and drought conditions in Indonesia and Australia (NOAA 2018). An El Niño year is one where the SST from five consecutive 3-month running means are anomalously high (above +0.5°C threshold) in the El Niño 3.4 region (Figure 1.3a). A La Niña is a similar scenario except the SST from five consecutive 3-month running means is anomalously low (below a −0.5°C threshold; Figure 1.3b) (www.climate .gov/enso). How can the walking-the-dog analogy be applied to El Niño region 4 in Figure 1.3b in terms of overall changes in seawater temperatures?

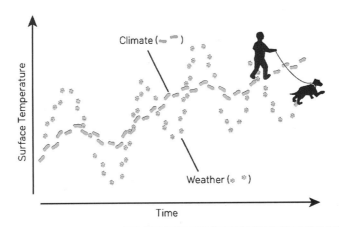

FIGURE 1.2   Conceptual model of weather versus climate, where the paw prints from the dog represent the weather being variable over the short term. The footprints from the human represents the climate or trend in the weather patterns over the longer term.

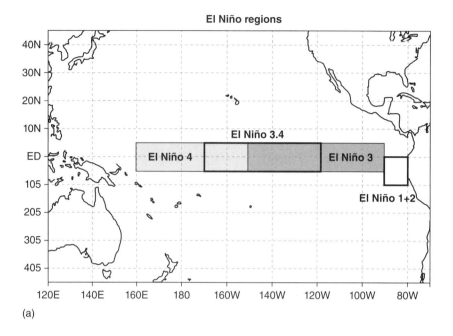

(a)

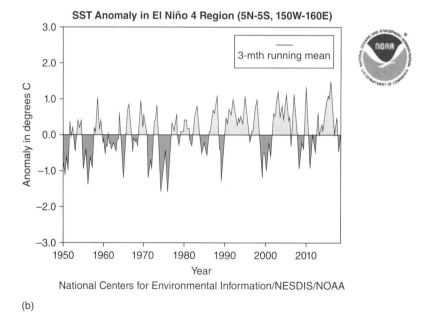

(b)

FIGURE 1.3 (a) Regions 1, 2, 3, and 4 where surface sea temperature (SST) data are being collected. Regions 3 and 4 are the prominent zones. (b) SST data since 1950 in El Niño region 4. (*From NOAA [2018].*)

Hydrogeologists may be aware that global climatic conditions are changing but fail to consider how or if these conditions may be scaled down to a local watershed. More typically, it is easy to become focused only on local phenomenon, such as water-level changes within a given valley or fluvial plane (Chapter 10). However, sometimes the locally observed water-level changes *are* locally derived. It is therefore best to keep an open mind until more data are collected to justify any interpretations. For example, are decreasing trends in hydrographs in wells that may be depleting local aquifer storage tied to water use in a particular valley, climatic drought conditions, or something else (Example 1.2)?

Example 1.2   Consider the hydrograph (water-level changes over time) shown in Figure 1.4. This is from a well at the north end of the Helena Valley (west-central Montana). Is the overall declining trend a result of increased drought conditions from climate change, changes in water use in the valley, or some other explanation? What do the long-term weather data indicate (climate change), what are the changes in water use (demands), or what other information might help explain what is being observed?

The geologic setting of this part of the Helena Valley is that the well from Figure 1.4 is in an area bounded by faulting (Chapter 2). This effectively reduces the extent of the aquifer from which pumping wells can extract water compared with other areas in the Helena Valley. Faulting results in localized boundary conditions (Chapter 3). Another factor appears to be from changes in well density (number of wells per parcel size) over time. The changes in well density can be attributed from actions taken by land-development teams. The original land development was at the 20-acre (8.1-ha) scale. Later, between the years 2000 and 2010, developer teams (attorney, engineer, realtor) have purchased and redeveloped (subdivided) these parcels into 5-acre (2-ha) plots, resulting in increasing the well density. One explanation, therefore, to the trend in Figure 1.4, is that the increase in well density has put a strain on the sustainability of the local aquifer. Knowing whether the local aquifer could sustain the water use at the original well density [20-acre (8.1 ha) scale] is unknown, but certainly the addition of a well for every 5 acres appears to be problematic.

In another, less fault-constrained geologic setting such a hydrograph response might be from climate change, where increasing drought conditions do not provide enough recharge to keep up with the current local demands.

Sometimes one can get too close to the subject being studied to be able to have the proper perspective to understand it. There is the story of the six blind men who came in contact with an elephant (Quigley 1959; Koukl 2013). Each described what they thought an elephant looks like. One felt the side of the elephant and exclaimed, "How smooth, an elephant is like a wall." One felt the trunk and exclaimed that the elephant "must be round like a snake." A third blind man felt the tusk and exclaimed, "How sharp! An

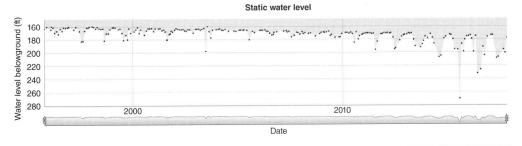

Figure 1.4   Static-water-level data from October 1995 to December 2017 from well M# 148259 from the northern part of the Helena Valley, Montana. See Example 1.2. (*http://mbmggwic.mtech.edu/*.)

elephant must be like a spear!" Yet the fourth blind man felt the leg and said an elephant "is tall and must be like a tree." Another felt the tail and said that "an elephant was like a rope." Finally, one felt the ears, so big and broad, and thought the elephant "must be like a fan." In their own way, each was right but presented only a part of the picture. Understanding the big picture can be helpful in explaining local phenomena.

1.2 Water-Budget Analysis

Most groundwater studies that a typical consulting firm may be involved with take place within a given watershed area (Figure 1.5). The hydrologic cycle is conceptually helpful, but a more quantitative approach is to perform a water-budget analysis, which accounts for all of the inputs and outputs to the system. It is based on a conservation of mass approach and can be expressed simply as Equation 1.1:

$$\text{INPUT} - \text{OUTPUT} = \pm\Delta\text{STORAGE} \tag{1.1}$$

The term ΔSTORAGE (change in storage) refers to any difference between inflow and outflow, resulting in a net increase or decrease in the system. An analogy with financial accounting may be useful to illustrate the concept. In a bank checking account, there are inputs (deposits) and outputs (writing checks or debits). Each month, if the sum of inputs placed into the account is greater than the sum of the outputs, there is a net increase in savings. If more checks are written than there is money in the checking account, one runs the risk of getting arrested. In a water-budget scenario, if more water is leaving the system than is entering, mining or dewatering of groundwater will take place (Figure 1.4). Dewatering may possibly cause permanent changes to an aquifer, such as a decrease in porosity, or compaction resulting in surface subsidence (USGS 1999). Areas where this has been significant include the San Fernando Valley, California;

FIGURE 1.5 Aerial view of a dendritic patterned watershed. This signifies that the underlying geology or soils are somewhat uniform.

Inflow	Outflow
Precipitation	Evapotranspiration
Surface water	Surface water
Groundwater flux	Groundwater flux
Imported water	Exported water
Injection wells*	Consumptive use
Infiltration from irrigation	Extraction wells

*Important for imported water.

TABLE 1.1 Major Inflow and Outflow Components

Phoenix, Arizona; Houston, Texas; and Mexico City. Several of the major components of inflow (sources) and outflow (sinks) are listed in Table 1.1.

It is often difficult to separate the phenomenon of transpiration from plants and evaporation from a water-surface body; therefore, they are combined together into a term called **evapotranspiration,** or **ET** (Chapter 11). In any area with a significant amount of vegetation close to the water table, there may be diurnal effects in water levels (Weight and Chandler 2010). Plants and trees act like little pumps that are active during daylight hours. During the day, ET is intense, and nearby water levels drop and then later recover during the night. Diurnal changes from plants and diel cycling can also cause changes in water quality in streams (Chapters 9 and 10). Accounting for all the components within water-budget analyses is difficult to put closure on, although it should be attempted. Simplifying assumptions can sometimes be helpful in getting a general idea of water storage and availability. For example, it can be assumed that over a longer period of time (e.g., >1 year). changes in storage are negligible. This approach was taken by Toth (1962) to form a conceptual model for groundwater flow by assuming that the gradient of the water table was uniform over a 1-year period, although the surface may fluctuate up and down. This type of model is also useful when performing back-of-the-envelope calculations for water availability.

Example 1.3 The Sand Creek drainage basin is located 7 mi (11.4 km) west of Butte, Montana (Figure 1.6). The basin covers approximately 30 mi^2 (7,770 ha). In 1992, the land was zoned as heavy industrial. In 1995, there were two existing factories with significant consumptive use. The author's phone rang one afternoon, and the local city manager calling from a meeting on a speaker phone wanted to know how much additional water was available for development. The question was posed by the author as to whether anyone was willing to pay for drilling a test hole so that a pumping test could be performed. After the laughter from the group subsided, they were informed that information to provide a quantitative answer was limited but that a number would be provided as a rough conjecture until better information could be obtained and that an answer would be forthcoming in a few minutes. Fortunately, there were some water-level data from which a potentiometric surface could be constructed (Chapters 3 and 4). From the potentiometric surface and a topographic map, a hydraulic gradient and an aquifer width were estimated (Chapter 4). A probable range of values was estimated for the hydraulic conductivity (Chapter 3), and a guess was given for aquifer depth. Darcy's law was used to estimate the volume of water moving through a cross-sectional area within the watershed per unit of time (Chapter 4). This quantity was compared with the water already being used by the existing industrial sources. It was reasoned that if the existing consumptive use was a significant portion of the Darcian flow volume (>20%), it wouldn't look like much additional development could be tolerated, particularly if the estimated

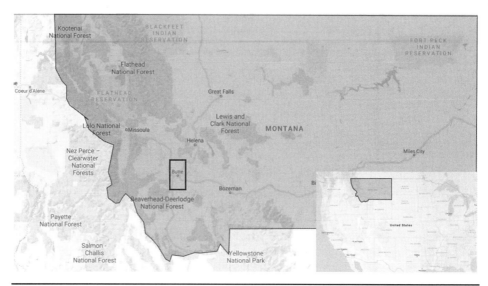

FIGURE 1.6 Silver Bow Creek drainage near Butte, Montana. See Example 1.3. [*Adapted from Burgher (1992).*]

contribution from precipitation did not look all that great. The local city manager was called back and provided with a preliminary rough guess of volume ranges. The caveat was that the answer provided was an extremely rough estimate; however, it did have some scientific basis. It was also mentioned that the estimate could be greatly strengthened by drilling test wells and performing additional studies.

Performing water-budget analyses is more difficult if there is significant consumptive use or if water is being exported. Sometimes there is a change in storage from groundwater occupying saturated media that end up in a surface-water body (Chapter 10). For example, in the Butte, Montana, area, short-term changes in storage can generally be attributed to groundwater flowing into a large mined-out open pit known as the Berkeley Pit (Burgher 1992). Water that was occupying a porosity less than 1% in granitic materials and greater than 25% in alluvial materials was being converted into 100% porosity in a pit lake.

Example 1.4 A variety of investigations have been conducted in the Butte, Montana, area as a result of mining, smelting, and associated cleanup activities (EPA 1995; Metesh and Madison 2004). It was desirable that a water-budget analysis be conducted in the upper Silver Bow Creek drainage to better manage the water resources available in the area. Field stations were established at two elevations, 5,410 ft and 6,760 ft (1,650 m and 2,060 m), to evaluate whether precipitation and evaporation rates varied according to elevation. Within the 123-mi^2 (31,857-ha) area, there were no historical pan evaporation data (Burgher 1992) (Figure 1.7). The period of study was from August 1990 to August 1991.

Part of the water balance required accounting for two sources of imported water from outside the area. One source was water from the Big Hole River, imported over the Continental Divide to the Butte public water supply system. Another source was water from Silver Lake, a mountain lake west of Anaconda, Montana, connected via a 30-mi (49-km) pipeline to mining operations northeast of Butte.

The water-budget equation (Equation 1.1) used was with sources and sinks (Table 1.1):

$$P + Q_{imp} - E_T - Q_{so} - Q_{uo} - Q_{exp} - \Delta S \pm n = 0 \qquad (1.2)$$

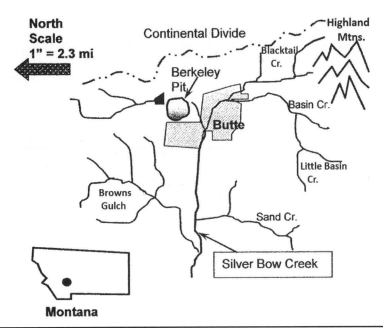

FIGURE 1.7 Sand Creek drainage basin, southwestern Montana, and field area detail. See Example 1.4. [*Adapted from Borduin (1999).*]

where P = precipitation from rain or snow

Q_{imp} = imported water from the Big Hole River and Fish Creek (in parentheses below)

E_T = evapotranspiration using the Penman method

Q_{so} = surface outflow at the western edge of the valley

Q_{uo} = estimated groundwater outflow

Q_{exp} = exported water through mining activities

S = change in storage in the system groundwater to surface water (Berkeley Pit)

n = error term, net loss or gain

Precipitation was higher at the upper site (13.35 in, 339 mm) compared to the lower site (10.5 in, 267 mm), while evaporation values were similar (23.79 in, 604 mm, and 23.46 in, 596 mm). All values in Equation 1.2 were calculated in units of millions of gallons per day (gpd), where the error term is used to balance the equation (Burgher 1992). The results are shown here:

$$112.73 + (9.18 + 5.20) - 113.34 - 12.25 - 0.15 - 1.40 - 5.32 + 5.35 = 0$$

Given the trend in population growth and increased pressure for water, one might ask, "Is more water within the region being used than is coming in?" Some areas have such an abundance of water that much development can still take place with little impact, while other areas are already consuming more water than their system can stand (Examples 1.2 and 1.4). An inventory of water use and demand needs to be taken into account if proper groundwater management is to take place.

Example 1.5 The Edwards Aquifer of central Texas is an extensive karstified system (Chapters 2 and 12) in Cretaceous carbonate rocks (Sharp and Banner 1997). Historical water-balance analysis shows that this aquifer receives approximately 80% of its recharge via losing streams (Chapter 10) that flow over the unconfined portions of the aquifer (Sharp and Banner 1997) (Chapters 3 and 12).

The amount of recharge has varied significantly over time and seems to be connected to the amount of stream flow (Figure 1.8; www.edwardsaquifer.org). The average recharge between 1938 and 1992 was 682,800 acre-ft/year (26.6 m³/s), reaching a maximum of 2,486,000 acre-ft/year (97 m³/s) in 1992 and a minimum of 43,700 acre-ft/year (1.7 m³/s) during 1956 (Sharp and Banner 1997). Other sources of recharge include leakage from water mains and sewage lines in urban areas and cross-formational flow where the aquifer thins, especially to the north (Figure 1.8).

Figure 1.8 indicates three different shades or patterns. The uppermost shaded area represents the drainage areas, where drainage basins collect surface runoff and direct it into stream channels that cross the recharge zone (middle shaded area) to the south (www.edwardsaquifer.org). The lowermost shaded area is known as the artesian zone, from which a number of springs discharge at the surface (www.edwardsaquifer.org). Spring discharge follows a subdued pattern of recharge, while pumping discharge indicates an increasing trend over time. Peaks in the trend in pumping rate are inversely proportional to the minima in recharge (Sharp and Banner 1997). Individual well yields are incredible. A single well drilled in San Antonio is reported to have a natural flow rate of 16,800 gallons per minute (gpm) (1.06 m³/s) (Livingston 1942), and another well drilled in 1991 is likely one of the highest-yielding flowing wells in the world at 25,000 gpm (1.58 m³/s) (Swanson 1991).

In the article by Sharp and Banner (1997), during the peak rate of population growth and development of the 1990s (https://www.census.gov/quickfacts/fact/chart/sanantoniocitytexas/PST120217#viewtop), the concern for management of the Edwards Aquifer was at an all-time high. For example, in 1996 the underdeveloped land north of Austin was being subdivided at a rate of 1 acre (0.40 ha) every 3 hours (Sharp and Banner 1997).

Complicating matters are the additional roles the Edwards Aquifer plays in supplying water for recreation areas; freshwater critical for nurseries in estuaries for shrimp, redfish, and other

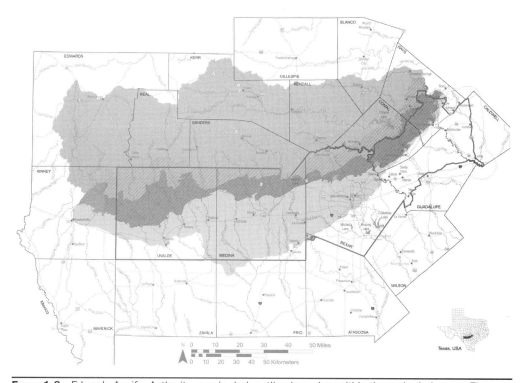

FIGURE 1.8 Edwards Aquifer Authority area in dark outline boundary within three shaded areas. The upper area is the drainage zone, the middle is the recharge zone, and the lower is the artesian zone. See Example 1.5. (*From www.edwardsaquifer.org.*)

marine animals; and spring water for threatened and endangered species that dwell in them (Sharp and Banner 1997). Fortunately, the Edwards Aquifer Authority (EAA), which now manages and monitors the Edwards Aquifer, became fully operational in 1996 (www.edwardsaquifer.org). The Edwards Aquifer and management by the EAA is a good example of how water-balance studies can assist in addressing the significant decisions that are continually badgered by special and political interests.

The previous discussion and examples point out that water-balance studies are complicated and difficult to understand (see next section). For this reason, many water-balance studies are being evaluated numerically (Anderson and Woessner 1992). A numerical groundwater-flow model organizes all available field information into a single system. If the model is calibrated and matched with historical field data (Bredehoeft and Konikow 1993), one can evaluate a variety of "what if" scenarios for management purposes (see also www.edwardsaquifer.org). For example, what would happen if recharge rates decreased to a particular level when stream flow also decreased? The areal effects of different management scenarios can be observed in the output (Erickson 1995).

Example 1.6   The author was involved in creating a three-dimensional (3D) numerical groundwater-visualization model for the North Plains Groundwater Conservation District (NPGCD) in the Texas Panhandle encompassing eight counties (Wood and Weight 2015). The conceptual model was developed from 16,000 well logs and from several related projects and geologic reports. The author guided a team of folks from Aquaveo Water Modeling Solutions (www.Aquaveo.com) as to how to select contacts between formations in the well logs. The hydrostratigraphy (Chapter 3) of the water-bearing and confining units of this visualization and characterization study consists of sedimentary rocks ranging from the distinctly bright red beds of Permian age at depth up through the coarser-grained units of the Tertiary Ogallala formation of Miocene age (Reeves and Reeves 1996).

To make the visualization model more universally available, 3D PDF versions of the model were created (Figure 1.9). Three-dimensional PDFs provide an application the district can deliver to staff, their board, and even the public to help educate and show the characteristics of the subsurface, aquifer units, water levels, cross sections, etc., included in the numerical model. Estimates of the available groundwater in the district were also made showing that groundwater volumes decreased steadily over time with a 28.6% decline in the Ogallala aquifer from pre-1973 levels to the 2014 water year (Wood and Weight 2015).

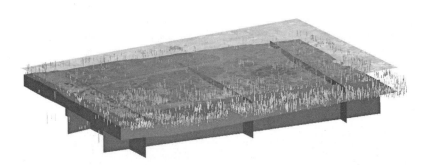

FIGURE 1.9   Three-dimensional PDF image from the visualization model for the North Plains Groundwater Conservation District (NPGCD) in the Texas Panhandle encompassing eight counties. See Example 1.6. (*Wood and Weight 2015.*)

1.3   Water-Budget Myth

In articles by John Bredehoeft (1982, 2002), a presentation is made about *the water-budget myth*. You might be wondering what this means and what is going on? The water-budget myth holds that natural aquifer recharge is the measure of the potential yield of groundwater in the aquifer to wells. It is a myth because it disregards the fact that, over time, the natural recharge to the aquifer determines the natural discharge *from* the aquifer. The natural recharge rate, however, is a fixed function of environmental and meteorological conditions and is independent of pumping. (Natural recharge is not affected by pumping.) Pumping therefore can reduce the rate of natural discharge from an aquifer only by an amount equal to the groundwater development. The idea of recharge being affected by pumping was argued to be a false idea many years ago by Theis (1940), who stated that groundwater development is based upon the principle that pumping causes a loss of water somewhere in the hydrologic system. Bredehoeft (2002) and Devlin and Sophocleous (2005) point out its persistence in the hydrogeology community.

Groundwater development requires a disruption of the balance or equilibrium between natural recharge and discharge. Groundwater development becomes sustainable only when hydrologic equilibrium is reestablished as groundwater withdrawal is eventually balanced by the capture of an equal amount of discharge from the aquifer. This may require long periods of time (tens to hundreds or even thousands of years) before a new equilibrium can be reached, depending on the distance of pumping from the discharge area and the aquifer properties (Kendy and Bredehoeft 2006). Many streams are only partially penetrating (they are in contact with only the upper portion of an unconfined aquifer; Chapters 3, 6, and 10), and therefore the stresses of pumping may be capturing groundwater discharge from the other side of the stream, a phenomenon known as **cross-river flow** (Gamache et al. 2003).

Hydrologic principles state that if wells *completely* intercept the natural recharge to an aquifer, the natural discharge from an aquifer to surface water will eventually end. If groundwater pumping exceeds natural recharge, both surface water and aquifer storage will be depleted over time. The degree of acceptable groundwater development depends on the amount of stream depletion that can be tolerated (Sophocleous 1997). What can be tolerated becomes a source of debate that eventually becomes policy that is managed by state agencies in the United States. The capture of natural discharge raises the controversy of stream depletion where pumping impacts stream flow that is already appropriated by downstream users and infringes on prior water rights. When the quantity of pumping is balanced by a reduction of stream flow (even though this may take a long time), a water right to withdraw groundwater from the well essentially becomes a water right to divert surface water from the stream at that same rate. In groundwater basins where surface-water allocations equal or exceed all of the groundwater discharge, groundwater development ultimately becomes a problem of stream overappropriation and must be managed by retiring existing water rights to augment or mitigate the effects of the requirements of new water rights. (A holistic approach examining the entire watershed is recommended for groundwater management.)

In groundwater basins where surface-water right allocations are less than the requirements for minimum stream flows or other defined purposes (defined by policies), there may be an opportunity for groundwater development. The quantity of how much additional groundwater discharge can be captured by a pumping well is known as **sustainable yield** (Bredehoeft 2002; Sophocleous 2005).

Sustainability

Sustainable development, as defined by Bredehoeft (2006), is the quantity of water available to meet present-day needs without compromising the needs of future generations. Here, we can learn from Native Americans because they evaluate most problems looking at least seven generations into the future (Medicine Head 2004; Brandon 2005). Current scientific thought is that sustainable yield should be evaluated using the tool of numerical groundwater modeling on a total basin scale over hundreds of years (Bredehoeft 2002, 2006; Hiscock et al. 2002; Alley and Leake 2004). If recharge rates are sufficiently high, some degree of stream depletion can be justified (Devlin and Sophocleous 2005). Watersheds where the concepts of recharge, discharge, and water consumption are poorly understood may result in disastrous outcomes.

Kendy (2003) describes an excellent example from the Hebei Province in the North China Plain of how management practices affect changes in the actual water being consumed in a watershed. Historical precipitation data showed a decrease over time from 54 cm/year pre-1980 to 46 cm/year from 1980 to 2000, while ET showed an increase over time from 46 cm/year to 66 cm/year (Kendy 2003). How could this be? There was a change in farming practices from dryland farming of a single "rain-fed" crop to an irrigated two-crop system, resulting in a disparity of 20 cm/year net loss. Irrigation waters, originating from surface-water diversions, were supplemented by pumpage to make up the deficit. Although some irrigation waters may return through infiltration, the rate of ET is proportional to the type and quantity of crops being farmed. Pumpage rates have been reduced but the water-table declines continue because of the large acreages of land being irrigated. The overall impact has been a steady decline in the water table (1 m/year), diminished surface-stream flows, and saltwater intrusion—truly a groundwater management nightmare.

Kendy (2003) emphasizes the point that hydrogeologists need not only consider what happens to groundwater once it is pumped from the ground but should also be concerned about what happens to that water once it reaches the surface, the water actually consumed. Therefore, trying to stabilize an unconfined aquifer through sound management practices means balancing the budget of the entire hydrologic system (a holistic approach) and not just the aquifer. This is the broader thought process that must be included by those involved in groundwater management.

Water as a Commodity

As the demand for freshwater increases, so does the need for its management. The supply and demand for water pushes it into the realm of a commodity (Southwest Hydrology 2004). The available water, both surface water and groundwater, are allocated through the water-right process, where the quantity of water is defined along with a specific beneficial use. Once the water right is defined with a title of ownership that can be transferable, it emerges as a commodity that may require a significant fee (Southwest Hydrology 2004). Rural agricultural lands are being converted into expanding urbanized areas that have a demand for water. Every product we consume requires water to produce it. As a result, many water laws and management practices continue to change in the western United States and will undergo more changes in the future.

Realization by state regulators that groundwater development negatively affects surface flow because they are hydraulically connected (Chapter 10) is relatively recent (since the late 1990s in most U.S. western states). There is a tremendous need to inventory, characterize, and rethink how management entities allocate water. The general trend is toward conservation and efficiency. For example, the old drainage ditch that conveys water over great distances (many miles or kilometers) may lose up to 75% of their flow from what flowed through a headgate (a diversion structure from a surface stream). Losses are not limited to downward seepage but are also attributed to evaporation and poor maintenance practices (Figure 1.10). Figure 1.10 represents an unmaintained irrigation ditch where over 0.75 cfs (21.2 L/s) is leaking through a makeshift dam constructed of carpet, covering a breach in the ditch approximately 7 ft (2.1 m) wide. The discovery occurred during an investigation by the author during a legal dispute. This and other breaches in this ditch result in a loss of over 50% of the original flow over a distance of approximately 330 ft (100 m). The quantity of flow lost through this ditch could supply several large-capacity irrigation wells.

Complaints from "old-time" surface-water users to fix and maintain their ditches will rend the air, but the reality is that water management has been lax, and stronger accountability must be part of the governing process.

Changes in Flow Systems

As water laws and management practices change, it is likely that changes in groundwater-flow systems will also result. For example, water seepage along an older irrigation ditch may be a significant source of groundwater recharge for relatively newer domestic wells. Should the irrigation ditch be retired from service? Annual recharge from ditch losses would result in a declining water table. This commonly occurs as rural farmland is converted into subdivisions and city expansion (Kuzara et al. 2012). Pre-irrigation groundwater flow systems, which were altered because of the addition of irrigation

FIGURE 1.10   Breached irrigation ditch using carpet as dam material approximately 7 ft (2.1 m) wide.

canals and ditches, will again readjust to the new post-irrigation ditch system and the water table will again decline to pre-irrigation levels. Changes in stresses result in groundwater-flow systems striving to reach a new equilibrium. Understanding the whole hydrologic system will be needed to manage a system of change. Basin-wide perspectives must be the guiding principle, or potential problems as experienced in the North China Plain (Kendy 2003) may result.

1.4 Groundwater Use

It is instructive to evaluate the requirements of water for different purposes. How much water is needed for domestic purposes? In the author's home, we had a water softener that had a built-in digital readout flow meter. When the shower was on, one could read the rate flowing in the pipe (4 to 6 gpm). It also provides an average total use per day. In our case, in the first quarter of 2000 the daily amount was 398 gal. This was for seven people and includes consumption in the form of showers, hand washing, toilet flushing, drinks of water, and water used for cooking and laundry. Or approximately 57 gpd per person. This does not account for outside water used in the yard from the hose. This seems to be fairly close with other estimates of daily water use in the United States in the range of 80 to 100 gpd (189 to 379 L) per person per day (https://water.usgs.gov/edu/qa-home-percapita.html).

In developing countries, the daily use requirements for water is approximately 10 gpd (40 L/day) when the source for water is close at hand (U.S. Army Corps 2000). The quantity for daily use drops off considerably as the water source becomes more distant (Figure 1.11). The volume of water for daily use drops to 15 L/day when sources exceed 200 m and drop further to 7 L/day when water supplies exceed distances of 1 km (U.S. Army Corps 2000). The recommended maximum distance to a water source is 500 m with a retrieval time of 15 minutes (www.squareproject.org).

For emergency situations like refugee camps or catastrophic disasters, there are recommended humanitarian standards for planning purposes (World Health Organization 2011), (www.squareproject.org). They recommend a minimum of 7.5 to 15 L/day, depending on whether the conditions are short term (2 weeks to 1 month) or long term (3 to 6 months) (Table 1.2; World Health Organization 2011).

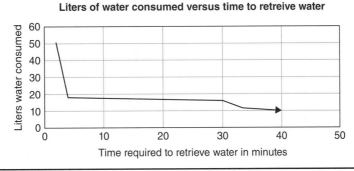

FIGURE 1.11 Image showing the relationship between consumed water in liters versus time to retrieve water in minutes. (*From www.squareproject.org.*)

Type of Need	Quantity (L/day)	Comments
Survival (drink and food)	2.5–3.0	Depends on climate and human physiology
Basic hygiene practices	2.0–6.0	Depends on social and cultural norms
Basic cooking needs	3.0–6.0	Depends on food type
Total	7.5–15.0	L/day

TABLE **1.2** Water Requirements for Emergency Planning Where 7.5 L/day Is Short Term and 15 L/day Is Long Term (Modified from World Health Organization 2011)

In Montana, domestic (exempt) well users can have a well drilled on their private property that yields up to 35 gpm (132 L/min) and does not exceed a volume of 10 acre-ft/year (12,300 m^3/year) without a special permit. Yields and volumes in other states will vary, but specific values are established through the legislative process. Wells used for irrigation or public water supply exceeding 100 gpm (378 L/min) or a volume greater than 10 acre-ft/year (12,300 m^3/year) require a water-use permit. As part of the technical evaluation, the well location (point of diversion), purpose, impacts to other existing users, demonstration of the water being there, and the well design (Chapter 15) are described by a hydrogeologist for evaluation by a state agency. As part of the process, a technical review is made of the application, and a notice is placed in the local newspaper to advise other users of this taking place. Any complaints must be filed and addressed. Once the water-use permit is granted, it is viewed as a water right for groundwater. In the western United States, all water rights are determined "a priori doctrine" using the following philosophy "first in time, first in right." A maximum rate and a maximum volume are applied for, conditioned on the water use. The rate is given in gallons per minute and the volume is given in acre-feet over a specified time.

Irrigation waters and water rights in the eastern United States are distributed following a Riparian doctrine. In this case, all water users in a given drainage basin share the allocatable water. If flow rates decrease by 10%, the users proportionally decrease water use by 10%. This works in areas where the precipitation is higher and streams are sustained by a fairly continuous source of base flow from groundwater discharge.

The use of groundwater for public water supply along with coordinating where to locate potential contaminating sources (sanitary landfills and industrial parks) falls under the area of groundwater resource management.

1.5 Groundwater Planning

As the demand for groundwater supply increases, there has been an improvement in management practices. This is being addressed in part by evaluating groundwater resources from the bigger picture (http://mbmggwic.mtech.edu/). Basins and watersheds are being delineated and the information stored as layers within databases in a downloadable format. Tables, spreadsheets, reports, and other sources of information can be stored and retrieved. Sources of information for these databases may come from drillers, landowners, or state agencies as groundwater characterization studies or aquifer vulnerability studies are performed.

A significant source of information within each state is being generated from source-water protection (SWP) area studies https://www.epa.gov/sourcewaterprotection (2017). All states have an SWP plan for all public water-supply wells. They vary by state, being adapted to each state's water resources and drinking-water needs (ibid.). Each source-water plan provides a wealth of information to each state's database and allows a more comprehensive management plan. Each state develops its own delineation and assessment process so that information is reported consistently and a minimum of technical requirements have been met.

Recharge areas to aquifers or areas that contribute water to surface waters used for public drinking water are delineated on a map. Geologic and hydrologic conditions are evaluated in the delineation study to provide physical meaning to the SWP areas. Inventories are made for contaminant sources within the area. Included may be businesses, surface or subsurface activities, or land uses within SWP areas where chemicals or other regulated contaminants are generated, used, transported, or stored that may impact the public water supply.

Source-Water Protection Studies

As a hydrogeologist, you may be hired to perform the technical assessment of the geologic and hydrologic information within an SWP area report. An example of the type of information needed is presented in the Montana Department of Environmental Quality Circulars PWS 5 and PWS 6 (Montana DEQ 1999). The following discussion is included for comparison with other state programs and to see what hydrogeologic topics are involved.

PWS 5 is a circular that helps an evaluator determine whether a public water supply has the potential to be directly influenced by surface water. In the circular, surface water is defined as any water that is open to the atmosphere and subject to surface runoff. Included as surface water are perennial streams, ponds, lakes, ditches, some wetlands, intermittent streams, and natural or artificial impoundments that receive water from runoff. Influence to groundwater supply can occur through infiltration of water through the streambed or the bottom of an impoundment (Chapter 10).

A scoring system of points is used during the preliminary assessment to determine whether the water supply can be classified as groundwater or whether there may be some influence from surface water. For preexisting wells, surface-water impacts are scored points by historical pathogenic or microbial organisms that have been detected in a well. Additionally, turbidity, distances from a surface-water source, depths of perforations below surface, and depth to the static water level (SWL) also contribute to points accumulated. In this case, like the game of golf, the fewer the points, the less likely a surface-water source impacts a given water-supply well.

If the number of points exceeds a limit, then further analysis is required. Springs or infiltration galleries immediately fall under the requirement of performing additional water-quality monitoring. A well may also be required to undergo intensive monitoring for 2 months following the completion of construction to determine its suitability as a public water-supply source. This would include weekly sampling for bacterial content and field parameters of temperature, turbidity, specific conductivity, and pH. The field parameters are also performed on the nearby surface-water source. A hydrograph of water-quality parameters versus time is plotted to compare similarities between surface water and groundwater. It is reasoned that groundwater parameters will not change much, while there may be significant variations in the chemistry of the surface water. Surface-water sources may be used for public water supply if the source water passes the biological and microbial tests.

If a microbial particulate analysis is required for a well, the client is required to conduct two to four analyses over a 12- to 18-month period, according to method EPA 910/9-92-029. The possibility of connection with surface water is indicated by the presence of "insects, algae, or other large diameter pathogens." A risk factor is also specified by the following bioindicators: giardia, coccidia, diatoms, algae, insect larvae, rotifera, and plant debris. The sampling method is performed using the following steps:

- Connect the sampling device as close to the source as possible.
- Assemble the sampling apparatus and other equipment *without* a filter in the housing to check whether the correct direction of flow is occurring.
- Flush the equipment using water from the source to be filtered for a minimum of 3 minutes. Check all connections for leaks. An in-line flow restrictor is desirable to reduce flow to 1 gpm (3.8 L/min).
- Filtering should occur at a flow rate of 1 gpm (3.8 L/min). During the flushing stage, the flow can be checked using a calibrated bucket and stopwatch.
- Shut off flow to the sampler. Put on gloves or wash hands and install the filter in the housing. Make sure a rubber washer or O-ring is in place between the filter housing and the base.
- Turn on the water slowly with the unit in the upright position. Invert the unit to make sure all air within the housing has been expelled. When the housing is full of water, return the unit to the upright position and turn on flow to the desired rate.
- Filtering should be conducted at a pressure of 10 pounds per square inch (psi). Adjustment of the pressure regulator may be necessary.
- Allow the sampler to run until 1,000 to 1,500 gal (3.78 to 5.67 m³) have been filtered. Mark the time when water was turned on and off.
- Disconnect the filter housing and pour the water from the housing into a ziplock plastic bag. Carefully remove the housing filter and place it in the bag with the water. Seal the bag, trying to evacuate all the air, and place it in a second bag and make sure neither bag leaks.
- Pack this into a cooler, making sure the filter does not freeze, as frozen filter fibers cannot be analyzed. Send the filter and data sheet to an acceptable laboratory within 48 hours.

If a public water supply source meets the criteria as an appropriate source, an SWP area must be delineated, as defined in the 1996 Federal Safe Drinking Water Act Amendments. In this process, recharge sources that contribute water to aquifers or surface water used for drinking water are delineated on a base map. Included is a narrative that describes the land uses within the delineated area along with a description of the characteristics of the community and nature of the water supply. Methods and sources of information used to delineate the SWP and potential contaminant sources should also be described. An example of the type of information is presented from PWS 6 (1999).

- **Introduction**—describes the purpose and benefit of the SWP plan.
- **Background**—includes a discussion of the community, the geographic setting, description of the water source, the number of residents to be served, well completion details, pumping cycles, the water quality, and natural conditions that may influence water quality at the public water system.
- **Delineation of water sources**—presents the hydrogeologic conditions (aquifer properties and boundaries), source-water sensitivity to contamination (high, moderate, or low), conceptual model based upon the hydrogeologic conditions, method of delineation, and model input parameters (Table 1.3).
- **Inventory of potential contaminant sources**—requires an inventory sheet for the control zone for each well. All land uses need to be listed. These may be classified as residential or commercial (sewered, unsewered, mixed), industrial, railroad or highway right-of-way, and agricultural (dryland, irrigated, pasture) or forest. Contamination sources are listed in Table 1.4.

Once the SWP area has been defined, the affected property owners are advised. If property owners do not like the outcome of the investigation, they may sue under a takings proviso (https://definitions.uslegal.com/t/taking-clause/). Takings issues are not something just for attorneys but something that should also be familiar to a hydrogeologist. Your greatest security in a courtroom situation is to be able to demonstrate that best possible practices have been used in the evaluation. Part of these practices include field studies and field data from which interpretations are made. As a field hydrogeologist, keep in mind the big picture and realize that your work may be revisited in court.

Input Parameter	Value(s) Used	Units	How Derived	Remarks
Elevation at well				
Static water level				
Transmissivity				
Thickness				
Hydraulic conductivity				
Hydraulic gradient				
Flow direction				
Effective porosity				
Pumping rate				
100-day total				
1-year total				
3-year total				

TABLE 1.3 Model Input Parameters for a Source-Water Protection Area

Agricultural	Commercial
Animal burial areas	Airports
Animal feedlots	Auto repair shops
Chemical applications (pesticides, fungicides, fertilizers, etc.)	Beauty parlors
	Boatyards
Chemical storage facilities	Car washes
Irrigation systems	Cemeteries
Manure spreading and pits	Construction areas
Industrial	Dry-cleaning establishments
Asphalt plants	Educational institutions (labs, storage)
Chemical manufacturing, warehouses, and distribution	Gasoline stations
	Golf courses (chemical applications)
Electrical and electronic products and manufacturing	Jewelry and metal plating
Electroplates and metal fabrication	Laundromats
Foundries	Medical institutions
Machine and metalworking shops	Mortuaries
Manufacturing and distribution of cleaning supplies	Paint shops
Mining and mine drainage	Photography establishments, printers
Paper mills	Railroad tracks and railyards
Petroleum products and distribution	Research laboratories
Pipelines (oil, gas, other)	Road de-icing activities (road salt)
Septic lagoons and sludge	Scrap and junkyards
Storage tanks	Storage tanks (above- and belowground)
Timber facilities	**Residential**
Toxic and hazardous spills	Fuel storage systems
Transformers and power systems	Furniture, wood strippers, refinishers
Wells (operating and abandoned)	Household hazardous products
Wood-preserving facilities	Lawns, chemical applications
Naturally Occurring	Septic systems, cesspools
Groundwater surface-water interaction	Water softeners
Iron and magnesium	Sewer lines
Natural leaching (uranium, radon gas)	Swimming pools (chlorine)
Saltwater intrusion	**Waste Management**
Brackish-water circulation	Fire training facilities
	Hazardous waste management units
	Municipal waste incinerators
	Landfills and transfer stations
	Wastewater and sewer lines
	Recycling reduction facilities

TABLE 1.4 Common Sources of Groundwater Contamination (U.S. EPA 1990)

1.6   Sources of Information on Hydrogeology

Information on hydrogeology can be gathered from direct and indirect sources. Direct sources would include specific reports where field data have been collected to evaluate the hydrogeology of an area (Todd 1983). Information may include water level (Chapter 4), geologic (Chapter 2), pumping test (Chapters 5 and 6), and groundwater flow direction (Chapter 4). An increasing number of these reports can be found in PDF form using Internet search engines on your phone. Indirect information are data sources that may be manipulated to be able to project data sources into desired areas with no information. For example, a few well logs placed on a geologic map and correlated with specific geologic units can be used to project target depths for drilling scenarios (Chapter 2).

In Montana, a Groundwater Information Center (GWIC) database was created, allowing public access to well logs, hydrogeologic reports, hydrograph data, and water chemistry (http://mbmggwic.mtech.edu/). All well drillers or licensed monitoring well constructors are required to submit a lithologic log, including well-completion details to a state agency within 60 days of completing the well. Sometimes this information is not reported, as some drillers are hesitant to report unsuccessful wells. If "no water" is found, sometimes the well logs do not get filed when in reality "no water" reported in a well log is helpful information. Once a well log is submitted, each well is given an "M number" (M#). Its location information, well depth, well yield, and a number of other parameters are also included. Wells can be located via the section, township, and range or by aquifer, subdivision, name or M#, drainage basin, or county. Once a well log is pulled up, there are options to locate it on a topographic map or color satellite imagery. Well logs may also show additional links to hydrographs of water-level data (Figure 1.4), water chemistry, or hydrogeologic assessments if they have been posted. This effort was pioneered by Montana Bureau of Mines and Geology researcher Tom Patton.

Since the first edition of this book, there has been an explosion of information accessibility through the Internet. It may be that if you need to find specific information about a given topic, it might be worthwhile to go to your local college or university. Universities buy the rights to access information databases. At a university, you could sit at a computer terminal and click on Databases A–Z, and a list of library-purchased databases would appear. A few of the most common ones are GeoRef, EarthWorks, GeoScience-World, Geofacets, and Compendex. During a topical search, specific references are listed with links to the full-text articles if the library has purchased the database.

The following list may give some ideas on where to start your search. It is by no means an exhaustive list, and no specific reason for the order presented is intended:

- Oil field logs or geophysical logs.
- Internet search engines.
- Peer-reviewed journals *Ground Water, Water Resources Research, Hydrogeology Journal,* and a host of other journals with "Water" in their titles (use the Database A–Z search at a local university library).
- Ground Water Online found at www.ngwa.org/gwonline/.
- NGWA: In addition they have a Daily Digest Groundwater Forum where members can pose any question and get responses from any other member about any topic.

- State-published or unpublished (incomplete) geologic maps.
- U.S. Geological Survey (on the Web at www.usgs.gov): This is a source of topographical maps and other published information plus a large section on water-supply information, including real-time surface-water stations and flows.
- U.S. Environmental Protection Agency (U.S. EPA 1990) (on the Web at www .epa.gov).
- The National Oceanographic and Atmospheric Association (NOAA) has a wealth of information about precipitation data (on the Web at www.noaa.gov).
- Topographic maps are helpful in locating wells (near building) and evaluating topography (for recharge or discharge areas) and making geologic inferences (by evaluating shapes of drainage patterns).
- Structure-contour maps that project the top or bottom of a formation and their respective elevations. See state geologic mapping programs for local in-state mapping efforts.
- State surveys and agencies are a great source for published information within a given state; additionally, they have well-log and water-quality databases and other production information (GWIC example described earlier).
- Libraries have a wealth of current and older published information. Sometimes the older publications are very insightful. Additionally, many libraries have search engines for geologic and engineering references (described earlier).
- Longtime landowners can give valuable historical perspectives and current observations.
- Master's theses and PhD dissertations have a variety of qualities, so glean what you can with a grain of salt. Some are very well done, while others should not have been published.
- Summary reports prepared by state agencies on geology, groundwater, surface water, or water use and demand.
- Consulting firms and other experienced hydrogeologists.
- U.S. Geological Survey water supply papers and open-file reports.
- County or city planning governments may have information in geographic information system (GIS) formats and have listings of wells, well ownership, or other useful information.
- Some authors or editors attempt to evaluate all references associated with groundwater and then group the references according to topics, such as done by van der Leeden (1991). This was a practice done before the 1990s and may be very helpful in identifying classic works.

In most states, a well log is required to be filed each time a well is drilled. Included should be the depth, lithologic description, perforation or screened interval, the SWL, and brief pumping or bailing-test information. This information can be used to evaluate depth to water and lithologies and to get a general idea of well yield in a given area. Experienced hydrogeologists learn how to combine a variety of geologic and well-log database sources into a conceptual model from which field decisions can be made.

1.7   Site Location for Hydrogeologic Investigations

As simple as it sounds, the first task is to know *where* the site is. As a hydrogeologist, you may be investigating a "spill," evaluating a property for a client who is considering buying the property, locating a production well, or participating in a construction dewatering project. It is imperative that you know where the site is so that you can assess what existing information there might be. If this is a preexisting site, there will be some information available; however, if you are helping to "site in" a well for a home-owner or a client for commercial purposes on an undeveloped property, you must know where the property is (see "Well Drilling" in Section 1.8). This is less problematic with GPS capabilities on a phone; however, part of knowing where a site is located comprises both its *geography* (surface conditions and access) and its *geology* (what the stratigraphic age is and what type of formations are present).

Many water-well drillers are successful at finding water for their clients without the help of a hydrogeologist. Either the geologic setting is simple and groundwater is gen-erally available at a particular depth, or they have experience drilling in a particular area. If they don't feel comfortable with the drilling location, they will always ask the client where *they* want the hole drilled so that they can't be held liable for problems when things go wrong or production is less than desired. The problem is that most homeowners don't have much of an idea about the geology of their site and choose a location based on convenience to their project rather than using field or geologic infor-mation (Example 2.5). The phone call to you, the hydrogeologist, generally comes after a "dry" hole or one with a disappointingly low yield has been drilled. The phone call may also come from the driller or the client. Before carefully looking into the situation, that is, answering questions about what happened and what the drilling was like, you must know where the location of the site is geographically and geologically. Unfortunately, sometimes you are given the wrong site-location information and you end up doing an initial geologic investigation in a place different from what you were led to believe. A personal example will help illustrate this view.

Example 1.7   A phone call came after drilling a 340-ft duster (dry hole). The drill site was chosen near an old, existing homestead cabin. The client (landowner) figured that since the homesteader had water and it was at an ideal location for his ranching operation, it would probably make a good site. The driller was shown the location by the client, and drilling commenced. After penetrating to a depth of 340 ft, with no water in sight, the driller decided to call and get some recommendations. The property consisted of more than 600 acres (243 ha) near a small town in western Montana. The section, township, and range of the location were provided to the author of where the hole was located by the driller. After an initial investigation in the library, geologic and topographic maps were located. The geologic information was superimposed onto the topographic map, and a couple of cross sections were constructed. The target zone would be a coarser-grained member of a Lower Cretaceous sandstone (Chapter 2). A meeting time and place were arranged. Within minutes of driving down the road from the meeting site, it became evident that the driller was heading to a location different from what was described (we were off the topographic map prepared beforehand). Instead of slightly undulating Cretaceous sedimentary rocks, the outcrops were basalt, rhyolite, and Paleozoic carbonates that had been tilted at a high angle. The dry hole had been drilled into a basalt unit, down geologic dip (Chapter 2). It becomes difficult to recommend anything when the structure and geologic setting are uncertain or unknown.

The purpose for well drilling was for stock watering. An initial design was recommended that would not require electrical power. A local drainage area could be excavated with a backhoe and cased with 24-in galvanized culvert material. A 3-in PVC pipe could be plumbed into the "culvert well"

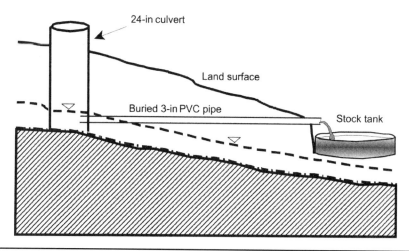

FIGURE 1.12   Gravity-feed stock tank design from Example 1.7 (not drawn to scale).

at depth and run to a stock tank (Figure 1.12). It was a gravity-feed design and would help keep cows at the far end of his property during the summer.

After this initial meeting, the rancher informed the author that they tried the backhoe method *and* found that the area had good aquifer materials but no water. Once the proper location was identified, we went back to the office and evaluated the geology. We determined that this drainage was situated on a large fault and that any infiltrating water collected would be lost quickly into the Paleozoic limestone formations at depth and would be found only by drilling a well several hundred feet below the surface.

In Example 1.7, an understanding of the geology was essential to locating a target for water development. Even if the geologic setting is known, modifications due to structural features may also need to be made. The geology of hydrogeology is further discussed in Chapter 2.

1.8   Taking Field Notes

A field notebook is your memory of events, locations, people, and figures. Without decent field notes, it will be difficult to reconstruct what happened in the field, let alone remember the important details that are necessary for billing out time and completing reports back at the office. Notebooks should be accessible, sturdy, and weatherproof, or at least you should write with a waterproof pen. Some people prefer to use a pencil (with something hard like a 6H lead), while others prefer pens. What is the problem with using a pencil? Pencils can smear as the pages chafe back and forth in a field bag. After a few times in the field, the information becomes hard to read. Writing with a pen can usually be clearly visible for years. After all, one should never erase in a field book. If your field notebook is needed in court, erase marks will be questioned and the book later thrown out as being tampered with. Don't use whiteout or some other cover-up to make your field book look prettier for the same reason. After all, it is a field book. Mistakes will be made, and when they occur, neatly draw a line through the mistake and continue on. Your notebook should not be a disorganized mess because it will be needed to help you remember what happened later. Some people have a "working" field book and a second field book that the day's work is translated into for neatness. It is a good idea to record or add information each evening or during slack time, or details will be missed. You take notes so that you don't *have* to remember everything.

Inside the cover, put your name, address, and phone number so that you can be reached in the event you lose your field book. Some prefer to staple a business card inside the front cover. The back of the field book should have some blank pages, reserved to write the names, addresses, and phone numbers of contacts or distributors of equipment, supplies, or parts frequently needed. Again, business cards can be stapled directly inside these back pages for ready access. For some people, it is also helpful to put conversion tables in the field book so that they are accessible for calculations. Your field book is your memory.

Although there are a host of topics that can be recorded in a field book, some examples of the more commonly encountered tasks are presented here.

Daily Information

There is some basic information that should appear in a field book each time one ventures out. The client, location, purpose, and objective of the day's work should be noted, for example, soil sampling along Skunk Creek for ACME remediation study. The day's date and time of departure can be very important. It is a good idea to record the times people arrive and leave a site as a daily itinerary. This could be significant in a lawsuit if claims are made about who was where and when. It is a good idea to record the people in the party who are present and their phone numbers and names (including the last names). Saying that Jim or Susan was there may not be very helpful a few years down the road. The basic who?, when?, where?, why?, and how? approach to field notes is a good way to remember what should be recorded.

Comments on the weather and how it changes during the day can also be significant in field interpretation. For example, a rainstorm moving in will likely reduce the air pressure. In confined aquifers, this may help explain why the water levels increased or slowed down during a pumping test (discussed in Chapter 5). Rainstorms may be the source of why the stage in a stream changed. If the weather is very cold, this may contribute to errors committed in the field from stiff fingers, trying to get the job done quickly, or trying to work with gloves on. Conversely, hot weather can also affect the quality of field data collected, as perspiration drops often land right where you need to record information. Equipment is also temperature sensitive, particularly water-quality equipment. This may contribute to data values drifting out of calibration. Of course, this can occur as one bounces down a bumpy road.

It is a good idea to record detailed observations and descriptions during slack time while your mind is still fresh. Field sketches and diagrams are useful. Many field books have grid lines on one side of the page so that well logs can be drawn vertically to scale. If your drawing or sketch is not to scale, say so in the field book. You may have a field book dedicated to a particular project or get a new one each year. It is advisable to organize or record information in a systematic way. For example, some people like to number the pages sequentially so that a table of contents can be made at the front of the field book. This is helpful if there are a variety of jobs over a broad area. Some of us just prefer to mark the beginning and end of a particular day's effort by noting, for example, "page 1 of 4," "page 2 of 4" and so on. Whatever system you use, strive to be consistent, complete, and neat.

Lithologic Logs

Most companies or agencies have their own forms for filling out lithologic information during drilling (Figure 1.13). This is desirable, but if for some reason you forgot your forms or none are provided, you may have to record lithologic log information in your field book. If you forgot your field book too, then, shame on you, you forgot your memory.

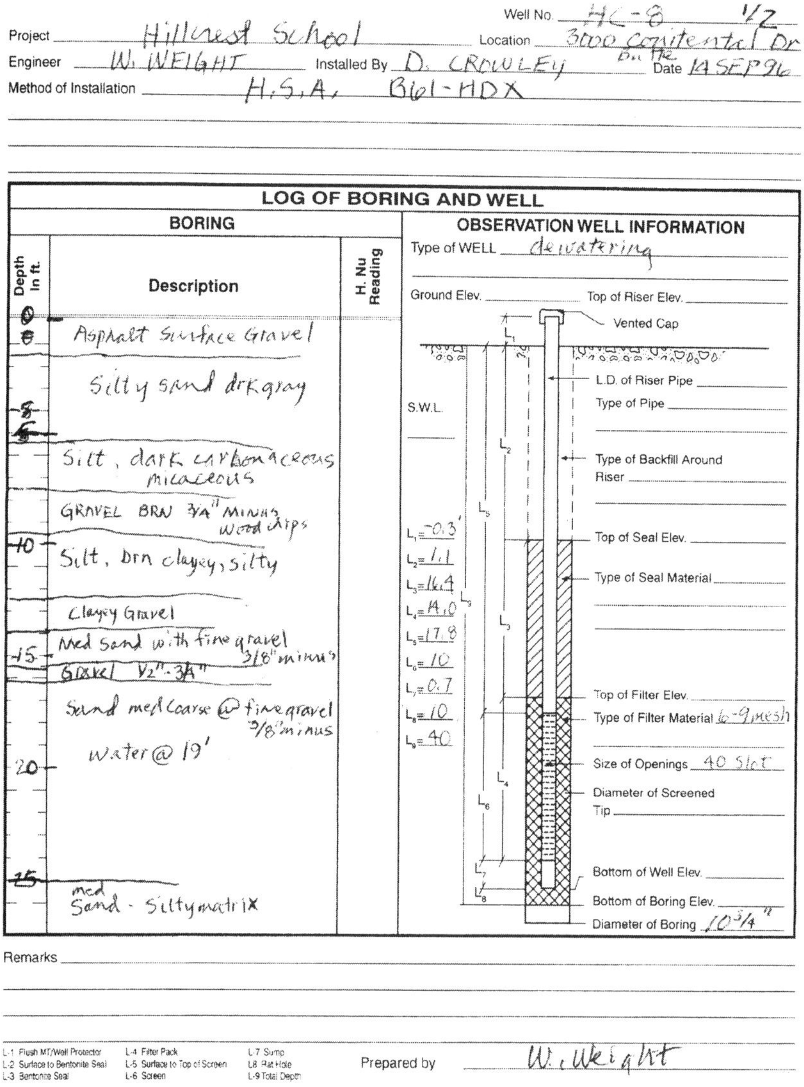

Project Hillcrest School
Well No. HC-8 1/2
Location 3000 Continental Dr
Engineer W. WEIGHT
Installed By D. CROWLEY
Date 14 SEP 96
Method of Installation H.S.A. B61-HDX

LOG OF BORING AND WELL

BORING

Depth In ft.	Description	H. Nu Reading
0	Asphalt Surface Gravel	
	Silty sand drk gray	
5	Silt, dark carbonaceous micaceous	
	GRAVEL BRN 3/4" MINUS wood chips	
10	Silt, brn clayey, silty	
	Clayey Gravel	
	Med Sand with fine gravel 3/8" minus	
15	Gravel 1/2"-3/4"	
	Sand med Coarse @ fine gravel 3/8" minus water @ 19'	
20		
25	med Sand - Silty matrix	

OBSERVATION WELL INFORMATION

Type of WELL dewatering
Ground Elev.
Top of Riser Elev.
Vented Cap
S.W.L.
I.D. of Riser Pipe
Type of Pipe
Type of Backfill Around Riser
Top of Seal Elev.
Type of Seal Material
Top of Filter Elev.
Type of Filter Material 6-9 mesh
Size of Openings 40 Slot
Diameter of Screened Tip
Bottom of Well Elev.
Bottom of Boring Elev.
Diameter of Boring 10 3/4"

L1 = -0.3'
L2 = 1.1
L3 = 16.4
L4 = 14.0
L5 = 17.8
L6 = 10
L7 = 0.7
L8 = 10
L9 = 40

Remarks

L-1 Flush MT/Well Protector
L-2 Surface to Bentonite Seal
L-3 Bentonite Seal
L-4 Filter Pack
L-5 Surface to Top of Screen
L-6 Screen
L-7 Sump
L8 Rat Hole
L-9 Total Depth

Prepared by W. Weight

Figure 1.13 Lithology and well-completion form from a well-drilling company. This example was for a monitoring well drilled using a hollow stem auger (HSA) drill rig, logged by the author.

It may be, for example, that you are making field descriptions of surface geology during a site characterization (Nielsen et al. 2006) or recording lithologic information from monitoring wells. (How to log a drill hole is discussed in Chapter 15, and a more detailed discussion of geologic information is presented in Chapter 2.)

Lithologic logs are generally recorded during the drilling process. The type of drilling affects the time that you have to record information (Chapter 15). If you are logging and bagging core, collecting chip samples, or doing a variety of other tasks, it can get tricky getting everything done while still recording a meaningful lithologic log.

Example 1.8 I can recall my first attempt at recording lithologic logs for drill holes in overburden (strata above a coal bed) on a coal property near Hanna, Wyoming. It was a forward rotary drilling rig (Chapter 15), and drillers could drill with air for the first 200 ft or so in relatively soft layered sediments. The mast was almost free-falling at a rate of approximately 20 ft/min. The drillers laughed as I frantically tried to record the geologic information. It was bewildering as the various lithologies changed in texture and color before my eyes.

Besides learning to work with your driller, it is helpful to have a shorthand set of descriptions for lithologies and textures (this is discussed further in Chapter 15). Many companies may have their own system, and these can later be incorporated into a software package for visualization. Rock or soil descriptions should be indicated by the primary lithology first, followed by a series of descriptors. Information recorded could include the following:

- Lithology name.
- Grain size and degree of sorting (is this actually crushed bedrock or broken sediments?).
- Color (can be affected by drilling fluids, such as mud).
- Mineralogy (HCl fizz test or mineral grains observed in the cuttings).
- Probable formation name, and etc. For example, an entry might say, "sandstone: very-fine to fine-grained, tan, with dark cherty rock fragments, the basal member of the Eagle sandstone."

You may notice things such as whether there were coarsening-upward or coarsening-downward sequences in sedimentary deposits, unconformities, or abrupt coloration changes, as would occur if the rocks changed from a terrestrial aerobic origin (red, yellow, or tan) to those of an anaerobic origin (gray, dark, or greenish) (Figure 1.14). The relative ability for rocks to react with dilute HCl can be useful in helping distinguish one formation from another or which formation member you may be in. Record dates and how samples were taken. For example, were the cuttings washed? (The fines would wash through a sieve, while the chips or coarse fraction would remain behind.) Were they grab, split-spoon, or core samples?

Another useful thing to note along with the lithologic log is the drilling rate. There may be a variety of similar rock types, but some may be well cemented and drill slower or faster than others. This may affect the ability for vertical groundwater communication between units, or it may be helpful for a blasting engineer who needs to know about a very hard sandstone or igneous unit that will require extra blasting agents in a mining operation. Drilling rates can be roughly compared to cone-penetrometer tests

FIGURE 1.14 Cross-sectional view of changing lithologies, showing a small-scale (several-meters-wide) thrust fault.

(Example 4.9; Figure 15.36). Some drilling may be smooth or "chatterier." For example, a hard sandstone may cause the drill string to chatter because the bit bounces somewhat as it chews up the formation, but the drilling will proceed slowly. Contrasted to this, a coal bed will "chatter" when it is soft and is ground up quickly. Notice how long it takes between the first "chatter" sound and the appearance of black inky water from the coal bed at the land surface. Use your senses (odors, for example) to notice changes. Additional comments about logging during drilling are made in Chapter 15.

Well Drilling

While performing well drilling, you need to record not only the lithologic information but also the well-completion information. Depending on the complexity and depth of the well, there may be other key people involved: other geologists, tool pushers, mud loggers, or engineers. Once again, write down the names, addresses, and phone numbers of the driller and these key people or, better yet, get their cards and staple them into your book. Hopefully, you have performed some background geologic work and have an idea about what formation you are drilling in and the targeted depth. What is the purpose of the well (monitoring, production, stock well, etc.)? Were there other wells drilled in the area? How deep did they drill for water or product contamination? At what depths were the production zones? What will the conditions likely be (hard, slow drilling, heaving sands, etc.)?

Record a detailed location of the well both the setting and access and GPS coordinates. The location may be need to be converted into state-plane coordinates, section township and range descriptions, or other identifiers. Will the elevations be surveyed or estimated from a topographic map? It should be noted that handheld GPS units cannot be relied on for accurate elevation data. More discussion about GPS systems and other location equipment is found in Chapter 4.

Write down the make and model of the rig and any drilling fluids that were used and when. Did you start with air or begin drilling with mud? It is a good idea to write down a summary of the work completed by your driller before you arrive. Quantities of materials billed by a driller should corroborate with your field notes to document errors. Big projects may need to be drilled continuously in shift-work format until the job is done.

> **Example 1.9** In western Wyoming in the early 1980s, some deep (1,200 ft, 370 m) monitoring wells were being placed down structural dip of a coal mining property. Each well took several days, so drilling and well completion took place in 12-h shifts by the geologists and drillers on a continuous basis until the wells were completed. A new drill crew and geologist would arrive every 12 h. It was critical that communication took place between the ones leaving and the ones arriving for continuity.

It is also important for billing purposes to write down incidents of slack time (standby), equipment breakdowns, or "runs" for water. The work submitted should match the field notes taken, or inquiries might be made. Each morning or during slack time, measure and record the static-water levels (Chapter 4). This is particularly important first thing in the morning after the rig has sat all night. The size and type of bit is also important for knowing the hole diameter. Any unusual or problematic situations or conditions should be recorded and discussed with your driller.

Well Completion

Once the hole is drilled, the well is completed from the bottom up. Your field book needs to contain sufficient information to construct well-completion diagrams back at the office or in the motel room at night. It is helpful to review the information on forms required by the local government so that all details are covered. For example, was surface casing used? If so, what was the diameter? What it grouted with bentonite? What about screen type or slot size? What are the diameters of the borehole and pump liner? The mechanics of well completion are discussed in Chapter 15; however, the following items should probably be included in your field book:

- Total depth drilled (TDD) and hole diameter (important for well hydraulics calculations) (Chapters 6 and 7).
- Indication of type and length of bottom cap (or was it left open hole?).
- Is there a sump or tailpipe (a blank section of casing above the bottom cap)?
- Interval of screened or perforated section, slot size and type or size of perforations, outer diameter (OD) and inner diameter (ID) of screen, and material type (schedule 80 PVC)? Is the screen telescoped with a packer, threaded, or welded on?
- Number of sections of casing above the screen, length of each interval, and height of stick up. Twenty-foot (6.1-m) pipe can vary in length. Fractions of a feet can add up when going several hundred feet into the ground.
- Was the well naturally developed, or was packing material used? If gravel packed, what volume was used, height above the screen [number of 100-lb (45.4-kg) or 50-lb (22.7-kg) bags]? Interval of packing material? This is helpful for planning future wells to be drilled.
- Was grouting material used? If so, over what interval?

- What kind of surface seal was used (concrete, neat cement, cuttings)?
- What kind of security system is there for the well cap?
- Was the well completed at multiple depths, and, if so, what are the individual screened intervals and respective grouting intervals?
- Was the well properly developed? What was the SWL before development?
- What method was used (pumping, bailing, lifting with air, etc.) and for how long? How many purge volumes or for what duration did development take place?

Pumping Tests

There are many details that need to be remembered during a pumping test, let alone many pieces of equipment to gather together. In addition to your field book, for example, it is imperative that you have well logs for all wells that will be included in the pumping test, with well-completion details. Your field book will play a critical role in remembering the details. One of the first items of business is to measure the static-water levels of all wells to be used during the test. Any changes will be important for interpretation. Ideally, a couple of days prior to the test, well sentinels or transducers will have been deployed to document any pretest regional trends (Chapter 5). The weather conditions may also affect the water-level responses. Is there a storm moving in? This can result in a low-pressure setting where water levels may rise. In confined systems, the author has observed changes measuring 0.5 to 1 ft (0.15 to 0.3 m) depending on the atmospheric pressure drop and confining conditions. Is there a river or stream nearby? Did the stage remain constant or raise or drop? During Montana Tech's hydrogeology field camps, there have been observed stage changes of 2 ft during a 24-h pumping test. How does this affect the results?

In October 1998, while attending a MODFLOW 98 conference in Boulder, Colorado, several groundwater modelers from around the world were informally interviewed and asked questions about what they thought should be included in a book like this. One resounding remark stood out: "Please tell your readers that they should always make backup 'hand' measurements in their field books on changing water levels in wells, during a pumping test! Don't just rely on the fancy equipment." Another comment was, "Plot the data in the field so that one knows whether the test has gone on long enough" (see Chapters 5 and 6). The data would be obtained from measurements recorded in field books or downloaded from software. Along this line, there should be a notation of the pumping rate during the test. Were there changes in discharge rate that took place during the test? Remember that theoretically the pumping rate is supposed to be constant (Chapter 6).

Troubles in the field need to be recorded. Sometimes individuals don't want to remember something goofy or embarrassing that happened in the field, but it may be the key to making a proper interpretation. Why didn't anyone get the first 5 min of the recovery phase of the pumping test? The reasons may range from being asleep and didn't notice the silence of the generator as the fuel ran out or perhaps off taking care of a personal matter or couldn't get the "darn data logger" to step the test properly. Perhaps a friendly Rottweiler powered through the equipment area and managed to knock out one of the connecting cables (Figure 1.15). I'm sure a documentary on the things that can go wrong during a pumping test would make fascinating reading.

FIGURE 1.15 Neighbor's Rottweiler dog disrupting vented transducer cables and harassing the author during a pumping test.

The following bulleted items will serve as a reminder of the most important general items to record but is by no means an exhaustive list. Additional discussion can be found in Chapter 5.

- Personnel on hand at the site, either assisting or present for observation.
- Static-water levels of all of the wells, prior to emplacement of any equipment.
- Sketch map showing the orientation and distances of observation wells to the pumping well; also include possible boundary effect features, such as steams, bedrock outcrops, other pumping wells, etc. that may affect the results.
- Weather conditions at the beginning, during, and at the end of the test.
- List of all equipment used: type of pump, riser pipe, discharge line, flow meters, or devices used to measure flow (e.g., bucket and stopwatch or flow meter?). Again, a sketch or photograph in the field is helpful.
- Exact time the pump was turned on and off for the recovery phase. (It is helpful to synchronize watches for this purpose or have a dedicated cell phone or stopwatch for the test.)
- Manual water-level measurements of observation wells. Include the well ID, point of reference (top of casing), method of measurement (e-tape or steel tape), and a careful systematic recording of times and water levels in a column format.
- Record of the pumping rate. How was this measured and how often during the test? (Times should be written 30 minutes or less.)
- Recovery phase of the test. Again, manual measurements, conditions, and times.
- Any other problems or observations that may prove helpful in the interpretation of the results.

Water-Quality Measurements

Any type of field measurement can have errors. Errors can be made by the equipment being out of calibration or bumped around in the field, or the operator may just get tired during the day. Maybe the weather conditions rendered the collecting of data fairly intolerable. Apparent errors may result from significant diel cycling of water-quality parameters over a 24-h period (Chapter 9). Honest observations in a field book can help someone remember the conditions under which the data were collected and help someone else reviewing the data or the interpretation. The field book can explain why the numbers look strange. For example, many parameters are temperature sensitive. Along with a pH measurement, there should be a knowledge of what the temperature is to set the calibration knob correctly (Figure 1.16). Many specific conductance meters have a temperature correction to 25°C internal within the device. Failing to take into account the temperature can significantly affect the results recorded. Chain of command and other details of the inherent problems that can occur in the field are discussed in greater detail in Chapter 9. The following items should be considered for field notes:

- What is the water sample type [e.g., surface water (lake, wetlands, or stream), domestic well, monitoring well]?

- What were the weather conditions throughout the day?

- What were the date and exact time when the sample was taken? (Many samples are required to reach a lab within 24 h.)

- Who were the personnel present (was the operator alone, or was there someone else helping)? Who was present at the respective sampling locations? Did the well owner come out? It may be important to be able to tell who was there.

- What were the methods of calibration and correction for drift?

Figure 1.16   Surface-water quality analysis (alkalinity and sampling) in the upper Big Hole River valley, Montana.

- What were the units? Were the data temperature corrected? Remember, many parameters are temperature dependent and temperature sensitive.

- If it is a water-well sample, what volume of water was evacuated before the samples were collected? What were the purge times and respective volumes? How was the sample taken? Was it collect via a bailer, a discharge hose from a pump, or a spigot or sink? Which parameters were monitored during the purging (pH, conductivity, temperature)? Where was the discharge water diverted to?

- How long did it take to reach a stable reading? Was the reading taken prematurely before stabilization took place? This can vary significantly between pieces of equipment. Was a stir rod going during the reading, or was the container just swished around every once in a while?

- What were the methods of end-point detection? Was a pH meter or color reagents used?

- Was the sample taken in situ via hydrasleave (www.hydrasleeve.com) or snap sampler (www.snapsampler.com)?

- Were the samples filtered or acidified? How?

- Was the laboratory protocol followed? Were the chain-of-command papers filled out?

- Units! Units! Units!

1.9   Summary

Hydrogeology with a field-application context is an invigorating subject. Each time you go into the field, each time you drill a new well or go into a new area, the geology and hydrogeologic conditions change. This is the fun and the challenge of it. Keep in mind the big picture of the hydrologic cycle while also paying attention to the detailed items, such as the geologic setting, weather patterns, diel water-quality changes in surface streams or storm fronts coming in during a pumping test, and recording these observations in your field book. By synthesizing the data, a conceptual model of a given area will emerge. The conceptual model will take shape as additional field data are collected.

Prior to heading for the field, check your sources of hydrogeologic information and make sure you have your field book, or you won't be able to remember the details of your daily experiences. Work hard, keep your wits about you, and be safe, and soon your personal experience database will transform you into a valuable team member.

1.10   Problems

1.1   Refer first to Figure 1.2. How can the walking the dog analogy be applied to El Niño region 4 in Figure 1.3b in terms of overall changes in seawater temperatures?

1.2   Select a relatively small drainage area (approximately 10 mi^2 or 25.9 km^2) near where you live and estimate the water budget in the area.

1.3   Go to the Edwards Aquifer Authority website (www.edwardsaquifer.org). Once there, click on the "Science and Maps" tab, then "Aquifer-Data," and finally "Charts and Graphs." There are at least two wells and two springs that have continuous water-level data over time. Do all four water-level sources agree? If not, determine why this may be the case.

1.4   How much water is needed for a refugee camp of 3,000 displaced people, including 800 school-age children and 20 relief agency staff?

1.5   Evaluate the source-water protection plan process in Section 1.5 and then compare with the state where you live.

1.11   References

Alley, W.M., and Leake, S.A., 2004. The Journey from Safe Yield to Sustainability. *Ground Water*, Vol. 42, No. 1, pp. 12–16.

Anderson, M.P., and Woessner, W.W., 1992. *Applied Groundwater Modeling—Simulation of Flow and Transport.* Academic Press, San Diego, CA, 381 pp.

Borduin, M.W., 1999. *Geology and Hydrogeology of the Sand Creek Drainage Basin, Southwest of Butte, Montana.* Master's Thesis, Montana Tech of the University of Montana, Butte, MT, 103 pp.

Brandon, Brenda. Personal Communication, November 2005, Spokane, Washington, Indian Reservation.

Bredehoeft, J.D., 2002. The Water Budget Myth Revisited: Why Hydrogeologists Model. *Ground Water,* Vol. 40, No. 4, pp. 340–345.

Bredehoeft, J.D., 2006. On Modeling Philosophies. *Ground Water,* Vol. 44, No. 4, pp. 496–499.

Bredehoeft, J., and Konikow, L., 1993. Ground-Water Models: Validate or Invalidate. *Ground Water*, Vol. 31, No. 2, pp. 178–179.

Bredehoeft, J.D., Papadopulos, S.S., and Cooper, H.H., 1982. Groundwater: The Water Budget Myth. In *Scientific Basis of Water-Resource Management*, Studies in Geophysics, pp. 51–57. National Academies Press, Washington, DC.

Burgher, K., 1992. *Water Budget Analysis of the Upper Silver Bow Creek Drainage, Butte, Montana*, Master's Thesis, Montana Tech of the University of Montana, Butte, MT, 133 pp.

Devlin, J.F., and Sophocleous, M., 2005. The Persistence of the Water-Budget Myth and Its Relationship to Sustainability. *Hydrogeology Journal*, Vol. 13, No. 4, pp. 549–554.

EPA, 1995. Superfund Record of Decision: Silver Bow Creek/Butte Area (O.U.3), Silver Bow/Deer Lodge, MT 9/29/1994. EPA/ROD/RO8-94/102. https://nepis.epa.gov/Exe/ZyPURL.cgi?Dockey=91000VHE.TXT

Erickson, E.J., 1995. *Water-Resource Evaluation and Groundwater-Flow Model from Sypes Canyon, Gallatin County, Montana.* Master's Thesis, Montana Tech of the University of Montana, Butte, MT, 69 pp.

Gamache, M., Screiber, R.P., and Weight, W.D. 2003. Estimating Induced Infiltration and Cross-River Flow from Numerical Modeling. In *Proceedings in MODFLOW and MORE 2003 Understanding through Modeling,* Vol. 1, International Ground Water Modeling Center, Colorado School of Mines, Golden, CO, pp.159–163.

Hayes, S.P., Mangum, L.J., Picaut, J., Sumi, A., and Takeuchi, K., 1991. TOGA-TAO: A Moored Array for Real-Time Measurements in the Tropical Pacific Ocean. *Bulletin of the American Meteorological Society*, Vol. 72, pp. 339–347.

Hiscock, K.M., Rivett, M.O., and Davison, R.M., eds., 2002. *Sustainable Groundwater Development.* GSL Special Publications 193. 344 pp. www.geolsoc.org.uk

Kendy, E., 2003. The False Promise of Sustainable Pumping Rates. *Ground Water*, Vol. 41. No. 1, pp. 2–4.

Kendy, E., and Bredehoeft, J.D., 2006. Transient Effects of Groundwater Pumping and Surface-Water-Irrigation Returns on Streamflow. *Water Resources Research*, Vol. 42, No. 8.

Koukl, G., 2013. The Trouble with the Elephant. www.str.org/articles/the-trouble- with-the-elephant#_ftn1

Kuzara, S., Meredith, E., and Gunderson, P., 2012. *Aquifers and Streams of the Stillwater and Rosebud Watersheds.* Montana Bureau of Mines and Geology Open-File Report 611, 65 pp.

Livingston, P., 1942. *A Few Interesting Facts regarding Natural Flow from Artesian Well 4, Owned by the San Antonio Public Service Company, San Antonio, Texas,* U.S, Geological Survey Open-File Report, 7 pp.

Medicine Head, Willis. Crow Tribal Elder, Personal Communication, January 2004, Crow Agency, MT.

Metesh, J.J., and Madison, J.P., 2004. *Summary of Investigation, Upper Silver Bow Creek, Butte, Montana.* Montana Bureau of Mines and Geology Open-File Report 507, 7 pp.

Montana DEQ, 1999. *Groundwater under the Direct Influence of Surface Water.* Montana Department of Environmental Quality Circular PWS 5, 1999 Edition, 30 pp.

Montana DEQ, 1999. *Source Water Protection Delineation,* Montana Department of Environmental Quality Circular PWS 6, 1999 Edition, 21 pp.

Nielsen D.M., ed., 2006. *Practical Handbook of Environmental Site Characterization and Ground Water Monitoring,* 2nd ed. CRC Press, Boca Raton, FL, pp. 35–205.

NOAA, 2000. *The El Niño Story.* http://www.pmel.noaa.gov, 4 pp.

NOAA, 2018. https://www.climate.gov/enso. Description of El Niño and La Niña and weather patterns (Figures 1.3a and 1.3b).

Philander, S.G.H., 1990. *El Niño, La Niña and the Southern Oscillation.* Academic Press, San Diego, CA, 289 pp.

Quigley, L., 1959. The Blind Men and the Elephant. Charles Scribner's Sons. New York.

Reeves, C.C., and Reeves, J.A., 1996. *The Ogallala Aquifer of the Southern High Plains.* Vol. 1, Geology. Estacado Books, Lubbock, TX, 360 pp.

Sharp, J.M., Jr., and Banner, J.L., 1997. The Edwards Aquifer: A Resource in Conflict. *GSA Today*, Vol. 7, No. 8, pp. 2–9.

Sophocleous, M., 1997. Managing Water Resources Systems: Why "Safe Yield" Is Not Sustainable. *Ground Water*, Vol. 35, No. 4, 541 pp.

Sophocleous, M., 2005. Groundwater Recharge and Sustainability in the High Plains Aquifer in Kansas, USA. *Hydrogeology Journal*, Vol. 13, No. 2, pp. 351–365.

Southwest Hydrology, 2004. Water as a Commodity [7 articles], Vol. 3, No. 2, pp. 12–31.

Swanson, G.L., 1991. Super Well Is Deep in the Heart of Texas. *Water Well Journal*, Vol. 45, No. 7, pp. 56–58.

Tarbuck, E.J., and Lutgens, F.K., 1993. *The Earth: An Introduction to Physical Geology.* Macmillan, New York, NY, 654 pp.

Theis, C.V., 1940. The Source of Water Derived from Wells: Essential Factors Controlling the Response of an Aquifer to Development. *Civil Engineer*, Vol. 10, pp. 277–280.

Todd, D.K., 1983. *Groundwater Resources of the United States.* Premier Press, Berkeley, CA, 749 pp.

Toth, J.A., 1962. A Theory of Ground-Water Motion in Small Drainage Basins in Central Alberta, Canada. *Journal of Geophysical Research*, Vol. 67, pp. 4375–4381.

U.S. Army Corps of Engineers, 2000. *Evaluation of Water Resources of Guatemala*. Mobile Office. http://www.oas.org/en/sedi/dsd/ELPG/DataBase/Guatemala.asp

U.S. EPA, 1990. *Guide to Groundwater Supply Contingency Planning for Local and State Governments*. EPA-440/6-90-003. US EPA Office of Groundwater Protection, Washington, DC, 83 pp.

USGS, 1999. *Land Subsidence in the United States*. U.S. Geological Survey Circular 1182, 177 pp.

USGS, 2012. https://water.usgs.gov/edu/earthwherewater.html. The breakdown of water on Earth.

van der Leeden, F., 1991. *Geraghty & Miller's Groundwater Bibliography*. 5th ed. Water Information Center, Plainview, NY.

Weight, W.D., and Chandler, K.M., 2010. Hydraulic Properties of Rocky Mountain First-Order Alluvial Systems and Diurnal Water-Level Fluctuations in Riparian Vegetation. *Journal of Environmental Science and Engineering*, Vol. 4, No. 9, pp. 12–23.

Whitlock, C., Cross, W., Maxwell, B., Silverman, N., and Wade, A.A., 2017. *2017 Montana Climate Assessment*. Montana State University and University of Montana, Montana Institute on Ecosystems, Bozeman and Missoula, MT, 318 pp. doi:10.15788/m2ww8w

World Health Organization, 2011. http://www.who.int/water_sanitation_health/publications/2011/tn9_how_much_water_en.pdf

Wood, T., and Weight, W.D., 2015. *North Plains Groundwater Conservation District. Stratigraphic Visualization Model*. Final report prepared for the North Plains Groundwater Conservation District, Dumas TX, commissioned by Aquaveo Water Modeling Solutions, Provo UT, 136 pp.

CHAPTER 2

The Geology of Hydrogeology

Before going into the field or performing field work, the hydrogeologist needs to have a general understanding and knowledge of subsurface geologic conditions. Engineers who work with geologic information may have forgotten some geologic terms or can't remember what they mean. This chapter is a reference on geologic topics and how they are specifically applied to hydrogeology.

Why is it important to understand the geologic setting? Perhaps the following questions will help. Is the area structurally complex (are the aquifers folded or faulted)? Are the rocks metamorphic, igneous, or sedimentary? What geologic mapping has been published or conducted in the area? Are there structural lineaments or other bedrock controls to groundwater flow in the area? Are the formations flat lying or tilted? What surface-water sources in the area pass over formational outcrops or **subcrops** (i.e., geologic formations exposed at the surface beneath alluvial or colluvial deposits that may be locations of recharge or discharge)? What are the topographic conditions like? Is the area flat lying, rolling hills, karst (Chapter 12), or mountainous? What is the climate like? Is it arid or humid? (This, for example, will make a big difference on evaluating recharge to the aquifer system.) What is the accessibility and location of the property (Chapters 1 and 4)? A few hours spent gathering this information will help in developing an appropriate conceptual model and assist in preventing mistakes one might make during field interpretations. Not understanding the general geology or field conditions before going into the field can be disastrous.

For any given project or field investigation, the hydrogeologist should put together a geologic model or conceptual geologic model that provides a physical framework of the groundwater-flow system. The geologic model identifies which units may dominate flow and which units or features may inhibit groundwater flow (including boundary conditions). The geologic stratigraphic units are grouped or divided into hydrostratigraphic units, or in other words, aquifers and confining units (presented in Chapter 3). Sedimentary rocks are not the only source of productive aquifers; metamorphic and igneous rocks may have zones within them that are productive or have significant localized fracture zones. It should also be noted that locations of significant water production may also be areas of concern for potential groundwater contamination, as they may become vulnerable (Baldwin 1997). The question might be asked, what is the vulnerability of this water supply source to contamination? Conceptual geologic models aid in predicting the direction of groundwater flow, its occurrence, its interaction with surface water (Chapter 10), and vulnerability. This chapter discusses the geologic properties of the three main rock types: igneous, sedimentary, and meta-morphic, and then provides a basic presentation of structural geology to help the reader think about components of a geologic model that may influence the direction of ground-water flow.

2.1   Geologic Properties of Igneous Rocks

Igneous (or fire origin) rocks are rocks that have formed from the cooling of melted geologic materials. They are an important source of water in some regions. Igneous rocks can be divided into two major categories: extrusive and intrusive. For the most part, **extrusive** igneous rocks (those that have erupted and formed on the land surface) have a greater capacity for water transmission and storage than do **intrusive** igneous rocks (those that formed and cooled beneath the earth's surface). Extrusive and intru-sive igneous rocks can generally be distinguished by their texture and mineral compo-sition. Extrusive igneous rocks will more commonly be finer grained and have fewer distinguishable (visible) minerals, because of their relatively rapid formation from cool-ing than intrusive igneous rocks. Both will vary in color and textural appearance according to mineral content, where those that are lighter in color are known as **felsic** (from a higher content of quartz and feldspar) rocks. **Mafic** rocks tend to be richer in iron and magnesium (ferromagnesian) mineral content (such as biotite, amphibole, and pyroxene), which tends to make these rocks darker in color.

For example, two common igneous rocks, rhyolite and granite, both tend to be lighter colored in appearance due to a high content of the light-colored mineral quartz and potassium feldspar. They have the same mineral composition; however, rhyolite will be a fine-grained textured rock from rapid cooling (typically resulting in mineral crystals too small to see with the naked eye). Granite, on the other hand, will be coarse grained through the process of slow cooling, and composed of larger, and visibly distin-guishable quartz, feldspar, and mica crystals commonly used as decorative rock. The naming of igneous rocks is built upon a classification scheme based upon mineral con-tent and whether they are intrusive (coarse grained) or extrusive (fine grained) (Streckeisen 1976; Figure 2.1). Figure 2.1 shows two ternary diagrams, with the upper half representing intrusive coarse-grained igneous rocks and the lower diagram extru-sive fine-grained igneous rocks. Major proportions of mineral content of the three end members are quartz, alkali feldspar (potassium or K-spar), and plagioclase (Na and Ca feldspar) are supplemented with increasing (right side of the diagram) or decreasing (left side of the diagram) mafic mineral content. The petrology of igneous rocks is beyond the scope of this chapter (Winter 2009), but the rock names shown in Figure 2.1 may help the reader with mineral context when reading geologic reports. (Additional discussion of the minerals in igneous rocks is found in the next section and in Section 2.7 under "Weathering").

2.2   Processes of Igneous Rock Formation

As discussed above, the classification of igneous rocks is based on the chemical compo-sition and the texture or size of the minerals present in the rock. The origin of magma (what melted to create it) determines the mineral *composition* of the rock, and the loca-tion (extrusive or intrusive) or where magma cools will determine the crystal size or *texture* of the minerals present. Those two properties (composition and texture) reflect

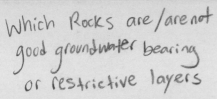

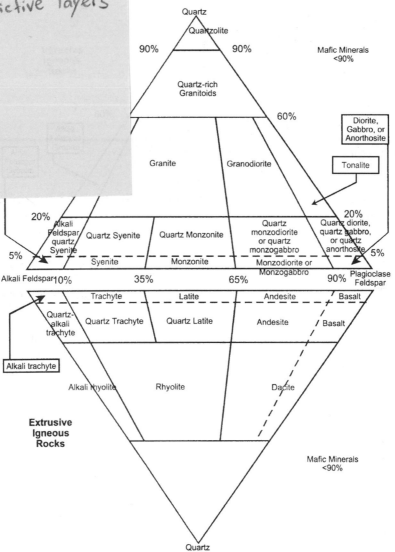

FIGURE 2.1 Summary classification scheme for the various igneous rocks. The upper ternary diagram corresponds to intrusive igneous rocks and the lower ternary diagram corresponds to extrusive igneous rocks. Note how the compositions of granite and rhyolite are the same. [*Modified from Streckeisen (1976).*]

the processes that led to the formation of the rock and can provide important geologic information about the tectonic setting in which it formed. Thus, the name of an igneous rock contains information about the tectonic processes that created it.

The crust and rigid part of the upper mantle (approximately 100 km thick) form the **lithosphere** that makes up the tectonic plates. Tectonic plates are made of two types of crustal material: oceanic crust (thinner, more flexible, dense, and darker in color with more mafic minerals) and continental crust (thicker, more rigid, less dense, and containing lighter-colored felsic minerals). Tectonic plates may contain both oceanic and continental crust. Beneath the lithosphere is a softer "plastic zone," also part of the upper mantle, known as the **asthenosphere** (between 100 and 700 km). Convection and heat flow in the asthenosphere cause the crustal plates to move around and relative to each other in a variety of ways. Active plate boundaries may be divergent (pull apart), convergent (move toward each other), and some move past each other as transform boundaries. When plates diverge, or pull apart, magma from the upper mantle moves up to the surface and may flow out as lava. When oceanic lithospheric crust converges with lighter continental crust on an adjacent plate, it bends and moves beneath the continental plate in a process known as **subduction**. All of these tectonic processes lead to the formation of magma. Molten magma derived dominantly from the mantle is more mafic in composition (more iron magnesium, with less silica). Magma that has formed through melting of continental crustal material will have more silica mixed in and give rise to more felsic minerals. A schematic diagram of igneous processes is shown in Figure 2.2 (Ridley 2010).

Because extrusive igneous rocks generally have more significant water-bearing zones, these will be discussed first.

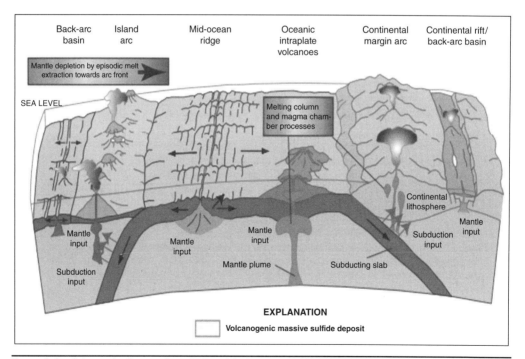

FIGURE 2.2   Schematic diagram from Ridley (2010) showing the principal components and processes involved in the production of island-arc and back-arc volcanics that are major lithostratigraphic units associated with volcanogenic massive sulfide deposits. Magma or lava rises to the surface at extensional locations and at convergent margins where subduction and partial melting occurs. Magma chambers that do not reach the surface provide places where intrusive igneous rocks form.

Extrusive Rocks

Extrusive rocks are also known as volcanic rocks. They have erupted to the surface through either volcanoes or extensional fissures. The major volcanic mountain ranges around the world form near the margins of plate boundaries. The Cascade Range in Washington and Oregon and the Andes Mountains of South America are forming where crustal plates are colliding. The mineralogically more mafic and denser oceanic crust is subducting (passing) under the lighter continental crust. At depths of 60 mi (100 km), temperatures and pressures are high enough to cause partial melting and mixing of crustal materials, forming molten rock under the surface known as **magma**. Magma is more buoyant than the surrounding solid rock and seeks a pathway to the surface where it emerges as **lava**. As magma rises, volatile gases within the fluid become less constricted and rapidly expand, resulting in an explosive eruption at the surface (Figures 2.3 and 2.4). Sometimes the volatile gases fracture the overlying rocks pneumatically into a **breccia pipe**. Magma chambers that do not reach the surface are places where intrusive igneous rocks form.

Other locations where igneous rocks commonly form are places where the crust is extending or pulling apart, known as **rifting**. Igneous activity has occurred, for example, in the Basin and Range province of the United States and in eastern Africa as a result of these forces. Extensional areas allow easier routes for deep-sourced magmas to rise to the surface. Most of these magmas are of basaltic composition. Sometimes rifting is initiated in a region and then suddenly ceases from changes in plate-movement

FIGURE 2.3   Mount St. Helens eruption, May 18, 1980. (*David Frank, USGS.*)

Figure 2.4 Weight (author) standing on angular pyroclastic rocks from Pacaya volcano eruption, near Guatemala City, Guatemala. Taken in 2010.

direction or heat-flow patterns in the asthenosphere. A schematic diagram of igneous processes is shown in Figure 2.2 (Ridley 2010).

Figure 2.2 (Ridley 2010) shows where magma or lava rises to the surface at extensional settings (arrows moving away from each other) in oceanic and continental crust and at convergent margins where subduction and partial melting occurs (mountain ranges and island arcs).

The most common extrusive igneous rocks are **basalt, andesite**, and **rhyolite**, listed in order of abundance and also increasing silica (SiO_2) content (Figure 2.1). Silica increases the viscosity of the igneous melt, much like cornstarch or flour thickens gravy. The greater the silica content, the more viscous or "sticky" the magma. Thicker magmas that are gas rich explode more violently, but do not flow very far. On the other hand, mafic low-silica magmas are less viscous and more "runny" allowing gas to escape easily rather than explosively. This often results in effusive lava flows as seen in Hawaii. In general, these effusive magmas are more likely to reach the surface making basalt the most common extrusive or volcanic rock to form.

Basalt

The most common extrusive igneous rock is basalt. It is the rock that constitutes oceanic crust and is common within the Pacific Rim in a series of island arcs known as the ring of fire (Figure 2.5). Basalts are common because they form by partial melting of the mantle and constitute the oceanic crust. In areas such as the mid-oceanic ridges and areas where the continental crust is extended, basaltic magmas rise quickly up their vents and flow onto the surface as a dark gray to black lava. These magmas are very

Figure 2.5 Stereo pair of steaming Redoubt Volcano, Chigmit Mountains, Alaska. The Chigmit Mountains are part of the ring of fire associated with subduction of the Pacific Plate at the Alaskan Aleutian Trench. (*Photos courtesy of Hugh Dresser.*)

fluid because of their low silica content (near 50%). Many basaltic lava flows, where they occur repetitively, are fairly thin, on the order of 15 to 40 ft (4.5 to 12.2 m) thick. At the top and bottom of these flows are scoria and **vesicular lava** zones of high porosity from burning vegetation and cooling processes (Figure 2.6). Vesicular lavas represent gas bubbles that expanded and were preserved as the lava cooled to a solid. Gas bubbles increase the porosity but tend to be poorly connected, and therefore do not increase permeability (Chapter 3). Some magmas literally expel fluids so quickly that they turn lava into a foamy froth that cools. This is **pumice**, which may have a porosity of 80%

Figure 2.6 Basalt cliff showing zones of dense basalt and more rubbly zones of gaseous vesicular basalt. Photo taken in southwestern Montana along the Big Hole River.

and will float on water because of the poor interstitial connections. The middle part of basalt flows can be quite dense (Figure 2.6). Some regions can have thicknesses of multiple lava flows that are in excess of thousands of feet (several hundreds of meters). Many of these are in areas where fissure eruptions have occurred. Fissure eruptions like those of the Deccan Plateau in India and the Oregon–Washington area may have resulted from meteorite impact (Alt and Hyndman 1995). Saturated, thinly layered basalt lava flows may result in some of the most prolific aquifers.

Example 2.1   In the eastern Snake River Plain, Idaho, is a very prolific aquifer known as the Eastern Snake River Plain Aquifer. As an undergraduate student, the author worked for Jack Barraclough of the U.S. Geological Survey, Water Resources Division, at the Idaho National Engineering Laboratory (INEL), while participating in several field geologic and hydrologic studies. [It is now known as the Idaho National Laboratory (INL)]. One project involved logging cores from deep drill holes (>1,000 ft, or 300 m). It was observed that basaltic lava flows ranged from 15 to 35 ft (4.5 to 10.7 m) thick. Most were clearly marked by a basal vitrophere or obsidian zones with some **scoria** (a high vesicular or bubbly zone) on the order of 1 to 2 ft (0.3 to 0.6 m) thick, above which was a dense basalt with occasional gas bubbles frozen into position as they had tried to rise to the surface (bubble trains). Each dense zone could be distinguished by its olivine content (a mineral characteristic from the mantle), which ranged from approximately 2% to 12%. The top was distinguished by another scoria zone. These layers extended from the surface down to the bottoms of the core holes, separated only by a few thin sedimentary layers, representing erosional hiatuses. It was obvious that there were abundant sources of connected permeability to allow free flow of groundwater.

  During the summer of 1979, a deep geothermal production well was attempted, known as INEL#1. Geologists thought they could identify the location of a caldera ring from surface mapping and geophysical efforts. A **caldera** is a collapsed magma camber characterized by ring faults and a crater rim more than 1 mi (1.6 km) across. The conceptual model was that at sufficient depth, fracturing and heat would allow the production of a significant geothermal production well. The proposed purpose of the well was for power generation and space heating. The drilling required a large oil rig with a 17-ft platform and 90-ft (27.4-m) drill pipe. Surface casing was set with a 36-in (91-cm) diameter and eventually "telescoped" down to an open hole 12¼ in (31.1 cm) in diameter at a total depth of 10,380 ft (3,164 m). The upper layered basalt flows are over 1,700 ft (518 m) thick at this location. These are underlain by dacitic and rhyodacitic rocks (Figure 2.1). Although the temperatures at depth were up to 325°F (149°C), there was poor permeability and water production.

    Some wells less than 150 m deep at the INL can produce in excess of 4,000 gallons per minute (gpm) (21,800 m³/day). High-capacity production wells start out with a pumping level approximately 1 ft (0.3 m) below static conditions that then recover back to static conditions (Jack Barraclough, personal communication, INL, July 1978)! Not all basaltic rocks can be thought of as potentially prolific aquifers. They can vary greatly in their ability to yield water. For example, Driscoll (1986) points out that incomplete rifting, about 1 billion years ago, in the central United States produced a massive belt of basalt extending from Kansas to the Lake Superior region. This massive belt contains relatively few fractures or vesicular zones, resulting in poor water production (only a few gpm, 10 to 20 m³/day). A wide range of hydrologic properties for basalts in Washington State has been reported by Freeze and Cherry (1979) in Table 2.1.

Andesite

Andesitic rocks are intermediate in silica composition (~60% silica) between rhyolite and basalt and are associated with the largest, most beautiful volcanoes in the world. Mount Fujiyama of Japan and Mount Rainier are examples (Figure 2.7). Andesitic rock typically forms where partial melting of oceanic and continental crust occurs together. The Andes Mountains represent a "type locality" for this kind of rock. Andesite is a rock with fine-grained matrix with other larger crystals sitting within them (Figure 2.8). Eruptions from andesitic magmas are usually violent, from explosively escaping volatile gases, followed

	Hydraulic Conductivity (cm/s)	Porosity (%)
Dense basalt	10^{-9}–10^{-7}	0.1–1
Vesicular basalt	10^{-7}–10^{-5}	5
Fractured basalt	10^{-7}–10^{-3}	10
Interlayered zones	10^{-6}–10^{-3}	20

TABLE 2.1   Range of Hydrologic Properties of Basalts in Washington State

FIGURE 2.7   Stereo pair of Mount Rainier, Washington, one of the volcanoes of the Cascade Mountain Range, part of a continental magmatic arc (see Figure 2.2). (*Photos courtesy of Hugh Dresser.*)

FIGURE 2.8   Photo of an andesite rock collected by the author from the Peruvian Andes Mountains. Note the large feldspar crystals sitting in a fine-grained matrix.

by lava flows. During the explosive part, ash and other ejecta spew out onto the flanks of the volcano and surrounding area. The ensuing lava flows cover parts of the ejecta, providing a protective blanket, or hydrogeologically form a semi-confining layer (Chapter 3). Subsequent eruptions result in a layering of ejecta and lava that builds up into large steep-sided volcanoes, hence the reason for the name composite volcano. Given the layered nature of andesitic terranes, there is a potential for water production among the layers.

> **Example 2.2**   Robert Schutt, a professional participant in the year 2000 Montana Tech hydrogeology field camp described drilling conditions near Guatemala City, where his company, Daho Pozos, drills water wells. A typical production well on the side of a volcano extends to a depth of approximately 800 to 900 ft (244 to 274 m) (Figure 2.9). A typical drilling sequence is as follows: from the surface to approximately 400 ft (122 m) is a dry unconsolidated zone of pyroclastics (originally hot angular rock materials), ash, and pumice followed by approximately 150 ft (45.7 m) of hard andesitic lava that serves to confine the next pyroclastic layer below. Once the hard andesite lava layer is penetrated, circulation is lost even when drilling with mud (Chapter 15). Drilling continues for an additional 250 ft (76.2 m) or so to provide sufficient water with yields up to 300 gpm (1,136 L/min).
>
>   Below is a well log and geophysical log for a typical production water well provided by Danny Lopez, a hydrogeologist with Daho Pozos (Figure 2.10). The well log indicates that the total well depth is 940 ft (286.5 m), the pump was set at 772 ft (235.3 m), and the static water level was at a depth of 621 ft (189.3 m) bgs. The achieved production after 24 hours of pumping was 215 gpm (814 L/min). A cement-grouting plug was placed between 380 and 400 ft (115.8 and 121.9 m) to seal off the lower production zone. A gravel pack was placed from 400 ft (121.9 m) to 940 ft (286.5 m). Below the well log is a photo of Danny Lopez standing near a live lava flow occurring on the Pacaya volcano near Guatemala City that erupted on August 19, 2006 (Figure 2.11).

Another source of water production can be associated with fractures from cooling. Liquid rock, as it cools, compacts to form polygonal-shaped columns, known as **columnar jointing** (Figure 2.12). Figure 2.12 was taken at Devil's Tower, Wyoming,

FIGURE 2.9   Ingersoll-Rand TH-75 drill rig drilling near the Hacienda Carmosa volcano near Guatemala City. (*Photo courtesy of Daho Pozos of Guatemala City, Guatemala.*)

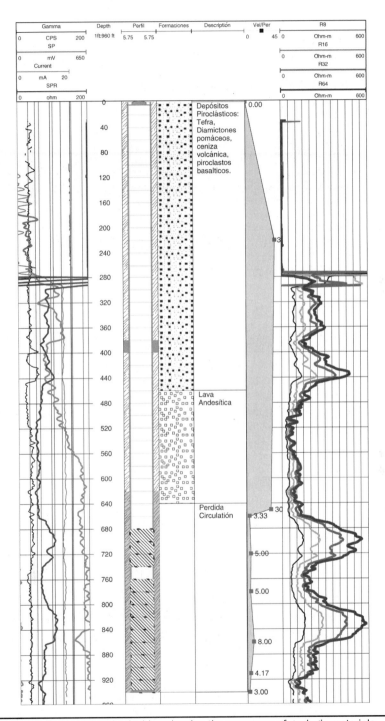

FIGURE 2.10 Well log and geophysical log showing the sequence of geologic materials (0 to 460 ft pyroclastics and tefra, 460 to 640 ft andesite lava, and 640 to 940 ft pyroclastic conglomerate—lost circulation). The pump depth was 772 ft and the static water level was at 621 ft. (*Courtesy of Daho Pozos, Inc., of Guatemala City, Guatemala.*)

FIGURE 2.11   Lava flow from an eruption from the Pacaya volcano near Guatemala City on August 2006. Arrow pointing to Danny Lopez standing near the vent. (*Courtesy of Daho Pozos, Inc., of Guatemala City, Guatemala.*)

(*a*)

FIGURE 2.12   (*a*) Columnar jointing at Devil's Tower, Wyoming, formed from cooling magma. Arrow pointing to rock climber. (*b*) Close-up of a single columnar joint, where columns are 3 to 4 m in diameter, also at Devil's Tower, Wyoming. Photo includes the Weight family from the 1980s for scale.

(b)

FIGURE 2.12   (Continued)

where the scale of the columnar jointing is especially spectacular (10 to 14 ft in diameter, 3 to 4 m). Spaces between the columnar jointing may provide significant permeability for groundwater flow (Chapter 3). The phenomenon is commonly observed in cooled lava flows of all compositions, but most notably with basalt.

Rhyolite

Rhyolite is the *least common* extrusive igneous rock. Compositionally, rhyolite is similar to granite (>70% silica), the *most common* intrusive igneous rock. Rhyolites generally form in areas of high heat flow within continental plates, for example, in Yellowstone National Park (a continental hot spot) or in New Zealand. Rhyolite eruptions are characteristically violent explosions followed by viscous lava flows that don't move far from the vent (Figure 2.13). Interstitial openings within rhyolitic tuffs (ashfall materials) and from openings between multiple layers may result in production wells in the tens of gallons per minute range (50 to several hundred cubic meters per day).

FIGURE 2.13   Stereo pair of rhyolite flow, north end of Mono Craters, California. Note how the flow did not move very far from the volcanic vent. (*Photos courtesy of Hugh Dresser.*)

Areas of high heat flow in geothermal settings may also cause the dissolution of minerals. As these waters rise to the surface, they cool and encounter cold-water recharge resulting in near-surface precipitation. This precipitation zone may create a confining layer that seals deeper aquifers from shallow aquifers or from interaction with surface water (Chapter 3).

Example 2.3   Near Gardiner, Montana, at the north end of Yellowstone National Park, the Yellowstone River flows northwestward into the Corwin Springs, a known geothermal area (Figure 2.14). In 1986, a production well on the west side of the Yellowstone River was drilled to a depth of 460 ft (140 m) and aquifer-tested in September of the same year (Sorey 1991). After pumping at a production rate of 400 gpm (25 L/s) for 13 h, La Duke Hot Springs on the east side of the Yellowstone River began to decrease in flow. This prompted a temporary moratorium on the drilling of production wells near Yellowstone Park (Custer et al. 1994). It is interesting that surface sealing from mineral precipitation

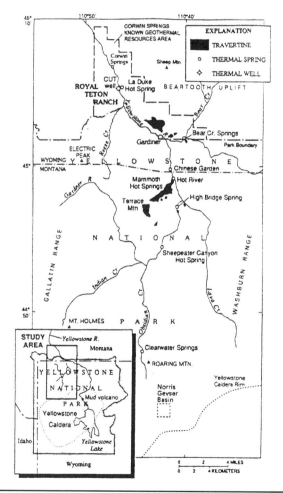

FIGURE 2.14   Corwin Springs, a known geothermal resource area near Yellowstone National Park from Example 2.1 where an aquifer test was conducted on Royal Teton Ranch that affected the LaDuke Hot Springs. (*Sorey 1991.*)

resulted in a separation of the shallow alluvial system of the Yellowstone River from the deeper rhyolitic rocks. A similar surface-sealing phenomenon occurs in the Rotorua area of New Zealand (Allis and Lumb 1992).

Intrusive Rocks

Intrusive igneous rocks are also known as plutonic rocks because they form large blob-like shapes underground that may have developed from partial melting of colliding plates, rifting, or melting above hot spots (Figure 2.2). They cool under the earth's surface. Characteristically, the texture of these rocks is a tight network of interlocking grains where minerals compete for space during the cooling process (Figure 2.15). Because cooling is slow and volatile gases such as water vapor are present, mineral growth is enhanced. The three most common intrusive igneous rocks whose extrusive counterparts have already been discussed (rhyolite, andesite, and basalt) are granite, diorite, and gabbro. Granite (Figure 2.16) is the most common intrusive rock because of its relatively high silica content (>70%). Its magma is so thick that it rarely reaches the surface as rhyolite. Diorite is a "salt and pepper" looking granite where approximately half of the minerals are white or light colored and the other half are dark or black (commonly biotite and amphibole). Gabbro is compositionally equivalent to basalt but has a coarse-grained texture. Since most intrusive rocks have similar hydrogeologic properties they will be discussed as a group.

Dormant magma chambers are given names according to their size. The largest of these cover a surface area greater than 60 mi^2 (100 km^2) and are known as **batholiths**. Smaller blob-like bodies are known as stocks. Sometimes molten igneous material intrudes (or is injected) into other older existing rocks that cut across them (discordant). Intrusive features that are discordant may do so at near vertical or subangled orientations from the horizontal, known as **dikes** (Figure 2.17). Concordant intrusive bodies (parallel to sedimentary layers) that are injected between layers are known as **sills**. Dikes and sills can be of any mineral composition. Figure 2.17a is a basalt dike that distinctively stands out more resistant to weathering than the material it intrudes into. Figure 2.17b shows where volatile gases escaped during the cooling process, along with the formation of columnar jointing (Figure 2.12). Each of these igneous features may influence groundwater-flow directions by obstructing flow paths or redirecting flow paths.

Figure 2.15   Interlocking crystals in weathering granite (hammer for scale).

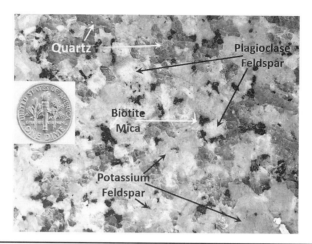

FIGURE 2.16 Close-up of granite showing K-feldspar, plagioclase feldspar, and quartz with arrows, US dime coin for scale.

(a)   (b)

FIGURE 2.17 (a) Columnar joining in a basalt dike more resistant to weathering and (b) close-up of same dike (see box area in 2.17a) showing where trapped volatile gases vented upward during cooling, Gros Ventre Range, Wyoming.

One of the hydrogeologic properties of rocks is porosity, where primary porosity is associated with the pore spaces at the time of formation (Chapter 3). The primary porosity of intrusive igneous rocks tends to be low, less than 1%, although granites with porosities greater than 1% are known (Fetter 1994). The ability of intrusive igneous rocks to produce water generally comes from pore spaces caused by secondary porosity where jointed fracturing and faulting have occurred. Contiguous joints and fractures enhance the permeability of a rock.

As molten magma chambers cool and are subject to the continued stresses involved in mountain building, the forces involved may produce fracture patterns that are oriented obliquely to the principal directions of stress. Fracturing tends to occur in a characteristic crossing pattern that becomes exposed to weathering that can be observed in the field (Figure 2.18). A list of porosity types and additional discussion of porosity and permeability are found in Chapter 3.

Minor drainages often develop in weaker fracture zones thus promoting recharge to granitic aquifers. Minor fractures can produce some water; however, larger sustained yields from production wells require more extensive fracture networks, such as fault zones. Large faults or fractures extend for distances of 1 mi (1.6 km) or more and are visible on aerial photographs. These longer fracture features are known as **lineaments** (Figure 2.19). Minor drainages that are controlled by these features tend to be abnormally straight, and thus can be recognized as potential target drilling areas. Another perspective of fracture zones being productive is that they are also the most vulnerable to surface contamination. This is important for wellhead or source-water protection issues (Chapter 1).

Example 2.4   In southwestern Montana, a large granitic body known as the Boulder Batholith, was once a magma chamber for a large volcanic system that has been stripped away by weathering, erosion, and uplift. Its associated mineralization is responsible for a significant amount of the colorful mining history of the Old West and current mining activity for precious metals. It extends from Butte to Helena, Montana, and is nicely exposed along Interstate 90 near Homestake Pass (Figure 2.20). This igneous body was formed from multiple intrusive events (Hamilton and Mayers 1974; Johnson et al. 2005) and is disrupted by numerous faults.

Figure 2.18   Jointed fracture patterns and spheroidal weathering in granite in the Laramie Range, Wyoming.

FIGURE 2.19 Stereo pair of the prominent Nez Perce lineament in southwestern Montana. (*Photos courtesy of Hugh Dresser.*)

A homeowner from Pipestone, Montana, on the east side of Homestake Pass called seeking the opinion of a hydrogeologist on where to drill a domestic well. The local state agency that keeps records of wells drilled in Montana was contacted to obtain information regarding any existing wells, their drilling depths, and production rates. After plotting this information on a topographic map, it was observed that the most productive wells are aligned with the significant canyons that have exceedingly straight lineament-like drainage patterns. The projection of these drainages extends westward into the Homestake Pass area. Production rates varied but yield from 8 to 20 gpm (43.6 to 109 m^3/day), while domestic wells that were located between drainages and away from or between major lineament patterns indicated production rates in the range of 2 to 3 gpm (10.9 to 16.4 m^3/day) or less. Static water levels in wells were approximately 30 to 70 ft (9.1 to 21.3 m) below ground surface.

This homeowner's property is located between the major lineaments, so prospects for a higher-producing well were not very good. The assumption was that away from major fracture zones, the smaller patterns were somewhat random. The hope was to drill a deeper well (400 to 500 ft, 122 to 152 m) and intersect sufficient minor fractures to yield a couple of gpm (10 m^3/day). This approach generated a well 450 ft (137 m) deep, producing 2 gpm (11 m^3/day). The static water was only 40 ft, so there was sufficient water from casing storage to yield significant quantities for a family.

FIGURE 2.20 Fracturing in granitic rocks near Homestake Pass, southwestern Montana. (*See also* Figure 2.21.)

Previously it was mentioned that intrusive igneous features such as dikes and sills can affect the flow patterns. Many magma chambers experience intermittent periods of activity, with multiple periods of intrusion. Older granitic bodies can be intruded by younger magmas of similar or differing composition (Johnson et al. 2005). Some fractures may be filled by younger intrusives, inhibiting groundwater flow. The Boulder Batholith in southwestern Montana is dominantly a granitic intrusion that is surrounded by as many as a dozen individually named peripheral intrusions of varying compositions (du Bray et al. 2012).

Example 2.5   Located at the southern end of the Butte, Montana, valley on the *west* side of Homestake Pass is a subdivision known as Terra Verde (Figure 2.21). Homeowners prefer to build at as high an elevation as possible for the best possible view of the nearby Highland Mountains (Figure 2.22). This results in having to drill deeper wells with a greater uncertainty of success, particularly if the locations they wish to place their homes on are not near any lineament features.

FIGURE 2.21   Location map of Terra Verde Heights subdivision (circled white) and Homestake Pass area, south of Butte, Montana.

FIGURE 2.22   Highland Mountains looking south, near Butte, Montana. The Highland Mountains exceed 10,000 ft (2,800 m).

In walking around the property from the prospective of a home builder, the author noticed that there were a series of large pine trees that grew in a straight pattern for at least 1/4 mi (400 m). It occurred to the author that this might represent a significant lineament even though it was not readily observable on an air photo. The author noticed (imagined?) a second such pattern that crossed the first on the prospective homeowner's property. The second observed or imagined lineament formed a subdued depression at the surface. Downslope there are several aplite (very fine-grained, light-colored) dikes. The logic was that perhaps these dikes resulted in inhibiting groundwater flow in the down-gradient direction from higher elevations, thus allowing greater storage capacity in the "up-gradient" direction. The recommendation was to drill in the lineament crossing pattern area a few hundred feet (several tens of meters) uphill from the dike's outcrop. This formed the reasoning for a best educated guess, presuming that the orientations of the intersecting fractures were nearly vertical. This illustrates the importance of understanding the fracture orientations (Section 2.8) and intersections.

Another recommendation made to the homeowner was that if no significant water was found after drilling 200 ft (60 m), then the probability of finding additional water-bearing fractures was even less likely (lithostatic pressures would tend to squeeze fractures closed). The well was drilled to a depth of 198 ft. At 185 ft, a significant fracture zone was encountered yielding 15 gpm (56.8 L/min).

This success story prompted another neighbor to call for advice. Unfortunately, the second property owner had already started the expensive process of blasting and constructing a basement for his home, located approximately 1/4 mi (400 m) downslope and to the east of his neighbor. The foundation work significantly limited the range of area available to recommend as drilling locations, because in Montana it is necessary to dig a trench from the wellhead to the basement to keep water lines from freezing. After performing a field investigation, surface lineaments were not observed and no other distinguishing features were observed. There was not much to go on. A recommendation was provided with a low but unknown probability for success. The hole was drilled 270 ft, and although the formation was getting softer, the homeowner told the driller to stop. Instead, a water witch was hired to locate the next location, where a 5-gpm (27-m³/day) well was drilled at a depth of 106 ft (32.2 m). This of course propagates wild ideas about the location and availability of water.

More about water witching is found in Section 15.7 of Chapter 15.

2.3   Metamorphic Rocks and Their Hydrogeologic Properties

Metamorphic rocks are those that have "changed form" (have become more compacted) through recrystallization from exposure to higher temperatures, pressures, and mineral-rich reactive fluids *without* melting. The original rock (**prototype**) may be igneous, sedimentary, or metamorphic. The elevated temperatures and pressures may occur during a mountain-building event or from being close to an intrusive igneous magma chamber (Figure 2.2). It is important to realize that this process occurs without melting, although fluids may be present to aid in the recrystallization process (**metasomatism**).

2.4 Processes of Metamorphic Rock Formation

There are two general groups of metamorphic rocks: **regional** and **contact** metamorphic rocks. The most common are regional metamorphic rocks (Figure 2.23). These formed from convergent plate-tectonic movement and tend to occupy large regional areas (e.g., Black Hills of South Dakota). Within convergent zones are also rising plutonic magma chambers (Figure 2.2). Magma being less dense than the solid country rock buoyantly seeks to rise to the surface. The "country rock" next to these magma chambers is subject to high temperatures and hot fluids, forming zones of contact metamorphic rocks. The temperature and pressure conditions and available fluids determine which minerals form, including ore minerals and explains why the country rock in some mining districts have such an "altered and baked" look. Fracturing of the overlying rocks from rising plumes may result in the injection of mineral-bearing fluids that harden into dikes and sills. Modern mining districts are often connected to some igneous source that was active in the geologic past.

Rocks away from magma chambers that are involved in plate convergence are subject to tremendous stresses. The pressure conditions are usually measured in the thousands of atmospheres range (1 atm = 101,325 Pascals), with units of **kilobars**. (One bar equals 10^5 Pascals; 1 kilobar equals 10^8 Pascals.) Temperatures are greater than 200°C. A general classification scheme for metamorphic rocks is shown in Figure 2.23. Regional metamorphic rocks are distinguished by the minerals present, which form in a characteristic temperature and pressure environment and therefore given a **facies** name. For example, lower temperature and pressure conditions are characteristic of the green-schist facies. The name comes from the alteration of basalts to chlorite, with its typical greenish color.

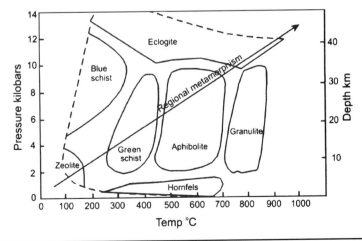

FIGURE 2.23 Classification of foliated metamorphic facies, showing path of regional metamorphism. Vertical axis is in kilobars of pressure and depth in kilometers, and the horizontal axis is in temperature °C. [*Modified from Turner (1968).*]

2.5   Foliated Metamorphic Rocks

The classification of metamorphic rocks is based upon specific mineral assemblages as these correlate with defined temperature and pressure conditions (Turner 1968; Winter 2009). The textural appearance of a metamorphic rock will also be described as being foliated or non-foliated. Rocks that are involved in colliding tectonic plates tend to have a foliated texture (Figure 2.24), whereas rocks forming near hot magma chambers (metamorphic changes mostly from heat) tend to be non-foliated.

Foliation occurs where platy minerals crystalize perpendicular to the stresses being applied (horizontal forces during collisional mountain building; Figure 2.25). Rocks that are deeply buried will metamorphose even though there may not be any relative tectonic movement (vertical downward forces from loading). Therefore, rocks relatively near to the earth's surface (2 mi, or 3 km) may indicate temperature and pressure conditions comparable to deep crustal rocks because they were involved in plate-tectonic collisions (Figure 2.2). Foliation is an important property in controlling the direction of groundwater flow and in identifying potential production zones. Having an understanding of the regional geology is critical to understanding the potential for water development in a local area. Measuring the orientation of foliation structures is discussed in Section 2.8.

FIGURE 2.24   Foliated rocks in Hoback Canyon, Wyoming. Professor Art Snoke on left with students on right.

FIGURE 2.25 An originally horizontal shale has been recrystallized perpendicular to tectonic forces during mountain building and converted into slate (the French Slate). Photo taken in the Snowy Range, southeastern Wyoming.

Example 2.6 Suppose that a shale, the most common sedimentary rock, is subject to tectonic stresses (Figure 2.26). As the conditions of temperature and pressure increase, the clay minerals will recrystallize perpendicular to the applied stresses forming a slate (Figure 2.27a). Slates are fine grained and have poor primary water-yielding capacities. However, if fractured, slates can yield sufficient water for most domestic purposes (Chapter 1).

If the process of increasing temperatures and pressures continues, the slate will first transform into a phyllite and then a schist (Figure 2.27b). Essentially, the clay minerals of the original shale will grow into sheet silicates of the mica group, resulting in a fabric known as **foliation**. Foliated fabric results in an anisotropy for fluid flow (Chapter 3). Flow parallel to foliation may be orders of magnitude greater than flow perpendicular to foliation. Continued increases in temperatures and pressures result in mineral separation into light and dark bands, forming a gneiss (Figure 2.27c).

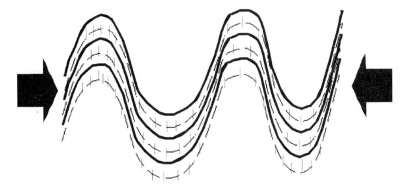

FIGURE 2.26 Schematic illustrating the reorientation of clay-mineral growth when a shale is subjected to horizontal stresses. (Heavy and dashed folded lines represent original bedding and clay orientation prior to applied stresses. Vertical markings represent foliation orientation of recrystallized minerals.)

(a)

(b)

FIGURE **2.27** Examples of foliated metamorphic rocks. (a) Example of slate (lowest temperature and pressure). (b) Muscovite-garnet schist—intermediate temperatures and pressures—taken in the Black Hills, South Dakota. (c) Example of gneiss where minerals have separated into light and dark bands. This example taken in Utah has been refolded from another tectonic event.

(c)

FIGURE 2.27   *(Continued)*

2.6   Non-Foliated Metamorphic Rocks

Non-foliated metamorphic rocks have typically been recrystallized from their original state from heat and pressure into a rock without foliated texture. All originated from a prototype of any rock type but examples will be given for the protoliths limestone and sandstone. Limestone will become changed into a marble and sandstone will become changed into a quartzite. In the case of limestone, it will not convert into a platy mineral no matter how much heat and pressure are applied; therefore, one cannot determine the original temperature and pressure conditions at the time of formation (geomaps.wr.usgs.gov, 2017). Many non-foliated metamorphic rocks occur from **contact metamorphism**. This is where pre-existing rock becomes baked and altered (changed) as a result of the intrusion of hot igneous rock without a change in pressure conditions (geomaps.wr.usgs.gov, 2017). It is possible that a protolith can have the characteristics of both foliated and non-foliated texture, but this condition is rare. For example, a conglomerate (lithified gravel) prototype can have the individual gravel clasts (next section) become metamorphized into quartzite and become stretched into a foliated fabric (Figure 2.28).

In the conditions described in Example 2.6 above, there is a tremendous competition for space during mineral growth, and the primary yielding capacity for metamorphic rocks tends to be very low. For example, Freeze and Cherry (1979) report intrinsic permeabilities in the range of 10^{-9} to 10^{-11} cm/s for metasediments of the Marquette Mining district in Michigan. A characteristic feature of metamorphic rocks and intrusive igneous rocks is that the hydrologic properties of porosity, permeability, and well yield decrease with depth (Davis and Turk 1964) (Figure 2.29).

FIGURE 2.28 Example of a gravel protolith where the individual gravel clasts were metamorphized into quartzite and then elongated and stretched in a tectonic shear zone. Photo taken in Idaho.

Note how well the physical properties of both rocks types in Figure 2.29 track each other in terms of changes with depth. This trend is believed by the author to be true of most rocks. Generally, igneous and metamorphic rocks are not known as big water producers without secondary porosity and permeability being created by faulting and fracturing.

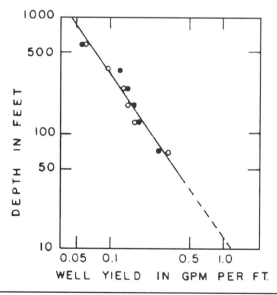

FIGURE 2.29 Yields of wells in crystalline rock of the eastern United States. Open circles represent mean yields of granitic rock based on a total record of 814 wells; black dots represent mean yields of schist based on a total record of 1,522 wells. [*From Davis and Turk (1964). Reprinted with permission of Groundwater, 1964.*]

2.7   Sedimentary Rocks and Their Hydrogeologic Properties

The most common water-bearing materials that produce potable water are sedimentary rocks. These can be consolidated or unconsolidated. Sedimentary rocks, by nature, tend to have high primary porosity and, depending on the depositional environment and particle size, they may have very high hydraulic conductivities. Sedimentary rocks are classified according to grain size and texture. The textural classes are **clastic**, or from fragments, and chemical, or formed from precipitation (chemical sedimentary rocks). Grain sizes are divided into gravel, sand, and mud according to a Wentworth-like classification scheme shown in Table 2.2 (Wentworth 1922; Folk 1966). Class sizes are determined by passing a sample through various sieve sizes (Chapter 15). The smallest size fraction is caught by the "pan" at the bottom, which is the mud portion. Mud includes all silt and clay-sized particles. The mud fraction is usually analyzed by a pipette or hydrometer method (Bouyoucos 1962).

Clastic sedimentary rocks are first formed by the breaking down of parent rock materials at the outcrop into fragments, called **clasts**, through the process of weathering. These clasts become transported by various media (wind, water, ice, or gravity). Later, the transported clasts come to rest as sediments in a depositional environment. The transport media impose a texture in the structure of the sediments that is observable. At this point it is important to realize that sediments are unconsolidated and not rock. Engineers refer to these materials as soils. Once the sediments become buried and subjected to increased temperature and pressures, they become lithified and converted into sedimentary rocks (a process known as diagenesis).

Class	Other Names	Particle Size (mm)	Particle Size (nm)	U.S. Sieve Size
Extremely coarse gravel	Boulders	>256		Wire mesh
Very coarse gravel	Cobbles	64–256		Wire mesh
Coarse gravel	Pebbles	16–64		Wire mesh
Medium gravel	Pebbles	8–16		Wire mesh
Fine gravel	Pea gravel	4–8		Wire mesh
Very fine gravel	Granules	2–4		10–5
Very coarse sand		1–2		18–10
Coarse sand		(1/2) 0.5–1.0	500	35–18
Medium sand		(1/4) 0.25–0.5	250	60–35
Fine sand		(1/8) 0.125–0.25	125	60–120
Very fine sand		(1/16) 0.0625–0.125	63	230–120
Very fine to coarse silt		(1/256) 0.0039–0.0625	3.9	<230
Clay		<0.0039	0.06–3.9	

TABLE 2.2   Sediment Classification Based upon Grain Size, after Wentworth (1922). Parenthesis values in the center column reflect minimum size in mm.

Weathering

Clastic sediment particles result from the weathering of igneous, metamorphic, and sedimentary rocks. The ease of weathering depends primarily on climatic conditions and rock type. Climates that are warm and moist produce the highest weathering rates (Figure 2.30). The composition of minerals in rocks is also a big factor. Rocks that crystalize at high temperatures and pressures tend to weather more quickly than minerals that form at lower temperatures and pressures. An example of this is known as Goldich's (1938) weathering series and is illustrated in Figure 2.31.

Figure 2.31 is essentially the inverse of Bowen's reaction series (1928). Bowen (1928) performed a series of laboratory experiments on igneous magmas to learn which silicate minerals form first from a molten state. For example, at high temperatures near 1,400°C to 1,500°C, olivine forms. As temperatures drop, the olivine is resorbed and pyroxene forms. This process continues with formation and resorption until the formation of quartz at about 600°C. The feldspar side of the diagram is a series of formation/resorption plagioclase minerals from highest calcium content to highest sodium content (anorthite, bytownite, labradorite, andesine, oligoclase, and albite), where albite has the highest sodium content (http://www.galleries.com/Feldspar_Group). The arrows between the calcium (Ca) and sodium (Na) plagioclase feldspars indicate a continuous solid-solution series because the ionic radii of these two are similar in size and can readily exchange in the crystal lattice. Potassium (K) feldspar (orthoclase or microcline) is distinguished from plagioclase because potassium's ionic radius is much larger than sodium or calcium. This makes these two minerals immiscible in the molten state.

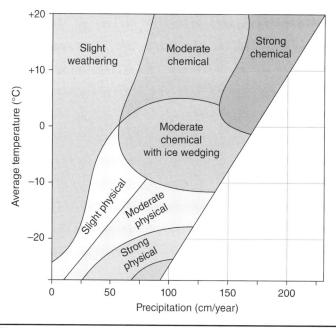

FIGURE 2.30 Diagram showing relationships between mean annual air temperature and precipitation, and primary mechanisms and intensities of chemical or physical weathering. [*Modified from Peltier (1950).*]

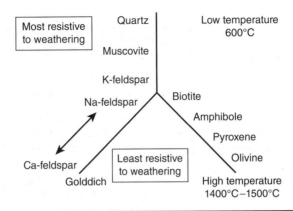

Figure 2.31 Goldich (1938) weathering sequence. (Illustrates which minerals are more or less resistant to weathering. Quartz is the most resistant and olivine and calcium plagioclase are most easily weathered.) Left side of the "Y" is a continuous series of plagioclase minerals and the right side of the "Y" is a discontinuous series of minerals that form according to temperature conditions. The inverse of this diagram is known as Bowen's (1928) reaction series.

The types of rocks that cool from a melt depend greatly on the original composition. Rocks that originate from the mantle form basaltic magmas, those that partially melt from a mixture of oceanic and continental crust are typically andesitic, and granitic rocks form from molten continental crust (near 600°C; Figure 2.2). Many metamorphic and sedimentary rocks are composed of the above mineral groups. Goldich (1938) determined that the reverse of Bowen's (1928) reaction series represented which minerals in rocks weathered the quickest. The only other minerals that weather more quickly than olivine and anorthite are those that are associated with chemical sedimentary rocks (carbonates and evaporites).

The most abundant minerals in the earth's crust, feldspars, weather into clay minerals. Clay minerals (like quartz and iron oxide minerals) are very stable when exposed to the atmosphere and form the basis of the most common sedimentary rock, **shale**. The other most common mineral in sedimentary rocks is quartz, a principal component in sandstone.

Weathering has two general categories: mechanical and chemical (Table 2.3). Mechanical weathering includes those processes that act to break down the larger rocks into smaller rocks without changing their physical properties. Another name could be disintegration. For example, a chunk of granite can be broken into smaller pieces during the process of frost wedging, where water enters a crack during the day and freezes during the night. Because water in a solid state occupies about 9% more space than in

Mechanical Weathering	Chemical Weathering
Frost wedging	Hydrolysis
Exfoliation	Oxidation
Heating and cooling	Dissolution
Plants and animals	

Table 2.3 Examples of Mechanical and Chemical Weathering

FIGURE 2.32   Talus slope of angular andesite rocks at the base of a cliff near the east entrance to Yellowstone Park, Wyoming.

the liquid state (water.usgs.gov, 2018), it will wedge or push the rock apart. Then gravity pulls the fragments downslope. This is why **talus** slopes form at the bottom of cliffs (Figure 2.32). When granitic rocks, which formed at higher pressures at depth, become exposed to lower pressures at the surface they tend to spall off in onion-skin-like layers (**exfoliation**) (Figure 2.33). Another reason for the onion-skin-like layers in granite is

FIGURE 2.33   Onion-skin-like layers of exfoliation in granite near Sentinel Dome in Yosemite Park, California.

from the exposure to temperature extremes at the outer surface of the rock. Temperatures of exposure to the outer edges of the rock may be high in the day and cool at night, while interior to the rock the temperature may be relatively constant, thus creating an opportunity for a zone of weakness.

Chemical weathering includes processes that change the chemical lattice structure and composition of minerals to reach equilibrium with surface conditions. Another name could be decomposition. Hydrolysis and oxidation are examples. Chemical weathering of rocks through exposure to water and carbon dioxide act to generates carbonic acid (Equation 2.1), which aids in the conversion of feldspars into clays:

$$H_2O + CO_2 \rightarrow H_2CO_3 \text{ (Carbonic Acid)} \tag{2.1}$$

Angular rocks are especially prone to weathering processes because of increased surface area. The result is the rounding of edges into a more spherical shape known as **spheroidal weathering** (Figure 2.18). Mechanical weathering processes break larger rocks into smaller ones, which produce geologic materials (clasts) with increased surface exposure to chemical weathering. The process is also important for enhancing the permeability of geologic materials and providing additional pathways of recharge into aquifer systems. Additionally, dissolution is important in adding dissolved minerals to groundwater and contributing to karstic conditions in carbonate rocks (Chapter 12). Smaller clastic fragments also become more available for sediment transport.

Transport of Sediment and Depositional Environments

Once sediments have been broken down and decomposed by the weathering process, they can be transported by wind, water, ice, or gravity into a depositional environment. Later, through **diagenesis**, unconsolidated sediments are compacted and cemented into sedimentary rock. Sedimentary rocks contain sedimentary structures that reveal the processes and mechanisms of transport. For example, ripples and certain cross-bedding types indicate **fluvial** or stream processes while large cross bedding up to 15 to 30 ft (4.6 to 9.1 m) high between bedding planes likely indicates **eolian** or desert wind conditions (Figure 2.34). As another example, glacial deposits transported by ice tend to be a mixture of large and small particles dumped together in no particular order (Figure 2.35). Having an understanding of depositional environments is necessary to produce a three-dimensional picture of the sediment distribution, and therefore a three-dimensional picture of the distribution of the hydraulic properties (Example 2.7). A conceptual groundwater model that incorporates interpreted cross sections, monitoring-well level data, and aquifer-test data will greatly assist in understanding groundwater flow.

Example 2.7   The author has been involved with numerical modeling of groundwater systems for many years. A proposal was made for a new water-bottling plant sourced from the deep Kalispell Aquifer in Montana. A group of concerned residents wanted to have a groundwater model built to show whether there would be impacts to their water rights associated with the shallow aquifer system. The area is geologically very complex with a mixture of depositional environments (glacial, fluvial, and lacustrine; Table 2.4). The author used over 220 published well logs to generate over 500 cross sections as a framework to create a multilayered numerical groundwater flow model (Weight 2017; Figures 2.36a–c). The model did demonstrate a connection as the confining unit separating the lower aquifer system from the upper aquifer system pinches out to the east toward the Swan Mountain Range (Figures 2.36a, b). There are several graphical user interface (GUI) groundwater-modeling packages on the market that contain powerful tools to visualize geological and numerical groundwater models.

FIGURE 2.34 Large cross bedding typical of an eolian environment. Taken in Zion National Park, Utah.

Sediments that have been transported by water and wind tend to be sorted and stratified. **Sorting** is a measure of the distribution of grain sizes. Sediments with a narrow range of grain sizes (all similar sized) are said to be well sorted. An example of a depositional environment where this occurs is a beach sand. Here, the wave action winnows the smaller grain sizes out to sea, leaving coarser sediments along the beach, in contrast to a well-sorted sedimentary unit is a glacial till where grain sizes ranging from boulders to clay-sized particles are mixed together (Figure 2.35). In this instance, the

FIGURE 2.35 Glacial till showing poorly sorted (well-graded) sediments near Leadville, Colorado.

Geologic Term	Environment	Comments
Alluvium	Streams, floodplains, or alluvial fans	Coarser-grained channel sediments surrounded by finer-grained sediments away from the channels, including silts and clays. Changes in lithology are commonly abrupt and laterally discontinuous.
Colluvium	Topographic slopes	Loose, incoherent coarse to fine-grained deposits collecting on slopes by gravity.
Drift	Glacial	Geologic materials deposited by ice or meltwater. Layered or stratified drift occurs from meltwater streams.
Eolian	Desert or pertaining to wind	Well-sorted fine to medium sands, with large cross bedding.
Fluvial	Streams, rivers, or stream action	Channels fine upward with fair to good sorting; a variety of cross bedding is visible.
Karst	Limestone dissolution	A topography formed by the dissolution of limestone, dolomite, or gypsum, creating caves, sinkholes, and underground drainage.
Lacustrine	Lake	Shales in thinly laminated beds.
	Beach	Well-sorted deposit that is longitudinally extensive along the oceanfront but laterally limited in the landward direction. If prograding, these can form sheet sand units.
Paludal	Marsh or swamp	Both are organically rich. Marsh sediments produce fibrous peat, and swamps produce woody peat, which can eventually become coal.
Pelagic	Ocean	Sediments originating in ocean water.
Playa	Ephemeral lake	The lowest part of an undrained basin receives intermittent water. Characterized by clay, silt, sand, and soluble salts.
Turbidite	Continental shelf, slope, or in a lake	Sediments well graded and laminated from moving downslope in a body of water.

Table 2.4 Listing of Geologic Terms, Depositional Environments, and General Characteristics

sediments are poorly sorted. Tills tend to become geologic units that are poor conductors of water.

Engineering literature uses the term **grading** instead of sorting. A well-graded unit is a soil mixture with a wide range of grain sizes, like the glacial till example above. The beach sand example would be described as being poorly graded (or not very many grain sizes represented). The terms are nearly opposite, so one must be careful of the descriptions and by whom. A geologist may be more comfortable using sorting terms, while an engineer may be more comfortable using graded sedimentary terms. Drill logs used for constructing cross sections may have been described by either or both professionals, so it is important to pay attention. It is recommended that a dictionary of geologic terms, such as that of the American Geologic Institute (Bates and Jackson 1984), be kept in the office library.

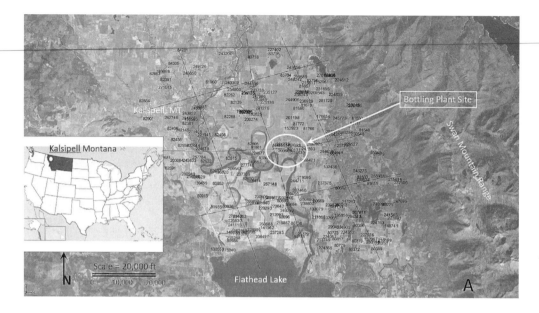

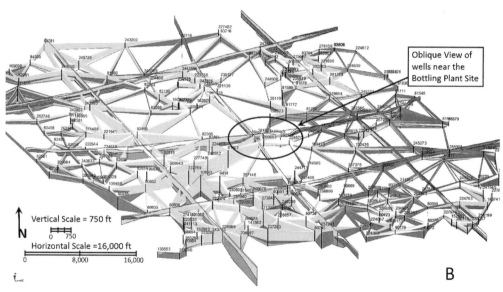

Figure 2.36 (a) Location map showing the location of the proposed water bottling plant, city of Kalispell, Montana; Flathead Lake, and other features near the controversial site. Black lines represent the locations of some cross sections and numbers represent the location names of published drill whole data.
(b) Distribution of over 500 cross sections constructed in the Groundwater Modeling Software (GMS, www.aquaveo.com) as a basis for a geologic model in numerical groundwater-flow model. The site of the bottling well and other nearby wells is indicated. The horizontal-to-vertical scale is about 4.5 to 1.0.
(c) Zoomed-in version of Figure 2.36b showing increased cross-section detail. The horizontal-to-vertical scale is about 18.5 to 1.0.

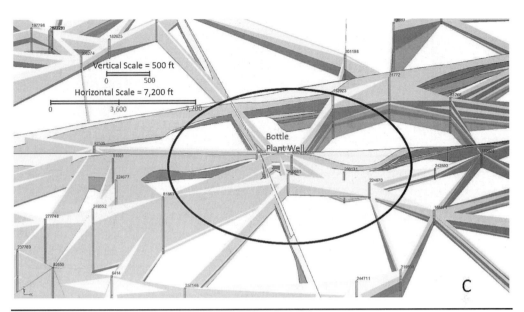

FIGURE 2.36   *(Continued)*

Stratigraphy

The hierarchy of names given to rocks is based upon the stratigraphic code. The breakdown of names can range from **group**, to **formation**, to **member**, to **bed** (Section 3.4; Figure 3.14). Groups represent a collection of formations. For example, the Colorado Group represents several marine shales that have been grouped together into one unit.

Formations are names given to mappable rock units that have formed in a similar depositional environment at a similar point in time (AAPG 2005). Generally, they consist of a certain lithologic type or combination of types (Bates and Jackson 1984). They are laterally extensive enough to be mapped and identified in the field. Formations are usually named from where they are best exposed and have been described in detail. For example, the Lahood Formation is a Precambrian marine fan deposit with turbidite sequences. This formation is best exposed near Lahood, Montana, where there are coarse boulders mixed with sandy turbidite beds. To the north, this formation grades into finer-grained, sandy turbidite beds (Table 2.4).

Formations are divided into members if there are distinctive characteristics that can be mapped or identified over significantly large lateral distances. For example, the Madison Limestone Member is divided into the Mission Canyon and Lodgepole members. Each member is distinguished by fossils and paleokarstic features. They are best exposed in the Little Rocky Mountains in northeastern Montana (Figure 2.37). More on how this nomenclature applies to hydrostratigraphic units is described in Section 3.4.

Further subdividing of formations from members to beds is done if they are particularly distinctive. For example, within the Fort Union Formation in eastern Montana and Wyoming is the Upper Tongue River, Lebo Shale, and Lower Tullock members (Brown 1993). Within the Tongue River Member are laterally extensive coal beds known as the Anderson, Dietz, and Monarch beds. These are important because they form regional aquifers (Figure 2.38). Having an understanding of the local stratigraphy and

FIGURE 2.37   Mission Canyon member of the Madison Limestone exposed in the Little Rocky mountains on the Fort Belknap Indian Reservation near Hayes, Montana.

rock formations is helpful in setting up the geologic framework for a groundwater-flow system.

Formation names often change at state boundaries, which can be confusing for regional studies. For example, the Madison Limestone is known as the Redwall Limestone in Arizona, where it weathers into a distinctive reddish wall exposed in the Grand Canyon.

FIGURE 2.38   Monarch coal bed from the Tongue River member of the Fort Union Formation taken at the Big Horn Coal Mine near Sheridan, Wyoming. Robert Weight, 6 ft 2 in, for scale.

Formation names are also given to distinct mappable units of igneous or metamorphic rocks. From the perspective of a hydrogeologist, formational units are grouped together based upon the hydraulic properties. Those with a similar enough hydraulic conductivity are combined together into one hydrostratigraphic unit (Section 3.4).

2.8   Structural Geology

Geologic formations may be folded, faulted, or tilted. These represent the response of geological materials to stresses. We live on a dynamic Earth. Plate-tectonic movements result in applied stresses that can cause geologic formations to be pulled apart, folded, or rumpled like carpet (Figure 2.39). These expressions of deformation are known as **strain**. During these processes, the physical properties of formations can change, affecting porosity and fluid permeability. The science of deciphering the physical orientation of rock relationships within an area is known as structural geology. The deformation and relative timing of tectonic stresses in large regional areas is the study of **tectonics**.

Disruptive changes in rock formations require well-developed observational skills. The surface expressions of geologic units are identified in the field, physically located (via GPS), and recorded on a map. Included are rock-type descriptions, observation of fossils that can be correlated with other locations, and the orientation of the strata relative to a flat plane. This is done by taking strikes and dips of the bedding planes in sedimentary rocks, foliation orientations in metamorphic rocks, or flow banding in igneous rocks. The locations (via GPS) and orientations of faults are also recorded. The structurally disturbed geological materials can control the direction of groundwater movement and therefore function as areas of higher or lower well yield. Not understanding the orientation of geological formations may result in drilling in the wrong location or missing a desired target.

Figure 2.39   Folded limestone beds of the Cretaceous Kootenai Formation in southwestern Montana. A reverse fault is visible near the right shoulder of the person in the photo.

Strike and Dip

A fundamental step in interpreting the orientation of formations in space is taking a strike and dip. The **strike** is the azimuth orientation of the intersection of a horizontal plane with any inclined plane or surface (Figure 2.40). This is usually taken with a geological compass. A bull's-eye bubble indicates when the compass is being held level. The intersection of two planes is a line and so the azimuth orientation of that line is read from the compass (Figure 2.41). The strike is indicated with a line drawn on the map in the azimuth orientation. In order for the orientation of the compass to match the orientation of a topographic map in the field, one must first correct for the declination of the earth's magnetic field. The declination is the horizontal angle between true north and magnetic north at a given location. This is always indicated at the lower left corner of a U.S. Geological Survey quadrangle topographic map.

The **dip** is the angle from a horizontal plane down to the inclined surface (Figures 2.40 and 2.42). The dip is read by holding the edge of the geological compass perpendicular to the strike line and moving a clinometer dial until its bubble is leveled and centered. Care must be taken to rotate the position of the compass back and forth slightly to obtain a maximum inclination for dip. Anything less than the maximum would be an apparent dip and not a true dip. The marking of the dip orientation is a short line drawn "teed" in the center of the strike line on the side of inclination with the dip angle written (Figure 2.40). For example, if a strike line is

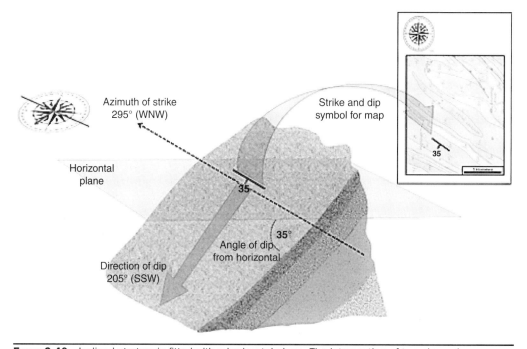

FIGURE 2.40   Inclined stratum is fitted with a horizontal plane. The intersection of two planes is a line—the strike. The angle of inclination from the horizontal surface to the surface of the strata is the dip. The strike and dip are marked on a geologic map as a long and a short perpendicular line with the dip angle indicated on the inclined side. (*Sourced from Patricia Heiser of Carroll College, Helena, Montana.*)

FIGURE 2.41 Strike reading taken with a Brunton compass. The compass is held flat with its edge against the field book. The bull's-eye bubble indicates when the compass is being held flat and level. The azimuth orientation is the strike reading.

FIGURE 2.42 Dip angle read from a clinometer on a Brunton compass. The compass is rotated perpendicular from the strike line (Figure 2.41) and a dial is moved to rotate the clinometer until the bubble in the level is centered. The maximum angle read is the dip angle.

drawn at 10° to 190° and the dip was 30° west, a line would be drawn on the map (say 1/2-in-long) and recorded in the field book with the GPS location, and the dip line (say 1/8-in-long) would be marked on the left side of the strike line with a 30° written on the left side.

A field geologist will take several strikes and dip readings around an area. In addition to strikes and dips, markings are also made of contacts between formations and faults that are observed. These are usually plotted on an aerial photo or on a topographic map. The different formations are given a distinctive pattern or color code (Geologic Time Scale, Section 2.9) and assembled into a geologic map. Geologic features that are covered (e.g., with colluvium or alluvium) or where faults are not observable but inferred or being projected are indicated with dashed lines.

The strike and dip of a formation can be estimated from a geologic map if the contacts are accurately drawn on a topographic map, and do not lie not in a straight line. This process is known as a **three-point problem**. Three-point problems can be solved with data that do not outcrop if they are from boreholes, mine shafts, or other subsurface information when the dip angle is uniform (Bennison 1990). The strike is determined by locating where the outcrop of a particular bed intersects the ground surface at the same elevation at two locations. A straightedge is used to draw a line connecting these two points. A perpendicular line is drawn to a third point where the outcrop elevation is known and the dip can be calculated or determined graphically. In Figure 2.43, the 2,200-ft contour is used to locate the strike line, and structure contours (projections of the bedding surface) parallel to the 2,200-ft contour extend down the planar surface of the bed. The contour spacing is determined from the 2,300-ft line (another outcrop of known elevation). Doing a three-point problem enables you to properly estimate how deep a target is when drilling, projecting inclined planes, or constructing cross sections.

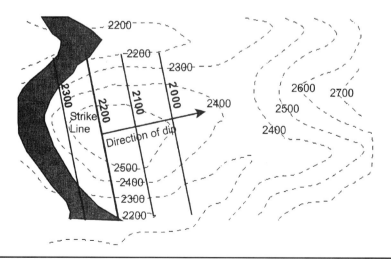

FIGURE 2.43 Illustration of a three-point problem used to determine the strike and dip from an outcrop and topographic map. The dashed lines are elevation contours. The thick black band is the outcrop of a sedimentary bed. The 2,200-ft structure contour line is oriented where the bed outcrop appears at two 2,200-ft elevation contours on either side of a hill. The same process is done for the 2,300-ft elevation contour to mark the spacing between structure contours. Additional structure contours are added in the direction of dip under the topographic surface.

Example 2.8   In 1997, summer students of the Montana Tech hydrogeology field camp were involved in logging a drill hole for the Meadow Village subdivision near the Big Sky Ski Resort south of Bozeman, Montana. The target zone was the basal Cretaceous Kootenai Formation at approximately 1,000 ft (300 m) below the land surface. It was hoped that sufficient permeabilities could produce a well in excess of 100 gpm (500 m^3/day). Since the geologic layers dip approximately coincident with the topographic slope, it was decided to see if the formations being drilled through would be exposed in the canyon walls near Ousel Falls in the south fork of the Gallatin River (Figure 2.44). The canyon cuts down the stratigraphic section. At the level of the river, the Cretaceous Thermopolis Shale, the unit just above the Kootenai Formation was exposed. It was helpful to compare the sedimentary layers exposed at the surface with the drill cuttings observed. A three-point problem was used to determine the strike and dip of the Thermopolis Shale from the elevation of two exposures in the canyon and the depth of intersection of this same unit at the drill hole. Unfortunately, tight cementation in the Kootenai Formation at depth limited the productivity of this well.

Fold Geometry

One form of strain is manifested when rocks subjected to stresses respond by folding. When strata become folded, there are some basic parts of the fold that define their geometry. A plane that bisects the structure is known as the axial plane (Figure 2.45a). On either side of the axial plane are the limbs. If the fold axis is oriented vertically, then the fold is symmetrical. If the fold axis is rotated, the fold is asymmetrical (Figure 2.45c). The orientation of the fold axis is the overall strike of the structure and the angle of the limbs relative to a horizontal surface represents the dip. Another feature associated with folds is the orientation of the fold trace. The fold trace is aligned at the top of the fold (Figure 2.45c). If the fold trace is inclined from the horizontal, then it is said to be plunging. The magnitude of plunge can be measured using a geological compass.

FIGURE 2.44   Cretaceous Blackleaf formation overlying the Cretaceous Thermopolis Shale exposed in the canyon wall of the south Fork of the Gallatin River near the Big Sky ski resort in southwestern Montana. (Example 2.8.)

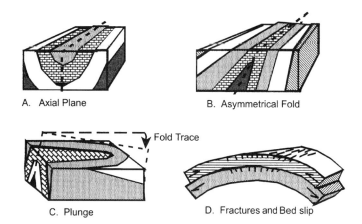

A. Axial Plane

B. Asymmetrical Fold

C. Plunge

D. Fractures and Bed slip

Figure 2.45 Fold geometry: (a) axial plane, (b) asymmetrical fold, (c) plunging fold, and (d) rotational fractures around axial plane and slip bed.

In Figure 2.45, two basic types of folds are illustrated. In Figure 2.45a, the fold is a syncline, and in Figures 2.45b and 2.45c anticlines are shown. In synclines the rocks in the middle are youngest, and the reverse is true of anticlines. Photographs of each are shown in Figures 2.46 and 2.47, respectively. The way to keep this straight is to try the exercise in Example 2.9.

Example 2.9   Take a paperback book or manual. If we use the convention of saying that page one is the oldest and that larger numbers represent younger beds, page one will be on the bottom. If the book is pushed together and the fold is anticlinal (arched), then page one will be to the inside, and the larger-numbered pages will be on the outside. This becomes more obvious if the book is cut horizontally along the crest of the fold (not recommended if you like this book). In folding the book into a syncline and cutting the book horizontally so that it has a flat surface, page one will now be to the outside. If the book is slightly rotated downward on one end so that it is plunging, the pattern in Figure 2.45c will be evident.

Faulting

Another form of strain is a brittle response where rocks rupture from the applied stresses. Rock masses that merely break into a pattern of fractures with no displacement are known as **joints** (Figure 2.18). When rock masses move relative to each other, faulting occurs. If a rock mass breaks into two blocks, then the plane separating the blocks is referred to as the fault plane. Typically, this plane is inclined. The block on top of the inclined fault plane is known as the **hanging wall**, and the lower block is known as the **footwall**. The term "hanging wall" originated from mining in mineralized areas where reverse faulting was common. The miner would stand on the footwall and hang his lantern on the hanging wall (Tingley and Pizzaro 2000).

Imagine a stack of sedimentary layers from A to I, with layer A being the oldest and on the bottom. A fault divides the layers into two rock masses. If the hanging wall moves vertically down relative to the footwall, the fault is said to be **normal** (Figure 2.48a). In this scenario, younger rocks still overlie older rocks (a normal relationship; layers D and C on the left are on top of layers A and B to the right). If the hanging wall has moved vertically up relative to the footwall, the fault is a **reverse**

FIGURE 2.46   Student standing in the axis of a synclinal fold in the Triassic Dinwoody Formation, southwestern Montana.

FIGURE 2.47   Stereo pair of the Big Sheep Mountain anticline, Bighorn Basin, Wyoming. (*Photos courtesy of Hugh Dresser.*)

fault (Figure 2.48b). Here, older rocks are overlying younger rocks (an "abby-normal" or reverse situation; layers C and D on the left are now on top of layers C and B). Normal and reverse faults tend to be at a relatively high angle (>45°). An example of the normal fault is shown in Figure 2.49 taken in the Nevada Ruby Range and a reverse fault at the Waterton Lakes National Park, Canada, is shown in Figure 2.50 (the rock mass on the right side is the hanging wall). When rock masses move horizontally relative to each other, faulting is known as a **lateral** or **strike-slip** fault, where the strike-slip or lateral motion is either right-lateral or left-lateral (Figure 2.48c). Figure 2.48c represents a right-lateral fault (looking across the fault plane the fence has moved to the right). If one

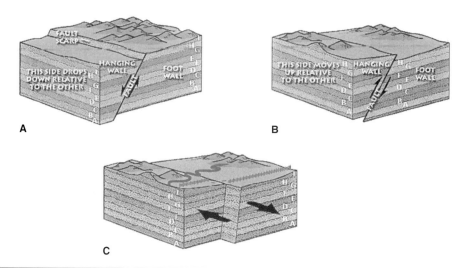

FIGURE 2.48 Fault Structures: (a) normal fault, layer A oldest, D youngest; (b) reverse fault, layer A oldest; (c) strike-slip or right-lateral fault. (*Modified from www.nature.nps.gov/geology/usgsnps/ deform/gfaults.html.*)

FIGURE 2.49 High-angle normal fault near the Ruby Mountains, northeastern Nevada.

FIGURE 2.50 Reverse faulting in Waterton Lakes National Park, Canada. The right side is the hanging wall.

crosses the fault plane to look back at the other block the motion appears to be the same (fence has still moved to the right). Most faults have some vertical and strike-slip motion.

In extensional areas, high-angle normal faulting is common near the point of separation where blocks slide or move downward and then the fault plane curves and becomes more horizontal with depth away from the separation point. When this occurs the fault is said to be **lystric** (sciencedirect.com, 2018). Additionally, in extensional areas, a series of blocks may move either down or up relative to each other to accommodate the widening additional space. When this occurs, a whole block may move downward bounded by two normal faults to form a **graben** valley (Figure 2.51). The adjoining uplifted blocks are known as **horsts**. A good example of where this has occurred is in the Basin and Range province in eastern Idaho, southwestern Montana, Nevada, Utah, and Arizona (Figure 2.52).

FIGURE 2.51 Stereo pair of a graben valley structure near Divide, Montana, in southwestern Montana. The valley has dropped down bounded by two normal faults on either side. (*Photos courtesy of Hugh Dresser.*)

FIGURE 2.52   Horst mountains and graben valleys of the Basin and Range province near Salt Lake City, Utah, looking west toward Nevada.

In compressional regimes, the rock masses tend to move into a more compact form. This is where reverse faulting is common (Figure 2.53). Low-angle reverse faults (commonly <15°) are known as **thrust** faults. Rock masses that are able to move as thrust packages tend have high fluid pressures involved to reduce the friction between the rock masses (Hubbert and Rubey 1959).

Faulting can occur within a thin zone (<1 m) or across a wide zone (tens of meters). This has important implications in forming boundary conditions for fluid flow (Figure 2.54).

FIGURE 2.53   Reverse faulting in the Kootenai Formation in Southwestern Montana. (*See also Figure 2.39.*)

Figure 2.54 Fault zone approximately 100 ft (30 m) wide at Double Springs, Idaho, taken 1 mo after the Borah Peak earthquake, October 1984.

A wide shear zone may serve as a conduit to allow confined aquifer waters to move upward and mix. If an aquifer is cut by a fault that brings a confining unit juxtaposed against a permeable unit, the fault may represent a barrier to flow.

Other Observations in Structures

As a geologist maps a structure, such as an anticline, there are other often additional smaller-scaled features that can be observed that contribute to understanding the larger structure. For example, one can observe smaller-scale folds within the larger structure known as parasitic folds (Figure 2.55). The strike of these will be similar to the larger structure; however, there is a systematic rotational component depending on which limb of the axis the minor structures are observed (Figure 2.45d). This same phenomenon is true of fractures. The outer edges of strata around a fold, whether the structure is an anticline or a syncline, are subject to extensional stresses and bed slip. The strata inside a fold are subject to compressional forces. This can greatly affect the fluid flow properties along strike of a structure.

Example 2.10 Within the Paradox Basin near Moab, Utah, along the Cane Creek anticline fracture permeabilities and associated enhanced flow conditions were described by Huntoon (1986). During the formation of the Cane Creek anticline, salt beds slowly migrated and bulged in the crest of anticlinal folds (Figure 2.45b). While bulging occurred, the overlying rocks at the crest of fold experienced extensional fracturing and normal faulting (Figure 2.45d) that very closely aligned with the strike of the Cane Creek anticline (Figures 2.56 and 2.57). Beds closest to the mine area are under compression.

It was decided that a solution-mining method would be employed to remove the salt by flooding the 150 mi (241 km) of mining cavities and then pumping this solution to the surface (Huntoon 1986). The Texasgulf 7 well was drilled in the crest of the anticline. The loss of circulation during drilling in the upper extensional beds and subsequent breakthrough into the mine cavity and rapid draining of the drilling fluid provide an interesting story.

FIGURE 2.55 Parasitic fold 1 m across within an anticlinal fold several hundred meters wide located in southwestern Montana.

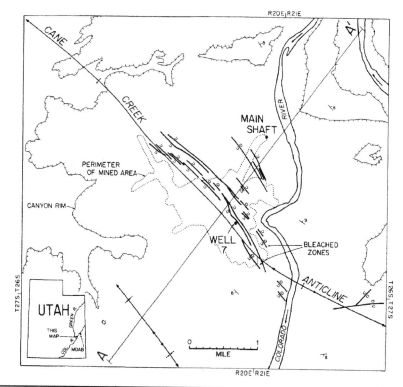

FIGURE 2.56 Extensional faults along the Cane Creek anticline near Moab, Utah. [*From Huntoon (1986). Reprinted with permission of the National Groundwater Association (1986).*]

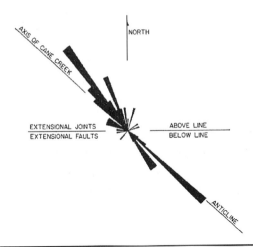

FIGURE 2.57 Strike relationship of 46 sets of extensional joints and 24 extensional faults near Moab, Utah. [*From Huntoon (1986). Reprinted with permission of the National Groundwater Association (1986).*]

Structural Effects in Karst Areas

Karst is a word that was derived from the word *kras* from the vicinity of Trieste, Italy, and adjacent Slovenia, which means "bare, stony ground" (Huntoon 1995; Quinlan et al. 1996). It does not have a universally accepted definition, but dissolution is the primary process in developing a distinctive surface topography, a topography characterized by sinkholes, caves, and underground drainage (Bates and Jackson 1984; Chapter 12). Karstic conditions often develop from dissolution of the host rock along fractures that become enlarged from millimeters to centimeters and even meters (Figure 2.58),

FIGURE 2.58 Keyhole caves developed in the upper 300 ft of the Mississippian Redwall Limestone, Marble Canyon, Arizona, illustrating triple porosity in one location. (*Photo courtesy of Peter Huntoon.*)

resulting in a triple-porosity system (Quinlan et al. 1996). This is significantly different from a porous media system, which tends to have enhanced permeability from secondary porosity features such as jointing and faults.

It is a mistake to assume that karstic groundwater flow systems develop only from enhanced secondary permeability features. Many untrained hydrogeologists or geologists in karstic systems may believe that cave systems and other karstic features developed where the fractures were largest (Ewers 2008). However, this is not necessarily true. In soluble host-rock karst systems, the greatest fluid flow occurs in the areas of steepest hydraulic gradients (Chapter 3) in concert with available carbon dioxide in the recharge area (Chapter 8). Here, fractures and other conduits are enlarged by dissolution of the host rock that develops in conduit systems parallel to fluid flow. There is an organizational hierarchy of dissolution tubes within a complex network that becomes progressively more organized in the "down-gradient" (downslope) direction that may develop into cave-sized conduits (Rahn and Gries 1973; Huntoon 1985; Huntoon 1995; Huntoon 1997; Ewers 2008) (Figures 2.59a and b). This development can ignore preexisting fracture systems developed by tectonic processes. Furthermore, karst systems change dynamically over time as a result of changes in stage levels and tectonic forces, which result in alternative flow paths being developed (Quinlan et al. 1996; Ewers 2008) that may even cross-cut older systems (Huntoon 1995).

The networks of conduit tubes within a karst aquifer network have multiple interconnections (Huntoon 1995). For this reason, there are numerous opportunities for

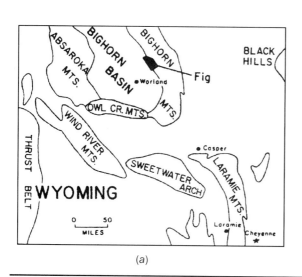

(a)

(b)

FIGURE 2.59   (a) Location of recharge areas on east side of Bighorn Basin. [*From Huntoon (1985). Reprinted with permission from the NGWA (1985).*] (b) P-bar cave dissolution cavity in the Big Horn Dolomite, Big Horn Mountains, Wyoming. Peter Huntoon is wearing the headband and the author is wearing a hard hat.

water to shunt across to other areas in the flow system, particularly in unconfined (Chapter 3) karst aquifers (Chapter 12). The direction of flow and multiple pathways appears to be stage dependent. Tracer studies (Chapters 13 and 14) reveal that although they are introduced at a single injection point, they can emerge at diverse exit points (Mills 1989; Huntoon 1992; Huntoon 1995; Ewers 2008). The scary thing is that at certain stage elevations, springs may be deemed safe because the water-quality data look clean. However, if the stage changes, then a once clean spring may become contaminated just because shunting within the flow system has shifted (Chapter 14).

Hydrogeologic field work in karst systems is markedly different from the traditional approaches used in non-soluble rock systems such as sedimentary rocks or igneous rock systems with significant secondary porosity. This is why a whole chapter is devoted to karst hydrogeology (Chapter 12), and additional insightful information is found in Chapters 13 and 14, where there is additional discussion on tracer tests.

2.9 Geologic Time

In Section 2.7 is a discussion of stratigraphic hierarchy where geologic units are given formation names or subdivided further into members or beds. Time units are also assigned with formation names such as the Mississippian Madison Formation or the Permian Quadrant Quartzite. The various geologic time units are divided into eons, eras, periods, and epochs in units of millions of years (Table 2.5; Kazemi et al. 2006; Geosociety.org, 2012).

Geologists have distinguished the different time units in the geologic time scale (Table 2.5) based upon absolute and relative dating methods including the fossil record. The Phanerozoic eon represents the eon of life. Eras (old life, middle life, and recent life) represent times of significant species of life forms. For example, trilobites lived during the Paleozoic era, but became extinct at the end of the Permian period, while dinosaurs existed only during the Mesozoic era. Periods are further divisions of time representing when significant species of animals appeared, such as the Devonian period (period of fishes) named after its type locale in Devonshire, England (ucmp.berkeley.edu, 2011). Epochs represent additional divisions of time devoted to most recent era (Cenozoic). The Pleistocene epoch is when the earth experienced its most recent ice ages and the Holocene represents time since the last glacial recession.

Absolute time is where exact times and dates can be given, for example, through counting tree rings or other measurable entities (Figure 2.60). Absolute dating methods typically include radiometric dating methods using unstable isotopes. Recall from basic chemistry that isotopes of a given element have the same number of protons (same atomic number) but differing numbers of neutrons, thus differing in atomic mass. Isotopes (Chapter 8) that are unstable will spontaneously emit radiation energy through a process known as decay. When one-half of the original parent material decays into its daughter product, one half-life is said to have occurred.

The various isotopes have fixed decay rates based upon the temperatures and pressures that exist in the outer layers of the earth. Tritium (^{3}H), for example, has a half-life of 12.43 years; therefore, it can be used to age date relatively young waters (Kazemi et al. 2006). Potassium (^{40}K)-Argon (^{40}Ar) ratios are used to date age old rocks because of the half-life of 1.25 billion years (USGS 2017). Radiometric dates in rocks are most commonly obtained from igneous rocks, or mineral clasts derived from igneous rocks. For example, detrital zircon grains found in ancient sandstones

Eon	Era	Period	Epoch	Millions of Years Before Present
Phanerozoic	Cenozoic	Quaternary	Holocene	0.01—present day
			Pleistocene	1.8–0.01
		Tertiary-Neogene	Pliocene	5.3–1.8
			Miocene	23.9–5.3
		Tertiary-Paleogene	Oligocene	33.7–23.9
			Eocene	54.9–33.7
			Paleocene	65.0–54.9
	Mesozoic	Cretaceous		144–65
		Jurassic		206–144
		Triassic		248–206
	Paleozoic	Permian		290–248
		Carboniferous-Pennsylvanian		323–290
		Carboniferous-Mississippian		354–323
		Devonian		417–354
		Silurian		443–417
		Ordovician		490–443
		Cambrian		543–490
Precambrian	Precambrian			
Proterozoic	Late			900–543
	Middle			1600–900
	Early			2500–1600
Archean	Late			3000–2500
	Middle			3400–3000
	Early			3960–3400
Hadean				4560–3960

Sources: www.geosociety.org/science/timescale; Kazemi et al. (2006); and GSA (2012).

TABLE 2.5   Geologic Time Scale

can be age dated. The relationship of time, parent material, daughter products, and decay rate are given in Equation 2.2 (Ristenen et al. 2016):

$$N = N_o \times e^{-kt} \tag{2.2}$$

where t = time for one half-life

k = appropriate decay constant

e = natural log

N = current number of grams or other quantity in the sample today

N_o = number of grams or other quantity in the original sample

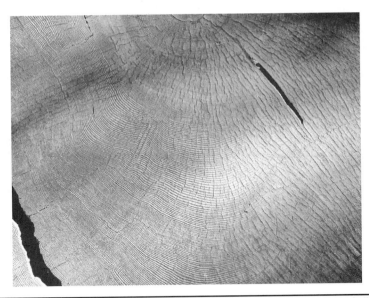

Figure 2.60   Tree rings in a 2,500-year-old sequoia tree in Sequoia National Park, California.

Example 2.11   If 10 g of Iodine-131 (half-life = 8.1 days) remain after 30 days, how large was the original sample?

Solution: The approach is to solve for the decay constant k after one half-life. Then use the formula to determine the number of grams in the original sample:

$$N = N_0 e^{-kt}$$

$$\frac{N}{N_0} = e^{-kt}$$

$$\ln \frac{1}{2} = -kt_{half}$$

$$-0.693 = -k(8.1 \, \text{days})$$

$$k = 0.0856 \, \text{days}^{-1}$$

$$N_0 = \frac{N}{e^{-kt}} = \frac{10g}{e^{-(0.0856 \, \text{days}^{-1})(30 \, \text{days})}} = 130g$$

Relative dating is where rock relationships and geologic principles are used to determine which rock unit is oldest in sequence without knowing its exact age. The most common geologic principles are listed in Table 2.6 and can be helpful in reconstructing the geologic history. Sometimes faulting or igneous intrusions are responsible for affecting the groundwater-flow paths, so understanding the geology is a must.

Unconformities (Table 2.6) are generally known by three types: (1) angular (where sedimentary layers meet at an angle to the erosional surface; Figure 2.62), (2) disconformity (where parallel layers have missing layers between younger and older layers), and (3) nonconformity (where sedimentary rocks come in contact with igneous or metamorphic rocks). Figure 2.61 represents an interesting case where one observes angular Cretaceous-age rocks exposed by the Shoshone River meeting younger surface rocks, Eocene in age (Willwood Formation about 55 million years), as an angular unconformity

Geologic Principle	Description
Principle of original horizontality	Geologic materials are basically laid out horizontally (lava flows, river deposits, etc.).
Principle of cross-cutting relationships	If a geologic unit cuts across another unit, it is younger than the one it cuts.
Principle of superposition	The oldest rocks will be on the bottom; successive layers are correspondingly younger.
Principle of lateral continuity	Geologic units are laterally continuous in all directions until they thin or pinch out or reach a boundary.
Principle of inclusions	The inclusions or pieces of any rock that become included in another rock are older than the rock they were included in.
Principle of unconformities	Unconformities represent hiatuses in time from erosion, non-deposition, or some other geologic activity.

TABLE 2.6   Geologic Principles (Modified after www.earth.rochester.edu)

(https://www.geowyo.com/heart-mountain.html). In the background one observes Heart Mountain, a detached Paleozoic-age rock mass that slid away from the Rocky Mountains that placed Ordovician- to Mississippian-age rocks (550 to 350 million years) on top of the Willwood Formation as a disconformity. This suggests that the slide event occurred after the formation of the Willwood Formation (younger than 55 million years).

A simple geologic history is shown in Figure 2.63. The unraveling of this history is left to the reader as one of the problems listed at the end of the chapter.

FIGURE 2.61   Example of unconformities. Cretaceous-age rocks are inclined or angled to the Shoshone River (north of Cody, Wyoming) and meet horizontal rocks, Eocene in age (55 million years), as an angular unconformity. In the background is Heart Mountain, composed of rocks Ordovician to Mississippian in age (300-million-year-older Paleozoic rocks) sitting on top of the younger horizontal rocks as a disconformity.

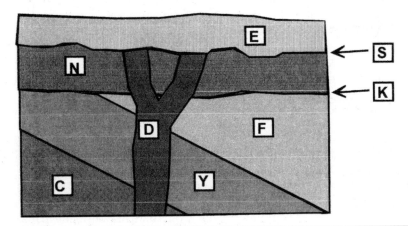

FIGURE 2.62 Geologic history, where units C, Y, and F are the oldest followed by uplift and tilting, then angular unconformity K, layer N, and then dike D. Erosional nonconformity S is followed by layer E, the youngest unit.

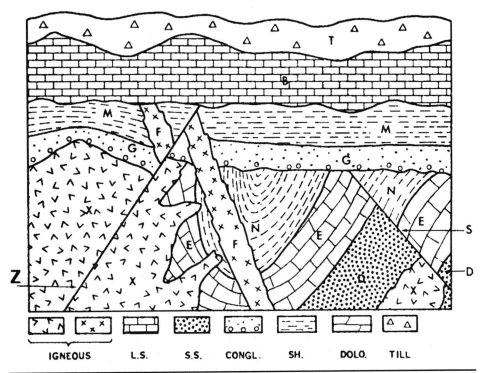

FIGURE 2.63 Geologically disturbed area. The letters represent the various lithologies, faults, and unconformities that influence what you see. (*Sourced from: http://www.science.smith.edu/ geosciences/intro/Exam1_11/section.jpg.*)

2.10   Using Geologic Information

The geology of hydrogeology could easily be the topic of many chapters or a whole book. Geologic concepts and examples are scattered throughout this book, because geology is inseparable from groundwater systems. The geologic features form the framework and boundary conditions for groundwater-flow systems. The more case histories and field examples one studies and experiences, the larger the context of mental library examples one can draw from to solve problems.

If you have avoided studying geology because of the inherent uncertainty in knowing what is occurring at depth, why study hydrogeology? It has the same types of uncertainties. The author's experience is that being able to know exactly what is occurring at depth is an uncomfortable area for many engineers, yet many will think nothing of using field data provided by geoscientists in a computer simulation model or design process. If you use geologic information, you have a responsibility to make sure that the constraints are also known. You are responsible for the information you use in a design or simulation. Take a course or two in geology or read up on it if it is a weakness in your life. It will make you a better professional.

2.11   Problems

2.1   What is the relationship between rhyolite and granite, andesite and diorite, and basalt and gabbro? Go to a college or university geology department where these most common igneous rocks might be available to view and handle and carefully compare the rock samples.

2.2   Evaluate the geophysical logs on Figure 2.10. Compare the lithologies reported and the gamma and resistivity geophysical responses and interpret what the lithologies might be present below 640 ft where they lost circulation (see Chapter 15 for concept of circulation).

2.3   Why is the porosity of extrusive igneous rocks greater than intrusive ones?

2.4   Describe the process of a shale becoming metamorphosed from a slate to a gneiss. Go to a college or university geology department where these most common metamorphic rocks might be available to view and handle and carefully compare the rock samples.

2.5   What is the difference between sediments and sedimentary rock? Carefully review the depositional environments listed in Table 2.4. How would sediments that were deposited in a eolian setting compare with sediments in a fluvial setting?

2.6   Lean a book against an object and measure its strike and dip using a geological compass.

2.7   Go to the following website: https://earthquake.usgs.gov/hazards/qfaults/ and click on the interactive fault map. Zoom into the San Francisco, California, area and determine what kinds of faults are present and what their movement rates are. How much movement should occur within the next 20 years?

2.8   How are karst areas different from other geological settings?

2.9   Create a stratigraphic column from oldest at the bottom to youngest at the top using the letters shown. Some of the letters might refer to faults or unconformities. Provide a brief geologic history of the area. http://www.science.smith.edu/geosciences/intro/Exam1_11/section.jpg.

2.10   The half-life of ^{137}Cs is 30 years. If there are about 5 curies left after 150 years, how many curies of Cs were there in the original sample?

2.12   References

AAPG, 2005. North American Stratigraphic Code. *AAPG Bulletin*, Vol. 89, No. 11 (November 2005), pp. 1547–1591. https://ngmdb.usgs.gov/Info/NACSN/05_1547. pdf

Allis, R.G., and Lumb, J.T., 1992. The Rotarua Geothermal Field, New Zealand: Its Physical Setting, Hydrology, and Response to Exploitation. *Geothermics*, Vol. 21, No. 1/2, pp. 7–24.

Alt, D., and Hyndman, D.W., 1995. *Northwest Exposures: A Geologic Story of the Northwest.* Mountain Press, Missoula, MT, 443 pp.

Baldwin, D., 1997. Hydrogeologic and Aquifer Vulnerability Investigation at Big Sky, MT. MS Thesis, Montana Tech of the University of Montana, Butte, MY.

Barraclough, J., 1978 (July). Personal communication while working at the INL about the productivity of basalt transmissivity.

Bates, R.L., and Jackson, J.A. (eds.), 1984. *Dictionary of Geological Terms, 3rd Edition.* American Geological Institute, Anchor Press, Garden City, NY, 571 pp.

Bennison, G.M., 1990. *An Introduction to Geological Structures and Maps.* Chapman and Hall, New York, 69 pp.

Bouyoucos, G.J., 1962. Hydrometer Method Improved for Making Particle Size Analysis of Soils. *Agronomy Journal*, Vol. 54, No. 5, pp. 464–465.

Bowen, N.L., 1928. *The Evolution of the Igneous Rocks.* Princeton University Press, Princeton, NJ.

Brown, J.L., 1993. *Sedimentology and Depositional History of the Lower Paleocene Tullock Member of the Fort Union Formation, Powder River Basin, Wyoming and Montana.* USGS Bulletin 1917, 142 pp.

Custer, S.G., Michels, D.E., Sill, W., Sonderegger, J.L., Weight, W.D., and Woessner, W.W., 1994. *Recommended Boundary for a Controlled Groundwater Area in Montana Near Yellowstone Park.* Water Resources Division, National Park Service, Fort Collins, CO, 29 pp.

Daho Pozos, http://dahopozos.com/en/who-we-are/

Davis, S.N., and Turk, L.J., 1964. Optimum Depth of Wells in Crystalline Rock. *Ground Water*, Vol. 2, No. 2, pp. 6–11.

Domenico, T.A., and Schwartz, F.W., 1990. *Physical and Chemical Hydrogeology.* John Wiley & Sons, New York, 824 pp.

Driscoll, F.G., 1986. *Groundwater and Wells.* Johnson Screens, St. Paul, MN, 1108 pp.

du Bray, E.A., Aleinikoff, J.N., and Lund, K., 2012. *Synthesis of Petrographic, Geochemical, and Isotopic Data for the Boulder Batholith, Southwest Montana.* USGS Professional Paper 1793.

Ewers, R., 2008. Practical Karst Hydrogeology with Emphasis on Ground-Water Monitoring. A short course at the North American Environmental Field Conference and Exposition, sponsored by the Nielson Environmental Field School, Inc., Tampa, FL.

Fetter, C.W., 1994. *Applied Hydrogeology, 3rd Edition.* Macmillan College Publishing Company, New York, 691 pp.

Folk, R.L., 1966. A Review of Grain-Size Parameters. *Sedimentology*, vol. 6. Elsevier, Amsterdam, pp. 73–93.

Freeze, A., and Cherry, J., 1979. *Groundwater.* Prentice Hall, Upper Saddle River, NJ, 604 pp.

Geomaps.wr.usgs.gov, 2017. https://geomaps.wr.usgs.gov/parks/rxmin/rock.html#igneous. Granite figure (New 2.16).

Geosociety.org, 2012. https://www.geosociety.org/GSA/Education_Careers/Geologic_Time_Scale/GSA/timescale/home.aspx

Goldich, S.S., 1938. A Study in Rock Weathering. *Journal of Geology*, Vol. 46, pp. 17–58.

Hamilton, W., and Myers, W.B., 1974. Nature of the Boulder Batholith of Montana. *GSA Bulletin*, Vol. 85, No. 3, pp. 365–378.

Hubbert, M.K., and Rubey, W.W., 1959. Role of Fluid Pressure in Mechanics of Overthrust Faulting: I, Mechanics of Fluid-Filled Porous Solids and Its Application to Overthrust Faulting. *Geological Society of America Bulletin*, Vol. 70, pp. 115–166.

Huntoon, W.P., 1985. Rejection of Recharge Water from Madison Aquifer along Eastern Perimeter of Bighorn Basin, Wyoming. *Ground Water*, Vol. 23, No. 3, pp. 345–353.

Huntoon, W.P., 1986. Incredible Tale of Texasgulf Well 7 and Fracture Permeability, Paradox Basin, Utah. *Ground Water*, Vol. 24, No. 5, pp. 644–653.

Huntoon, W.P., 1992. Hydrogeologic Characteristics and Deforestation of the Stone Forest Karst Aquifers of South China. *Ground Water*, Vol. 30, No. 2, pp. 162.

Huntoon, W.P., 1995. Is It Appropriate to Apply Porous Media Groundwater Circulation Models to Karstic Aquifers? In *Groundwater Models for Resources Analysis and Management*, Aly I. El-Kadi (ed.), CRC Press, Boca Raton, FL, 339–358 pp.

Huntoon, W.P., 1997. The Case for Upland Recharge Area Protection in the Rocky Mountain Karsts of the Western United States. In *Karst Waters and Environmental Impacts*, Gunay, G., and Johnson, A.I., (eds.), A.A. Balkema, Rotterdam/Brookfield.

Johnson, B.R., Ihinger, P.D., Mahoney, J.B., and Friedman R.M., 2005. Re-Examining the Geochemistry and Geochronology of the Late Cretaceous Boulder Batholith, MT. GSA National Meeting, Denver, CO, October.

Kazemi G.A., Lehr, J.H., and Perrochet, P., 2006. *Groundwater Age*. John Wiley & Sons.

Mills, J.P., 1989. *Foreland Structure and Karstic Ground Water Circulation in the Eastern Gros Ventre Range, Wyoming*. Master's Thesis, University of Wyoming, Laramie, WY, 101 pp.

Quinlan, J.F., Davies, G.J., Jones, S.W., and Huntoon, P.W., 1996. The Applicability of Numerical Models to Adequately Characterize Ground-Water Flow in Karstic and Other Triple-Porosity Aquifers. In *Subsurface Fluid-Flow (Ground-Water and Vadose Zone) Modeling*, ASTM STP 1288, Ritchey, J.D., and Rumbaugh, J.O., (eds.), American Society for Testing and Materials, 115–133 pp. https://www.astm.org/DIGITAL_LIBRARY/STP/PAGES/STP38382S.htm

Peltier, L., 1950. Geographic Cycle in Periglacial Regions as It Is Related to Climatic Geomorphology. *Annals of the American Association of Geographers*, Vol. 40, No. 3, pp. 214–236.

Rahn, P.H., and Gries, J.P., 1973. Large Springs in the Black Hills, South Dakota and Wyoming. *South Dakota Geology Survey Report of Investigation 107*, 46 pp.

Ridley, W.I., 2010. Petrology of Associated Igneous Rocks. https://pubs.usgs.gov/sir/2010/5070/c/Chapter15SIR10-5070-C-3.pdf

Ristenen, R.A., Kraushaar, J.J., and Brack, J., 2016. *Energy and the Environment*, 3rd Edition. John Wiley & Sons. New York, 334 pp.

Sciencedirect.com, 2018. https://www.sciencedirect.com/topics/earth-and-planetary-sciences/listric-fault

Sorey, M.L. (ed.), 1991. *Effects of the Potential Geothermal Development in the Corwin Spring's Known Geothermal Resources Area, Montana, on the Thermal Features of Yellowstone National Park*. U.S. Geological Survey Water-Resources Investigations Report 91-4052, pp. A1–H12.

Streckeisen, A.L., 1976. To Each Platonic Rock Its Proper Name. *Earth Science Review*, Vol. 12, pp. 1–34.

Tingley, J.V., and Pizarro, K.A., 2000. *Traveling America's Loneliest Road: A Geologic and Natural History Tour*. Nevada Bureau of Mines and Geology Special Publication 26, Nevada Bureau of Mines and Geology, 132 pp.

Turner, F.J., 1968. *Metamorphic Petrology*. McGraw-Hill, New York, 366 pp.

Ucmp.berkeley.edu, 2011. http://www.ucmp.berkeley.edu/paleozoic/paleozoic.php

USGS, 2017. https://geomaps.wr.usgs.gov/parks/gtime/ageofearth.html

Water.usgs.gov, 2018. https://water.usgs.gov/edu/density.html

Weight, W.D., 2017. Pre-Filed Expert Testimony Weight for Objectors; Regarding the Montana Artesian Water Company Application for Proposed Water Bottling Plant in Kalispell Montana. 37 pp.

Wentworth, C.A., 1922. A Scale of Grade and Class Terms for Clastic Sediments. *Journal of Geology*.

Winter, J.D., 2009. *Principles of Igneous and Metamorphic Rock Petrology, 2nd Edition*. Prentice Hall, Englewood Cliffs, NJ, 20 pp.

Web Resources

Excellent YouTube Channel with numerous short videos illustrating and explaining everything from rock identification to faulting to glaciers and groundwater: https://www.youtube.com/channel/UCtQfVk8PDyHU6e9q_1cEY0Q

Main page: https://geology.usgs.gov/

Water resources data page:

https://www.usgs.gov/products/data-and-tools/real-time-data/water

Good site with a lot of photos of different rock types to help with visual identification: https://geology.com/rocks/

Grains size and rock textures: http://faculty.chemeketa.edu/afrank1/rocks/sedimentary/sedtexture.htm

History of geologic time—life on earth:

http://www.ucmp.berkeley.edu/exhibits/historyoflife.php

Geologic time scale:

https://www.geosociety.org/GSA/Education_Careers/Geologic_Time_Scale/GSA/timescale/home.aspx

American Geoscience Institute—Great Image Bank:

http://www.earthscienceworld.org/

https://www.nature.nps.gov/geology/usgsnps/deform/gfaults.html

http://www.galleries.com/Feldspar_Group. Examples of feldspars.

CHAPTER **3**

Aguifer Properties

It is important to be able to translate the geology (Chapter 2), when it becomes saturated, into hydrogeology. The physical properties of geologic materials control the storativity and ability of fluids to move through them. Rock units that do not allow fluids through them become barriers to fluid flow and in turn change the direction of groundwater movement (Chapter 4). Other features such as fault zones may serve as conduits to fluid flow or act as barriers. In this chapter, the physical properties of saturated geologic materials are presented to provide a basic understanding of aquifers, confining layers, and boundary conditions as a basis for understanding groundwater flow presented in Chapter 4. Boundaries are often determined directly through drilling (Chapter 15), pumping tests (Chapters 5 and 6), or geophysical methods.

3.1 From the Surface to the Water Table

When precipitation reaches the land surface, some water enters the soil horizon. This process is known as **infiltration**. Water that accumulates on the surface faster than it can infiltrate becomes **runoff** (Chapter 1; Figure 1.1). The rate at which water infiltrates or runs off is a function of the physical properties of the surficial soils. Some of the important factors appear to be soil thickness, clay content, moisture content, and intrinsic permeability of the soils' materials (Baldwin 1997). (Additional discussion on intrinsic permeability is given in Section 3.2.) Infiltrating water that encounters soils with higher clay content tends to clog the pores, causing precipitation to mound up and run off, unless they are exceedingly dry (Stephens 1996). Sandier soils promote infiltration and exhibit less vegetative growth (Figure 3.1), while soils with a higher clay content appear to promote plant growth. Glaciated areas provide an example of an environment where many soil types can be found. Glacial sediments deposited via moving water become stratified or layered and tend to be well drained; examples include outwash deposits, kames, and eskers. Sediments transported by ice that accumulate along the sides and end of a glacier are poorly sorted and contain a higher content of clay and silt; examples include lateral and end moraines, which are poorly drained (Figure 3.2). Once infiltration occurs, any groundwater that descends below the rooting depth eventually reaches the regional water table as recharge. This has serious implications for dissolved chemicals that accompany the descending waters.

Between the soil horizon and the regional water table is an area referred to as the vadose zone (Figure 3.3). The ability of the vadose zone to hold water depends upon the moisture content and grain size. Wells completed in the vadose zone will have no water in them, even though the geologic materials appear to be wet, while wells completed in

FIGURE 3.1   Gravel channel exposed in road cut near Sheridan, Wyoming. Coarser sands lack vegetation and have a rougher appearance, while finer-grained upper and lateral floodplain deposits support more vegetative growth.

FIGURE 3.2   Terminal moraine in southwestern Montana, indicating poorly sorted sediments.

saturated fine-grained soils will eventually collect groundwater. Chapter 11 is devoted to the vadose zone and its properties and field methodologies.

Another part of the vadose zone immediately above the regional water table is the capillary fringe. The capillary fringe is essentially saturated, but groundwater is being held against gravity under negative pressure (Chapters 4 and 11), a phenomenon known as capillarity. In groundwater applications atmospheric pressure is referenced

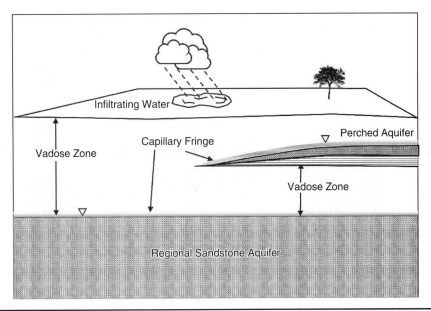

FIGURE 3.3 Schematic of the vadose zone, infiltrating water, and the capillary zone.

as being zero pressure head. The water table, for example, is at atmospheric pressure, while below the water table, pore water is under a pressure greater than atmospheric. Water in the capillary fringe and the rest of the vadose zone is under a pressure less than atmospheric. The phenomenon of capillarity is observed when one puts a paper towel into a pan of water. The water is attracted to the surfaces of the towel fibers being drawn up through very small pore tubes between the fibers. Similarly, within the capillary fringe of aquifer materials, groundwater seeks to wet the surfaces of geologic materials with an attraction greater than the force of gravity. The thickness of the capillary fringe is grain-size dependent. The finer grained the material, the higher the capillary fringe rises above the water table because of smaller-sized pore throats, increased surface area, and surface tension (Chapter 11).

Example 3.1 An example here provides context for understanding the role of capillarity. One of the largest Superfund sites in the United States is associated with over 100 miles (161 km) of streamside tailings cleanup along the Clark Fork River (http://deq.mt.gov/Land/fedsuperfund/sst). This resulted from a 300- to 500-year flood event (Chapter 10) on the Clark Fork River in 1908 that spread mine tailings from Butte, Montana, mining operations to Milltown Dam near Missoula, Montana, into the floodplain. Capillarity from the water table reaches the floodplain surface resulting in a constant flow of moisture through metals-laden materials (Figure 3.4). Once at the surface, soluble metal salts precipitate out. When subsequent rain events occur, these metal salts quickly go back into solution resulting in fish kills and other negative impacts to aquatic life on the river; hence, the need for Superfund cleanup efforts.

When drilling wells or installing monitoring equipment, one must also be careful that the *first* water encountered is actually the regional water table and not a **perched aquifer**. Perched aquifers represent infiltrating groundwater that accumulates over confining layers of limited areal extent above the regional water table (Figure 3.3).

FIGURE 3.4   Streamside tailings along the headwaters of the Clark Fork River, Montana. Note the stage of the river below and the darkened sediments from capillary moisture rising upward. Metal salts at the surface are being observed by the students.

Perched aquifers may be capable of sustaining enough water for several residences, but generally not enough for many residences or long-term production. Obtaining multiple water levels in wells in the same area would help one determine whether a perched water table exists or not. Aquifers are defined and discussed in Section 3.4.

Example 3.2   A consulting company was evaluating the drilling depths for production wells for a proposed subdivision. Drilling estimates were being made based upon existing wells in the area. The evaluator did not realize that there were wells completed in a local perched aquifer and also in a regional unconfined aquifer. In the end, he averaged the well depths to estimate drilling costs. This resulted in bidding the project way too low and the consulting company lost serious money. The crux of the problem was in failing to understand the hydrogeology system and that the differences in well depths represented two separate aquifers.

3.2   Porosity and Aquifer Storage

The volume of water that an aquifer can take into or release from storage for a given change in head is often determined by its porosity. The porosity of earth materials is a function of size, shape, and particle arrangement or packing. The ability of water to move through an aquifer is described by its permeability or hydraulic conductivity (see Section 3.3).

Porosity

The **porosity** is represented as the nonsolid fraction of geologic materials. This is where fluids can be held. In the vadose zone, the porosity or open spaces is filled with air and water (Section 3.3 and Chapter 11). The total porosity of geologic materials expressed as a percent is represented by:

$$\eta\% = \frac{V_V}{V_T} \times 100\% \qquad (3.1)$$

where η = porosity
V_V = volume of the void
V_T = total volume

Porosity is often broken down into primary and secondary porosity. Primary porosity is the void space that occurred when the rock or geologic material formed. Secondary porosity refers to openings or void space created after the rock formed (Figure 3.5). Examples of these are given in Table 3.1.

Figure 3.5 Secondary porosity from faulting. A spring is emanating from the fault zone.

Primary Porosity	Secondary Porosity
Vesicles	Faults
Intergranular pores	Fractures
Interlayer partings and unconformities	Solution channels
	Stylolites
Spaces between lava flows	Enhanced pathways from plants and animals
Lava tubes	
Intercrystalline pores	

Table 3.1 Example of Primary and Secondary Porosity

Generally, each aquifer or confining layer is numerically modeled using a single overall porosity unless its lateral changes are known through a distribution of cores. In fractured rock, such as a fractured sandstone or granite, a "dual-porosity" model may be more appropriate. A dual-porosity model assigns a porosity to the fracture zone (secondary porosity) and to the geologic block materials (primary porosity) (see Chapter 11). Carbonate karstified systems may have a triple porosity where there are microscale and two levels of macroscale pore spaces (larger fractures and caves, see Section 2.6 and Chapter 12). The larger-scale fractures may lead to misleading interpretations of unlimited supply.

Example 3.3   When John Sonderegger (co-author of the first edition of this book) worked for the Alabama State Survey in the United States, he heard about a pumping test on a Huntsville, Alabama, municipal well [pumped at approximately 2,000 gallons per minute (gpm), 7.57 m³/min] that ran for a week with only a couple of feet (0.5 m) of drawdown. The test was continued to 10 days and ran out of water on the 8th or 9th day with a sustainable pumping rate of only a couple of hundred gpm (0.76 m³/min) or less. The larger macroscale spaces had become depleted.

Of the total porosity of geologic materials, there is a portion that will drain freely by gravity and an amount retained in the geologic materials. The volume of water that will drain by gravity per unit drop in the water table per unit volume of aquifer is referred to as the **specific yield** (S_y). The water that remains clinging to the surfaces of the solids in the interstices is referred to as **specific retention** (S_r). Although they are strictly different entities, the specific yield is often used as an estimate for the **effective porosity** (η_e), a term used to describe the porosity available for fluid flow. It should be noted that specific yields are estimated from vertical lab-drainage tests (Roscoe Moss Company, 1990), while effective porosities are generally used for horizontal flow calculations, such as to calculate groundwater velocities (Chapter 4). Although one may generally assume that the specific yield and the effective porosity are the same for a coarser-grained aquifer, in some soils there may be a high content of soil aggregates and water may become tied up in dead-end pore spaces. This results in the effective porosity and specific yield being significantly different (Cleary et al. 1992). Specific yield and specific retention make up the total porosity, expressed in the following relationship.

$$\eta = S_Y + S_R \tag{3.2}$$

Example 3.4   Suppose you have a 5-gal bucket full of oven-dried sandy material. You add just enough water until the sand becomes completely saturated at the 5-gal mark (Figure 3. 6). Assume, also that it takes exactly 1 gal to fill all of the pores. From Equation 3.1, the percent porosity equals:

$$\eta = \frac{1 \, \text{gal}}{5 \, \text{gal}} \times 100\% = 20\% \tag{3.3}$$

If an opening was placed in the bottom of the bucket and the water was allowed to drain into a container (in a humidity-controlled lab setting) until it stops, the specific yield and specific retention could be estimated. Let us assume the amount of water that drained was 0.9 gal. By definition, the specific yield would be:

$$S_Y = \frac{0.9 \, \text{gal}}{5 \, \text{gal}} = 0.18 \tag{3.4}$$

From Equation 3.4, the effective porosity would therefore be estimated to be 18%. The specific retention would be 0.02 or 2% of the total volume. In most sand and gravel

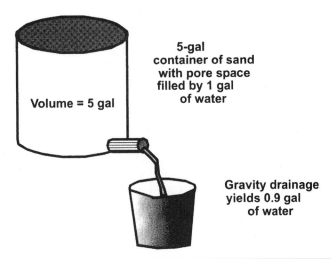

Figure 3.6   Illustration of the concept of specific yield. A 5-gal bucket of sand is filled to the top with 1 gal of water; 0.9 gal drains out by gravity, while 0.1 gal is retained in the pore spaces.

aquifers, the specific retention is quite low (<5%). As the grain size decreases, the total porosity increases (Table 3.2; Figure 3.7), but the specific retention increases as well. The smaller the grain size, the greater the surface area for water to cling to.

One notices from Table 3.2 that the porosity of sand and gravel together (porous media) is less than that of either material separately. This is a function of the packing of grains and sorting. Sorting and grading was described in Chapter 2 and is an expression of ranges of grain sizes. The packing of grains is a function of the size, shape, and arrangement of grains. If one takes spheres of equal size and arranges them, there are two end-member packing arrangements that represent the most porous packing (cubic packing) and the least porous (rhombohedral packing) (Figure 3.8). Cubic packing is where grains are stacked vertically on top of one another with their edges touching. This yields a porosity of 47.6%. In rhombohedral packing the spheres are arranged together in their most compact form, yielding a porosity of 25.9%. Notice in Figure 3.8 how porosity is independent of grain size (r) regardless of packing style. When grains

Unconsolidated Materials	η%	Consolidated Rock	η%
Clay	40–70	Sandstone	5–35
Silt	35–50	Limestone/dolomite	<1–20
Sand	25–50	Shale	<1–10
Gravel	20–40	Crystalline rock (fractured)	<1–5
Sand and gravel	15–35	Vesicular basalt	5–50

Table 3.2   Ranges of Porosities in Typical Earth Materials [Modified after Freeze and Cheery (1979), Driscoll (1986), and Roscoe Moss Company (1990)]

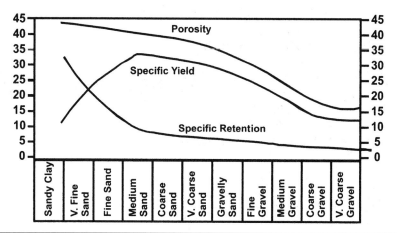

FIGURE 3.7 Relationship of porosity, specific yield, and specific retention with grain size. [*Modified after Scott and Scalmanini (1978).*]

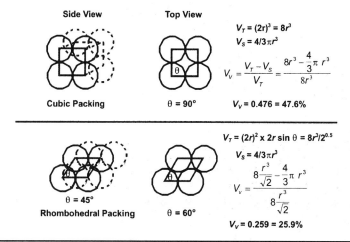

FIGURE 3.8 Diagram of cubic and rhombohedral packing and determination of porosity (V_T = total, V_S = volume solid, V_v = void).

of differing sizes are mixed together, the smaller grains fill in between the grains of the larger ones, thus reducing the overall porosity:

$$V_T = (2r)^3 = 8r^3 \quad \text{and} \quad V_v = (V_T - V_s)/V_T$$

In the saturated zone below the water table, the porous openings are *completely* filled with water. In many porous media aquifers, the total porosity can also be calculated from lab work and estimating the particle density in the following relationship:

$$\eta\% = 100\left[1 - \frac{\rho_b}{\rho_p}\right] \tag{3.5}$$

where η = porosity

ρ_b = dry bulk density (m/L^3)

ρ_p = particle density (m/L^3)

Equation 3.5 is solved using values obtained from cores taken in the field. The volume of saturated material is measured and then weighed (the wet weight). The geologic materials are then placed in a container and heated at 105°C until dry. The new dry weight compared to the volume represents the **dry bulk density** (ρ_b). The **particle density** (ρ_p) is the density of the solid materials. A common value for aquifer materials is that of quartz, 2.65 g/cm^3. The density of groundwater is close to 1.0 g/cm^3 depending on the water temperature (Appendix B). The ratio of the particle density to the density of water is known as the **specific gravity**.

Example 3.5 A core of sandy materials of approximately 80 cm^3 was taken from the field and weighed. The net soil material minus the tare weight was 166 g (wet weight). The soils were saturated and reweighed with a value of 172 g. After placing the sample in an oven at 105°C for about 24 hours, the dry sample was weighed at 148 g (dry weight). The water temperature in the field was 12°C, and the density is 0.99950 g/cm^3 (Appendix B).

The dry bulk density is

$$\rho_b = 148 \text{ g}/80 \text{ cm}^3 = 1.85 \text{ g/cm}^3$$

If the particle density can be assumed to be mostly quartz at 2.65 g/cm^3, the porosity is estimated using Equation 3.5 to be

$$\eta = 100 \left[1 - \frac{1.85 \text{ g/cm}^3}{2.65 \text{ g/cm}^3} \right] = 30.2\%$$

Another way the porosity can be estimated is by evaluating the volume occupied by the water in the core at saturation minus the dry weight:

$$\text{Saturation weight} - \text{dry weight} = 172 \text{ g} - 148 \text{ g} = 24 \text{ g}$$

This occupies a volume of

$$V = 24 \text{ g}/(0.99950 \text{ g/cm}^3) = 24.01 \text{ cm}^3$$

Using Equation 3.1,

$$\eta = 100 \frac{24.01 \text{ cm}^3}{80 \text{ cm}^3} = 30.0\%$$

The discrepancy in values in Example 3.5 can be attributed to an assumed particle density in the first case or because of volume errors. The above example was artificially created to illustrate the concept of porosity. In reality, the core must be tapped into a container in the laboratory. During the extraction process, differential compaction results in field conditions being lost. It is always more difficult to place a field core into a container of equal volume. Porosity and specific yield are often estimated in a laboratory.

Anderson and Woessner (1992) summarized the results of their findings for specific yield in Table 3.3. It is interesting how the arithmetic means of unconsolidated materials follow the trends indicated in Figure 3.7. When the reported arithmetic means differ significantly from the midpoint of the range values, this indicates that the distribution is skewed. Notice also that the differences between the unconsolidated sedimentary materials, such as fine and medium sand, compared to their lithified counterparts, fine

Material Class	Material	No. of Analyses	Range	Arithmetic Mean
Sedimentary	Clay	27	0.01–0.18	0.06
	Silt	299	0.01–0.39	0.20
	Sand (fine)	287	0.01–0.46	0.33
	Sand (medium)	297	0.16–0.46	0.32
	Sand (coarse)	143	0.18–0.43	0.30
	Gravel (fine)	33	0.13–0.40	0.28
	Gravel (med)	13	0.17–0.44	0.24
	Gravel (coarse)	9	0.13–0.25	0.21
	Siltstone	13	0.01–0.33	0.12
	Sandstone (fine)	47	0.02–0.40	0.21
	Sandstone (medium)	10	0.12–0.41	0.27
	Limestone	32	0–0.36	0.14
Wind deposits	Loess	5	0.14–0.22	0.18
	Eolian sand	14	0.32–0.47	0.38
Metamorphic	Schist	11	0.022–0.033	0.026
Igneous	Tuff	90	0.02–0.47	0.21

TABLE 3.3   Ranges of Values of Specific Yield [Adapted from Anderson and Woessner (1992)]

and medium sandstone, are different because of the volume occupied by cementing agents. Also note that the specific yield is *always* less than the total porosity.

In a study conducted at the U.S. Geological Survey laboratory reported by Morris and Johnson (1967), differing earth materials were tested and evaluated for the physical and hydrologic properties. An example from this study illustrating the grain-size distributions of water-laid sandy materials and their physical and hydrologic properties is shown in Tables 3.4A and 3.4B. It is interesting to note the similarities of specific gravity regardless of the grain-size distribution and the range of grain-size distributions, dry bulk densities, and hydraulic conductivities. It is apparent from the grain-size

Location	Clay <0.004 mm	Silt 0.004 0.063 mm	V. Fine sand 0.063– 0.125 mm	Fine sand 0.125– 0.25 mm	Med. sand 0.25– 0.5 mm	Coarse sand 0.5–1.0 mm	V. Coarse sand 1–2 mm	V.F. gravel 2–4 mm	Fine gravel 4–8 mm	Med. gravel 8–16 mm
Brunswick, GA	10.4	19.6	57.2	7.2	3.4	1.6	0.6	—	—	—
McCurtain Co., OK	2.8	8.8	19.2	64.6	4.4	0.2	—	—	—	—
Gallaway Co., KY	4.0	0.2	1.1	19.4	73.4	1.8	0.1	—	—	—
Arapahoe Co., CO	0.3	0.0	0.0	0.2	3.5	24.3	40.5	23.7	6.8	0.7

TABLE 3.4A   Grain-Size Distribution of Differing Water-Laid Soils [From Morris and Johnson (1967)] Notice How Sums of Numbers along Rows Add to 100.0

Location	Depth (ft)	Specific Gravity (g/cm³)	Dry Bulk Density (g/cm³)	Specific Retention (%)	Specific Yield (%)	Total Porosity (%)	Hydraulic Conductivity (ft/day)
Brunswick, GA	496–497	2.71	1.58	22.8	18.9	41.7	1.7
McCurtain Co., OK	—	2.65	1.90	0.7	27.6	28.3	11
Gallaway Co., KY	—	2.67	1.48	1.0	43.6	44.6	53
Arapahoe Co., CO	33.5–34	2.61	1.67	2.9	33.1	36.0	802

TABLE 3.4B Physical and Hydrologic Properties of Water-Laid Soils [From Morris and Johnson (1967)]

distributions that there is a correlation between grain size and hydraulic conductivity (intrinsic permeability; Section 3.3).

Example 3.6 Drought conditions have resulted in reducing water levels an average of 2.3 ft (0.7 m) in wells within a 10.2 mi² (2641.8 ha) area. A specific yield of 0.21 is believed to be representative of the fine-grained sandstone aquifer the wells are completed in. (1) How much water is represented by this change in water levels in acre-feet and in gallons? (2) How long could this volume of water supply a community of 90 residents, in days and years, if it is assumed that each resident uses approximately 50 gal/day?

Solution 1 The first step is in calculating the total volume of aquifer (V_T). Once this has been determined, V_T should be multiplied by the specific yield (0.15) to obtain the usable water volume for part 2:

$$V_T \times 0.15 = 10.2 \text{ mi}^2 \times \frac{(5{,}280 \text{ ft})^2}{(1 \text{ mi})^2} \times 2.3 \text{ ft} \times 0.15 = 2.342 \times 10^7 \text{ ft}^3 \times \frac{1 \text{ acre-ft}}{43{,}560 \text{ ft}^3} = 538 \text{ AF}$$

$$2.342 \times 10^7 \text{ ft}^3 \times \frac{7.48 \text{ gal}}{1 \text{ ft}^3} = 1.752 \times 10^8 \text{ gal avaliable}$$

Solution 2 Determine the volume required per day by the community and divide the gallons available by this number:

$$90 \text{ residents} \times \frac{50 \text{ gal/day}}{1 \text{ resident}} = 4{,}500 \text{ gal/day}$$

$$\frac{1.752 \times 10^8 \text{ gal}}{4{,}500 \frac{\text{gal}}{\text{day}}} = 38{,}936 \text{ days or } 106 \text{ years}$$

Storativity

The amount of water an aquifer can take into or release from storage is known as the **storage coefficient**. The numerical value assigned to an aquifer is the **storativity**, a dimensionless value determined from pumping tests (Chapters 5 and 6). In a pumping test, the storativity represents the storativity from the saturated thickness contributing to the well bore. Storativity is defined according to Equation 3.6:

$$S = S_Y + S_S \times b \tag{3.6}$$

The first term has already been defined as the specific yield (S_y). The second term is the **specific storage** (S_s), dimensioned 1/L, and multiplied by the saturated thickness (*b*) to become dimensionless. The specific yield is approximately equal to the storativity value for most unconfined aquifers. The usual range of storativity values in unconfined aquifers is 0.03 to 0.3 (Fetter 1994). The value of S_s in unconfined aquifers is practically negligible, as it is about 1,000 times less (Robson 1993), unless there are sections within the aquifer where the grain size is very small (interbedded clay lenses). In confined aquifers (Section 3.4), the water released from storage is a function of the compressibility of the aquifer materials and the compressibility of water (Equation 3.7):

$$S_S = \rho \times g(\alpha + \eta\beta) \qquad (3.7)$$

where ρ = fluid density (m/L^3)
g = gravitational force (L/t^2)
α = compressibility of mineral skeleton [1/(M/Lt^2)]
β = compressibility of water [1/(M/Lt^2)]

Specific storage is also known as elastic storage (Robson 1993; Section 7.4). This is from the flexible nature of mineral skeletons when changes in stress are "felt," such as during a pumping or slug test. Aquifer dewatering causes the mineral skeleton to be stressed sufficiently that compression of the mineral skeleton takes place, with or without gravity drainage. This phenomenon can also occur during a significant seismic event, such as an earthquake; however, this usually results in permanent structural changes to the mineral skeleton. Permanent structural changes in the mineral skeleton are not restorable. Compressional stresses on the mineral skeleton are also enhanced by overpumping. Earth materials within an aquifer most susceptible to permanent changes are the finer-grained materials (Table 3.5). The larger the compressibility value, the easier the material is to compress.

In confined aquifers, storativities range from 10^{-3} to 10^{-6} (0.001 to 0.000001). Because of the compressibility of water, storativity values less than 10^{-6} are not possible in porous media aquifers. When evaluating a report, be aware that the calculated storativities from a software package may be reported much lower than is possible in real life.

Material	Compressibility (m²/N)	Compressibility (ft²/lb)
Clay	10^{-6}–10^{-8}	10^{-5}–10^{-7}
Sand	10^{-7}–10^{-9}	10^{-6}–10^{-8}
Gravel	10^{-8}–10^{-10}	10^{-7}–10^{-9}
Fractured rock and sedimentary rocks	10^{-8}–10^{-10}	10^{-7}–10^{-9}
Igneous and metamorphic rock	10^{-9}–10^{-11}	10^{-8}–10^{-10}
Water	4.4×10^{-10}	2×10^{-9}

TABLE 3.5   Range of Compressibility Values [Adapted from Freeze and Cherry (1979) and Roscoe Moss Company (1990)]

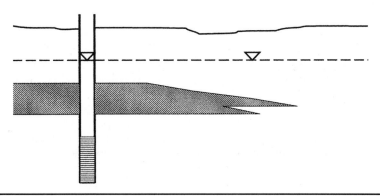

FIGURE 3.9 Schematic of a semiconfined aquifer where a screened interval occurs below a clay layer that pinches out laterally.

So now the question comes up: what about storativities that occur between 0.001 and 0.03? These fall into the leaky confined or semiconfined range (Chapter 6). The closer values are to 0.001, the more semiconfining the aquifer is and the harder it is for leakage to occur. Contrasted to this, storativities nearer to 0.01 reflect a fairly leaky system (Chapter 6). Figure 3.9 is a schematic of a semiconfined system.

In Figure 3.9, the clayey horizon semiconfines the aquifer during the initial stages of pumping. As the cone of influence extends outward, water from above the clay layer eventually contributes water from the pumping stress. In a leaky aquifer, the clayey horizon would more likely be a silty layer that continues to extend across the diagram. As the hydraulic head becomes reduced from pumping in the lower zone, water from above the silty layer is induced through the silty layer as **leakance**. (See Chapter 6 for type curves associated with these conditions in aquifer tests.)

3.3 Movement of Fluids through Earth Materials

Now that we have considered the ability of earth materials to store water, another important element refers to the ability of fluids to move through them. The term most commonly used is **hydraulic conductivity (K)**. It encompasses the ability of a material to transmit fluids under a *unit* hydraulic gradient considering the dynamic viscosity in units of length over time (L/t) (Fetter 1994). **Viscosity** is indicative of the resistance of fluid to flow. It has units of newton-seconds over meters squared (N × s/m²). Thick fluids, such as tar or cold molasses have high viscosity, whereas alcohol is an example of a low-viscosity fluid. **Permeability** is another term commonly used to express the ability of fluids to pass through earth materials, and both hydraulic conductivity and permeability are often used interchangeably in groundwater studies, although it should not be done. This needs an explanation.

The conductivity properties of fluids are related to the **specific weight (γ)** and the **dynamic viscosity (μ)** of the fluid (Appendix B). The specific weight represents the gravitational driving force of the fluid. The ability of fluids to move is also inversely proportional to the resistance of fluids to shearing (Fetter 1994). This is expressed in Equation 3.8:

$$K = k_i \frac{\gamma}{\mu} = k_i \frac{\rho g}{\mu} \qquad (3.8)$$

where k_i = intrinsic permeability, darcy (9.87×10^{-9} cm²)
γ = specific weight = density $\times$ gravity
ρ = density (m/L³)
g = force due to gravity (L/t^2)
μ = dynamic viscosity (m/Lt)

The **intrinsic permeability** (k_i) represents the physical flow properties of the geologic materials. It is essentially a function of the pore-size openings only (no fluids are included). The larger the pore opening, the larger the intrinsic permeability. This relationship can be seen in Tables 3.4A and 3.4B. The specific weight (γ) indicates how a fluid of a given density will be driven by gravity. The dynamic viscosity (μ) indicates that if a fluid is less resistive to fluid flow, the earth material will be more conductive. Those that equate hydraulic conductivity to intrinsic permeability consider most fresh groundwater to have insignificant changes in specific weight and dynamic viscosity; therefore, the ability of groundwater to move is proportional mainly to the intrinsic permeability alone. This should not be done because the hydraulic conductivity will change by a factor of 3 simply by changing the water temperature between 2°C and 30°C. This also becomes problematic when contaminants interact with groundwater with different fluid properties, such as nonaqueous phase liquids (NAPLs) or with denser saline waters.

Example 3.7 What is the hydraulic conductivity of water at 2°C, 16°C, and 30°C if the intrinsic permeability is 0.2 darcy?

Solution We start by seeing what a darcy is [modified from Fetter (1994)]:

$$1\,\text{darcy} = \frac{1\,\text{cP} \times 1\,\text{cm}^3/\text{s}/1\,\text{cm}^2}{1\,\text{atm}/1\,\text{cm}} = 9.87 \times 10^{-9}\,\text{cm}^2 \tag{3.9}$$

where cP = centipoise, a unit of viscosity = 0.01 dyn · s/cm²
atm = atmosphere = 1.10132×10^6 dyn/cm²

From Appendix B one can obtain the density and viscosity values as a function of temperature. An example calculation for water will be done at 2°C with the results for the other two temperatures following.
At 2°C for water,

$$\rho = 0.99994\,\text{g/cm}^3$$

$$\mu = 0.01673\,\text{g/s} \cdot \text{cm}$$

Acceleration from gravity is given as $-g = 980$ cm/s²:

$$0.2\,\text{darcy} \times 9.87 \times 10^{-9}\,\text{cm}^2 = 1.97 \times 10^{-9}\,\text{cm}^2$$

From Equation 3.8, the hydraulic conductivity of water at 2°C is

$$K = K_i \frac{(\rho g)}{\mu} = 1.97 * 10^{-9} \times \frac{0.99994\,\text{g/cm}^3 \times 980\,\text{cm/s}^2}{0.01673\,\text{g/s} \cdot \text{cm}}$$

$$= 1.16 \times 10^{-4}\,\text{cm/s}$$

Similarly, substituting the appropriate values from Appendix B for density and viscosity at temperatures of 16°C and 30°C, respectively (0.99894 and 0.99565 g/cm³ for density and 0.01111 and 0.00801 g/s · cm for viscosity), we obtain hydraulic conductivity values for water:

$$\text{At } 16°C, K = 1.74 \times 10^{-4} \text{ cm/s}$$

$$\text{At } 30°C, K = 2.4 \times 10^{-4} \text{ cm/s}$$

Hydraulic conductivity values for earth materials range over 12 orders of magnitude (Figure 3.10). The distribution of hydraulic conductivity is log normally distributed, so any averaged values should be from geometric means or some other transformation. Obtaining precise values for hydraulic conductivity is unlikely, so when they are reported as such they should be viewed with a jaundiced eye. This leads to a caution when reporting hydraulic conductivity values. Just because your calculator or spreadsheet gives a number with many decimal places, your responsibility as a hydrogeologist is to report only that level of precision that you feel is justified. Typically, this would be with a maximum of two significant figures (Table 3.6).

Hydraulic conductivities are reported in a variety of different units in the literature and in software packages. A hydrogeologist needs to be able to convert back and forth from one unit system to another using unit conversions and by memorizing a few key simple conversion factors (Appendix A). It may be that you are out in the field with only your field book and a calculator and you have to make a decision based upon some numbers.

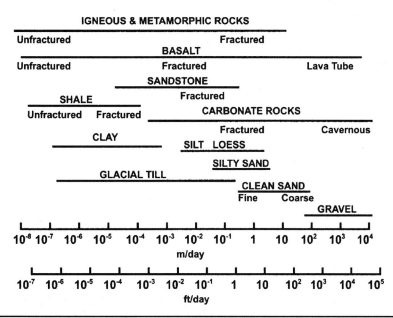

FIGURE 3.10 Ranges of hydraulic conductivity values for earth materials. [*Adapted from Anderson and Woessner (1992).*]

Material Type	Material	Hydraulic Conductivity cm/s
Unconsolidated	Unweathered marine clay	8×10^{-11}–2×10^{-7}
	Clay	1×10^{-11}–4.7×10^{-7}
	Silt, loess	1×10^{-7}–2×10^{-3}
	Fine sand	2×10^{-5}–2×10^{-2}
	Medium sand	9×10^{-5}–5×10^{-2}
	Coarse sand	9×10^{-5}–6×10^{-1}
	Gravel	3×10^{-2}–3
	Till	8×10^{-10}–2×10^{-4}
Sedimentary rocks	Shale	1×10^{-11}–2×10^{-7}
	Siltstone	1×10^{-9}–1.4×10^{-6}
	Sandstone	3×10^{-8}–6×10^{-4}
	Limestone, dolomite	1×10^{-7}–6×10^{-4}
	Karst and reef limestone	1×10^{-4}–2
	Anhydrite	4×10^{-11}–2×10^{-6}
	Salt	1×10^{-10}–1×10^{-8}
Crystalline rocks	Unfractured basalt	2×10^{-9}–4.2×10^{-5}
	Fractured basalt	4×10^{-5}–2
	Weathered granite	3.3×10^{-4}–5.2×10^{-3}
	Weathered gabbro	5.5×10^{-5}–3.8×10^{-4}
	Fractured igneous and metamorphic rocks	8×10^{-7}–3×10^{-2}
	Unfractured igneous and metamorphic rocks	3×10^{-12}–2×10^{-8}

TABLE 3.6   Ranges of Hydraulic Conductivity Values for Various Earth Materials [Adapted from Domenico and Schwartz (1990)]

Example 3.8   A value of hydraulic conductivity of 2×10^{-4} cm/s was estimated for a fine sandy material. The number needed to be converted into ft/day to perform a calculation for average linear velocity during a tracer test (Chapters 13 and 14). The only resources you have are a writing utensil, a field book, and a calculator:

$$K = 2 \times 10^{-4} \frac{\text{cm}}{\text{s}} \times \left(\frac{1 \text{ in}}{2.54 \text{ cm}} \right) \times \left(\frac{1 \text{ ft}}{12 \text{ in}} \right) \times \left(\frac{60 \text{ s}}{1 \text{ min}} \right) \times \left(\frac{1,440 \text{ min}}{1 \text{ day}} \right)$$

$$= 0.57 \text{ ft/day}$$

The average linear velocity was estimated to be

$$V_{\text{ave}} = \frac{K}{\eta_e} \times \frac{\delta h}{\delta l} = \frac{0.57 \text{ ft/day}}{0.26} \times \frac{3 \text{ ft}}{145 \text{ ft}} = 0.045 \text{ ft/day}$$

The hydraulic conductivity can be estimated for sandy materials where the **effective grain size** (d_{10}) is between 0.1 and 3.0 mm (Hazen 1911), where d_{10} represents the smallest 10% of the sample. (It is important to pay attention to the limits over which this is applicable.) The effective grain size is determined from a grain-size distribution plot from passing an oven-dried field sample through a series of sieves (see Chapter 15; Example 3.9). In a grain-size plot, the d_{50} is the median grain size (Table 2.2). Grain-size plots are helpful in determining the sorting or ranges of grain sizes present (Section 2.3). The sorting is estimated with the **uniformity coefficient** (C_u) expressed in Equation 3.9:

$$C_u = \frac{d_{60}}{d_{10}} \tag{3.9}$$

Values less than 4 are considered to be well sorted, and values greater than 6 are considered to be poorly sorted (Fetter 1994). The Hazen (1911) equation relating hydraulic conductivity to effective grain size and a sorting coefficient is shown in Equation 3.10. The most common error made by users of this equation is to forget to convert the grain-size parameters from millimeters to centimeters:

$$K = C(d_{10})^2 \tag{3.10}$$

where K = hydraulic conductivity (cm/s)
  d_{10} = effective grain size (cm, not mm!)
  C = Hazen (1911) sorting and grain-size coefficient (1/cm/s)

The Hazen (1911) coefficient C, which is different from the uniformity coefficient, is assigned according to sorting *and* grain size (Table 3.7). The grain size is determined by evaluating the median grain size (d_{50}) (see Table 2.2) from a grain-size distribution curve (Example 3.9). Values that are poorly sorted and finer grained receive smaller Hazen (1911) coefficient numbers. I recommend that the coefficients in Table 3.7 be estimated only to the nearest value of 10.

Shepard (1989) also evaluated the data from published studies relating grain size to hydraulic conductivity by plotting hydraulic conductivity (in ft/day) versus median grain size (d_{50}) on log-log paper. Various plots were made based upon sediments from different depositional environments (Chapter 2), each forming a straight-line plot. The slope of the plot was related to an exponent (Equation 3.11). The values of the exponent range between 2.0 for glass spheres of equal size and 1.5 for poorly

Description	Coefficient
Poorly sorted to well-sorted very fine sand	40–80
Poorly sorted to moderately sorted fine sand	40–80
Moderately sorted to well-sorted medium sand	80–120
Poorly to moderately sorted coarse sand	80–120
Moderately sorted to well-sorted coarse sand	120–150
Moderately sorted to well-sorted very coarse sand	150–200

TABLE 3.7 Hazen Equation Coefficients in (cm/s) Based on Sorting and Grain Size

Deposit. Environ. (Shape Factor)	0.0l mm K ft/day	0.1 mm K ft/day	1.0 mm K ft/day	10.0 mm K ft/day	Exponent
Glass beads (40,000)	4.0	400	40,000	—	2
Eolian dunes (5,000)	1.0	70	5,000	—	1.85
Beach deposits (1,600)	0.5	28	1,600	—	1.75
Channel deposits (450)	0.23	10	450	20,100	1.65
Sedimentary rock (100)	0.1	3.2	100	3,160	1.5

TABLE 3.8 Hydraulic Conductivity (K in ft/day) Related to Shape Factors (C) Based on Depositional Environment and Median Grain Size (mm) [Adapted from Shepard (1989)]

sorted unconsolidated materials. Example shape-factor C_F values and exponents "i" are shown in Table 3.8:

$$K = C_F \times d_{50}^i \qquad (3.11)$$

where K = hydraulic conductivity (ft/day)

$\quad C_F$ = shape factor (based upon depositional environment) in units that convert mm² to ft/day

$\quad d_{50}$ = median grain size in mm

$\quad i$ = exponent (between 2.0 and 1.5) = slope on log-log plot

Example 3.9 A grain-size distribution based upon water-laid deposits from McCurtain County, Oklahoma (triangles), and Arapahoe County (boxes), Colorado (Morris and Johnson 1967), is plotted in Figure 3.11. The respective hydraulic conductivities are calculated based upon the Hazen (1911) method and the relationship developed by Shepard (1989). These are compared with the hydraulic conductivities determined in the laboratory by Morris and Johnson (1967). The values used in the equations are given in Table 3.9.

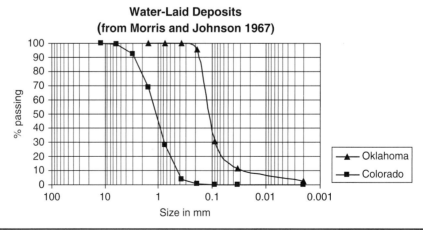

FIGURE 3.11 Plot of grain size versus percent finer by weight. Triangles represent McCurtain County, Oklahoma, and boxes represent Arapaho County, Colorado.

Location	d_{10} mm	d_{50} mm	d_{60} mm	C_u	C	K_{hazen} ft/day	C_F	Exp	$K_{shepard}$ ft/day	K_{lab} ft/day
Oklahoma	0.06	0.17	0.18	3.0	90	9	1,000	1.7	54	11
Colorado	0.67	1.5	1.8	2.6	150	2,080	1,200	1.75	3,360	802

TABLE 3.9 Summary of Calculations to Evaluate Hydraulic Conductivities in ft/day for Two Water-Laid Sand Deposits

Equations 3.9 and 3.10 and Table 3.7 are used to evaluate the hydraulic conductivity using the Hazen (1911) method, and Equation 3.11 and Table 3.8 are used to evaluate the hydraulic conductivity using the Shepard (1989) relationship. Both are compared to the coefficient of permeability reported by Morris and Johnson (1967). The results are shown in Table 3.9 with an example calculation below.

Although the sand from Oklahoma has an effective grain size less than 0.1 mm, a well-sorted sample still allows the calculations to be usable. The uniformity coefficient indicates a well-sorted sand with a d_{50} in the fine sand range (between 0.125 and 0.25 mm; Table 2.2). Converting the median grain size into centimeters and using a Hazen (1911) C factor of 90 yields

$$K_{Hazen} = 90(0.006 \text{ cm})^2 = 3.2 \times 10^{-3} \frac{\text{cm}}{\text{s}} \left(\frac{86,400}{1 \text{ day}} \right) \left(\frac{1 \text{ ft}}{30.48 \text{ cm}} \right)$$

$$= 9.2 \text{ ft/day}$$

For the Shepard (1989) method it is noticed that the sands are water laid and well sorted, a shape factor C_F between beach and channel deposits was selected, with an exponent "i" of 1.7 to yield the following hydraulic conductivity:

$$K_{Shepard} = 1,000 \times (0.18 \text{ mm})^{17} = 54 \text{ ft/day}$$

All values were within one order of magnitude, but the Hazen (1911) method appears to have better agreement with the laboratory-calculated value.

Example 3.7 illustrates how estimates for hydraulic conductivity vary based upon grain-size analysis. It is the author's experience that similar discrepancies also occur when performing pumping tests (Chapters 5 and 6) and slug tests (Chapter 7). Conversions from one system to the units of another are also illustrated in the example. Table 3.10 provides useful conversions for the more commonly used hydraulic conductivity values.

Transmissivity

Hydraulic conductivities (K) can be estimated based upon grain-size relationships or directly in the field by performing slug tests (Chapter 7). Hydraulic conductivities are also estimated from pumping tests (Chapters 5 and 6), by first obtaining a **transmissivity** (T) value. Transmissivity represents the ability of fluid flow through earth materials of a given saturated thickness (b). The transmissivity is directly related to the hydraulic conductivity by the following relationship:

$$T = K \times b \qquad (3.12)$$

where K = hydraulic conductivity
b = saturated thickness

Multiply	By	To Obtain
1 gal/day/ft²	0.1337	ft/day
1 cm/s	2,835	ft/day
1 m/day	3.2808	ft/day
1 gal/day/ft²	4.72×10^{-5}	cm/s
1 ft/day	3.53×10^{-4}	cm/s
1 m/day	1.16×10^{-3}	cm/s
1 gal/day/ft²	4.07×10^{-2}	m/day
1 cm/s	864	m/day
1 ft/day	0.3048	m/day
1 cm/s	21,203	gal/day/ft²
1 ft/day	7.481	gal/day/ft²
1 m/day	24.54	gal/day/ft²

TABLE 3.10   Conversion Factors for Commonly Used Hydraulic Conductivity Values

During a pumping test, the estimated saturated thickness contributing to the screened interval varies depending on the time of pumping, geology, and whether an aquifer is confined or unconfined. More discussion on this topic is found in Section 6.7. The common approach is to assume that the saturated thickness (b) in an unconfined aquifer represents the thickness of saturated materials from bedrock to the water table and in a confined aquifer, b is simply the thickness of the aquifer between confining units. One must realize that this assumes there is no low hydraulic conductivity barrier layers within the aquifer and/or that the entire thickness is screened. This typically occurs only in thin aquifers.

3.4   Aquifer Concepts

An aquifer is a formation, a part of a formation, or a group of formations that contains *sufficient* saturated permeable material to yield *significant* or economic quantities of water to wells or springs. Water within the zone of saturation is at a pressure greater than atmospheric pressure (Section 3.1). The definition is useful because of its flexible application, depending on water use within an area. For example, in a dry, semiarid region, any aquifer that can produce 10 gpm from a well used for stock watering purposes would be considered sufficient and significant (Figure 3.12). Contrasted to this, a center-pivot irrigation production must produce a minimum of 500 gpm for a 160-acre piece (quarter section) of land to be considered sufficient and significant (Figure 3.13). In the latter case, a 10-gpm domestic well in the same area as the pivot well may not be considered very significant.

FIGURE 3.12   Well discharge for stock watering.

FIGURE 3.13   Pivot irrigation system in southwestern Montana.

An aquifer can be made up of a single formation, such as the Dakota Sandstone, where a name followed by a lithology is typically given (AAPG 2005). Other aquifers that consist of groups of formations are probably better known as aquifer systems. Well-known examples are the Floridian Aquifer System (Miller 1986) and the High Plains Aquifer System (Gutentag et al. 1984), both of which consist of units of varying lithologies and ages, where no specific lithology is attached (AAPG 2005). Some aquifers may consist of only one of several members within a formation. For example, within the Fort Union Formation, exposed in eastern Montana and Wyoming are three members: the Tongue River Member, an aquifer that overlies two other members; the Lebo Shale Member, a confining unit; and the underlying Tullock Member, which is also considered an aquifer (USGS 2002). The lowest hierarchy in nomenclature is bed and flow (for lava flow). Within the Tongue River Member of the Fort Union Formation are several thick subbituminous coal beds. The Anderson, Dietz, and Monarch coal beds extend regionally and are important aquifers uses for the watering of stock and have been exploited to extract coal-bed methane (Flores 2004). Aquifer units are separated by confining units of varying physical properties. A summary of the stratigraphic nomenclature is shown in Figure 3.14.

Unconfined Aquifers

When water infiltrates into the ground (Figure 3.3), gravity pulls it downward until it encounters a confining layer. Because flow is inhibited, the zone of saturation literally piles up above this confining layer. The saturated thickness represents the height of the pile, and since the porosity of the overlying materials is connected to the land surface, it is referred to as an **unconfined aquifer**. It can be viewed as an open system. The top of this fully saturated material is called the **water table** and is located where the pressure is atmospheric (Figure 3.15). It also represents a physical moving boundary that fluctuates according the amount of recharge or discharge. Groundwater beneath the water table is under a pressure greater than atmospheric, because of the weight of water above a given point of reference. This concept will be explained in more detail in Chapter 4.

Confined or Artesian Aquifers

Geologic materials that outcrop at the surface at a relatively high elevation (recharge area) receive water by infiltration recharge (Figures 3.16 and 2.20). Materials of sufficient transmissivity that become saturated and are overlain and underlain by confining layers become aquifer units with trapped groundwater. The elevation of the zone of saturation propagates a pressure head throughout the system from the weight of the

```
Group
   Formation
      Member - Lens or Tongue
         Bed(s) or Flow(s)
```

FIGURE 3.14 Lithostratigraphic units to be considered aquifers or confining units based upon the North American stratigraphic code. (*www.nacstrat.org/north-american-stratigraphic-code.*)

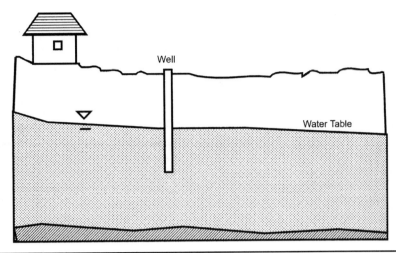

FIGURE 3.15   Unconfined or water table aquifer with pore spaces openly connected to the surface.

FIGURE 3.16   Mountain recharge in southwestern Montana.

water (Chapter 4). When this occurs, there is sufficient pressure head within the system to lift the water within cased wells above the top of the confined aquifer (Figure 3.17). When this occurs, by definition, this is a **confined** or **artesian** aquifer. Both terms refer to the same condition.

Generally, most confined aquifers receive recharge through vertically leaking geologic materials above or below them (Freeze and Cherry 1979). When water levels in

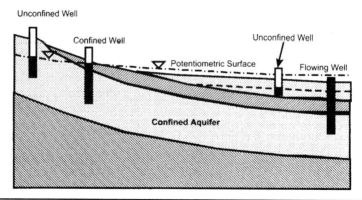

FIGURE 3.17 Confined aquifer and surficial unconfined aquifer setting. Three wells completed in the lower aquifer, two of which are confined. Overall the head in the upper unconfined aquifer is less than the lower confined aquifer.

cased wells rise to the land surface and become flowing wells, they are still considered to be from a confined aquifer (Figure 3.18). The spatially distributed height to which water levels in cased wells rise forms a surface known as the **potentiometric surface** (Figure 3.17). Notice in Figure 3.17 that the confined aquifer and unconfined aquifer have different potentiometric surfaces. In an unconfined aquifer the potentiometric surface is the water table. The slope of this surface is known as the **hydraulic gradient**.

FIGURE 3.18 Artesian flow in an 8-in (20-cm) borehole at a depth of 190 ft (58 m) in fractured bedrock at the Moonlight Basin Ranch near the Big Sky Ski Resort, Montana. Photo taken by David Potts.

The slope of the hydraulic gradient is proportional to the hydraulic conductivity of the geologic materials. The lower the hydraulic conductivity, the greater the slope of the hydraulic gradient (Section 4.1). Each aquifer has its own potentiometric surface from which groundwater flows and other interpretations are made (Chapter 4); therefore, it is important to know stratigraphically where each well is completed (Chapters 4 and 15). It is also important to construct separate potentiometric surfaces for each aquifer to evaluate vertical connectivity. If the potentiometric surfaces match, then they are vertically connected.

Confining Layers

Aquifers yield significant quantities of water, while **confining layers**, although they store water, generally inhibit fluid flow through them. Some prefer to use separate terminology for varying types of confining layers. For example, layers that pass water slowly may be referred to as **aquitards**, while those that are considered to be impermeable are called **aquicludes** (Freeze and Cherry 1979; Fetter 1994). Some substances may have *very* low hydraulic conductivity, but nothing is really impermeable, including geotextiles, the most compacted clay, and the tightest bedrock. Hydraulic conductivities that are six orders of magnitude (a million times) less permeable than another substance may *just* appear to be impermeable.

For practical purposes, confining layers can be defined as those geologic materials that are more than two to three orders of magnitude less permeable than the aquifer above or below it. Those less than two orders of magnitude different are more likely hydraulically connected enough to be considered a single aquifer unit. Using this as a basis, formational stratigraphy, discussed previously, is regrouped and reevaluated in terms of hydrostratigraphy.

Hydrostratigraphy

When one considers the definitions of aquifers and confining layers in performing a site characterization or defining the hydrogeology of an area, it is also important to define the **hydrostratigraphy**. This involves the combining or separation of units with similar hydraulic conductivities into aquifers or confining layers. It may be, for example, that the Ordovician and Mississippian formations (group of formations) collectively form one thick Madison Aquifer, while the Cretaceous Muddy Sandstone (part of a formation) forms another. These may be isolated from aquifers above or below by a part of a formation, a single formation or a group of formations or units. The hydrostratigraphy is used along with structural data to build a **conceptual model** of a flow system (Chapter 4).

Example 3.10 On the Hualapai Plateau, Arizona, it was desired to drill wells for water development rather than depending upon temporary stock tanks behind earthen dams (Huntoon 1977). The earthen dams were successful only if flash floods occurred, which only trapped relatively poor-quality water. The target zone for the drill project was the lowermost Rampart Cave Member of the Muav Limestone, some 1,500 ft below the canyon rim, Cambrian in age. The Cambrian section was originally described by Nobel in 1914 and 1922 and then in detail by McKee and Dresser in 1945. They honored the nomenclature designated by Nobel (1922) but assigned all shale units as part of the Bright Angel Shale and the limestone units as part of the Muav Limestone. This posed a significant problem because the depositional environment was that of the transgressive sea (with typical back-and-forth progression) where the two formations interbed and lithologic subunits extend for miles. The lowest three members of the Muav Limestone are interbedded with shale units that are "unnamed" (Figure 3.19). The hydrogeologist who was hired to define the target zones said that if they drilled past the limestone

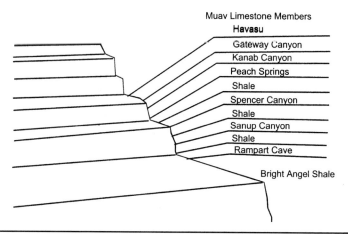

FIGURE 3.19 Detailed Cambrian section of the lower part of the Muav Limestone on the Hualapai Plateau, Arizona. [*Adapted from Huntoon (1977).*]

into the upper part of the shale, they would reach maximum production (Huntoon 1977). The problem was that drilling took place through the Peach Springs Member of the Muav Limestone (Figure 3.19) and the driller terminated in the uppermost "unnamed" shale. The hydrogeologist failed to inform the driller that there were two other limestone and shale units that needed to be penetrated before the target zone would be encountered. At most of the drill locations, the drillers had unwittingly stopped short of the target by 150 to 250 ft (Huntoon 1977).

A given conceptual model of groundwater movement depends somewhat on scale. There are regional basin-wide groundwater studies that may include several aquifers and confining units, contrasted to local groundwater studies, that may be a small as at a construction site. The hydrostratigraphy of a small residential site (e.g., 5 acres, 2.0 ha) would likely be defined by local shallow drill-hole information.

3.5 Boundary Concepts

The hydrostratigraphy, structural changes, and spatial distribution of earth materials all affect groundwater flow and how one separates the saturated materials into aquifer units. Each aquifer will have its own potentiometric surface and a hydraulic gradient corresponding to its hydraulic conductivity. If there are multiple aquifers in a particular area, it is important to identify the potentiometric surface of each aquifer separately. Figure 3.20 shows the potentiometric surface of four aquifers: an unconfined bedrock aquifer associated with well D, unconfined sand, and gravel aquifer (dashed line associated with well A), and two other confined aquifers (solid lines, wells B and C). Note how the potentiometric surface of the intermediate layer (well B) is lower than the others, but the potentiometric surface of the bottom layer (well C) is higher than that of wells A and D. The slope of the potentiometric surfaces is toward or away from the reader, except in the case of the bedrock aquifer, which slopes toward the alluvial system. Because groundwater always moves from areas of higher total head to areas of lower total head, the groundwater from the lowest system (well C) wants to move upward through a confining layer toward the intermediate aquifer. Additionally, the

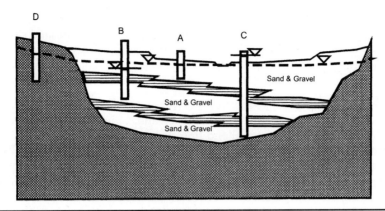

FIGURE 3.20 Schematic illustrating four aquifer units: well A, an unconfined aquifer connected to the river; well B, an intermediate confined aquifer; well C, completed in a confined aquifer; and well D, completed in bedrock.

groundwater from the surficial unconfined aquifer wants to move downward toward the intermediate (well B). Groundwater will always move as long as there is a head difference from one location to another, thus creating a hydraulic gradient. There may be a horizontal component to groundwater flow (within aquifers) and a vertical component (between aquifers and in recharge or discharge areas) of groundwater flow. Groundwater continues to move until equilibrium is reached. The water level observed at well C is above the land surface in the cross-sectional view of Figure 3.20, because "up valley" the bottommost confining unit extends far enough up gradient that recharge waters enter the lowest aquifer layer at an elevation above the stage of the stream at the position of the cross section.

Figure 3.21 is an example that shows how structural influences and variations in aquifer hydraulic properties create other types of aquifer boundaries. In the diagram

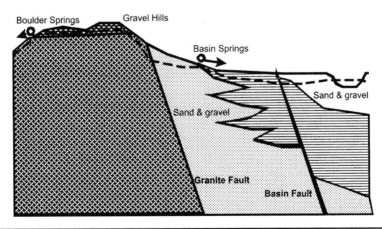

FIGURE 3.21 Diagram illustrating aquifer boundary conditions. Boundary conditions are caused by varying hydraulic conductivities of rock units and structural offsets. The horizontal pattern signifies shale.

there are two springs: Boulder and Basin. Boulder Spring is a contact spring where a unit of higher hydraulic conductivity (surface gravels) intersects the land surface where the granite becomes exposed. Questions regarding the occurrence of Basin Spring and other features in Figure 3.21 are posed for the reader at the end of the chapter.

Homogeneity and Isotropy

There are two terms that are often confused and/or ignored in hydrogeologic studies that are related to the physical and flow properties of an aquifer: **heterogeneity** and **anisotropy**. Heterogeneity refers to the lateral and vertical changes in the physical properties of an aquifer. For example, the grain-size arrangement or packing, thickness, porosity, cementation, and hydraulic conductivity at one location (site A) can be compared with the same physical properties at another location (site B) (Figure 3.22). If the variability of the physical properties relative to the volume of the system of interest is small, the aquifer is said to be **homogenous**. A **heterogeneous** aquifer will have a different saturated thickness, porosity, or some other physical property that is different from location to location. For example, if a sandstone becomes finer grained laterally (as in a facies change) or thins or thickens laterally, it is *not* homogenous.

Anisotropy refers to differences in hydraulic properties with respect to the direction of flow. If a given porous media were made up of spheres in cubic packing, the permeability of the media would be the same in all directions. In most depositional environments, the sediments are laid down in a stratified manner, so that the vertical hydraulic conductivity is *very* different from the horizontal hydraulic conductivity. According to the author's experience, this difference can be from percentages to a couple of orders of magnitude or more. In considering a Cartesian coordinate system, where Z represents the vertical dimension, within the X-Y plane there is a preferential flow in one direction over the other, where the maximum ability is typically assigned to the X-direction (Figure 3.23). Hydraulic conductivity (K) is a tensor; therefore, the X-direction (K_{xx}) represents a maximum value and any other oblique path would have a lesser magnitude. A summary of heterogeneity and anisotropy concepts in the X-Y plane is shown in Figure 3.24. Faults, fractures, bedding planes, and other features are potentially important causes of anisotropy and heterogeneity, and therefore have an influence on fluid flow.

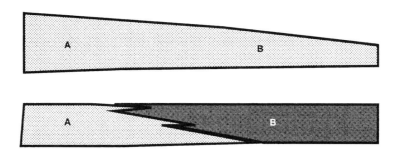

FIGURE 3.22 Schematic showing heterogeneity. In the upper diagram, the thickness changes from point A to point B. In the lower diagram, the grain size and packing change from point A to point B.

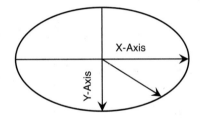

Figure 3.23 Plan view of *X-Y* plane flow properties. Since hydraulic conductivity is a tensor, the net flowability in the oblique direction is less than the magnitude of the *X*-direction alone.

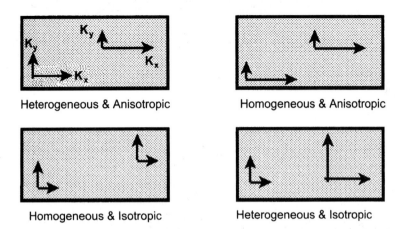

Figure 3.24 Combinations of heterogeneity and isotropy. [*Adapted from Freeze and Cherry (1979).*]

3.6 Springs

Springs are an important source of hydrogeologic information. They occur because the hydraulic head in the aquifer system intersects the land surface (Figure 3.25). By being observant to their distribution, flow characteristics, and water qualities, much valuable information can be derived without the need of drilling a single well. Springs can sometimes provide more information than wells.

Perhaps the most common types of springs are contact springs. In an area of layered sedimentary rock, springs may develop if recharge enters a higher hydraulic conductivity unit overlain by a lower hydraulic conductivity unit where the contact is exposed. In the field, these are observed as places where trees and other vegetation flourish in an otherwise dry area. Also look for game trails that lead into the vegetation, as local animals know where to find water. Springs sourced from confined aquifers may occur in structurally complex areas where the recharge elevation is high, relative to the location of the rock outcrops lower in the system. They occur either from folded beds, depositional contacts, or fault planes. Importantly, they represent places of discharge where minimum hydraulic head occurs locally within an aquifer system. A summary of the physical characteristics of springs and their significance is listed in Table 3.11.

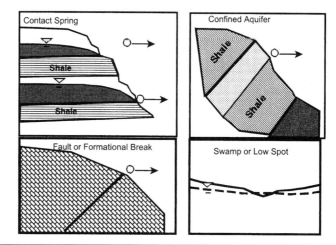

Figure 3.25 Cross-sectional views of a variety of springs: contact spring, confined aquifer (could be a fault or depositional plane), and a swamp or low spot.

Characteristic	Comments
Provides elevation of minimum head	As single points or a string of points on a topographic map
Hydrostratigraphy defined	Vegetation at discharge points separated by areas lacking in vegetation
A point of maximum hydraulic conductivity	Either coarser grained or more fractured
Discharge type, fractures, porous discharge	Type of aquifer permeability observed
Discharge rate	Volume and frequency may be steady or vary
Discharge time	Steady flow indicates a larger system than one that stops flowing after a short time
Type of aquifer	Confined or unconfined
Groundwater quality	General chemistry of aquifer
Place to age date	Provides a place of maximum residence time in the aquifer
Thermal data	Circulation depths or thermal influences
Periodicity	Is there a regular discharge interval?
Can be used to understand structural relationships	Discharge areas can be correlated with geologic structures

Table 3.11 Physical Characteristics of Springs

3.7 Summary

The subsurface geology once it becomes saturated is divided into hydrostratigraphic units consisting of aquifers and confining units. The physical properties of aquifers include their ability to store and transmit water while confining layers tend to inhibit flow and serve as barriers to flow. Porosity represents the volume of pore space that can be occupied by groundwater. Porosity can be estimated from cores taken in the field and then tested for their water content. The effective porosity is that part of porosity available for fluid flow. It is often estimated from values of specific yield and is an important part of estimating average linear groundwater velocities.

Hydraulic conductivity is a measure of the transmissive properties of saturated earth materials. It can be estimated based upon grain size or by stressing the aquifer through pumping tests. The rate at which water can move through an aquifer is reflected by the slope or hydraulic gradient of the potentiometric surface; high hydraulic conductivity materials have a comparatively flatter hydraulic gradient than low hydraulic conductivity units.

Aquifers may be unconfined (open systems), confined, leaky confined, or semiconfined. Each aquifer needs to be evaluated for its own potentiometric surface. The hydrostratigraphy is placed within the geologic framework to create a conceptual model of groundwater flow. Fault zones and bedrock may behave as barriers and influence the distribution of hydraulic head. An often overlooked approach in deciphering the structure of an area and determining aquifer characteristics is taking time for observations at springs. Springs in an area provide direct quantitative information about the aquifer and its chemistry.

3.8 Problems

3.1   Assume you have a container with inside dimensions of 10 by 10 by 10 cm into which you place enough 1-cm-diameter spheres to fill the box. For this problem, let us assume that the packing of the spheres is cubic packing. Answer the following:

   a. What is the pore volume in cm^3 in the box? (this is not porosity; it is a volume).

   b. What is the total surface area of the spheres exposed to the pore volume in cm^2? (assume that the area at the "touch" points are negligible).

   c. What happens to the numbers in "a" and "b" when the diameter of the spheres are changed to 0.02, 0.2, 2, and 5 cm? (you should make a table).

   d. Are there any relationships or patterns that emerge? Explain.

3.2   You have a fractured rock with uniformly spaced joints that are orthogonal to each other (every joint is at 90° angles, like little boxes). The rock is cubic, 10 m on a side and the typical joint separation is 1 mm in each direction. What is the pore volume and surface area exposed for the following joint spacing?

   a. 1 m   b. 0.5 m   c. 0.1 m

An aquifer has a specific yield of 0.19. During a drought period, the following drops in water level were noted. How much water is represented by the drop?

Area	Size	drop (ft)
A	8.6 mi²	1.40
B	12.3 mi²	2.44
C	14.9 mi²	3.35
D	8.1 mi²	2.49

3.3 Convert the following with unit conversions you could remember during an exam or in the field without a reference book:

   a. 122 acre-ft/year to gal/min (gpm)

   b. 37,880 gal/day-ft² (gpdf²) to centimeters per second (cm/s).

   c. 9×10^{11} cm³ of water to (1) liters and (2) gallons

From the table below, plot the grain-size distribution for footage 185 to 195 and 200 to 205 (all values are in grams) and estimate the hydraulic conductivity using the Hazen method.

Footage	Clay <0.004 mm	Silt 0.004– 0.063 mm	V. Fine sand 0.063– 0.125 mm	Fine sand 0.125– 0.25 mm	Med. Sand 0.25– 0.5 mm	Coarse Sand 0.5– 1.0 mm	V. Coarse Sand 1–2 mm	V. F. Gravel 2–4 mm	Fine Gravel 4–8 mm	Med. Gravel 8–16 mm
180–185	0.6	1.1	7.6	16.8	23.4	20.1	4.4	1.5	0	
185–190	0	0.3	1.7	1.8	2.9	9.3	30.9	21.0	9.9	6.2
190–195	0	3.4	5.2	8.2	14.4	34.4	31.1	19.6	10.5	
195–200	0	0.5	1.7	3.2	8.5	14.0	21.6	15.2	10.9	8.6
200–205	0	1.1	3.4	4.7	7.0	12.4	33.2	32.8	17.1	26.5

TABLE 3.4A   Grain-Size Distribution of Differing Water-Laid Soils [From Morris and Johnson (1967)]

3.4   A sedimentary rock has a permeability of 0.58 darcy and a porosity of 24%. What is the hydraulic conductivity (ft/day) for pure water at the following temperatures (use Appendix B to help you):

   a. 5°C   b. 22°C   c. 40°C

In each case of Problem 3.4, calculate the average linear velocity (ft/day) assuming a hydraulic gradient of 90 ft/mi and the distance that a nonreactive contaminant would travel in 1 year.

3.5   In considering the boundary conditions of Figure 3.21, the granitic bedrock and the clay layers separate the three sand and gravel aquifers from each other, evidenced by the differing water-level elevations. Given the low porosity of the granite, long-term yields from well D would be expected to be what compared to well C? What would the water levels in wells be like if the bedrock were siltstone instead of granite? Think of the physical properties of siltstone compared to granite.

3.6   In considering Figure 3.22, why is Basin Spring there? What effect is the Basin Fault having on the system?

3.7   Notice that the river is losing water. Are the slopes of the hydraulic gradients within the different units appropriate? What would you suggest?

3.9   References

AAPG, 2005. North American Stratigraphic Code. *AAPG Bulletin*, Vol. 89, No. 11 (November 2005), pp. 1547–1591. https://ngmdb.usgs.gov/Info/NACSN/05_1547.pdf

Anderson, M.P., and Woessner, W.W., 1992. *Applied Groundwater Modeling: Simulation of Flow and Advective Transport*. Academic Press, San Diego, CA, 381 pp.

Baldwin, D.O., 1997. *Aquifer Vulnerability Assessment of the Big Sky Area, Montana*. Master's Thesis, Montana Tech of the University of Montana, Butte, MT, 110 pp.

Clark Fork River Cleanup. http://deq.mt.gov/Land/fedsuperfund/sst

Cleary, R.W., Pinder, G.F., and Ungs, M.J., 1992. *IBM PC Applications in Ground Water Pollution & Hydrology*. Short course notes. National Ground Water Association, San Francisco, CA.

Domenico, P.A., and Schwartz, W.W., 1990. *Physical and Chemical Hydrogeology*. John Wiley & Sons, New York, 824 pp.

Driscoll, F., 1986. *Groundwater and Wells*. Johnson Filtration Systems, St. Paul, MN, 1108 pp.

Fetter, C.W., 1994. *Applied Hydrogeology, 3rd Edition*. Macmillan, New York, 691 pp.

Flores, R.M., 2004. *Coalbed Methane in the Powder River Basin, Wyoming and Montana: An Assessment of the Tertiary-Upper Cretaceous Coalbed Methane Total Petroleum System*. https://pubs.usgs.gov/dds/dds-069/dds-069-c/REPORTS/Chapter_2.pdf

Freeze, A., and Cherry, J., 1979. *Groundwater*. Prentice Hall, Upper Saddle River, NJ, 604 pp.

Gutentag, E.D., Heimes, F.J., Krothe, N.C., Luckey, R.R., and Weeks, J.B., 1984. *Geohydrology of the High Plains Aquifer in Parts of Colorado, Kansas, Nebraska, New Mexico, Oklahoma, South Dakota, Texas, and Wyoming*. U.S. Geological Survey Professional Paper 1400-B.

Hazen, A., 1911. Discussion: Dams on Sand Foundations. *Transactions of the American Society of Civil Engineers*, Vol. 73, 199 pp.

Huntoon, P.W., 1977. Cambrian Stratigraphic Nomenclature and Ground-Water Prospecting Failures on the Hualapai Plateau, Arizona. *Ground Water*, Vol. 15, No. 6, pp. 426–433.

McKee, E.D., and Dresser, C.E., 1945. *Cambrian History of the Grand Canyon Region*. Carnegie Institute, Washington Publication 563, 168 pp.

Miller, J.A., 1986. *Hydrogeological Framework of the Floridian Aquifer System in Florida and Parts of Georgia, Alabama, and South Carolina*. U.S. Geological Survey Professional Paper 1043-B.

Morris, D.A., and Johnson, A.I., 1967. *Summary of Hydrologic and Physical Properties of Rock and Soil Materials, as Analyzed by the Hydrologic Laboratory of the US Geological Survey 1948–1960*. U.S. Geological Water-Supply Paper 1839-D.

Nobel, L.F., 1914. *The Shinumo Quadrangle, Grand Canyon District, Arizona*. U.S. Geological Survey, Bulletin 549, 100 pp.

Nobel, L.F., 1922. *A Section of the Paleozoic Formations of the Grand Canyon at the Bass Trail*. U.S. Geological Survey Professional Paper 131, pp. 23–73.

Robson, S.G., 1993. *Techniques for Estimating Specific Yield and Specific Retention from Grain-Size Data and Geophysical Logs from Clastic Bedrock Aquifers*. U.S. Geological Survey, Water-Resources Investigations Report 93-4198, 19 pp.

Roscoe Moss Company, 1990. *Handbook of Ground Water Development*. John Wiley & Sons, New York, 493 pp.

Shepard, R.G., 1989. Correlations of Permeability and Grain Size. *Ground Water*, Vol. 27, No. 5, pp. 633–638.

Stephens, D.B., 1996. *Vadose Zone Hydrology*. Lewis Publishers, CRC Press, Boca Raton, FL, 347 pp.

USGS, 2002. *Water Quality and Environmental Isotopic Analyses of Ground-Water Samples Collected from the Wasatch and Fort Union Formations in Areas of Coalbed Methane Development—Implications to Recharge and Ground-Water Flow, Eastern Powder River Basin, Wyoming.* Water-Resources Investigations Report 02-4045. U.S. Geological Survey, Cheyenne, WY.

CHAPTER 4

Groundwater Flow

Willis D. Weight

Professor of Engineering, Carroll College, Helena, Montana

Gregory P. Wittman

Geoscience Associates, LLC, Kalispell, Montana

4.1 Groundwater Movement

Groundwater is generally always moving. Movement occurs from higher hydraulic head in **recharge** areas (natural or artificial), where precipitation is generally higher, to **discharge** areas of lower hydraulic head (wells, springs, rivers, lakes, and wetlands). The reason groundwater moves is because there always seems to be a "change in head" somewhere in a groundwater system. This "change in head" results in a slope (hydraulic gradient) in the potentiometric surface (Chapter 3). Gravity is the driving force that moves water. Infiltrating groundwater moves downward until it reaches a horizon with low enough hydraulic conductivity to begin accumulating. Groundwater moves so slowly in porous media (ft/year to ft/day) that accumulated water in recharge areas may build up for months or years before their effects in the system can be observed. If you raised up one corner of a bathtub full of water, gravity would cause the water to move toward the low corner. If the bathtub was filled with saturated sand and you raised up a corner (you would be very strong!), water would still move, to the lower corner but much more slowly. The quantity of groundwater movement through porous media is defined through a governing flow equation known as **Darcy's law** (Figure 4.1).

Darcy's Law

In the mid-nineteenth century, a French engineer Henry Darcy (1803–1858) systematically studied the movement of water through sand columns (Darcy 1856). He was able to show that the volumetric rate (Q) of groundwater was proportional to the intrinsic permeability (see Chapter 3) of the porous media (k) and the change in head (hydraulic gradient) over the length of the sand column (Figure 4.1). This can be expressed as Darcy's law:

$$Q = K\frac{\partial h}{\partial l}A = KiA = Tiw \tag{4.1}$$

131

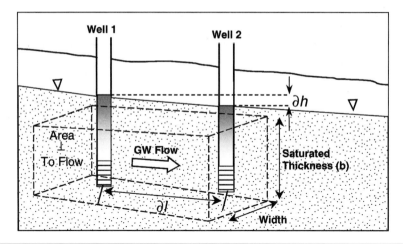

FIGURE 4.1 Schematic of Darcy's law showing change in head (δh) over the flow path length (δl) and the area perpendicular to flow (aquifer width times thickness (b) for horizontal flow).

where Q = volumetric discharge rate (L^3/t)

$\partial h/\partial l = i$ = hydraulic gradient (L/L) or slope of the potentiometric surface

$A = b \times w$, the cross-sectional area perpendicular to flow (L^2), (in horizontal flow, the saturated thickness (b) multiplied by the width of the aquifer (w) or a part of the width of an aquifer)

K = hydraulic conductivity (L/t)

$$T = K \times b \; (L^2/t)$$

There are other governing flow equations that have a similar form as Darcy's law (Equation 4.1); one is for the flow of electricity (Ohm's law), and another is for heat flow or heat transfer (Fourier's law). The governing principles behind all of these (Sert 2012) include:

- Conservation of mass
- Conservation of linear momentum (Newton's 2nd law)
- Conservation of energy (1st law of thermodynamics)

Georg Ohm (1789–1854), a German physicist, also did a series of experiments like Darcy did and came up with the following equation:

$$i = -K\frac{V_A - V_B}{L}A = -K\frac{dV}{dx}A \qquad (4.2)$$

where i = current

K = electrical conductivity

$K = 1/\rho$, where ρ = resistivity

V = voltage, or voltage difference

L = distance

A = area

It is also written as $V = IR$, where $R = \rho(L/A)$.

Jean-Baptiste Joseph Fourier (1768–1830), a French mathematician and physicist, also conducted several experiments and published his work on heat flow (Fourier 1822). He formed the following equation (Jones 2012; Smith 2017):

$$q = -K\frac{T_A - T_B}{dx}A = -K\frac{dT}{dx}A \tag{4.3}$$

where q = rate of heat flow
$\quad K$ = thermal conductivity
$\quad T$ = temperature, or temperature difference
$\quad x$ = distance
$\quad A$ = cross-sectional area

In all three of these cases, there is movement of water, current, or heat from higher or larger amounts to lower values.

There is a significant difference between the hydraulic conductivity (K) and the intrinsic permeability (k), although they are frequently used interchangeably in practice because the fluid properties of most uncontaminated groundwaters are fairly similar at similar temperatures (Chapter 3; Example 3.7). Waters with total dissolved solids (TDS) less than 5,000 mg/L also have similar enough viscosities and specific weights (Fetter 1994).

Darcy's law can also be viewed as

$$Q = q \times A \tag{4.4}$$

where

$$q = K\frac{\partial h}{\partial l} = Ki \tag{4.5}$$

q = the Darcian velocity, also known as the specific discharge

Even though the Darcian velocity or **specific discharge** (q) has units of velocity (L/t), it does not represent a true groundwater velocity. This is because the porous media occupy some of the area perpendicular to flow. To account for this, one must include the effective porosity (n_e) of the porous media. The n_e represents the porosity available for fluid flow. This velocity is actually an average linear velocity (V_{ave}) as the water moves around the grains from one point (high head) to another (low head). Picture in your mind's eye a water molecule moving as a star-fighter pilot dodging and maneuvering through the porous media in its tortuous path down the hydraulic gradient; V_{ave} is also known as the **seepage velocity** (V_s), expressed as

$$V_s = \frac{K}{n_e}\frac{\partial h}{\partial l} = \frac{q}{n_e} \tag{4.6}$$

where all terms have been defined previously (Section 3.3).

Example 4.1 The seepage velocity (V_s) or average linear velocity can be used to estimate the time of travel of groundwater for a nonreactive dissolved contaminant as it is being advectively carried along by groundwater moving down hydraulic gradient. The general relationship is

$$\text{Time of travel} = \frac{\text{Distance traveled}}{V_s} \tag{4.7}$$

Bizarre Application   Franco Horowitz escaped from a low-security work farm and dumped a highly soluble deadly contaminant "methyl death" down his ex-wife's domestic well. He was sent to prison after being convicted for assault from chasing her lover, the town mayor, with a tire iron. The town mayor has a domestic well 200 ft down gradient and Franco was seeking revenge. We are going to assume it takes one day for "methyl death" to begin a direct flow path toward the mayor's well. If the hydraulic gradient is 45 ft/mi (1 mi = 5,280 ft), the effective porosity is estimated to be 0.22, and the hydraulic conductivity is approximately 20 ft/day, how long will it take "methyl death" to reach the mayor's well?

  We apply the above relationship after using Equation 4.6 to solve for V_s, the groundwater velocity of "methyl death."

$$V_s = \frac{K \times \partial h}{n_e \times \partial l} = \frac{20 \text{ ft/day} \times 45 \text{ ft}}{0.22 \times 5,280 \text{ ft}} = 0.775 \text{ ft/day}$$

$$\text{Time of travel} \sim \frac{200 \text{ ft}}{0.775 \text{ ft/day}} = 260 \text{ days}$$

Discussion   This relationship is used to make a back-of-the-envelope "first approximation" estimate for time of travel. Since V_s represents the average linear velocity, the leading edge of the plume for "methyl death" may arrive much sooner than 260 days. As the mayor uses his well, the cone of depression from pumping will intermittently steepen the slope of the hydraulic gradient. Therefore, to report a number more precise than 260 days is discouraged because most of the parameters used in the estimate for time of travel were also estimated.

  Darcy's law is valid in the saturated zone, in multiphase flow, and in the vadose zone, although adjustments to Equation 3.1 have to be made. For example, in the case of unsaturated flow in the vadose zone, hydraulic head is due to matrix suction (tension) and elevation head and hydraulic conductivity ($K(\theta)$) becomes a function of the moisture content (θ) (Fetter 1993; Section 13.4).

Hydraulic Head

In Chapter 3, the concepts of potential, hydraulic head, and total head are mentioned. These will be developed here.

  Whenever water moves from one place to another, it takes work (W) to accomplish this. The work performed in moving a unit mass of fluid is a mechanical process and is known as the fluid potential (Φ) (Freeze and Cherry 1979). The forces moving the fluid have to overcome the resistant frictional forces of the porous media. The work calculation represents three types of mechanical energy (Freeze and Cherry 1979). The total fluid potential (Φ_T) (the mechanical energy per unit mass) is expressed as the sum of the three components of the Bernoulli equation, which are velocity potential, elevation potential, and pressure potential (the dimensions of work and energy are the same: ML/t^2):

$$\Phi_T = \frac{\upsilon^2}{2} + gz + \frac{P}{\rho} \tag{4.8}$$

where υ = fluid velocity (L/t)
   g = acceleration due to gravity (L/t^2)
   z = elevation above a datum (L)
   P = pressure = Force/A, (ML/t^2)/L$_2$ = ML/t^2
   ρ = fluid density, (M/L^3)

M. King Hubbert (1940) showed that the fluid potential at any point is simply the total hydraulic head multiplied by the acceleration due to gravity (g).

$$\Phi_T = gh_t \tag{4.9}$$

Since g is very nearly constant near the earth's surface, fluid potential and hydraulic head are nearly perfectly correlated (Freeze and Cherry 1979). If the components of the Bernoulli equation (Equation 4.8) (expressed as energy per unit mass) are divided by gravity, we obtain the three components of total hydraulic head expressed as fluid potential on a unit weight basis.

$$h_t = \frac{v^2}{2g} + Z + \frac{P}{\rho g} \tag{4.10}$$

Dimensionally, these are all measured in terms of length (feet or meters). The first term is known as the *velocity head* (h_v), the second term is called the *elevation head* (h_z), and the last term is called the *pressure head* (h_p).

Since groundwater moves so slowly (ft/year to ft/day), the velocity head (h_v) is considered to be negligible (see Example 4.2). Therefore, the total head (h_t) at any point in a groundwater system (confined or unconfined) can be expressed in terms of the elevation head (h_z) and the pressure head (h_p). Equation 4.10 can be simplified to:

$$h_t = h_z + h_p \tag{4.11}$$

Example 4.2   The typical hydraulic conductivity of a fine to silty sand is approximately 0.3 ft/day (1×10^{-4} cm/s). Is the velocity head negligible?

Solution   The hydraulic conductivity must be converted to a velocity using Equation 4.6 by making some assumptions. If the effective porosity can be assumed to be 25% and a gradient of 0.008 ft/ft is used, then the velocity can be calculated.

By using the "first" head term in Equation 4.10, one can calculate the relative velocity head.

$$v = \frac{[(0.3\ \text{ft/day})(0.008\ \text{ft/ft})]}{0.25} = 9.6 \times 10^{-3}\ \text{ft/day}$$

$$\frac{v^2}{2g} = \frac{[(0.0096\ \text{ft/day})(1\ \text{day}/86{,}400\ \text{s})]^2}{2(32.2\ \text{ft/s}^2)} = 1.92 \times 10^{-16}\ \text{ft}$$

This appears to be a negligible value.

Another view point that may help the reader understand what hydraulic head is can be viewed from the perspective of pore-water pressure at a static position within the porosity of geologic materials below the ground surface (GS) (hydrostatics). In the context of the energy equation (Bernoulli) the position being considered is referenced from some datum [usually above mean sea level (AMSL)]. The total head at the position of interest then would include the elevation and pressure head.

Earlier in Section 3.1 we considered the water table to be at atmospheric pressure or referenced as zero pressure head, it is also known as the phreatic or piezometric surface. In the vadose zone the pressure head is less than atmospheric (negative pressure head) because of capillarity and tension or matric suction, and is a function of the degree of saturation or percent of water content (Mitchell 1960; Kohler 2014). Below the water

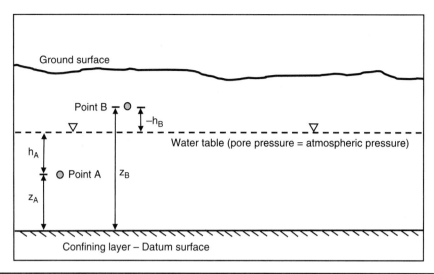

FIGURE 4.2   Diagram showing total head at point A and point B and respective elevation and pressure head values of each. [*Modified after Veissman and Lewis (2003)*.]

table, where the porosity of geologic materials is fully saturated, groundwater is at a pressure head greater than atmospheric according to the depth below this surface.

The hydrostatic head conditions at a particular point in the aquifer or vadose zone can be referred to by its **pore-water pressure** or gauge pressure (u) (Mitchell 1960; Kohler 2014). These concepts are illustrated in Figure 4.2. It can be expressed as

$$u = h_a \gamma \tag{4.12}$$

where u = pore-water pressure or gauge pressure (lbs/ft² or kN/m²)
  h_a = vertical head measured from a point relative to the water table (ft or m)
  γ = specific weight of water (roughly 62.4 lb/ft³, or 9.81 kN/m³)

In Figure 4.2 at point A the pore-water pressure (u) is equal to the vertical distance from point A to the water-table surface (h_A), caused by the weight of the water above it. In a swimming pool if one swims below the surface to a depth of 3 meters, they feel the hydrostatic pressure head of 3 meters of water acting upon them. Whether in a small pool or a large lake it is the depth below the surface that the swimmer feels, not the horizontal dimension of the water body. In the case of point B, tension or matrix suction of the water clinging to the surfaces of the geologic materials against the force of gravity is what causes the hydraulic pressure head to be negative ($-h_B$). From an energy equation point of view, the negative pressure head combined with the elevation head may produce a net total head equivalent to the elevation of the water table above a datum. The relationship between degrees of pore-water pressure above and below the water table, including the capillary fringe, is shown in Figure 4.3. The pore-water pressure conditions in the vadose zone are more complicated because of the variable water degree of saturation that may occur.

The total head in a given aquifer, whether unconfined or confined, is reflected by the elevation of the static water level (SWL) in a cased well. When these elevations are

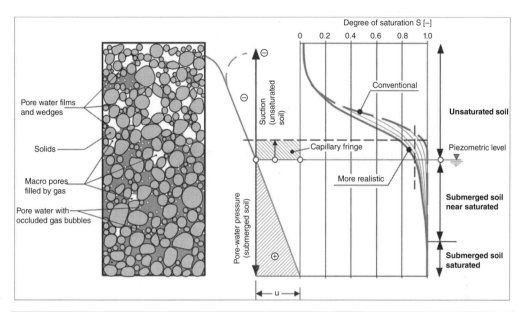

FIGURE 4.3 Relationship of pore water (*u*) and magnitude above and below the water table, also showing the conditions in the capillary fringe. (*Used by permission from Taylor and Francis Publishing.* http://blog.hj-koehler.de/boden-unter-wasser.)

contoured, a surface is generated known as the **potentiometric surface**. The potentiometric surface is a surface representing total head (elevation head, h_z; and pressure head, h_p). In an unconfined system, where the pressure head is zero everywhere (atmospheric), it is the water table.

A piezometer can be thought of as a mini well with a very short screen, where the length of openings in the screen or perforations in the casing is in terms of centimeters (Chapter 5; Section 15.6). Therefore, the water level observed in a piezometer essentially reflects the pore-water conditions at a "vertical point" within the aquifer. In all cased wells the water level observed reflects the total head occurring at the midpoint of the screen. The longer the screen length in a particular well the greater the range of vertical pore-water pressures represented (more on this in the next section, where vertical flow is discussed).

In a confined aquifer the total head is also represented by water levels in cased wells. For an aquifer to be confined, by definition the water level rises above the physical top of the aquifer (Chapter 3). This occurs because confined aquifers are pressurized. Pressurization results from an aquifer being confined from above and below by materials of lower hydraulic conductivity, while recharge waters entering the system (at some position laterally) do so at a significantly higher elevation than the elevation of the aquifer at the position of the cased well (Figure 3.17; Chapter 3). Since water levels in confined aquifers rise proportionally to the pressure head within the aquifer system, their elevations represent points of total head in the potentiometric surface. When these points are contoured one can determine the slope of this surface and estimate the direction of groundwater flow and make calculations using Darcy's law. The potentiometric surface can also be conceptualized as the total head

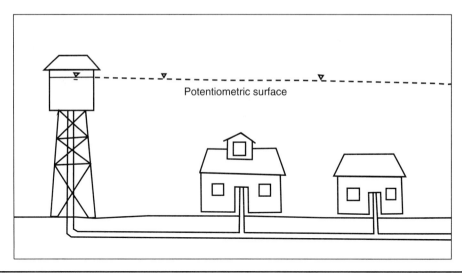

FIGURE 4.4 Potentiometric surface associated with a water tower showing head losses in the direction of water delivery.

in a water-supply system (Figure 4.4). The water tower provides pressure head in the system. Head losses occur from frictional interaction of water with pipes as it moves down through the system.

Examples 4.3 and 4.4 and Figure 4.5 illustrate the concept of total head in an unconfined and confined aquifer relative to a common datum. Again, the most commonly used datum is mean sea level (MSL).

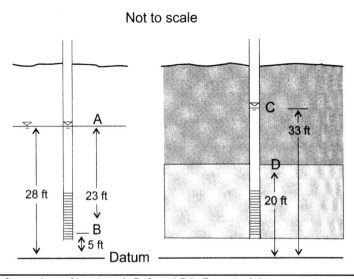

FIGURE 4.5 Comparison of heads at A, B, C, and D in Example 4.4.

Example 4.3 A well completed in a unconfined aquifer (left) and a confined aquifer (right) are compared at two points within the well to show that the static water level in cased wells represents the total head (Figure 4.5). Note now the water level at point C is above the top of the aquifer (point D) indicating that this aquifer is confined.

Solution Equation 4.11 can be used to calculate the heads at A, B, C, D.

$$h_A = h_z + h_p = 28 \text{ ft} + 0 \text{ ft} = 28 \text{ ft}, \qquad h_B = h_z + h_p = 5 \text{ ft} + 23 \text{ ft} = 28 \text{ ft}$$

$$h_C = h_z + h_p = 33 \text{ ft} + 0 \text{ ft} = 33 \text{ ft}, \qquad h_D = h_z + h_p = 20 \text{ ft} + 13 \text{ ft} = 33 \text{ ft}$$

Hydraulic Head and Darcy's Law

Between the Bernoulli equation and Darcy's law, most groundwater problems associated with fluid flow can be solved. Groundwater models are designed to solve for head distributions, from which Darcy's law is applied to determine the quantity of flow. Where most people have trouble, however, is in the proper interpretation of the Bernoulli equation and the correct application of Darcy's law.

Remember that the total head is composed of the elevation head and the pressure head (velocity head is negligible Example 4.2). The pressure head is created by the weight of the water above a datum. Pressure is often measured in pounds per square inch (psi). There is a simple relationship between pressure head and equivalent hydraulic head (Equation 4.13):

$$1 \text{ psi} = 2.31 \text{ ft of H}_2\text{O} \tag{4.13}$$

Let's see where this came from. Recall from the Equation 4.10 that the pressure head can be expressed as

$$h_p = \frac{P}{\rho g} \tag{4.14}$$

Pressure (P) is equal to force/area, where force (F) is equal to mass times acceleration due to gravity (g). The denominator essentially represents a force over a volume (vol). Therefore, there is an inherent g in both the numerator and the denominator:

$$\frac{P}{\rho g} = \frac{mg/A}{\rho g} = \frac{mg/A}{mg/\text{vol}} \tag{4.15}$$

If it can be assumed that the density of water (ρ) is approximately 62.4 lb/ft³ and the area (A) is converted from square inches to square feet, then

$$1 \text{ psi} = \frac{1 \text{ lb}/1 \text{ in}^2}{62.4 \text{ lb}/\text{ft}^3} \times \frac{144 \text{ in}^2}{1 \text{ ft}^2} = 2.31 \text{ ft} \tag{4.16}$$

where gravity in the numerator and denominator (Equation 4.15) has already canceled out. The relationship in Equation 4.13 can also be conceptualized by taking a piece of 1-in inner-diameter (ID) PVC pipe and filling it with water. The weight of the water would be one pound.

This means that if flowing wells are capped and equipped with a pressure gauge, the pressure read in psi could be converted into feet of water. If psi readings are small (a few psi), then a standpipe can be used to get a true reading of total head (Figure 4.6).

FIGURE 4.6   Well with a standpipe where the total head can be measured above the ground surface.

Example 4.4   A well completed in Death Valley, California, is capped and has a pressure-gauge reading of 23 psi. The ground surface (GS) elevation is −234 ft (using MSL as a datum) and the well casing is capped at 2.5 ft above the GS. What is the total head at this location that would be used as a point on a potentiometric surface?

Solution   Using Equation 4.12, we obtain the equivalent pressure head, and using Equation 4.11, we determine the total head:

$$h_p = 23 \text{ psi} \times 2.31 \text{ ft}/1 \text{ psi} = 53.1 \text{ ft of pressure head}$$

$$h_z = -234 + 2.5 = -231.5 \text{ ft of elevation head}$$

$$h_t = h_p + h_z = 53.1 + (-231.5) = -178.4 \text{ ft, relative to mean sea level}$$

Now that you have a basic understanding of hydraulic head and Darcy's law, Examples 4.5, 4.6, and 4.7 illustrate some applications.

Example 4.5   As a graduate student in Laramie, Wyoming, I lived in a small house with a basement apartment. In the springtime, the basement apartment would flood from seeping groundwater. The landlord and his son were down there mopping the water up and wringing it into buckets, which they would throw out onto the lawn. After observing them make a couple of trips to the yard, I took pity and suggested they design a sump-pump system to manage the groundwater seepage from the high water table. We will use this as an example of using Darcy's law for vertical flow. The walls are assumed to be impermeable and that all the water is seeping through the concrete slab of the basement. We can use this information to estimate the hydraulic conductivity of the concrete slab. Suppose the following data were gathered after a brief field excursion, which included placing a few piezometers around the house (Figure 4.7):

Rate of water seepage is 20 gal/h.
Average head outside the house is 3 ft above the basement floor.
Basement slab is 4-in (1/3-ft) thick.
Basement floor area is 24 × 32 ft.

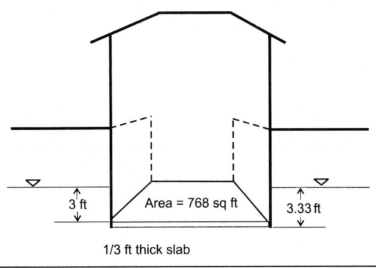

1/3 ft thick slab

Figure 4.7 Basement flooding problem. Example 4.5.

Solution If it can be assumed that the physical properties of the concrete slab are homogenous and isotropic, Equation 4.1 can be manipulated algebraically to obtain an estimate for the hydraulic conductivity (K):

$Q = 20$ gal/h $\times 1$ ft^3/7.48 gal $\times 24$ h/1 day $= 64.2$ ft^3/day
$A = 24$ ft $\times 32$ ft $= 768$ ft^2, the area perpendicular to upward vertical flow
$\partial h = 3.0$ ft
$\partial l = 1/3$ ft, the distance over which the head change occurs
$i = (3.33 - 0.33)$ ft/0.33 ft $= 9$ ft/ft

$$K = \frac{Q}{Ai} = \frac{(62.4 \text{ ft}^3/\text{day})}{(768 \text{ ft}^2)9 \text{ ft/ft}} = 9.0 \times 10^{-3} \text{ ft/day}$$

Example 4.6 (An example from Dr. Bruce Thompson from New Mexico) Suppose that along the Rio Grande River there is a canal that runs parallel to the river. Because of flooding, the surficial materials are fine grained and clayey. However, there is a horizontal connection through a sandy layer between the canal and the river that averages 2-ft thick (Figure 4.8). Suppose that the stage of the river is 2,110 ft and the stage of the canal is 2,113 ft. If the hydraulic conductivity of the sandy material is 10 ft/day and the average distance from the canal to the river is 35 ft, how much **leakance** or seepage loss occurs per mile of canal back to the river?

Solution Equation 4.1, Darcy's law, can be used to calculate the result.
Change in head is $\partial h = 2,113 - 2,110 = 3$ ft.
Flow path length over which the head changes is 35 ft.
Area perpendicular to flow is 1 mile long $\times 2$ ft wide $= 5,280$ ft $\times 2$ ft $= 10,560$ ft^2.
Leakance $Q = (10$ ft/day$)$ $(3$ ft/35 ft$)$ $(10,560$ ft$^2) = 9,050$ ft^3/day per mile of canal.

Example 4.7 One of the best ways to visualize Darcy's law and gain a perspective of the area perpendicular to flow is by imagining that the room you are in is a confined aquifer. Pretend that two wells were drilled from the roof and completed on opposite sides of the room. Suppose that the water levels in the two wells are above the height of the ceiling and the difference in head of the two wells is 20 cm. Water moves which way? What is the hydraulic gradient? If you assigned a hydraulic conductivity to the aquifer materials, can you determine the amount of flow through the room?

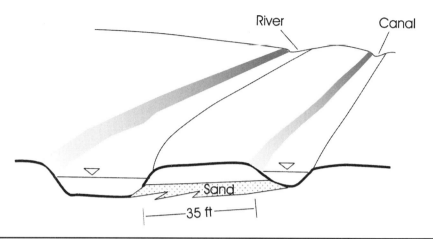

FIGURE 4.8   Leakage from canal with a higher stage than that of the nearby river through a connecting sand horizon. Example 4.6.

Solution   Since flow is horizontal from one side of the room to the other, the distance from one well to the other along the flow path is the flow length (∂l) over which the 20 cm of head change is (∂h). The $\partial h/\partial$l is the hydraulic gradient. The area perpendicular to flow (A) is the height of the room from floor to ceiling multiplied by the width of the wall. The height of the room represents the saturated thickness (b).

4.2   Flow Nets

One of the ways to graphically represent a two-dimensional view of groundwater flow, either planar (x,y) or cross-sectional (x,z), is through a **flow net**. The net consists of two sets of lines, flow lines and **equipotential lines**, or contours of equal hydraulic head. In the simplest case flow lines are parallel to groundwater flow, and the equipotential lines are perpendicular to flow (homogeneity and isotropy; Section 3.5). The spacing of the equipotential lines reflects the hydraulic gradient (Figure 4.9). When creating a flow net, one starts by contouring the heads (making equipotential lines). Next, flow lines are added perpendicular to the equipotential lines in such a way that the spacing of the flow lines and potential lines roughly form an equidimensional box. This gives the appearance of a net. The quantity of flow (**Q**) between any two flow lines is the same based on the principle of conservation of mass. If there are sources (recharge contributions) or sinks (discharge areas), the effect would result in the bending (divergence for injection or convergence for discharge) of flow lines. This will be discussed in more detail later on. In a given flow net, the areas of highest transmissivity have the flattest gradients. Some excellent examples on how to construct flow nets, with wonderful drawings and examples related to flow beneath concrete or earthen structures, are given by Watson and Burnett (1993). It is my experience that flow nets can be easily created using a variety of software (Groundwater Modeling Systems, Waterloo Hydrogeologic, Groundwater Vistas, and others).

The steepening of the hydraulic gradient in Figure 4.9 suggests that the transmissivity of the aquifer has changed. Heath and Trainer (1981) provided two explanations: one is attributed to a subregion of lower hydraulic conductivity (sand instead of sandy

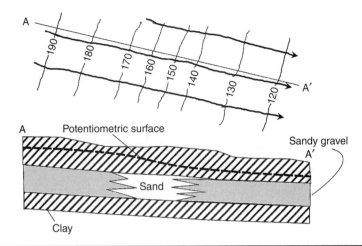

Figure 4.9 Plan view and cross-sectional view of a potentiometric surface map in a confined aquifer. In this example the steepening of the potentiometric surface is coincident with the sand unit compared with the coarser-grained sandy gravel up and down gradient. [*Modified after Heath and Trainer (1981).*]

gravel) and the other a thinning of the aquifer thickness (b). The sand or thinner sandy gravel both provide reasons the hydraulic gradient would become steeper.

Three examples (Examples 4.8, 4.9, and 4.10) illustrating how hydraulic gradient and hydraulic conductivity are related will be given from litigious hydrocarbon cases. In Example 4.8 a gasoline spill cleanup took place without investigating possible contamination sources up gradient (a bad idea). In Figure 4.10 of the example is an oblique

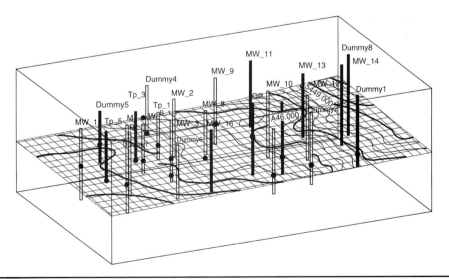

Figure 4.10 Oblique view from a groundwater model of a water-table surface from wells completed in a shallow unconfined aquifer in western Montana. MW stands for monitoring well. The scale of the grid spacing is approximately 10×10 ft (3×3 m). Example 4.8.

view of the potentiometric surface of a shallow water table. One can observe the steeper gradient on the right side of Figure 4.10, where finer-grained (siltier) sediments prevail. On the left side of the figure the hydraulic gradient flattens, corresponding to a coarser-grained sandy aquifer material.

Example 4.9 takes place at a railroad fueling site at a town in north-central Montana near the Canadian border. The site was under investigation as a likely source for diesel and halogenated hydrocarbons contamination of the shallow aquifer in a residential area. The technical personnel representing the railroad assumed that the hydrogeology at the railroad property prevailed northward into the residential area. Additional drilling and correlation analysis of the available drillhole data indicated a marked coarsening of aquifer sediments in the residential area. The coarser-grained sediments included a paleochannel that aided in distributing the contaminants throughout the neighborhood (Weight 2004). The flattening of the hydraulic gradient in Figure 4.11 directly correlates to the sediments being coarser grained.

In Example 4.10, a tanker truck carrying unleaded gasoline with a pup trailer (second smaller trailer) was traveling north to Kalispell, Montana, along the east side of

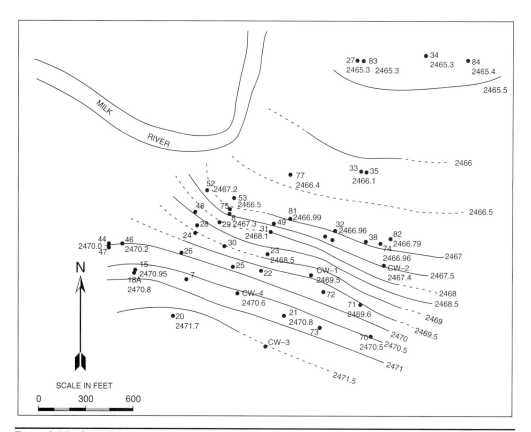

FIGURE 4.11 Potentiometric surface map of a shallow aquifer near the Milk River in north-central Montana. Contour interval is ½ ft (0.15 m). The steeper hydraulic gradient is where the finer-grained sediments occur at the railroad property compared to the flatter hydraulic gradient associated with the coarser-grained floodplain and paleochannel deposits. Example 4.9.

FIGURE 4.12   Tanker truck with a crushed pup trailer that leaked 6,800 gallons of unleaded gasoline onto the side of the road and seeped under five homes near Flathead Lake, Montana. Example 4.10.

Flathead Lake. The tire of the pup trailer got off of the roadway and slid into metamorphic bedrock, emptying its contents into the roadway ditch. The gasoline quickly seeped into the soils and moved downslope toward the lake under residential homes (Figure 4.12). Figure 4.13 represents the potentiometric surface generated from the water-level data from monitoring wells completed in bedrock and the sediments near the lake. The bedrock, with lesser hydraulic conductivity, matches the steeper part of the potentiometric surface.

Example 4.8   In a city in western Montana was a business that stored and sold gasoline from the 1950s through the 1980s. During the 1980s several leaks and spills were documented, resulting in contamination of the local soils. A brief investigation funded by a state agency during the early 1990s revealed a BTEX (Benzene, Toluene, Ethyl Benzene, and Xylene) plume in the shallow aquifer (<20 ft or 6.1 m). Although some of the up-gradient backhoe pits, which extended down to the water table, indicated other possible sources (evidenced by surface sheens and odors), this was not followed up on. In an effort to cleanup the site, soils were excavated followed by the installation of a soil-vapor extraction system. This effort "appeared" to cleanup the vadose zone to a zero-detection limit, at least temporarily. The insurance company representing the business was approached by the state agency that financed the cleanup to recover operating expenses plus interest.

   A review of the documents and field data associated with the case indicated a strong possibility that at least one other contamination source existed up gradient of the site. Additional monitoring wells and sampling points were installed to evaluate the extent of off-site sources. The additional data points, after being sampled, revealed at least one additional up-gradient source of BTEX chemicals. Water-level data from all wells were used to construct a water-table surface map (potentiometric surface; Figure 4.10). Figure 4.10 represents an oblique view of the water table taken from a numerical model.

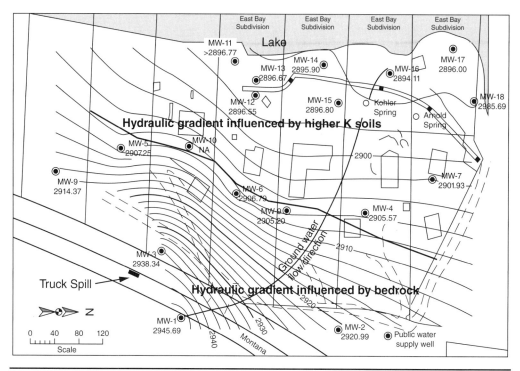

Figure 4.13 Potentiometric surface showing hydraulic gradient differences in bedrock (steep) versus unconsolidated sediments (much flatter). Unleaded gasoline contaminated five residences along Flathead Lake (Example 4.10).

The data were contoured using a ½ ft (0.15 m) interval, with every third contour marked in bold. (The original contaminated soils were thought to be between MW 1 and MW 4 in the figure and MW 2 was considered to be up gradient during the early 1990s). The drill cuttings from monitoring wells MW 10 through MW 14 (on the right-hand side of Figure 4.10) showed a predominance of siltier materials compared to wells at the original site near MW 1, MW 3, and MW 4 at the left of the diagram showing a predominance of medium to coarse sand. Notice how the slope of the potentiometric surface corresponds well to the geologic materials of the aquifer.

Example 4.9   Along the Montana hi-line (a highway that runs east to west across northern Montana just south of the Canadian border) in north-central Montana is one of several railroad towns. The railroad site had been a fueling facility since the 1880s and was converted from steam to diesel power in the 1940s. Since that time a sizable pool of free product (approximately one-half million gallons, 1.9 million liters) had built up on the water table. The direction of groundwater flow is north to northeast through a residential area. A class-action law suit from the local residents resulted in an investigation of all existing data, followed by new drilling with subsequent water-quality sampling and aquifer analysis.

The claim by technical folks representing the railroad was that hydrogeologic conditions at the fueling site continued northward into the residential area without any significant change. The drillhole data and the potentiometric surface maps contradicted this interpretation (Weight 2004). Figure 4.11 represents a mutually agreed upon interpretation of water-level contours of the local shallow aquifer. The contour interval is ½ ft (0.15 m). Using lithologic log descriptions from drillhole data and considering the depositional environment of the area, it became clear that a paleochannel extends, from where the Milk River bends northward, eastward across the residential area (Figure 4.11). Notice how the hydraulic gradient is very flat east of the Milk River (representing the paleochannel and associated floodplain deposits), while the tighter contours represent the clayey silt materials occurring at the railroad property.

Example 4.10   There are two highways that head northward to Kalispell, Montana, from the south. The western route is much longer, although safer with wider roadways. The eastern route is more direct but much narrower, especially for larger vehicles. A tanker truck towing a pup trailer was winding along the eastern highway when the right back tire slid off the roadway, pulling the pup trailer quickly off the road. The tank of pup trailer overturned and tore open, spilling its 6,800 gallons (25,750 L) of unleaded gasoline into the roadway ditch (Figure 4.12). The gasoline seeped into the surface soils and then moved downslope under five homes. The residents had to be evacuated. A series of monitoring wells were drilled into the bedrock and the near-lake sediments to determine the extent of free product and vapor-phase BTEX constituents. An interpreted potentiometric surface map from the water-level data was constructed (Figure 4.13). There is a clear distinction in the equipotential contours of those in the lower hydraulic conductivity bedrock (steeper hydraulic gradient) compared to the higher hydraulic conductivity sediments bordering Flathead Lake.

In the previous three examples the potentiometric surfaces could be correlated with the underlying aquifer materials. In *every* case where the hydraulic conductivity of the aquifer materials was lesser, because of consisting of finer-grained sediments or bedrock, the equipotential contours were markedly steeper. This was because the hydraulic gradient had to be steeper for gravity to accommodate and move groundwater through those materials, a fundamental relationship of Darcy's law.

This brings up another principle when applying Darcy's law. When one uses flow nets to estimate flow, it is important to make calculations between flow lines where the hydraulic gradient is similar, rather than averaging the hydraulic gradient over a larger region with varying gradients. Calculations using Darcy's law are *only* valid where the hydraulic conductivity or transmissivity is constant. Where this is true, the flow net will show a uniform hydraulic gradient corresponding to the hydraulic properties of the aquifer.

Getting back to Figure 4.9 the hydraulic gradient between contours 190 and 170 is uniform; therefore, the hydraulic properties between these contours are similar. Between contours 170 and 140 the hydraulic gradient is also similar but its slope has changed (it is steeper, the contours are closer together) indicating a change in aquifer properties (finer grained or the aquifer becoming thinner). The widening of the hydraulic gradient between contours 140 and 120 indicates the aquifer has changed once again (became coarser grained or the aquifer became thicker). Hence, when performing calculations using Darcy's law, one must be sure the hydraulic gradient ($I = dh/dl$) is matched with the corresponding aquifer hydraulic properties of K or T.

If the location at hand is not in a recharge or discharge area, the flow is generally considered to be horizontal (horizontal flow). This can occur if the material is relatively permeable in flat terrain. Here the equipotential lines in the third dimension are assumed to be more or less vertical. As a practical matter, this means that wells completed at different depths within the same aquifer at the same location will also have the same water-level elevations. In plan or map view flow nets and in two-dimensional groundwater modeling, however, the vertical distribution of horizontal conductivities (and other characteristics) are averaged over the entire saturated thickness.

Rarely, within a given aquifer unit, is the distribution of horizontal hydraulic conductivity the same. This phenomenon is a function of the depositional environment of the sediments (Chapter 2). There are usually lenses of varying hydraulic conductivity or fining upward or downward sequences within an aquifer (Figure 3.1). The coarser units tend to have faster relative velocities than the finer-grained units. This contributes

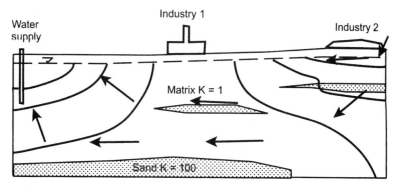

FIGURE 4.14   Equipotential lines and flow arrows for a recharge-discharge area with two potential contamination sources affecting a public water supply.

to varying contaminant arrival times than would be predicted using the average linear velocity from Equation 4.6 (Figure 4.14). Figure 4.14 is a cross-sectional view of an area where there are two industrial plants. Industrial Plant 1 is located where horizontal flow prevails, but industrial Plant 2 is located in a recharge area, where a downward component to groundwater flow exists. If the water supply on the left side of Figure 4.14 is in a place where groundwater discharges, there will be an upward component to groundwater flow. If the water supply becomes contaminated, which plant is most likely the source?

Vertical Groundwater Flow

In discharge areas, recharge areas, or undulating topography, there is always a significant vertical component to groundwater flow (Figure 4.15). This means that water levels in wells completed (screened) at different depths within the same aquifer unit

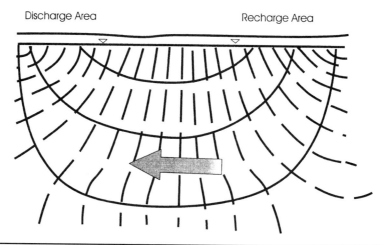

FIGURE 4.15   Flow net showing recharge and discharge areas with no vertical exaggeration. [*Modified from Toth (1962; 1963).*]

will be at different elevations. The best way to determine this is to complete two or more wells at varying depths within the same location as a grouped set. There should be a trend of downward or upward changes in water levels with depth, depending on whether the location is a recharge or discharge area. For example, wells completed in the central region of Figure 4.15 will have similar water levels, whereas wells completed deeper in the discharge area will show higher relative heads than wells completed nearer to the surface. The reverse will be true in a recharge area. This becomes significant later on in the interpretation of water-level data (Section 4.4).

The water level observed in any well is reflective of the head at the midpoint of the screened interval. Piezometers (Section 10.2) with screen lengths less than ½ ft (0.15 m) provide strategic vertical point estimates of hydraulic head. Monitoring wells with screen lengths longer than 10 ft (3 m) may not provide a sufficiently accurate distribution of hydraulic head to understand the vertical nuances of groundwater flow. Depending on the goal of the project, however, it just may be a matter of scale.

At a given contamination site, it is important to know if there is a significant component to vertical flow and which way that component is. For example, wells with long screened intervals may have components of groundwater flow entering the well and components leaving the same well (McIlvride and Rector 1988). For example, in a discharge area, the higher equipotential contours are intercepted at the lower part of the well. This results in water entering the well in the lower portion of the well screen and becoming captured (Figure 4.16). In the upper portion of the well screen, the head is higher than the head in the nearby lower equipotential contours. This results in water

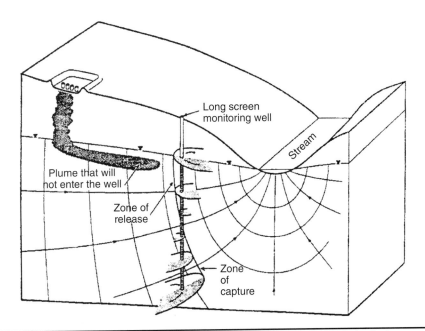

FIGURE 4.16 Water circulation patterns within fully screened monitoring wells in discharge areas. When the well screens are long the head varies greatly in the vertical dimension causing a groundwater mound to occur at the surface, thus causing dissolved contaminants in groundwater to bypass the monitoring well. [*From McIlvride and Rector (1988).*]

in the upper portion of the well moving outward from the well. Any floating light-phase contamination that would potentially become intercepted in the upper portion of the well may go around the well and may not be detected.

Gaining and Losing Systems

Many perennial streams receive groundwater discharge (gaining streams) as baseflow along their courses that sustain their flow even during the driest months. For example, in Montana, USA, over two-thirds of the **base-flow index** (% of total annual stream-flow) is contributed by groundwater (Montana DNRC 2014). In this case, the river represents the lowest point of hydraulic head in a groundwater surface-water system. The equipotential contours in a flow net bend such that the flow lines converge on the stream (they "V" upstream) (Figure 4.17, location "D"). In Figure 4.17 near location D the water table slopes toward the stream, whereas at location A the stage of the stream represents a local maximum and the slope of the water table is away from the stream (the equipotential contours "V" downstream). In a two-dimensional flow net in plan view all of the equipotential contours appear to be vertical, thus giving the impression of horizontal flow everywhere. However, whenever there are discharge areas (e.g., gaining streams; Figure 4.18a) or recharge areas (e.g., losing streams; Figure 4.18b) the equipotential contours actually dip or tilt representing a three-dimensional flow system. The corresponding flow lines in the flow net are actually curvilinear (Wampler 1998) (Figures 4.18a and 4.18b). A given stream valley can have both components of losing stretches and gaining stretches as shown in Figure 4.17.

In the arid west of the United States, many mountain streams are sustained by base-flow because the associated alluvial sediments are relatively thin and bounded at depth by low hydraulic conductivity bedrock. These same streams lose a significant quantity or all of their water into alluvial fans or the thicker sediments of larger valleys. This is especially true in areas with low precipitation. Here stream losses occur because the water table slopes away from the river, and may be several feet below the river stage.

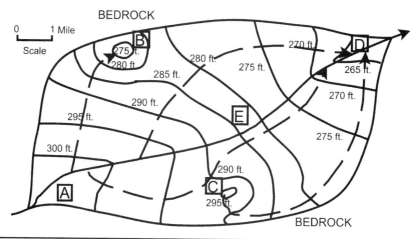

Figure 4.17   Schematic of groundwater/surface water flow in a valley surrounded by bedrock: losing stream (A), discharge area (B), recharge area (C), gaining stream (D), and neither gaining nor losing conditions (E). [*Adapted from Davis and Dewiest (1966).*]

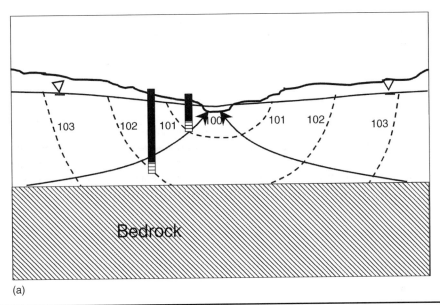

(a)

FIGURE 4.18a Schematic of a flow net of a gaining stream or groundwater discharge area. Notice how the deeper well has a higher head than the shallow well indicating upward flow. Not to scale.

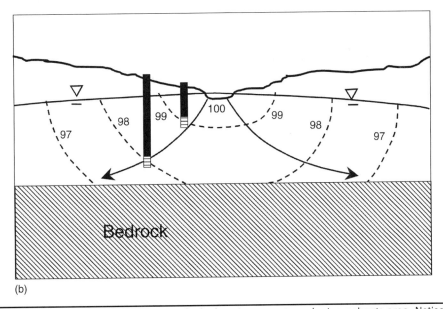

(b)

FIGURE 4.18b Schematic of a flow net of a losing stream or groundwater recharge area. Notice how the deeper well has a lower head than the shallow well indicating downward flow. Not to scale.

In desert areas after a significant precipitation event the stream is likely perched above the highest margin of the unconfined aquifer and quickly loses all its water over a matter of several days.

Refraction of Groundwater Flow

As water passes from porous materials of a certain hydraulic conductivity into a differing hydraulic conductivity, the flow lines will bend or refract according to the principles described by Snell's law. This phenomenon is observed in a clear glass of water with a straw sticking out of it. The straw seems to bend at the air/water interface. This is due to the refraction of light as light rays pass from the less dense medium of air into the denser medium of water. Snell's law applied to groundwater flow is described by

$$\frac{K_1}{K_2} = \frac{\tan(\sigma_1)}{\tan(\sigma_2)} \qquad (4.17)$$

This phenomenon is illustrated in Figure 4.19. Groundwater flow lines (lines with arrows) move vertically or subvertically through low-conductivity layers (at wider spacing) and somewhat horizontally through higher-conductivity layers.

In terms of pipe flow, if

$$Q_1 = Q_2 = v_1 A_1 = v_2 A_2 \qquad (4.18)$$

then the closer-together (tightly spaced) groundwater flow lines represent areas with horizontally faster velocities (moving through flatter hydraulic gradients), and the wider-spaced streamlines represent areas with slower vertical velocities (moving through steep hydraulic gradients), because gravity needs a steeper gradient with more cross-sectional area to move the same quantity of water. This is a function of conservation of mass.

Since groundwater flow lines will refract when they encounter geologic materials with differing hydraulic conductivities, it is extremely difficult to predict where these streamlines will go. For example, many depositional environments will create lenses of materials with differing hydraulic conductivities. This becomes significant when one

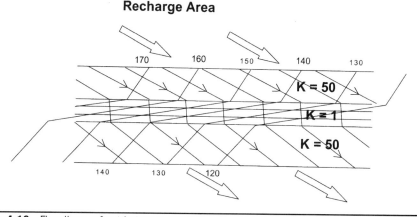

FIGURE 4.19   Flow lines refract between or at the boundary of dissimilar materials of differing hydraulic conductivity.

considers locating a landfill or attempts to predict where a source of contamination came from. Depending on the geology and the flow system, a contamination source that appears to be obviously close may not be the source in question at all, but may have come from some more distant source (Figure 4.15).

4.3   Level Measurements in Groundwater Wells

The basis for determining the direction of groundwater flow is based upon static water-level data from the same aquifer. Level measurements in groundwater wells are one of the most fundamental tasks performed at any site. It is one of the first things done when arriving to a site and performed on a routine basis. Water levels can be collected manu-ally with a portable field device or rigged to collect continuous data with a data logger sentinel. It is possible to telemetrically send these data back to the office, powered by a solar panel providing the ability to have real-time access to critical monitoring points. Water-level data are of little value without knowing the well-completion information and what the relative elevations are. For example, vertical flow interactions between multiple aquifers are evaluated from water-level data completed in different aquifers. Grave errors can result if one assumes that all water-level data for a given area are col-lected from the same aquifer unit, ignores recharge and discharge areas, and fails to survey the elevation of the measuring points (MPs)!

The purpose of this section is to present the practical aspects of obtaining level-measurement data and to describe the most commonly used devises to obtain them. It is instructive to present the most common sources of error from field mistakes and idiosyncrasies of the equipment to give the reader a better understanding of the pitfalls one can fall into. The consequences of misinterpreting water-level data are discussed in Section 4.4.

Access to Wells

Taking level measurements in wells is straightforward, provided that access to wells is known in advance. In a site where level measurements are being taken from wells con-structed by a variety of contractors, there may be an assortment of security devices and locks. Each may require its own key or some special way to remove the cap. For example, monitoring wells that are constructed in playgrounds or in paved roadways may have a flush-mount security plate (Figure 4.20). In this case, a socket or wrench is required to remove the surface metal plate before a subsequent lock can be reached. A bucket with whisk brooms and an array of screwdrivers can be invaluable in removing sediment trapped in the spacing between the cap and its housing and in the mount bolts. This flat well-completion method allows vehicles to run over the well locations or individuals to run around without tripping.

Removing a Well Cap

To save on costs for more regional studies, monitoring networks are sometimes expanded to include domestic, stock-water, or other existing wells. Nothing is more frustrating in field work than not to have access to a well. This may be something as simple as forget-ting the keys to a gate or well caps. Many stock wells or domestic wells will typically have a well cap attached by three bolts (Figure 4.21). These are usually more than "finger tight" to prohibit animals from knocking off the well caps, and to discourage the casual

FIGURE 4.20 Flush-mount monitoring well covering, so that vehicles or other traffic can pass over the monitoring site. Two bolts hold the flush mount in place.

FIGURE 4.21 Typical well cap on domestic well with security bolts being loosened by a multitool.

passerby from intruding. An indispensable device is a crescent wrench or the handy multitool purchased at most sporting goods or department stores. These should be equipped with pliers that can be used to loosen the bolts and free the cap. This is a good excuse to tell your spouse that you need one; it makes a great safety award.

Well caps that become wedged over a long season from grit, changes in air temperature, or moisture can usually be loosened by taking a flat stick or your multitool and tapping upward on the sides of the well cap until it loosens up. This process will eventually allow access to the well. Tight-fitting polyvinyl chloride (PVC) well caps can be modified with slits cut on the sides of the well cap to allow more flexibility. Another important point to remember is to place the well cap back on and resecure any locking device the way you found it. Well owners can get really upset about such neglect and may not allow a second opportunity for access.

Example 4.11   In areas where gas migration takes place (coal mines or landfill areas), gas pressures may build up underneath well caps if they are not perforated with a "breathe hole." One field hydrogeologist was measuring water levels in a coal-mining region and had a premonition to kick the well cap. When the cap blew 150 ft (46 m) into the air he contemplated what would have happened to his head. This resulted in a company policy of drilling "breathe holes" in well caps of monitoring wells or in the casing to bleed off any methane gas pressures. Pressure changes in any monitoring well from changing weather conditions may also affect the static water level; therefore, drilling a small hole below the well cap in PVC casing can alleviate pressure effects.

Water-Level Devices

There are a host of water-level devices used to collect level-measurement data. Don't be fooled into thinking one is best; the different devices have their own unique applications. Manual devices, such as steel tapes and electric tapes (E-tapes), are discussed first, followed by those with continuous recording capabilities, such as chart recorders, transducer/data logger combinations, and self-contained mini loggers.

Most level measurements are taken under static conditions. Dynamic conditions occur during a pumping test. If the intention is to take a static water-level measurement and the well is in use or was recently used, it is important to allow recovery to equilibrium conditions before taking the measurement. For example, a stock well may be on all the time or the pump activated by a timer. The pump can be shut off for a while (until there are no changes between readings). Then take a reading and turn the pump back on or reset the timer. Once again, if timers are not reset or pumps turned back on, then you may not be given a second chance.

Water-level measurement techniques have evolved over time from basic steel tape and E-tape measurements recorded in the field to modern self-contained mini loggers that can be connected to telemetric systems to provide remote real-time data. Long-term water-level recording devices range from pen-to-paper-chart recorders (older technology) to modern enhanced downloadable telemetric systems. This section outlines each of these devices.

Manual Water-Level Devices

How to Take a Level Measurement

The basic procedure of taking a level measurement involves gaining access to the well, removing the well cap, and lowering a device down the well to obtain a reading. Normally, this procedure takes only a matter of a few minutes depending on the well depth, device used, and whether the well is a dedicated monitoring well or a domestic well being used as a monitoring well. As simple as this procedure sounds, if one is not aware of the weaknesses and idiosyncrasies of the equipment, serious errors in readings can result that affect the quality and interpretation of the data.

Steel tapes and E-tapes are still the basic and essential tool for water-level measurement. These tools have been used for years and provide basic water-level readings for monitoring surveys and in setting up pressure transducers for long-term monitoring. E-tapes are essential to have on any groundwater monitoring project as it is prudent to collect backup water levels if your pressure transducers and data loggers fail.

Steel Tapes

Most data collected before the late 1970s were likely collected with a steel tape. Steel tapes represent the "tried-and-true" method that still has many important applications today. Water levels from the 1940s on were all measured this way until electric tapes (E-tapes) became the norm. The proper way to use a steel tape is to apply a carpenter's chalk to approximately 5 to 10 ft (1.5 to 3 m) of tape before lowering it into the well. The tape is lowered to some exact number next to the MP, for example, 50 ft (15 m). The tape is retrieved to the surface, where the "wet" mark on the tape, accentuated by the carpenter's chalk, is recorded (Figure 4.22). A reading like "50 ft minus 4.32 ft" would yield a reading of 45.68 ft depth to water below the MP. There is usually a historical knowledge of the approximate depth to water to guide this process. Otherwise it is done by trial and error. Usually a good reading can be made even on a hot day, with a wet chalk mark, otherwise the tape dries quickly. Getting a clean reading may require more than one trip with the tape, hence its loss in popularity.

FIGURE 4.22 "Wet" mark on a steel tape after lowering a known depth of tape into the well from the measuring point. The SWL is the known depth minus the length of the wet portion.

Sometimes the *only* way to get a water level *is* with a steel tape. In many pumping wells, with riser pipe and electrical lines, or wells with only a small access port, the only way to get a water level is with a steel tape. Steel tapes are thin, rigid, and tend to not get hung up on equipment down the hole. A 300-ft (100-m) steel tape is an important part of any basic field equipment. It can also be used to measure distances between wells or other useful tasks.

Electrical Tapes (E-Tapes)

By the early 1980s, E-tapes began to dominate the market and are currently considered as basic field equipment. The first devices were usually marked off in some color-coded fashion every 5 ft (1.5 m), with 10-ft markings and 50- or 100-ft markings in some different color. E-tapes now are usually marked every 1/100th of a foot (3 mm) (Figure 4.23) and come in a variety of probe shapes and sizes (Figure 4.24). All E-tapes function with

FIGURE 4.23 E-tapes with probes marked every one-hundredth of a foot on a flat-style (left) and round-style tape (right).

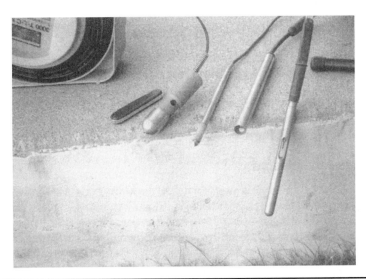

FIGURE 4.24 E-tape probe sensors come in a variety of shapes and sizes. Swiss army knife for scale.

the same basic principle. A probe is lowered into the water, which completes an electrical circuit, identified by a buzzer sound, an activated light, or both. The signal from the water level is transmitted up the electronic cable of the E-tape to the reel where the signal occurs. Each has its own appropriate function. During a pumping test with a generator running, it may be hard to hear a buzzer sound, or you may have a hearing impairment from hanging around drill rigs too long without wearing hearing protection. In this case a light is helpful. In very bright sunlight, your light indicator may be hard to detect and the buzzer can be more helpful. It should be noted that if oils or other floating organics are present in the well, they will not be detected or may provide a false water-level reading.

Most E-tapes have a sensitivity knob from the conductivity sensor in the probe. The sensitivity knob is usually a turn dial with markings from 1 to 10. Probe sensitivity is needed for a variety of water qualities. Low sensitivity (1 to 3) will give a clear signal in high-TDS waters, such as those found in Cretaceous marine shales. The higher the TDS in the water, the lower the sensitivity needed to detect a water level. Relatively pristine or low-TDS waters require that the turn dial be adjusted to a higher sensitivity setting (8 to 10) to get a clear signal. A good general rule is to put the dial in midrange (4 to 6) and lower the probe until a buzz/light signal is detected. If little attention is paid to the sensitivity setting, the hydrogeologist may obtain false readings. This may be one of the first questions asked to a field technician when trying to explain why a water-level reading does not make sense. It is usually handy to have a couple of different depth capabilities. It is noted here that there are a number of multiparameter probes that will measure water-quality parameters in addition to static water levels.

To take a water-level reading using an E-tape, turn the device on by moving the sensitivity knob dial to midrange. Lower the probe until the light goes on or the buzzer sounds. This gets you close to where the reading will be. At this point, lift the E-tape line above the depth that triggered the buzzer and gradually lower the probe until the buzz sound is repeated. Hold this spot on the E-tape line with your thumbnail or with a pointer, like a pencil (Figures 4.25a and 4.25b). The E-tape line should be held away from the MP toward the center of the well and shaken lightly to remove excess water and the process repeated. If the light signal is clear and the buzz is crisp, the reading is most likely accurate.

(a)   (b)

FIGURE 4.25   (a) Taking a water-level reading using E-tape by pinching the position where the buzzer sounded. (b) Student with thumb positioned on E-tape and the exact location pointed at with a pencil to mark a clear water-level reading.

False readings are fairly common if one does not pay attention to the sensitivity knob setting. In some wells, scraping of the probe into condensation inside the casing will trigger a false reading. The sound of the buzzer may not be clear or the light may flicker instead of providing a clear, bright reading. In this case, the sensitivity should be turned down to the lowest setting that yields a good signal. This will be variable, depending on the water quality involved. Cascading water from perforated sections above the actual water level can also be problematic. This can occur during an aquifer test, where one is trying to get manual readings in the pumping well that has been perforated at various depth intervals. In pristine waters, it may be difficult to get a reading at all. In this case, the lower sensitivity knob settings may not be discerning enough to detect when the water table has been reached. Detection may or may not sound off until the probe has been submerged well under the water surface. In this case, turn the dial to the high sensitivity range (8 to 10) and the buzzer or light should give a clearer signal. It is always a good idea to lightly shake the line and repeat the process until a clear reading is obtained.

Water levels taken in monitoring wells tend to be uninhibited by riser pipe and electrical lines, which are typically found in pumping wells. In this case the only concern, as to probe size, may be the diameter of the monitoring well. Pumping wells, however, may present some more daunting challenges. Here, probes may wind around equipment or get hung up in wiring extending up from the pump or in spacers designed to hold the riser pipe into the central part of the well. Getting hung up can be a real problem, not to mention being costly and annoying. This is where steel tapes or cheaper E-tapes can be helpful, particularly those with a very narrow probe size. In some cases, a service call to a well driller may be cheaper than paying for a new E-tape. It has been the author's experience that lowering and raising an E-tape *slowly* in a pumping well increases the likelihood of avoiding getting a probe stuck.

Chart Recorders

Continuous water-level readings were recorded for many years using chart recorders. Although the use of chart recorders has declined over the past decade, they are still available and are often encountered in the field. One of the more commonly used chart recorders is Steven's Recorder (www.stevenswater.com/products/sensors/hydrology/level/type-f). These devices use a drum system onto which a chart is placed. The position of the water level in the well is tracked by having a weighted float connected to a beaded cable that passes over the drum and is connected to a counterweight (Figure 4.26).

As the drum turns forward or rolls backward with the movement of the water level, a stationary ink pen marks the chart. The pen is only allowed to move horizontally, which corresponds with the time scale set by a timer. Timers can be set for a month or up to 3 months. Chart paper is gridded where each column line usually represents 8 hours. The row lines mark the vertical water-level changes. Since the turning of a drum can be significant, control of movement is through setting the ratio of float movement to drum movement of the chart paper. This is known as setting the gear ratio. For example, a 4:1 gear ratio means that the float will move four times the distance the drum would turn forward or backward. This helps keep the pen in contact with the chart paper. Once the old chart is replaced with new paper, the information from the previous chart has to be "reduced" or converted into numbers. A technician often logs numbers from a chart into a file for data analysis or hydrograph plotting. Corrections for drift are made by

FIGURE 4.26 Chart recorder in the field, resting on a platform with a protective cover that can be locked.

noting the time and date at the end of the chart and then applying a linear correction factor from the beginning time. For example, if the final marking on the chart is 12 hours short (because the timer setting is off) of the supposed time, a correction factor is applied to "stretch" the data to match up with real time. Reduced data are easily adjusted on a spreadsheet. Frequently, wells located near surface-water streams are fitted with a chart recorder to compare the groundwater and surface-water elevations or in an area where water-level fluctuations are significant. Other reasons may be for use during winter months or in remote areas when access is difficult. The chart recorder is positioned directly above the well on some kind of constructed platform (Figure 4.26). The platform sits above the well casing, with the float attached to a beaded cable lowered down to the water surface. A hole cut in the platform allows the beaded cable to pass up over the drum and then through another hole cut in the platform for the counterweight. The top of the well casing and the recorder can be encased in a 55-gal drum or some other container that can be locked.

Errors and malfunctions can occur in a variety of ways. For example, the batteries can go dead on the timer or the pen can become clogged. The beaded cable occasionally comes off the drum from lack of maintenance or from rapid water-level changes. Replacement floats can be jury-rigged with plastic 1- or 2-L soda bottles weighted with sand, if one cannot find a replacement float in a timely way. Chart recorders tend to require occasional maintenance and need to be enclosed in a structure at the wellhead. The chart recorder is being phased out and replaced by enhanced telemetry systems (ETSs) or well sentinels, but this section was included for those still using them or for those who wonder what they are.

Transducers and Data Loggers

Transducers are pressure-sensitive devices that "sense" the amount of water above them. The "sensing" is performed by a strain gauge located near the end of the transducer, which measures the water pressure. Transducers are placed into a well at a selected depth connected with a cable to a data logger at the surface near the well.

There are two types of pressure transducers used in the field. The first type is a vented (gauged) type of pressure transducer that is attached to a cable that allows barometric compensation for atmospheric changes at the head of the well. The vented pressure transducer uses a cable with a very small vent tube that runs down the length of cable from the surface and terminates behind the pressure transducer. The vent tube acts as a conduit for barometric pressure changes occurring at the surface and allows the barometric pressure on the water column to be "canceled out" by the air pressure transmitted within the tube. The vented gauge reference is thus always open or connected to atmospheric pressure. The main advantage of this type of level-reference gauge is that over time, since pressure changes occur regularly with changing atmospheric conditions, zero readings become self-compensated. This allows the device to always read zero when the pressure is released no matter how much the atmospheric conditions vary. Care must be taken not to kink a vented cable when installing the pressure transducer in the well. The usual field technique is to create a loop in the cable secured with duct or Gorilla Tape that can rest on and be secured to the top of the casing. Pressure transducers must be connected to a surface data logger as the unit has no internal data storage capabilities.

Water-level changes accumulated by a vented pressure transducer are recorded at the surface by an attached data logger (Figure 4.27). Data loggers can be single channel for a single well or have multiple channels to record the information from several wells.

Figure 4.27 Vented data loggers in use in a field setting during an aquifer test.

The second type of pressure transducer is an absolute pressure transducer that is sealed and connected to the surface with an electrical wire connection. This pressure transducer senses and records all the pressure changes it encounters: both water pressure (hydraulic head) and air pressure (barometric or atmospheric pressure). Barometric corrections for water-level data collected with an absolute pressure transducer need to be adjusted from data collected by a self-contained barologger transducer, mini data logger placed at or near the wellhead. Only one barologger transducer is needed to correct data from other absolute pressure transducers as long as they are within a several miles or kilometers of the barologger. Suppliers of both vented and absolute pressure transducers normally provide software and a communication package with options to adjust the atmospheric data with groundwater-level readings.

Surface mounted data loggers that connect to pressure transducers are often located in well pump houses recording water level and/or flow volumes. Data loggers with multiple channels were available for recording water-level changes from multiple wells during pumping tests in the 1990s and early the 2000s. However, the multichanneled surface data loggers are becoming less common since self-contained systems with barologgers have been developed.

Self-Contained Pressure Transducer: Data Logger

The most commonly used equipment currently used is the self-contained data logger pressure-transducer combination or well sentinel. These systems consist of a pressure transducer and an on-board computer chip to record data. The computer chip is programmable and has various memory capabilities depending on the model. The typical size for well sentinels is 1.83 cm (0.78 in) in diameter and 21.6 cm (8.5 in) in length. They typically can store up to 130,000 records and up to 50 different logs. Well sentinels can be programmed similar to surface-mounted data loggers. The computer chip allows for programming log formats of linear, fast linear, linear average, event, step linear, and true logarithmic. Units of measure can be set to psia, kPa, bar, mbar, mm Hg, inches Hg, and cm H_2O. Most well sentinels also include temperature measurements in either Celsius or Fahrenheit. The operating temperatures are generally between −20°C and 80°C (−4°F to 176°F); however, uses of these well sentinels have been observed working in wells with significantly higher temperatures such as hot springs or geothermal wells. Well sentinels can be placed into wells to record data over a long time period, similar to chart recorders. For example, In Situ Inc. has one called the Troll or mini Troll. Solinst Inc. has one called the Levellogger (Figure 4.28) and Schlumberger has one called the Diver. Both vented and absolute well sentinels are available from most equipment suppliers.

Vented well sentinels are connected to the surface by a vented cable the same as a vented pressure transducer. The vented cable allows for barometric correction and access to the data while the well sentinel is operational. The absolute well sentinel can be connected to the surface by a wired cable or capped for some applications. A capped absolute well sentinel may be used in applications where the pressure transducer is required to be sealed in a specific section of a well, such as in a packer test. Well sentinels have an advantage over pressure transducer and surface-mounted data loggers in that the unit can be placed in the well with a cable and simply enclosed in a locked wellhead. Well sentinels also eliminate the need to protect vented cables, running over the surface connected to a multichannel surface data logger, during aquifer tests. A monitoring well with a deployed well sentinel is shown in Figure 4.29. Note that the extra vented cable can be coiled in the well casing and securely locked.

Figure 4.28 A small-diameter well sentinel that measures level and temperature over time.

Figure 4.29 Monitoring well with a well sentinel in place and the extra vented cable to be coiled inside the well casing. [*Photo by GP Wittman.*]

Figure 4.30   Example of an in-line battery for a Troll well sentinel system (In-Situ, Inc., Fort Collins, CO). (*Used by Permission.*)

Well sentinels are programmed by using either laptop computers, tablets, or cell phones with software and mobile handheld field units provided by the well sentinel supplier. The handheld field devices have preloaded apps to program and store information recorded by the well sentinel and are ideal for field use in all weather conditions.

Well sentinels contain batteries that require replacement after long-term use. Batteries in well sentinels can last up to 10 years depending on the frequency of data collection. Changing batteries usually needs to be performed by the manufacturer; therefore, the authors suggest that acquiring newer equipment be considered. There may be the option of using external batteries placed "in-line" with the cable system for some well sentinels, as a means of extending battery life. An example of an in-line battery is shown in Figure 4.30.

Pressure Ranges of Pressure Transducers

The water pressure can be converted into feet of head above the strain gauge by Equation 4.19, repeated here for convenience as:

$$1 \text{ psi} = 2.31 \text{ ft of H}_2\text{O} \tag{4.19}$$

The above relationship is important in knowing the maximum pressure or how deep the transducer can be placed below the static water surface. Typical ranges of pressure transducers are 5, 10, 15, 20, 30, 50, 100, 200, 300, and 400 psi. For example, a 10-psi transducer should not be placed more than 23 ft below the static water surface. If it is, the transducer may become damaged. Another problem may be that the data logger may not record any meaningful data. Personal experience indicates that the data logger will give level readings of zero if a transducer/data logger is placed below the pressure depth range. During a monitoring event one should anticipate the range of water-level changes that may occur during the testing period. Table 4.1 lists the practical recordable ranges for commonly used absolute and vented data loggers provided by the manufacturers.

During a pumping test, one should anticipate the potential range of water-level changes that will take place. The transducer placed in the pumping well is usually equipped with the highest variable range followed by lower water-level ranges for observation wells located at assorted distances away. If the transducer is to be placed in an observation well, it is a good idea to lower the transducer to the appropriate depth and let it hang and straighten out before defining an initial level measurement. A common error is to forget to establish the base level of each well as a last setting in the data logger before the beginning of the test. For example, if base level is set immediately after the transducer is placed into the well, especially with the placement of the pump and riser pipe, the water levels may not be properly equilibrated. It is also a good idea to select a base level other than zero (such as 100 or some other easily referenced number). Fluctuations of the water levels may occur above the initial base-level reading, resulting

Absolute (nonvented)	Range (m)	Range (ft)
30 psia	11	35
100 psia	60	197
300 psia	200	658
500 psia	341	1,120
1,000 psia	639	2,273
Gauged (vented)		
5 psig	3.5	11.5
15 psig	11	35
30 psig	21	69
100 psig	70	231
300 psig	210	692
500 psig	351	1,153

TABLE 4.1 Practical Depth Ranges for Specific Data Logger Pressure Settings Provided by the Manufacturers

in negative numbers for drawdown. This can be problematic when trying to plot the data in log cycle on a spreadsheet. Strategic placement of the transducer in the pumping well just above the top of the pump is what the authors recommend. The transducer should be secured with electrician's tape, with additional tape being added approximately every 5 ft (1.5 m) along the cable as the pump and riser pipe are being lowered into the well (Figure 4.31). It is sometimes useful to attach the transducer to above the pump with bailing wire in the case of warmer water.

Logging Methods

Data loggers and well sentinels have the capability to record data at various rates depending on the application. The following is a list of log types for recording data and their descriptions available depending upon the capabilities of the instrument.

Logging Methods for Recording Data for Long-Term Monitoring

Linear   The linear log type of setting measures and records data at a user-defined fixed interval of half a minute to several hours or more. This method is commonly used for long-term studies, landfill monitoring, stream gauging, tidal studies, and background water levels prior to aquifer testing. Fixed intervals can be recorded in days, hours, minutes, or seconds.

Linear Average   Linear average log type provides the option of smoothing out anomalous highs and lows that may occur in a data set, for example, if a water wave passes over the instrument. Each stored measurement represents the rolling average of several rapid measurements. This method is used for long-term studies, stream gauging, tidal and open-water studies where trends are more important than accuracy. Intervals are measured in days, hours, minutes, or seconds.

FIGURE 4.31   John D lowering a 1-HP pump down a 4-in (10.2-cm) production well to be aquifer tested. Notice how the pressure transducer is attached with electrician's tape just above the pump on the riser pipe.

Event   Linear event log type combines basic fixed-interval logging of specified parameters with the ability to log data at a faster interval when a single-parameter event condition is present. For example this is often used to record storm surge or flood crest events.

Logging Methods for Aquifer Testing

True Logarithmic   True logarithmic log type captures early-time water-level data during aquifer testing. Measurements are very closely spaced at the start of the test (40 measurements per second) followed by a logarithmically decaying schedule of longer time intervals as the test progresses. There are typically 20 to 40 measurements per log decade. This log type is commonly used for step-drawdown pump tests (Chapter 5), constant-rate pump tests (Chapter 6), and slug tests (Chapter 7).

Fast Linear   Fast linear log type measures and records at a user-defined fixed interval of 1 min or less. The interval is small (seconds to milliseconds), and the test is usually of short duration due to the volume of data logged and the impact of very fast sampling on battery life and storage memory.

Step Linear Step linear log type measures and records data according to a number of user-defined elapsed time intervals or "steps" within a schedule. Both the elapsed time and the number of measurements within each step can vary. After completing the elapsed time for each step, the schedule will automatically move to the next step. Some data loggers allow up to 10 separate defined steps. This can be used during a step test; however, coordination of the timing of steps must be carefully planned and executed (Chapter 5).

Setup Recommendations

Proper installation of well-monitoring equipment is critical for good data collection for both pressure transducers and well-sentinel applications. Care must be taken in the field to make sure the cable running down the well is as straight as possible and undamaged. Sometimes, even though a proper connection is made, the cable line from the transducer to the surface may be kinked or damaged, thus prohibiting accurate data collection. Cables connected to vented pressure transducers have a hollow tube that allows the strain gauge to sense changes in air pressure. The typical field method is to make a loop greater than 1 in (2.5 cm) with duct or Gorilla Tape that is attached to the well casing, also with tape. The "tape-loop" may slip in hot weather, so a dowel or similar object can be placed to prevent kinking. If using vented pressure transducers connected to a surface data logger, the vented cable can be protected in heavy-traffic areas by placing two parallel boards on the ground taped together at high traffic areas. The cable should be placed under the juncture of where the two boards touch. Non-vented (absolute) pressure transducers are commonly connected with wire connections to allow data access. In this case kinks in the cable are not usually an issue but it is advisable to ensure the cable is straight. It is important to keep the transducer or well sentinel a few inches above the well bottom when deploying installing as sediment may be present. Dropping the sensing end of the transducer into sediment could hinder the operation of the unit.

Cables connected to the pressure transducer or well sentinel have various coating materials. Polyester elastomer is found on many cables and is applicable for most monitoring applications. It is advisable to used cables with Teflon coatings when monitoring contaminated groundwater. The Teflon cables tend to be easier to decontaminate (Figure 4.32).

Desiccant Options

Some of the suppliers of well sentinels ship the equipment with a desiccant attached to the data logger. There are two types of desiccants normally available. The first type is used for shipping the data logger and should not be used during deployment of the instrument. The second type of desiccant is to be used in the field and can be attached in-line with the pressure transducer. Desiccants are very useful when working in more humid environments. Desiccants can come in various sizes. Standard-sized desiccants are best suited for low-humidity environments or deployments where maintenance occurs regularly. Extra-large desiccant sizes are available and are utilized in high-humidity environments or deployments where maintenance occurs infrequently. Figure 4.33 shows an in-line desiccant attached to a vented cable.

Figure 4.32   Steam cleaning of slug-testing equipment. Similar decontamination applications can be used to clean well sentinels; such as a simple Alconox-water solution and a brush.

Figure 4.33   Desiccant (right) with connection to a vented cable (left), which helps keep the equipment dry under humid conditions (In-Situ, Inc., Fort Collins, Colorado). (*Used by permission.*)

Additional Data Logger Options

Well sentinels may also be equipped with a variety of water-quality probes in addition to a pressure transducer (level measurement). Popular options include measuring specific conductance, pH, dissolved oxygen, and temperature. One probe recently introduced includes multiparameter processers that continuously measure 12 water parameters in a variety of applications. Standard suite of six sensors measure actual and specific conductivity, salinity, total dissolved solids, resistivity, density, optical dissolved oxygen, ORP, pH, temperature, water level, and water pressure. These can be very helpful in monitoring changes in water quality over time. For example, during a pumping test, changes in water quality may provide helpful information that aids in interpretation. A case example occurred during a pumping test in mining impacted waters near Butte, Montana. A steady-state head condition (water levels stabilizing) was being approached during the test. It was obvious to see that a recharge source was contributing cleaner water to the pumping well, indicated by a higher pH and lower specific conductance, possibly from a nearby stream or from a gravel channel connected to a nearby stream. The water-quality data nicely augmented the interpretation of the pumping test.

Most field technicians are encouraged to obtain manual backup readings to the data logger readings using E-tapes in case an automatic system fails. This is also recommended to students collecting data during field exercises or collecting data for their theses.

Well sentinels can record can store multiple events or tests. It is advisable to back up the stored information from the well sentinel or data logger to a computer and free up the available memory at the start of any water-level recording program.

Enhanced Telemetry Systems

The most recent changes in groundwater monitoring have been the development of telemetric monitoring systems. These systems have the ability to transmit recorded or live data from the well's data logger to an on-site or off-site computer or to a website. When monitoring conditions change rapidly, wells are located in remote sites, or require real-time data, ETS's may be a cost-effective way to go. These systems utilize wireless signal technology, which provides greater control over the data, and has the benefit of reducing the number of site visits (travel and manpower costs reduced). The basic layout is shown in Figure 4.34.

Remote telemetric systems connect dedicated well sentinels in monitoring wells to a central data collection center through various communication systems. Telemetry options include telephone, radio, cell (smartphone), or satellite connections. Telephone-modem data transmission systems are used with remote water-monitoring systems that are accessible to telephone lines. Radio systems are used for relatively short-range transmission of remote water-monitoring system data with a range of approximately around 15 miles (24.1 km). Remote water-monitoring systems that use cellular data transmission require sites that are well covered by cellular transmission towers. These remote water-monitoring system sites will generally be closer to developed areas. The satellite data transmission method was designed for remote water-monitoring systems in areas where power is not available, there are no telephone lines, or cellular coverage is nonexistent and that are far enough away from the data collection point that a radio system is impractical. Power options include

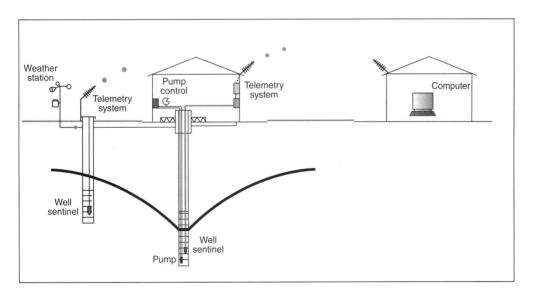

FIGURE 4.34 Schematic of an enhanced telemetry system (ETS) designed by Summit Envirosolutions, Inc., St. Paul, Minnesota. The schematic shows the components of a monitoring system that can transmit real-time drawdown-level data from a monitoring well to a computer at a desired location. (*Used by permission.*)

FIGURE 4.35   Remote *solar-powered* monitoring well with a cell phone telemetric system allowing real-time data to be retrieved from a remote location. (*Courtesy of Summit Evirosolutions, Inc. Used by permission.*)

either 100–240 VAC line power or solar-charged battery power. An example of a remote telemetric system at a monitoring well is shown in Figure 4.35 with a view of the control box in Figure 4.36.

Information collected from the remote monitoring wells is sent to a central computer that can be accessed through the Internet. This allows the data to be controlled yet observed at multiple locations. This is a great application for monitored sites where multiple individuals or stakeholders need to view or share the information.

Telemetric well-monitoring technology has led to the development of software and apps that allow the user to monitor an entire well field in real time. For example, some software applications may continually recontour the changes of groundwater levels in real time. The software is often joined with a Supervisory Control and Data Acquisition (SCADA) system allowing for control of pumping rates in an active well field in Minnesota, Oklahoma, and Alabama. (John Dustman, personal communication, September 2018).

FIGURE 4.36 Control box for the remote telemetric system shown in Figure 4.35. (*Courtesy of Summit Envirosolutions, Inc. Used by permission.*)

Telemetric sensors will continue to become smaller, less expensive, and capable of detecting more parameters at lower concentrations. Wireless communication improvements such as Long Range (LoRa) developed by Semtech will improve data accuracy with low-cost chips. Continuous data streaming will become the norm to monitor Earth's processes and groundwater conditions in particular. The subsequent advancements in modeling, neural networks, genetic algorithms, and kriging all bode well to better understand groundwater mechanics, particularly in fractured media.

Software and Apps

Groundwater-monitoring equipment suppliers usually provide software or apps to communicate with the data-logging equipment. The software is compatible with laptops, tablets, and smartphone applications. The software allows for exporting the collected data in various formats to be compatible with software packages for analyzing or plotting the water-level records. The data from the data loggers of well sentinels are normally uploaded to the computer via a cable; however, some software is available via a Bluetooth connection. The software provided by the suppliers will also allow for reading the data being collected as the pressure transducer or well sentinel is actively recording. Many of the telemetric systems in use today also include other sensor logging options such as specific conductance, pH, dissolved oxygen, and temperature.

Defining Level Measurements

For a given project area or data set, it is imperative that there be some consistency among those who collect the data. One of the first considerations is where the common datum is. A common datum is MSL. This would mean that all data would be reduced to elevations above MSL. For some projects, an arbitrary relative datum may become the base datum. For example, you are out in the field and there is no convenient way to tie into a benchmark of known elevation. You still need to evaluate the relative elevations or may have security reasons for keeping your database as arbitrary elevations. Keeping exact elevations from being known by others may be important if a public presentation is being conducted in a politically sensitive area. Whatever ends up being used as the base station needs to be a relatively permanent feature that is not likely to be disturbed. It would be disastrous to select a large rock or drive a stake that is later excavated and removed during construction or by curious children.

Surveying the well locations and MPs is critical to proper interpretation of the data. Surveying can be done by simple level surveys; by total station systems that give northing, easting, and elevation; or by a global positioning systems (GPS). For example, level surveys provide a relative vertical positioning, but are not capable of providing northing and easting positions. GPS is very useful for widely spaced monitoring locations, distributed farther than is practical with traditional surveying equipment. Depending on the scale of the project, reasonable estimates of location and elevation can be estimated using software such as Google Earth.

The handheld GPS units are widely used by drillers and geologists with X and Y accuracies within 10 ft (3 m) if the receivers have Wide Area Augmentation System (WAAS) capabilities. The WAAS provides greater accuracy and precision by incorporating ground stations in the United States to correct for ionospheric disturbances and timing of the 24 Department of Defense satellites and corresponding satellite orbit errors (Garmin 2006). Without the WAAS, GPS units are typically accurate to within 33 ft (10 m). The GPS satellites circle the earth twice a day, in six earth-centered orbital planes (www.faa.gov). A signal is being continuously transmitted that GPS receivers "lock" unto. With three satellites a latitude and longitude (2D) location can be calculated, with more satellites vertical estimates can also be made. Again handheld units are not very accurate in the vertical (approximately 50 ft or 15 m or so). This is hardly accurate enough to use for constructing potentiometric surface maps. Additional discussion of GPS, including links to Europe's European Geostationary Navigation Overlay Service (EGNOS) and Japan's Multi-functional Satellite Augmentation System (MSAS) systems can be found at https://water.usgs.gov//osw/gps.

To obtain survey-grade vertical GPS accuracy a more expensive system that includes a base station (positioned at a known surface elevation) configured with a rover unit must be used. This type of system can bring vertical accuracies within less than a centimeter. To accomplish this, a tripod is positioned over a point of known vertical elevation, with the base station unit and rover unit coupled together at the exact same elevation (Figure 4.37). Software in the base unit is given instructions via a handheld

Figure 4.37 GPS base station and rover unit during initial setup. The rover unit is attached to a rod and taken from location to location to determine X, Y, Z global positioning.

keypad and the rover unit is correlated with the base station. Once this initial setup has taken place the rover unit may need to be held continuously out the window of a vehicle between stations, so as not to lose the connective signal. A bull's-eye bubble is held against the rover unit to maintain as near vertical position as possible and the portable keypad is activated to mark each way point (Figure 4.38). The farther away from the base station (some units are limited to a distance of 14 mi or approximately 20 km, depending on unit cost), the greater the drift in vertical accuracy. Also, the battery life of some units may only be for 8 to 10 h, so closure of the survey should be accomplished before the batteries lose power, although this isn't absolutely necessary. Closure is ideally completed by bringing the rover unit back to the base station and coupled with the unit once again for a final reading. A software package that accompanies the GPS unit is used to reduce the handheld keypad data and perform any corrections.

Measuring Points

Once a datum has been established, data collected in a fieldbook are brought back to the office. All persons collecting data should clearly indicate whether data were collected from top of casing (TOC) or GS, or some other point of reference. For example, a well casing cut off below the floor of a shed may be recorded from the shed floor surface. The distance that the casing extends above the GS is referred to as the "stick up." Proper measuring and recording of reference points become important later on when a level survey is conducted and relative elevations need to be established.

Identifying MPs may not be as straightforward as would seem. For example, suppose several monitoring wells are installed in a short period of time. Simple annoyances, such as not cutting the casing off horizontally, can occur. It may be that a piece of

FIGURE 4.38 GPS rover unit being held vertical at a way point. A bull's-eye bubble is attached to the rod to make sure the positioning is held straight up and down.

steel casing was cut off with a torch, or that a section of PVC casing was cut diagonally. This is not only a problem with placement of a well cap, but also creates a myriad of problems when different individuals go out into the field to take measurements. Where do you take the measurement? At the lowest part of the cut, the top of the cut, or somewhere in between? You may not be responsible for the well completion, but still need to collect meaningful data.

It is suggested that markings be clearly made on the casing indicating where to take a measurement, such as a black mark on PVC casing (Figure 4.39). On steel casing, it is helpful to take a hacksaw or file and make three nick marks close together. A dark permanent marker can also be used to make an upward arrow pointing to the three marks or marked MP. This works best if all markings are made *inside* the casing. Markings on the outside, even those made with bright spray paint, tend to fade quickly in the weather. If a name plate is not with the well, it is also helpful to indicate well name and completion information, such as total depth drilled (TDD) and screen interval, on the inside of the well cap.

Retrieving Lost Equipment

When pumping wells are being used to obtain level-measurement readings, there is always a risk of getting the equipment stuck. This may be during a pumping test, where an operator is trying to follow the pumping level in a well, with an E-tape or simply trying to obtain a level measurement from a domestic well. Extending upward from the top of the pump is the riser pipe and electrical lines to power the motor (Figure 4.40). If the pump is fairly deep, plastic centralizers are placed to keep the riser pipe and wiring near the middle of the casing.

Figure 4.39   The identification of a measuring point on a PVC casing with permanent marker.

Figure 4.40   Pump and riser pipe being tightened with aluminum pipe wrenches. Aluminum is much lighter and easier to handle in the field.

In cold climates, **pitless adapters** are used to make a connection from a waterline pipe placed in a trench from the house to the well. An access port is welded to the casing, approximately 6 to 8 ft below the GS to prevent freezing of waterlines. An elbow connector from the riser pipe fits into the pitless adapter. This way the riser pipe elbow can be removed from the pitless adapter and the pump can be pulled from the well for servicing. (This practice eliminates the need for a well house with a rickety ladder leading down to the location of the top of the well surrounded by creepy spider webs, Figure 4.41.) Nevertheless, E-tapes or other level-measuring probes can still become stuck down a well.

When an E-tape becomes stuck, the first task is to try and free the probe. Depending on how far down the probe is, one can try different methods. Sometimes success occurs from loosening the electronic cable and striving to jiggle the probe free. If this doesn't work and the probe is within 20 ft (6.1 m), a "fishing" device can be made from one or more sections of 1-in PVC electrical conduit pipe, typically 10 ft (3 m) long. In the bottom piece, a slit or notch about 6 in (15.2 cm) long is cut. The level-measuring cable is hooked into the notch. By sliding the conduit pipe toward the probe, there is more leverage available to move the probe around. In the event the well probe really becomes stuck, it may be necessary to pull the pump and riser pipe to retrieve the E-tape. This can be a real pain, particularly in a production well that has a lot of heavy piping. If you want to save the E-tape, it is cost-effective to call in a service truck with a cable-and-hoist capability to pull up the piping and pump. The cost of service is a minor percentage of the cost of a new E-tape.

Figure 4.41 View looking down the inside of a well house, with ladder access, down to the lower floor boards, just above the top of well casing (TOC). Also shown is the pressure tank which provides water supply and pressure to a home. SWL may be referenced from upper or lower floor boards.

Opening Tight-Strung Gates and Negotiating Fences

Another obstacle in gaining access to wells in the western United States is barbed-wire fences. After one obtains permission to obtain level measurements in rural areas, it is important to leave all fences and gates the way they were found. Some gates are tough to open and are tougher to close. There is actually a technique to opening and closing a tightly strung gate. The gate is usually held at the bottom and the top by a looped wire attached to the fence post. The trick is to place the gate post into the bottom loop and pull toward the top loop (sort of like a headlock in the sport of wrestling), with the fence post braced against your shoulder for more leverage (Figure 4.42). It is embarrassing to have a coworker or support vehicle wait long periods of time for you to close a gate. Merely pushing the gate post toward the fence results in safety problems and frustration. The headlock method works much better. In some cases, in extremely tight fences a "cheater bar" will be hanging there along the fence post. In this case, you again place the gate post into the lower loop and then use the "cheater bar" to pull the fence close enough to attach the upper loop. Both apply the same principle.

Some fences may have an electrically hot top wire to keep livestock in. In this case if you touch the wire, you will get shocked. The wire looks distinctively different from the other wires and is located near the top of the fence. Test the wire by gently tossing something metal at the suspected wire. Having rubber boots on can be helpful. Hot fence wires can be especially problematic for fences keeping buffalo in. Network mogul Ted Turner has a fence in Montana that puts out 5,000 V (although the amperage is low) to the unwary animal. It is shocking enough to keep a buffalo away but also strong enough to knock a person down and stop their heart from beating for a couple of minutes.

FIGURE 4.42 How to properly open or close a barbed-wire gate, using the headlock method to apply sufficient leverage to loosen the gate, a critical technique in gaining access to a well.

Decontaminating Equipment

If an operator is to perform a level-measurement survey of a series of monitoring wells, it is important to be prepared to decontaminate, or DECON, the equipment. Merely going from one well to the next with the same device may cause cross-contamination, especially at a hazardous waste site. DECON practices vary from company to company. Examples; such as, rinsing equipment with an Alconox-water solution rubbed with a brush, submergence bath, or steam cleaning (Figure 4.32) may be required. A small amount of Woolite into a bucket of water makes a great DECON solution. It depends on the nature of the site and scope of the project. If the purpose for monitoring is for trends in the hydrographs of water levels, a simpler DECON step would be taken, compared to monitoring wells completed in a plume of toxic organics. In the case of a hazardous waste site, and the task is water-quality sampling, it may be better to use dedicated bailers tied onto and then left inside each well.

Level-measuring devices may be designed to have the electronic reel built as a module that can be removed for DECON purposes or the electronics be encased in an epoxy coating for easier decontamination. It is significant to realize that certain chemicals may adsorb onto probes (Fetter 1993). Most probes are constructed of stainless steel because this tends to be less reactive than other materials, except in mining-impacted waters where pH values are low and dissolved metals are high. It is also helpful to have clean-looking equipment when you go knock on a landowner's door.

Level Measurements in Flowing Wells

When a well is drilled into a confined aquifer whose recharge area is high in elevation, the pressure head at the wellhead may be sufficiently high to result in artesian flow. The volume of flow at the surface is a function of hydraulic head and the casing diameter. Predicting the actual rate of flow at the surface is difficult to do, even if the shut-in pressure is known. From field observations two wells with identical total hydraulic heads but different casing diameters will flow differently at the surface as a function of casing diameter. For example, a 4-in ID well with a shut-in pressure of 5 psi (11.6 ft of head) may flow at 20 gpm (1.26 L/s), while a 2-in ID well nearby, also with shut pressure of 5 psi (11.6 ft of head), flows at 5 gpm (0.32 L/s). If enough stand pipe is added to each well the result is that each has the exact same static water level. A reduction of casing diameter at the surface results in a reduction in the rate of flow.

Flowing artesian wells can be capped and equipped with a pressure gauge so that the pressure head at the cap can be read directly (Figure 4.43). In this case the elevation of the cap would be surveyed and the pressure gauge reading converted into feet of head and added to the elevation head. If the pressure head is less than 5 psi (11.6 ft of head) then a well can be fitted with a standpipe sufficiently high to keep the well from flowing and direct level-measurements made (Figure 4.6). During a pumping test if the pressure head is too great, a self-contained pressure-transducer data logger can be plumbed directly into the well casing. A section on running pumping tests on flowing wells is given in Chapter 11.

Example 4.12   In Petroleum County, Montana, tens to hundreds of wells have been drilled into several confined aquifers, such as the Cretaceous Basal Eagle Sandstone, and the 1st, 2nd, and 3rd Cat Creek (Brayton 1998). Many of these wells had been flowing for decades until a rehabilitation program took place to control free-flowing wells. Some of the wells were controlled by installing flow reducers, while others were redrilled and completed with new casing and insulation and located below the frost zone to keep from splitting from freezing (Figure 4.44). This resulted in saving millions of gallons of water and the recovery of water levels on the order of 5 ft (1.5 m) in just 1 year (Brayton 1998; Weight et al. 1999).

FIGURE 4.43 Well cap pressure gauge in a capped artesian well. The gauge is read in psia.

Summary of Level-Measurement Methods

Level-measurement devices are necessary to obtain SWL and dynamic level data. Each has its own application and design. It may be helpful to have a variety of pieces of equipment with a range of depth capabilities.

FIGURE 4.44 Artesian well completed below the frost zone inside a corrugated culvert well house. The water line to the house T's with the well.

Summary of Manual Methods

1. Establish a common datum, such as MSL. It may be necessary to define an alternate elevation above MSL. In addition, consideration of the different scales of the maps, the data will be used for may be important.

2. Make sure that the chosen base station is a relatively permanent feature, unlikely to be moved or removed later on.

3. Don't forget your keys, wrenches, and multitool to remove the well cap.

4. Establish an MP physically on the well casing (Figure 4.39) or the location from which all water-level measurements will take place (floor of the well house; Figure 4.41). A sketch or reference photo will help document the MP.

5. To prohibit effects from weathering, make all markings on the inside of the well or under the well cap.

6. Use a carpenter's chalk with a steel tape for a clearer reading.

7. E-tapes should have their sensitivity dials adjusted as appropriate to the water quality to get a crisp, clear buzz, or light indication.

8. Always move the E-tape cable line to the middle of the well, shake lightly, and repeat, to make sure the reading is the same.

9. Record the depth to water from TOC or GS. The "stick up" should also be measured and recorded. Remember to put the well cap back on after the reading and secure the well.

10. Do not use your E-tape to sound well depths.

11. Make sure your field notes are clear, because getting back to some wells again may be difficult.

If wells are locked or inside locked gates, one must remember the appropriate keys or tools necessary to gain access. Multitools are handy to remove tightened bolts and loosen wedged well caps. Steel tapes are helpful to gain access to small openings in well caps and can also be used to measure distances to objects in the field. E-tapes are the most common level-measuring device and come in a range of probe-size designs and depth capabilities.

Chart recorders are still helpful in monitoring continuous hydrograph data. However, chart recorders are being replaced by ETS or a network of well sentinels. ETS are the way to go for obtaining "real-time" data, while the data from well sentinels can be uploaded at any convenient or necessary moment in time.

Pressure transducers and data loggers are indispensable for rapid succession readings needed during pumping and slug tests. These are more expensive but are easier to manipulate during the data reduction and analysis stage. Pressure transducers may also be helpful in obtaining levels on flowing wells.

Summary of Automated Methods

1. Lower vented pressure transducers down the well to allow the cable to straighten and then spool each transducer cable to the data logger. Allow a 1-in loop in the cable (to facilitate air-pressure changes), secured with duct or Gorilla

Tape and then tape-attach to the well so that it doesn't move. Place a dowel in the loop.

2. Check the psi range of the transducer or well sentinel and make sure it isn't lowered into a water depth that exceeds its pressure capacity.

3. Attach the transducer or well sentinel in the pumping well above the pump with electrician's tape and secure the cable to the riser pipe every 5 ft (1.5 m) to the surface (Figure 4.31).

4. Make sure each vented pressure transducer is connected to the data logger. (This can also be checked from the data logger or from an APP on the cell phone.)

5. Establish the base level as a last item before starting the pumping test and use a value greater than zero.

6. Back up any automated system with manual E-tape level measurements.

Getting level-measurement devices stuck down a well is a real possibility in pumping wells. Fishing techniques may be used as a first step, but it may be necessary to call a service rig to free an E-tape probe. This would only be a percentage of the cost of a new probe. Decontamination of level-measuring devices is essential to prevent cross-contamination. Simple rinsing or brushing using an Alconox-water solution or steam cleaning can do a reasonably good job. Having clean equipment is important when showing up at the door of a landowner.

4.4 Misinterpretation of Water-Level Data

The misinterpretation of water-level data is a relatively common occurrence among those just getting started in field hydrogeology. It takes time and experience to think through certain questions to put together a reasonable conceptual groundwater-flow model. Some of the following reasons for making errors in interpreting water-level data can be grouped as topics. The following list is by no means exhaustive but will hopefully get the hydrogeologist thinking:

- Are there vertical components of flow?
- Misunderstandings of the difference between static water levels in wells and the local elevation of the water table
- Combining the level data from shallow and deep wells completed in the same aquifer
- Combining the level data from wells completed with long and short screen lengths in the same aquifer (previous section on vertical groundwater flow)
- Combining level data for wells completed in different aquifers vertically or horizontally
- Combining level data from different times, seasons, or well-completion dates.

Shallow and Deep Wells

The first three topics can be essentially discussed together. Even if the geology of an area is fairly uniform, there is usually a vertical component to flow emanating from recharge to discharge areas (see Figure 4.16). It should be kept in mind that the flow path of water particles is generally rectilinear (Wampler 1998). This means that very few water molecules actually move along the top surface of the water table; rather they take a rectilinear path among local, intermediate, or regional flow systems (Example 4.13). The development of the different scaled flow systems is a function of the dimensions of the basin. Deeper basins tend to have a greater variety of flow systems developed (Fetter 1994). In humid areas where there is a fairly constant source of precipitation recharge, increased topography areas also affect the local flow regimes. For example, more undulations in the surface topography result in a greater number of local recharge and discharge areas (Fetter 1994). This must be taken into account when deciding on the depth, placement, and completion of wells. It also will affect one's interpretation of water-level data. For example, it may mean that wells completed at different depths may be screened within different parts of a local or intermediate system.

When one combines level-measurement data from wells completed shallow and deep an improper interpretation of groundwater flow may result (Saines 1981). The question arises, how deep must the well completions be before they represent a different flow system? Can I use level measurement data from 30 ft (9.1 m) wells, 100 ft (30.5 m) wells, and 300 ft (91.4 m) wells and still be in the same groundwater-flow system? The answer is complicated by the proximity of wells to recharge or discharge areas, variable screen lengths, hydrostratigraphy (Chapter 3), and physical dimensions of the valley. Rather than drilling more wells it may be that the number of wells needed may be reduced by incorporating water-quality data (Chapters 9 and 10). It may be that water-quality signatures are typical of water from certain flow systems. The following three examples will help illustrate the above issues.

Example 4.13   A study was conducted along Blacktail Creek in Butte, Montana, to investigate why local residents were experiencing basement flooding (Figure 4.45). The geology in the immediate vicinity of Blacktail Creek consists of fine-grained sediment lenses mixed with sandy units. The sandy units appear to be laterally connected, with the fine-grained sediments generally constrained within the floodplain area of the creek. Nested wells completed at depths of 15 and 30 ft, respectively,

FIGURE 4.45   Well drilling along Blacktail Creek, Butte, Montana, as part of an investigation as to why basement flooding was occurring. Example 4.13.

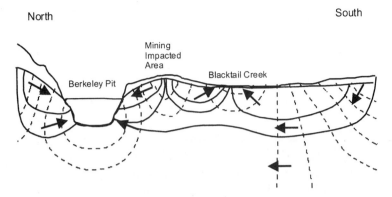

North   South

FIGURE 4.46 Schematic of Example 4.13 interpretation, showing the localized flow systems of Blacktail Creek and the Berkeley Pit and a more regional flow system from south to north (see also Example 4.14). Dashed lines represent equipotential head contours. Not to scale; however, from the south, the margin to the Berkeley Pit is approximately 8 mi (12.8 km).

within 100 ft (30.5 m) of the creek showed water-level differences within 1 ft (0.3 m) between wells. Deeper wells indicated a water level approximately 1 ft (0.3 m) higher than shallow wells, indicating an upward gradient. Differences between nested pairs indicated that if all data were combined, a confused groundwater-flow interpretation would result.

The question exists whether all groundwater within the residential area comes from more than one flow system. Water-quality samples from the area indicate the possibility that waters from mining-impacted areas may be influencing waters discharging into the area. A schematic of this interpretation is presented in Figure 4.46.

Example 4.14   Butte, Montana, is nestled in an extensional valley surrounded by mountains including those of the North American continental divide. Like many extensional intermontane basins of southwestern Montana and southeastern Idaho, the basement rocks of the valley are structurally tilted to the east, resulting in thicker sedimentary deposits on this side. In this example, the eastern sediments are approximately 900-ft thick compared to only about 100- to 200-ft thick on the western margin. Sediments are generally coarse grained, sourced from weathered granite, with lesser discontinuous micaceous clay lenses that locally have a confining effect. The Butte Valley slopes to the north in a gradual gradient as does groundwater flow. Shallow wells that penetrate to "first" water indicate a gradient of 15 ft/mi (or 0.003) (Botz 1969).

Two of the local drainages, Little Basin Creek and Blacktail Creek, initiate their headwaters in the Highland Mountains to the south and flow northward into the Butte Valley. The streams are bedrock controlled as they flow out onto the weathered granitic valley-fill materials, losing most of their water within about 0.25 mi (400 m), which results in basin recharge (downward vertical component to groundwater flow). Deep circulation occurs in the valley, and then stream flow increases once again as groundwater begins to discharge about 4 mi to the north. Wells completed near the discharge area about 5 mi (8 km) to the north at depths of 120 ft (36.5 m) and 35 ft (10.7 m), respectively, show SWL differences of 2 ft (0.61 m) between wells indicating an upward gradient, while nested wells in the middle part of the valley show equal water-level elevations, indicating only horizontal flow. This essentially is the scenario depicted in Figure 4.17 and in cross-sectional view in Figure 4.46.

Example 4.15   A watershed committee in southwestern Montana was concerned about the drilling of new production wells during a period of extended drought. They wanted to know what the impacts of new production wells would have on the Beaverhead River (Figure 4.47). The Quaternary fluvial deposits and anastomosing sloughs associated with the floodplain of the Beaverhead River

FIGURE 4.47 View of Albers Slough and the Beaverhead River floodplain taken from the Tertiary bench from the west side looking east. (Example 4.15)

are approximately 3 mi (5 km) wide in the study area. Flanked on both sides of the floodplain to the east and to the west are Tertiary sediments that extend to the surrounding mountains. Additionally, on the west side, and to the north, are Tertiary volcanics from which irrigation-well production rates may exceed 1,000 gpm (63 L/s). Precipitation recharge is augmented by leakage from a distribution of canals originally sourced from the Beaverhead River many miles upstream. The general groundwater-flow system originates from the mountains and flows toward the floodplain of the Beaverhead River, a fairly complex groundwater flow system.

The financial budget for this project was very low, so new drilling was out of the question. To better understand the groundwater-flow system(s), a network of more than 40 wells was developed by seeking permission from property owners. These were mostly livestock and domestic sources. These were measured on a bimonthly basis over a period of 2 years (Weight and Snyder 2006). Well depths ranged from approximately 30 ft (10 m) to several hundred feet (>100 m). Hydrographs were created of the level-measurement data, coupled with a study of stream flow and precipitation data. Survey grade GPS elevations were determined for each of the monitoring well and river MPs (Figures 4.37 and 4.38). The following observations were determined from the level-measurement data:

- Level-measurement data within the floodplain correlated well with stage elevations of the Beaverhead River indicating the floodplain is well connected with the river.

- Level data from wells less than 100 ft (30 m) on the west side appear to be influenced by recharge from surface-water canal irrigation return flows.

- A 50- to 100-ft (15- to 30-m) thick clay layer on the west side of the valley separates the shallower level elevation data (wells completed at depths <100 ft, 30 m) from "deep" wells completed greater than several hundred feet (>100 m). The shallower wells have an average SWL elevation level 50 ft (15 m) higher than the deeper wells.

- The Tertiary volcanics also appear to be a source of recharge to the Tertiary sediments on the west side. They show an average level elevation 80 to 100 ft (24 to 30 m) higher than all other west side wells. Thus, illustrating the problem with combining level from different aquifers.

- East side Tertiary sediments have a greater thickness, are coarser grained, and have a higher average transmissivity than sediments on the west side.

- Level-measurement data on the east side show a cyclical pattern in all well hydrographs, reflecting the influence of irrigation pumping and the timing of precipitation infiltration recharge from the nearby mountains.

Water-quality data indicated a clear distinction from Beaverhead River surface-water sources and groundwater irrigation return flows. For example, there was an inverse relationship between specific conductivity (SC) readings and pH data. For every 0.5 decrease of pH there was a 1.5 times increase in SC. Surface water and shallow wells within the floodplain showed higher pH values with lower SC values. Groundwater influenced by irrigation return flows showed lower pH and higher SC values (Weight and Snyder 2006).

Short versus Long Screen Lengths

Another source of interpretation errors may come from wells completed with varying lengths of screen.

The level measurements in a well represent the average head at the midpoint of the screened interval, including elevation and pressure head. The longer the screen length, the more vertical head changes that may be included in the average. The hydrogeologist needs to be careful in considering the geology, the hydrogeologic setting, and the well completion when coming up with a conceptual groundwater-flow model. This is particularly important if one is in a recharge or discharge area (Section 4.2).

If a well is completed in a recharge or discharge area, the average head may be below or above the water table. Consider the schematic in Figure 4.48. Static water levels in paired shallow and deep wells at a given location in a recharge area will show a downward gradient, while paired wells in a discharge area will show an upward gradient from SWL data. Why?

Combining Different Aquifers

An example of the misinterpretations of combining SWLs from different aquifers was discussed in Example 4.15. As another example, a woman from a conservation district in eastern Montana called one day to say she had monthly water-level and water-quality data for 24 months. Could I please make some sense out of it and tell whether

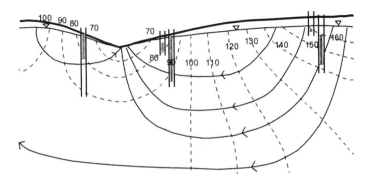

Figure 4.48 Water levels in a recharge area and discharge area, showing an average water level for a longer screen length, shaded.

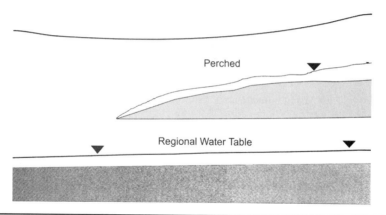

Figure **4.49**   Schematic of perched and regional aquifers from Example 4.16.

there were any water-quality trends? My first question to her was, could I please obtain a base map and well-completion information, including lithologic logs? One must know how many aquifers and confining units may be involved before proper interpretation can be performed. Understanding this is crucial to interpreting the direction of groundwater flow and the possible direction of contaminant movement.

It may be that there is a significant depth difference between sedimentary aquifers because of substantially different elevations at recharge areas. For example, deeper aquifers may outcrop at higher elevations than younger, shallower formations because of mountain uplift (Figure 3.17). If this is the case, the SWL variations between shallow and deeper aquifers may be dramatically different. Static water-level differences between shallow and deeper aquifers may be something as simple as comparing a localized perched aquifer with a deeper regional aquifer.

> **Example 4.16**   A consulting engineer was estimating drilling costs based upon water-level data within a subdivision area of a couple of square miles (5.2 km²). Within the area was a perched aquifer and a deeper regional aquifer with wells completed in both (Figure 4.49). The engineer did not understand the geologic setting and averaged the well-completion depths for both aquifers to obtain an estimate for total drilling footage. This resulted in a poor estimate of total drilling costs (financial disaster) and surprises in the field when drilling began.

4.5   Summary

Groundwater is generally always moving from higher to lower hydraulic head. The slope of the "head" surface or potentiometric surface of an aquifer is known as the hydraulic gradient. The total head at each location on the potentiometric surface is determined using components of the Bernoulli equation, where the velocity head term was shown to be negligible from most applications. The quantity of groundwater movement moving through a defined cross-sectional area is solved using applications of Darcy's law. Two-dimensional applications of Darcy's law can be performed by constructing flow nets. Recharge areas are defined where flow lines diverge and move downward from a given area, whereas flow lines converge and move upward in groundwater discharge areas.

As groundwater encounters hydrogeologic units of differing hydraulic conductivity the flow lines refract or bend, thus changing the flow path. Flow paths through finer-grained units tend to be somewhat vertical, while groundwater flow through coarse-grained units is more horizontal. Level measurements in wells define the potentiometric surface at a given location. The various principles and problems associated with taking level measurements were discussed. The point was made that each level measuring device has its own application.

There are many reasons groundwater-flow systems are poorly understood. The misinterpretation of level measurements was suggested as a key factor. Ignoring vertical flow, combining shallow and deeply completed wells, and combining wells with short and long well screens contribute to the difficulty of interpreting level-measurement data. Care must be taken to make sure the level-measurement data are from a known aquifer. Mixing level-measurement data from different aquifers leads to misinterpretations of groundwater flow and may result in a host of problems.

4.6 Problems

4.1 We begin by evaluating water-level data to determine the direction of groundwater flow. When there are water-level elevations from wells the direction of groundwater flow can be estimated by drawing a flow line (dashed in Figure 4.50) perpendicular to the equipotential contour (or lines of equal head). This is a first approximation to the direction of groundwater flow. In areas of recharge, flow lines diverge from an area, while flow lines converge in a discharge area (Figure 4.50).

Please visit the potentiometric surface (Kinnaman and Dixon, 2011) of the Upper Floridian Aquifer at http://pubs.usgs.gov/sim/3182/pdf/sim3182.pdf.

Place approximately 10 to 15 flow lines from Polk City on this map showing your understanding of groundwater flow (large recharge area with 120 ft contour southwest of Orlando)

Zoom into various areas of the Upper Floridian Potentiometric Surface Map and identify at least three natural recharge areas and three natural discharge areas.

4.2 Take the following water-level elevation data in Figure 4.51 and use a 2-ft contour to establish the direction of groundwater flow. Then, place two flow arrows on the diagram showing your understanding of the direction of flow.

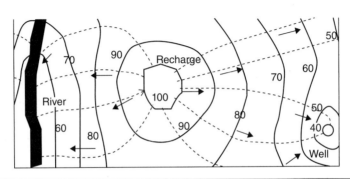

Figure 4.50 Schematic of recharge (diverging flow lines) and discharge areas with converging flow lines.

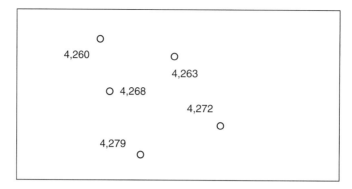

Figure 4.51   Water-level elevations for five wells.

4.3   Darcy's law can be used to evaluate the quantity of groundwater moving through an area. The hydraulic gradient used in the calculations is reflective of the conditions occurring in the field. In Figure 4.9, it will be assumed for problem (a) that the flow (Q) through the aquifer is 64,000 ft³/day. For all problems the saturated thickness "b" is assumed to be 120 ft thick.

a)   Use the information in Figure 4.9 to determine the hydraulic conductivity of the sand *and* the sandy gravel.

_____ sand _____ sandy gravel

b)   If the hydraulic conductivity of the sandy gravel was changed to be 175 ft/day, calculate the flow (volume of water per unit time) through a part of the aquifer 2,600 ft wide. _____ Q

c)   In Equation 4.1, we also note that Darcy's law can be applied using the transmissivity (*T*), the width of the aquifer (*w*) and the hydraulic gradient. Once again considering Figure 4.9 and b = 120 ft, apply the results you determined in part (a) to answer this question:

   1.   What are the transmissivities of the sand and sandy gravel portions of the aquifer (use cross-sectional view)? _____ sand, and _____ sandy gravel.

   2.   Assuming an effective porosity of 0.22 now estimate the seepage velocity of the water moving through the _____ sand, and _____ sandy gravel.

   3.   How long does it take groundwater to move through the sandy gravel from the "A" edge to the beginning of the sand? _____ How long does it take to move through the sand? _____.

4.4   You have been hired to evaluate the production potential within the area indicated in Figure 4.52. Apparently, all of the land is for sale and you are to select the "best" 20% of the land area. "Best" is being defined as best potential for water development. Figure 4.52 indicates where data from well logs and water-level data were derived. A summary of the findings is shown in Table 4.2. Please respond to tasks a to g listed below Table 4.2.

a)   Contour a copy of the water level data in the area. (Hand contouring is best.)

b)   On a separate piece of paper create an isopach map of the aquifer unit.

c)   Pumping-test data at well "D" and well "I" indicate transmissivities of 10,000 ft²/day and 2,520 ft²/day, respectively. What range of hydraulic conductivities might the area have? Explain.

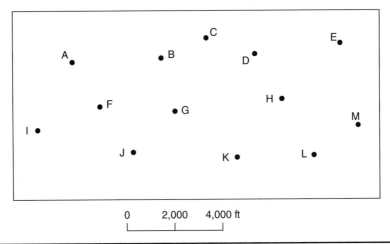

FIGURE 4.52 Information associated with problem 4.4. Lettered dots represent the location of wells. (*The original diagram was constructed on engineering graph paper.*)

d) Assuming porous media what lithologies might the K-values represent and what is their aerial distribution? What evidence did you use to explain your answer?

e) Calculate the daily volume of water passing between wells C and J.

f) Is the aquifer confined or unconfined? Please explain.

g) What is the "best" 20% of the land area and what would you recommend to your client?

Well	Surface Elevation	Water-Level Elevation	Top of Aquifer	Bottom of Aquifer
A	671	633	607	571
B	677	631	605	545
C	681	630.5	605	553
D	678	630	604	554
E	675	628	600	557
F	669	631.5	605	561
G	670	630.5	604	549
H	667	629	601	546
I	666	633	606	579
J	662	630.5	603	560
K	665	629.5	601	548
L	664	628	599	539
M	665	627	598	543

TABLE 4.2 Well-Log Data Indicating Aquifer Thickness and Static Water-Level Elevations

A 48-h pumping test is planned for an unconfined aquifer. There are only three wells to be monitored for this test. The pumping well (PW-1) is 600 ft deep. SWL is measured at 25 ft below ground surface (bgs). Step tests the previous day indicate the well can be pumped at 300 gpm with a predicted drawdown of 400 ft. The pump in the well is located at a depth of 575 ft. The first monitoring well (MW-1) is 50 ft from the pumping well and is 450 ft in depth with a SWL at 24 ft bgs. The second monitoring well is 1,000 ft from the pumping well that is 250 ft in depth with a measured water level at 22 ft bgs. Vented well sentinels are to be used to record drawdown during and recovery after the test.

4.5a   What are the depth ranges for the transducers to be used and at what depths should they be placed in each well?

4.5b   Which data-logging format should the well sentinels be programmed at to record the drawdown and recovery?

4.7   References

Botz, M.K., 1969. *Hydrogeology of the Upper Silver Bow Creek Drainage Area, Montana.* Montana Bureau of Mines and Geology Bulletin 75, 32 pp.

Brayton, M.D., 1998. *Recovery Response from Conservation Methods from Wells in the Basal Eagle Sandstone, Petroleum County, Montana.* Master's Thesis, Montana Tech of the University of Montana, Butte, MT, 68 pp.

Davis, S.N., and Dewiest, R.J.M., 1966. *Hydrogeology.* John Wiley & Sons, New York, 463 pp.

Fetter, C.W., 1993. *Contaminant Hydrogeology.* Prentice Hall, Upper Saddle River, NJ, 500 pp.

Fetter, C.W., 1994. *Applied Hydrogeology, 3rd Edition.* Macmillan, New York, 691 pp.

Freeze, A., and Cherry, J., 1979. *Groundwater.* Prentice Hall, Upper Saddle River, NJ, 604 pp.

Heath, R.C., and Trainer, F.W., 1981. *Introduction to Groundwater Hydrology.* Water Well Journal Publishing Company, Worthington, OH, 285 pp.

Hubbert, M.K., 1940. The Theory of Ground-Water Motion. *Journal of Geology*, Vol. 48, pp. 785–944.

Kinnaman J.L., and Dixon J.F., 2011. Potentiometric Surface of the Upper Floridan Aquifer in Florida and Parts of Georgia, South Carolina, and Alabama, May–June 2010. https://pubs.usgs.gov/sim/3182/pdf/sim3182.pdf

McIlvride, W.A., and Rector, B.M., 1988. Comparison of Short- and Long-Screen Monitoring Wells in Alluvial Sediments. *Proceedings of the Second National Outdoor Action Conference on Aquifer Restoration, Ground Water Monitoring and Geophysical Methods*, Vol. 1, Las Vegas, NV, pp. 375–390.

Saines, M., 1981. Errors in Interpretation of Ground-Water Level Data. *Ground Water Monitoring and Remediation*, Spring Issue, pp. 56–61.

Toth, J.A., 1962. A Theory of Groundwater Motion in Small Drainage Basins in Central Alberta. *Journal of Geophysical Research*, Vol. 67, pp. 4375–4381.

Toth, J.A., 1963. A Theoretical Analysis of Groundwater Flow in Small Drainage Basins. *Journal of Geophysical Research*, Vol. 68, pp. 4795–4811.

Veissman, W., and Lewis, G.L., 2003. *Introduction to Hydrology, 5th Edition.* Prentice Hall, Upper Saddle River, NJ, pp. 333–334.

Wampler, J.M., 1998. Misconceptions about Errors in Geoscience Textbooks, Problematic Descriptions of Ground-Water Movement. *Journal of Geoscience Education*, Vol. 46, pp. 282–284.

Watson, I., and Burnett, A.D., 1993. *Hydrology: An Environmental Approach*. Buchanan Books, Cambridge, 702 pp.

Weight, W.D., 2004. Integrating Site Characterization Data into a Numerical System: A Case Study from Northern Montana. *Abstract in Proceedings of the 2004 North American Environmental Field Conference and Exposition*. Nielson Environmental Field School, Tampa, FL.

Weight, W.D., Brayton, M.D., and Reiten, J., 1999. Recovery Response from Conservation Methods in Wells from the Basal Eagle Sandstone, Petroleum County, Montana. *1999 GSA Abstracts with Programs*, Vol. 31, No. 4, pp. A60.

Weight, W.D., and Snyder, D., 2006. Basin Analysis of Groundwater Changes in the Northern Dillon, Montana area. *Abstracts in Proceedings of the AWRA Montana Section Annual Meeting*, Polson, MT.

CHAPTER 5

Pumping Tests

As young people in high school, my friends and I would come to a stop at a red light, put the vehicle in park, open the doors, run around the vehicle, and then get back in before the light turned green. This silly activity was fun because of the chaotic flurry of confusion that took place, reminiscent of the Keystone cops, plus the bemused expressions of other drivers as they looked on. It was always a challenge to get everyone around the vehicle and back in so that traffic was not held up. This activity will be referred to as a wacky fire drill (WFD). The first few times one runs a pumping test, the chaotic flurry at the beginning of the test of synchronizing watches, obtaining manual readings every few tens of seconds, the management of a fairly long list of equipment, and the stress of being responsible to obtain meaningful data is much like a WFD.

Example 5.1 The author still remembers his first experience in being involved with a pumping test back in the fall of 1980 in northern Wyoming. My job was to continue manual readings for the night shift and to radio in if there was any trouble. At about 2 o'clock in the morning the pump started making a terrible noise and the radio had gone faint because the vehicle battery had run low from not restarting the engine to keep things charged up. I felt foolish and scared and wondered how long it would be before anyone found me. To my rescue came the hydrology supervisor, Jim Bowlby, who immediately realized the water level had drawn down to the pump intake, creating a potential cavitation situation. He also helped give me a jump-start to get the truck and the much needed heater going. We attempted to salvage the test, but the data were messed up. This was a WFD gone bad.

5.1 Why Pumping Tests?

There are many methods of obtaining hydraulic information from aquifers, but perhaps the most common and best is the pumping test or aquifer test. Note that there is an industry-wide error in referring to these as "pump tests." Hopefully it will be the aquifer we are testing, *not* the pump. Without an estimate of the hydraulic properties of an aquifer, calculations of groundwater movement and contaminant transport cannot be performed with any level of confidence. Saying that pumping tests are the best method of obtaining hydraulic information presumes that the test has been designed properly and was run for an adequate period of time. This also includes setting the pump discharge at a sufficient rate to stress the aquifer and supposes that there is at least one observation well to obtain storativity values. By stressing the aquifer for a sufficient amount of time, a glimpse of the aquifer properties away from the pumping well can be obtained, including boundary conditions that may affect groundwater flow. Although less expensive point estimates can be obtained from specific capacity tests and slug tests, the insights gained from pumping tests can be far more meaningful. This, of course, depends on the objectives of the study. (Specific capacity tests and the evaluation of aquifer-test data are discussed in Chapter 6, and the field methodologies for slug testing and analysis are presented in Chapter 7.)

Perhaps the most important reason for conducting an aquifer test is to conceptualize the general hydraulic properties of an aquifer. Well-hydraulic theory assumes that an aquifer is homogenous and isotropic (Chapter 6) when in fact this is rarely the case (Section 5.2). Another assumption that is often violated is that there are no boundary conditions, meaning that the aquifer is of infinite areal extent. To satisfy the condition of horizontal radial flow, the aquifer is supposed to be screened over the entire thickness of the aquifer. Geologic conditions and well-completion costs often make this condition not practical. How can a pumping test be designed that will yield meaningful results? How can one discover the spatial variations in aquifer properties and the effects of boundary conditions either perceived or not? This is the topic of the next section.

5.2   Pumping-Test Design

Pumping-test design begins with establishing the objectives and conditions of the study. The objectives are usually to obtain hydraulic information; however, conditions may also correspond to circumstances that may exist in the field, for example:

- Is the pumping test being conducted near or within a contaminant plume? If so, where will the discharge water be stored and treated and how much will be produced? Or will the pumping test affect the configuration of a plume?

- Will pumping affect or impact other existing senior water users' wells?

- Will pumping occur near a construction site or downtown area where dewatered sediments may result in compaction and foundation stabilization problems?

- What size of pump will be necessary to adequately stress the aquifer? Will a sampling pump be adequate, or will a submersible or even a turbine pump be required? Think of equipment and costs.

- Is this well being tested for production purposes, such as for irrigation or public water supply, that would require long-term pumping? What impacts, if any, are expected to occur, and what additional personnel and equipment might be necessary?

- Is the aquifer being tested to check the requirements for a subdivision or to perform dewatering estimates at a construction site or mining property?

- Is the test going to take place near a recharge or discharge area (Chapter 4)? Are vertical gradients or other components of groundwater flow important?

- Are multiple aquifers being affected?

- What are the potential boundary conditions in the area? Are there streams, faults, constructed barriers, or confining layers or discontinuities that may affect the shape of the cone of depression (Chapter 6)?

Sometimes there aren't any special conditions; it's just that a pumping test is needed to estimate the hydraulic properties for use in calculations for a groundwater-flow model or transport model. Maybe initial estimates are needed to estimate distance-drawdown relationships.

Geologic Conditions

When designing a pumping test, it is imperative that a general background of the geology be understood. Will this be a porous media test in unconsolidated or lithified sediments? What depositional environments are probably represented? Will this take place in fractured rock or karst conditions? Is this a dual or triple porosity system? For a general presentation on geologic conditions, see Chapter 2, and see Chapter 12 for karst conditions.

General geologic information can be obtained from other wells drilled in the area. State agencies have well logs that are accessible by section, township, and range (Chapter 1). For larger production wells, a small-diameter pilot hole is advisable to be drilled first to anticipate the well construction design. In a monitoring-well field, it may be that all wells will basically be constructed the same. The pipe and screen will be fairly uniform. Details on monitoring-well construction and design are found in Chapter 15.

For the purposes of pumping tests and the analysis of results, the following geologic conditions are helpful to notice in advance of the pumping test:

- Do the sediments fine upward or downward? This may affect the rate of vertical contribution. Layered sediments will have an effect on the results in terms of potential delayed yield (Chapter 6) or other sources of recharge. These may also constrain the lateral contribution of water from course-grained sediments that may be confined or semi-confined above and below. This is especially important in determining the saturated thickness contributing water to partially penetrating wells during the pumping test (Chapter 6). To check the vertical connection of sediments, wells completed at various depths will be needed to observe any changes.

- What is the potential for lateral boundary affects? Are there surface water bodies, such as lakes, streams, or ponds, that are hydraulically connected with the groundwater system (Chapter 10)? Or, could the aquifer be expected to thin or thicken or become coarser or finer grained in a particular direction? Are either recharge or barrier conditions anticipated? If so, it is advisable to locate an observation well away from the pumping well in the direction of anticipated boundary-condition impacts.

- Is the aquifer likely of low or high transmissivity? This will affect the spacing and distance of the observation wells. Cones of depression are steep in low-transmissivity aquifers and do not extend as far as in higher-transmissivity sediments.

- In fractured systems, what are the orientations of the major fracture zones? Observation wells drilled nearby may show little or no response if not connected to the main system. In fracture systems the anisotropy can be so great that flow is 60 degrees contrary to the hydraulic gradient (Marek Zaluski, personal communication, September 2018). This is especially important in a forced gradient tracer test if a nearby monitoring well is to be used as the source well (Chapters 13 and 14).

Example 5.2 During the 1992 Montana Tech field camp a pumping-test site was selected with five observation wells (Figure 5.1). The students were to perform a 24-h pumping test and interpret the results of a preexisting site. The pumping well was sounded to check for depth and was found to

FIGURE 5.1 Pumping test with five observation wells. Application described in Example 5.2.

be silted in over the entire length of the 10-ft screened zone. The slot size of the screened interval during well completion was too coarse (Chapter 15). Through a bailing process the pumping well was cleaned out. Each of the three existing observation wells was also cleaned out, and two new ones were drilled and completed. The configuration of the aquifer test is shown in Figure 5.2. One thing that surprised the students was the rapid response of an observation well located much farther away (Ob-3 well) than wells completed closer (Ob-2 well), until they considered the geologic conditions. After some thought, they realized that the completion depth and lithologies of the more distant well were very similar to the pumping well (Figure 5.3). Wells completed closer to the pumping well were completed more shallowly and screened in finer-grained sediments. Partial penetration effects were also a factor (Chapter 6).

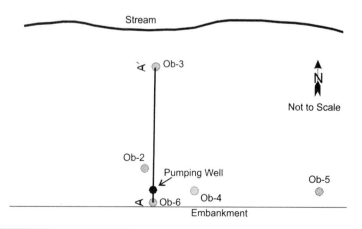

FIGURE 5.2 Plan view of 1992 field camp pumping test layout. (Wells Ob-3 and Ob-5 are about 49 and 53 ft, respectively, from the pumping well.)

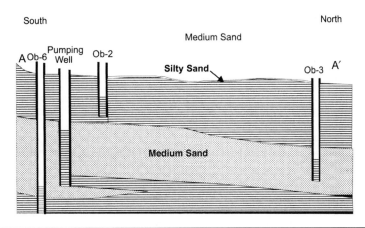

FIGURE 5.3   Cross-sectional view of Figure 5.2, from south to north. Well Ob-3 showed a drawdown response before well Ob-2.

Observation well Ob-3 in Figures 5.2 and 5.3 responded sooner than Ob-2. Both the pumping well and northernmost observation well were completed at approximately 30 to 35 ft depths (9.1 to 10.7 m), while Ob-2 was only completed at a depth of 12 ft (3.7 m) in finer-grained sediments. In time, all wells were impacted by drawdown. A schematic cross section is shown in Figure 5.3.

Distance-Depth Requirements of Observation Wells

The depth of completion of observation wells should reflect the conditions of the aquifer to be developed or the hydraulic properties of the aquifer where calculations will be applied for the intended purpose of the study. If there are additional objectives, such as vertical connections or gradients between shallow and deep systems, then nested pairs or other well-completion options are desirable (Chapter 15).

The distance of observation wells from the pumping well should also reflect a basic understanding of the geologic conditions already mentioned above and whether the aquifer is confined or unconfined. Confined aquifers, in some cases, can have observation points literally miles (kilometers) away. It will also be a function of the time of pumping. Short-duration tests may require that monitoring wells be placed closer than those being tested at longer pumping times (>72 h). If delayed yield is expected in the time/drawdown response curve (Chapter 6), an observation well close enough to the pumping well will be necessary to observe this. This is usually on the order of 10 to 20 ft (3 to 6 m) or so, but distances in excess of hundreds of feet have been observed by the author. Once again it depends on the duration of pumping and aquifer properties.

Example 5.3   In the Little Bitterroot Valley in northwestern Montana is an elongated north-to-northwest-trending intermontane basin (Figure 5.4). Before homesteaders arrived in the valley, water was only utilized at springs (Donovan 1985). The first developable water came from the Lone Pine Aquifer.

The Lone Pine Aquifer consists of very permeable unconsolidated glaciofluvial gravels and sands of early Pleistocene age. These are overlain by about 200 ft (61 m) of silts and clays from Pleistocene Lake Missoula. A shallow surficial aquifer consisting of Pleistocene sand and gravel deposits and Holocene fluvial terrace gravels sits on top of the lacustrine sediments (Donovan 1985). (See Chapter 2 for explanation of geologic terms.)

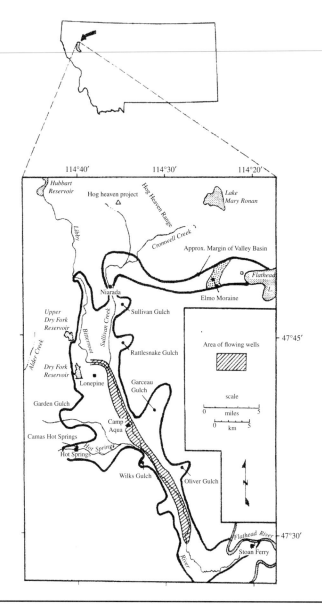

FIGURE 5.4   Location map of the Little Bitterroot Valley, northwestern Montana. [*Reprinted with permission from the MBMG (1985).*]

Drilling was accomplished through a jetting technique that cuts through the soft lacustrine deposits but was able to penetrate only a few feet into the Lone Pine gravels. The pressure head within the aquifer was sufficient to cause flowing artesian wells, with yields in excess of 1,000 gpm (3,790 L/min). The main use of developed water was for irrigation of approximately 3,000 to 3,500 acres, although some was used for domestic and stock watering purposes (Donovan 1983). Controversy arose over declines in flow that resulted from an increase in the number of wells. Additionally, the valley has warm water 25 to 53°C over approximately 600 acres of the Camp Aqua area (Figure 5.4). This resulted in problems with irrigation applications and with bathhouses being developed (Donovan 1985).

Pumping tests were conducted between 1980 and 1983 to determine the aquifer characteristics of the Lone Pine aquifer. Measurable drawdown was observed in observation wells as far away as 9 mi (14 km)!

Ideally, any pumping test should have (at least) two observation wells. The two wells should be placed at different radial distances from the pumping well and at different azimuth orientations. The closer well should detect near-well conditions and be sure to pick up drawdown, and the other well is helpful in evaluating the distance the cone of depression expands in a particular azimuth direction. It is also extremely important to have an understanding of the direction of groundwater flow. Drawdown in wells in the up-gradient or down-gradient direction is typically more pronounced compared to wells lateral to the pumping in a sloping potentiometric surface at a given time t. This is because of the influences of gravity but also may be a function of anisotropy within the aquifer and from concepts associated with capture zones (Harter 2002). It may be the case that observation wells not detect any drawdown regardless of how long pumping takes place (Figure 5.5). This concept is significant to groundwater capture zones and remediation design.

Figure 5.5 represents six schematics of cones of depression occurring in a southward-sloping hydraulic gradient from pumping wells over the same time t. In each case there are two observation wells with one placed close to the pumping well and one located farther away. However, in cases C, E, and F, one or more wells did not detect drawdown at a given time t. It appears that the duration of pumping would have little effect on whether drawdown would be detected or not. In case B, although both wells would detect drawdown, little information would be gained as to changes occurring east or west of the orientation line of the two wells. There are not many advantages to a design

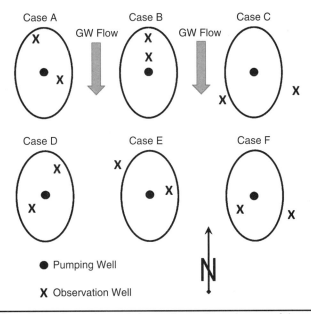

FIGURE 5.5 Configurations of pumping and observation wells and cones of depression with direction of groundwater flow at some time t. (Scenarios A, B, and D will detect drawdown in all wells, and Scenarios C, E, and F will not detect drawdown in one or more wells at time t.)

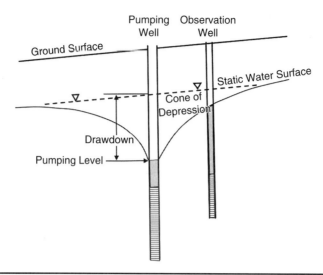

FIGURE 5.6   Cross section showing a cone-of-depression in a sloping potentiometric surface (in this case the aquifer is unconfined).

like this. A cross-section view of a pumping well and an observation well in a sloping potentiometric surface is depicted in Figure 5.6.

In Figure 5.6 the sloping potentiometric surface is exaggerated for emphasis. It is the author's experience that a sloping water table often yields an asymmetrical cone of depression, where the greatest degree of drawdown occurs in the up-gradient direction. This is likely from inhomogeneity and anisotropy within the aquifer. This is important in having a conceptual model of the aquifer and plays a significant in a capture-zone analysis and also likely an important component in pumping-test design.

Example 5.4   Near Dillon, Montana, in the southwestern part of the state, a rancher wanted to obtain a water-use permit to supplement his surface-water rights for irrigation. A 12-in production well was drilled and completed, capable of producing 1,800 gpm (114 L/s). This required a 72-h aquifer test be conducted to estimate aquifer properties, demonstrate well yield, and estimate the impacts to the aquifer at the end of the irrigation season (Chapter 6). The beauty of this test was that there were four possible observation wells at four different azimuth orientations from the pumping well (A, B, C, and D). A 74-h pumping test was performed in July 2004 by the 2004 Montana Tech hydrogeology field camp and pumping continued to the end of the irrigation season (78 days later). The author monitored the drawdown at 20 days and at the end of the irrigation season with the data shown in Table 5.1. Drawdown was observed in all four wells at 74 h and beyond.

The four observation wells used to monitor drawdown (Table 5.1) are shown in Figure 5.7. The data from all three times (3, 20, and 78 days) were used to estimate a plan-view shape of the cone of depression over a full irrigation season (April to October each year), in this case 188.6 days (Figure 5.7). A 2-ft drawdown cone was depicted (dashed line) along with the full radius of influence (solid line) to show that no other neighboring wells would experience any drawdown during irrigation activities. The direction of groundwater flow is to the northwest. Note the asymmetry

Well	Distance "r" in feet	Comment	dd @ 3 days	dd @ 20 days	dd @ 78 days
A – shop	2,112	Down-gradient	0.12	0.87	3.52
B – trailers	3,511	Directly east	0.04	0.55	2.83
C – test well	4,224	Up-gradient	0.23	1.2	4.64
D – swale	2,620	South—up-gradient	0.70	1.49	4.22

TABLE 5.1 Drawdown Data (in feet) from Observation Wells Surrounding the Pumping Well in Example 5.4, with Location of Observation Wells A, B, C, and D Shown in Figure 5.7

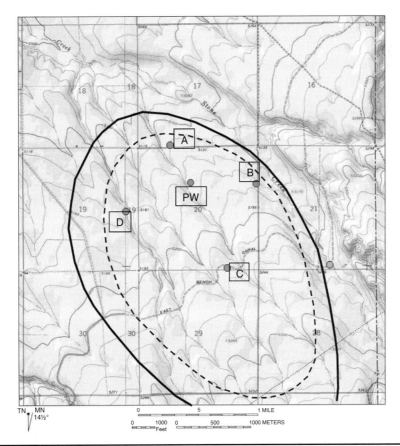

FIGURE 5.7. Plan-view delineation of cone of depression at 188.6 days from data used in Table 5.1 (solid line) and delineation of 2-ft drawdown (dashed line). Wells A, B, C, and D have drawdown data measured from the field. The stock well, marked in lighter color outside the solid line east of well C showed no drawdown.

of the cone of depression and compare with the schematics shown in Figure 5.6. Asymmetry can be attributed to decreasing grain size in the down-gradient direction (inhomogeneity) and lateral grain-size distribution perpendicular to the direction of flow (anisotropy). Conceptually, groundwater is flowing from the direction of the sediment source, where coarser-grained materials (with corresponding greater K-values) are found.

A second production well was aquifer tested to the northwest and its transmissivity was approximately one order of magnitude lower (550 ft²/day) than the production well shown in Figure 5.7 (5800 ft²/day), thus it was easier for the cone of depression to extent up gradient.

5.3   Step-Drawdown Tests

In order for pumping tests to effectively stress the aquifer, a proper pumping rate must be established. Even though the wells may be properly placed and developed (Chapter 15), if the pumping rate is too low, a small cone of depression will result and the drawdown in observation wells may not be detected. Conversely, if pumping rates are too high, then the test will not run very long because the pumping level will reach the pump, possibly causing cavitation (Example 5.1). Another cause of cavitation occurs when the suction pipe is of too small a diameter, producing turbulence and creating air bubbles. This is not a problem with submersible pumps unless the pumping level has reached the pump.

The procedure of changing pumping rates over a consistent time interval in a deliberate manner is known as a **step-drawdown test**. This is a single-well test where the initial pumping rate is lower than the maximum expected rate. After the drawdown stabilizes, the rate is increased, or stepped up, to a higher rate for the same amount of time, usually 30 min to 2 h each (Clark 1977; Kruseman and de Ridder 1990; Mathias and Todman 2010). The key is to run each step for the same amount of time and a minimum of least three steps. This procedure is also done to check well performance. Step tests were first devised by Jacob (1947) to check what would happen to the drawdown if the pumping rate varied during a pumping test.

Well-hydraulics theory is based on the concept of laminar (nonturbulent) flow for groundwater (Chapter 6). In the near vicinity of a well bore, velocities may increase to the point that turbulent conditions result. If the conditions are laminar, then the drawdown is proportional to the pumping rate (Q). If turbulence occurs, there are other well losses that also contribute that result in a nonlinear relationship, where the Forchheimer equation can be used (Mathias and Todman 2010). Collectively these can be represented by the following relationship presented in Equation 5.1 (Roscoe Moss Company 1990) and illustrated in Figure 5.8:

$$s = ds + ds' + ds'' + ds'''$$ (5.1)

where s = total drawdown measured in the pumping well
ds = head loss in the aquifer (formation loss)
ds' = head loss in the damage zone (skin effect)
ds'' = head loss in filter pack
ds''' = well loss from water entering the screen

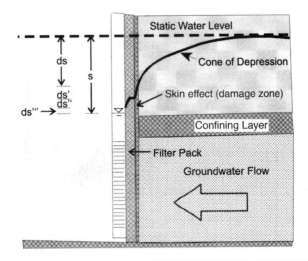

Figure 5.8 A schematic illustrating the individual contributing head losses to the total head loss observed at the pumping level. [*Modified after JAWWA (1985).*] Head losses are described in Equation 5.1.

The total drawdown observed within a pumping well is the difference between the static water level and the pumping level, where the pumping level is the level inside the well casing (Figure 5.6). The actual water level that exists outside of the casing varies according to a group of additional head losses that occur from other effects (Figure 5.8). Included are head losses that occur as water moves through the aquifer (ds) and head losses as the groundwater encounters a skin effect or damage zone (ds'). The skin effect is a result of fine drill cuttings or films from fluids that remain on the borehole wall from drilling. Groundwater then passes through the filter pack material (ds''). Naturally developed wells may not have significant skin effects but will have the natural gravel packing materials. As groundwater enters the well screen and changes direction from horizontal to vertical flow toward the pump, additional wells losses take place (ds''').

Well Efficiency

Before continuing the discussion, this is an appropriate place to present the idea of **well efficiency**. If the water level outside the casing matches the pumping level inside the casing, the well efficiency is said to be 100% (Figure 5.6). This is one of the criteria assumed for aquifer hydraulic theory (Chapter 6). In reality the pumping level is usually greater than the level outside the casing. The concept is illustrated in Figure 5.9 and defined in Equation 5.2, where K represents the depth to the pumping level from static conditions and L represents the depth to the water level outside the casing from static conditions:

$$\% E = L / K \times 100\% \tag{5.2}$$

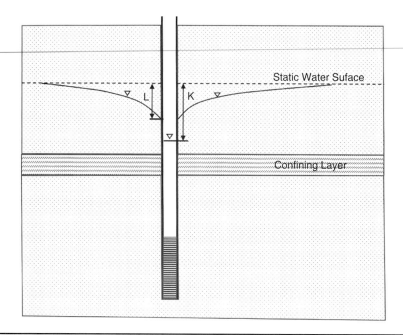

Figure 5.9 Concept of well efficiency, where K is the pumping level and L is the water level outside the casing and where all terms for % efficiency are presented in Equation 5.2.

Laminar versus Turbulent Flow

Laminar flow in a perfectly efficient well in a confined aquifer will yield the following relationship described as the modified nonequilibrium equation by Driscoll (1986):

$$h_o - h = \frac{264Q}{T} \log\left(\frac{0.3T\,t}{r^2 S}\right) \tag{5.3}$$

where $h_o - h$ = drawdown
$\quad$ Q = pumping rate (gal/min)
$\quad$ T = transmissivity (gal/day-ft)
$\quad$ t = time (day)
$\quad$ r = radial distance of observation point (ft)
$\quad$ S = storativity (dimensionless)

This is also okay for unconfined aquifers if the drawdown $(h_o - h)$ is small (the saturated thickness is within 90% of the original thickness) relative to the saturated thickness (b). The next step is to represent the laminar term by (B):

$$B = \frac{264}{T} \log\left(\frac{0.3T\,t}{r^2 S}\right) \tag{5.4}$$

Combining Equations 5.3 and 5.4 yields the following simple relationship:

$$h_o - h = BQ \qquad (5.5)$$

If all of the additional well loss terms shown in Equation 5.1 are combined together into a nonlinear turbulence factor (C) and added to the laminar term, we have a crude representation of the total drawdown (Equation 5.6):

$$h_o - h = BQ + CQ^2 \qquad (5.6)$$

A more rigorous analysis of these processes and how they can be solved is presented by Clark (1977), Kruseman and de Ridder (1990), and Mathias and Todman (2010). A method on how to interpret a variable-rate pumping test is also described by Birsoy and Summers (1980). However, a more simplified approach was presented by Walton (1970), which is useful for field approximations of maximum pumping rates and for selecting pumps. Walton (1970) provides a procedure where the turbulence factor (C) can be estimated from a step-drawdown test (Figure 5.10; Example 5.5a). The process involves conducting a step-drawdown test with successive increases in Q and plotting the results on a

Figure 5.10 Students conducting a step-drawdown test during a hydrogeology field camp.

time-drawdown graph (Cooper-Jacob plot; Chapter 6). The first step should be within about 75% of the targeted Q followed by step increases in Q on the order of 15% to 20%. Each step appears as a straight-line segment followed by a marked drop in drawdown between steps (Example 5.5b). Initially, during laminar flow, the slopes of the segments are approximately the same. As well efficiencies decrease and turbulence effects increase, the slopes of the straight-line segments will steepen. A rough estimate of well efficiency can then be calculated, and an appropriate maximum pumping rate can be estimated.

The turbulence factor can be approximated by

$$C \approx \frac{\left(\dfrac{\Delta(h_o - h)_n}{\Delta Q_n} \right) - \left(\dfrac{\Delta(h_o - h)_{n-1}}{\Delta Q_{n-1}} \right)}{\Delta Q_n + \Delta Q_{n-1}} \qquad (5.7)$$

where C = turbulence factor, in ft/(ft^3/s)^2
$\Delta(h_o - h)_n$ = the change in drawdown from step $n - 1$ to n, in ft
ΔQ_n = the change in pumping rate from step $n - 1$ to n, in ft^3/s

Values of C also provide a qualitative measure of well efficiency. Walton (1970) gives the following groupings in Table 5.2.

The methodology will be illustrated with two examples: one from Walton (1970) with an efficient high-yield irrigation well followed by a less-than-ideal example of how a step test should be done. The second example is useful in trying to make sense out of some actual field data, when the C-values don't make sense. Sometimes the equations don't work out and applying principles in the field is your best tool.

Example 5.5a A 24-in-diameter (0.61-m) irrigation well was tested and the data are summarized in Table 5.3.

The respective C values are determined from Equation 5.7:

$$C_{1,2} \approx \frac{\left(\dfrac{1.59 \text{ ft}}{0.62 \text{ ft}^3/\text{s}} \right) - \left(\dfrac{5.40 \text{ ft}}{2.22 \text{ ft}^3/\text{s}} \right)}{2.22 \text{ ft}^3/\text{s} + 0.62 \text{ ft}^3/\text{s}} = 0.046 \text{ ft}/(\text{ft}^3/\text{s})^2$$

$$C_{2,3} \approx \frac{\left(\dfrac{0.72 \text{ ft}}{0.27 \text{ ft}^3/\text{s}} \right) - \left(\dfrac{1.59 \text{ ft}}{0.62 \text{ ft}^3/\text{s}} \right)}{0.62 \text{ ft}^3/\text{s} + 0.27 \text{ ft}^3/\text{s}} = 0.115 \text{ ft}/(\text{ft}^3/\text{s})^2$$

Turbulence Factors C	Comments
<5	Great
5–10	Good
10–40	Fair to poor
>40	Bad

TABLE **5.2** Qualitative Values for Turbulence Factor C, in ft/(ft^3/s)^2 [From Walton (1970)]

Step	Q (gpm)	Δ Q (gpm)	Δ Q (ft³/sec)	Δ s (ft)
1	1,000	1,000	2.22	5.40
2	1,280	280	0.62	1.59
3	1,400	120	0.27	0.72

TABLE 5.3 Step-Drawdown Data from a 24-in Irrigation Well [From Walton (1970)]

By using Equation 5.6, one can determine the well losses from the turbulent factor at 1,280 gpm (0.081 m³/s) and 1,400 gpm (0.088 m³/s), respectively:

$$(h_o - h)_{1280} = 0.046 \text{ ft}/(\text{ft}^3/\text{s})^2 \, (2.84 \, (\text{ft}^3/\text{s}))^2 = 0.37 \text{ ft}$$
$$(h_o - h)_{1400} = 0.115 \text{ ft}/(\text{ft}^3/\text{s})^2 \, (3.12 \, (\text{ft}^3/\text{s}))^2 = 1.12 \text{ ft}$$

The relative % losses can now be calculated:

$$\% \text{ loss}_{1280} = 0.37 \text{ ft}/(5.4 \text{ ft} + 1.59 \text{ ft}) \times 100\% = 5.3\% \text{ very efficient!}$$
$$\% \text{ loss}_{1400} = 1.12 \text{ ft}/(5.4 \text{ ft} + 1.59 \text{ ft} + 0.72 \text{ ft}) = 14.5\% \text{ still efficient}$$

At 1,400 gpm (0.088 m³/s), the well is still approximately 85% efficient.

Example 5.5b Near Three Forks, Montana, an irrigation well was drilled and completed and the drillers wanted to know what pumping rate to use for the constant-discharge test. They were optimistic and felt the well could yield 500 gpm (31.6 L/s). The author suggested doing a step-drawdown test, starting near 200 gpm (12.6 L/s) and increasing the pumping rate approximately 50 gpm (3.2 L/s) for the next five steps and each step be run for a period about 45 min each. There was approximately 200 ft of available drawdown. When I arrived the day after the step test the drillers had done their own thing and the data in Table 5.4 were given to me.

From this table one can see the rates were variable and not run for the same length of time. The data were plotted so that the drawdown response per step could be evaluated (Figure 5.11). It can be noted from Table 5.4 and Figure 5.11 that the data stabilized after each step until the pumping rate was increased to 500 gpm (31.6 L/s), where the slope is notably steeper. It is apparent from the data that 400 gpm (25.2 L/s) would be appropriate for the constant-rate pumping test and the starting rate for the step test should have been 250 gpm (15.8 L/s). A constant-rate pumping test was subsequently conducted at a constant rate of 450 gpm (28.4 L/s) and had to be "valved" back to a constant rate of 375 gpm (23.7 L/s) after 2 h for the duration of the 72-h test. It was evident from the constant-rate discharge test that the step-discharge test was providing reasonable guidance as to what the starting pumping rate should have been (375 gpm).

It is not always necessary to conduct a formal step-drawdown test prior to performing a pumping test. However, it can be particularly helpful if one is sizing a pump for production purposes and wishes to make a purchase. After one gains some experience, the appropriate pumping-discharge rates for a constant-discharge pumping test can be estimated by observing the drawdown in the pumping well after several minutes. Control of the discharge can be made over a fairly good range by constricting the outflow. However, one must start with a pump size that is within the appropriate range. Relying solely on the driller's estimate of what a well can make can get you into trouble. Another issue occurs if well development (Chapter 15) has not adequately taken place. In this case sediment may be produced during the step-drawdown test and well production and efficiency actually improve!

Q in gpm	Time, min	Drawdown, ft	Q in gpm	Time, min	Drawdown, ft
170	1	60.78	400	55	126.61
170	2	65.4	400	60	130.08
170	3	65.4	370	65	130.08
170	4	65.4	370	70	130.08
170	5	65.4	370	80	130.08
250	6	79.26	370	90	130.08
250	7	81.57	500	91	150.87
250	8	83.88	500	92	162.42
250	9	83.88	500	93	167.04
250	10	83.88	500	95	171.66
250	15	83.88	500	98	173.97
250	20	86.19	500	99	179.75
250	25	88.5	500	100	178.59
250	30	88.5	480	105	179.75
250	35	88.5	480	110	180.9
250	40	88.5	480	120	181.48
250	45	88.5	480	125	182.05
250	50	88.5	480	135	183.21
400	51	106.98	480	145	184.94
400	52	111.6	480	155	185.52
400	53	116.22	480	165	186.65
400	54	123.15	480	175	187.25

TABLE 5.4 Step-Test Data from the Jones Irrigation Well, Three Forks, Montana

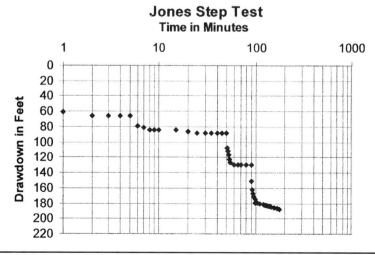

FIGURE 5.11 Step-drawdown test with four steps conducted on an irrigation well near Three Forks, Montana. The first step from 170 gpm (10.7 L/s) to 250 (15.8 L/s) is very subtle. Note that the last time step at 500 gpm has a steeper gradient in the plotted data indicating too high of a sustainable pumping rate.

5.4   Setting Up and Running a Pumping Test

Setting up a pumping test can be frustrating if something is left off the equipment list and one must retrieve something back at the office or shop. This section discusses the typical equipment that one should bring to have a successful experience. Discussion of the pitfalls that may occur is also presented. A discussion of the information that should be recorded in field books is given in Chapter 1.

The following is a list of equipment needed for a pumping test (items to bring are indicated in italics):

- *Power supply* (usually a generator, equipped with outlets that will allow the pump control box to be plugged in).
- *Extra fuel cans* for generator.
- *Control box* and pump.
- *Data logger*—for storing data and user's manual.
- *Transducers* (vented or well sentinels with barologger)—anticipate the drawdown expected in each well and have a transducer or well sentinel within range. Don't forget the *jumper cables* that connect the vented transducer to the data logger (Chapter 4).
- *E-tapes*—to check conditions prior to setting up the test and for backup readings in case equipment fails.
- *Laptop computer* or *cell phone* preloaded with the appropriate processing software or app, with a flash drive as a backup.
- *Discharge system*—this may consist of a *garden hose* for small discharge rates or *riser pipe, elbows,* and *connectors* (1¼ in for flows about 30 gpm to 4 in for flows at hundreds of gpm), depending on the rate, to convey the discharge away from the site. A means of measuring discharge is needed, either a *flow meter* or a *calibrated container* (buckets), and a *stopwatch.* An in-line flow constrictor, such as a *gate valve* (preferred) or ball valve, is needed to put back pressure on the pump and control the discharge. *Teflon tape* may be needed for the threads when connecting the discharge lines and riser pipe.
- *Pipe wrenches*—to assemble the riser pipe and discharge line (aluminum ones are much lighter).
- *Duct tape* or the newer *Gorilla Tape* (the most versatile tool in the box).
- *Electrical tape.* Always handy.
- *Miscellaneous tools* (screwdrivers, wrenches, electrical tester, etc.).
- *Field books, field forms* to record data, and *pens or pencils* (with refills).
- *Well logs* and *field map.*
- A way to plot the data for each observation well. (Data should always be plotted in the field to observe data behavior; Chapter 6.)
- *Cell phones* or *kitchen timers* (for timing readings).
- *Keys* (for gate locks or well access).
- *Shovel* and *rope.*

- Miscellaneous field equipment (such as water-quality equipment to check changes in pH, temperature, or specific conductivity).
- Hat, sunscreen, bug spray, and personal items.
- Rain clothing, tarp, and/or tent.

Power Supply and Pumps

In considering the appropriate power supply for a pumping test, one must have a knowledge of the pump and electrical feed requirements. Because many pumping tests are remote and a generator is often used, buying the cheapest generator may be a big mistake. To run a small sampling pump system, it is necessary to provide a steady feed of voltage. Cheaper units will often sputter and surge. Unless the power feed is steady, trying to use these sensitive sampling pump systems will be a problem. For example, good results for small pumps [up to 1.5 horsepower (hp)] have been obtained from Honda generators (Figure 5.12).

These generators run smoothly and are relatively quiet. Some manufacturers over-rate their products, so it is necessary to make sure there is sufficient capacity to start and run a pump. Other pumping systems may have their own recommended generator systems, so one should consult the user's manual.

In starting a generator, particularly if it has sat idle for several months, it may be necessary to activate a warm-up switch for 15 to 30 s before switching the generator on. Changing the oil once a year is also a good idea, regardless of how little it may have been used. When the oil level gets low, many generators will shut down. It has been the author's experience that when the oil level has not been checked and is low, the generator will make a couple of changes in sound (hiccups) from the normal purr of the engine approximately 1 h before it suddenly sputters and stops. Newer generators have a safety feature that shuts the engine off when the oil level is low.

Figure 5.12 Pumping test conducted with a steady-feed voltage Honda generator.

Figure 5.13   Jump-starting a generator battery after cleaning the battery cables.

Generators may require diesel or unleaded gasoline. No particular preference is recommended. Another maintenance item is the battery. It may be that everything else is working fine, but the battery cables may become corroded. Sometimes after cleaning, one can give the generator a jump start, and then the battery will become charged during the pumping test (Figure 5.13). It is helpful to test the equipment before taking it out into the field.

Generators are usually distinguished by the number of kilowatts they are capable of. Pumps require a boost of electrical power to get started. A general rule of thumb is that a *kilowatt is required for every horse power)*. This is not entirely true but will get one in the ballpark. For example, a 20-kW generator might get a 25-hp submersible pump started but would require 220 V and three-phase power. A 25-hp pump may need a 440-V source. As another example, one driller used a 150-kW generator for a 75-hp pump at 440 V producing 500 gpm (31.6 L/s). He commented that the generator could probably keep a 100-hp submersible pump going, but he wasn't sure whether it would start one. This is part of knowing what your electrical feed requirements are. Submersible pumps are wired to control boxes that are then plugged into the generators with 110-, 220-, or 440-V outlets. These pumps also come with quite a range of horsepower and electrical requirements, such as how many phases and amperage. Smaller pumps (up to 1 hp) are usually single phase and have simple wiring requirements (wires are white, red, or black, and green for ground; Figure 5.14). From 1.5 hp on, submersible pumps may require three-phase configurations. *Unless you are very adept at electrical wiring, this should be done by a professional.* This can cause all sorts of problems, such as the pump impellers going the wrong way or causing damage to the pump. The higher the horsepower rating, the greater the voltage needed. For example, in the 10- to 20-hp range, the electrical configurations may require a shift from 220 to 440 V. Pumps requiring 440 V will yield hundreds of gpm (thousands of m³/day) discharge range. At this level, it may be advisable to use a turbine pump system (Figure 5.15).

FIGURE 5.14 Wiring a small 1-hp pump in the field. White wire goes to white wire, black wire to black wire, and green is to ground. Described in Example 5.6.

Example 5.6 A new 1-hp pump was purchased for use during a field camp. The wiring was done on the tailgate of a pickup truck by students (Figure 5.14). The pumping test was set up, and no water would come out. The pump was retrieved and placed in a 55-gal drum to test and see if it could pump water. It was determined that the wiring was done such that the impellers turned the wrong way. Once the wires were switched, the pump worked properly. Testing the pump in a drum was a quick way to see if the wiring is correct.

On another occasion, a sampling pump would not work. The impellers were seized up from sand (it is not a good idea to use a sampling pump for well development! Chapter 15). In this case, the pump impellers were cleaned and then put back in upside down. Again, no water would come out and the pump got warm. After a careful inspection and comparison to diagrams in the user's manual, the impellers were placed properly and the pump worked once again (Figure 5.16).

The wire to the pump is usually spooled on a drum. The pump and wire can be lowered down the well if a broom handle or something like it is placed through the

FIGURE 5.15 Turbine pump yielding 375 gpm. Turbine pumps are used for high-yield pumping. Pump is mounted on a transferable trailer.

FIGURE 5.16 Fixing the impellers on a small sampling pump. Application described in Example 5.6.

drum for unspooling. Pumps can be very heavy, and a security cable is recommended for retrieval in case it becomes hard to maneuver (Figures 5.17 and 5.18). For a sampling pump, the riser pipe or hose is already attached to the pump. In a submersible pump, lengths of riser pipe must be fitted into the top of the pump, extending upward.

FIGURE 5.17 Installing a heavy 5-hp pump, with a security cable attached to prevent loss down the well.

Figure 5.18 Lifting a 50-hp submersible pump using a hydraulic hoist in southwestern Montana.

Each piece should be tightened with wrenches, with Teflon tape wrapped reverse style on the threading so it won't "bunch up." Once the appropriate pumping depth is achieved, a 90° elbow fitting is needed to extend the discharge line away from the test site, preferably beyond the expected extent of the cone of depression or to a drain, if available (Figure 5.19).

Some pumps must be primed or have a chamber filled with water so they will produce water (Figure 5.20). This brings up a topic known as net positive suction head (NPSH). On the suction side of a pump, the NPSH is determined by subtracting the negative pressures (suction lift, friction loss, and vapor pressure) from the positive pressures (a function of atmospheric pressure) (Figure 5.21). The atmospheric pressure is a function of altitude (Table 5.5). If the distance between the intake and the pump chamber is too great, the pump will never produce water, no matter how large the motor is. The vapor pressure requirement is not needed for temperatures greater than 60°F (Jacuzzi 1992). The pump design varies by manufacturer.

Data Loggers, Transducers, and Well Sentinels

Data loggers, transducers, and well sentinels are a vast improvement over manual methods but can cause some field workers to become complacent and careless. It is

FIGURE 5.19 Using a 90° elbow on a discharge line directed away from the well beyond the expected reach of the cone-of-depression to prevent recharge affects.

FIGURE 5.20 Sump pump requiring priming before successful function.

always a good idea to collect manual backup measurements. One of the blessings of the data logger/transducer configuration is best illustrated by the considering effort it used to require to collect data in a single-well test. In the "good old days," the trick was to finagle an E-tape down past the riser pipe and pump wiring in search of the pumping level. This would often result in hanging up the E-tape along with missed data. The test would often have to be repeated several times with no guarantee of success.

Hydrogeologists will likely use data loggers and transducers or well sentinels to collect pumping-test data (Chapter 4). Numerous companies have rental arrangements; however, if several pumping tests are expected to be run on a regular basis, it is wise to consider buying some equipment. The payback on the investment will take place in

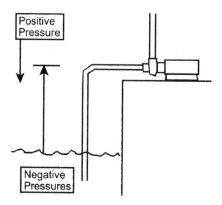

FIGURE 5.21 Schematic illustrating the concept of net positive suction head (NPSH). If the distance between the intake and the pump chamber is too great, the pump will never produce water no matter how large the motor is.

Units	Sea level	2,000 ft	4,000 ft	6,000 ft	8,000 ft	10,000 ft	12,000 ft
lb/in^2	14.7	13.7	12.7	11.8	10.9	10.1	9.3
in·Hg	30.0	28.0	25.9	24.1	22.3	20.6	19.1
ft water	34.0	31.7	29.4	27.3	25.2	23.4	21.6
kPa	1013.3	942.1	875.1	812.0	752.6	696.8	644.4

TABLE 5.5 Variation of Atmospheric Pressure as a Function of Altitude [From Jacuzzi (1992)]

short order. A significant discussion and description of data loggers, transducers, and well sentinels can be found in Chapter 4 (Section 4.3); however, the context of using the equipment in aquifer tests is included here.

Vented transducers can come with a cable spool and extra cable lines greater than 350 ft (107 m) (Figure 5.22). The transducer or well sentinel should be lowered into the well within the depth range for the transducer and below the expected depth of total drawdown. It is noted here that the greater the transducer or well sentinel rating, the less sensitive (lower precision) they become. For example, it doesn't make much sense to place a 100-psia rated transducer or well sentinel in a well that may only encounter 1 ft (0.3 m) of drawdown. Here a 5-psia rating transducer would be more precise. An example of typical ranges is shown in Table 5.6. Compare also with the information presented in Chapter 4. Drawdown in the hundreds of feet range may occur in mine shafts and large production wells. The principle here is to select a transducer or well sentinel based on depth below the water surface also factored by the expected drawdown.

Once a vented transducer is lowered into a well, a loop (bite) greater than 1 in (2.5 cm) should be made to make sure that the cable's air line is not kinked. The cable should then be wrapped around the well casing one or more times and then secured in place with duct tape (Figure 5.23). The duct-tape loop may slip in hot weather, so a dowel or similar object can be placed within the loop to prevent kinking. The cable then is extended to the data logger and connected. It is a good idea to lower the transducer down the well early in the setup phase to allow for cables to stretch out and straighten.

Figure 5.22 Shown are vented transducers with hollow cabled lines exceeding 300 ft (100 m). These are connected to the data logger on the trainer bed with "jumper cables."

Transducer Range in psia	Range in ft	Range in m
5	11.5	3.5
10	23.1	7.0
15	34.6	10.6
20	46.2	14.1
30	69.3	21.1
50	115.5	35.2
100	231.0	70.4
200	462.0	140.7
300	693.0	211.1
400	924.0	281.4

Table 5.6 Transducer Ranges in psi Referenced from Atmospheric Pressure

The transducer in the pumping well should be attached with electrician's tape to the riser pipe or with duct tape approximately 1 ft above the pump, with additional tape added every 5 ft (1.5 m) to the top of the well (Figure 5.24). If the cable must cross a road or a heavy traffic area, it is advisable to pass it under the road, if a culvert is available. Otherwise, two boards can be loosely duct-taped together with the cable placed in a space between them, allowing for traffic to drive over the boards without impacting the cable directly.

The point of preventing a vented transducer cable from getting kinked is to enable any changes in barometric pressure to be sensed by the transducer. The cable is hollow

FIGURE 5.23 Transducer with vented cable with greater than a 1-in loop for airflow. Placing a dowel or similar object prevents kinking if the securing tape allows slippage in the hot sun.

FIGURE 5.24 Taping a vented transducer cable to a riser pipe in a pumping well.

and air-pressure conditions must be allowed to move freely through the line. Well sentinels on the other hand are self-contained and placed at depth down the well to "sense" changes in water level over time. If significant changes in barometric pressure are occurring at the surface, these will not be detected by the well sentinel. For this reason most manufacturers of well-sentinel equipment also make "barologgers," whose purpose is to keep track of the barometric changes at the surface (Chapter 4). Barologgers are placed in a protected place at the surface or a few feet down the well bore. The accompanying software or app will pull in the information from the well sentinel and the barologger and simultaneously adjust the data at a command prompt.

Well sentinels are communicated with via a cable with a USB connection to a laptop to facilitate the timing of when data are gathered and subsequently stopped. The parameters of the test are communicated through software to the sentinel unit including the option to synchronize with the time clock on the laptop computer or set up via cell phone using an app, particularly if wireless coverage is available. One must be careful to make sure the clocks of any devices are current. This can happen if you have changed time zones and forgot to make adjustments.

Example 5.7 While performing a pumping test in southwestern Montana, a client's production well was monitored while another pumping test was being conducted on a neighboring property. The question was whether the neighbors well would impact the client's well. A starting time and synchronization was established with an older laptop with the appropriate software loaded. It was not noticed that the clock of the laptop was 4 years earlier than the existing time. Therefore, the well sentinel would not have started collecting data until 4 years hence from the date of the actual test! Fortunately, manual readings had been taken before and during the test to confirm no impact had taken place. Without the manual readings the effort would have been completely wasted. This was also inspiration to retire the older laptop.

Data loggers are usually capable of collecting data from multiple vented transducers. In this case each transducer needs to be registered and enabled. Transducer parameters, such as scale, offset, linearity, type, and ID number, are required. The reference level is usually the last thing keyed in before starting the test. Extra care should be taken to ensure that the "jumper cables" with vented transducers are well connected to the data logger (note that these are not the cables used to charge a dead vehicle battery). Checking for a proper connection can be done by using the data logger to take a reading from the transducer. If "no signal" occurs or the reading is zero, check the connection and try again. The connections generally work with only one orientation. Look at the prongs of male and female connections to see if any wires may be bent. Many transducers have a strain gauge near the probe tip that has a screw-on protective cap.

It is a good idea to unscrew the cap and make sure this area is clean. The following suggestions may explain why a vented transducer may be giving an improper signal or no signal at all:

- Check all connections.
- Check for cable damage—an unsecured cable can fall out of a truck and become damaged on the highway.
- Check the strain gauge end—strange field readings have been cleared up by unscrewing the end cap and washing the cap and strain gauge. This can happen if there is some muddy or silty material in the bottom of the well that is not cleaned off.

- Make sure the transducer is above the bottom of the well and free hanging.

- Vented transducers have "breather lines" in their cables to sense changes in air pressure, therefore transducer cables should have a loop of at least 1-in diameter, so that the cable does not become kinked. Kinked cables can result in strange or no readings at all. Well sentinels just need to be hung in the well with a barologger to keep track of pressure changes.

- The transducer or well sentinel may be in water deeper than the range of the transducer—this will result in a reading of zero and may harm the strain gauge.

- Check for damage to the probe—it may need to be shipped off to the manufacturer for recalibration.

Once vented transducers are connected to the data logger and all the transducer information has been entered, it is time to set the reference level. This can be an elevation or an arbitrary point. It is *not* a good idea to use 0.0 as the reference level, since negative readings may occur. This does not bode well when plotting on log or semilog graphs (since negative log values are undefined in a spreadsheet program). The user has the option of referencing all data from a positive downward or positive upward position. For example, top-of-casing (TOC) readings are referenced so that when water levels drop (like drawdown in a pumping test), the changes are recorded as positive downward. Surface referenced readings increase upward as they are often used in stage measurements to sense the increase in stage as levels rise such as in surface water or tidal areas (Chapter 10). Readings can be set to log-cycle or linear-time scales, which are described in more detail in Chapter 4.

In addition to pressure transducers, many companies provide additional probe types for collecting other kinds of data. For example, water-quality probes are available to collect pH, specific conductance, dissolved oxygen, oxidation-reduction potential (ORP), temperature, or specific ion data (Chapter 4). Well sentinels often come with at least temperature and level capabilities. This can be helpful, along with the drawdown data, to detect sources of recharge or for other interpretations. Some self-contained units have multiple capabilities of level and water-quality data all in one. These can be programmed directly by attaching a cable to a laptop computer or programed via an app on a cell phone. Access to the data during the pumping test is helpful to be able to observe the behavior of the time-drawdown data or check if water-quality parameters have changed (Figure 5.25).

E-Tapes

Electrical water-level meters or E-tapes have many uses at a pumping-test site. They are usually needed to take level measurements in all of the wells prior to lowering the transducers or well sentinels and the pump. They can be used to sound well depths for wells with missing well logs and even measure the distances between wells if a regular measuring tape is not available. One can determine the appropriate psia range for transducers from well logs and the depth to water. E-tapes are essential for backup readings. Backup readings are needed because they can "save" a test if the equipment malfunctions or the power supply unexpectedly shuts down (Example 5.6). E-tape measurements also help keep the field person focused and occupied. Kitchen timers are useful to alert the field person when the next reading should be taken. A discussion of E-tapes and their use and design is presented in Chapter 4.

FIGURE 5.25   Evaluation of real-time data on a laptop computer. Similar applications can be done with cell phones and apps (Chapter 4).

Discharge System

The riser pipe provides a pathway for pump discharge. Once it reaches the wellhead and is connected with a 90° elbow to convey the water away from the pumping well, it becomes the discharge line. The discharge line should extend far enough to not affect or influence the cone of depression. Sometimes steady-state conditions in unconfined aquifers are reached because the discharge line is too close to the pumping well. In this case, the pumped water may be recharging the aquifer by infiltrating back from the surface.

For smaller sampling pumps, the discharge line can be a garden hose. Additional hoses can be attached to extend the line outward. Control of discharge can be made by attaching a spigot to the end of the hose. The spigot acts as a mini gate valve to control discharge. Pumps are designed to deliver a certain amount of flow per head loss. It is a good idea to have a pumping rate that is less than the maximum pump capacity. Back pressure can be placed on the pump by partially closing a gate valve or ball valve while maintaining a steady discharge rate. As drawdown continues during the test, the lift required by the pump may increase. Discharge can be maintained by opening the valve slightly. *There is much more control during this process with gate valves than with ball valves.*

One of the assumptions required for well hydraulic theory is that the discharge (Q) be constant (Chapter 6). One of the requirements of a pumping test is to configure the discharge system so that a constant discharge rate can be maintained. This requires regular checks of pumping rates throughout the test. The first 15 min or so of the test are especially critical, until the drawdown stabilizes and valve adjustments are less frequent. For discharge rates up to approximately 100 gpm (545 m³/day) or so, a container and a stopwatch can be used. This can get the field person very wet if the container is too small. For higher discharges, a U.S. 5-gal (18.9-L) bucket will be too small. It is noted here that just because the bucket is supposed to be 5 U.S. gallons (18.9 L), that doesn't mean that it is. From experience these have been observed to range from 4 to 5.5 gal

Figure 5.26 Students filling a 6-gal bucket, timed with a cell phone to monitor discharge during a pumping-test exercise.

(15.1 to 20.8 L). It is important to calibrate a bucket with a mark for every gallon, or some other desired volume, using a permanent marker. This way, there is more confidence in the readings. Time can be measured with a stopwatch on a cell phone (Figure 5.26). The greater the discharge, the greater the precision needed for time. Table 5.7 may be helpful in comparing time to particular discharge rates in gpm and L/s.

Another method to measure discharge is with an in-line flow meter. These are usually installed in the discharge line within 10 ft (3 m) or so of the pumping well. The flow meter is also placed between a control valve and the well (Figure 5.27). The control valve creates a back pressure and fulfills the requirement of the pipe being full flowing for readings to be accurate. In-line flow meters may have impellers that rotate to calculate a velocity and are calibrated to read a Q in gpm. They are accurate within a small percentage of the flow and are convenient because all you have to do is read the gauge. In addition, flow meters record the total flow so that at the end of the test one can determine the average pumping rate (Figure 5.28).

For larger flow volumes, such as those from a high-capacity submersible or turbine pump, a different arrangement may be used. A circular orifice weir is fitted at least 6 ft from the gate valve within the discharge line (Driscoll 1986). An access nipple is fitted to the discharge line at midpipe, 2 ft (0.6 m) toward the gate valve from the orifice plate. To the nipple a manometer tube is attached that extends upward to be measured on a scale (Figure 5.29). The manometer measures the pressure (head) in the pipe and should be made of clear tubing so that its height can be observed next to a calibrated stick.

Time to Fill 5 gal in s	Discharge Rate in gpm	Discharge Rate in L/s	Time to Fill 3 gal in s	Discharge Rate in gpm	Discharge Rate in L/s
2	150	9.47	1	180	11.36
3	100	6.31	2	90	5.68
4	75	4.73	3	60	3.79
5	60	3.79	4	45	2.84
6	50	3.16	5	36	2.27
7	42.9	2.71	6	30	1.89
8	37.5	2.37	7	25.7	1.62
9	33.3	2.10	8	22.5	1.42
10	30	1.89	9	20	1.26
12	25	1.58	10	18	1.14
14	21.4	1.35	12	15	0.95
16	18.8	1.19	14	12.9	0.81
18	16.7	1.05	16	11.3	0.71
20	15	0.95	18	10	0.63
22	13.6	0.86	20	8.2	0.52

TABLE 5.7   Reference of Time to Fill a 5-gal Bucket in Seconds to Discharge Rates in gpm and L/s

FIGURE 5.27   A discharge line for a pumping test equipped with an orifice weir (in the pipe) and an in-line flow meter with bubble read for flow. Notice that the gate valve to the right allows for better control of variable back pressure to maintain a constant discharge.

FIGURE 5.28 An in-line flow meter used to monitor flow in gpm (needle pointing at about 300 gpm) and total flow that has occurred (numbers on the meter are multiplied by 0.001 acre-ft). Other meters, for example, may record total flow data in thousands of gallons.

Duct Tape

Duct tape is one of those essential items that is helpful in setting up a pumping test. Duct tape is used to attach transducers to the well casing. It can be used to hold braces together to support discharge lines, patch things together, and do a host of other jobs. Any serious user of duct tape should take a look at *The Duct Tape Book* by Berg and

FIGURE 5.29 Manometer tube setup with circular orifice weir to measure flow.

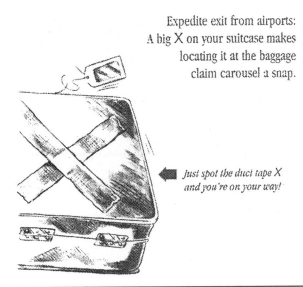

Expedite exit from airports:
A big X on your suitcase makes
locating it at the baggage
claim carousel a snap.

*Just spot the duct tape X
and you're on your way!*

FIGURE 5.30 Hydrogeologists are always on the go. Place a nice silver X on your travel bag to readily spot your luggage. [*After Berg and Nyberg. Used with permission from Pfeifer-Hamilton Publishers (1995).*]

Nyberg (1995). As an example pertinent to hydrogeology work, consider Figures 5.30 and 5.31. Since the first edition of this book, Gorilla Tape (www.gorillatough.com) has come out, and it has similar applications; however, it is even stronger. For example, it can hold a plastic bumper together for over 2 years, through two Montana winters (Figure 5.32).

FIGURE 5.31 Duct tape holding a drill rig cab seat together.

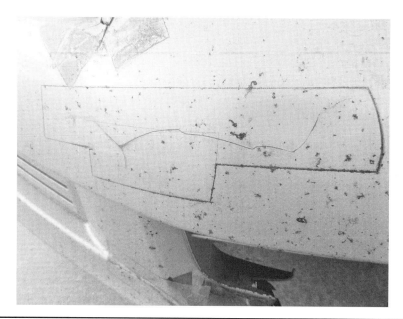

Figure 5.32 Clear Gorilla Tape can hold together a cracked plastic bumper together for over 2 years though two cold Montana winters. Note even the acids from the bugs have not affected its functional abilities.

Setup Procedure

Now that the essential equipment has been described, it may be useful to list the tasks generally associated with pumping tests. It is assumed at this point that all observation wells have been drilled, completed, and developed, and that permission to be on the property has been obtained. One should have a lamp or source of light (cell phone) that can be plugged into the generator for lighting and flashlights (with extra batteries) for taking manual readings during the night. Making the appropriate arrangements for food, supplies, and restroom facilities or designated areas will help make the time go smoother. A portable field table with chair and a tarp for inclement weather conditions are also recommended. Here is the list:

1. Gather together the equipment listed at the beginning of this section. Obtain extra fuel.

2. Check the weather forecast and bring any essential clothing for the anticipated conditions.

3. Measure and record the static water levels in all wells to be used in the test. Monitoring of wells for a few days prior to the test (pretest levels) is helpful to determine if any regional trends are occurring (general decreases or increases in water levels).

4. If there are any nearby surface streams, establish the stage with a physical marker so that it can serve as a gauging point should changes in stage occur (Figure 5.33).

Figure 5.33 A metal fence post was driven into the streambed of Warm Springs Creek, north-central Montana, during an aquifer test occurring in an irrigation pit. A maximum stage from a rainstorm was marked with red flagging to compare changes in stream flow with any observed water-level changes occurring in the irrigation pit.

5. Offload the equipment and begin placing the transducers or well sentinels in the wells, securing them in place with duct or Gorilla Tape or clamps.

6. Place the pump and riser pipe in the well along with the transducer designated to measure levels in the pumping well. This transducer is needed to determine how close the pumping level is to the pump intake.

7. Attach and support the discharge line so that the discharge water is a suitable distance from the pumping well (Figure 5.34). Configure the discharge line for the appropriate discharge volume (e.g., in-line flow meter or manometer and scale) or get the discharge measuring device (calibrated bucket) and place it near the discharge point if another measuring system is not being used.

8. Attach all vented transducers to the data logger with connecting cables. The data logger should be placed fairly close to the generator, so that they can both be activated simultaneously at start time. Or, in the case of well sentinels one can use a linear timescale, say every minute, and have everything in place long before the aquifer test begins, thus capturing pretest data, data during the test, and recovery data. In this case, a barologger will be needed to correct for atmospheric pressure changes.

FIGURE 5.34   Discharge line away from pumping well. Manual backup readings with E-tape being collected during an aquifer test.

9. Set up the test in the data logger by giving the test a name and enter all the vented transducer information. Make sure there is a good connection with each transducer by taking a reading of each one at the data logger. Do the readings make physical sense? If not, then check out why. The case of using well sentinels was described in number 8.

10. Run a step-drawdown test or test the pumping rate so that the gate valve or discharge control valve can be closed back to the appropriate discharge rate. Once there, leave it in the desired position.

11. Prepare field books for frequent readings (listed below) and synchronize all watches (Chapter 1, Section 1.8).

12. Prepare personnel with E-tapes for manual backup readings; one E-tape dedicated for each well (or two) is advised. If you are working alone, have an E-tape next to each well for ready use.

13. Make sure any safety issues are taken care of (listed below).

14. Prepare a signal so that all parties can begin taking readings at the same time. The most critical is activating the data logger and the pump at the same time. A "1, 2, 3, go!" system while dropping an arm works well. One designated cell phone can be used to monitor total time and other cell phones can be used to measure specific time intervals.

15. Once the test has begun, frequent manual readings are necessary. If there are two of you, it is helpful to have one call out the readings and the other keep track of time and record the numbers. The pumping rate in the calibrated bucket should also be measured or the discharge read from the flow meter (Figure 5.28) or manometer tube (Figure 5.29). If manual readings coincide with data-logger recordings "splash effects" may distort the data-logger value.

Time Since Pump is Turned on or off, in min	Length of Time Interval, in min	Comments
0–10	0.5	Readings can be staggered when observation wells are being measured.
10–15	1.0	
15–30	2.0	
30–60	5.0	
60–90	10.0	All later readings should be within 1 min of the appropriate time interval.
90–120	15.0	
120–360	30.0	
360–1,440	60.0	
1,440–end of test	180.0	

TABLE 5.8 Recommended Time Intervals for Manual Readings in a Pumping Test

16. When the time spacing between readings gets longer (15 to 30 min), plot the data or graph it to see what is happening. One should never stop a test based solely on a specific time frame. Look at the data and see whether boundary conditions or delayed yield or other effects are indicated in the data. When sufficient data have been collected, prepare for recovery mode.

17. If the data logger has a step function to go directly into recovery log mode, all that needs to be done is to synchronize the step-function button and deactivate the generator or pump. Some data loggers require that another test be defined and prepared beforehand to activate the recovery phase. By starting the "new" test, the "old" one is stopped automatically. If a linear scale has been set on the well sentinels this is not an issue.

18. Once again, field books need to be prepared for frequent readings (listed in Table 5.8) and all watches synchronized.

19. The signal is given and the data logger is "stepped" while the pump is shut off. In the new silence immediate recovery-mode manual readings are taken and recorded.

20. Make sure to plot the data during the test when time allows. There is a tendency to stop the recovery phase too soon, although it can be projected in a spreadsheet program (Chapter 6). It is an excellent additional source of information, so it should be collected. Plan accordingly and be patient.

Frequency of Manual Readings

The majority of drawdown takes place during the first hour or so of the test. Continued drawdown does occur, especially if a barrier condition is encountered (Chapter 6); however, changes are usually not as dramatic. The same is true for recovery mode. The initial hour after pumping stops is when recovery occurs most rapidly. At these times, if extra personnel can be on hand to help with manual readings, it can make a big difference in reducing stress. The flurry of activity at the beginning of the pumping test and the beginning of recovery is similar to a WFD.

The frequency of readings depends on the response of drawdown in the observation wells. Wells farther away may not "see" a response for several minutes or even hours. This depends on the completion depth of the wells and the geology (Example 5.2). It also depends on whether confining or semi-confining conditions are present. Small (confined) storativities correspond to a rapid expansion of the cone of depression (Chapter 6). The key is to watch the changes in the data. Usually, the first observed static water level (SWL) difference of 0.03 to 0.05 ft (0.91 to 1.5 cm) from starting conditions is a fair indication that the drawdown is real. When drawdown changes between readings drop to approximately 0.02 ft (0.6 cm) per reading, one should advance to the next time interval (Table 5.8). As a general guide, the manual reading schedule shown in Table 5.8 is what the author recommends, and then it should be adjusted according to what is observed.

Safety Issues

As with any activity, there are always safety issues. Common sense will generally help one to be alert to most problems. A few specific items will be mentioned based on experiences by the author:

- Make sure the fuel containers are placed away from the exhaust of the generator. During one test, it was observed that part of the plastic fuel container was beginning to melt!

- Make sure vehicle exhaust is away from where people are working. Notice the wind direction and park in a safe orientation.

- Make sure the generator and data logger are protected from rain and placed on a surface with good drainage so that puddles of water do not accumulate.

- Secure the cables from vented transducers so that tripping hazards are avoided.

- Secure the area, if possible, to keep children and animals away from the working area (Figure 5.35).

- Use a funnel when refueling the generator during a test. Splashing fuel can be a problem.

- Have lighting at night and a first-aid kit on site.

- Don't panic.

- Don't work alone. This way someone can go for help if serious problems arise.

Once again, common sense will help one to avoid most problems. The buddy system is always advisable for keeping each other alert or making runs for supplies. It is terrible to have a problem with no one around to help (Example 5.1).

5.5   Things That Affect Pumping Test Results

If none of the field conditions change from the beginning to the end of a pumping test, you are very fortunate. During the course of most pumping tests, some type of condition change may occur that affects the interpretation of the results. These changes should be recorded. If one is not paying attention to changes during the test and writing them down, they may not be accounted for during the interpretation phase (Chapter 6).

FIGURE 5.35 If possible, children and animals should be kept away from the work area. In this case, a group of 12 large Angus bulls was fenced away from the pumping and discharge area during an aquifer test.

Weather and Barometric Changes

Consider the following scenario. At the beginning of a pumping test, it is sunny and clear. About 4 h into the test, a storm front moves through. The sky is dark and a steady rain falls. During the night, the storm passes over and the sky clears again. A nearby stream increases in stage 1 ft (0.3 m) higher than at the beginning of the test and later drops 2 ft (0.61 m) over the rest of the test.

Another condition to consider is that if an aquifer is confined to semi-confined, the water levels in wells will likely be affected by barometric changes. Sunny and clear weather accompanied by high barometric pressure causes the ambient air to press down harder on the water surface in wells, thus decreasing the relative water elevation. When a storm comes into the area, the relative air pressure decreases. Thus, the ambient air is not pressing as hard on the water surface, allowing the water level to rise. Personally, the author has observed water levels rise as much as 1 ft (0.3 m) when a storm has come in. During a pumping test, the water levels are supposed to be decreasing if they do not encounter any recharge sources. It may be that a recharge source is being interpreted from a data plot resulting from a decrease in drawdown under a confined aquifer setting, when it is really from a storm front that reduced the ambient air pressure. This is where field notes come in handy.

Other Apparent Sources and Sinks

Other physical phenomena occurring in the field may affect the interpretation of pumping-test results. Nearby wells that were pumping and then stop during a pumping test may reduce the drawdown results. Since overlapping cones of depression are additive (Chapter 6), the drawdown rates are accelerated when additional nearby wells are turned on. In this case the interpretation of the transmissivity may be smaller than the actual value. When a nearby well stops pumping, the time/drawdown curve will flatten thus the transmissivity may appear to be greater or the curve may suggest that a source of recharge has been encountered.

Example 5.8   A pumping test was being conducted to see whether a mining company supply well would impact other concerned local residents. The geologic setting is depicted in Figure 5.36. The pumping well was completed within the granitic bedrock (granodiorite), and the residential wells were completed in both the granodiorite and the overlying sediments. The sediments ranged in thickness from 30 to 55 ft (9.1 to 16.8 m) in the pumping-test area. In this instance, a preliminary distance-drawdown analysis (Chapter 6) indicated that any resident within 2,000 ft (610 m) of the pumping well could have their wells monitored. Approximately seven residential wells were monitored by students during the 12-h aquifer test (the time frame the mining company proposed using their well each day during the 5 mo each year of use, thus allowing 7 mo of recovery). The instructions given to the residents were that no water use should take place during the test, which would begin at 8 a.m. on a Saturday morning. During the test, one of the observation wells seemed to be recovering over the duration of the test. In addition, none of the wells completed within the overlying sediments indicated any response except this well. After conferring with the well owner, it was discovered that someone used the bathroom earlier that morning (before the start of the test) and what was being observed was a recovery response from pumping. Wells completed within the granodiorite within 1,100 ft (335 m) *did* show a drawdown response.

Another factor that can affect the drawdown response is a change in stage of a local stream. At the beginning of the test, the stage of the stream should be marked so that its stage is known. If changes in stage occur during the test, they can be noted (Figure 5.33).

Did it rain during the test? This could be a potential source of recharge in a shallow unconfined aquifer. Is the discharge line too close to the pumping test site? The author

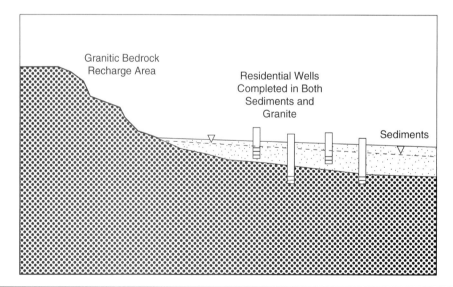

Figure 5.36   Geologic setting for Example 5.8. Residential wells are completed in both the granitic bedrock and the overlying sediments. (Not to scale.)

has observed a test reach steady-state conditions merely because someone failed to make sure discharge was not beyond the cone of depression.

Another consideration is the proximity of railroads and heavy traffic areas to shallow confined systems. A documented case in Long Island, New York, showed a hydrograph response near a train station (Jacob 1940). The weight of the train compressing the aquifer resulted in water-level rise, and when the train departed, the water level would drop once again. Be alert to physical changes that may occur during a pumping test. Write them down and use your field notes during the interpretation phase. Additional discussion on the interpretation of pumping-test results is presented in Chapter 6.

5.6 Summary

Pumping tests are valuable field methods in estimating the hydraulic properties of aquifers. The time and effort required to perform a pumping test can provide information about the aquifer that cannot be obtained from slug tests (Chapter 7) or specific capacity tests (Chapter 6). Pumping tests can be stressful because they require a fairly large list of equipment and many things can go wrong. A checklist is useful to remind field personnel what needs to be brought and tasks that need to be done.

It is important to have an understanding of the geology so that observation wells can be oriented appropriately and completed at the proper depths. Appropriate pumping rates are needed to stress the aquifer so that drawdown can be observed. Transducers and well sentinels should be selected that will be compatible with the water depths and the expected drawdown ranges. Discharge systems should be designed to maintain a constant discharge rate and obtain accurate information. The discharge line should be configured with a constricting device, so that it can be carefully opened to accommodate more flow. This would be needed should drawdown conditions cause the flow rates from pumping to decrease. Extra personnel are helpful to have around at the beginning of a pumping test to help set up equipment and collect the early-time manual readings. This is also true at the beginning of the recovery phase. Changing weather and other field conditions that occur during the test should be recorded in field books for use during the interpretation phase. The flurry of activity that can occur during a pumping test reminds the author of a WFD.

5.7 Problems

5.1. Shown in Figure 5.37 is a plan-view image of the Mustang Camp complex near Gardiner, Montana. The black dots represent existing wells along with surface elevations. All wells have been drilled to a depth of approximately 80 (24.4 m) to 100 ft (30.5 m). The stage of the Yellowstone River is approximately 5,100 ft above mean sea level (ANSL). You have been asked to design a pumping test for a water-supply well. As part of your design develop a plan that would help you do the following:

  a) Determine aquifer properties
  b) Consider boundary conditions and other possibly important features that may need to be considered in the interpretation of the data
  c) Deal with logistical challenges such as where to put the discharge water and other ideas

5.2. Take the data from Table 5.1 and Figure 5.7 and estimate the shape of the cone of depression at 3, 20, and 78 days.

Figure 5.37   Plan-view map (oriented north to south)) of the Mustang camp complex near Gardiner, Montana. Yellowstone River is to the west. All wells are approximately the same depth (approximately 100 ft deep). The numbers represent surface elevations in feet AMSL. Associated with Problem 1.

5.3.   Take the drawdown data from Table 5.1 after 3 days of pumping compared with 13.9 ft of drawdown in the pumping well. If one plots the data in Excel (spreadsheet program) using a log scale for distance (r) on the x-axis and uses a linear scale for drawdown as the y-axis (scale *in reverse order* so that 0 ft is at the upper left corner) and assumes that the "r" distance for the pumping well is 1 ft, what is the approximate well efficiency (see Equation 5.2 and Figure 5.9)? Knowing that the maximum drawdown did not change over time did the well efficiency change?

5.4.   Take the step-drawdown data from Table 5.4 from time 51 to 90 min and plot this in Excel. Fit the data with a straight line and project the data out to 5,000 min (use a log scale for the x-axis). Knowing that the total available drawdown is 220 ft would the well be able to sustain a constant-discharge aquifer test at 370 to 400 gpm for 5,000 min? What would you suggest for a pumping rate to be used in a 72-h aquifer test?

5.5.   An aquifer test was performed in 2013 at the pumping test site from Problem 1. It was a 72-h test and the pumping rate was 130 gpm. Considering the information in Section 5.4, what kind and size of pump should be used (in hp) and what size of generator would be needed, along with the appropriate voltage requirements? What other equipment should be considered?

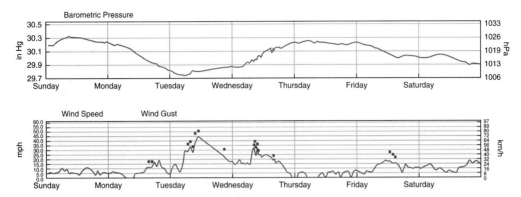

FIGURE 5.38   Barometric and wind speed data over a week from the Browning, Montana, weather station.

5.6.   Read the short document about net positive suction head found at http://www.pumpschool. com/applications/NPSH.pdf. Why are most recommendations for ultimate drawdown in centrifugal pumps limited to approximately 20 ft?

5.7.   Take the step-drawdown data from Table 5.4 and determine the weighted-average pumping rate over the 175-min test.

5.8.   You are involved in a pumping test where the pumping well and two observation wells are being monitored. Consider that the aquifer test coincides with the timing of data in Figure 5.38. The test started Monday morning and ended Thursday afternoon. What issues with your time-drawdown data might arise from the weather changes that have taken place? Identify the major weather events and briefly describe what happened during the test.

5.8   References

Berg, J., and Nyberg, T., 1995. *The Duct Tape Book*. Pfeifer-Hamilton, Duluth, MN, 64 pp.

Birsoy, Y.K., and Summers, W.K., 1980. Determination of Aquifer Parameters from Step Tests and Intermittent Pumping Data. *Ground Water*, Vol. 18, pp. 137–146.

Clark, L., 1977. The Analysis and Planning of Step Drawdown Tests. *Quarterly Journal of Engineering Geology and Hydrogeology*, Vol. 10, pp. 125–143.

Donovan, J.J., 1983. *Hydrogeology and Geothermal Resources of the Little Bitterroot Valley, Northwestern Montana*. Montana Bureau of Mines and Geology, Memoir 58, 60 pp. and two sheets.

Donovan, J.J., 1985. *Hydrogeologic Test Data for the Lonepine Aquifer, Little Bitterroot Valley, Northwestern Montana*. Montana Bureau of Mines and Geology, Open-File Report 162, 9 pp.

Driscoll, F.G., 1986. *Groundwater and Wells*. Johnson Screens, St. Paul, MN, 1089 pp.

Fetter, C.W., 1993. *Contaminant Hydrogeology*. Prentice Hall, Upper Saddle River, NJ, 500 pp.

Gorilla Tape source: www.gorillatough.com.

Harter, T., 2002. *Delineating Groundwater Sources and Protection Zones*. Edited by L. Rollins. University of California Agricultural Extension Service and the California Department of Health Services, 25 pp.

Heath, R.C., and Trainer, F.W., 1968. *Introduction to Ground Water Hydrology*. Water Well Journal Publishing, Worthington, OH, 285 pp.

Jacob, C.E., 1940. Correlation of Groundwater Levels and Precipitation in Long Island, NY. *Transcripts of American Geophysical Union*, Vol. 24, Pt. 2, pp. 564–573.

Jacob, C.E., 1947. Drawdown Test to Determine Effective Radius of Artesian Well. *Transactions of the American Society of Civil Engineers*, Vol. 112, Paper 2321, pp. 1047–1064.

Jacuzzi, 1992. *Jacuzzi Brothers Hydraulics Training Seminar Water Systems Workbook*. Jacuzzi, Chino Hills, CA, 40 pp.

Kruseman, G.P., and de Ridder, N.A., 1990. *Analysis and Evaluation of Pumping-Test Data*. 2nd ed. Publication 47. International Institute for Land Reclamation and Improvement, Wageningen, the Netherlands, 377 pp.

Mathias, S.A., and Todman, L.C., 2010. Step-Drawdown Tests and the Forchheimer Equation. *Water Resources Research*, Vol. 46, W07514. doi:10.1029/2009WR008635.

Reilly, T.E., Franke, O.L, and Bennett, G.D. 1984. *The Principle of Super Position and Its Application in Ground-Water Hydraulics*. U. S. Geological Survey Open-File Report 84-459, 36 pp.

Roscoe Moss Company, 1990. *Handbook of Ground Water Development*. Wiley, New York, 493 pp.

Walton, W., 1970. *Groundwater Resource Evaluation*. McGraw-Hill, New York, 664 pp.

Williams, D.E., 1985. Modern Techniques in Well Design. *Journal of American Water Works Association*, Vol. 77, No. 9, pp. 68–74.

CHAPTER 6

Aquifer Hydraulics

One of the critical pieces of information needed to solve problems in the field of hydrogeology is the hydraulic properties of an aquifer. Questions, such as, how long will it take water from the recharge area to reach a production well? How long will it take a contaminant to move from point A to point B? If this production well is activated, how far will the cone of depression reach? And how many other wells will be affected? What distribution of hydraulic properties should be assigned to the layers within this groundwater-flow model? In order to answer such questions, it is necessary to measure the drawdown aquifer responses during pumping tests (Chapter 5) and slug tests (Chapter 7) in wells over time. This chapter begins by discussing the traditional analytical methods followed by a discussion of applications when the data do not fit the ideal case. Unfortunately, the latter case is the more common scenario found in the field. It is beyond the scope of this chapter to provide an in-depth discussion of aquifer hydraulics, but it is hoped that the presentation here will be useful. There are excellent discussions in Hantush (1964), Bear (1979), Lohman (1979), Heath (1983), Driscoll (1986), and Kruseman and deRidder (1991). A wonderful collection of references pertinent to aquifer tests and slug tests has been put together by Duffield (2018). Slug-test analytical methods and discussion are presented in Chapter 7.

The Theis (1935) method is presented first to show what time-drawdown data look like. Although there are more complicated analytical methods that may be appropriate, any pumping-test data set should probably be evaluated with a Theis (1935) analysis. If the graphical image does not fit or make sense, alternative methods may be consulted (Duffield 2018). Pumping-test data are also easily evaluated using the Cooper-Jacob (1946) method, particularly if they are single-well tests (discussed later in Section 6.3); however, more assumptions are required. Additional topics covered in this chapter include distance-drawdown relationships, image-well theory, derivative analysis, and a discussion of the application of pumping-test data to aquifer properties. Explanations for deviations from the Theis (1935) curve are offered along with field examples. Dual porosity and fracture-flow data sets have their own characteristic response. These are also presented with field examples. Some of the groundwater-flow equations relative to the applications in this chapter can be found in Appendix E.

6.1 Wells

Wells provide a point of access to the water-bearing materials and are one of the most important features of groundwater studies. When properly designed, constructed, and developed (Chapter 15), measured level-change responses in wells provide useful

Figure 6.1   Student properly measuring a static water level during a pumping test. A second student is recording the time-drawdown data.

information about the characteristics of an aquifer. Characterizing the physical properties of aquifers is the study of **aquifer hydraulics**.

The water level in a well, when no pumping is occurring, is known as the **static water level** (SWL). The SWL reflects the total head at the midpoint of screened intervals of a well (Chapter 4). (It is recommended that the reader refer to Section 4.4 in the interpretation of water-level data section.) If the well is an open-hole completion (open at the bottom with no screen or casing perforations), it is considered to be representative of the head at in the aquifer at the bottom of the well. Typically, SWLs are measured from the top of the casing (TOC) (Figure 6.1). Level surveys are then performed to determine the elevation of the TOC and the ground surface level (GSL) (Chapter 4). Once the reference level has been defined, the change in water level with time can be measured during a pumping or aquifer test to evaluate the hydraulic properties of an aquifer. The difference measured between the nonequilibrium water level at a particular time "t" and the initial SWL is referred to as **drawdown** (Driscoll 1986) (Figure 6.2; see also Figure 5.6).

Cone of Depression

When pumping begins, the water level in the pumping well is lowered. This lowering (measured as drawdown) induces a gradient or slope all the way around the wellbore known as the **cone of depression**. Flow toward the well proceeds radially, much like the spokes of a bicycle converging on the hub. The cone of depression extends outward until it can capture enough water to meet the demands of the pumping rate (Q). The shape of the cone of depression is estimated by observation wells that also show a drawdown response. If these are located in different azimuth orientations away from

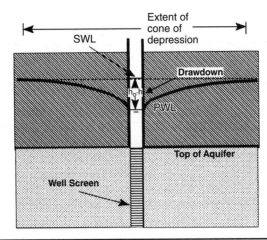

FIGURE 6.2 Schematic illustrating drawdown and cone of depression in a confined aquifer. Note the cone of depression occurs in the potentiometric surface above the top of the aquifer while the aquifer remains fully saturated.

the pumping well (Chapter 5), one can evaluate anisotropy by plotting the drawdown at a specified time "t" and contouring it (see Figures 5.6 and 5.7).

The cone of depression in a confined aquifer (Figure 6.2) occurs within the potentiometric surface above the top of the aquifer. This means that the saturated thickness (b) of the aquifer is always maintained. Since confined aquifers have smaller storativity values (Chapter 3) their cones of depression extend farther away from the pumping stress. Recall also that the potentiometric surface is a surface that reflects both the pressure head and the elevation head (Chapter 4).

The cone of depression in an unconfined aquifer represents a physical draining of the porous materials near the pumping well, creating a dewatered depression in the water-table surface (Figure 6.3). The extent is indicated by the respective drawdowns

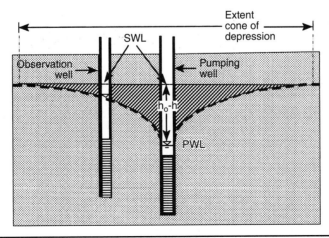

FIGURE 6.3 Schematic illustrating drawdown and cone of depression in an unconfined aquifer, with an observation well.

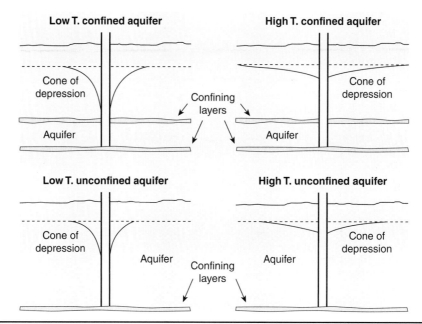

FIGURE 6.4   Schematic of the differences in the shapes of cones of depression based on aquifer properties. The confined aquifers have a larger cone of depression because of lower storativity values and higher sensitivity to pressure-head changes.

observed in nearby observation wells. The extent of the cone of depression from the pumping well is also referred to as the **radius of influence**. The shape and coverage of the cone of depression depends on the pumping rate (Q) and the nature of the geologic materials. Finer-grained materials produce a steeper cone of depression because a steeper hydraulic gradient is required for gravity to move groundwater toward the pumping well. Pumping within transmissive materials produces a flatter cone of depression (Figure 6.4) that may extend further out. The relationship between the steepness of the cone of depression and the transmissivity is a proportional relationship between the hydraulic conductivity and the hydraulic gradient expressed in Darcy's law (Chapter 4). Note how the size of the respective cones of depression is smaller for the unconfined cases because of their relatively higher storativities (Chapter 3). Context for how this can be applied is illustrated in Example 6.1.

Example 6.1   Within the Butte, Montana, area in southwestern Montana are a number of Superfund operable units associated with mining operations. During the remediation design and subsequent construction phase, 1.3 million cubic yards of mine tailings were to be excavated and removed to another location. It was required that dewatering of the mine tailings and the underlying alluvial materials take place to perform the excavations.

  Adjacent to the excavation operation is another Superfund operable unit associated with "organics" contaminated groundwater resulting from chemicals used in a timber-treating process. The chemicals were used in treating support timbers used as underground mine supports. In this operable unit, the shallow groundwater system became contaminated with a mixture of diesel and pentachlorophenol (PCP) (Figure 6.5) (http://www.buttectec.org/?page_id=77). The concern with dewatering the mine tailings operable unit was that if the hydraulic conductivity was high, then the cone of depression from dewatering the tailings would extend sufficiently to the southeast, potentially mobilizing the organic plume.

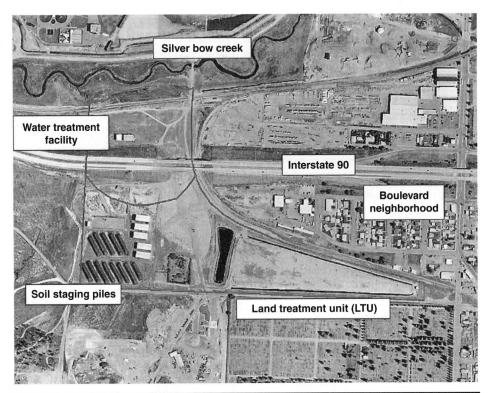

FIGURE 6.5 Plan view of the Montana Pole Treatment Plant Superfund site in Butte, Montana. The view is oriented north-south. The mine tailings described in Example 6.1 were excavated where Silver Bow Creek is located, north of Interstate 90. The Montana Pole Plant is situated south of Interstate 90 above and below where the soil staging piles.

The crux of the problem was centered in the estimated value for the hydraulic conductivity of the alluvium and its influence on the extent of the cone of depression. The hydraulic conductivity for the alluvial materials was believed to be approximately 80 ft/day (24.3 m/day). This was based upon expensive numerical groundwater models that could not reach calibration unless hydraulic conductivity numbers for the alluvium were at this magnitude. In the model, the alluvium was represented by a relatively thin package of sediments over impermeable bedrock. This layer could not allow sufficient water through it unless the hydraulic conductivity value was increased over values estimated from pumping tests. Personnel from a local state agency (Montana Bureau of Mines and Geology) argued that the pumping tests suggest that the alluvial materials had a lower hydraulic conductivity. They pointed out that the required volume of water *could* move through the area using a lower hydraulic conductivity if an additional saturated thickness consisting of weathered granitic bedrock also be used. The matter was settled by conducting a pumping test screened in the alluvial material. In the results, a hydraulic conductivity of 20 to 30 ft/day (6.1 to 9.1 m/day) was calculated for the alluvium, and a hydraulic conductivity for the weathered granite was estimated to be 1 to 5 ft/day (0.3 to 1.5 m/day). By adding an additional thickness of weathered bedrock layer to the model, calibration was reached with the lower hydraulic conductivity value more reflective of the local geology and the pumping-test results. This permitted the dewatering activities to take place (John Metesh, personal communication, Butte, Montana, September 1995).

Figure 6.4 illustrates that a lower transmissivity (hydraulic conductivity time thickness) results in a steeper, less extensive cone of depression, whereas a higher hydraulic conductivity results in a flatter, farther-reaching cone of depression. In Example 6.1, the flatter cone from the higher estimated hydraulic conductivity would potentially reach the organic plume, mobilizing the contaminants. During the actual excavation and dewatering process in 1997, it was found that the lower hydraulic conductivity was a better estimate.

Comparison of Confined and Unconfined Aquifers

The size of the cone of depression is also affected by the storage coefficient or storativity and aquifer transmissivity. Typical ranges of storativity for a confined aquifer are between 10^{-3} and 10^{-6}. Unconfined aquifers have storativities that range between 0.03 and 0.30 (Fetter 1993). Semiconfined or leaky-confined aquifers fall somewhere in between. The significance of the smaller storativity value is that the cone of depression in a confined aquifer will extend faster and farther than in an unconfined aquifer. This is shown in Figure 6.4 and illustrated by the Example 6.2.

Example 6.2 While performing the technical analysis for a source-water protection plan in Ramsay, Montana, a pumping test was conducted on the local public water supply well (Figure 6.6) (O'Connell and Smith 1993). There are two production wells plumbed into a water tower pressurizing system. The wells were drilled in 1917, and the well logs are nonexistent. It was unclear whether the wells were drilled into younger layered sediments or deeper into the underlying Tertiary Lowland Creek Volcanics. The pump in each well could be activated manually or controlled by an automatic system.

Figure 6.6 Water tower for the Ramsay, Montana, public water supply. Site description is in Example 6.2.

The well closest to the water tower (the south well) was used as the pumping well, and the other well located 510 ft (155 m) to the northwest was used as the observation well. During the test, the pumping well was activated manually after instrumenting the wells with drawdown measuring equipment. The pumping rate was 210 gallons per minute (gpm) (13.2 L/s). The north observation well experienced 0.05 ft (1.5 cm) of drawdown within 30 s of pumping. This indicates that the cone of depression extended at a rate of at least 17 ft/s (5.2 m/s), which is at the rate of a very fast person running toward the well at full speed! The rapid expansion of the cone of depression implied the aquifer is confined!

Drawdown in a confined aquifer is from response to compression of the mineral skeleton, decompression of the water, and an instantaneous release of water from storage, and this occurs quickly. The cone of depression in an unconfined aquifer is a physical depression in the water table from drainage of the pores, is of more limited extent, and develops more slowly.

Example 6.3 To obtain a mental picture of what is occurring in a confined aquifer, imagine taking a household cleaning sponge (flat rectangular shapes). If this sponge is saturated and sandwiched between two pieces of plywood, then sealed with silicone around the sides, it could theoretically represent a confined aquifer. The sponge represents the aquifer materials, and the plywood represents the confining layers. A hole is drilled through the top plywood piece into the sponge, and a straw is inserted to represent a well (Figure 6.7). If the sponge becomes saturated the water level in the straw may be visible above the top plywood piece. Any slight compression of the "aquifer" results in water rising quickly up the straw. The amount of compression would be almost imperceptible, even though a dramatic expulsion of water takes place. Lifting your hand off of the upper plywood piece results in the water level immediately dropping back to static conditions. This represents the elastic-like response of aquifers. A similar concept is illustrated by encased beverages that come sealed with a straw; such as CapriSun (http://parents.caprisun.com). The straw is inserted into the beverage and when the container is slightly squeezed, fluid emerges from the straw; however, in this case there are no "aquifer" materials.

The equations describing nonequilibrium drawdown with time were first attributed to Theis (1935). In analyzing pumping-test data sets, the usual first step is to perform a Theis analysis to see what the data look like.

6.2 Traditional Pumping-Test Analytical Methods

In the early 1930s, Theis noticed that there was a relationship between drawdown and well yield with time. He approached a friend named Lubin, a mathematician at the University of Cincinnati, and described his problem (Bob Cleary, personal communication,

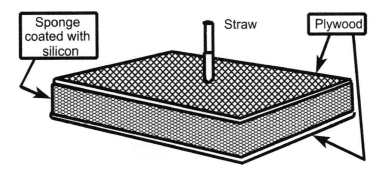

FIGURE 6.7 Model of a confined aquifer made with a sponge sandwiched between plywood and sealed with silicon calking. A hole is drilled and a straw inserted to represent a well.

San Francisco, CA, 1993). Lubin stated that he had a solution that was well known from the heat-flow literature. The analog to a well in an aquifer was compared to a wire within a cube of steel, heated by a battery. Lubin said that the solution of the radial heat-flow equation (Equation 6.1) being integrated was dependent upon the following assumptions:

- No edge effects (aquifer is of infinite areal extent).
- Uniform thickness (pumping well fully penetrates and receives water from the full thickness of the aquifer).
- Constant heat source (constant pumping rate).
- Homogenous and isotropic (aquifer is uniform in character and the hydraulic conductivity is the same in all directions).
- No sources or sinks (water is discharged instantaneously from storage and not from external sources either adding or removing water).

Additional assumptions inherent in the Theis equations listed in Driscoll (1986) are the following:

- Pumping well is 100% efficient.
- Laminar flow (Darcian) flow prevails throughout the well and aquifer.

The equation related to the heat-flow literature has the following form:

$$h_o - h = \frac{Q}{4\pi T} \int_u^\infty \frac{e^{-u}}{u} du \tag{6.1}$$

where Q = pumping rate (L^3/t)
$\quad\quad$ T = transmissivity (L^2/t)
$\quad$ $h_o - h$ = drawdown (L), sometimes denoted in the literature as "s" but kept as "$h_o - h$" to keep this term from being confused with Storativity (uppercase "S").

The solution is known as the Theis equation:

$$h_o - h = \frac{Q}{4\pi T}\left[-0.5772 - \ln u + u - \frac{u^2}{2 \times 2!} + \frac{u^3}{3 \times 3!} - \frac{u^4}{4 \times 4!} + - \cdots \right] \tag{6.2}$$

where [ ] (part in brackets of Equation 6.2) is known as the **well function $W_{(u)}$**, or simply

$$h_o - h = s = \frac{Q}{4\pi T} W_{(u)}$$

Note: The units need to be worked out correctly to get the appropriate drawdown in units of length! For example, gallons and feet will need to be converted appropriately. The well function is a function of u, where

$$u = \frac{r^2 S}{4Tt} \tag{6.3}$$

Note: Also, u is dimensionless!

where r = radial distance of the observation well or point of measurement to the pumping well (L) (Problem 6.6)

S = storativity or storage coefficient (unitless)

T = transmissivity (L^2/t)

t = time (in days or minutes)

Interestingly enough, Theis (1935) published the solution of a partial differential equation (PDE) that did not exist until C.E. Jacob, a graduate student of his at the University of New Mexico, came up with it in 1941 (Bob Cleary, personal communication, San Francisco, CA, January 1993).

Considering the assumptions given, radial flow toward a well can be thought of as water moving toward the well in a polar coordinate system. In the case where it is assumed that the aquifer properties are isotropic and homogenous (Chapter 3), however, with sources or sinks, the radial flow equation developed by Jacob (1950), derived in Appendix E, can be expressed as

$$\frac{\partial^2 h}{\partial r^2} + \frac{1}{r}\frac{\partial h}{\partial r} + \frac{W}{T} = \frac{S}{T}\frac{\partial h}{\partial t} \tag{6.4}$$

where r = radial distance from the well (L)

t = time (t)

h = hydraulic head (L)

S = storativity (dimensionless)

T = transmissivity (L^2/t)

W = source or sink term (L/t)

Instead of coming up with Equation 6.2 and solving it numerically, Theis (1935) solved it graphically. The method involves superimposing two curves. The first curve, laid on the bottom, called the **normal type curve**, is a plot of the well function "$W_{(u)}$" (the bracketed part of Equation 6.2 along the ordinate) versus values of "u" (along the abscissa) given in Equation 6.3. Corresponding values for $W_{(u)}$ and "u" are given in Appendix D (see Example 6.4 and Figures 6.8a, 6.8b, and 6.8c). Note how the plot of the normal type curve, shown in Figure 6.8a is in log-log scale. The second curve (Figure 6.8b) represents the time-drawdown data from the pumping test, which is laid on top of the normal type curve (or superimposed). This plot, also in log-log scale, puts drawdown ("$h_o - h$") expressed either in feet or in meters (increasing upward in the y-axis) versus r^2/t along the x-axis. It is imperative that the log-log scales of the two plots be of the same scale without distortion. If holding the two plots up to a window or light table, care is taken to make sure the x- and y-axes are kept square (both x- and y-axes are parallel to each other) as the data plot is slid along the normal type curve, to obtain a "best" fit of all the data. Software packages accomplish the superposition of both plots and provide results from the calculations.

A convenient match point is usually selected where both $W_{(u)}$ and u equal 1.0. This simplifies the calculations greatly. The corresponding match point on the data plot is determined for drawdown ($h_o - h$) and r^2/t to solve for transmissivity from Equation 6.2 and then for storativity by rearranging Equation 6.3. An example showing a fit of the pumping-test data and how to make the calculations is given in Example 6.4.

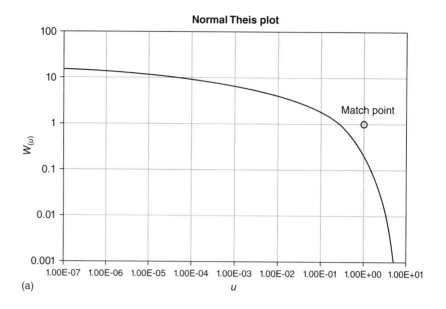

(a)

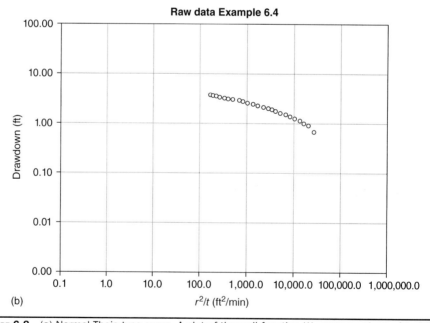

(b)

Figure 6.8 (a) Normal Theis type curve. A plot of the well function $W_{(u)}$ versus values of "u." The open circle is the match point where both $W_{(u)}$ and "u" (E + 00) equal 1. (b) Plot of field data (from Todd 1980) at the same log-cycle scale as the type curve (Figure 6.8a). (c) Image showing the field data (Figure 6.8b) juxtaposed on top of the normal Theis type curve (Figure 6.8a) to determine where the match point impacts the field data (drawdown and r^2/t).

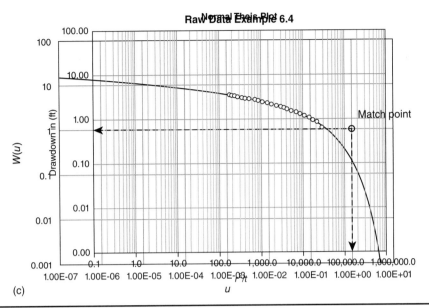

Figure 6.8 *(Continued)*

Example 6.4 A 4-h aquifer test was conducted to obtain a quick estimate of aquifer properties from Todd (1980). The pumping rate was approximately 450 gpm and the data in Table 6.1 are from an observation well located approximately 200 ft away. The Theis normal type curve is shown in Figure 6.8a, and the raw data from Table 6.1 are plotted in Figure 6.8b. The raw data superimposed on top of the normal Theis type curve are shown in Figure 6.8c. In order to appropriately superimpose the data onto the type curve the scales of both the data and the type curve *must* match. The match point on the type curve was selected at $Wu = 1.0$ and $u = 1.0$. The arrows indicate where the match point coincides with the raw data plot (Figure 6.8b). The match on the raw data indicates that the drawdown is at $h_o - h = 0.6$ ft and $r^2/t = 227{,}000$ ft^2/min.

Using Equation 6.2 the transmissivity is

$$T = \frac{450 \text{ gpm} \times 1 \text{ ft}^3/7.48 \text{ gal}}{4 \times \pi \times 0.6 \text{ ft}} \times 1.0 = 7.98 \text{ ft}^2/\text{min} = 11{,}500 \text{ ft}^2/\text{day}$$

Using Equation 6.3, one solves for the S (storage coefficient or storativity) using the transmissivity "T" just calculated. The storativity suggests that the aquifer is confined:

$$u = \frac{r^2 S}{4 \times t \times T} \rightarrow S = \frac{u \times 4 \times T \times t}{r^2} = \frac{1.0 \times 4 \times 7.98 \text{ ft}^2/\text{min}}{227{,}000 \text{ ft}^2/\text{min}} = 0.00014$$

The industry standard used by hydrogeologists today is a **reverse** type Theis curve (hereafter referred to as the Theis curve) (Figure 6.9). Both the type curve and the field-observed data curve are "reversed" by taking the inverse of the abscissa (*x*-axis) information. The reverse type curve is a plot of $W_{(u)}$ versus $1/u$, both on log-log scale. The data curve is a plot of drawdown ($h_o - h$) versus t/r^2, also on log-log scale. The reason

Time (min)	$h_o - h_{o'}$ ft	r^2/t, ft^2/min
0	0.00	0.0
1	0.66	40,000.0
1.5	0.89	26,666.7
2	0.98	20,000.0
2.5	1.12	16,000.0
3	1.21	13,333.3
4	1.35	10,000.0
5	1.48	8,000.0
6	1.57	6,666.7
8	1.74	5,000.0
10	1.87	4,000.0
12	1.97	3,333.3
14	2.07	2,857.1
18	2.20	2,222.2
24	2.36	1,666.7
30	2.49	1,333.3
40	2.66	1,000.0
50	2.79	800.0
60	2.95	666.7
80	3.05	500.0
100	3.15	400.0
120	3.28	333.3
150	3.41	266.7
180	3.51	222.2
210	3.61	190.5
240	3.67	166.7

TABLE 6.1 Pumping-Test Data from Todd (1980). The Data are Plotted in Figure 6.8b and Matched with the Type Curve in Figure 6.8c.

for using the reverse Theis curve is because any given observation well will have a constant radial distance (r). This means that the drawdown can be plotted more conveniently with corresponding points in time (usually plotted in minutes). Plots of more than one observation well *on the same plot must use* the calculated t/r^2 values for the respective distances (r). The transmissivity is calculated from Equation 6.5,

$$T = \frac{Q}{4\pi(h_o - h)} W_{(u)} \tag{6.5}$$

where T = transmissivity (L^2/t)

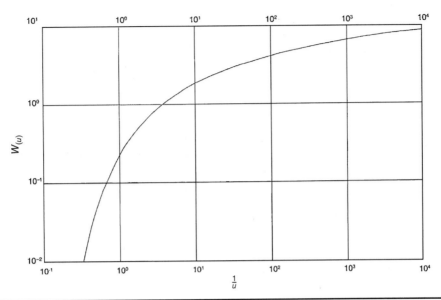

FIGURE 6.9   Plot of $W_{(u)}$ versus $1/u$ resulting in a reverse Theis curve. [*Modified after Lohman (1979).*] The field data are plotted in drawdown versus time to simplify the calculations.

and the calculation for storativity,

$$S = \frac{u4Tt}{r^2}$$

(6.6)

where S = storativity (dimensionless).

Remember to make sure all your units are consistent! Many errors occur in the calculations from not being careful with units! Calculations of transmissivity are usually in L^2/t, ft^2/min, ft^2/day, m^2/day, or gal/day-ft (day × feet). Not being careful with your calculations can lead to erroneous interpretations and poor design. More applications will be given in Section 6.4. An example showing a data plot fit to a reverse type curve with corresponding calculations is shown in Example 6.5.

Example 6.5   A pumping test for a municipality was conducted for 70 days. The pumping rate was 350 gpm (22 L/s) and the data from the observation well located 450 ft (91 m) away were collected over the testing period. The time-drawdown data superimposed on the reverse type curve yielded a drawdown of 0.4 ft with time at 5 min at the match point ($W_{(u)} = 1/u = 1.0$). Calculations of transmissivity and storativity were performed using Equations 6.5 and 6.6. The storativity suggests that the aquifer is confined:

$$T = \frac{350 \text{ gpm } (1 \text{ ft}^3/7.48 \text{ gal})}{4\pi(0.4 \text{ ft})} = 9.3 \text{ ft}^2/\text{min} = 13,400 \text{ ft}^2/\text{day}$$

$$S = \frac{4(9.3 \text{ ft}^2/\text{min})(5 \text{ min})}{(450 \text{ ft})^2} = 9.2 \times 10^{-4}$$

Example 6.6   One of the important requirements during a pumping test is that the pumping discharge rate be kept constant. This can affect the shape of the Theis curve. During a legal conflict, a homeowner and a well driller were at odds over drilling a well too deep (deep-holing). The total depth of the well was 325 ft (99 m).

At 200 ft (61 m) the well driller told the well owner the well was yielding between 4 and 5 gpm (13.1 and 16.4 L/min) but they were going to "chase it a bit." They chased it to 325 ft while the owner was at work and declared the well a success with a yield of 8 gpm (30 L/min). An argument led to the disagreement. The well owner was advised by the author that if another driller came in and pulled the casing, backfilled the hole with gravel, and retested the well, achieving the same production yield, it would indicate the additional footage was unnecessary. The well owner was cautioned that there was great risk in doing this because he may lose the well from materials caving into the well. The well owner was angry enough to go ahead with the suggested idea anyway. Another driller was hired, the casing was pulled, the well was backfilled with gravel to 212 ft, and then plugged with bentonite before reinstalling the well screen.

A pumping test was conducted by the other driller to see what kind of yield the well would make. The pump was set at 187 ft (57 m) and the pumping commenced at 10:00 a.m. The data collected are presented in Table 6.2.

During the jury trial the expert witness representing the driller used the data from Table 6.2 to generate a time-drawdown "Theis plot." The shape of the drawdown curve was disjointed, caused by variations in the pumping rate, and the departure from the "type curve" was explained using fracture-flow theory to impress an uneducated jury. The author thought it prudent to visit the site and conduct his own brief pumping test at 8 gpm (30.2 L/min). A very characteristic Theis curve resulted when the pumping rate was constant, to the dismay of the other expert.

Time	Elapsed Time (min)	Water Level (depth)	Pumping (gal/min)
10:00	0	18.83	7
10:25	25	61.42	7
10:30	30	65.83	7
10:35	35	69.66	7
10:40	40	73.75	5
10:45	45	79.0	7
10:50	50	85.0	7
10:55	55	89.83	7
11:00	60	93.75	7
11:05	65	98.58	7
11:15	75	105.67	6.5
11:25	85	115.0	7
11:45	105	120.0	5.5
11:55	115	140.0	8.17
12:00	120	144.0	7.5
12:12	132	164.0	7.2
12:20	140	170.0	7.2
12:35	155	174.0	7.91
12:40	160	175.0	8.3
12:45	165	175.0	8.6
13:45	225	175.0	6

TABLE 6.2   Pumping-Test Data and Well Yield from Example 6.5

Cooper-Jacob Straight-Line Plot

Cooper and Jacob (1946) recognized that if "u" in the Theis (1935) equation was sufficiently small, the nonequilibrium equation could be modified to a logarithmic term instead of an infinite series well function (Equation 6.7). Agreement as to when u is sufficiently small is not standard. Values range from 0.05 (Driscoll 1986; Fetter 1993) to 0.005 (Fletcher 1997). The software package AQTESOLV (Duffield 2018) indicates u is sufficiently small at 0.01. The smaller u is, the less error is involved. The author and John Sonderegger (co-author of the first edition of this book) believe 0.02 is sufficiently small. With "u" sufficiently small, Equation 6.2 can be rewritten as

$$h_o - h = \frac{Q}{4\pi T}[-0.5772 - \ln(u)] \tag{6.7}$$

By substituting in the parameters from Equation 6.2 for "u" and finding a natural log value for $e^{0.5772}$, we obtain

$$h_o - h = \frac{Q}{4\pi T}\left[-\ln(1.78) - \ln\left(\frac{r^2 S}{4Tt}\right)\right] \tag{6.8}$$

remembering that $-\ln A = \ln(1/A)$ and $\ln B \times C = \ln B + \ln C$:

$$h_o - h = \frac{Q}{4\pi T}\left[+\ln\left(\frac{1}{1.78}\right) + \ln\left(\frac{4Tt}{r^2 S}\right)\right] \tag{6.9}$$

Factoring out the natural log function and combining terms, we obtain

$$h_o - h = \frac{Q}{4\pi T}\ln\left[\frac{(2.25\,Tt)}{r^2 S}\right] \tag{6.10}$$

Since the relationship between $\log_{10} X$ and $\ln_e X$ is $\log_{10} X = 0.43429 \ln_e X$, we obtain

$$h_o - h = \frac{2.3\,Q}{4\pi T}\log\left(\frac{2.25\,Tt}{r^2 S}\right) \tag{6.11}$$

where all units are assumed to be consistent
$h_o - h$ = drawdown (L)
   S = storativity (dimensionless)
   T = transmissivity (L^2/t)
   t = time (t)
   Q = pumping rate (L^3/t)

When one plots drawdown (arithmetic scale on the "y" axis, with drawdown increasing downward (reverse order in a spreadsheet) versus time (log scale on the "x" axis) a straight-line results if the conditions of "u" being small are met along with the other Theisian conditions (Figure 6.10). By evaluating Equation 6.3 one can see that "u" is small when "r" (the distance of the observation point) is small or when the time "t" (time since pumping began) is large. This eliminates having to work with the well function

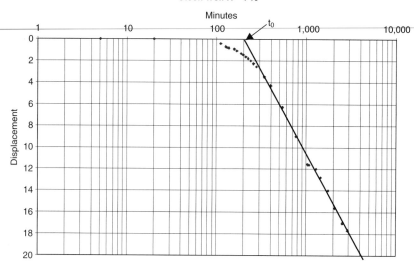

Figure 6.10 Cooper-Jacob plot showing straight-line behavior in drawdown in feet versus time. Zero drawdown ($h_o - h = 0$) occurs at time $t_o = 200$ min. Context for the figure is from Example 6.7a.

($W_{(u)}$). It should be noted that for a given pumping rate (Q) and observation well, located at a constant distance (r), drawdown ($h_o - h$) and time (t) are the only variables. By separating the variable (t) from the constant terms in Equation 6.11:

$$h_o - h = \frac{2.3\,Q}{4\pi T} \log\left(\frac{2.25\,T}{r^2 S}\right) + \frac{2.3\,Q}{4\pi T} \log(t) \tag{6.12}$$

Cooper and Jacob (1946) noticed that this could be compared to the equation for a line:

$$h_o - h = C_1 + C_2 \log(t), \quad \text{or} \quad y = b + mx \tag{6.13}$$

By selecting two times t_1, t_2, where $t_2 > t_1$, the drawdown at time t_1 is described by:

$$(h_o - h)_1 = \frac{2.3\,Q}{4\pi T} \log\left(\frac{2.25\,T t_1}{r^2 S}\right) \tag{6.14}$$

and the drawdown at time t_2 is described by:

$$(h_o - h)_2 = \frac{2.3\,Q}{4\pi T} \log\left(\frac{2.25\,T t_2}{r^2 S}\right) \tag{6.15}$$

The difference of the two drawdowns is expressed as:

$$(h_o - h)_2 - (h_o - h)_1 = \frac{2.3\,Q}{4\pi T} \log\left(\frac{2.25\,T t_2}{r^2 S}\right) - \frac{2.3\,Q}{4\pi T} \log\left(\frac{2.25 T t_1}{r^2 S}\right) = \frac{2.3\,Q}{4\pi T} \log\left(\frac{t_2}{t_1}\right) \tag{6.16}$$

By selecting time over one log cycle (a factor of 10), the (t_2/t_1) term simplifies to 1.0. The transmissivity is calculated using consistent units by the familiar expression in Equation 6.17:

$$T = \frac{2.3\,Q}{4\pi\Delta(h_o - h)} \tag{6.17}$$

where $\Delta h_o - h$ is the drawdown over one log cycle.

The data, then, can be plotted in the field using a spreadsheet program with $h_o - h$ (arithmetic scale as the ordinate, y-axis) versus time [abscissa in log scale (x-axis), usually in minutes; Figure 6.10].

Example 6.7a The context for Figure 6.10 will be used to illustrate the use of the Cooper-Jacob (1946) straight-line method to estimate the transmissivity. A farmer/rancher near Dillon, Montana, in southwestern Montana was seeking a water-use permit to irrigate his acreage. A 24-h test was conducted on a 345-ft (105 m) deep production well at 800 gpm in 2003. The production well experienced about 144.5 ft of drawdown after 24 h and a nearby stock well ($r = 740$) was monitored throughout the test to estimate the aquifer properties (T and S). It showed a drawdown at about 14 ft after 24 h. In comparing the straight-line fit in Figure 6.10 over one log cycle, an estimate of 19.8 ft of drawdown was occurring at 4,000 min and a drawdown of 6.2 ft was occurring at 400 min, or 15.6 ft of drawdown over one log cycle. Equation 6.17 was used to estimate T:

$$T = \frac{2.3Q}{4\pi\Delta(h_o - h)} = \frac{2.3 \times 800 \text{ gpm} \times \left(\dfrac{1 \text{ ft}^3}{7.48 \text{ gal}}\right)}{4 \times \pi \times (15.6 \text{ ft})} = \frac{1.25 \text{ ft}^2}{\text{min}} = 1,800 \text{ ft}^2/\text{day}$$

How do we get the storativity? The straight-line plot is extended to where the zero-drawdown axis occurs (Figure 6.10). The time where drawdown equals zero is known as "t_0" (T-naught). Cooper-Jacob (1946) noticed that Equation 6.11 could be used to solve for the storativity. The first step is to move the constant portion to the other side:

$$\frac{(h_o - h)4\pi T}{2.3\,Q} = \log\left(\frac{2.25\,Tt_0}{r^2 S}\right) \tag{6.18}$$

By setting drawdown $(h_o - h) = 0.0$ (value at t_0) and taking the exponent of both sides, we obtain

$$10^0 = \frac{2.25\,Tt_0}{r^2 S} \tag{6.19}$$

which simplifies to

$$1.0 = \frac{2.25\,Tt_0}{r^2 S} \tag{6.20}$$

The storativity can be calculated by Equation 6.21:

$$S = \frac{2.25\,Tt_0}{r^2} \tag{6.21}$$

where T, r = as described previously

t_0 = straight-line projection of the time-drawdown curve up to where it intersects the zero-drawdown axis.

Example 6.7b   Revisiting Example 6.7a and Figure 6.10 one observes that the straight-line projection up to zero drawdown occurs at about 200 min. By including the newly calculated $T = 1,800$ ft^2/day and the radial distance of the stock well, being 740 ft, Equation 6.21 can be used to estimate S:

$$S = \frac{2.25Tt_0}{r^2} = \frac{2.25 \times 1,800 \text{ ft}^2/\text{day} \times 200 \text{ min} \times 1 \text{ day}/1,440 \text{ min}}{(740 \text{ ft})^2} = 1.0 \times 10^{-3}$$

When the results from the Theis curve are compared to the Cooper-Jacob method, the results are generally comparable; however, the Cooper-Jacob method is easier to plot in the field and does not require curve matching. If the conditions for "u" are not met, the Theis method is still viable for interpretation.

Example 6.8   A pumping test was conducted in July 1979 with a pumping rate of 1,200 gpm (75.7 L/s). The data are from an observation well located 850 ft (259 m) away. The data are shown in Table 6.3 and plotted using the Theis (1935) and Cooper-Jacob (1946) methods, as shown in Figures 6.11a and 6.11b.

Notice the characteristic Theisian fit in Figure 6.11a plotted using the Aqtesolv software (Duffield and Rumbaugh 1991). The calculated values for both methods are comparable; however, the Cooper-Jacob straight-line fit is a least-squares best fit. When using software programs and you seek a straight-line fit of the data, it may do what you see in Figure 6.11b. Where is a straight-line fit characteristic of the aquifer? To identify where the data form a straight line the author advises that you hold the Cooper-Jacob plot away from you at a low angle. If you try this, you will note that the data after 180 min form a straight line. Another approach is to lay the plot on a table. Now get your face close to the table to evaluate the straightness of the data. The appropriate fit is slightly steeper than the best-fit line indicated on the plot and will result in the transmissivity value being slightly smaller than indicated but will better represent the data at later time.

Time	Minutes	Water Level	Drawdown
0700	0	40.0	0
0715	15	40.13	0.13
0730	30	40.5	0.5
0800	60	41.19	1.19
0830	90	41.7	1.70
0900	120	42.11	2.11
1000	180	42.78	2.78
1100	240	43.32	3.32
1300	360	44.01	4.01
1500	480	44.52	4.52
1700	600	44.89	4.89
1900	720	45.23	5.23
2100	840	45.55	5.55

TABLE 6.3   Data from July 1979 Pumping Test

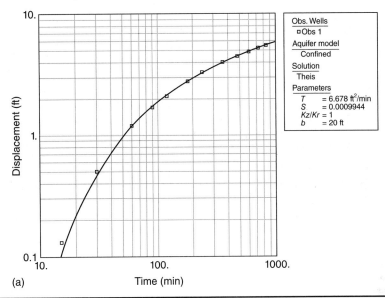

(a)

FIGURE 6.11a A typical Theis fit. The data are from an observation well located 850 ft away from the pumping well collected in July 1979. The data are shown in Table 6.3 and described in Example 6.8.

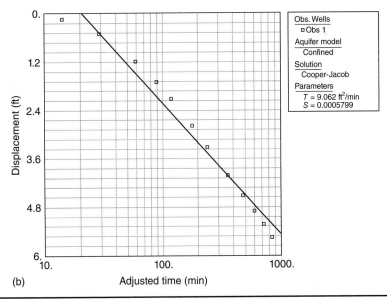

(b)

FIGURE 6.11b A least-squares best-fit match using the Cooper-Jacob (1946) method. The data after 180 min form a straight line. (Straight line incorrectly fit by software.) The data are from Table 6.3 and correspond to the same data set as Figure 6.11a.

Example 6.8 brings up a caution about aquifer testing software. These programs are wonderful for plotting the data but tend to remove the "thinking part" of data analysis. If one accepts a best-fit line to the data using the Cooper-Jacob (1946) method, and there are multiple slopes resulting from boundaries and other conditions, the interpretation will be wrong. Spreadsheet programs are useful for plotting data, especially using the Cooper-Jacob (1946) method; however, plots within spreadsheet programs using the Theis method are more complex. Errors result if the log cycles of the data plot and the log cycles of the well function characteristic curve are not at the same scale dimensions when curve matching is attempted. Both should have the same log-log spacing (boxes of log scales should overlap) when superimposing the field data on the type curve and selecting a match point. This was not an issue during the days of using 3-by-5 log cycle graph paper (https://www.printablepaper.net/category/log or other options on the Internet).

Once the Theis equation (Equation 6.5) has been used to obtain a transmissivity (*T*) and Equation 6.6 has been used to solve for storativity (*S*) the aquifer properties are assumed to prevail throughout the entire aquifer (Theis 1935). Therefore, *T* and *S* are assumed to be constants and can be used to evaluate drawdown for *different* pumping rates (*Q*), radial distances (*r*), and different times of pumping (*t*). This is what the "nonequilibrium" part of the equation name means. This also assumes there are no boundary conditions (one of the assumptions of the Theis equation at the beginning of this section).

How is this done? First a new value must be recalculated for "*u*" (Equation 6.3). The new "*u*" must be adjusted with any changes in the observation point "*r*" and any new time "*t*." One then can use Appendix D to look up a corresponding value for $W_{(u)}$, that is applied to Equation 6.2, along with any changes in pumping rate *Q*. The new drawdown can then be calculated. Example 6.8 provides an illustration using the outcomes from Example 6.7.

Example 6.9   Assume that the aquifer properties of transmissivity (*T*) and storativity (*S*) are 6.68 ft²/min and 9.9×10^{-4}, respectively (obtained from Example 6.7). We want to estimate the drawdown at a distance 2,000 ft away, after increasing the pumping rate to 1,500 gpm (94.7 L/s) over a time period of 5 days. We first adjust the transmissivity into units of ft²/day (to match the time scale):

$$6.68 \text{ ft}^2/\text{min} \times 1{,}440 \text{ min}/1 \text{ day} = 9{,}620 \text{ ft}^2/\text{day}$$

Next, we use Equation 6.3 to recalculate a new value for "*u*":

$$u = \frac{r^2 S}{4Tt} = \frac{(2{,}000 \text{ ft})^2 (9.9 \times 10^{-4})}{4(9{,}620 \text{ ft}^2/\text{day})(5 \text{ days})} = 0.021$$

From Appendix D the corresponding value for $W_{(u)}$ is 3.31. This is substituted into Equation 6.5, along with the new pumping rate and proper conversions (Appendix A) to make the units work out correctly:

$$h_o - h = \frac{Q}{4\pi T} W_{(u)}$$

$$= \frac{(1{,}500 \text{ gal/min})\,(1 \text{ ft}^3/7.48 \text{ gal})\,(1{,}440 \text{ min}/1 \text{ day})}{4\pi(9{,}620 \text{ ft}^2/\text{day})} (3.31) = 7.9 \text{ ft}$$

Distance-Drawdown Relationships

Back in 1906, Thiem developed a method of estimating transmissivity and storativity after steady-state conditions have been achieved (i.e., the cone of depression had stabilized).

Once steady state occurs, time becomes inconsequential, and the hydraulic properties of an aquifer estimated from drawdown become a function of distance.

The Cooper-Jacob (1946) time-drawdown straight-line approach was modified to plot drawdowns at various radial distances from the pumping well at a chosen time "t." These plots are known as **distance-drawdown plots** (Jacob 1950). By graphing drawdown (arithmetic scale) versus distance (on a log scale), a straight-line fit results (Jacob 1950), if the aquifer is homogenous and isotropic (Chapter 3). Estimates of the hydraulic properties can also be made, if the assumption of u being sufficiently small (0.02, previous section) can be made. The intercept where the straight-line fit crosses the zero drawdown represents the *range of influence* of the cone of depression (Figures 6.2 and 6.3). In other words, the range of influence is the radial distance from the pumping well to where zero drawdown occurs (for practical purposes 0.01 ft of drawdown). Since "t" is constant, the variables now are drawdown "$h_o - h$" and radial distance "r."

There is a powerful relationship between a Cooper-Jacob (1946) time-drawdown plot and a distance-drawdown plot (Jacob 1950). The slope of the straight-line fit (drawdown per log cycle of distance) is precisely twice that of the time-drawdown plot (drawdown per log cycle of time). The reason for this can be seen by evaluating Equation 6.11. Within the "log" term, notice that time (t) is to the first power while radial distance (r) is to the second power. The "double slope" results from the following relationship:

$$\log(r^2) = 2 \times \log(r) \tag{6.22}$$

This ratio is a fixed relationship. The approach in performing a straight-line fit (drawdown per log cycle of distance) analysis is to choose a particular point in time (4 h, 8 h, 1 day?), where you have drawdown data from nearby observation wells from an aquifer test (Problem 6.10). This "fixes" the shape of the cone of depression at the chosen time "t." If one plots the drawdown (ordinate in arithmetic scale, zero drawdown at the top, values *in reverse order* in a spreadsheet) verses distance "r" along the abscissa in log scale, a straight-line fit will occur with the distance-drawdown data *if* the aquifer is homogenous and isotropic. The equations used to calculate transmissivity and storativity are analogous to Equations 6.17 and 6.21 using consistent units (note the difference in the denominator for transmissivity):

$$T = \frac{2.3Q}{2\pi\Delta(h_o - h)} \tag{6.23}$$

$$S = \frac{2.25Tt}{r_0^2} \tag{6.24}$$

where T, t are as described previously, and r_0 = range of influence, straight-line projection of the distance-drawdown curve up to where it intersects the zero-drawdown axis as described above.

Example 6.10   A pumping test was conducted where the pumping rate was 400 gpm (25.2 L/s). Drawdowns were measured in three observation wells, A ($r = 40$ ft, $h_o - h = 6.5$ ft), B ($r = 300$ ft, $h_o - h = 3.9$ ft), and C ($r = 900$ ft, $h_o - h = 2.5$ ft), after 12 h. Estimate the T and S of this aquifer using the distance-drawdown method.

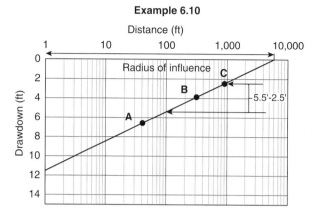

Figure 6.12 Distance-drawdown plot from Example 6.10. The drawdown from three wells at time 4 h were plotted versus distance in feet.

Solution Set up three columns in a spreadsheet program with the headings: Well, Distance, and Drawdown. Enter the data as described in the problem. Insert an *x-y* scatter plot using Distance along the *x*-axis and Drawdown down the *y*-axis. Right-click on the *y*-axis and select *values in reverse order* and change the *maximum drawdown* to around 15 ft. Right-click on the *x*-axis and select *logarithmic scale* with the *maximum distance value* at 10,000 ft. Make a straight-line fit to the three points. The straight-line fit should reach zero drawdown at about 6,000 ft and the drawdown is about 11.5 ft at a distance of $r = 1.0$ ft (Figure 6.12).

From the plot estimate the change in drawdown over one log cycle (5.5 ft − 3.5 ft = 2.0 ft). Convert the pumping rate Q = 400 gpm into ft³/day to match the time units:

$$Q = 400\text{ gpm} \times \frac{1\text{ ft}^3}{7.48\text{ gal}} \times 1{,}440\text{ min/day} = 77{,}005\text{ ft}^3/\text{day}$$

Now use Equation 6.23 to calculate *T* and then Equation 6.24 to calculate *S*.

$$T = \frac{2.3Q}{2\pi\Delta(h_0 - h)} = \frac{2.3 \times 77{,}005\text{ ft}^3/\text{day}}{2 \times \pi \times 3.0\text{ ft}} = 9{,}400\text{ ft}^2/\text{day}$$

$$S = \frac{2.25Tt}{r_o^2} = \frac{2.25 \times 9{,}400\text{ ft}^2/\text{day} \times 0.5\text{ day}}{(6{,}000\text{ ft})^2} = 2.9 \times 10^{-4}$$

If one can estimate the aquifer values *T* and *S* then distance-drawdown analysis can be applied in a variety of circumstances.

Example 6.11 A mining operation producing railroad ballast required water to provide dust control for their operations (Figures 6.13a and 6.13b). They spent 2 years preparing for the mining operations and had ignored the water issue until the end. A wildcat well drilled over 570-ft (173-m) deep near their operations failed to produce water. Approximately 0.8 mi (1 to 1.5 km) to the south is a small community of approximately 60 persons, including a recreational vehicle park (Good Sam's). Each home or business has its own well and septic tank. The geology consists of approximately 50 ft of sediments overlying granitic rocks (see Figure 5.36). The production zones for wells are in fractures in granite that parallel Pipestone Creek flowing down from the mountains to the west and from the upper sandy sediment units. The mining company made a deal with a local resident in the small

(a)

FIGURE 6.13a   Railroad ballast loadout facility for the mining operation described in Example 6.11.

(b)

FIGURE 6.13b   Dusty conditions for the mining operation described in Example 6.11. The water from the well was needed as a dust suppressant.

community to drill a well on his property that could pump water up the hill to the north. When the drill rig was set up in the yard, the other residents started asking questions.

The author was asked to represent the mining company in answering a "few questions" at a public meeting. Unaware of the sensitive nature of the problems to be discussed, he agreed to be present. Before the meeting, the author obtained lithologic logs for each of the wells in the area. By evaluating the production testing performed on each well recorded on the well log, an estimate of the transmissivity was made (Specific Capacity, Section 6.4). The storativity was estimated based upon the geologic setting. By rearranging Equation 6.24 an estimate of the range of influence "r_o" for the proposed pumping rate could be performed. The transmissivity was estimated using a specific capacity equation for fracture flow (Section 6.4), and the storativity of the granite layer was estimated to be 0.001. A time of 100 days of continuous pumping was used as a worst-case scenario.

$$r_0 = \sqrt{\frac{2.25(13\,\text{ft}^2/\text{day}) \times 100\,\text{days}}{0.001}} = 1{,}710\,\text{ft}\,(520\,\text{m})$$

At the meeting, the room was filled with angry residents. By the time the author was given the floor to speak, everyone was at peak irritation, similar to a hissing group of geese. (Ironically, the lady who was most upset and called the governor did have geese on her property that attacked us every time we took a water level in her well; Figure 6.14). The before-meeting distance-drawdown analysis was used to tell all residents who lived more than 2,000 ft from the proposed well that they could go home. During the meeting, the author committed the mining company to having a pumping test performed and a numerical groundwater flow model be constructed to evaluate pumping and recovery scenarios. This solution appeased the residents and ultimately led to a practical solution. The actual range of influence observed in a 12-h aquifer test was about 1,200 ft.

Predictions of Distance-Drawdown from Time-Drawdown Plots

As was mentioned, a powerful predictive tool results from the slope relationships between time-drawdown and distance-drawdown graphs. If a short-duration pumping test (say 24 to 72 h) is conducted and a straight-line relationship develops, the straight

Figure 6.14 Gaggle of geese that attacked the author during the aquifer test described in Example 6.11.

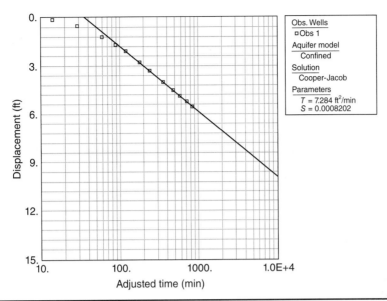

Figure 6.15 Proper fit of the data from Figure 6.11a, showing the Cooper-Jacob straight-line fit and solution. The slope of the data was extended out to 10,000 min (Example 6.12).

line of the time-drawdown graph can be extended to longer periods of time (many days or weeks) to estimate impacts at the longer-time conditions, provided there are no boundary conditions. Once the drawdown at the longer period of time has been estimated, a distance-drawdown graph (using twice the slope of the time-drawdown graph) can be used to estimate the range of influence of the cone of depression. This is a useful exercise when evaluating well-interference effects (Section 6.5) of a potential cone of depression on other senior water users in the area.

Example 6.12   Suppose the data from the 14-h aquifer test in Example 6.8 are to be used to estimate the range of influence of the cone of depression after 1 week (7 days ~ 10,000 min). Figure 6.11b is refitted to the proper aquifer response (Figure 6.15) and extended out to 10,000 min and the drawdown at the observation well ($r = 850$ ft) was used to estimate the radius of influence at 1 week.

At 10,000 min the drawdown is approximately 9.8 ft and the slope of the straight-fit line is 7.1 ft (at 2,000 min) minus 3.0 ft (at 200 min) or 4.1 ft per log cycle for the time-drawdown data. The estimated aquifer parameters are $T = 7.28$ ft^2/min or 10,500 ft^2/day and a storativity of 8×10^{-4}. This information is then used to plot the new drawdown with the slope at 8.2 ft per log cycle to estimate the range of influence (Figure 6.16).

Figure 6.16 indicates that after 14 h of pumping (original data) the range of influence was about 4,000 ft (upper line). Using the time-drawdown data projected out to 10,000 min it shows a drawdown at $r = 850$ ft (259 m) of about 9.8 ft (3 m). Applying twice the slope of the time-drawdown per log cycle to the drawdown at $r = 850$ ft the range of influence is estimated to be about 12,000 ft (3,110 m).

As a check one can use once a rearranged version of Equation 6.24 to solve for r_o. In using the values obtained from Figure 6.15 ($T = 10,500$ ft^2/day and $S = 8.2 \times 10^{-4}$) the reader is invited to estimate r_o. As a comparison using the values from Figure 6.11a ($T = 9,620$ ft^2/day and 1×10^{-3}) one obtains another estimate. The numbers estimated by plotting and by calculation give an idea of what could happen. It looks like the range of influence is expected to extend roughly a couple of miles.

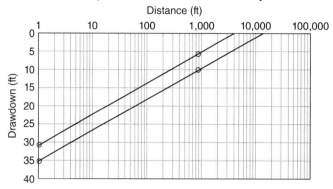

Example 6.12 Distance-drawdown analysis

FIGURE 6.16 Spreadsheet plot of drawdown from an observation well located 850 ft from the pumping well after 14 h (840 min) and the projected drawdown (Figure 6.15) at 10,000 min showing the range of influence for both conditions (Example 6.12). The slope of the time-drawdown data was doubled when placed on the distance-drawdown plot.

Suppose you have no pumping-test data and you are trying to get a permit to drill a well, you can use Equation 6.24 to estimate the range of influence r_0 for a particular time of your proposed well, if you can come up with an estimate a value for T and S (Example 6.13).

Example 6.13 Suppose there is a need to irrigate 30 acres (12.1 ha). You need to find out from your local state agency if any wells exist near the proposed well site. The well logs may give an idea of how deep it is to water and what lithologies the wells are completed in. The well logs may also indicate how much yield in gpm they make. Since $T = K \times b$ (Chapter 3), if an estimate for hydraulic conductivity can be obtained, and the saturated thickness (b) is known or can be projected (Chapter 2), a value for T can also be estimated (your approximation of T should be conservative). The storativity can be estimated from the lithologic conditions (Chapters 2 and 3) and whether the aquifer is confined ($S \sim 10^{-3}$ to 10^{-6}) or unconfined ($S \sim 0.03$ to 0.30). In this example, a range for T and various times should be used to evaluate impacts to other wells (Table 6.4). By estimating the drawdown in the pumping well and the range of influence (distance to zero drawdown) one can perform a distance-drawdown plot to evaluate the impacts to neighboring wells. To produce Table 6.4, it was necessary to first use Equation 6.3 to calculate u. It was assumed for the pumping well, that the effective radius was 1 ft (0.3 m). A corresponding value for W_u was looked up in Appendix D. The drawdown was calculated

Transmissivity ft²/day	1/3 day		30 days		60 days		180 days	
	Draw-Down at 1 ft	Radius of Influence, ft	Draw-Down at 1 ft	Radius of Influence, ft	Draw-Down at 1 ft	Radius of Influence, ft	Draw-Down at 1 ft	Radius of Influence, ft
1,000	16.3	71	24.9	671	26.2	949	28.3	1,643
3,000	6.1	122	9.0	1,162	9.4	1,643	10.1	2,846
5,000	3.9	158	5.6	1,500	5.9	2,121	6.3	3,674
10,000	2.1	224	2.9	2,121	3.1	3,000	3.3	5,196

TABLE 6.4 Comparison of Drawdowns, Transmissivity, and Time

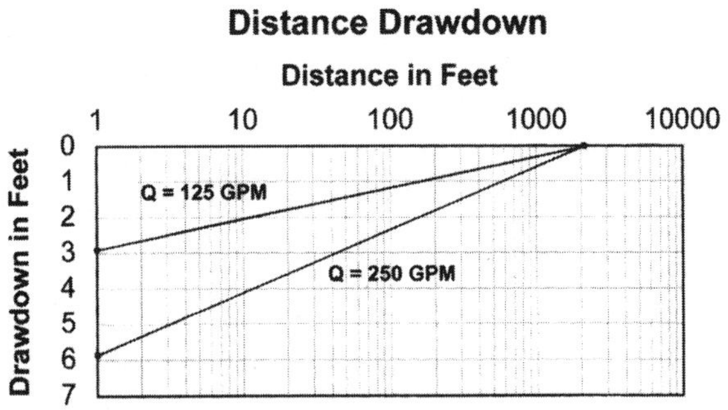

Distance Drawdown

FIGURE **6.17** Relationship of pumping rate to range of influence. Pumping rate does not affect the range of influence but does influence the amount of drawdown.

using Equation 6.2. As a sample calculation, suppose the transmissivity is estimated to be 10,000 ft²/day and the storativity is estimated to be 0.15, the range of influence after 30 days from rearranging Equation 6.24 is:

$$r_0 = \sqrt{\frac{2.25(10,000 \text{ ft}^2/\text{day}) \times 30 \text{ days}}{0.15}} = 2,121 \text{ ft } (647 \text{ m})$$

From Example 6.13, notice that the pumping rate Q was *not* used to evaluate r_0. The range of influence is dependent on T, S, and time (t). Since, the pumping rate does *not* affect the range of influence (see Equation 6.24), what is observed then is how the pumping rate affects the slope of the distance-drawdown plot. For a particular time, t, the r_0 is fixed, but the slope of the distance-drawdown line drawn back to the vicinity of the pumping well depends on the pumping rate (Figure 6.17). Notice that the drawdown in Figure 6.17 at approximately 160 ft (49 m) is expected to be 1.0 ft (0.3 m) for a pumping rate of 125 gpm (425 L/min). For most practical problems, the distance (r) at the pumping well is the **(effective radius, r_e)**, or the distance out to where the formation was disturbed during drilling (damage zone; Figure 5.8). The author typically uses 1 ft (0.3 m). But for practical applications one must also consider an estimate of the well efficiency (Figure 5.9).

The Cooper-Jacob (1946) method of time-drawdown and the Jacob (1950) method of distance drawdown are a modification of the Theis equation and assume that u is small. The Theis method is always a valid "first approach" to evaluating pumping-test data, and the Cooper-Jacob (1946) and Jacob (1950) methods are valid when the time (t) is sufficiently large or radial distance (r) is sufficiently small. It is useful to evaluate data using as many methods as applicable, a real positive aspect of software packages (Duffield 2018). Each may help provide a different perspective and aid in a better interpretation.

6.3 Non-Ideal Pumping-Test Analytical Methods

Leaky-Confined and Semiconfined Aquifers

Storativities between 0.001 and 0.03 suggest that the aquifer could be semiconfined or leaky-confined. The Theis method, as previously described, theoretically represents a confined aquifer where the aquifer thickness always remains fully saturated and that all water derived from pumping comes instantaneously from storage. The type curve (Figure 6.9) becomes a kind of standard for comparison to which all other time-drawdown

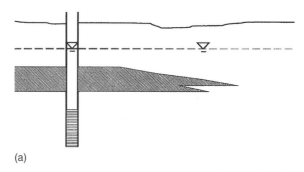

(a)

FIGURE 6.18a   Pumping well positioned below a clayey unit that pinches out to the right. The clay semiconfines the lower production zone.

solutions are compared. Consider once again Figure 3.9 (reprinted here as Figure 6.18a). In Figure 6.18a there is a confining clayey unit separating the screened portion of the well from the upper unconfined sediments. After the pump is engaged the preliminary drawdown response in the cone of depression is confined, just like the Theis method already described. This early time-drawdown response follows a Theis type A curve (comparable to Figure 6.9 and discussed soon). However, once the cone-of-depression extends past the restricting influence of the confining unit, the pumping stress is able to draw from a thicker unconfined saturated thickness (*b*) (Figure 6.18a). This is observed in the time-drawdown data as a recharge response, where the plotted data take a flatter trajectory below the Theis curve. This is a semiconfined hydrogeologic setting connected with an unconfined aquifer.

Consider a second scenario as portrayed in Figure 6.18b. Suppose the clayey part of Figure 6.18a is all replaced by a thin silty layer **aquitard** (of lower hydraulic conductivity compared to the aquifer materials above and below) that extends in all directions. In this situation the early time-drawdown data also follow the theoretical Theis curve (type A). After time passes and the cone of depression extends outward, the hydraulic head in the aquifer below the aquitard becomes significantly less than the hydraulic head in the aquifer materials above the aquitard resulting in a head differential. Darcy's law tells us that groundwater moves from higher hydraulic head to lower hydraulic head; therefore, groundwater in the upper aquifer is "induced" downward through the silty materials as **leakance**. The leakance effect is also observed as a flattening of the time-drawdown data as a recharge response. This is one example of a leaky-confined aquifer setting.

The scenario depicted in Figure 6.18b is where the upper aquifer is unconfined. Hantush and Jacob (1955) developed a mathematical relationship and solution for an aquifer being pumped in a leaky-confined setting. The conceptual model is that of a confined two-aquifer system separated by a relatively thin aquitard, where all sources of recharge are moving through the aquitard. The situation is depicted in Figure 6.19. The solution to the Hantush-Jacob equation (1955) is of similar form to Equation 6.2; however, they use the function "*u*" (identical to Equation 6.3) and a dimensionless parameter r/B:

$$h_o - h = \frac{Q}{4\pi T} W\left(u, \frac{r}{B}\right) \quad \text{and} \tag{6.25}$$

$$B = \sqrt{\frac{Tb'}{K'}} \tag{6.26}$$

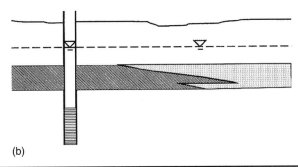

(b)

FIGURE 6.18b   Figure 6.18a changed to an extensive silty unit that extends in all directions. The silty unit with lower hydraulic conductivity constrains the rate of recharge from the upper unit after pumping in the lower unit.

where $W(u, r/B)$ = the leaky well function (dimensionless)

r = radial distance from the pumping well to the observation well (L)

T = hydraulic conductivity of the pumped aquifer $K \times$ thickness b of the pumped aquifer (L2/T)

b' = thickness of the aquitard (L)

K' = vertical hydraulic conductivity of the aquitard (L/T)

Values of $W(r/B)$ versus $1/u$ have been plotted into a family of type curves (Heath 1983) (Figure 6.20), so the field data, as previously described can be plotted, superimposed, or matched accordingly. Hantush (1960) published a modified solution that takes into account effects of leakage from the aquitard (discussed later) (http://hydrogeologist-swithoutborders.org/wordpress/1979-english/chapter-8). Neuman and Witherspoon

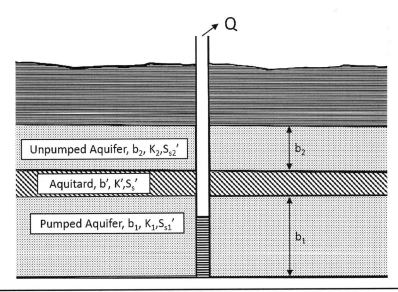

FIGURE 6.19   Schematic of a two-aquifer system separated by a leaky aquitard. The pumped aquifer may have leakance through the aquitard without the aquitard contributing water from storage. [*After Hantush and Jacob (1955).*]

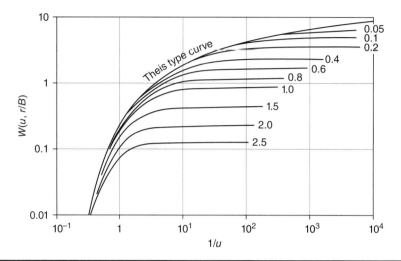

Figure 6.20   Family of leaky type curves of $W(u, r/B)$ = the leaky well function (dimensionless) (after Hantush and Jacob 1955). The smaller the r/B value is, the closer the curve is to the reverse Theis type. (*Used by permission from Prentice Hall Publishing, Upper Saddle River, New Jersey.*)

(1969) successfully provided a more complete solution of the scenario depicted in Figure 6.19 by including the release of water from storage of the aquitard and drawdowns in the unpumped aquifer. Because of the relative simplicity of using the type curves developed by Hantush (1960) they are often used for the case shown in Figure 6.18b (Heath 1983). A more rigorous mathematical derivation of the expressions of pumping-induced leakage from pumping is presented by Butler (2003).

Both semiconfined and leaky-confined aquifers initially follow the Theis type A curve and then flatten or "curve" below the curve, indicating a recharge-response condition. The hydraulic conductivity of the aquitard would normally be up to a couple of orders of magnitude lower than the aquifer unit(s) (Figure 6.19) (Heath 1983). Earlier it was mentioned that aquifer tests need to be conducted long enough to observe the behavior of the data. In the case of the family of curves of Figure 6.20 (Hantush 1960) the horizontal nature of the field data associated with the later-time flat type curves suggests that the data would continue horizontally even if pumped a long time. This response reflects the relatively lower hydraulic conductivity control of the aquitard.

When the drawdown data over time fall below the Theis curve on a log-log plot but *remain curved* rather than becoming flat (Figure 6.21), it is likely that the aquitard is releasing water from storage. This occurs because the aquitard is saturated and has a head higher than the head in the aquifer being pumped. This head differential, therefore, causes the aquitard to release water from storage, in an attempt to reach equilibrium (Hantush 1956, 1960). The solution uses the function "u" (identical to Equation 6.3) and a dimensionless parameter Beta (β) (see Equation 6.27). The various Beta (β) curves represent different combinations of hydraulic properties of the aquifer and the confining units (Heath 1983). The approach is to fit the field time and drawdown data ($h_o - h$, or s) to one of the beta (β) curves and select a convenient match point to obtain values for $H(u, \beta)$, $1/u$, time, and drawdown ($h_o - h$).

$$h_o - h = \frac{Q}{4\pi T} H(u, \beta) \qquad (6.27)$$

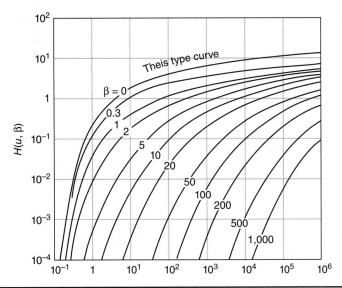

Figure 6.21 Family of Hantush (1960) type curves of $H(u, \beta)$ = leaky well function where the aquitard contributes water from storage. The various β's represent different combinations of hydraulic properties of the aquifer and the confining units (Heath 1983).

where the aquifer properties are determined using the following equations:

$$T = \frac{Q}{4\pi(h_o - h)} H(u, \beta) \quad \text{and} \tag{6.28}$$

$$S = \frac{4Tt(h_o - h)}{r^2} \tag{6.29}$$

Example 6.14 Near Three-Forks, Montana, in southwestern Montana, a 72-h aquifer test was conducted in January 2006 as part of an application for a 216-ft (66-m) deep irrigation well. The surface sediments were clayey and fine grained and the production zone was in coarse sands and fine gravels about 30 ft thick at depth. The source aquifer thins to the east. The production well is 10-in (25-cm) in diameter and a 6-in (15-cm) well, located 24 ft (7.3 m) to the east, was used as an observation well. The SWL was 16 ft and the total drawdown was 183 after 72 h with an average pumping rate of 363 gpm (22.9 L/s) (Figure 6.22).

Commercial software was used to evaluate the time-drawdown data from both wells. The Theis method was applied to the 10-in production well and the 6-in well fit the Hantush (wedge-shaped) solution (Duffield 2018) to obtain T and S. The results of T are very similar (370 ft^2/day and 355 ft^2/day, respectively) and the solutions seem to be a good match with the field data (Figures 6.23a and 6.23b).

In a semiconfined aquifer scenario, aquitard units may pinch out laterally to encounter another part of the aquifer that is actually unconfined. The field time-drawdown data are plotted and the flattening of the data below the Theis curve reflects a time delay of water from the aquifer yielding to the well as an apparent source of recharge or from an increased saturated thickness (Figure 6.18a).

Figure 6.22 Discharge from an aquifer test conducted near Three Forks, Montana, described in Example 6.14. The weighted-average discharge was 363 gpm (22.9 L/s).

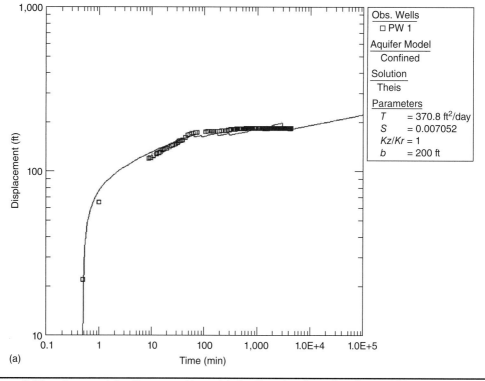

Figure 6.23a Commercial software used to match the time-drawdown data from a 10-in production well using the Theis solution. From Example 6.14.

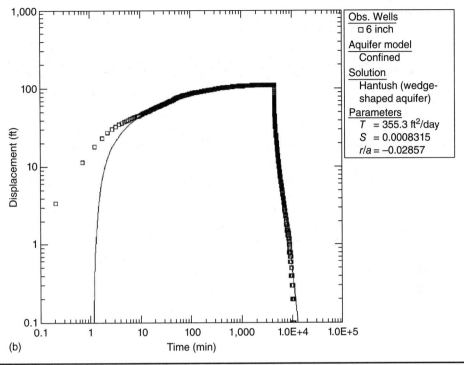

(b)

FIGURE 6.23b Commercial software used to match the time-drawdown and recovery data from a 6-in observation well using the Hantush wedge-shaped solution (Duffield 2018). From Example 6.14.

Unconfined Aquifers

Analysis of aquifer-test data in unconfined aquifers present a different set of challenges because the water-table surface is also a boundary condition. A cone-of-depression develops from pumping and the sediments within the cone change from fully saturated to partially saturated conditions from draining (Figure 6.3; Chapter 11). Jacob (1950) has shown that the Theis methods described above are applicable as long as the saturated thickness (b) is approximately 90% of the original saturated thickness. If the saturated thickness drops below 90% a correction factor must be applied to the field data before plotting the time-drawdown data for analysis. The relationship is shown in Equation 6.30.

$$(h_o - h)' = (h_o - h) - \frac{(h_o - h)^2}{2b} \tag{6.30}$$

where $(h_o - h)'$ = corrected drawdown
$(h_o - h)$ = observed drawdown
b = saturated thickness

Observation wells within the cone-of-depression, however, will show a Theis-like confined drawdown response early on, followed by a recharge response, and finally by a later-time Theis-like delayed response. This is best explained by describing the shape of the time-drawdown data as three segments. First the data follow the Theis type

A confined response from water being instantaneously released from storage. In the middle segment a recharge response is observed where the data curve below the type A curve and then flatten (somewhat like in Figure 6.20). This can be explained by water draining from sediments in the descending cone-of-depression by gravity drainage at a rate that is less than instantaneous. Finally, the aquifer becomes sufficiently stressed to follow a Theis curve once again (in this case the type B curve). The "climbing" of the drawdown data back onto the Theis type B curve at a later time is known as **delayed yield** (Boulton 1963; Neuman 1972, 1975, 1979, 1987).

Analysis of the three-segment unconfined case was pioneered by Boulton (1963, 1973) where he produced a family of type curves showing the three segments, assuming the response was due to an aquifer empirical constant (http://hydrogeologistswithoutborders.org/wordpress/1979-english/chapter-8). Additional research by Neuman et al. (Streltsova 1973; Gambolati 1976) led to solutions conforming to the three observed segments without requiring theoretical aquifer constants. Neuman (1975) provided a solution using a beta (β) parameter (a function of r from the pumping well and the "averaged" drawdown). This has been simplified here as the unconfined well function $W(u_A, u_B, \beta)$ (Equation 6.31) and the publishing of a family of type curves (Neuman 1975; Figure 6.24).

$$h_o - h = \frac{Q}{4\pi T} W(u_A u_B, \beta) \qquad (6.31)$$

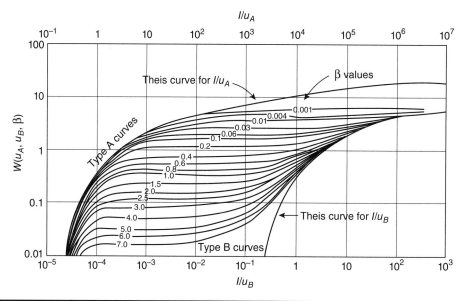

FIGURE 6.24 Family of type curves $W(u_a, u_b, \beta)$ = delayed yield responses after Neuman (1975). There are three segments to the curves: (1) an early-time confined response (type A) from instantaneous release of water from storage, (2) a recharge segment where water from gravity drainage (less than instantaneous) causes the time-drawdown data to flatten below a Theis type curve for a given β function (a function of distance r from the pumping well and the "averaged" drawdown), and (3) the aquifer becoming sufficiently stressed to "climb back onto a second Theis type curve (type B). The scale for $1/u$ for type A curves runs along the top, and the scale for the type B curves runs along the bottom. (*Used by permission from Prentice Hall Publishing, Upper Saddle River, New Jersey.*)

where the early time time-drawdown field data are fit to the type A curve and a match point is selected using the unconfined well function $W(u_A, u_B, \beta)$ and $1/u_A$ (along the top of Figure 6.24). T and S are calculated by

$$T = \frac{Q}{4\pi(h_o - h)} W(u_A, \beta) \quad \text{and} \tag{6.32}$$

$$u_A = \frac{r^2 S_y}{4Tt} \tag{6.33}$$

The late-time data are fit to the Theis type B curve and a match point is selected using the unconfined well function $W(u_A, u_B, \beta)$ and $1/u_B$ (along the bottom of Figure 6.24). T and S are calculated by:

$$T = \frac{Q}{4\pi(h_o - h)} W(u_B, \beta) \quad \text{and} \tag{6.34}$$

$$u_B = \frac{r^2 S_y}{4Tt} \tag{6.35}$$

Fitting the later-time data results in roughly a three-order of magnitude increase in the calculated value of storativity [shift of match point $W(u_A, u_B, \beta) = 1.0$ and $u_A = 1.0$ for early-time field data to match point $W(u_A, u_B, \beta) = 1.0$ and $u_B = 1.0$ for the late-time data]. This puts the results closer to the unconfined range.

The series of curves between the two Theis curves (A and B) represent β values (a function of r from the pumping well and the "averaged" drawdown). The β value of the bottom-most curve is 7.0 and the β value becomes smaller with successive curves in the upward direction toward the confined (type A curve). A curve with a β value of 7.0 indicates that recharge from gravity drainage (or an increased saturated thickness in Figure 6.18a) was detected almost immediately, while a β value of 0.3 indicates that this process occurred much later in the test. The β value is defined and discussed further in the next subsection. The response observed is a function of the geologic setting, the position of the observation well and the pumping rate.

The author suggests that a similar delayed-yield shape response can occur from at least another couple of other geologic settings. The geology of the first was described in Example 6.1. In this example, an alluvial aquifer was overlain by silty (mine tailings) materials. The hydraulic head of the alluvial aquifer is approximately 6 in (15.3 cm) above the top of the alluvial materials (or within the silty materials zone). Pumping of the alluvial aquifer results in lowering the head below the confining silty materials (see Example 6.15), thus allowing gravity drainage to occur (Figure 6.25). Another type of geologic setting is described in Section 6.6 (Fracture-Flow Analysis). Geologic materials that are fractured release water from the larger fractures more "instantaneously" and induce a stress on the matrix materials. Over time, with increased stress from pumping, the smaller fractures and matrix materials of some rocks also release water from gravity drainage, reflected as a recharge response (flattening of the data in the time-drawdown plot). Once the stress from pumping "catches up" with the recharge release a delayed-yield-like response occurs in the data with a resurgence of increased drawdown.

Example 6.15 A 22-h pumping test was conducted near Miles Crossing west of Butte, Montana (Figures 6.26a and 6.26b) in a fluvial setting impacted by mine tailings from flooding. The pumping test

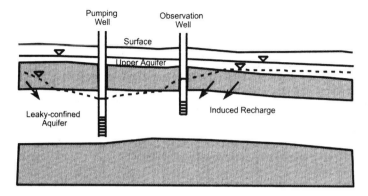

Figure 6.25 Pumping test conducted in a leaky-confined aquifer, with induced recharge through a silty layer, separating the upper and lower aquifer. Near the pumping well the cone-of-depression has dropped below the silty layer and is undergoing gravity drainage. The cone of depression is indicated by a dashed line, thus explaining a delayed-yield response in the time-drawdown data.

was conducted during the Montana Tech 1997 hydrogeology field camp. Well MT97-2 is the pumping well, and the other three are observation wells mapped to scale (Figure 6.26a). The pumping rate was approximately 4.5 gpm (17 L/min). The results of the log-log plot of the pumping well MT97-2 and observation well MT97-3 are shown in Figures 6.27a and 6.27b, respectively (responses in the other two observation wells are very similar). The direction of groundwater flow is generally from east-northeast to west-southwest (Figure 6.26a).

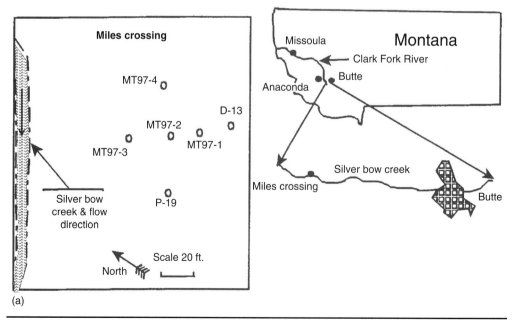

(a)

Figure 6.26a Location map for the Miles Crossing aquifer test site (Example 6.15). The streamside tailings were deposited in a fluvial setting.

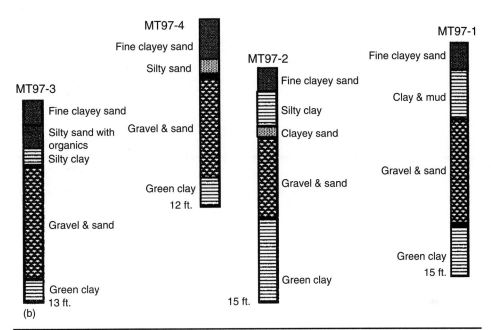

FIGURE 6.26b Oblique view of the lithologies of the four wells installed by students at the Miles Crossing research site. Well MT97-2 is the pumping well and all others are observation wells (Example 6.15).

One notes the immediate confined-like response followed by a recharge event (flattening of the curves in Figures 6.27a and 6.27b). During early time (first few minutes), the SWL is up in the finer-grained silty materials (still confined-like). This is followed by a draining of the silty materials (flat segment). Near the end of the test, it is apparent in the pumping well and observation wells that delayed

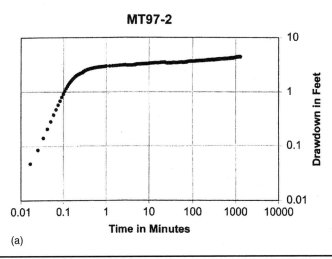

FIGURE 6.27a and b Time-drawdown plots for the pumping well MT97-2 and an observation wells, from top to bottom, respectively. Each indicate the beginnings of a delayed-yield effect at time >800 min (Example 6.15).

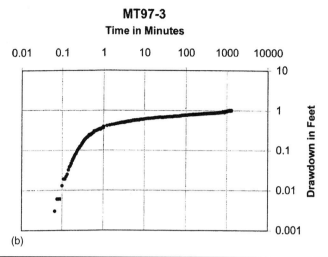

(b)

FIGURE 6.27a and b *(Continued)*

yield is just beginning at later time (>800 min) indicated by the upward trend (especially evident in Figure 6.27b).

It is interesting to see the same phenomenon with a different perspective in a Cooper-Jacob plot (Figure 6.28). Figure 6.28 represents the same time and drawdown data for observation well MT97-3, but here in semi-log form. You will recall that a Theis-like response results in a straight line in the data plot. The straight line is projected upward to zero drawdown to determine a t_o, used in the calculation of storativity S. In this case two straight lines are present (Figure 6.28). The first straight line observed represents the confined early-time response corresponding with the type A curve, followed by the second straight line observed from the delayed-yield

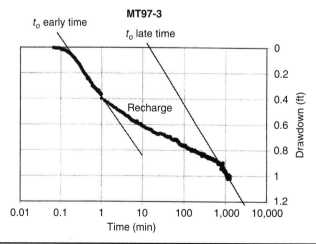

FIGURE 6.28 Cooper-Jacob plot of observation well MT97-3 in Miles Crossing pumping test (compare with Figure 6.27b). Note how delayed yield is displayed as the second straight-line slope.

response corresponding with the type B curve. You will notice that this shifts the second projected t_o to an appreciably later time. This results in a significant change in the estimate of storativity. Delayed yield did not occur until approximately 800 min. Had the pumping test been terminated at 600 min, this phenomenon would not have been observed.

The question then becomes, is this a leaky-confined aquifer or really a semiconfined aquifer with delayed yield? This question can be answered only by running the pumping test sufficiently long (2 to 3 days, depending on the geologic setting). If the semiconfining layer extends far enough away, then the recharge effect will remain flat as in Figure 6.20, while one that pinches out within the reach of the cone of depression will show delayed yield if pumped long enough.

Understanding the significance of the potential changes in storativity values is important in evaluating aquifer conditions and the proper application of distance-drawdown analysis, such as the impacts of pumping wells on neighboring wells (Table 6.4). The radial distance from the pumping well to zero drawdown or the *range of influence* happens to be a sensitive parameter depending on the storativity value.

Reporting that an aquifer is confined or unconfined when the storativity falls between confined and unconfined values suggests a lack of understanding of the aquifer system, miscalculations, or errors in interpretation (Example 6.16).

Example 6.16 The following personal story may help illustrate an example of improperly reporting storativity values. Back in the mid-1980s the author participated in writing up a report of aquifer test data results on the Eocene Wasatch Formation for a coal gasification study (Borgman et al. 1986). Dr. Shlomo P. Neuman of the University of Arizona was one of the reviewers, and his pen bled freely where I had incorrectly said that a particular unit was confined. In retrospect, the unit was semiconfined. A particular carbonaceous shale unit was confining a sandy aquifer unit where it existed; however, there were places where the shale unit pinched out and thickened once again, allowing the aquifer unit to become connected to another unconfined sandy unit above. Had the author been more experienced or aware, the data would have properly been interpreted. Interestingly enough this story has come full circle as the author had the opportunity to work with Dr. Neuman once again in 2017 on a dispute over a water-bottling plant and the interpretation of the use of one of his models in projecting impacts on senior water users (Example 2.7).

Specific Yield and Vertical Hydraulic Conductivity from Unconfined Aquifer Tests

Unconfined aquifer tests lend themselves to making additional calculations of aquifer properties. As mentioned previously, time-drawdown data have a tendency to follow a three-segment pattern including two Theis type curves (both A and B). Because of this there is an opportunity to evaluate the early-time data, the late-time data, and obtain a beta (β) value for the data. In addition to calculating transmissivity values, one can also evaluate changing values of storativity with time. Furthermore, one can use the β value to calculate specific yield and estimate the vertical hydraulic conductivity. How this is done is presented in Example 6.17.

In Chapter 3 the property of aquifer storativity is discussed. The low storativity of confined aquifers (10^{-3} to 10^{-6}) results in observing an immediate change in head with barometric pressure changes or when a pumping stress is imposed. Pumping stresses within a confined aquifer decreases the pore-water pressure (Chapter 4), resulting in increasing the compressive stresses acting on the aquifer. Therefore, during the very beginning of *any* aquifer test a confined-like response from changes in compressive stresses acting on the mineral skeleton accounts for the drawdown data following the

Theis type A curve (Weight and Wittman 1999). As gravity drainage begins during pumping of an unconfined aquifer, the middle segment (specific yield), a recharge response (flattening of the curve) is observed in the time-drawdown data plot (Example 6.17). Once gravity drainage "catches up" with the pumping stress, the time-drawdown data begin to "climb" up on the Theis type B curve once again. This means that even a very coarse-grained aquifer will display an initial confined-like response followed by a delayed-yield pattern, even if the confined response is limited to the first minute of the aquifer test (Chapter 7).

When evaluating the early time-drawdown data of an unconfined aquifer test the transmissivity and storativity are calculated using Equations 6.32 and 6.33 using the match point from the Theis type A curve (Figure 6.24). As the time-drawdown data drop below the type A curve they will follow a more flattened curve (middle segment), with a particular β value. The latest time-drawdown data should begin to follow the Theis type B curve (Figure 6.24). Equation 6.34 is used to calculate the transmissivity (T) and Equation 6.35 is used to calculate the storativity (S) using the second match point with late time-drawdown data following the Theis type B curve. Recall that at the second match point one conveniently uses the point where $W(u_B, \beta)$ and $1/u_B = 1.0$. If the aquifer is anisotropic ($K_h \neq K_v$) the relationship for β is used to obtain an estimate of the vertical hydraulic conductivity according to the following relationship (Fetter 1994):

$$\beta = \frac{r^2 K_v}{b^2 K_h} \tag{6.36}$$

where r = radial distance to the observation well
  b = saturated thickness

and K_h and K_v are the horizontal and vertical hydraulic conductivities, respectively
  Recall from Equation 3.12 that $T = K_h \times b$ and rearranging Equation 6.36 we obtain a relationship to estimate the vertical hydraulic conductivity (Equation 6.37).

$$K_v = \frac{K_h b^2 \beta}{r^2} \tag{6.37}$$

Example 6.17 A pumping test was conducted to estimate the aquifer properties for dewatering purposes at a mining operation in Florida. The relative coordinates for the pumping well (PW) and three observation wells (Obs1, Obs2, and Obs3) are shown in Table 6.5. The geologic setting is a Quaternary sand aquifer, where each well was approximately 50 ft deep and the saturated thickness averaged 43 ft. The pumping rate was 82 gpm (5.2 L/s) and the test was run for 24 h.

Well	Easting, ft.	Northing, ft.	Radial dist., ft.	Transm., ft²/day	Storativity	Specific yield	Beta
PW	307.7	315.1					
Obs1	292.3	317.9	15.6	960	0.0015	0.14	0.058
Obs2	323.0	359.7	47.2	970	0.0005	0.025	0.075
Obs3	293.4	212.3	106.6	940	0.0007	0.085	3.26

TABLE 6.5 Locations of the Pumping Wells and Observation Wells for a Pumping Test Conducted in Florida

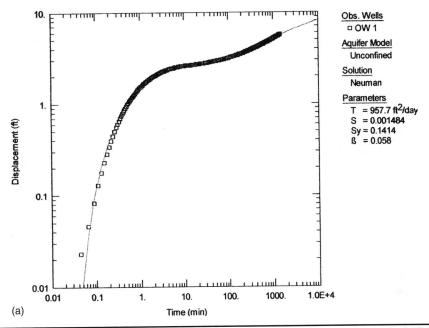

(a)

FIGURE 6.29a Florida pumping test at 82 gpm. Data are from observation well 1 located 15.6 ft (4.8 m) south (Example 6.17).

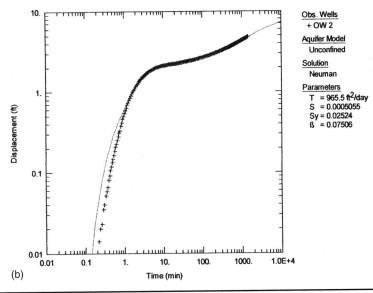

(b)

FIGURE 6.29b Florida pumping test at 82 gpm. Data are from observation well 2 located 47.2 ft (14.4 m) north-northwest (Example 6.17).

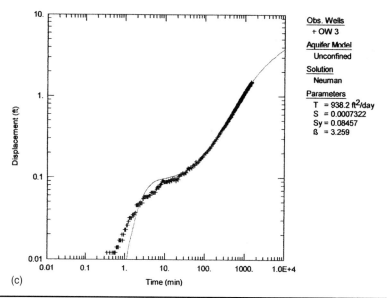

FIGURE 6.29c   Florida pumping test 82 gpm. Data are from observation well 3 located 106.6 ft (32.5 m) southeast of the pumping well (Example 6.17).

The time-drawdown data were plotted with Aqtesolv (Duffield 2018), where each curve was fit using the Neuman method as a solution (Figures 6.29a, 6.29b, and 6.29c). Equally good fits can also be obtained using the Moench (1996) solution. The results are also listed in Table 6.5.

The transmissivities are remarkably uniform and all of the time-drawdown data plots exhibit a delayed-yield response. Obs1 and Obs3 are south and southeast of the pumping well, respectively, and indicate specific yields typical of an unconfined aquifer. (Note how one uses the S_y rather than S for the aquifer properties because of its dominance.) Obs2 is located north-northeast of the pumping well and indicates a more semiconfined response, with a value at the low end of the unconfined range ($S_y = 0.025$). From the relationship of saturated thickness (b) and transmissivity (T) above (Equation 3.12) the horizontal hydraulic conductivity (K_h) is estimated to be 22 ft/day (6.7 m/day). Substituting the β values from Table 6.5 and the other known quantities into Equation 6.37 we can estimate the vertical hydraulic conductivity (K_v). An example calculation is shown for well Obs1.

$$K_v = \frac{22 \text{ ft/day} \times 43 \text{ ft}^2 \times 0.058}{(15.6 \text{ ft})^2} = 0.44 \text{ ft/day}$$

The K_v values for Obs2 and Obs3 are left for the reader. How do the K_v values for all three wells compare? Do they help one interpret why the S_y values are different and help in formulating a conceptual model?

Recovery Plots

If you are running a pumping test, it makes a lot of sense to also collect the **recovery data** once the pump is turned off. The data are collected over a period of time until full recovery or near full recovery takes place. An example of how the data appear in an arithmetic scale drawdown versus arithmetic time plot is shown in Figure 6.30. (It contains the same data as in Figure 6.23b.) Note how drawdown occurs most dramatically during the early time of the drawdown phase and also at the beginning of the recovery

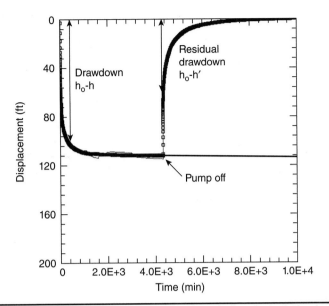

FIGURE 6.30 An arithmetic scale drawdown versus arithmetic time for both the drawdown and the recovery phases. The data are the same as in Figure 6.23b in Example 6.14. Drawdown occurs most dramatically during the early time of the drawdown phase and also at the beginning of the recovery phase.

phase. (This is why when collecting data in the field the frequency in Table 5.8 was recommended.) **Residual drawdown** $(h_o - h)$ is the amount of drawdown measured directly after the pump is turned off until full recovery occurs.

So how are the aquifer properties determined using recovery data? This type of plot is known as a recovery plot, where the x-axis (abscissa) is the ratio of time since pumping started (T) over time since pumping stopped (T') in log scale and the y-axis (ordinate) is the residual drawdown $(h_o - h')$ in arithmetic scale. This produces a relationship where the early-time data are indicated by large numbers (first points on the right-hand side) and the late-time data (data to the left-hand side) become smaller as the ratio of T/T' gets closer to 2.0 more or less. If a straight-line relationship develops, then projecting the straight line to where the residual drawdown becomes zero represents the time at which full recovery will take place. Theoretically, this should occur at the projection to where the ratio of T/T' is 2.0. In a homogenous isotropic aquifer with no recharge sources, the time it took to generate drawdown is the same time it would take to reach full recovery. Ratios less than 2.0 indicate slower recovery suggestive of poorer aquifer recharge or storage, and ratios greater than 2.0 indicate more rapid recovery, suggestive of better aquifer recharge (Example 6.19). The transmissivity can be calculated similarly to the equation used in the time-drawdown Cooper-Jacob plot (1946).

$$T = 2.3 * Q'/4\pi\Delta(h_o - h')$$ (6.38)

where Q' = weighted-average pumping rate L^3/T
$\Delta(h_o - h')$ = residual drawdown per log cycle

Q in gpm	Time pumped, min	Total, gal
450	65	29,250
430	15	6,450
415	100	41,500
390	120	46,800
375	1,300	487,500
360	300	108,000
375	1,100	412,500
330	1,328	438,240
	4,328 total minutes	**1,570,240** total gallons

TABLE 6.6 Pumping Rates and Quantities Needed to Estimate a Weighted-Average Pumping Rate

Once the pump is shut off, theoretically an imaginary injection well with an injection rate Q' representative of the weighted-average pumping rate Q used during the pumping phase will cause the aquifer to recover. If the pumping rates vary during the drawdown phase the imaginary injection well will reflect a weighted-average constant rate calculable from the data.

Example 6.18   In Example 6.14 is a description about a 72-h pumping test that was conducted near Three Forks, Montana, for a new irrigation well. The desired pumping rate was 400 gpm (25.2 L/s), but over time the pumping rate decreased. The pumping rates and minutes pumped at each rate are given in Table 6.6. It was desired that a weighted average Q' be calculated as an input parameter to evaluate the aquifer properties.

The total gallons divided by the total minutes pumped yields a weighted, average pumping rate for the imaginary recovery well.

Example 6.19   Just east of Judith Gap in central Montana a 72-h pumping and recovery test was conducted to evaluate the irrigation potential of some unmapped gravels. The pumping rate was kept constant at 425 gpm (26.8 L/s); however, after the first 200 min of the drawdown phase there was a "breakthrough" in well development and the well recovered approximately 18 ft (5.5 m) before continuing to drawdown. The recovery data resulted in a smooth unambiguous data set that was used to estimate the aquifer transmissivity (Figure 6.31) using Equation 6.38.

If for whatever reason one cannot collect all of the recovery data to complete recovery it is possible to project when full recovery would occur. This is done by projecting the later-time T/T' data to where zero residual drawdown occurs. Recall that $T =$ time since pumping started and $T' =$ time since pumping stopped. The relationship between the time for projected full recovery can be expressed as Equation 6.39.

$$T/T' = \frac{(PT + X)}{X} \qquad (6.39)$$

where $PT =$ time during which pumping occurred, where $PT + X$ represents the total time since pumping started
$X =$ time required for full recovery

For example, suppose the projected $T/T' = 1.33$ with $PT = 24$ h. Solving for X one obtains $X = 72$ h. This suggests that it would take three times as long to fully recover for

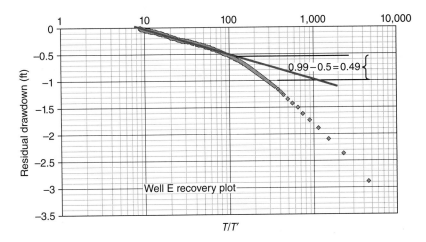

Figure 6.31   Recovery plot of T/T' (time since pumping started divided by time since pumping stopped) versus residual drawdown ($h_o - h'$) for a pumping test conducted near Judith Gap, Montana (central Montana). The pumping rate was 425 gpm (26.8 L/s) and the $\Delta h_o - h'$/log cycle is shown. Equation 6.38 is used to estimate the transmissivity T (Example 6.19).

a given pumping time PT. This is does *not* bode well for making a case that there is plenty of water.

The time it takes for recovery becomes extremely important if you are evaluating pumping and recovery scenario impacts. Suppose, for example, that your distance-drawdown plot shows impact to other senior water users, wells. It may be that by pumping and then letting the well recover before pumping again would provide a reasonable compromise to everyone's water-use needs. In the example just described it may be that pumping could occur for 24 h if the waiting time were an additional 72 h. The author's general experience is that many unconsolidated aquifers take 1½ *times* the pumping rate time to fully recover and that fractured bedrock aquifers may take up to *four times* the pumping rate time to fully recover unless there are additional sources of recharge. In Figure 6.31 full recovery occurred at $T/T' = 8.5$ suggesting that recharge recovery is very good!

Clearly, more observation wells in different azimuth orientations and varying distances from the pumping well would improve the understanding of the "true" distance-drawdown effect (Figure 5.5). A comparative plot of drawdown versus distance can be made from several different observation wells during a pumping test to see if a straight-line plot results. If this is true, it indicates that the cone of depression is symmetrical around the pumping well. If this is not true, there is likely some anisotropy to the shape of the cone (Figure 5.7). This is one reason data should be collected from all of the observation wells to provide a suite of recovery values for interpretation.

There are a number of errors that can be made in evaluating recovery data. The common one is the tendency for newcomers to recovery data sets to simply plot the recovery (residual drawdown) versus time in minutes similar to on a regular Cooper-Jacob (1946) plot (Figure 6.10). This does not work out. Instead the *residual drawdown* versus ratio of T/T' needs to be plotted.

Once the pump and generator are shut off, the data logger should be activated or "stepped" to the next data collection phase (Chapter 5) to capture the recovery data. This can be simplified if a linear time scale is selected in a well sentinel and established at times before the test began and continues until the postrecovery time period. If recovery data are being collected manually, a frequency of data collection similar to Table 5.8 should take place. As one might expect the early-time recovery data change quickly at first and then slow down over time (Figure 6.30). Practically speaking one must be careful to make sure that there is a check valve in the pump *or* that water from the discharge line and riser pipe will drain back into the well turning the impellers backward and giving a "false" pulse of recharge.

6.4   Other Analysis Considerations

Derivative Plots

In addition to time-drawdown and recovery plots it is also useful to consider the derivative of pressure versus the natural logarithm of time (Spane and Wurstner 1993). In this type of analysis the extra sensitivity of pressure derivatives can greatly assist with the recognition of other subtle occurrences in the data such as wellbore storage (next section), well inefficiencies (Chapter 5), boundaries (Section 6.4), and when radial flow has been established (Section 6.4). This type of analysis has been widely applied in the petroleum industry (Bourdet et al. 1983; Ehlig-Economides 1988); however, its applications in aquifer testing are increasing (Renard et al. 2009; Duffield 2018).

The concept of radial flow simply means that flow to the well has gotten past well-bore-storage and formation inefficiencies and is representing the infinite aquifer case. There are characteristic shapes in the pressure-derivative plots corresponding to confined, leaky-confined, and unconfined aquifers (Figure 6.32). Notice in Figure 6.32 the characteristic "wellbore-storage hump" in the early-time pressure-derivative data (Spane and Wurstner 1993) and that radial flow occurs whenever the pressure-derivative data trend horizontal. The other responses in the confined setting reflect boundary conditions (Figure 6.32). In the unconfined condition with the three-segment delayed yield curve (described in the previous section) there is a U-shape until the second Theis curve, where the data trend horizontal once again. An example of a horizontal radial-flow infinite-aquifer response is shown in Figure 6.33 using the data from Table 6.3 (wellbore storage is not evident). The pressure-derivative data curve over and trend horizontal while the time-drawdown data follow the Theis curve. The reader is encouraged to refer to the articles mentioned or visit the Aqtesolv website (Duffield 2018) for additional information and references. The author recommends using professional software to include this additional type of analysis.

Example 6.20   A controversial water-bottling case (Example 2.7) where local residents using water in an unconfined aquifer system were worried about impacts from a water-bottling production well. The production well was touted as being completed in a deep confined aquifer but the behavior of the time-drawdown and pressure-derivative data appears to be closer to leaky-confined (Figure 6.34). The pressure derivative (diamond pattern) becomes somewhat scattered without a smoothing algorithm (Renard et al. 2009; Duffield 2018). The dip in the scattered pressure-derivative area is somewhat unconfined-like and the time-drawdown data (boxes) curve over like a leaky-confined aquifer that receives leakance from storage (Figure 6.21).

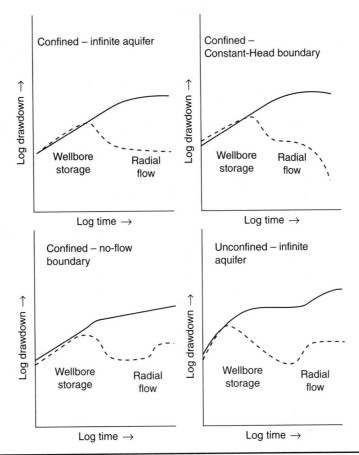

Figure 6.32 Examples of pressure-derivative plots versus natural log of time. The pressure derivatives can greatly assist with the recognition of other subtle occurrences in the data, such as wellbore storage, boundaries, and when radial flow has been established (horizontal conditions).

Single-Well Tests

Sometimes physical circumstances do not allow conducting pumping tests with observation wells. One reason may be that the aquifer is very deep (Belcher et al. 2001). In irrigation settings, such as in the Ogallala Formation of Kansas and Nebraska, wells are deliberately spaced far apart to reduce overlapping cones of depression. In this case neighboring wells do not make good observation wells.

In a study by Halford et al. (2006), 628 radial MODFLOW (McDonald and Harbaugh 1988) simulations of single-well pumping tests were conducted to see how well choices by analysts, using the Cooper-Jacob (1946) method compared with known transmissivities. Hydraulic properties were limited to plausible ranges. Transmissivities ranged from 10 to 10,000 m^2/day. Transmissivities of less than 10 m^2/day were excluded because here slug tests are more practical than pumping tests. A single specific storage

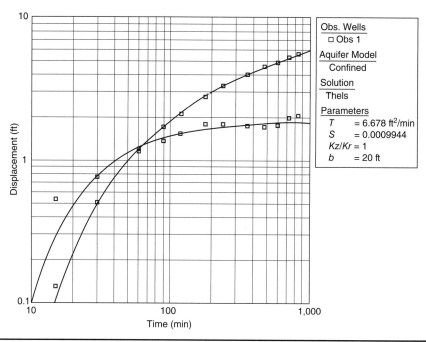

Figure 6.33 Example of infinite aquifer case with horizontal radial flow in the pressure-derivative data, while the time drawdown data follow the Theis curve. The data are from Table 6.3 and wellbore storage is not evident (description of the data is in Example 6.8).

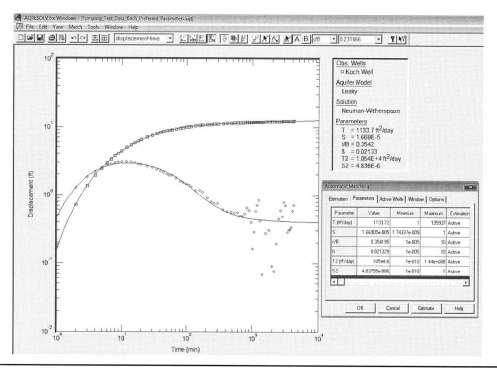

Figure 6.34 Commercial software plot of time-drawdown (upper plot, with boxes) showing a Hantush leaky-aquifer response (Figure 6.21) and pressure derivatives (diamond pattern) that become scattered and dips at about 1,000 min (compare with Figure 6.32 unconfined case) (Example 6.20).

of 5×10^{-6} m^{-1} was assigned to all aquifers because transmissivity estimates were insensitive to specific storage. Specific yields ranged from 0.05 to 0.3. Vertical anisotropies ranged between 0.02 and 0.2 and were assumed to represent a sedimentary system. Partial penetration (Section 6.5) effects in confined and unconfined aquifers were also included with penetrations ranging from 10% to 100% of the aquifer. All simulations were conducted for 2 days.

Pumping wells were simulated in column 1 and aquifer material was simulated with columns 2 through 99. Column 2 was 0.02-m wide and the remaining columns were 1.15 times the width of the previous column. The 100-m thick aquifers were uniformly subdivided into 100 rows. All models had no-flow lateral boundaries, initial heads of 0 m, and a single stress period of 50 time steps. The first time step was 0.1 s in duration. Each successive time step was 1.3 times greater than the previous one.

Cooper-Jacob transmissivity was estimated for each single-well test by a mechanistic approach (set up in an Excel spreadsheet) and separately by six analysts, including the three authors and three volunteers. A semi-log slope was defined by the drawdowns at 0.5 and 2 days after pumping started. A minimum time of 0.5 day was selected as a compromise between avoiding the early-time complications due to well-bore storage, partial penetration, and water-table effects and assumed measurement sensitivity for actual aquifer tests. Experience guided analysts' best fit of semi-log slopes were performed.

Cooper-Jacob (1946) transmissivity estimates in confined aquifers were affected minimally by partial penetration, vertical anisotropy, and interpretative technique. Transmissivities of 100 m^2/day or greater were estimated within 10% of their value in all but one case (Halford et al. 2006). A steady additional drawdown from partial penetration and vertical anisotropy was established before 12 h of pumping had elapsed. Confined aquifer test results were unambiguous and transmissivity estimates varied little among analysts.

Transmissivities of unconfined aquifers were overestimated with a mechanistic application of Cooper-Jacob (1946). More than 75% of known transmissivity values between 10 and 1,000 m^2/day were overestimated. Estimates averaged twice known transmissivity values in this range. About 80% of known transmissivity values between 1,000 and 10,000 m^2/day were estimated within a factor of two (Halford et al. 2006). Transmissivity estimates were not improved by interpreting results with an unconfined analytical solution instead of using the Cooper-Jacob (1946) method.

Hydraulic conductivity of confined aquifers was unambiguously determined as the transmissivity estimate divided by aquifer thickness, rather than the screen length, whenever transmissivity was 100 m^2/day or greater. At low transmissivity values (<10 m^2/day) hydraulic conductivity estimates using the screen length were believed to be appropriate. Hydraulic conductivity estimates ranged from 1.6 to 8 times known values where transmissivity estimates were divided by screen length (see Section 6.5).

Casing Storage

Schafer (1978) pointed out that early-time pumping-test data may lead to an erroneous interpretation of transmissivity. If the casing is of large diameter and the pumping rate is small, most of the water will be coming from the casing (emptying the bucket, so to speak). For example, pumping an 8-in (20.3-cm) well at a rate of 5 gpm (18.9 L/min) will not result in much stress of the aquifer or create much drawdown. In a single-well test, this can be especially important. A plot of the time-drawdown data may indicate a relatively steep-sloped pattern during the first 10 or 20 min (emptying the casing, Figure 6.32), followed by a flattening of the slope as water from the aquifer begins to come in through the screened interval. Fitting a slope to the early-time data to calculate transmissivity will result in under-estimating the appropriate value. Schafer (1978) suggested that a critical time (t_c) be calculated to indicate when casing storage is no longer contributing to the yield of a well. The t_c can be calculated by using Equation 6.40, sourced from Driscoll (1986):

$$t_c = \frac{0.6(d_c^2 - d_p^2)}{Q/s} \qquad\qquad t_c = \frac{0.017(d_c^2 - d_p^2)}{Q/s} \qquad (6.40)$$

where t_c = time in minutes, when casing
    storage is negligible
d_c = inside diameter, in inches, of
    well casing
d_p = outside diameter, in inches, of
    pump riser pipe
Q/s = specific capacity (next section)
    of the well in gpm/ft at time t_c

where t_c = time in minutes when casing
    storage is negligible
d_c = inside diameter, in
    millimeters, of well casing
d_p = outside diameter, in
    millimeter, of pump riser pipe
Q/s = specific capacity (next section)
    of the well in m³/day/m at
    time t_c

Being able to determine when t_c has occurred becomes an iterative process, because Equation 6.40 assumes that the drawdown at t_c is known. To begin one must start by selecting a given drawdown and inserting this into Equation 6.40 to calculate t_c in an iterative fashion. The real drawdown at the calculated t_c is used to recalculate a new t_c. If the new t_c does not change much from the previous value, then the "true" t_c has been determined.

Example 6.21   A pumping test was conducted at 6 gpm in a 6-in (15.2-cm) domestic well near Yellowstone National Park. The time-drawdown data are shown in Figure 6.35. The inside diameter of the casing is 6 in and the outside diameter of the riser pipe was 1.5 in (3.8 cm). An iterative method to determine time critical was used to determine when the effects of casing storage were negligible.

The first guess for t_c was at a drawdown of 7 ft. Using Equation 6.40:

$t_c = \dfrac{0.6(6^2 - 1.5^2)}{6/7} = 23.6$ mm. The drawdown at 23.6 min is 7.6 ft. We substitute again this drawdown into Equation 6.40 and recalculate t_c.

$t_c = \dfrac{0.6(6^2 - 1.5^2)}{6/7.6} = 25.7$ mm. The process is repeated again and we obtain 26.1 min, which is a small change. The data before 26 min are not used in calculating the transmissivity.

6.5 Specific Capacity

One "first cut" approach to estimating transmissivity from drill logs is to use the **specific capacity** (S_c) relationship. Specific capacity is defined as:

$$S_c = \frac{Q}{h_o - h} \qquad (6.41)$$

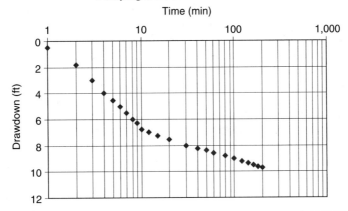

Figure 6.35 Time-drawdown data in semi-log from a pumping test conducted near Yellowstone National Park. The steeper straight line in the early-time data represents influences from wellbore storage (Example 6.21).

where Q = gpm

$h_o - h$ = drawdown in feet after at least 1-h pumping

From the specific capacity equation, a rough rule of thumb derived from distance-drawdown relationships yields the following relationship for transmissivity (Driscoll 1986).
For confined aquifers:

$$T = S_c \times 2,000 \tag{6.42}$$

For unconfined aquifers:

$$T = S_c \times 1,500 \tag{6.43}$$

where T is in gal/day-ft (day times feet).

A specific capacity approach applied to fractured bedrock aquifers was developed experimentally by Huntley and Steffey (1992), yielding the following relationship for bedrock aquifers:

$$T = 38.9 \, (S_c)^{1.18} \tag{6.44}$$

where T is in ft²/day.

It should be noted that the units of the numerical factors 2,000, 1,500, and 38.9 in Equations 6.42, 6.43, and 6.44, respectively, convert the transmissivities into the specified units. For example, suppose that the lithologies in the well log from a well drilled 300 ft (91.4 m) deep indicate several clay horizons above the production zone, suggesting that the aquifer may be confined. The production zone is screened over a 25-ft (7.6-m) thick sandstone. Suppose that the SWL is at 20 ft (6.1 m) below ground surface. Through bailing, the production rate appears to be about 50 gal/min. Suppose further, that the measured drawdown is approximately 24 ft (7.3 m) from static conditions. This yields a specific capacity of 50 gpm/24 ft = 2.08 gpm/ft. The transmissivity would be estimated to be 4,200 gal/day-ft (560 ft²/day) using Equation 6.42.

Something should be said about the drawdown reported in well logs by drillers. On many report forms the method of production is usually listed as pump, bailer, or air. When a pump is used the confidence in the pumping rate and drawdown is good. Estimates of production rates and drawdown using bailers are also pretty good. However, one should know what production by air means. The driller will first report the SWL, such as at 20 ft (6.1 m) below the surface. To produce water by air using a rotary rig (Chapter 15) means that the drill pipe is lowered to a depth near the screened interval, with the drill bit removed, and then injected with air to lift the water to the surface. The quantity of water discharging at the surface is estimated to be the production rate. In this case the reported drawdown then is the difference between the SWL and the depth to the end of the drill pipe. This leads to very low and strange specific capacity values. If all the wells drilled in a certain area are developed and tested in this way a relative transmissivity can be obtained. An additional comment is that Equations 6.42, 6.43, and 6.44 traditionally assumed that the measured drawdown associated with specific capacity estimates was after 24 h of pumping (Driscoll 1986), when in most cases the development time for most domestic wells is 1 to 2 h. Jim Butler (personal communication, 2007) found that relatively short developmental times still yield reasonable estimates for specific capacity. In either case, having a "first approximation" estimate for transmissivity is useful to compare with other analysis methods.

In general, the usefulness of the specific capacity can be summarized as follows:

- Gives a rough estimate of T (even with lousy data)
- Data usually available online (free)
- Can be applied to confined, unconfined, or fractured bedrock aquifers
- Gives a rough idea of well completion (should the well yield more water than it is getting?)

If the specific capacity value is grossly smaller than expected, the following reasons could explain the results:

- Poorly completed, not screened in the right zone, or not screened at all.
- Well screen is clogged, either poor well development, precipitation of minerals in the screen, or biofouling.
- All well development was performed by air.
- Contractor fraud? It may be that if there is a pattern to poorly completed wells in an area where specific capacity values are noticeably higher in adjacent wells. If the poor wells were all completed by the same contractor, you could be suspicious of fraud.

Equation 6.44 was used in Example 6.11 to perform the distance-drawdown relationship needed for the public meeting.

6.6   Well Interference and Boundary Conditions

Suppose there are two or more wells whose cones-of-depression overlap. In this case if two or more wells are spaced closely enough the drawdown from pumping one well will result in drawdown in the other adjacent well (Figure 6.36a). How does one estimate

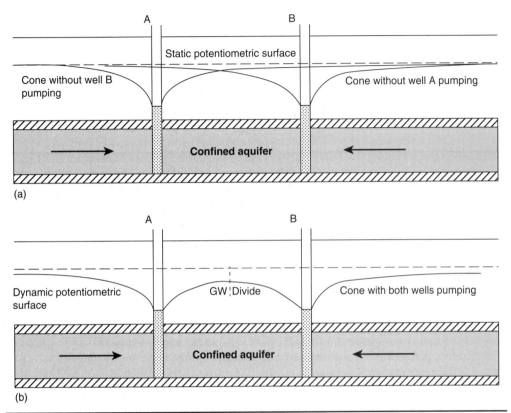

FIGURE 6.36 (a) Case showing two cones-of-depression without the influence of the other well. In this case the cones-of-depression overlap. (b) The influence of the drawdowns is additive so this case represents the added effects of both pumping wells into a single potentiometric surface. [*Modified after Heath (1983)*.]

the impacts? The net effect is the *sum of the drawdowns* from all of the wells (Figure 6.36b) at a given point based upon the principle of superposition. Superposition comes from the fact that linear systems of equations are added together to form composite solutions (Reilly et al. 1984). For example, if the natural drawdown at the groundwater divide in Figure 6.36b is 3 ft from the pumping of well A and 2.5 ft from the pumping of well B, the total drawdown at the groundwater divide will be 5.5 ft from the influence of both wells.

This can be determined either graphically, using distance-drawdown methods, or mathematically, using Equation 6.2 or 6.11. Equation 6.2 can be used anytime, whereas Equation 6.11 is appropriately used only if u is small (i.e., <0.02). It is the author's opinion that using Equation 6.2 after using Equation 6.3 to obtain a value for "u", and looking up a value for W_u in Appendix D is the easiest way.

Drawdown can be positive (downward, $h_o - h$) or negative (with groundwater mounding, $-h_o - h$). When there are injection or recharge wells (the pumping term becomes negative, $-Q$), instead of creating a cone of depression, there is a **cone of impression**. A cone of impression can be thought of as an upside-down cone of depression, creating a mound in the potentiometric surface at the location of the injection well

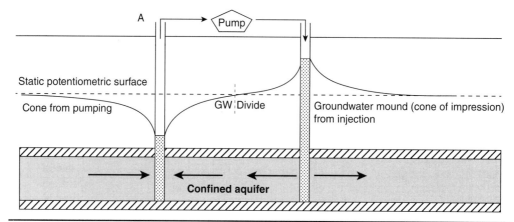

FIGURE 6.37 Schematic of a cone of depression and a cone of impression (showing mounding). [*Modified after Heath (1983).*]

(Figure 6.37). The net drawdown effect of pumping and injection wells can be evaluated independently and then added together at a particular point of interest (Example 6.22).

Example 6.22 Suppose we wish to evaluate the net drawdown at point A from the effect of three wells. Two of these wells are pumping wells, and one of the wells is an injection well (Figure 6.38). The pumping rates (500 and 600 gpm) and injection rate (-450 gpm) in Figure 6.38 are shown, and the duration of pumping is 1 year (365 days). The transmissivity is 5,000 ft²/day, and the storage coefficient is 0.20. The steps are to calculate u with the different r and t values and then look up a corresponding value for $W_{(u)}$ in Appendix D and apply Equation 6.2. The author suggests setting up a table similar to Table 6.7.

The net value is about 10.0 ft of drawdown. The values for W_u are given in Appendix D. A sample calculation for the first row in Table 6.7 follows:

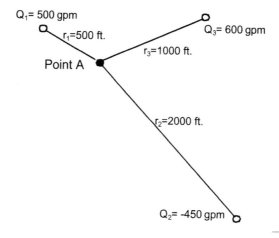

FIGURE 6.38 Schematic for Example 6.22. Two wells are pumping (+Q), and one well is injecting (−Q).

Well	r (ft)	u	$W_{(u)}$	Q gpm	Drawdown (DD, ft)
1	500	0.00685	4.413	500	6.76
2					
3					
				Net DD	9.97 (10 ft)

TABLE 6.7 Parameters Associated with Drawdown Calculations

$$h_o - h = \frac{Q}{4\pi T} \times W_u$$

$$= \frac{500\,\text{gal/min}}{4\pi(5,000\,\text{ft}^2/\text{day})} \times \left(\frac{1\,\text{ft}^3}{7.48\,\text{gal}} \times \frac{1,440\,\text{min}}{1\,\text{day}}\right) 4.413 = 6.76\,\text{ft}$$

Graphically, a distance-drawdown graph would be created separately for each pumping or injection well in Example 6.22 (Figure 6.39). The injection well would be treated the same way as the pumping well except that the drawdown would be negative. The drawdowns (plus or minus) at the respective distance from the point of interest (point A) would be picked from the distance-drawdown graph and then summed up.

In the case of well 1 in Example 6.21 one can use Equation 6.24 and the given T and S to calculate the radius of influence (r_0). The drawdown at a distance of 1 ft (0.3 m) would represent the position of a given pumping or injection well. With $r = 1$ ft, calculate "u" using Equation 6.3 and look up the corresponding value for $W_{(u)}$ from Appendix D. Finally, calculate the drawdown at a distance of $r = 1$ ft using Equation 6.2. A graph is made at $r = 1$ ft and at r_0 using the two drawdowns (Figure 6.39). This process is repeated for every pumping or injection well. The results should be pretty much the same.

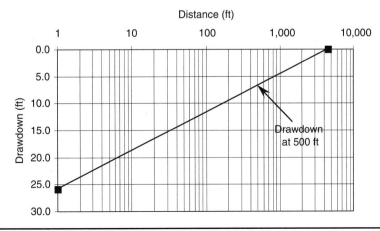

FIGURE 6.39 Distance-drawdown graphic for well 1 in Example 6.22, where $Q = 500$ gpm, $r_0 = 4,531$ ft using Equation 6.27, and $u = 2.74\text{E-08}$, $W_u = 16.84$ and the drawdown at $r = 1.0$ ft is 25.8 ft. Compare the drawdown at 500 ft with the calculated value in Table 6.7.

This brings up a point regarding the power of a numerical groundwater model because the drawdown is determined not only at a single point A, but at every cell in a grid. If the aquifer is not of infinite aerial extent, boundary conditions must also be considered.

Aquifer Boundary Conditions

Up to this point, we have been considering the aquifer to have infinite areal extent (radial flow in a pressure-derivative plot; Section 6.3). This means that the cone of depression continues outward forever. The cone of depression will stop if it:

- becomes large enough to capture enough water demanded by the pumping rate from recharge and leakage, thus reaching steady-state conditions, or
- reaches a hydrogeological recharge boundary.

In reality, most hydrogeologic settings rarely have aquifers that are of infinite areal extent; rather they change laterally in grain size, shape, or lithology (heterogeneity; Chapter 3). These changes affect the shape of a time-drawdown curve in characteristic ways. Some boundaries that are encountered are less permeable than the aquifer unit being pumped. These are known as barrier boundaries. Time-drawdown data are affected by a *steepening* of the straight-line plot in the Cooper-Jacob (1946) method and a rising of the drawdown data *above* the reverse Theis (1935) curve (Figure 6.40a). How this is simulated is through image-well theory (next section, where real wells and image wells produce different drawdown influences). Examples of **barrier boundaries** include:

- Decrease in aquifer thickness from thinning or erosion
- Decrease in aquifer permeability from decreasing grain size (e.g., facies change; Chapter 2)
- Encountering a fault plane, bedrock contact, or some other barrier
- Startup of a nearby pumping well, or the effect of another nearby pumping well

Some cases occur where the boundaries encountered are more permeable than the aquifer unit being pumped. These are known as recharge boundaries. The time-drawdown data are affected by a flattening of the straight-line plot in the Cooper-Jacob (1946) method and the drawdown data dropping below the reverse Theis (1935) type curve (Figure 6.40b). Examples of **recharge boundaries** include:

- Increase in aquifer thickness (thus an increase in T)
- Increase in aquifer permeability from increasing grain size (e.g., facies change)
- Encountering a recharge source, such as a lake, stream, or gravel channel
- Shutdown of a nearby pumping well or the effect of another nearby pumping well shutting off
- Leakance from adjacent aquifers

The effects of boundaries are usually seen when the drawdown is plotted, either by the Theis (1935) or the Cooper-Jacob (1946) method (Figures 6.24 and 6.25). Notice how in a barrier boundary the drawdown increases with time, and in a recharge boundary the drawdown decreases with time. The way boundaries are accounted for in drawdown calculations is by applying both well-interference and **image-well theory**.

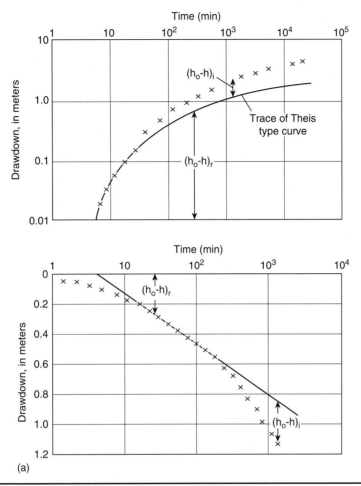

(a)

Figure 6.40a Time-drawdown data are affected by rising of the drawdown data *above* the reverse Theis curve and a *steepening* of the straight-line plot in the Cooper-Jacob (1946) method (Heath 1983).

Image-Well Theory

The concept of superposition also applies to boundary conditions within or near the pumping well. If a boundary exists, the observation well closest to the boundary will pick up the change reflected in the time/drawdown data first. To simulate the net effect of the pumping well *and* the influence of the boundary on drawdown (positive or negative), its effects are simulated by use of an image well. Perhaps it would be better to explain with an example. Suppose a pumping well begins pumping. After 100 min, the cone of depression encounters a bedrock barrier boundary. The effect on a time-drawdown plot is a steepening of the slope in a Cooper-Jacob plot (Figure 6.40a) because the aquifer has become depleted in that direction; thus, the drawdown is now increasing more rapidly with time. To simulate this effect, an *imaginary well* is placed at an equal distance on the other side of the boundary, at the same distance the pumping well is to the boundary, and at the same pumping rate (Figure 6.41) (this is like a mirror image). Since the boundary is a barrier boundary, the imaginary well is also acting as a

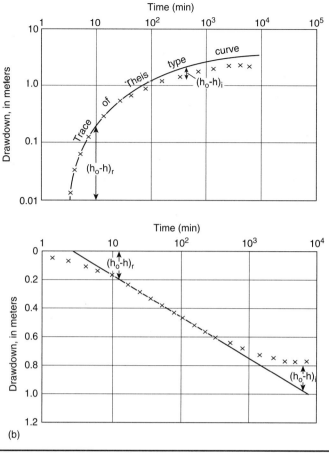

FIGURE 6.40b Time-drawdown data are affected by the drawdown data dropping below the reverse Theis (1935) type curve and a flattening of the straight-line plot in the Cooper-Jacob (1946) method (Heath 1983).

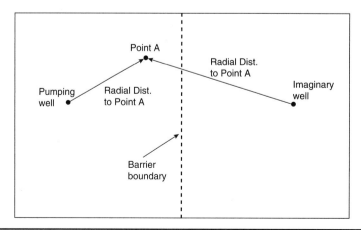

FIGURE 6.41 Schematic of pumping well and associated image well and distance from a barrier boundary (note the distances to the barrier boundary are equal). Also shown are the respective distances "r" to a point of evaluation of drawdown (Point A).

pumping well (barrier boundaries result in the pumping rates of the image wells having the *same* sign as the *real* well) and since the influences of their respective cones of depression overlap, from superposition the drawdowns ($h_o - h_r + h_o - h_i$) are added (Figure 6.42a). The drawdown effects at the boundary, or anywhere within the radius of influence of the imaginary well, are added together with the effects of the real pumping well according to the principles explained previously.

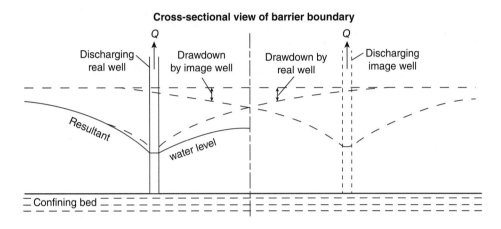

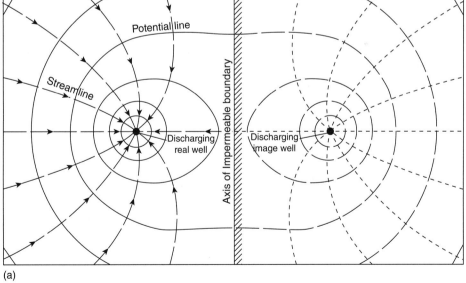

(a)

Figure 6.42a Cross-sectional view (upper part) showing what happened when the cone-of-influence from a pumping well reaches a barrier boundary. Its effect is simulated by placing an image well equidistant from the barrier (on the other side) and the pumping rate of the image well has the *same* sign, thus creating its cone of depression. The water level shows the added effect of both wells. Shown below is how the two wells (real and imaginary) are simulated in plan view (Heath 1983).

If the boundary is a recharge boundary, the image wells are located in the same way except that the "sign" of pumping for the imaginary well is *opposite* that of the real well. The drawdown from pumping becomes negated by the recharge effect from the imaginary well with the opposite sign (Figures 6.37 and 6.42b). If two boundaries are encountered, the resulting effect would be to have an image well for each boundary. You would likely have to have more than one observation well during a pumping test to "see" two boundaries in the time-drawdown data.

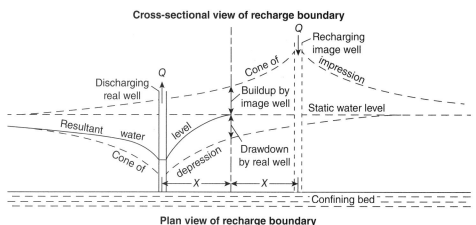

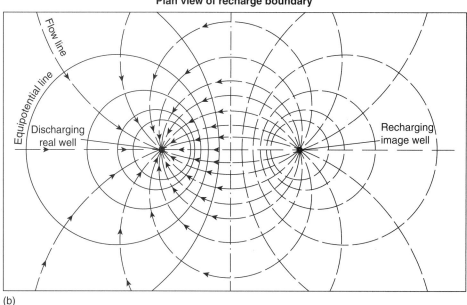

(b)

FIGURE 6.42b   Cross-sectional view (upper part) showing what happened when the cone-of-influence from a pumping well reaches a recharge boundary. Its effect is simulated by placing an image well equidistant from the barrier (on the other side) and the pumping rate has the *opposite* sign, thus creating its cone of impression. The resultant water level shows the added effect of both wells. Shown below is how the two wells are simulated in plan view (Heath 1983).

6.7 Partial Penetration of Wells and Estimates of Saturated Thickness

Wells that fully penetrate the aquifer and are fully screened have radial (horizontal) flow toward the wellbore during pumping. If the well only partially penetrates the aquifer, the flow paths have a vertical component to them (Figure 6.43). The flow paths are, therefore, longer and converge on a shorter well screen, resulting in increased head losses and decreased well efficiency.

Many times it is not possible or desirable to design a well that fully penetrates an aquifer. This is particularly true in an unconfined aquifer that may be hundreds of meters thick (Figure 6.43); hence, these wells are partially penetrating. In the Helena, Montana, Valley, many domestic wells only have an "open-hole" completion or have a few circular saw slots in PVC casing or torch-cut slots in steel casing while partially penetrating in an aquifer that may be hundreds of feet (meters) thick. These limited openings provide the only access to the aquifer to retrieve water. In an unconfined aquifer, it is practical if possible to place screen in the lower one-third of the aquifer to allow maximum drawdown in the production zone. Observation wells for pumping tests should be placed far enough away from the pumping well to avoid partial penetration effects.

When is an observation well in a pumping test far enough away from the pumping well to avoid the effects of partial penetration? Hantush (1964) indicates that if the observation well fully penetrates the aquifer *or* if it is partially penetrating and meets the minimum requirements of Equation 6.45 for minimum distance ($Dist_{min}$) away from the pumped well, the effects are negligible.

$$\text{Dist}_{min} = 1.5b \, (K_h/K_v)^{0.5} \tag{6.45}$$

where b is the saturated thickness, and K_h and K_v are the horizontal and vertical hydraulic conductivities, respectively (discussed earlier).

The vertical hydraulic conductivity may be an order of magnitude or less than the horizontal hydraulic conductivity (Example 6.23).

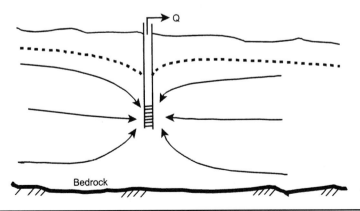

FIGURE 6.43 Schematic of partial penetration effects illustrating horizontal and vertical flow components.

Example 6.23 Layered sedimentary units may be somewhat homogenous within layers, with a production screen that spans over more the one layer. In this example we consider four sedimentary layers that contribute to a production well. Each layer has a known horizontal hydraulic conductivity (K_h) and the vertical hydraulic conductivity (K_v) is assumed to be a factor of 15 less than K_h. In this case the averaged horizontal (K_{hAve}) and vertical (K_{vAve}) hydraulic conductivities are computed using Equations 6.46 and 6.47.

$$K_{hAve} = \sum_{i=1}^{n} \frac{K_{hi} \times b_i}{b} \tag{6.46}$$

and

$$K_{vAve} = \frac{b}{\displaystyle\sum_{i=1}^{n} \frac{b_i}{K_{vi}}} \tag{6.47}$$

where K_{hi} = the horizontal hydraulic conductivity of the ith layer (length/time)
b_i = the thickness of the ith layer (length)
K_{vi} = vertical hydraulic conductivity of the ith layer (length/time)
n = number of layers
b = total aquifer thickness (thickness)

The described four-layer scenario is depicted in Figure 6.44. The four contributing sedimentary layers from top to bottom are Unit A, limestone, 8 m thick, with K_h = 5 m/day; Unit B, siltstone, 15 m thick, with K_h = 0.5 m/day; Unit C, medium-grained sandstone, 5 m thick, with K_h = 10 m/day; and Unit D, conglomerate, 3 m thick, with K_h = 65 m/day.

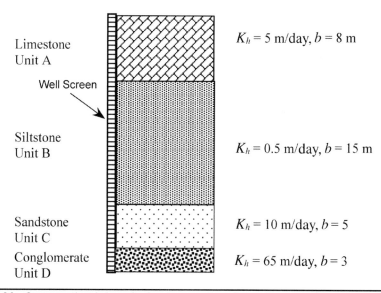

FIGURE 6.44 Cross-sectional view of a four-layer sedimentary unit spanned by a production well screen (Example 6.23).

Equation 6.46 is used to estimate the average horizontal hydraulic conductivity:

$$K_{hAve}$$

$$= \left[\frac{(5\,\text{m/day}) \times 8\,\text{m} + (0.5\,\text{m/day}) \times 15\,\text{m} + (10\,\text{m/day}) \times 5\,\text{m} + (65\,\text{m/day}) \times 3\,\text{m}}{31\,\text{m}} \right]$$

$$= 9.4\,\text{m/day}$$

One can observe that the relative weighting from each unit is 13.7%, 2.6%, 17%, and 66.7%, respectively. Notice how Unit D contributes two-thirds, even though it is the thinnest of all the units. The averaged vertical hydraulic conductivity is found using Equation 6.47:

$$K_{vAve}$$

$$= \left[\frac{31\,\text{m}}{(8\,\text{m}/5\,\text{m/day}) + (15\,\text{m}/0.5\,\text{m/day}) + (5\,\text{m}/0.5\,\text{m/day}) + (3\,\text{m}/65\,\text{m/day})} \right]$$

$$= 1\,\text{m/day}$$

Estimates of the Contributing Saturated Thickness in Unconfined Aquifers

In the preceding discussion, there is a big question about the saturated thickness b, especially for unconfined aquifers. In this section we explore the practical applications of estimating b for confined and unconfined aquifers and explore an appropriate method of assignment of the horizontal hydraulic conductivity K. A common characterization method for estimating horizontal hydraulic conductivities is by deriving transmissivity estimates from pumping tests and then dividing by the saturated thickness b. There is a tendency in practice to assume that the saturated thickness is equal to the screen length, particularly with engineers seeking approval of a subdivision. This is appropriate if the screen fully penetrates the aquifer and horizontal flow prevails. However, this ideal situation is not often accomplished because most pumping wells are completed with partially penetrating screen settings or "open holes," where an open hole is just that: a non-capped casing with no perforations. Groundwater flowing toward the screened interval or open hole has vertical components, resulting in an effective saturated thickness that can differ significantly from the screened (or open hole) interval. Depending upon the length of time of pumping and the existing geologic conditions, horizontal hydraulic conductivities can be significantly overestimated. Rules of thumb for estimating "effective saturated thickness" were proposed as a function of pumping time (Figure 6.45) and the characterization of vertically heterogeneous geology (Figure 6.46) in the first edition of this book (Weight and Sonderegger 2001). The two general principles are presented once again followed by results from a numerical analysis of the problem.

Time

When a pumping test is being conducted in relatively homogenous sediments and time is short (to be defined later), the contributing saturated thickness b' is approximately the screen length (L) times a multiplication factor (MF) (Figure 6.45). When the pumping

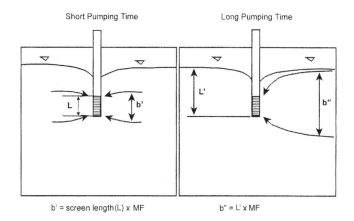

Short Pumping Time   Long Pumping Time

b' = screen length(L) x MF   b" = L' x MF

FIGURE 6.45 Contributing thickness of an aquifer depending on time for a relatively homogeneous hydrogeologic setting. MF = multiplication factor. [*After Weight et al. (2003).*]

time is longer (24 h or more) the reference thickness (L') in an unconfined aquifer becomes the distance from the water table to the bottom of the screen (Figure 6.45). The estimated effective saturated thickness is L' times MF. In the first edition (Weight and Sonderegger 2001) the MF was suggested to be 1.3 based upon our years of experience performing pumping tests.

Geology

When lithologic units are layered or interbedded with units of significantly differing physical properties in the vertical direction, the transmissivity assigned to the saturated thickness must be evaluated according to the contributing hydrogeologic units. If the

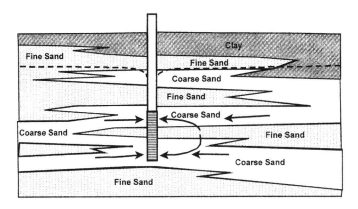

FIGURE 6.46 Pumping test in variable hydrogeologic units. In short-duration tests, the actual contributing thickness may be less than the screen length. Dashed line represents the potentiometric surface under pumping conditions. Lower K units may take a shorter pathway to nearby higher K units (arrows) rather than moving horizontally to the well screen.

hydraulic conductivity of a hydrogeologic unit is one order of magnitude or greater than other units, the majority of water will be produced from the higher K unit (Example 6.22). Water seeks a path of least resistance. It will be easier for water from the lower hydraulic conductivity units to find a pathway up or down to a higher hydraulic conductivity unit rather than take the relatively longer pathway to the well screen (Figure 6.46). This may result in water being primarily produced from an aquifer thickness *less than* the screen length in shorter-duration pumping tests in heterogeneous aquifers.

Numerical Analysis

After additional feedback from other professionals it made sense to evaluate the original rules of thumb numerically as a cost-effective method of analysis. The numerical scenarios were conducted using USGS finite-difference model MODFLOW (McDonald and Harbaugh 1988), supplemented with commercially available pre- and post-processing software. The MODFLOW model grid was 1,000 ft by 1,000 ft with 10-by-10-ft grid spacing. Twenty layers of equal thickness were needed to evaluate vertical flow without significant "boundary effects." The top of the model domain was at an elevation of 100 ft and the bottom datum at 0 ft. A pumping well was placed in the exact center of the grid, spanning layers 4 and 5 (elevations 75 to 85 ft) to represent a screened section of 10 ft. Transient conditions were established over one stress period with 30 time steps. An accelerator parameter of 1.2 was used to spread out the influences of pumping over a time frame of 5 days (Weight et al. 2003).

Pumping rates were varied from 50 to 200 gpm. A horizontal hydraulic conductivity (K_h) of 200 ft/day was fixed throughout all of the simulations to represent coarse sand to sandy gravels, representative of public supply wells. The vertical hydraulic conductivity (K_v) was varied from 200 to 20 ft/day to represent a typical range of stratification-caused anisotropy, and storage coefficient values were alternated between 15% and 30%.

Drawdowns, heads, and streamlines from each simulation run were evaluated. All drawdown values were contoured at an interval of 0.1 ft. It was reasoned that 0.1 to 0.2 ft of drawdown could be used to define "significant hydraulic influence" and thus the vertical extent of significant flow to the pumping well. For each run, six numerical values were visually estimated to establish hydraulic response curves over the first 2 days of pumping. The two "depth" results represent the model elevations, extended reach by the 0.1' and 0.2' drawdown contours, below the bottom of the well screen, and the two "width" results are the 0.1' and 0.2' drawdown contours corresponding to radial distances from the pumping well. The two "top" results are the model elevations, above the top of the well screen, where the 0.1' and 0.2' contours breached the surface. The maximum value for each of these results was determined for each time step and plotted. The curves were evaluated to determine the sensitivity of simulated hydraulic responses to variations in the basic parameters and to reevaluate the two rules of thumb discussed above and what the appropriate MF value should be. A summary of the modeling scenarios and results are shown in Table 6.8.

Graphical depictions of each modeling scenario were performed to look for trends and patterns, used in reevaluating the rules of thumb. An example is shown in Figure 6.47.

Run	$K_h : K_v$ Ratio	Pumping Rate (gpm)	Storage Coefficient (%)	Maximum Drawdown (ft)	Time (h) for Drawdowns (0.1' and 0.2') to Reach Water Table or Certain Depths					
					Water Table		1.5 × Screen		1.5 × Well Bottom	
					0.1'	0.2'	0.1'	0.2'	0.1'	0.2'
PP5	1	100	15	0.9	0.6	1.0	0.2	0.4	0.6	1.7
PP6	5	100	15	1.6	2.1	4.0	0.3	0.4	2.0	5.0
PP7	10	100	15	2.0	4.9	8.9	0.3	0.6	3.5	9.0
PP8	5	200	15	3.2	1.7	2.1	0.1	0.3	1.0	2.0
PP9	10	200	15	4.0	3.3	4.0	0.2	0.3	2.0	3.5
PP10	5	200	30	3.0	3.3	4.0	0.3	0.5	2.1	3.9
PP11	10	200	30	3.8	6.0	8.9	0.5	0.7	4.0	7.3
PP12	5	100	30	1.5	4.0	7.3	0.5	0.9	4.0	10.8
PP13	10	100	30	1.9	8.9	15.7	0.7	1.3	7.3	19.0
PP14	5	50	30	0.7	7.3	22.9	0.9	2.6	10.8	>48
PP15	10	50	30	0.9	15.7	47.9	1.2	3.3	19.0	>48

TABLE 6.8 Summary of Results from Modeling Scenarios and Times in Hours to Reach the Water Table or Specified Depths in an effort to determine a multiplication factor (MF). [From Weight et al. (2003).]

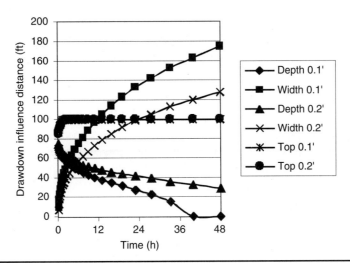

Figure 6.47 Graphical representation of one modeling scenario from Table 6.8. [*After Weight et al. (2003).*]

Two columns in Table 6.8 indicate when the influence of pumping reached or exceeded 1.5 times the screen length. Most of these results are less than an hour, suggesting that an MF factor of at least 1.5 multiplied by the screen length could be applied to the contributing saturated thickness b' with aquifer tests of duration 1 h or more. Additionally, the last two columns of Table 6.8 indicate the time required for detecting the influence of water movement within 1.5 times the distance from the water table to the bottom of the screen. This is achieved in less than half a day, except for scenarios with the lowest pumping rate simulated (50 gpm). This suggests that using the screen length could significantly overestimate "b" and dramatically underestimate K_h. The rules of thumb suggested by Weight and Sonderegger (2001) should be updated to at least 1.5 based on the numerical results.

The issue concerning the saturated thickness b is an important one that is not given enough attention. It has been the author's experience that many hydrogeologists or engineers working for consulting companies are mainly concerned with obtaining a transmissivity (T) value during the pumping test and then divide T by the screen length to obtain a hydraulic conductivity value without regard to whether the K value matches the lithology. Worse yet is the case where hydraulic conductivity values obtained this way are used in calculations. Based upon basic geologic principles their reported transmissivity values from pumping tests and the calculated saturated thicknesses make no sense, whatsoever! It is as though the pumping tests were performed because they were necessary to estimate the hydraulic properties of an aquifer, and then in the quantitative analysis of whatever application, the hydraulic conductivity didn't come out as expected so the field results were ignored.

Example 6.24 While performing a technical analysis of a consultant's report of a flow and transport model for a landfill, the author noticed that a very careful analysis was performed by geologists during the site-characterization phase. A significant amount of expense, time, and effort was made to establish a conceptual model based upon drill-hole information and constructed cross sections. Additionally, pumping tests were performed with both transmissivity and hydraulic conductivity values reported that were used in the numerical model. The hydraulic conductivity values used in the groundwater-flow model were inconsistent with the lithologic units and reported transmissivities. (Also, the length of time of pumping was short, on the order of a few minutes to a few hours.) The respective saturated thicknesses were calculated to be 50 to 60 ft (15.2 to 18.3 m) thick, when the screen length was only 10 ft (3 m). Based on the hydraulic conductivities used in their model, coupled with the true saturated thickness based on the geology, the model layers should have extended 80 ft above the ground surface! It was as though a list of field exercises were performed to fulfill certain requirements, with little effort made to be consistent when it came time for the numerical model. Calibrating the model with constant head boundaries to yield a particular flow scenario seemed to be more important than reality.

6.8 Fracture-Flow Analysis

The previous section has been primarily concerned with pumping tests in porous media. Pumping tests in fractured porous media have a very different time-drawdown behavioral responses. The physical setting is one where high conductivity fractures drain quickly, while smaller fractures and block media drain at later time. Diagrammatically there is a different flow-path length for groundwater in fractures compared to porous media. The flow-path length compared to lateral distance is known as the **tortuosity** factor. In fractured media, the tortuosity factor can be significantly less (Figure 6.48).

The distances over which groundwater moves can be more direct and therefore may result in significant velocities (Ewers 2008). As the velocity of groundwater flow increases, it begins to gain kinetic energy and become turbulent. Turbulent flow in fractured aquifers is a common phenomenon (Carriou 1993). Once turbulent flow dominates, then Darcy's law is no longer valid and a different equation for hydraulic gradient is used (Bear 1979).

$$\frac{dh}{ds} = A_L q + B_T q^n \tag{6.48}$$

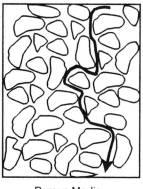 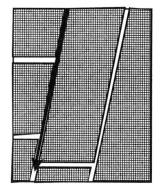

Porous Media,
High Tortuosity

Fractured Media,
Low Tortuosity

FIGURE 6.48 Differences in tortuosity (flow-path length) in porous versus fractured media.

where dh/ds = hydraulic gradient (L/L)
$\quad\quad q$ = specific discharge $(L/t) = K \times dh/ds$
$\quad\quad A_L$ = laminar flow constant (t/L)
$\quad\quad B_T$ = turbulent flow constant (t/L)
$\quad\quad n$ = turbulent flow exponent
$\quad\quad K$ = hydraulic conductivity (L/t)

Turbulence increases the drawdown within the pumping well by a factor of $B_T q^n$, where n increases with the degree of turbulence, which has experimentally been found to range from 1.6 to 2.0 (Bear 1979) (see also Chapter 5).

The dynamics of groundwater flow to a pumping well in fractured media was first considered by Barenblatt et al. (1960). They gave the following equation for describing flow in a fractured aquifer to a pumping well:

$$\Delta s_f = \frac{Q}{4\pi T} \int J_0(xr) \left[1 - \exp \frac{ktx^2}{1 + x^2 \eta_f} \right] \frac{\partial x}{x} \quad\quad (6.49)$$

where $(h_o - h)_f$ = drawdown in the fractured aquifer (L)
$\quad\quad Q$ = pumping rate (L^3/T)
$\quad\quad T_f$ = transmissivity of fractured aquifer (L^2/t)
$\quad\quad J_0$ = Bessel function of the first kind, or zero order
$\quad\quad r$ = radial distance from the pumping well (L)
$\quad\quad t$ = time since pumping began
$\quad\quad \kappa$ = ratio of fractured aquifer transmissivity to aquifer storativity,
$\quad\quad\quad$ or $T_f/S_p (L^2/t)$
$\quad\quad \eta$ = fissure characteristic (dimensionless)
$\quad\quad Sp$ = porous block storativity (dimensionless)
$\quad\quad d_x$ = differential $\times$ operator (radial distance from center of well to the limit
$\quad\quad\quad$ of integration (L))

The part after the integral sign in Equation 6.49 is the well function associated with a discretely fractured aquifer. The type curve from Barenblatt et al. (1960) was plotted by Carriou (1993) and is shown in Figure 6.49. Notice the linear early-time portion of

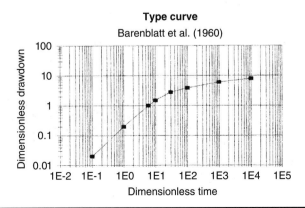

FIGURE 6.49 The type curve from Barenblatt et al. (1960) for discrete fractures (Carriou 1993).

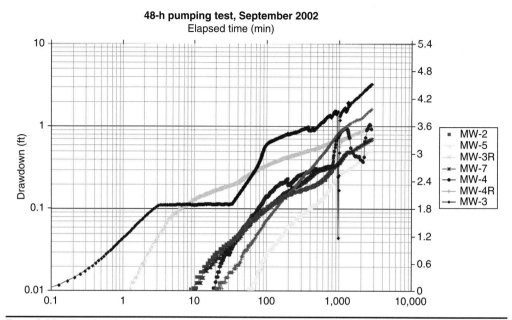

Figure 6.50   Time-drawdown data from a 48-h pumping test conducted in a bedrock aquifer with multiple observation wells in Pennsylvania. The characteristic linear angled pattern is typical of a fracture-flow response.

the curve. This represents water yield from discrete fractures crossing the wellbore coming from fracture storage (Carriou 1993). As more water is withdrawn, the hydraulic gradient within the bedrock induces water from the bedrock and smaller fractures to move to the larger fractures, thus transitioning to a more Theisian-like or even a delayed-yield type of response at later time (next section). It is the author's experience that time-drawdown data exhibit a characteristic diagonal trend on a log-log plot (normally used in a Theis analysis; Figure 6.50), more typical of the type curves shown in Streltsova (1976). Figure 6.50 represents the time-drawdown data from a 48-h pumping tested conducted in a bedrock aquifer with multiple observation wells in Pennsylvania. Field studies in petroleum settings indicate that fractures greater than 0.05 mm are large enough to contribute significantly to fluid extraction volume while contributing very little to overall aquifer block storativity (Gringarten et al. 1974; Streltsova 1976).

The analytical approach for pumping tests in discrete fractures after the Barenblatt et al. (1960) method is to plot field values of drawdown (s) versus time (t) in log-log scale. This plot is overlain on top of the appropriate type of curve for dimensionless fracture drawdown (s_{df}) versus dimensionless time, also in log-log scale (Figure 6.49). The axes are kept square similar to the Theis curve-matching approach (Section 6.2), except that both the linear pattern (representing fracture dewatering) and the late time are both included in the Barenblatt et al. (1960) solution. This means that there is no need to emphasize early- or late-time in the "fit." The transmissivity and storativity are solved using Equations 6.50 and 6.51 from Barenblatt et al. (1960).

$$T_f = \frac{Qs_{df}}{4\pi\Delta s}$$

(6.50)

$$S_p = \frac{4T_f t}{t_d r^2} \qquad (6.51)$$

where Q = withdrawal rate (ft³/min)

S_{dr} = dimensionless fracture drawdown from match point

T_f = fracture transmissivity (ft²/min)

S_p = aquifer storativity (dimensionless)

t = time from match point (minutes)

t_d = dimensionless time from match point

r = radial distance from pumping well to observation well (feet)

Once again the author recommends to the reader the use of a variety of fracture-flow solutions available in commercial software (Duffield 2018).

Example 6.25 A subdivision was proposed in an area in south-central Montana that was already challenged with adequate yields from wells and springs. A state agency conducted a 48-h aquifer test in a production well completed in a sandstone to estimate the aquifer properties. The production well was located in the middle of the subdivision while the observation well at a distance of 937 ft (286 m) was positioned near an unmapped fault in Lot 8. The time-drawdown data shown in log-log scale in Figure 6.51a and the Cooper-Jacob plot used by the state agency for aquifer properties is shown in Figure 6.51b. Note how similar the shape of responses is in Figures 6.51a and Figure 6.50. Full

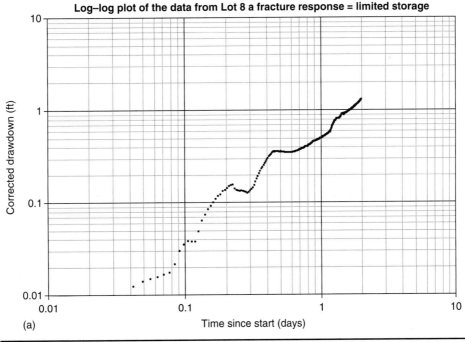

FIGURE 6.51a Time-drawdown data from an observation well (Lot 8) at a distance of 937 ft (286 m) from the pumping well in log-log scale, during a 48-h aquifer test conducted in a fractured sandstone aquifer. Developers were seeking to setup a new subdivision (Example 6.25).

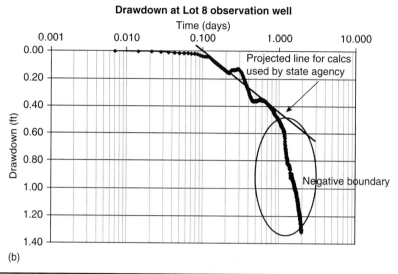

Figure 6.51b   Time-drawdown data from an observation well (Lot 8) at a distance of 937 ft (286 m) from the pumping well this time in semi-log scale (Cooper-Jacob 1946), during a 48-h aquifer test conducted in a fractured sandstone aquifer. The negative boundary of an unmapped fault is circled (Example 6.25).

recovery to background levels took 9 days. The negative boundary pointed out by the author, probably constrained by faulting, and lack of storativity prospects resulted in the subdivision being denied.

Additional fracture responses are possible from fractured media depending on the physical properties of the block materials. For example, if there is a fractured sandstone being stressed by pumping, both the fractures and the blocks will yield water in a dual-porosity model. The dual-porosity model is similar to a delayed-yield response in porous media (Neuman 1979), where the fractures behave linearly; once the blocks begins to yield water, a recharge effect (flattening of the drawdown curve, middle segment; Section 6.2) occurs. Later on when the blocks become stressed, another curvilinear response occurs, indicated by increased drawdown once again.

Pumping Tests in Flowing Artesian Wells

Pumping tests can also be conducted in flowing (artesian) wells if the "pumping well" can be controlled with a shutoff valve. At start time, the pumping well is opened up, allowing the free flow of water. A flowing observation well can be plumbed with a transducer to measure the reduction in head over time (Figure 6.52). If the flow does not decrease more than 10% or so, a reasonable fit can be made to the pumping-test data. "Pumping" times should be similar to conventional pumping tests (Chapter 5).

Example 6.26   In Petroleum County, in central Montana, are several flowing wells from differing Cretaceous hydrogeologic units. The flowing wells were abandoned exploration oil wells from the 1960s. In a study performed by Brayton (1998), pumping tests were conducted in the basal member of the Cretaceous Eagle Formation. A flowing pumping well was shut down, while the flowing observation well was plumbed with a pressure transducer (Figure 6.52). The pumping test was conducted by opening the valve on the "pumping well" and allowing the data logger to record the decrease in pressure over time. Decreased pressure was converted to drawdown in feet over time. An example plot of the data is shown in Figure 6.53, indicating a dual-porosity fracture-flow response.

Figure 6.52 A plumbed-in vented transducer in a flowing well used as an observation well during an aquifer test. The vented transducer detects any decreasing pressure as the valve for the pumping well is opened.

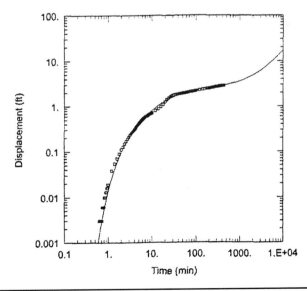

Figure 6.53 Dual-porosity fracture-flow response from a pumping test in the Cretaceous Eagle Formation (sandstone), central Montana. [*After Brayton (1996).*]

6.9   Summary

One of methods of evaluating the hydraulic properties of aquifers is through conducting pumping tests. How to conduct a pumping test is described in Chapter 5, while helpful suggestions on what to record in your field book are found in Chapter 1. This chapter addressed several of the analytical methods used to analyze pumping-test data. The most common approach is to perform a Theis analysis or do a Cooper-Jacob plot *in the field during the test* to see how the data behave. Boundary conditions and other effects are observed directly in the plot. Care must be taken to ensure that the pumping rate was kept constant, or strange patterns may occur that are an artifact of varying pumping rates.

One must also make sure the pumping test was conducted long enough when a recharge response was observed early on. Does the recharge response reflect a semiconfined or leaky-confined setting, or is delayed yield occurring? This can only be resolved through recording additional time-drawdown responses out for at least a couple of days.

Direct relationships exist between time-drawdown and distance-drawdown plots. These can be used to evaluate the impacts of longer-duration pumping scenarios on neighboring wells with senior water rights.

The analysis of pumping-test data is closely connected with having an understanding of the geologic setting and a knowledge of the well-completion details. Pumping tests yield an estimate for transmissivity (T) and storativity (S). These are two-dimensional parameters in which the vertical dimension is averaged. When estimating the hydraulic conductivity (K) from the transmissivity value, it is important to consider test duration and geology, particularly in a partial penetrating aquifer scenario.

Fracture-flow responses have their own characteristic behavior. Type curves for discrete-fracture or dual-porosity models are very different. A common aspect of both is the early-time linear-like response on a log-log plot of drawdown versus time. Care must be taken to make sure the analysis method, field observations, and geologic setting all make sense.

6.10   Problems

6.1.   Take the data from Table 6.3 and manually plot these data on 3-by-5 log cycle paper. It should look like a part of a Theis type curve. Find a 3-by-5 cycle Theis curve or create one from Appendix D and make sure the scale of the data and the scale of the type curve match. Go to a window and overlay the field data onto the type curve and make a dot with a pencil where the match point coincides with the field data sheet. Read off the values for drawdown ($h_o - h$) and time (t). Use Equation 6.2 to calculate T. Use Equation 6.3 and the time from the data sheet to calculate S. Recall that the match point Wu and $1/u$ are equal to 1.0. The pumping rate was 1,200 gpm and the observation well was 850 ft away.

6.2.   Enter the time-drawdown data from Table 6.1 (Todd 1980) into a spreadsheet and create a Cooper-Jacob plot. Determine the slope of the straight-line fit and determine the change in drawdown per log cycle. Estimate T using Equation 6.17. Project the straight-line fit to where zero drawdown occurs to obtain t_o. Use Equation 6.21 to estimate S. The pumping rate was 450 gpm and the observation well was 200 ft away.

6.3.   Enter the time-drawdown data from Table 6.2 onto a spreadsheet and see how it looks. How did varying pumping rates during the test affect the intended straight-line result if the data under a constant pumping rate behaved Theisian?

6.4. Take the results of you aquifer properties (*T* and *S*) from Problem 6.1 and estimate what the drawdown would be if time = 60 days and the pumping rate was changed to 600 gpm and the new observation point was at a distance of 330 ft.

6.5. A 72-h aquifer test site was set up to test aquifer properties for the Davis family in a small pivot operation (10 acres). A plan view schematic in Figure 6.54a indicates the location of the pivot well, the house well, and two shallow stock wells (SWs) (<15 ft deep). Pipestone Creek flows through the property as shown. The static water levels were measured within minutes of each other. Both plots of time-drawdown data for the two wells are shown in Figures 6.54b and 6.54c. No changes in water levels were observed in the SWs during the test.

a) Calculate transmissivity and storativity of the aquifer.
b) What kind of aquifer was tested? Unconfined, leaky, or confined? Explain.

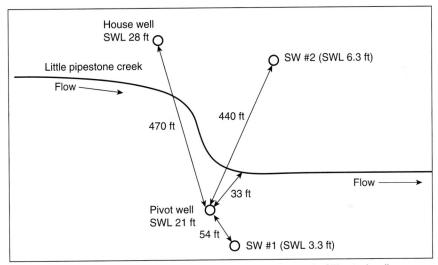

Problem 5. Plan view schematic of wells on the Davis Property, not to scale. SW = stock well

Davis-House well

From	To	Description
0.0	5.0	Topsoil
5.0	28.0	Sand
28.0	35.0	Sand & clay
35.0	42.0	Clay
42.0	50.0	Clay
50.0	60.0	Sand & clay
60.0	68.0	Sand & clay

Pivot well

From	To	Description
0.0	2.0	Topsoil
2.0	25.0	Clay
25.0	30.0	Hard black basalt
30.0	50.0	Clay
50.0	62.0	Sandy clay
62.0	75.0	Coarse sand and clay
75.0	78.0	Fine sand with trace of coarse sand
78.0	84.0	Sandy clay
84.0	91.0	Fine sand with trace of coarse sand
91.0	114.0	Course sand with pea gravel

(a)

FIGURE 6.54 (a) Plan view schematic of wells on the Davis property, lithologic log of the Davis house well, and lithologic log of the pivot well (PW). Problem 5. (b) Time-drawdown plot of PW. The data were effected during initial pumping at 100 gpm to fill the pivot system (about 7 min), and stabilization of pumping occurred at 60 gpm at about 30 min. (c) Time-drawdown plot of the house well (about 470 ft to the northwest). Somebody probably flushed the toilet at about 900 min.

Time-drawdown data from the pivot well

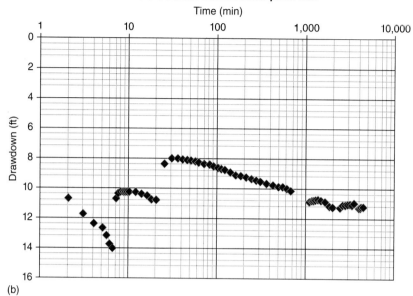

(b)

House well plot
May 30–June 2, 2004

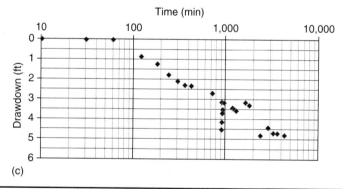

(c)

FIGURE 6.54   *(Continued)*

c)   What interpretations can you offer about the nature of the aquifer and its impact on Little Pipestone Creek, the SWs, or the productivity of the aquifer?

6.6.   A local retailer outlet (Big Mart) tested a production well for its new store pumping at a rate of 1.5 cubic meters per minute. The aquifer is believed to be homogenous and isotropic. The time-drawdown data have been plotted for an observation well located 30 m away directly west of the pumping well (Figures 6.55 and 6.56).

a)   Calculate transmissivity and storativity for the observation well. Explain your results.
b)   Comment on how long it would take for the cone of depression to reach the next nearest business located 400 m directly east of the production well. Hint: Use Equation 6.21 to solve for time (*t*).

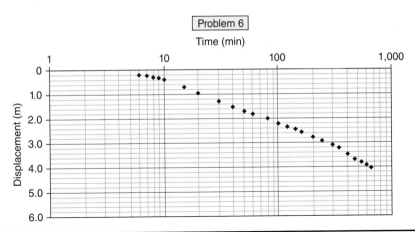

FIGURE 6.55 Time-drawdown data for an observation well located about 30 m away from the Big Mart production well (Problem 6).

c) Does the other existing business have any reason to be concerned? Explain your answer.

d) What kind of aquifer is this? Unconfined, leaky, or confined? Explain.

e) What is the transmissivity using the specific capacity method?

6.7. Consider the diagram below. A real estate broker wants to know if there is a significant impact from two pumping wells at point Q. The aquifer is assumed to have a transmissivity of 2,500 ft²/day, the storativity is 0.004, the time is 3 days, Q_1 is 150 gpm, Q_2 is 200 gpm, and the grid spacing is 300 by 300 ft. Assuming the boundary is a barrier type No, what would you tell her?

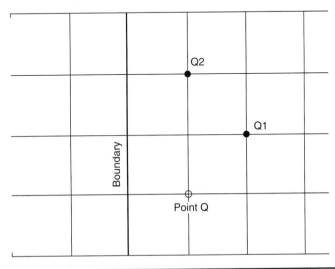

FIGURE 6.56 Evaluation of drawdown at point Q from two other pumping wells and a boundary (Problem 7).

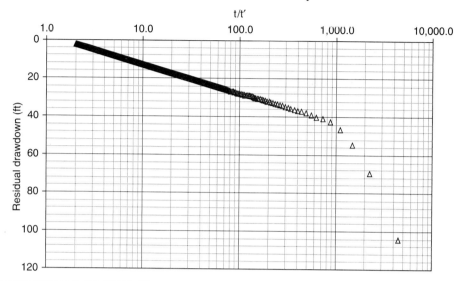

Figure 6.57   Recovery data from the Carroll transfer well after a 72-h pumping test. The recovery data show residual drawdown plotted versus t/t' (Problem 8).

6.8.   An aquifer test was conducted at the transfer well at Carroll College at a pumping rate of 159 gpm. Plotted in Figure 6.57 is the recovery data of residual drawdown versus t/t' (Figure 6.57). Pumping occurred for 72 h. The well was drilled to a depth of 500 ft (15 points).

a)   What is the transmissivity?
b)   When did the well fully recover?
c)   Comment on the data and its potential as an irrigation well.

6.9.   The data in Table 6.9 (Kruseman and deRidder 1991) are from an observation well located 90 m away from the pumping well. The pumping rate was 873 m³/day. Plot the data and fit the early-time data with the Theis type A match point. Find an appropriate beta curve and fit the late-time data with the second Theis type B match point. Compare the transmissivity and storativity values (Figure 5.58).

6.10.   Refer to the diagram below. The drawdown in the pumping well (radius of 0.4 ft) is 10 ft. The pumping rate is 200 gpm and the drawdown observed in wells A, B, and C are 6 ft, 4 ft, and 3.8 ft, respectively. Pumping has occurred for 48 h.

a)   Is the aquifer homogenous and isotropic? Explain.
b)   Sketch the approximate outline (shape) of the cone of depression where 4 ft of drawdown is at 48 h.
c)   Estimate the transmissivity(-ies) of the aquifer.

Time (Min)	$h_o - h$ (m)
1.17	0.004
1.34	0.009
1.7	0.015
2.5	0.03
4	0.047
5	0.054
6	0.061
7.5	0.063
9	0.064
14	0.09
18	0.098
21	0.103
26	0.11
31	0.115
41	0.128
51	0.133
65	0.141
85	0.146
115	0.161
175	0.161
260	0.172
300	0.173
370	0.173
430	0.179
485	0.183
665	0.182
1,340	0.2
1,490	0.203
1,520	0.204

TABLE 6.9 Time-Drawdown Data from Kruseman and deRidder (1991) (Problem 9)

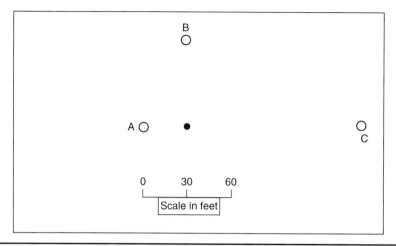

FIGURE 6.58 Diagram for Problem 10. The drawdown in the pumping well (radius of 0.4 ft) is 10 ft. The pumping rate is 200 gpm and the drawdown observed in wells A, B, and C are 6 ft, 4 ft, and 3.8 ft, respectively.

6.11   References

Barenblatt, G.I., Zheltov, I.U.P., and Kochina, I.F., 1960. Basic Concept in the Theory of Seepage of Homegenous Liquids in Fissured Rocks. *Journal of Applied Mathematical Mechanics*, Vol. 24, No. 5, pp. 1286–1303.

Bear, J., 1979. *Hydraulics of Groundwater*. McGraw-Hill, New York, 569 pp.

Belcher, W.R., Elliot, P.E., and Geldon, A.L., 2001. *Hydraulic-Property Estimates for Use With a Transient Ground Water Flow Model of the Death Valley Regional Ground-Water Flow System, Nevada and California*. USGS Water-Resources Investigations Report 01-4120, 28 pp.

Borgman, L.E., Weight, W.D., and Quimby, W.F., 1986. *Geological and Hydrological Characteristics of the Hoe Creek Underground Coal Gasification Site*. Technical Report 86-2. Prepared for the United States Environmental Protection Agency EMS Laboratory, Las Vegas, NV, 71 pp.

Boulton, N.S., 1963. Analysis of Data from Non-Equilibrium Pumping Tests Allowing for Delayed Yield from Storage. *Proceedings Institution of Civil Engineers*, Vol. 26, No. 3, pp. 469–482.

Boulton, N.S., 1973. The Influence of the Delayed Drainage on Data from Pumping Tests in Unconfined Aquifers. *Journal of Hydrology*, Vol. 19, pp. 157–169.

Bourdet, D., Whittle, T.M., Douglas, A.A., and Pirard Y.M., 1983. A New Set of Type Curves Simplifies Well Test Analysis, *World Oil*, May 1983, pp. 95–106.

Brayton, M., 1998. *Recovery Response from Conservation Methods in Wells Found in the Basal Eagle Sandstone, Petroleum County, Montana*. Master's Thesis, Montana Tech of the University of Montana, Butte, MT, 68 pp.

Butler, J.J., 2003. *Mathematical Derivations of Expressions for the Total Pumping-Induced Leakage Entering an Aquifer*. Kansas Geological Survey Open-File Report 2003–6, 13 pp.

Carriou, T.M., 1993. *An Investigation into Recovering Water Levels in the Butte, Montana, Bedrock Aquifer*. Master's Thesis, Montana Tech of the University of Montana, Butte, MT, 160 pp.

Cleary, B., 1993. Personal Communication, San Francisco, CA.

Cooper, H.H., Jr., and Jacob, C.E., 1946. A Generalized Graphical Method for Evaluating Formation Constants and Summarizing Well-Field History. *Transactions of the American Geophysical Union*, Vol. 27, pp. 526–534.

Driscoll, F.G., 1986. *Groundwater and Wells*. Johnson Screens, St. Paul, MN, 1108 pp.

Duffield, G., 2018. http://www.aqtesolv.com/aquifer-tests/aquifer-testing-references.htm

Duffield, G.M., and Rumbaugh, J.O., III, 1991. AQTESOLV, Aquifer Test Solver Software. Version 1.00, Geraghty and Miller Modeling Group, Reston, VA, 90 pp.

Ehlig-Economides, C., 1988. Use of the Pressure Derivative for Diagnosing Pressure-Transient Behavior, SPE 18594. *Journal of Petroleum Technology*, October 1988, pp. 1280–1282.

Ewers, R., 2008. Practical Karst Hydrogeology with Emphasis on Ground-Water Monitoring, a short course at the North American Environmental Field Conference and Exposition, sponsored by the Nielson Environmental Field School, Inc., Tampa, FL.

Fetter, C.W., 1993. *Applied Hydrogeology, 3rd Edition*. Macmillan, New York, 691 pp.

Fletcher, F.W., 1997. *Basic Hydrogeology Methods: A Field and Laboratory Manual with Microcomputer Applications*. Technomic, Lancaster, PA, 310 pp.

Gambolati, G., 1976. Transient Free Surface Flow to a Well: Analysis and Theoretical Solutions. *Water Resources Research*, Vol. 12, pp. 27–39.

Gringarten, A. C. et al., 1974. Unsteady-State Pressure Distributions Created by a Single Infinite-Conductivity Vertical Fracture. *Society of Petroleum Engineers Journal*, Vol. 14, No. 4, pp. 347.

Halford, K.J., Weight, W.D., and Schreiber, R.P., 2006. Interpretation of Transmissivity Estimates from Single-Well Pumping Tests. *Ground Water*, Vol. 44, No. 3, pp. 467–471.

Hantush, M.S., 1956. Analysis of Data from Pumping Tests in Leaky Aquifers. *Transactions of the American Geophysical Union*, Vol. 37, pp. 702–714.

Hantush, M.S., 1960. Modification of the Theory of Leaky Aquifers. *Journal of Geophysical Research*, Vol. 65, pp. 3713–3725.

Hantush, M.S., 1964. Hydraulics of Wells. In Chow, V.T., ed., *Advances in Hydroscience*, Vol. 1, Academic Press, New York, pp. 281–432.

Hantush, M.S., and Jacob, C.E, 1955. Non-Steady Radial Flow in an Infinite Leaky Aquifer. *Transactions of the American Geophysical Union*, Vol. 36, No. 1, pp. 95–100.

Heath, R.C., 1983. *Basic Ground-Water Hydrology*. USGS Water-Supply Paper 2220, 85 pp. https://pubs.er.usgs.gov/djvu/WSP/wsp_2220.pdf

Huntley, D.R., and Steffey, D., 1992. The Use of Specific Capacity to Assess Transmissivity in Fractured-Rock Aquifers. *Ground Water*, Vol. 30, No. 3, pp. 396–402. http://hydrogeologistswithoutborders.org/wordpress/1979-english/chapter-8/

Jacob, C.E., 1950. Flow of Ground-Water. In Rouse, H., ed., *Engineering Hydraulics*, John Wiley & Sons, New York, 321–386 pp.

Kruseman, G.P., and deRidder, N.A., 1991. *Analysis and Evaluation of Pumping Test Data, 2nd Edition*. Publication 47. International Association for Land Reclamation and Improvement, Wageningen, Netherlands, 377 pp.

Lohman, S.W., 1979. *Ground-Water Hydraulics*. U.S. Geological Professional Paper 708, Washington, DC, 70 pp.

McDonald, M.G., and Harbaugh, A.W., 1988. *A Plea Modular Three-Dimensional Finite-Difference Ground-Water Flow Model*. USGS Numbered Series. 586 pp.

Metesh, J., 1995. Personal Communication, in Butte, Montana. Techniques of Water-Resources Investigations 06-A1.

Mishra, P.K., and Kuhlman, K.L., 2013. Unconfined Aquifer Flow Theory—From Dupuit to Present. https://arxiv.org/pdf/1304.3987.pdf

Moench, A.F., 1996. Flow to a Well in a Water-Table Aquifer: A Improved Laplace Transform Solution, *Ground Water*, Vol. 34, No. 4, pp. 593–596.

Neuman, S.P., 1972. Theory of Flow in Unconfined Aquifers Considering Delayed Response to the Water Table. *Water Resources Research*, Vol. 8, pp. 1031–1045.

Neuman, S.P., 1975. Analysis of Pumping Test Data from Anisotropic Unconfined Aquifers Considering Delayed Gravity Response. *Water Resources Research*, Vol. 11, pp. 329–342.

Neuman, S.P., 1979. Perspective on "Delayed Yield." *Water Resources Research*, Vol. 15, No. 2, pp. 989–998.

Neuman, S.P., 1987. On Methods of Determining Specific Yield. *Ground Water*, Vol. 25, No. 6, pp. 679–684.

Neuman, S.P., and Witherspoon, P.A., 1969. Applicability of Current Theories of Flow in Leaky Aquifers. *Water Resources Research*, Vol. 5, pp. 817–829.

O'Connel, W.T., and Smith, T.C., 1993. *Aquifer Characterization for the Well-Head Protection Plan for Ramsay, MT*. Senior Design Project, Department of Geological Engineering, Montana Tech of the University of Montana, Butte, MT.

Reilly, T.E., Franke, O.L., and Bennett, G.D., 1984. *The Principle of Super Position and Its Application in Ground-Water Hydraulics*. U.S. Geological Survey Open-File Report 84–459, 36 pp.

Renard, P., Glenz, D., and Mejias, M., 2009. Understanding Diagnostic Plots for Well-Test Interpretation. *Hydrogeology Journal*, Vol. 17, pp. 589–600. doi:10.1007/s10040-008-0392-0

Schafer, D.C., 1978. Casing Storage Can Affect Pumping Test Data. *Johnson Drillers' Journal*, Third Quarter.

Spane, F.A., Jr., and Wurstner S.K., 1993. DERIV: A Computer Program for Calculating Pressure Derivatives for Use in Hydraulic Test Analysis. *Ground Water*, Vol. 31, No. 5, pp. 814–822.

Streltsova, T.D., 1973. On the Leakage Assumption Applied to Equations of Groundwater Flow. *Journal of Hydrology*, Vol. 20, No. 3, pp. 237.

Streltsova, T.D., 1976. Hydrodynamics of Groundwater Flow in a Fractured Formation. *Water Resources Research*, Vol. 12, No. 3, pp. 405–414.

Theis, C.V., 1935. The Lowering of the Piezometer Surface and the Rate and Discharge of a Well Using Ground-Water Storage. *Transactions of the American Geophysical Union*, Vol. 16, pp. 519–524.

Thiem, G., 1906. *Hydrologische Methoden*. Gebhardt, Leipzig, 56 pp.

Todd, D.K., 1980. *Groundwater Hydrology, 2nd Edition*. John Wiley & Sons, New York, 552 pp.

Weight, W.D., Schreiber R.P., and Gamache, M., 2003. Numerical Evaluation of the Effective Thickness in Pumping Tests, in *Proceedings MODFLOW and MORE 2003 Understanding through Modeling*, vol. 2, International Ground Water Modeling Center, Colorado School of Mines, Golden, CO, pp. 702–706.

Weight, W.D., and Sonderegger, J.L., 2001. *Manual of Applied Field Hydrogeology*, McGraw-Hill, New York, 608 pp.

Weight, W.D., and Wittman, G.P., 1999. Oscillatory Slug-Test Data Sets: A Comparison of Two Methods. *Ground Water*, Vol. 37, No. 6, pp. 827–835.

CHAPTER 7

Slug Testing

Slug testing has been used over the years to obtain a cost-effective quick estimate of the hydraulic properties of aquifers. Slug testing has gained popularity since the 1980s since it can be used to obtain hydraulic property estimates in contaminated aquifers where treating pumped waters is not desirable. Slug-testing analysis methods were first developed during the 1950s (Hvorslev 1951; Ferris and Knowles 1954). The simple method devised by Hvorslev (1951) led to its wide use for both confined and unconfined aquifers as an estimate of the hydraulic conductivity within the screened interval. Improvements of the methods by Ferris and Knowles (1954) were made by Cooper et al. (1967) and Papadopulos et al. (1973) for confined aquifers. Later, Bouwer and Rice (1976) and Bouwer (1989) developed a method for analysis in confined, semi-confined, or unconfined aquifers that takes into account aquifer geometry and partial penetration effects. Campbell et al. (1990) summarizes the various methods used and their relative merits with an extensive list of references. Additional references to articles on slug testing and aquifer testing have been compiled by Duffield (2018). An article by Butler et al. (1996) describes the protocol that should be used in the field to reduce errors of estimation, with a newer field guide prepared by Butler et al. (2011). The book by Butler (1998) presents a fairly comprehensive summary on slug-test methodology and analysis. If this chapter does not satisfy the reader on the topic of slug testing, the book by Butler (1998) is highly recommended.

Constant-discharge pumping tests are a better method of analyzing the hydraulic properties of aquifers (Chapter 5) since their influence goes beyond the immediate vicinity of the borehole, although in high-transmissivity aquifers the influence can be detected several hundred feet away (Bredehoeft et al. 1966). Slug tests are especially useful in locations where handling the discharge of contaminated pumping waters is prohibited yet a rough estimate of the hydraulic properties of the aquifer is needed. Another great thing about slug tests is that they can be conducted relatively quickly, so that several point estimates of hydraulic conductivity can be collected within a day's field work. The chances of a hydrogeologist needing to perform slug tests are pretty high. The intent of this chapter is to discuss the methodology of performing slug tests in the field followed by methods used to analyze the data. This is done for both the overdamped and the underdamped cases. Discussion of the problems that can occur in the field are also be presented along with the typical mistakes that are made when performing the analysis.

7.1   Field Methodology

Slug testing involves lowering a "slug" (a cylinder of known volume, usually constructed of PVC pipe or some other suitable material, filled with sand or gravel that is capped at both ends) attached to a rope or cable, down a well bore to displace the static water with an equivalent volume (W_o in Figure 7.1). A pressure transducer, previously placed below the slug level, senses the water-level responses, and the data are recorded in a data logger (Figure 7.2). The slug remains in place until the water level equilibrates (static water level, SWL). Once at equilibrium, the data logger is activated and the slug is briskly retrieved above the SWL (Figure 7.3). The displaced water (W_o) then recovers to the original static level. The time required for static conditions to occur once again is proportional to the hydraulic conductivity of the aquifer materials. Graphical methods are used to obtain parameters that are used to calculate the hydraulic conductivity values.

In the pre-1990s, water-level changes during slug tests were measured using steel tapes or electrical tapes (E-tapes). This only allowed slug tests to be conducted on relatively low hydraulic conductivity materials (Lohman 1972). Therefore, this was applicable when water levels slowly recovered over a matter of minutes to several hours. With the current wide use of high-speed data loggers, pressure transducers, and well sentinels (Chapter 4), slug-test analysis may be conducted on materials over a wide range of hydraulic conductivities.

How to Make a Slug

It is important that the size of the slug be appropriate to the amount of water desired to be displaced. For example, 1¼-in PVC works well for 2-in wells, and 3-in PVC works well in 6-in wells (larger-diameter slugs can be used but become heavy and difficult to handle). The lengths should be designed so that at least 1 ft (0.3 m) or more

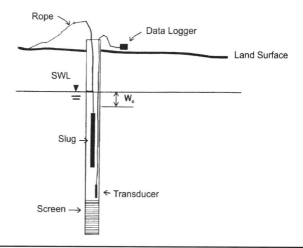

FIGURE 7.1   Slug-testing schematic. The slug displaces the SWL by an amount W_o or h_o. When lowering the slug, the water is displaced by W_o above the SWL. When removing the slug, the water is displaced by W_o below the SWL. [*Weight and Wittman (1999). Used with permission of Groundwater Publishing Co.*]

Figure 7.2 Using the data logger in the field to gather slug-test data. Shown is a vented transducer cable and data logger.

Figure 7.3 Lowering a slug down a well. The slug is shown suspended above the 2-in well with the vented transducer cable secured into place with duct tape and attached to a data logger.

of displacement can occur. One can calculate the approximate volume of the slug ($\pi r^2 h$, where h = length) before going into the field. Typical slug lengths are 3 ft (1 m), 6 ft (1.8 m), or 10 ft (3 m). The general procedure for constructing slugs is described next:

- Take a blank piece of PVC casing and cut it to the desired length, allowing for the top and bottom caps to contribute to the total length.

- The bottom of the casing is capped and either glued with PVC glue or screwed together or both. Preferably, this is reinforced with duct or Gorilla Tape.

- Sufficient sand or gravel is added so that when the slug is capped at the top, it will sink below the SWL surface.

- A hole is made in the top cap, large enough for the rope or cable to thread through. A large knot, like a figure eight, can be tied to keep the rope or cable from pulling back through.

- The top cap is then secured into place. Once again, this should have duct or Gorilla Tape wrapped *around the cap and rope* or cable for additional security.

- Make a variety of sizes and have additional materials with you in the vehicle to make additional slugs or in case modifications are needed.

Example 7.1 While performing several slug tests in the Beaverhead Valley, near Dillon, Montana (Figure 7.4), in 1996, a supposedly secure top cap separated from the slug and we helplessly watched the rest of the slug drop out of sight. It had been held on with PVC glue. Reinforcing each slug, top and bottom, with duct tape became standard procedure ever since and even after several tens of more

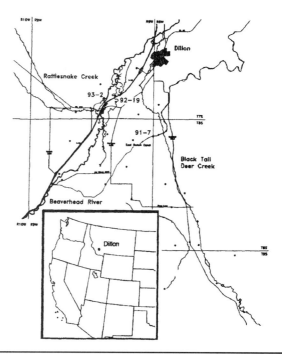

FIGURE 7.4 The Beaverhead Groundwater Project Area (BGPA) slug-test area, Beaverhead Valley, Dillon, Montana. [*From Wittman (1997).*]

tests, this security step proved itself again and again. Securing slogs with Gorilla Tape provides an added measure of security (Chapter 5).

In another setting, a heavy pump was to be lowered down a well via a braided stainless-steel cable. A loop through the top of the pump was used to link the cable, and a purchased clamp that was wrenched tight was supposed to hold the loop. Fortunately, after raising the pump, the faulty clamp slipped, revealing its inability to hold the cable before it was positioned over the well. A new approach of tying a knot and duct-taping the knot, in addition to the clamp, worked beautifully. It is much easier to cut away duct tape than "fish out" lost equipment. Again, Gorilla Tape is even stronger for this application.

Performing a Slug Test with a Vented Transducer

Performing a slug test is fairly straightforward and should take less than half an hour per well, unless the hydraulic conductivity is very low. The following steps are presented to help the field person remember what to do:

- Obtain general information about the well, such as well-completion information (inside radius of well casing and screen and diameter of the borehole), total depth of the well, location of and diameter of the screen, packing material, and height of the water column above the screen. This is important to select an appropriately sized slug and to know what transducer pressure rating is appropriate (Chapters 4 and 5).

- Remove the well cap and take a SWL measurement. This should be checked a few times before starting the test to see if changes or trends are occurring.

- Select the proper transducer and lower it *below* the location of where the bottom of the slug will be placed (Figure 7.1). Note that if the transducer is rated at 10 psi and if it is lowered into more than 23 ft of water, it may become damaged and not take a reading at all (Chapters 4 and 5). Allow the transducer to stretch and hang for a few minutes, so that vertical movement is not occurring during the test.

- Create a loop (bite) of at least 1-in diameter in a vented transducer cable and secure it with duct tape or electrical tape. The loop rests on the top of the well casing, and an additional wrap of the transducer cable is made around the well casing. This is all secured into place with duct tape or Gorilla Tape so that the transducer doesn't move during slug retrieval (Figure 7.3).

- Connect the vented transducer to the data logger, establish the transducer parameters, and prepare for the starting of the test. Before lowering the slug and setting the reference level, take a reading from the data logger to see if the water depth looks reasonable. Reasons for strange readings on transducers are discussed in Chapter 5. It is import to record the timing of testing in your field book.

- Lower the slug with a rope or cable below the water level and tie it off until the water level equilibrates. Note that if the slug floats, it needs to be retrieved and filled with more sand and gravel to make sure it sinks below the water surface.

- Once the water level has equilibrated, the test is ready. Recheck the data logger and establish the reference level to be some value greater than the maximum expected displacement (a value of 10 or 100 is usually what the author uses).

- While holding the line, carefully untie the slug rope and be ready for the signal to retrieve the slug. Start the data logger in log mode, and with only a second or two delay, briskly retrieve the slug. One way to think of it is to count to four and on three start the data logger.

- Check the data logger to see whether the water level has equilibrated and stop the test. This process should be repeated three times to make sure the data behave similarly.

- Decontaminate the slug and make sure well caps are replaced and the site is left the way you found it.

What was just described is a **rising-head test**. Theoretically, one could get the same results by suspending the slug above the SWL and lowering the slug into the water. This would cause the water level to be displaced upward (W_o) and gently return down to the original level (a **falling-head test**). Another way to perform a falling-head test is to pour a known volume of water down a well and record the "fall" in water level to static equilibrium. This whole process assumes that adequate well development has occurred (Chapter 15). The author prefers the rising-head test because the water that is displaced must come from the aquifer rather than pushed out into the aquifer as is done in a falling-head test.

The reason for turning the data logger on just a few seconds before the slug is "instantaneously" retrieved is to make sure that the slug clears the water surface. If the slug hasn't cleared the water surface, "splash effects" and an erroneously large reading will occur (Table 7.1). One must look for the maximum displacement (h_o or w_o in Figure 7.1) in the data. This is presumed to occur at time zero. All other data (h) are adjusted and referenced from the maximum displacement point. In Table 7.1, the measurement at time 0.0133 min represents where the slug did not break the water surface yet, thus creating an erroneously large displacement value. The subsequent oscillation downward represents the "splash" effect. Time zero has been adjusted from 0.0266 min. Displacement versus time plots are used to estimate the hydraulic properties. Slug testing is also performed pneumatically by plumbing the well so that a "blast" or pulse of air instantaneously displaces the water surface (Butler 1998). The data logger records the displaced water-level data over time. The collected data are plotted and form the basis for estimating aquifer hydraulic properties.

Slug Tests Using Well Sentinels

Well sentinels are uniquely suited to obtaining slug-test data. The parameters of the well sentinel are established prior to the test communicated via a USB cable and laptop computer or smartphone. The author has found it convenient to set the well sentinel to collect data in linear mode every 0.5 s. This is usually sufficient to capture the aquifer response. The protocol differs only slightly from what was described in the previous section. A brief synopsis of the appropriate steps to take is presented.

- After the SWL has been measured, attach the well sentinel to a secure cable on a spool (Figure 7.5). The parameters should have previously been established with the laptop and the sentinel may already be collecting data.

- Lower the sentinel well below the "active depth" of where the slug will be raised and lowered, but not below the psi rating of the sentinel. The cable can be secured in place with a strong spring-tensioned clip or some other method.

Time (min)	Displacement (ft)	Time (min)	Displacement (ft)
0.0	99.984	0.12	101.218
0.0033	99.984	0.14	101.155
0.0066	99.984	0.16	101.91
0.01	100.095	0.18	101.033
0.0133	107.954	0.20	100.947
0.0166	101.392	0.30	100.767
0.02	101.423	0.40	100.623
0.0233	101.55	0.50	100.509
0.0266	101.566	0.60	100.442
0.03	101.534	0.70	100.378
0.0333	101.518	0.80	100.316
0.0366	101.502	0.90	100.269
0.04	101.487	1.00	100.236
0.05	101.455	1.20	100.175
0.06	101.408	1.40	100.148
0.07	101.376	1.60	100.116
0.08	101.329	1.80	100.092
0.09	101.297	2.00	100.079
0.10	101.265	2.50	100.042

TABLE 7.1   Slug Test Illustrating a Slug Not Clearing the Water Surface with Initial Reference of 100 ft (3-ft Slug in 2-in Well)

FIGURE 7.5   Well sentinel attached to a galvanized cable on a spool. The cable can be held in place to the casing with a heavy spring-loaded clamp.

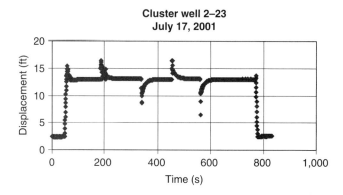

FIGURE 7.6   Data set showing the water-level response captured on a well sentinel after lowering and retrieving a slug several times and finally lifting the well sentinel back to the surface. A data set like this would be cut and pasted into separate spreadsheet pages for individual analysis.

- Lower the slug into the water and tie it off for 10 min or so (depending on the expected recovery time) then briskly retrieve the slug out of the well and wait for another 10 min or so (once again depending upon the expected time needed for recovery). The well sentinel should have collected the first falling-head and rising-head test.

- Repeat the process twice more to obtain three falling-head tests and three rising-head tests (Figure 7.6).

- The well sentinel will need to be reconnected with the laptop or smartphone to upload the data and reconfigured before moving on to the next well.

- Any decontamination should also be performed before transferring the well sentinel to another well.

7.2   Common Errors Associated with Performing Slug Tests

There are two topics the author would like to emphasize about the hydrogeologic setting where slug tests are typically performed. One is performing slug tests in monitoring wells. The other topic has to do with equipment problems. Most of this information was taken from a short course taught by the author at a field conference in Florida (Weight 2006).

Monitoring Well-Completion Issues

Everything mentioned in the previous section about the slug-testing procedures sounds fine until one considers a few things about monitoring wells. Monitoring wells tend to be installed quickly and methodically, ideally to establish a monitoring network to estimate the spatial distribution of hydraulic head and of water-quality constituents. Hopefully, by the time the wells are installed the connection with the aquifer is good enough to believe those data. Monitoring wells are usually not drilled *or* developed with the objective of estimating aquifer properties. Slug testing, therefore, can be seen as an afterthought or as "Hey let's estimate the aquifer properties too while we are collecting field data." Nevertheless, slug testing can be performed to obtain useful relative information. To explore some of the problems with monitoring wells it is useful to consider the many ways monitoring wells are drilled and completed (Chapter 15). Some of the following questions are worth considering:

- What is being tested: the well completion or the aquifer? What kinds of screen and sand packing materials were used?
- Is the well an "open-hole" completion? Or is the bottom cap missing?
- Does the well screen or perforated interval extend above the SWL?
- What is the position and length of the screen?
- Is sand packing material present or is the well naturally developed? If so over what length?
- Which drilling methods were used and were any problems encountered during drilling? (This is often the reason of a "skin" effect; Chapter 5.)
- How long and to what extent was well development performed? This is another explanation for there being a "skin effect."
- Was grouting used and if so where?
- If the well is surged using a slug does it produce sediment? When doing this, does surging develop or clog the sand packing?
- Was slug testing performed multiple times on each well?

The list above is by no means exhaustive but serves to get the reader thinking. There are some general principles that can be derived from posing the above questions. First of all if the screen and packing materials are of a smaller diameter than the aquifer materials the hydraulic properties of the aquifer are *not* what will be tested. Many times "in-stock" monitoring screens of a given slot size or packing materials are used on every monitoring well, regardless of the aquifer's physical properties. In this case monitoring wells should *only* be used for static head and water-quality parameters.

Slug-testing analysis procedures require that one know the radius, length, and diameter of the screen, the distribution of packing materials, the inside diameter of the casing, the diameter of the well boring, and the moment in time of maximum water-level displacement. Water in wells with screens that extend above the static level surface is being displaced within the casing *and* the packing materials. Screens and packing materials that are *not* wet should not be included in the calculations. As obvious as this statement sounds the author has seen this occur many times.

Skin effects represent places where the borehole wall is disturbed or damaged and not well connected with the aquifer. This can occur from a sudden gush of water entering a hollow-stem drill string once a "knockout plate" is removed (Chapter 15). During this event, muddy water in the drill hole "slimes" or coats the borehole wall, making it difficult to remove the fines during development. Another example of borehole wall damage occurs when the drill bit encounters a larger cobble and wobbling of the drill string extends the disturbed zone out farther than usual. Any of these mishaps may render the monitoring well useless, even when extra time is being used during well development. When problems like the ones just explained happen during drilling and well completion it may be more appropriate to abandon the well and start another one.

Equipment Problems

While performing slug tests there are some unique problems that may occur in the field that may affect the data. As one arrives in the field and removes the well cap the first step should be to measure the SWL. This should be done a few times before starting any test. Without knowing the SWL there is no reference point to know when equilibrium has been reached. If one is sensing water-level changes with vented pressure transducers using cables with "breathable" (hollow) barometric lines, care must be made to watch for kinks in the line. Kinks may result in the data not being accurately transmitted to the data logger. Usually a minimum-sized loop of 2.5 cm (1 in) in the line is enough to prevent kinking (Figure 7.2). Allow 10 to 15 min of time to allow the cables to be stretched straight so they aren't moving with time. Duct tape or Gorilla Tape is usually a good medium to secure cables to the well bore. It should be noted that in the hot sun duct tape may "loosen its grip" and slippage or kinking of the line may result.

Another equipment problem results from idiosyncrasies associated with pressure transducers. For example, one of the last steps taken before one lowers the slug is to set the vented transducer reference level. All subsequent levels recorded by the data logger will be added or subtracted from the established reference level. If this is set before one allows cables to stretch or the transducer is secured, adjustments in the displacement data will be required. If one is performing a series of slug tests, one must reset the reference position for each well, as it will likely vary from well to well and contribute to confusion in interpretation of the data from several wells later on. Well sentinels may help alleviate this problem especially if one has taken good field notes.

Also, before performing a slug test using vented pressure transducers, one should "read" the level sensed by the transducer at the data logger. If the unit was not lowered sufficiently below the water surface the slug will probably disturb it. If the pressure transducer level reads zero, then there may be a couple of explanations. First, check the cable connections with the data logger. Are any of the small wires in the connecting junctions bent? Was the transducer or down-hole data logger stressed beyond its capacity? If you lower a 10-psi transducer into 35 ft of water, you may damage the unit and most likely obtain a reading of "zero feet of head" at the data logger (this is the voice of experience here).

With down-hole data loggers or well sentinels, the pressure-level sensor and data logger are contained within the same unit (Chapter 4). In this case the unit is programmed by a USB connection attached to a laptop computer to be programmed prior to lowering the unit down the well. With the well sentinel secured in place, a common scenario may be to program the time readings in linear scale to every 0.5 or 1.0 s and then perform three falling-head tests and rising-head tests. When the data are read back into the laptop one can tell when the slug displaced the SWL and reached equilibrium and was subsequently retrieved, repeating the process three times (Figure 7.6). The question comes up, how long does one wait between tests? This can be evaluated by measuring the SWL with an E-tape (this needs to be noted in a field book for possible level corrections, as the E-tape may disturb the displacement data). In high-transmissivity aquifers a linear time setting of 0.5 or 1.0 s may allow too much time to pass between measurements to provide sufficient detail needed for underdamped analysis (Section 7.4). In this case the slug test would need to be redone with log cycle time settings and the down-hole data logger reread or reprogrammed each time.

7.3 Analyzing Slug Tests—The Damped Case

Slug-testing methods and other topics in well hydraulics presume that Darcy's law is valid (Chapter 4). Specifically, this means that viscous forces or frictional effects during groundwater flow predominate over inertial forces, which are usually considered to be negligible (Chapter 3). The viscous forces allow sufficient friction to keep the water column from oscillating. This is known as the damped case. When the water column is displaced to the maximum (h_o or W_o), recovery occurs quickly at first (often up to 50% to 80%) and then tapers off. This is viewed in a data plot as a flattening in the slope. In high-transmissivity aquifers, water-level responses recover quickly and may have sufficient momentum to overcome the viscous forces and oscillate as an underdamped spring. What to do in this situation is discussed in Section 7.4.

This section presents how to analyze slug tests using the most common methods of slug-test analysis, the Hvorslev (1951) and Bouwer and Rice (1976) methods, followed by the author's modification. Frequently these are used for analysis in unconfined aquifers, but they have been used in analyzing partially penetrating and confined aquifers if the screen is well below the confining layer (Fetter 1994). Both methods provide an estimate of the hydraulic conductivity but have no means for estimating storativity. Papadopulos et al. (1973) provide a set of type curves that data can be matched to calculate transmissivity and storativity for confined aquifers. The data are plotted on semi-log paper as the ratio of displacement (h) over maximum displacement (h_o) as h/h_o (log scale) versus time in arithmetic scale. A description of this method is summarized in Fetter (1994). Additional discussion and methodologies are presented by Butler (1998) in his book.

Hvorslev (1951) Method

One of the most popular methods of analyzing slug-test data is known as the **Hvorslev** (1951) method (pronounced horse-loff). It is probably the quickest and simplest method. It is restricted to estimates of hydraulic conductivity within the screened zone of a well and can be applied to confined and unconfined aquifers. The maximum displacement value (h_o) is identified in the data. This is assumed to occur at time zero. The ratio of all other displacement data (h) to h_o or h/h_o are plotted on the ordinate (y-axis) in log scale, and time is plotted arithmetically along the abscissa (x-axis). There are no type curves with which to match the data. The plot is needed to identify the parameter (T_o) or the time it takes the water level to rise or fall to 37% of the initial maximum, h_o. This 37% is derived from the inverse of the natural log e. The value e is 2.7182818 ..., and the inverse of this number is 0.368 or roughly 0.37, hence the 37%. [It will be shown after the section "Bouwer and Rice Slug Test" how to analyze these data without using a ratio approach to obtain T_o (author's modification); there is an easier way.] Hvorslev (1951) developed several equations to meet certain conditions, but the most commonly used equation is

$$K = \frac{r^2 \ln (L/R)}{2LT_o} \tag{7.1}$$

where K = hydraulic conductivity
   r = inside radius of the well casing
   R = radius of borehole (i.e., effective radius)
   L = length of well screen plus filter packing if it extends above the top of the screen
   T_o = time to reach 37% of h_o

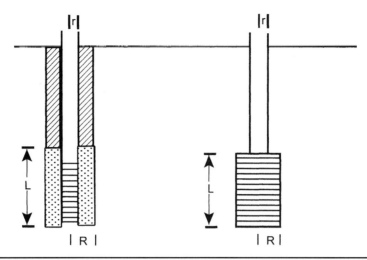

FIGURE 7.7 Well-completion information used to obtain parameters used in the Hvorslev (1951) method in Equation 7.1. The diagram on the right indicates a naturally developed well without a gravel pack. [*Modified from Fetter (1994).*]

Equation 7.1 is appropriate for any case where $L/R > 8$. This is easily satisfied unless the well screen and packing interval are very short. The other parameters are derived from the well-completion information (Figure 7.7; Fetter 1994). The radius of the well casing "r" is straightforward unless the displacement water level drops within the range of the screened interval. In monitoring wells that are completed to monitor light non-aqueous-phase liquids (LNAPLs) such as gasoline, the screen typically extends above the SWL (Figure 7.8, left side of the figure). In this case, the casing radius becomes

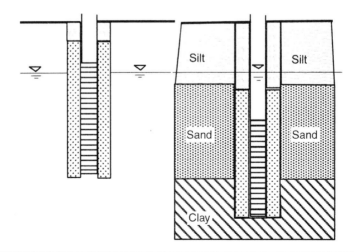

FIGURE 7.8 Examples of well-completion showing gravel pack above the water table and the screen and gravel pack partially completed in clay. The logic for extending the well screen above the SWL was to make sure there was well screen present between the range of high and low seasonal water-level changes.

an effective casing radius. If the gravel-packing material also extends above the top of the screen, the porosity of the gravel-packing material is taken into account to obtain an adjusted casing radius (r_A) by using Equation 7.2 (Bouwer 1989). The parameter η represents the porosity of the packing materials (Chapter 3):

$$r_A = [(1.0 - \eta)r^2 + \eta R^2]^{1/2} \tag{7.2}$$

Example 7.2 As an example to this adjustment for r_A, suppose the radius of the well screen is 0.167 ft (5.1 cm) and there is a gravel-packing material extending an additional 0.25 ft (7.6 cm) to the edge of the borehole with a porosity of approximately 25% in the packing materials around the well screen. The radius of the borehole is 0.42 ft (12.7 cm). The adjusted radius of the casing using Equation 7.2 would be

$$r_A = [(1.0 - 0.25)(0.17 \text{ ft})^2 + 0.25(0.4167 \text{ ft})^2]^{1/2} = 0.253 \text{ ft}$$

If the water level is influenced within the packing materials, then the height or magnitude of displacement is proportionally less than would be seen if all displacement changes occurred solely within the casing. The parameter R is the effective radius of the borehole. (The diameter of the borehole may vary, according to problems encountered during drilling.) If the borehole is damaged or coated with drilling fluids, then a "skin effect" will separate the borehole from the aquifer and the hydraulic conductivity may reflect the "skin" and not the aquifer. The parameter L *includes* the gravel-packing materials. It is a common error during analysis to use only the screen length. The full *saturated* gravel-pack length is capable of allowing water from the aquifer to enter the packing material and thus the screen. One must properly apply the well-completion information to perform the calculations. For example, Figure 7.8 shows an example of a well completed with screen above the SWL and an example of a well partially screened within a clay unit. It doesn't make much sense to extend the parameter L above the water table or into the clay unit if they are not contributing water. Example 7.2 shows how to adjust for the radius of the casing when displacement levels drop within the screen interval.

The author is aware of cases where the driller insists on completing a monitoring well with packing materials at a lower hydraulic conductivity than the aquifer. If this is true and the well is supposed to be used for estimating hydraulic properties, then the driller should be "slapped up the side of the head." In this case the hydraulic parameters estimated will only reflect the gravel pack and not the aquifer. It is the hydrogeologist's responsibility to make sure that well completion reflects the purposes of the well.

Example 7.3 The data plotted in Figure 7.9 comes from a well completed in a streamside mine tailings area in southwestern Montana. Monitoring wells were installed to learn about the flow characteristics in the impacted area. The raw data are found in Table 7.1. The adjusted time zero is at 0.0266 min. All data previous to this time are presumed to be attributed to "splash effects" and are not used. The reference of 100 ft (30.5 m) is subtracted from all the displacement data before plotting.

If one looks at the data from a low angle (now get your face near the table surface and look), roughly three straight-line segments can be identified. The first break is at approximately 0.25 min (15 s) and the second break is after 1.0 min. Approximately 50% of the total recovery took place during the first straight-line segment. The second and third segments are collectively referred to by the author as a "tailing effect" as a tapering recovery takes place. This is described in the paper by Bouwer (1989). From the diagram, a T_o of approximately 0.36 min is identified and used in Equation 7.1. The inside diameter of the casing is 2 in (5.1 cm), $r = 0.083$ ft, and the diameter of the borehole is approximately

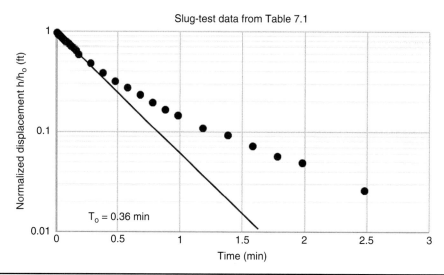

FIGURE 7.9 Slug test illustrating the Hvorslev slug test method. Flattening of the curve shows the tailing effect. See Example 7.3. (*Data are from Table 7.1.*)

8 in (22.8 cm), radius of $R = 0.33$ ft. The length of the screen (L) is 5 ft (1.5 m) and the packing materials are 7 ft (2.1 m):

$$K = \frac{r^2 \ln(L/R)}{2LT_o} = \frac{(0.083 \text{ ft})^2 \times \ln(7 \text{ ft}/0.33 \text{ ft})}{2 \times 7 \text{ ft} \times 0.36 \text{ min}}$$

$$= 4.2 \times 10^{-3} \text{ ft/min} \times 1{,}440 \text{ min/day} = 6 \text{ ft/day}$$

Bouwer and Rice Slug Test

Another commonly used slug-test method was developed by Bouwer and Rice (1976) that would estimate the hydraulic conductivity of aquifer materials from a single well. It was designed to account for the geometry of partially penetrating or fully penetrating wells in unconfined aquifers (Bouwer 1989). Many software packages provide the Hvorslev (1951) and Bouwer and Rice (1976) methods as the only choices for slug-test analysis. Having more than one method to calculate the hydraulic conductivity is helpful. It is especially useful for cross-checking values for comparison purposes. If the results of one method differ significantly from the other, then this is helpful in detecting calculation errors.

The slug-test data are collected, as before, except that Bouwer and Rice (1976) use a slightly different notation. They prefer to call the maximum displacement y_o instead of h_o and refer to the other displacement values as y_t instead of h. There is also *no* ratio of the data from y_o. Instead, y_o is chosen from the plot at time zero, unless there is a double-line effect (discussed later on in this section). The geometry of the variables used in the analysis is shown in Figure 7.10.

The variables are defined as follows:

h = depth from the water table to bedrock
L_w = length from bottom of screen to the water table
L_e = length of well screen plus any packing materials

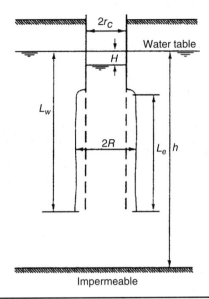

FIGURE 7.10   Geometry and parameters for a partially penetrating well screen in an unconfined aquifer. [*Bouwer (1989). Used with permission of Ground Water Publishing.*]

R_w = effective radius of the well bore or distance the displaced water dissipates away from the well bore

y = displacement interval (shown as H in Figure 7.10)

r_c = radius of the casing where the rise of the water level is measured

As was mentioned earlier, Bouwer (1989) suggested that if the water level rises within the open or screened portion of the well, an adjusted r_c value needs to be calculated that accounts for the porosity of the packing material. This adjustment is not necessary if the movement of the water level back to SWL conditions is restricted within the well casing. The hydraulic conductivity is calculated from

$$K = \frac{r_c^2 \ln(R_e / R_w)}{2L_e} \frac{1}{t} \ln \frac{y_o}{y_t} \qquad (7.3)$$

where y_o = the maximum displacement at time zero

y_t = displacement at a chosen time t

$\ln(R_e/R_w)$ = dimensionless ratio that is evaluated from analog curves (R_e is the effective distance over which the head displacement dissipates away from the well)

The analog data in Bouwer and Rice (1976) are used to fit one of two equations: one for the case that $L_w < h$ (i.e., partial penetration) and one for $L_w = h$ (i.e., fully penetrating). The equation for the partial penetration case is

$$\ln \frac{R_e}{R_w} = \left[\frac{1.1}{\ln (L_w / R_w)} + \frac{A + B \times \ln [(h - L_w)R_w]}{L_e / R_w} \right]^{-1} \qquad (7.4)$$

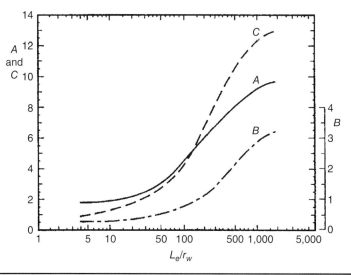

FIGURE 7.11 Parameters A, B, and C (dimensionless) plotted as a function of L_e/R_w used in Equations 7.4 and 7.5. [*Bouwer (1989). Used with permission of Ground Water Publishing.*]

and the equation for the fully penetrating case is

$$\ln \frac{R_e}{R_w} = \left[\frac{1.1}{\ln(L_w / R_w)} + \frac{C}{L_e / R_w} \right]^{-1} \qquad (7.5)$$

where A, B, and C are dimensionless numbers plotted as a function of L_e/R_w.

The plots of A, B, and C are shown in Figure 7.11. One of the possible benefits of the Bouwer and Rice (1976) method is that the aquifer geometry is taken into account. A problem arises in using this method if the depth to bedrock is not known or cannot be estimated.

Another phenomenon that may be observed in pumping tests is what is known as the double straight-line effect (Bouwer 1989). During initial displacement, the water in the gravel-packing material drains very quickly into the well, which may be indicated by a steep slope over a very short time period in the displacement data (Figure 7.12). When the water level in the packing materials equals the water level in the casing, the water level slows to reflect the contribution of the aquifer materials, thus forming a second straight line. Any deviations after the second straight line may be attributed to the "tailing effect" described earlier in the Hvorslev (1951) method. In the author's opinion the "tailing effect" can be attributed to well-completion issues, such as a skin effect. The steep first-line portion of the curve usually takes place over a matter of a few seconds. In Example 7.3, one may argue that a double-line effect is observed; however, the first straight line takes place over 15 s, and the calculated value is consistent with the lithology (fine to medium sand). It is the author's experience that steep first-line segments reflecting drainage from the gravel pack occur over a couple of seconds or a fraction of the total time of recovery.

Since the first (very steep) straight line is representative of the gravel-packing materials, the second straight line is used in performing the calculations. Notice that y_o is

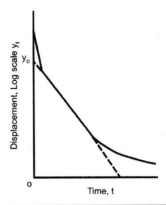

FIGURE 7.12 Displacement of head in a well versus time, showing the double straight-line effect. The first segment represents drainage of the gravel pack, the second segment the aquifer response, followed by a curved tailing effect. [*Modified from Bouwer (1989).*]

projected back to time zero (Figure 7.12). The straight line reflecting the aquifer response is projected below the tailing effect. Any time "*t*" can be chosen along the aquifer response line for the calculations. In Equation 7.3, the selected time "*t*" is matched with a corresponding displacement value y_t that is used when calculating the hydraulic conductivity. Since there are no requirements concerning 37% of the recovery, as in the Hvorslev method, it is convenient to select a time and displacement that are most clearly read from the plot.

Example 7.4   Three students were sent out to a tailings pond site (mining property) to conduct slug tests. Figure 7.13 shows a cross section of the setting. Each student selected a slug that would displace the same amount of water (in feet) as the well number. All students evaluated the data using the Bouwer and Rice (1976) method. Notes on the lithologic materials were misplaced or lost. Implications concerning the nature of the tailings were to be evaluated from the slug-test results. The data are shown in Table 7.2, and the plots and calculations are shown below.

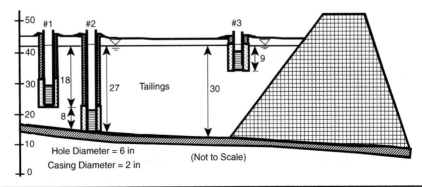

FIGURE 7.13   Cross-section view of the tailings pond with monitoring-well-completion information described in Example 7.4.

Well 1		Well 2		Well 3	
Time (s)	Displacement (ft)	Time (s)	Displacement (ft)	Time (s)	Displacement (ft)
0	1.00	0	2.00	0	3.00
1	0.78	1	1.76	1	2.10
2	0.61	2	1.54	3	1.37
3	0.48	3	1.34	5	1.05
4	0.37	4	1.18	7	0.98
5	0.29	5	1.00	9	0.90
6	0.22	6	0.88	14	0.78
7	0.17	7	0.78	19	0.69
8	0.13	8	0.68	24	0.60
9	0.10	9	0.60	29	0.54
10	0.081	10	0.54	34	0.48
11	0.062	11	0.48	39	0.43
12	0.049	12	0.41	44	0.38
13	0.039	14	0.35	49	0.34
14	0.03	16	0.30	59	0.30
		18	0.28	69	0.28
		20	0.263	79	0.26
		22	0.248	89	0.24
		24	0.236	99	0.22
		26	0.224		
		28	0.213		
		30	0.205		

TABLE 7.2 Slug-Test Data from a Tailings Pond Site

Well 1 Since well 1 is partially penetrating, Equation 7.4 is used, where

$$L_e = 9 \text{ ft}, L_w = 18 \text{ ft}, r = 0.083 \text{ ft}, R_w = 0.25 \text{ ft}, h = 26 \text{ ft},$$
$$A = 2.6, B = 0.45, t = 9 \text{ s}, y_t = 0.1 \text{ ft}, y_o = 1.0 \text{ ft}$$

$$\ln \frac{R_e}{R_w} = \left[\frac{1.1}{\ln (18 \text{ ft}/0.25 \text{ ft})} + \frac{2.6 + 0.45 \times \ln [(26 \text{ ft} - 18 \text{ ft})/0.25 \text{ ft}]}{9 \text{ ft}/0.25 \text{ ft}} \right]^{-1}$$
$$= 2.683$$

$$K = \frac{(0.083 \text{ ft})^2 (2.683)}{2(9 \text{ ft})} \frac{1}{9 \text{ s}} \ln \frac{(1.0 \text{ ft})}{(0.1 \text{ ft})} = 2.63 \times 10^{-4} \text{ ft/s} = 23 \text{ ft/day}$$

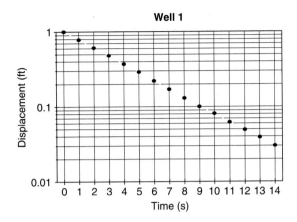

Well 2   Since well 2 is fully penetrating, Equation 7.5 is used, where

$$L_e = 8 \text{ ft}, \, L_w = 27 \text{ ft}, \, r = 0.083 \text{ ft}, \, R_w = 0.25 \text{ ft}, \, h = 27 \text{ ft}, \, C = 2.2$$

$$t = 5 \text{ s}, \, y_t = 1.0 \text{ ft}, \, y_o = 2.0 \text{ ft}$$

$$\ln \frac{R_e}{R_w} = \left[\frac{1.1}{\ln(27 \text{ ft}/0.25 \text{ ft})} + \frac{2.2}{8 \text{ ft}/0.25 \text{ ft}} \right]^{-1} = 3.29$$

$$K = \frac{(0.083 \text{ ft})^2(3.293)}{2(8 \text{ ft})} \frac{1}{5 \text{ s}} \ln \frac{(2.0 \text{ ft})}{(1.0) \text{ ft}} = 1.98 \times 10^{-4} \text{ ft/s} = 17 \text{ ft/day}$$

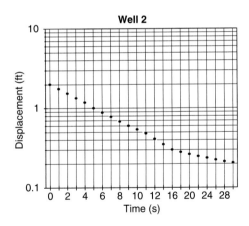

Well 3   Since well 3 is partially penetrating, Equation 7.4 is used, where

$$L_e = 9 \text{ ft}, \, L_w = 9 \text{ ft}, \, r = 0.083 \text{ ft}, \, R_w = 0.25 \text{ ft}, \, h = 30 \text{ ft}$$

$$A = 2.6, \, B = 0.45, \, t = 9 \text{ s}, \, y_t = 0.9 \text{ ft}, \, y_o = 1.12 \text{ ft}$$

$$\ln \frac{R_e}{R_w} = \left[\frac{1.1}{\ln(9 \text{ ft}/0.25 \text{ ft})} + \frac{2.6 + 0.45 \times \ln[(30 \text{ ft} - 9 \text{ ft})/0.25 \text{ ft}]}{9 \text{ ft}/0.25 \text{ ft}} \right]^{-1} = 2.3$$

$$K = \frac{(0.083 \text{ ft})^2(2.30)}{2(9 \text{ ft})} \frac{1}{9 \text{ s}} \ln \frac{1.12 \text{ ft}}{0.9 \text{ ft}} = 2.14 \times 10^{-5} \text{ ft/s} = 2 \text{ ft/day}$$

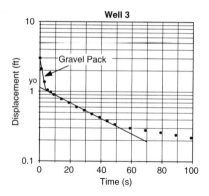

Even though the lithologic logs were lost, one can make some inferences concerning the hydraulic and lithologic properties of the tailings sediments from the location of the wells and the calculations. It is evident that the hydraulic conductivity decreases toward the dam. This is something that might be expected. As the tailings are slurried toward the dam, the coarsest fraction would likely settle out first. The finer materials would continue toward the dam. It appears that the coarsest fraction lies near the bottom with a fining upward sequence to the sediments (Chapter 2). This suggests that higher hydraulic conductivities may be found nearer the bedrock in a direction away from the dam.

Common Errors Made in Analyzing Slug-Test Data

There are some common ways that individuals make errors in performing calculations of hydraulic conductivity from slug-test data. The most routine ones are presented below with an example with discussion following (Example 7.5):

- One uses the diameter of a casing or borehole instead of the radius in the calculations. All of the methods require the *radius* instead of the *diameter*.

- Forgetting to include the gravel-packing materials above and below the screen as the contributing length L or Le. Many have tendency to use only the screen interval, which leads to less accurate results.

- Be careful not to average the data to make a single "best fit" line. If the data are roughly in a straight line, then this is no problem, but including the tailing effect portion along with the aquifer response data will result in erroneously underestimating the hydraulic conductivity. Get your face close to the table to evaluate the plot to detect the appropriate straight-line section from which to draw the line. The aquifer response will likely reflect more than 50% of the recovery data. Those who fit the "tailing effect" are most likely fitting the last 25% or so of the recovering data. Ask the question, which part of the data reflects where most of the recovery is taking place?

- Don't forget to square the casing radius value. In both methods described above, the radius of the casing is squared in the calculations. In the Bouwer and Rice (1976) method, it is important to take the inverse of the bracketed quantity in Equations 7.4 and 7.5.

- In the Bouwer and Rice (1976) method, in the partial-penetration case, make sure that the constants A and B are picked from the appropriate scales (the B scale is on the opposite side of the graph as the A scale).

Example 7.5 The author has observed numerous examples of "poor fits" from data plots performed by consultants, resulting in hydraulic parameters that make no physical sense compared to the aquifer materials. A typical example of one of these plots is shown in Figure 7.14. The displacement data are in feet and the selected time scale is in days. Notice that the maximum displacement occurs at about 4.2 ft (1.3 m) and much of the recovery data (down to about 2 ft is a nice straight line) are followed by a curved tailing effect. The presumed aquifer response was fit to the tailing effect (between 0.8 and 0.4 ft), where less than 10% of the recovery data occur (most likely associated with well-completion issues). The time scale of days was a poor choice, as it compressed the data where the aquifer response was occurring. The scale is also confusing because each tick mark on the x-axis represents 0.008 days (not exactly a convenient scale to work with). From the data-fit the consultants arrived at an estimate of hydraulic conductivity of approximately 0.03 ft/day (0.001 m/day) for fine sand. If one fits the steep slope of the curve, one obtains a value closer to 2 ft/day (0.6 m/day). The author did not have confidence in the consultants, work and redid the slug tests obtaining values of approximately 10 ft/day (3 m/day). It was unknown whether the consultants wished to skew the data to the lower side, as they had contamination issues and did not want to show much impact.

How to Analyze Slug Tests Using Both Damped Methods from a Single Plot

Now that the Hvorslev and Bouwer and Rice methods have been presented, it is useful to see how to perform the calculations using both methods from a single plot. The simplest way of dealing with slug-test data is to plot displacement versus time, as in the Bouwer and Rice (1976) method. Recall that in the Hvorslev (1951) method the data were normalized in ratio format with h/h_o so that at time zero the value would be near 1.0. From this plot, the time at 37% was selected as T_o to be used in the calculations. Any

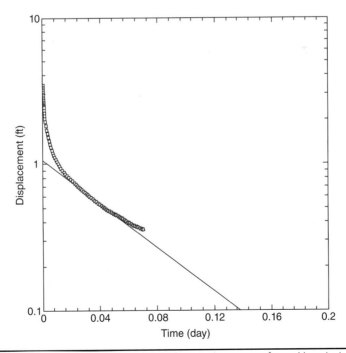

Figure 7.14 Example of a "poor fit" of the data from a slug test performed by a hydrogeologic consulting company. The data should have been plotted with time in minutes or seconds, not days.

displacement versus time plot can be used for the equation in the Hvorslev (1951) method (Equation 7.1). One only needs to determine what the maximum displacement (h_o or y_o) is and then take 37% of this maximum value for the corresponding T_o value.

For example, in well 3 in Example 7.4, if the projected h_o in the tailings pond is 1.12 ft, then 37% of this is 0.41 ft. T_o would be approximately 40 s. This would be used in Equation 7.1 to obtain

$$K = \frac{(0.083\ \text{ft})^2\ \ln(9\ \text{ft}/0.25\ \text{ft})}{2(9\ \text{ft})\ 40\ \text{s}} = 3.4 \times 10^{-5}\ \text{ft/s} = 3\ \text{ft/day}$$

In comparing the calculated values between the two damped case methods, the Hvorslev (1951) method yielded a larger number, although both values are smaller than those of the other two wells. The reader is invited to make a comparison of well 1 and well 2 using the Hvorslev (1951) method (taking 37% of y_o to determine T_o) to recalculate the hydraulic conductivity. Similar trends to the Bouwer and Rice (1976) method should be observed; however, the numbers are not exactly the same. This is typical of aquifer hydraulics (Chapter 6); in some cases, one method may yield a higher or lower value than another method, but should yield comparable trends.

The author noticed that when the well was fully penetrating (the case of well 2) the numbers from the two methods were almost the same. The Hvorslev (1951) method produces a higher hydraulic conductivity value for wells 1 and 3. In both of these cases, the wells are partially penetrating. The Bouwer and Rice (1976) method takes into account partial-penetration effects by including the aquifer geometry and may more accurately reflect the conditions. This is a subject for additional study and debate.

7.4   Analyzing Slug Tests—The Underdamped Case

Unlike the previous sections where the methodologies for the damped case have been presented, there are occasions in high-transmissivity aquifers where the water-level response during slug tests behaves like an underdamped spring. The water level literally undulates back and forth above and below the static equilibrium level in an oscillatory motion. This behavior is unexpected by many field personnel, since slug tests were originally developed for low-transmissivity aquifers (Lohman 1972). Although different methods for analyzing these data sets exist, it is the author's experience that many individuals who encounter oscillatory water-level data during slug tests either deem the data "confusing or unsolvable" or simply assign a "high" hydraulic conductivity. Analytical methods for oscillatory data in the literature are mathematically complex and challenging to use (van der Kamp 1976; Uffink 1984; Kipp 1985; McElwee and Zenner 1998; Butler and Zhan 2004; Audouin and Bodin 2007).

Simplifying methods for practitioners have been developed for spreadsheets for the van der Kamp (1976) method (Wylie and Magnuson 1995) and for the Kipp (1985) method (Weight and Wittman 1999). One potential drawback is that both methods assume there is no skin effect (nonlinear head losses), although the equations in the Kipp (1985) method do allow for skin effects. It is known that both positive and negative skins can occur (Yang and Gates 1997; Butler 1998); however, the impact of skin effects is often evaluated with lower hydraulic conductivity formations and can be significantly reduced with proper well development (Chapter 15). The author believes that a skin effect is a "fudge factor" associated with poor well completion and is difficult to quantify.

A relatively small group of researchers and practitioners have published information on water-level responses that indicate inertial effects resulting in oscillations. van der Kamp (1976) presented the first significant paper leading to the analysis of oscillatory data using a sinusoidal approximation method. In 1985, Kipp expanded the theory presented by Bredehoeft et al. (1966) to produce a series of type curves. Butler et al. (2003) have developed a spreadsheet method for slug tests performed on partially penetrating unconfined and confined aquifers, respectively, which has been adopted into software packages (Duffield 2018; McCall and Shipley 2018). There are other underdamped methodologies that have been developed by Uffink (1984), Springer and Gelhar (1991), McElwee et al. (1992), and Butler and Garnett (2000). Some of these are also similar to the method by Kipp (1985) in that they also estimate a damping parameter (ζ) and the effective water-column height (L_e) and have been tested in the field. Slug-test data have also been numerically processed in the Fourier-frequency domain (Crump 1976; McElwee and Zenner 1998) in a manner that takes into account the inertia of the water column, turbulent head losses in the well (Chapter 5), and disregards aquifer storage to make a quasi-steady-state interpretation (Audouin and Bodin 2007).

Nature of Underdamped Behavior

The first significant explanations of underdamped oscillatory behavior were offered by Bredehoeft et al. (1966). They were intrigued by a couple of examples that occurred in Florida and Georgia. In Florida, a 12-in well completed in a cavernous limestone aquifer (transmissivity of approximately 120,000 ft^2/day) responded in an underdamped oscillatory fashion when a float was periodically raised and lowered. This behavior was picked up by a pressure transducer within the well. In Georgia, a recorder located 200 ft from a city supply well showed an oscillatory response every time the city well's pump kicked on (Bredehoeft et al. 1966). Additional geologic field settings are described by Butler (1998). According to Bredehoeft et al. (1966), systems are said to be

- overdamped or damped, where no oscillations occur following an initial disturbance, indicates that viscous forces or Darcian conditions dominate the system (Chapter 4);
- underdamped, where oscillations occur as in a damped sine wave, indicating that inertial forces are significant; and
- critically damped, where the system is in transition between the two.

Bredehoeft et al. (1966) first studied inertial effects by simulating an oscillatory system using an electric analog but did not develop a method of slug-test analysis. They present a relationship between transmissivity and what they term as the "effective column height" (roughly the distance from the midpoint of the screen to the SWL surface) for a given transmissivity and storativity (Figure 7.15). The fitted line in Figure 7.15 represents where the critically damped transition takes place. A sinusoidal approximation to the underdamped oscillatory slug-test response was presented by van der Kamp in 1976. Another sinusoidal approximation, using a pneumatic procedure, is described by Uffink (1984) in Kruseman and de Ridder (1990) and by McCall and Shipley (2018). Kipp (1985) was later able to develop a method of slug-test analysis with type curves. The van der Kamp (1976) approximation method is easily performed using a spreadsheet solution developed by Wylie and Magnuson (1995). The type curves of the Kipp (1985) method along with a spreadsheet solution were developed by Weight and Wittman (1999).

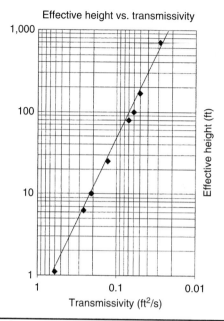

FIGURE 7.15 Relationship of transmissivity to effective water column height. The area above the plotted line represents the underdamped case, whereas the area below the plotted line represents the damped case, and the fitted line represents the critically damped transition zone. [*Modified after Bredehoeft et al. (1966).*]

Example 7.6 During the 1988 Hydrogeology Field Camp of Montana Tech of the University of Montana, underdamped oscillatory responses from slug tests were observed in wells west of Butte, Montana (Manchester 1990). Matching oscillations were noticed from two wells located 40 ft apart, although the observation well was significantly dampened. Follow-up pneumatic slug tests were performed, including one in a well "completed" in the swimming pool on the Montana Tech campus to simulate "infinite" transmissivity. A spectacular oscillatory response was observed (Manchester 1990).

Another "infinite" transmissivity example was observed in a 12-in production well after a blasting charge was set off. The 12-in well was supposed to be drilled into a mine adit that was successfully penetrated by two adjacent wells A and B (Figure 7.16) that *were* completed in the mine adit. The adit was believed to be missed by only 1 to 3 ft. A blasting charge was set off to breach the connection (Figure 7.17). A pressure transducer in well B was destroyed during the process so no data were collected; however, a reflected beam of light from a mirror (Figure 15.41) revealed a clear oscillatory response for over 7 min! The production well was subsequently tested at 200 gpm with 0.6 ft of drawdown observed, indicating that the blast was successful at connecting the well with the adit, resulting in an excellent production well.

Weight and Wittman (1999) describe oscillatory water-level responses while collecting slug-test data from wells scattered within the Beaverhead Groundwater Project Area (BGPA) in southwestern Montana (Figure 7.4). Point estimates of hydraulic conductivity values within the BGPA were needed to provide additional hydraulic control within a large area that was to be numerically modeled (Wittman 1997). These data would augment other hydraulic parameter estimates from pumping tests and

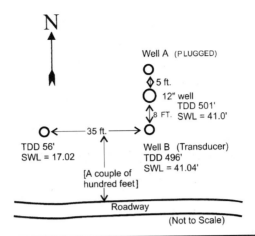

FIGURE 7.16 A 12-in production well missed the void space of a mine adit that was successfully encountered by wells A and B. A blasting charge successfully breached the adit wall and an oscillatory behavior occurred in the water level for over 7 min.

FIGURE 7.17 Constructing explosives pack by a local expert.

previous slug tests. Of the 30-plus wells that were tested, over half of the wells showed an underdamped oscillatory response. It became necessary to analyze these data or have significantly less hydraulic control within the model domain.

Example 7.7 Perhaps one of the more spectacular oscillatory data sets collected by the author occurred west of Butte, Montana, in July 1996, during the Hydrogeology Field Camp at Montana Tech of the University of Montana. The slug test was conducted prior to a constant discharge rate test (Figure 7.18). The normalized oscillatory response data are shown in Table 7.3 and plotted in Figure 7.19. The effective column height was very tall, approximately 190 ft (58 m).

FIGURE 7.18   Constant-rate discharge pumping test, Hydrogeology Field Camp, July 1996.

Effective Stress and Elastic Behavior

Physically, the oscillatory behavior phenomenon can be partially explained by the concept of effective stress and inertia, resulting in an elastic wave propagation. The effective stress is the difference between the total stress and the fluid pressure (Chapter 3). Within an aquifer, the total load from the water and geologic materials is "felt" above a given plane (Fetter 1994). The load carried by the mineral skeleton is somewhat offset by the fluid pressure, resulting in an effective stress:

$$\sigma_t = \sigma_e + P_f$$

(7.6)

where σ_t = total stress
σ_e = effective stress
P_f = fluid pressure

In confined aquifers, changes in pressure can occur with very minute changes in actual thickness (Fetter 1994). In this condition, there is very little change in the total stress; however, the effective stress increases as the fluid pressure decreases and vice versa. The exchange back and forth in stresses produces an elastic-like response in a confined aquifer from being under pressure. The inertial effects of the displaced water column eventually balance out with the frictional forces of the geologic media and well-completion materials (van der Kamp 1976).

The phenomenon of elastic behavior does not appear to be restricted to confined aquifers only. It is the author's observation that *any* aquifer that is being stressed during the first few seconds to a minute or so produces water from compression of the mineral skeleton or expansion of water (Chapter 6), even in unconfined aquifers (Prickett 1965; Neuman 1979; Moench 1993). This essentially describes the specific storage (S_s) or elastic storage viewed on a type A Theis (1935) curve (or the early-time data; Chapter 6).

Time (s)	Displacement (ft)	Time (s)	Displacement (ft)
4.0	−0.019	25.0	−1.471
4.5	−0.529	26.0	−1.377
5.0	−1.011	27.0	−1.112
5.5	−1.458	28.0	−0.718
6.0	−1.846	29.0	−0.252
6.5	−2.158	30.0	0.223
7.0	−2.384	31.0	0.642
7.5	−2.529	32.0	0.954
8.0	−2.583	33.0	1.121
8.5	−2.548	34.0	1.134
9.0	−2.438	35.0	0.995
9.5	−2.249	36.0	0.734
10.0	−1.997	37.0	0.390
10.5	−1.682	38.0	0.012
11.0	−1.320	39.0	−0.344
11.5	−0.923	40.0	−0.636
12.0	−0.510	41.0	−0.832
12.5	−0.095	42.0	−0.907
13.0	0.315	43.0	−0.863
13.5	0.696	44.0	−0.712
14.0	1.039	45.0	−0.476
14.5	1.335	46.0	−0.189
15.0	1.571	47.0	0.107
15.5	1.741	48.0	0.375
16.0	1.845	49.0	0.576
16.5	1.880	50.0	0.693
17.0	1.849	51.0	0.715
17.5	1.754	52.0	0.639
18.0	1.603	53.0	0.485
18.5	1.398	54.0	0.277
19.0	1.152	55.0	0.044
19.5	0.872	56.0	−0.186
20.0	0.573	57.0	−0.375
21.0	−0.044	58.0	−0.510
22.0	−0.621	59.0	−0.570
23.0	−1.081	60.0	−0.551
24.0	−1.373		

TABLE 7.3 Normalized Oscillatory Response Data, Sand Creek Well, West of Butte, Montana

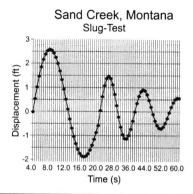

FIGURE 7.19   Slug test illustrating oscillatory behavior, described in Example 7.7. (*Data from Table 7.3.*)

Higher-transmissivity aquifers tend to propagate perturbations within an aquifer faster, indicating significant inertial effects over lower-transmissivity aquifers. An example of the quick transition of confined to unconfined conditions is presented in Example 7.8.

Example 7.8   Near Kalispell, in northwestern Montana, is a glaciofluvial setting where a shallow coarse gravel aquifer (<30 ft or 9.1 m) overlies lacustrine clays. A constant-discharge pumping test at 350 gpm (1,908 m³/day) was conducted in a 10-in (25 cm) well to evaluate the hydraulic properties of the aquifer (King 1988). A plot of the data from an observation well located 37 ft away from the pumping well is shown in Figure 7.20. No transducer was on hand to collect the earliest time data (<1 min); however, delayed yield is evident within a couple of minutes (Chapter 6). Delayed yield indicates that the aquifer is unconfined and that the elastic response of the aquifer must have occurred within a minute or less, the approximate time of a slug test in a high-transmissivity aquifer. The transmissivity calculated was very high for a porous media aquifer, about 70,000 ft²/day.

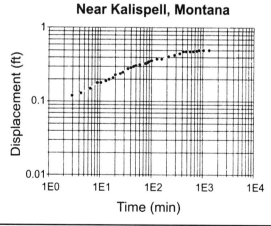

FIGURE 7.20   Displacement (ft) versus time (min) with pumping rate of 350 gpm in a pumping test near Kalispell, Montana in 1988. Note that a delayed-yield effect is already under way, and any confined response occurred previous to 30 s. The data were all collected via an e-tape (Example 7.8).

During an instantaneous removal of a slug, the fluid pressure suddenly decreases, causing an increase in the effective stress. The mineral skeleton has to take more of the load, and this squeezes the aquifer slightly. Water from the aquifer then gushes into the well bore past the original "pre-slug" static equilibrium position resulting in an upward oscillation. The inertial effects, that is, rate of change of momentum of the water column, pulled downward by gravity, gushes back outward into the aquifer, increasing the fluid pressure once again during the downward oscillation. This can occur because the water column height is sufficiently high or the transmissivity is sufficiently large to overcome the critically damped transition zone (Bredehoeft et al. 1966). This process repeats itself until the aquifer materials and well-completion materials finally dampen the process. It can be thought of as an elastic response created by a pressure wave propagating radially from the well. The higher the transmissivity, the easier it is for the pressure wave to propagate outward. Wylie and Magnuson (1995) draw the analogy of a mechanical system of a spring in a viscous medium, where the water column is the mass and the aquifer is the spring. In the "infinite" aquifer scenario of a cavernous limestone, swimming pool, or mine adit, eventually the well-completion materials and boundary effects of the cavity cause damping to finally occur. In the above discussion, it is assumed that there were no skin effects.

Two methods were previously presented on how to analyze data that behave in an underdamped fashion in the previous editions of this book: the van der Kamp (1976) and the Kipp (1985) methods. Each was derived from the spreadsheet methods developed by Wylie and Magnuson (1995) and Weight and Wittman (1999). However, since user-friendly commercially available software options exist [www.aqtesolv.com/slug_test_analysis.htm from Duffield (2018) and https://geoprobe.com/slug-test-analysis-software-new-v20 from McCall and Shipley (2018)], the previous edition's detailed examples have been omitted. Nevertheless, it is worth reviewing the theory and ideas behind underdamped methods.

van der Kamp Method

Garth van der Kamp (1976) developed a sinusoidal approximation method to under-damped slug-test data. A brief theoretical development is presented, followed by the use of a spreadsheet analog with an example. The geometry needed to discuss the analysis of the well response is shown in Figure 7.21.

The effective length of the water column (L), as defined by Cooper et al. (1967) with the terms in the equation shown in Figure 7.21, is

$$L = L_c + 3/8L_f \tag{7.7}$$

The equation for the balance of forces within the well filter, including inertial effects, is given in Equation 7.8:

$$\frac{d^2w}{dt^2} + \frac{g}{L}w = \frac{g}{L}h_f \tag{7.8}$$

where w = transient water level in the well (L)
$\quad g$ = acceleration due to gravity (L/T^2)
$\quad h_f$ = hydraulic head of water in the filter (L)
$\quad L$ = length of the effective water column (L)

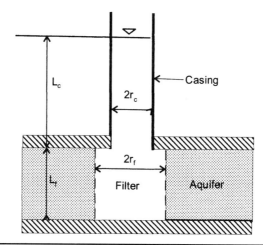

FIGURE 7.21   Geometry of slug test as described in the van der Kamp slug-test method. [*Modified from van der Kamp (1976).*]

If the transient part of the hydraulic head can be assumed to be governed by the standard equation for radial flow and h_f is assumed to be the same everywhere inside of the filter pack, an approximate solution can be used to simulate an exponentially damped cyclic fluctuation (Wylie and Magnuson 1995):

$$\omega(t) = \omega_o e^{-\gamma t} \cos(\omega t) \tag{7.9}$$

where γ = damping constant (gamma) (T^{-1})
$\quad\quad\omega$ = angular frequency of the oscillation (omega) (T^{-1})
$\quad\quad\omega_o$ = initial water-level displacement from slug removal (L)
$\quad\quad T$ = time from the start of the test (T)

The field data are fitted to the approximation equation to determine values for γ and ω. It is presumed that storativity (S) and the filter radius (r_f) are known or can be estimated.

Example 7.9   An example of a sequential fit to the data from Table 7.3 is shown in Figure 7.22. As in the Kipp (1985) method, the storativity values are somewhat sensitive to the transmissivity outcomes. A discussion about this follows the presentation of the Kipp (1985) method.

Kipp Method

Another method that can be used to estimate transmissivity values for aquifers was developed by Kipp (1985). He was the first to expand the earlier theory developed by Bredehoeft et al. (1966) that takes into account both well-bore storage and inertial effects and produces a series of type curves that allow for oscillatory responses to be analyzed (Figure 7.23).

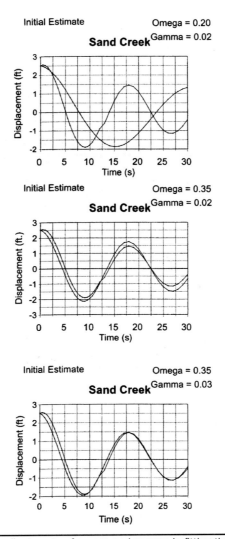

FIGURE 7.22 Sequential adjustments of omega and gamma in fitting the Sand Creek slug-test data (Table 7.3; Figure 7.19) using the sinusoidal approximation spreadsheet algorithm by Wylie and Magnuson (1995).

A summary of the basic assumptions given by Kipp (1985) was summarized by Manchester (1990) with additional discussion by Weight and Wittman (1999):

- Confined aquifer, bounded on the top and bottom
- Uniform aquifer thickness
- Well fully penetrates the aquifer
- Aquifer of infinite areal extent

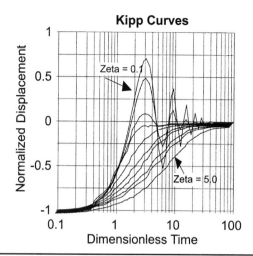

FIGURE 7.23 Type curves for normalized oscillatory slug-test data sets. Damping parameters (z) are 0.1, 0.2, 0.5, 0.7, 1.0, 1.5, 2.0, 3.0, 4.0, and 5.0, with the smallest value representing the most oscillatory responses. [*Weight and Wittman (1999). Used with permission from Groundwater Publishing Co.*]

- Homogenous porosity (η) and matrix compressibilities
- Flat potentiometric surface
- Delayed yield not considered (Chapter 6)
- Constant water density in the well bore and a constant compressibility in the aquifer

Kipp (1985) was able to generate dimensionless variables and parameters that were plotted into a series of type curves (Figure 7.23). For example, there is a dimensionless inertial parameter (β) that can be held constant while different values of the storage parameter (α) are used to produce the dimensionless damping parameter (ζ). The tabled values from the Kipp (1985) paper were entered into a spreadsheet (Weight and Wittman 1999).

Methodology

The steps of the Kipp (1985) methodology is described from the paper by Weight and Wittman (1999):

1. The uploaded data should first be converted into time units of seconds (*x*-axis) and displacement values in meters (*y*-axis) to match the type curves. Read step 2 before this step is completed.

2. Evaluate the raw data and determine the reference point that the oscillations eventually dampen out to. This should compare closely to the original SWL or transducer reference point. For example, if the reference point in the raw data is 10.0 ft (3 m) and the data eventually dampen out to 10.019 ft, then 10.019 ft becomes the reference point. This also becomes the zero-displacement reference level (*w*). All other displaced data are referenced to this final damped

value so that differences will oscillate above and below the zero-displacement line.

3. Determine the maximum displacement (w_o). (Note: Sometimes there is some confusion here if the data logger is turned on *before* the slug has been pulled; Section 7.1.) The author has seen cases where the data logger records data as the slug is being pulled from the well bore and has not broken the water surface yet. This results in an abnormally large maximum displacement value (Table 7.1). If this occurs, use the next largest displacement value as w_o. This is also the reason for activating the data logger a few seconds before pulling the slug. Furthermore, it is important to also rescale the time value back to zero at the chosen w_o value.

4. Calculate $-w/w_o$ values for the displaced data. The minus sign is there because the type curves are viewed from a negative displacement perspective (i.e., rising-head test). Graph time in seconds (x-axis in log scale) versus $-w/w_o$ (y-axis in arithmetic scale).

5. Match the type curves with the graphed data. As you do this you will determine three values: the damping parameter (ζ), the first peak $\hat{t}$, and the subsequent inflexion point t. Highly oscillatory systems, with parameters smaller than 0.5, should be matched to the first largest inflexion point or peak to select time $\hat{t}$. Time t is picked at the second inflexion point (where the displaced data first change from concave downward to concave upward, usually near where it crosses the displacement axis; see the examples that follow). For oscillatory systems with a parameter (ζ) greater than 0.7, use the lower inflexion point (where concave upward just begins, again consult the example).

6. Calculate the effective column length (L_e) using $\hat{t}$ and t using Equation 7.9 (Kipp 1985):

$$\hat{t} = \frac{t}{(L_e / g)^{1/2}} \qquad (7.10)$$

where L_e = effective static water length (m)
g = acceleration of gravity (m/s²)
$\hat{t}$ = dimensionless time
t = time

7. Calculate the dimensionless storage parameter (α). This takes into account well-bore storage by considering the radius of the casing. It is presumed that storativity (S) is known or can be estimated (Kipp 1985):

$$\alpha = \frac{r_c^2}{2 \times S \times r_s^2} \qquad (7.11)$$

where α = dimensionless storage parameter
r_c^2 = radius of the casing (m)
S = storativity
r_s^2 = radius of the screen (m)

8. Solve for the dimensionless inertial parameter (β) (Equation 28 in Kipp 1985). The critical damping parameter (ζ) is selected or estimated from the fit of the data to the type curves. There is additional discussion following step 9:

$$\zeta = \frac{\alpha(\sigma + 1/4 \times \ln(\beta))}{2 \times \beta^{1/2}} \tag{7.12}$$

where ζ = dimensionless damping factor
α = dimensionless storage parameter
β = dimensionless inertial parameter
σ = dimensionless skin factor

9. Calculate the transmissivity (T) in m²/s, using the previously derived parameters (Kipp 1985):

$$\beta = \frac{L_e}{g}\left(\frac{T}{S \times r_s^2}\right)^2 \tag{7.13}$$

Equation 7.12, used to solve for β, is most easily derived by setting up an iterative solver on a spreadsheet. The reader is encouraged to use the data from Tables 7.3 and 7.4 to evaluate an underdamped data set in a spreadsheet program. A few comments about the data from Table 7.3 are presented in Example 7.10 and followed by comments about the data from Table 7.4 in Example 7.11. It is also recommended that these data be tested in commercially available software.

Example 7.10   Input in the data from Table 7.3 into a spreadsheet. The displacement values in feet need to be converted to meters (multiply by 0.3048 m/ft); the time data in seconds are already in the appropriate format.

Evaluate the data to determine a reference point where the data dampen to. In this case, assume this is at 0.0 ft since the time/displacement data stop before full dampening takes place.

The maximum observable displacement w_o is 0.787 m (2.583 ft). The data previous to 4 s are clouded by splash effects; this makes it difficult to determine where to rescale time zero. The author subtracted 6 s from the time data to rescale the data back to time zero.

By setting up a column on the spreadsheet of $-w/w_o$ values, by plotting time versus $-w/w_o$ values the reader should see a plot like is shown in Figure 7.24.

Example 7.11   To illustrate an example of a larger damping parameter (ζ) (Kipp 1985), here is a data set from well 93-2 south of Dillon, Montana, along Blacktail Deer Creek (Figure 7.4). The data are shown in Table 7.4.

Input in the data from Table 7.4 into a spreadsheet. The displacement values in feet need to be converted to meters (multiply by 0.3048 m/ft); the time data in seconds are already in the appropriate format.

Evaluate to reference point where the data dampen to. In this case, dampening takes place near 0.0 ft (0.003 ft).

The maximum observable displacement w_o is 0.417 m (1.367 ft). (The data previous to 1.8 s are clouded by splash effects.) The author subtracted 1.8 s from the time data to rescale the data back to a new time zero.

By setting up a column on the spreadsheet of $-w/w_o$ values, by plotting time versus $-w/w_o$ values the reader should see a plot like is shown in Figure 7.25.

Time (s)	Displacement (ft)	Time (s)	Displacement (ft)	Time (s)	Displacement (ft)
0.20	0.213	4.8	0.980	20.0	0.194
0.4	2.261	5.2	0.948	21.0	0.171
0.6	0.546	5.6	0.914	22.0	0.149
0.8	0.371	6.0	0.875	23.0	0.133
1.0	0.476	6.4	0.844	24.0	0.114
1.2	1.082	6.8	0.812	25.0	0.102
1.4	1.183	7.2	0.780	26.0	0.089
1.6	1.316	7.6	0.749	27.0	0.076
1.8	1.367	8.0	0.72	28.0	0.067
2.0	1.348	8.6	0.679	29.0	0.054
2.2	1.297	9.2	0.638	30.0	0.048
2.4	1.266	9.8	0.600	31.0	0.041
2.6	1.234	10.4	0.565	32.0	0.035
2.8	1.202	11.0	0.533	33.0	0.029
3.0	1.186	12.0	0.479	34.0	0.026
3.2	1.158	13.0	0.431	35.0	0.019
3.4	1.129	14.0	0.387	36.0	0.016
3.6	1.113	15.0	0.346	37.0	0.013
3.8	1.085	16.0	0.311	38.0	0.007
4.0	1.066	17.0	0.276	40.0	0.007
4.4	1.037	18.0	0.247	42.0	0.003
		19.0	0.219	44.0	0.003

TABLE 7.4   Data Set from Well 93-2 South of Dillon, Montana

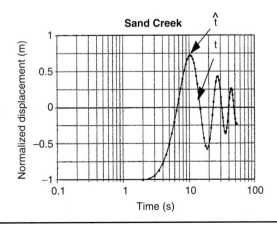

FIGURE 7.24   Plot of normalized displacement $-w/w_o$ versus time, illustrating how $\hat{t}$ and t are estimated in the Kipp (1985) method for the Sand Creek data (Table 7.3; Example 7.10).

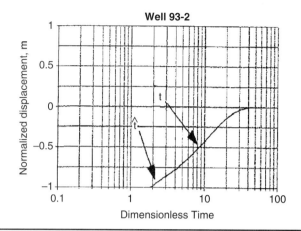

FIGURE 7.25 Plot of normalized displacement $-w/w_o$ versus time, illustrating how $\hat{t}$ and t are estimated in the Kipp (1985) method for well 93-2 in the Beaverhead, Montana, area (Table 7.4; Example 7.11).

Generally, the more oscillatory the data, the smaller the damping parameter (ζ) and the higher the transmissivity. Wells with higher water columns also tend to have higher transmissivities (Figure 7.15). Butler (1998) points out that the water-column length is usually measured from the midpoint of the screen to the SWL. Wells with estimated damping parameters greater than 1.0 can probably be analyzed using the Hvorslev (1951) method or the Bouwer and Rice (1976) method (Weight and Wittman 1999).

Discussion of Storativity

Both methods discussed earlier require an estimate of storativity. If the aquifer is confined, values between 10^{-3} and 10^{-5} are typical (Chapter 3). However, as was discussed earlier with Example 7.8, both spreadsheet algorithms can be applied to leaky-confined or even unconfined aquifers (Weight and Wittman 1999). In all cases, it is necessary to use a confined storativity value or the results will not match the lithologies represented. Wells that are more shallowly completed and have less confining materials should begin evaluation using storativities of 0.001, and wells completed deeper or have semi-confining materials present should be evaluated using a storativity of 0.0001 (Weight and Wittman 1999). Storativity values can be adjusted by a factor of 5 to see where both underdamped methods agree best. One should use the same storativity when comparing methods for a particular data set.

At the beginning of Section 7.4 it was mentioned that Audouin and Bodin (2007) came up with a model that eliminates the high-frequency oscillations and produces a signal that can be interpreted using the McElwee and Zenner (1998) model that accounts for the inertia of the water column and nonlinear head losses at the well, and neglects aquifer storage as a quasi-steady-state approximation. They present the results from approximately 100 slug tests from wells drilled into a fractured-limestone setting in southern France. Their paper describes how they interpret the high-frequency oscillations, how they are filtered out, and then apply the McElwee and Zenner (1998) model. The author highly recommends perusing their paper.

7.5   Other Observations

It is the author's experience that the results from slug testing tend to underestimate hydraulic conductivity values compared with constant-rate pumping tests. This is based upon field experience where the author has observed data from wells that have been tested by both pumping tests and slug tests and from discussions with other colleagues. The amount of underestimation may range from approximately 30% to 100% to over an order of magnitude. This may be a function of well development and, therefore, skin effects and a reduced effective radius of influence for slug testing.

Within contaminated plume areas, where monitoring wells are installed, wells are not designed to produce much water and are not likely to be well developed. Monitoring wells completed in high-transmissivity aquifers are installed with the intention of providing hydraulic head values and estimates of hydraulic properties. It is in high-transmissivity aquifers where monitoring wells that are slug tested may also show underdamped behavior. The methods described in this chapter can be used to evaluate slug-test data.

7.6   Problems

7.1   It is necessary to make a slug for a 3-in (7.6 cm) ID monitoring well. The displacement of the water level is expected to be contained inside the casing only. You want to be able to displace the water about 1.5 to 2.0 ft (0.5 m). Use Table 7.5 to estimate the dimensions of your slug and what materials you will use.

7.2   It is your responsibility to order three new well sentinels to run a series of slug tests and aquifer tests. The slug-test wells are typically 30 ft (10 m) with a SWL about 8 to 10 ft below ground surface (bgs). Some of the pumping tests might see pumping levels (Chapter 6) at 75 to 80 ft (23 to 25 m) below SWL. What do you recommend and why?

7.3   A slug test was conducted on a 6-in well in northern Montana, where the diameter of the borehole was 10 in and the length of the gravel pack was 5 ft. The data are shown in Table 7.6. The material below the well completion is clayey. For context on the location of this site, see Example 4.9 in Chapter 4. Input the data into Excel and consider the following:

(a) What is the estimated hydraulic conductivity (K) using the Hvorslev and the Bouwer and Rice methods? How well do they compare and which equation for the Bouwer and Rice method (7.4 or 7.5) is the most appropriate before using Equation 7.3?
(b) Compare the results with Figure 7.12. Is a double straight-line response evident? Where is the aquifer response?
(c) Change the x-axis scale maximum out to 150 s. Convert the time scale into minutes and replot the data. How does scale influence the interpretation of the results?
(d) Is this a rising-head or falling-head test? Explain.

7.4   Consider using the author's combined method where displacement versus time is plotted and the 0.37 of the maximum displacement is used to determine T_o in the Hvorslev equation. How well do the results compare with number 3?

7.5   Suppose the length of the well screen in Table 7.2 (well 1) is 5 ft even though the gravel pack is 9 ft. Recalculate a new estimate of K if L_e is changed to 5 ft. Was the change significant?

7.6   Recalculate the K values for Example 7.4 (Table 7.2) using the Hvorslev method (Equation 7.1) and compare with the results shown for the Bouwer and Rice method. Would the same interpretation

Nominal Pipe Size (in)	Outside Diameter (in)	Minimum Wall Thickness (in)	Inside Diameter (in)
Schedule 40 PVC pipe			
0.50	0.84	0.109	0.622
0.75	1.05	0.113	0.824
1.00	1.315	0.133	1.049
1.25	1.66	0.14	1.38
1.50	1.9	0.145	1.61
2.00	2.375	0.154	2.067
2.50	2.875	0.203	2.469
3.00	3.5	0.216	3.068
4.00	4.5	0.237	4.026
Schedule 80 PVC pipe			
0.50	0.840	0.147	0.546
0.75	1.050	0.154	0.742
1.00	1.315	0.179	0.957
1.25	1.660	0.191	1.278
1.50	1.900	0.200	1.500
2.00	2.375	0.218	1.939
2.50	2.875	0.276	2.323
3.00	3.500	0.300	2.900
4.00	4.500	0.337	3.826

TABLE 7.5 Table Schedule 40 and Schedule 80 PVC Pipe Dimensions, Modified after www.petersenproducts.com

of the nature of the tailings result? (Are the patterns the same even if the numbers aren't exactly the same?)

7.7 Consider the relationship shown in Figure 7.15 about transmissivity and effective column height. Why is a taller effective column height more prone to produce and underdamped (oscillatory) response?

7.8 Do the recommended activities of the Kipp method described in Example 7.10. Follow the example shown in Figure 7.24 and select a $\hat{t}$ and t to do your best to estimate a transmissivity. Also consult Figure 7.23 for an estimated ζ value. The well and screen are both 6 in in diameter.

7.9 Do the recommended activities of the Kipp method described in Example 7.11. Follow the example shown in Figure 7.25 and select a $\hat{t}$ and t to do your best to estimate a transmissivity. Also consult Figure 7.23 for an estimated ζ value. The well and screen are both 6 in in diameter.

Time (s)	Head (ft)	Displacement (ft)	Time (s)	Head (ft)	Displacement (ft)
0	4.46	2.45	15	6.6	0.31
0.5	4.99	1.92	15.5	6.61	0.3
1	5.22	1.69	16	6.62	0.29
1.5	5.34	1.57	16.5	6.63	0.28
2	5.44	1.47	17	6.63	0.28
2.5	5.52	1.39	17.5	6.64	0.27
3	5.58	1.33	18	6.64	0.27
3.5	5.69	1.22	18.5	6.65	0.26
4	5.76	1.15	19	6.65	0.26
4.5	5.84	1.07	19.5	6.66	0.25
5	5.9	1.01	20	6.67	0.24
5.5	5.97	0.94	20.5	6.66	0.25
6	6.04	0.87	21	6.67	0.24
6.5	6.1	0.81	21.5	6.67	0.24
7	6.15	0.76	22	6.68	0.23
7.5	6.2	0.71	22.5	6.69	0.22
8	6.25	0.66	23	6.69	0.22
8.5	6.3	0.61	23.5	6.69	0.22
9	6.34	0.57	24	6.7	0.21
9.5	6.38	0.53	24.5	6.7	0.21
10	6.43	0.48	25	6.7	0.21
10.5	6.46	0.45	25.5	6.7	0.21
11	6.48	0.43	26	6.71	0.2
11.5	6.5	0.41	26.5	6.71	0.2
12	6.54	0.37	27	6.71	0.2
12.5	6.54	0.37	27.5	6.71	0.2
13	6.56	0.35	28	6.71	0.2
13.5	6.57	0.34	28.5	6.72	0.19
14	6.58	0.33	29	6.72	0.19
14.5	6.6	0.31	29.5	6.72	0.19
			30	6.73	0.18

TABLE 7.6 Rising-Head Slug-Test Data from Northern Montana

7.10 Take the displacement data from Table 7.4 without doing the adjustments for $-w/w_o$ but rather just plot displacement versus time and see what K value you obtain.

7.11 Compare your efforts using the Kipp method in Problem 7.8 with the van der Kamp method.

7.12 Read the paper by Audouin and Bodin (2007) and describe what they did to interpret high-frequency oscillation slug-test data, how they filtered the slug-test data, and how they applied their interpretation to a model.

7.7 References

Audouin, O., and Bodin, J., 2007. Analysis of Slug-Tests with High-Frequency Oscillations. *Journal of Hydrology*, Vol. 334, pp. 282–289.

Bouwer, H., and Rice, R.C., 1976. A Slug Test for Determining Hydraulic Conductivity of Unconfined Aquifers with Completely or Partially Penetrating Wells. *Water Resources Research*, Vol. 12, pp. 423–428.

Bouwer, H., 1989. The Bouwer and Rice Slug Test—An Update. *Ground Water*, Vol. 27, No. 3, pp. 304–309.

Bredehoeft, J.D., Cooper, H.H., Jr., and Papadopulos, I.S., 1966. Inertial and Storage Effects in Well-Aquifer Systems: An Analog Investigation. *Water Resources Research*, Vol. 2, No. 4, pp. 697–707.

Butler, J.J., Jr., 1998. *The Design, Performance, and Analysis of Slug Tests*. Lewis Publishers CRC Press, Boca Raton, FL.

Butler, J.J., Jr., Duffield, G.M., and Kelleher D.L., 2011. *Field Guide for Slug Testing and Data Analysis*. Midwest Geosciences Group Press, Carmel, IN. http://www.midwestgeo .com/fieldtools/slugtesting.php?item=101

Butler, J.J., Jr., and Garnett, E.J., 2000. *Simple Procedures for Analysis of Slug Tests in Formations of High Hydraulic Conductivity Using Spreadsheet and Scientific Graphics Software*. KSG Open-File Report 2000-40. http://www.kgs.edu/Hdro/Publications/ OFR00_40/index.html

Butler, J.J., Jr., Garnett, E.J., and Healey J.M., 2003. Analysis of Slug Tests in Formations of High Hydraulic Conductivity. *Ground Water*, Vol. 41, No. 5, pp. 620–630.

Butler, J.J., Jr., McElwee, C.D., and Liu, W., 1996. Improving the Quality of Parameter Estimates Obtained from Slug Tests. *Ground Water*, Vol. 34, No. 3, pp. 480–490.

Butler, J.J. Jr., and Zhan, X., 2004. Hydraulic Tests in Highly Permeable Aquifers. *Water Resources Research*, Vol. 40, W12402, pp. 12.

Campbell, M.D., Starrett, M.S., Fowler, J.D., and Klein, J.J., 1990. Slug Tests and Hydralulic Conductivity. In *Proceedings of Petroleum Hydrocarbons and Organic Chemicals in Groundwater*, October 31–November 2, Houston, TX, NWWA, Dublin, OH, pp. 85–99.

Cooper, H.H., Jr., Bredehoeft, J.D., and Papadopulos, I.S., 1967. Response of a Finite-Diameter Well to an Instantaneous Charge of Water. *Water Resources Research*, Vol. 3, pp. 263–269.

Crump, K.S., 1976. Numerical Inversion of Laplace Transforms Using a Fourier Series Approximation. *Journal of ACM*, Vol. 28, No. 1, pp. 89–96.

Duffield, G.M. 2018. HydroSOLVE.INC. http://www.aqtesolv.com/aquifer-tests/ aquifer-testing-references.htm

Ferris, J.G., and Knowles, D.B., 1954. *Slug Test for Estimating Transmissibility*. U.S. Geological Survey, Note 26, 7 pp.

Fetter, C.W., 1994. *Applied Hydrogeology, 3rd Edition*. Macmillan, New York, 691 pp.

Hvorslev, M.J., 1951. *Time Lag and Soil Permeability in Ground Water Observations*. U.S. Army Corps of Engineers Waterway Experimentation Station, Bulletin 36.

King, J.B., 1988. *Hydrogeologic Analysis of Septic-System Nutrient-Attenuation Efficiencies in the Evergreen Area, Montana*. Master's Thesis, Montana College of Mineral Science and Technology, Butte, MT, 91 pp.

Kipp, K.L., Jr., 1985. Type Curve Analysis of Inertial Effects in the Response of a Well to a Slug Test. *Water Resources Research*, Vol. 21, No. 9, pp. 1397–1408.

Kruseman, G.P., and de Ridder, N.A., 1990. *Analysis and Evaluation of Pumping-Test Data*. International Institute for Land Reclamation and Improvement, Publication 47, 377 pp.

Lohman, S.W., 1972. *Well Hydraulics*. U.S. Geological Survey Professional Paper 708.

Manchester, K.R., 1990. *Oscillatory Responses Due to Inertial Effects and Observation Well Water Level Fluctuations Induced by Pneumatic and Vacuum Slug Test Methods*. Master's Thesis, Montana College of Mineral Science and Technology, Butte, MT, 85 pp.

McCall, W., and Shipley, G., 2018. Geoprobe. https://geoprobe.com/slug-test-analysis-software-new-v20

McElwee, C.D., Butler, J.J., Jr., and Bohling, G.C., 1992. *Nonlinear Analysis of Slug Tests in Highly Permeable Aquifers Using a Hvorslev-Type Approach*. Kansas Geological Survey, Open-File Report 92–39.

McElwee, C.D., and Zenner, M.A., 1998. A Nonlinear Model for Analysis of Slug-Test Data. *Water Resources Research*, Vol. 34, No. 1, pp 55–66.

Moench, A.F., 1993. Computation of Type Curves for Flow to Partially Penetrating Wells in Water-Table Aquifers. *Ground Water*, Vol. 31, No. 6, pp. 966–971.

Neuman, S.P., 1979. Perspective on "Delayed Yield." *Water Resources Research*, Vol. 15, pp. 899–908.

Papadopulos, I.S., Bredehoeft, J.D., and Cooper, H.H., Jr., 1973. On the Analysis of "Slug-Test" Data. *Water Resources Research*, Vol. 9, pp. 1087–1089.

Prickett, T.A., 1965. Type-Curve Solution to Aquifer Tests under Water-Table Conditions. *Ground Water*, Vol. 3, No. 3, pp. 5–14.

Springer, R.K., and Gelhar, L.W., 1991. *Characterization of Large-Scale Aquifer Heterogeneity in Glacial Outwash by Analysis of Slug Tests with Oscillatory Responses, Cape Cod, Massachusetts*. U.S. Geological Survey Water Resource Investigation Report 91–4034.

Theis, C.V., 1935. The Relation between the Lowering of the Piezometric Surface and the Rate and Duration of Discharge of a Well Using Ground-Water Storage. *Transactions of the American Geophysical Union*, Vol. 16, pp. 519–524.

Uffink, G.J.M., 1984. *Theory of the Oscillating Slug Test*. National Institute for Public Health and Environmental Hygiene, Bilthoven, Netherlands. Unpublished research report, 18 pp.

van der Kamp, G., 1976. Determining Aquifer Transmissivity by Means of Well Response Tests: The Underdamped Case. *Water Resources Research*, Vol. 12, No. 1, pp. 71–77.

Weight, W.D., and Wittman, G.P., 1999. Oscillatory Slug-Test Data Set: A Comparison of Two Methods. *Ground Water*, Vol. 37, No. 6, pp. 827–835.

Weight, W.D., 2006. Field Performance and Errors in the Analysis of Slug Tests. Conference Proceedings of the 2006 North American Environmental Field Conference and Exposition, Tampa, Florida. Presented by the Nielson Environmental Field School, 15 pp.

Wittman, G.P., 1997. Computer Simulated Flow Model of the Groundwater Resources of the Beaverhead Valley in the Dillon Area, Beaverhead County, Montana. Master's Thesis, Montana Tech of the University of Montana, Butte, MT, 101 pp.

Wylie, A., and Magnuson, S., 1995. Spreadsheet Modeling of Slug Tests Using the van der Kamp Method. *Ground Water*, Vol. 33, No. 2, pp. 326–329.

Yang, Y.J., and Gates T.M., 1997. Wellbore Skin Effect in Slug-Test Data Analysis for Low-Permeability Geologic Materials. *Ground Water*, Vol. 35, No. 6, pp. 931–937.

CHAPTER 8

Water Chemistry: Theory and Application

Chris Gammons

Professor of Geological Engineering, Montana Technological University, Butte, Montana

The purpose of this chapter is to discuss different major and trace solutes that dissolve into natural waters, to outline common ways to present and interpret water quality data, to briefly review geochemical thermodynamics and how it can be used to solve problems related to water chemistry, and to discuss some of the fundamental processes that influence water chemistry in the hydrosphere. An introduction to the application of stable H- and O-isotopes of water is included at the end. A companion chapter (Chapter 9) deals with practical considerations related to project planning, field sampling, analysis, and interpretation of water-quality data with a focus on quality control.

It is not possible to cover all aspects of the geochemistry of natural waters in a single chapter. For more depth, good choices for students or entry-level professionals include Drever (1997), Kehew (2001), and Eby (2004). More advanced textbooks include Stumm and Morgan (1996), Langmuir (1997), and Clark (2015). Fetter (1999) and Bedient et al. (1999) two of the many books that deal with contaminant transport of inorganic and organic compounds. Useful information on the geochemistry of natural waters, arranged by element, can be found in Hem (1985), which is available online.

8.1 What's in Water?

Water is a remarkable solvent due to its high dielectric constant (a measure of charge polarity in the H_2O molecule) which allows it to dissolve oppositely charged ions such as Na^+ and Cl^-, the components of common salt. Uncharged molecules, such as silica, gases, and many organic compounds can also dissolve in water to varying degrees. After a review of units and notation, some of the most important major, minor, and trace solutes in water are discussed.

Units of Measurement

The most common units of measurement for major and trace solutes dissolved in water are mg/L (ppm or parts per million) and µg/L (ppb or parts per billion). It is also common to report concentrations in units of molar or molal concentration. Molar (M) concentration is the number of moles of solute per L of solution, whereas molality (m) is the number of moles of solute per kg of H_2O (i.e., the solvent). In dilute waters up to about 0.1 m, the difference between molar and molal units is very small and can be ignored. At higher salinity, conversion between molarity and molality factors in the density of the solution. Since the density of water changes with temperature, molar concentrations change with temperature. In contrast, molal concentrations are independent of temperature, which is one reason that molal is the concentration unit of choice in thermodynamic calculations.

> **Example 8.1**   A water sample contains 12.5 mg/L of Na^+. What is the molar concentration of Na^+? To answer this question, divide the mg/L Na value by the atomic mass of Na in mg:
>
> $$m Na^+ = (12.5\ mg/L) \times (1\ mole\ Na/22{,}990\ mg) = 5.44 \times 10^{-4}\ moles/L.$$
>
> Because molar or molal units are typically very small numbers, analytical data are sometimes presented in units of mmol/L (millimole per L = molar $\times 10^{-3}$) or µmol/L (micromole per L = molar $\times 10^{-6}$). Thus the present example would contain 0.544 mmol/L Na^+ (544 µmol/L).

Confusingly, the concentration in mg/L of certain solutes depends on how the compound is written. For example, water that contains 12 mg/L of nitrate expressed as NO_3^- will contain only 2.7 mg/L of nitrate expressed as N. This is because NO_3 has a gram formula weight of 62.007 g/mole whereas N weighs 14.007 g/mole. Most labs specify their units by writing "2.7 mg/L NO_3 as N" or "12 mg/L NO_3 as NO_3." In contrast to mg/L units, the concentrations expressed as mol/L of NO_3 or as mol/L of N are the same. Other solutes that can lead to potential confusion of this sort include nitrite (NO_2 vs N), ammonium (NH_4 vs N), phosphate (PO_4 vs P), sulfate (SO_4 vs S), and dissolved silica (SiO_2 vs Si vs H_4SiO_4).

Dissolved versus Colloidal versus Suspended Solids

Groundwater samples are usually filtered in the field to remove suspended solids. In surface-water sampling, it is common to collect both a filtered and a nonfiltered sample. Both samples are analyzed in the lab, and the difference between the two measurements gives the concentration present in the suspended solids fraction of the water column. It is common to use filters with a 0.45-µm nominal pore size, although other pore sizes are available (e.g., 0.2 or 0.1 µm). In contrast, most dissolved solutes have diameters in the range of 0.001 to 0.01 µm. Particles that have diameters greater than 0.01 µm but less than the filter pore size are often termed colloids. Colloidal particles can become important for Fe, Al, and other trace metals. In addition, some bacteria and all types of virus will pass through a 0.45-µm filter. Colloids can be removed from a water sample by ultrafiltration, although this is rarely done in practice.

Major Solutes

Most of the total dissolved solids (TDS) in a water sample can be accounted for by a relatively short list of solutes, including the major cations, Na^+, K^+, Ca^{2+}, and Mg^{2+}; the

Median Concentration of Major Solutes,[1] mg/L			Median Concentration of Dissolved Trace Elements in Groundwater,[2] µg/L			
Solute	Surface Water	Groundwater	Element	Median	Element	Median
HCO_3^-	58	200	Al	3.0	Mn	7.0
Ca^{2+}	15	50	As	0.79	Mo	1.0
Cl^-	7.8	20	Ba	54	Ni	1.1
K^+	2.3	3	B	35	Pb	0.07
Mg^{2+}	4.1	7	Co	0.17	Se	0.34
Na^+	6.3	30	Cr	1.2	Sr	270
SO_4^{2-}	3.7	30	Cu	1.0	U	0.52
$SiO_2(aq)$	14	16	Fe	7.9	V	1.4
TDS	120	350	Li	6.0	Zn	4.8

Sources: [1] Langmuir 1997; [2] USGS NAWQA groundwater samples, 1999–2003 (Ayotte et al. 2011).

TABLE 8.1 Median Concentrations of Major Solutes and Dissolved Trace Elements in Surface Waters and Groundwaters

major anions Cl^-, HCO_3^- (bicarbonate), and SO_4^{2-} (sulfate), and a few molecules that are present as uncharged species, such as dissolved silica [expressed as Si, $SiO_2(aq)$, or H_4SiO_4], dissolved CO_2 [expressed as $CO_2(aq)$ or H_2CO_3], and dissolved O_2 (DO). Table 8.1 summarizes the median concentrations of major solutes in surface waters and groundwaters (Langmuir 1997). Also shown is the median value of TDS, which is traditionally measured by evaporating a known volume of water to dryness and weighing the solid residue. Although their concentrations are often well above the ppm range, dissolved CO_2, N_2, and O_2 are not included on this list, as they volatilize on evaporation and therefore do not contribute to TDS. Average groundwater has higher TDS than average surface water. This is due to the longer residence time of groundwater in the subsurface and the greater opportunity to dissolve minerals out of soil and bedrock.

Trace Elements

Table 8.1 also summarizes median concentrations of dissolved trace elements in U.S. groundwater. Some of the more abundant elements include barium (Ba), boron (B), iron (Fe), lithium (Li), manganese (Mn), strontium (Sr), and zinc (Zn). Trace element concentrations can vary over many orders of magnitude, and they are often highly sensitive to changes in pH or redox state. Groundwaters with low DO concentration may contain dissolved Fe and Mn at mg/L levels. Some of the trace elements listed in Table 8.1 are toxic to humans and/or aquatic organisms, and therefore have maximum contaminant levels (MCLs) that are enforceable by state or federal agencies. For example, arsenic is highly toxic to humans, and the U.S. drinking water standard for As is 10 µg/L. Although there is no drinking water standard for zinc, concentrations of Zn > 100 µg/L can be toxic to fish (Sprague 1971).

Some trace elements can exist in aqueous solution in a number of different chemical forms, each of which may have a different chemical behavior, mobility, and toxicity to humans and other forms of life. For example, arsenic has two dominant oxidation states, arsenate (As^{+5}) and arsenite (As^{+3}), each of which has different hydrogeochemical behavior (Welch et al. 2000). Another example is mercury, which has three oxidation states, Hg^{2+}, Hg^+, and Hg^0. Both of these elements are highly toxic to humans. In strongly reducing environments, Hg^{2+} can combine with methane gas to form methyl mercury, CH_3Hg^+. Methyl mercury is highly poisonous and tends to biomagnify up the food chain (Wolfe et al. 1998). A detailed discussion of all of the possible trace elements and their different chemical species is beyond the scope of this chapter. The main point is to realize that a conventional chemical analysis only gives the total concentration of a given trace element in water. Other analytical tests are needed to speciate trace elements into different oxidation states and organic versus nonorganic forms. Geochemical programs, such as PHREEQC (Parkhurst and Appelo 2013), Visual Minteq (Gustafsson 2013, an updated version of the original USEPA program of Allison et al. 1991), or Geochemist's Workbench (Bethke and Yeakel 2015), can solve hundreds of equations simultaneously to predict the exact chemical speciation of a given trace element.

Nutrients

The category of "nutrients" includes phosphate (PO_4^{3-}) and the nitrogen compounds *nitrate* (NO_3^-), *nitrite* (NO_2^-), and *ammonium* (NH_4^+). With the exception of nitrate, these compounds rarely exceed 1 mg/L in uncontaminated natural waters. A further discussion of N-compounds in groundwater is given in Section 8.4.

Aquatic plants (including algae) uptake C, N, and P to make biomass in a ratio that is usually close to 106:16:1 (mole scale), the so-called **Redfield ratio**. Microbial biomass in soil is more P-enriched, with an overall stoichiometry of 60:7:1 (Cleveland and Liptzin 2007). If water has a dissolved N/P ratio much greater than the Redfield ratio, then that water is said to be "phosphorus limited" with respect to growth of plants. If the N/P ratio is low, then nitrogen is the limiting nutrient. In the calculation of N/P, "N" includes nitrate (NO_3^-), nitrite (NO_2^-), and ammonium (NH_4^+), but not $N_2(g)$, since the latter is biologically unavailable to most plants. Phosphorus exists in aqueous solution mainly as *soluble reactive phosphate* (SRP), which includes the compounds $H_2PO_4^-$ (dominant at pH < 7.2) and HPO_4^{2-} (dominant at pH > 7.2). These are the products of the first and second dissociations of phosphoric acid (H_3PO_4). Trivalent PO_4^{3-} does not become important until pH > 12, so its concentration is negligible in most waters. A certain (usually minor) amount of N and P can also be associated with dissolved organic carbon (DOC) compounds such as humic and fulvic acids that contain amine (NH_3) and phosphate (PO_4) functional groups. The sum of organic-N and ammonium (NH_4^+ + NH_3) is referred to as *Kjeldahl nitrogen*.

Dissolved Organic Carbon

The term "dissolved organic carbon" is not well defined in the literature. One definition would be any compound that contains C-H bonds. A few examples include methane (CH_4) and other alkanes (e.g., C_2H_6, C_3H_8), dissolved carboxylic acids (e.g., acetate, CH_3COO^-), alcohols (e.g., CH_3OH, C_2H_5OH), and aromatic compounds that contain C_6 rings (e.g., phenols, benzene). In natural waters, DOC includes all of the above, as well as the large and structurally complex humic and fulvic acids. Humic and fulvic acids mainly form by the microbial decay of plant matter, and they may contain as many as 30 or 40 carbon atoms linked in branching chains and rings, with attached functional groups including carboxylic acid (COOH), phosphate (PO_4), and amine (NH_3) molecules.

By definition, fulvic acids are soluble at all pH values, whereas humic acids separate from aqueous solution at pH < 2. Natural DOC compounds are mainly formed by microbes in soil or in organic-rich sediment as is found in wetlands and bogs. Most groundwaters and surface waters contain DOC at concentrations in the range of 0.1 to 10 mg/L. Rivers and lakes with high concentrations of DOC become tea-colored ("black water"), owing to photochemical reactions involving DOC that adsorb sunlight. DOC is important in hydrogeochemistry because it can form strong aqueous complexes with trace metals (e.g., Fe, Al, Cu). At pH values above about 4, most DOC molecules carry a negative charge, due to deprotonation of carboxylic acid functional groups. Although often not analyzed, DOC can be a significant contributor to the charge balance of aqueous solutions owing to its negative charge. DOC is also the "food" for many types of bacteria that perform redox reactions, including iron-reducing and sulfate-reducing bacteria.

Human-made organic compounds can also contribute to DOC, especially in waters draining industrial, urban, or agricultural areas. The list of possible compounds is exceedingly long and is beyond the scope of this chapter to detail. More information can be found in Chapter 9 or in textbooks that focus on organic contaminants in surface water and groundwater (e.g., Fetter 1999; Weiner 2015).

8.2 Presentation of Water Quality Data

It is useful to review some of the common ways that water chemistry data are graphically presented before jumping into the discussion on thermodynamics.

Piper Diagrams

One of the most common methods for summarizing groundwater chemistry data (but less common for surface water) is the Piper diagram (Piper 1944). The Piper diagram is a trilinear diagram that displays the relative amounts of each of the common cations ($Na^+ + K^+$, Ca^{2+}, Mg^{2+}) and anions ($HCO_3^- + CO_3^{2-}$, SO_4^{2-}, Cl^-) in a given sample. With experience, the pattern of sample data on a Piper diagram reveals much information about water subtypes and possible trends due to water mixing or chemical reaction. Many licensed software packages will create Piper diagrams. The USGS offer the program "GW Chart," which can be downloaded at: https://water.usgs.gov/nrp/gwsoftware/GW_Chart/GW_Chart.html.

The first step in plotting a sample on a Piper diagram is to recalculate the concentration of each ion into milli-equivalents per liter (meq/L), which is simply the millimolal concentration (mmol/L) multiplied by the charge of the ion. The sum of the meq/L of Na^+, K^+, Ca^{2+}, and Mg^{2+} gives the total meq/L of common cations. Likewise, the sum of the meq/L of HCO_3^-, CO_3^{2-}, SO_4^{2-}, and Cl^- gives the total meq/L of common anions. The percentage of the total cations or anions for each ion in the sample is then calculated and plotted on the respective cation or anion triangle in the Piper diagram. For example, the water sample plotted in Figure 8.1 has a cation composition (based on meq/L) of 50% Mg^{2+}, 10% ($Na^+ + K^+$), and 40% Ca^{2+}, and an anion composition of 50% Cl^-, 22% ($HCO_3^- + CO_3^{2-}$), and 28% SO_4^{2-}. It takes some practice to learn how to plot a sample on a trilinear diagram, so be sure that you see how this was done for the example. The way in which the tick labels are inclined helps to see which set of grid lines to use. To plot the sample on the top diamond field, simply use a ruler to extend the data point parallel to the left edge of the cation triangle and parallel to the right edge of the anions triangle. The datum plots where the two lines intersect.

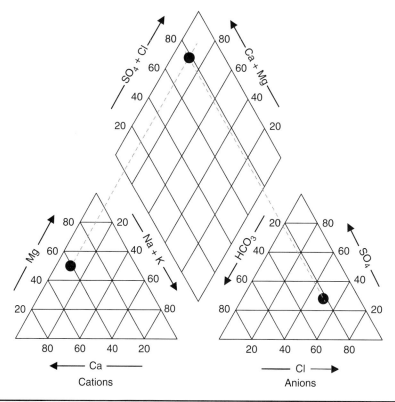

FIGURE 8.1   Construction of a Piper diagram. See text for explanation.

Figure 8.2 shows another Piper diagram with each hydrogeochemical facies labeled. Also shown are some example groundwater data for samples collected from three different aquifers in central Montana (Gammons et al. 2010). The water samples range from "Ca-type" to "Mg-type" in terms of their cations, and from "bicarbonate-type" to "sulfate-type" in terms of their anions. Although they may be present at significant concentrations, other cations and anions such as Fe^{2+}, Mn^{2+}, or NO_3^- are not included in a Piper plot. As well, a Piper plot tells us nothing about the absolute concentration of solutes in the water, only their relative concentrations.

Stiff and Schoeller Diagrams

Two other common ways that chemical analyses are graphically displayed include the Stiff diagram (Stiff 1951) and the Schoeller diagram (Schoeller 1962). In a Stiff diagram (Figure 8.3c), each major cation or anion is arranged vertically and their concentrations are plotted on a horizontal axis in units of meq/L. By convention, the cations are always on the left, in the following order from top to bottom: ($Na^+ + K^+$), Ca^{2+}, Mg^{2+}. Anions are plotted on the right in the order Cl^-, ($HCO_3^- + CO_3^{2-}$), and SO_4^{2-}. The data points are connected by line segments that create a two-dimensional polygon. The shape of the polygon instantly conveys the type of water present. For example, if the polygon has a fat waist and narrow top and bottom, this would be "calcium-bicarbonate type" water. Stiff

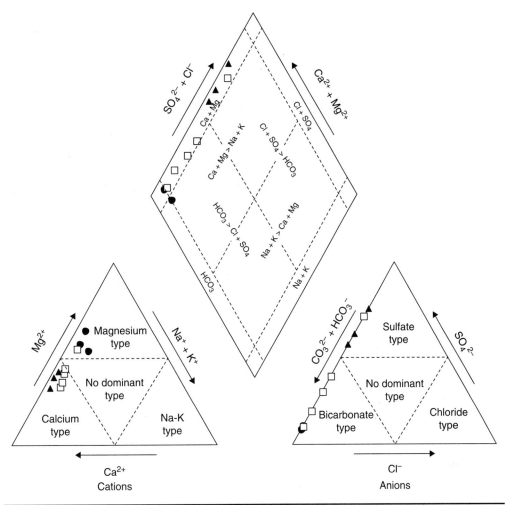

FIGURE 8.2 Piper diagram with hydrogeochemical facies labeled. Data points show compositions of groundwater in three different aquifers in central Montana. [*Data from Gammons et al. (2010).*]

diagrams are well suited for display on maps of the field area, where a polygon can be clipped and pasted next to each sample location. The width of the polygon can be adjusted to scale to the total concentration of solutes. A scale is typically given for at least one of the polygons so that absolute concentrations can be conveyed. Additional solutes could be added to a Stiff diagram if desired. For example, $(Fe^{2+} + Mn^{2+})$ and NO_3^- are sometimes added as another row at the bottom.

A Schoeller (1962) diagram is the easiest to construct and, in some ways, conveys the most information. Solute concentrations are plotted on a log-concentration line graph, with a different line for each sample. A traditional Schoeller diagram arranges the major ions in the order: Mg^{2+}, Ca^{2+}, $(Na^+ + K^+)$, Cl^-, SO_4^{2-}, HCO_3^-. However, it is easy to add as many additional solutes as desired on the plot. Also, the logarithmic y-axis scale makes it possible to simultaneously display major ions and trace elements.

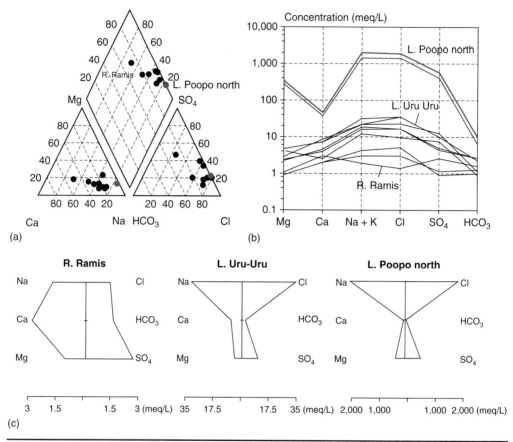

FIGURE 8.3 Surface water samples from the Lake Titicaca-Lake Poopo region, Peru and Bolivia, plotted up on (a) a Piper diagram, (b) a Schoeller diagram, and (c) Stiff diagrams for three selected samples. R. Ramis is located at the headwaters; L. Uru-Uru is located near the end of the watershed but has a small outlet; L. Poopo is a terminal playa lake at the end of the watershed. [*Data from Guyot et al. (1990).*]

Be careful not to get carried away though, as your diagrams will get messy pretty quickly if you are plotting too many samples and elements on the same graph!

Example 8.2 A set of river and lake samples in the arid drainage basin of lakes Titicaca and Poopo (Peru-Bolivia) was collected by Guyot et al. (1990). The samples ranged from tributary rivers draining the Andes Mountains (Ramis, Illave, Coati) upstream of Lake Titicaca to an evaporated playa lake (Lake Poopo) near the terminus of the watershed in the Bolivian Altiplano. Table 8.2 summarizes the data obtained and Figure 8.3 presents the data graphically using a combination of Piper, Schoeller, and Stiff diagrams. The Piper diagram shows how the waters evolved from "no dominant type" to "sodium-chloride type" with distance traveled. The Schoeller diagram shows how solute concentrations increased two to three orders of magnitude through the system from evapoconcentration. The Stiff diagrams show a progressive change in shape from headwaters to the playa lake, with Ca^{2+}, SO_4^{2-}, and HCO_3^- becoming less important relative to Na^+ and Cl^-. Most likely the concentrations of Ca^{2+} and SO_4^{2-} were limited by precipitation of gypsum ($CaSO_4 \cdot 2H_2O$), whereas bicarbonate would have been limited by precipitation of calcite. Note the change in scale for each Stiff diagram.

Sample ID	T(°C)	pH	TDS	Na	Ca	Mg	K	HCO$_3$	SO$_4$	Cl
R. Ramis	14.7	8.46	310	41	58	13	3	95	125	48
R. Coata	16.8	9.65	397	91	40	11	8	59	42	175
R. Ilave	19.4	9.29	324	66	38	11	8	73	57	107
R. Des P. Int	19.4	9.64	953	256	48	51	23	53	360	367
R. Des-Cal	16.3	8.99	1,830	510	165	36	17	68	600	795
R. Des P. Jap	23.2	8.84	1,240	355	77	28	21	158	220	600
R. Des Chuq	19.7	9.08	1,290	404	90	26	19	160	240	604
R. Des P. Esp	16.4	8.29	2,360	469	143	55	30	135	475	1,220
L. Uru-Uru	15.3	8.43	2,340	739	143	55	30	134	500	1,215
L. Poopo north	23.0	8.27	107,000	44,700	880	3,990	2,100	606	27,500	66,100
L. Poopo south	23.6	8.48	82,600	32,300	770	3,025	1,650	387	19,000	49,430

TABLE 8.2 Chemical Composition of Surface Water in the Lake Titicaca Region, Peru-Bolivia. All Data in mg/L. [Data from Guyot et al. (1990)]

Eh-pH Diagrams

Many elements in the Periodic Table are redox-sensitive, meaning that they form different sets of compounds that have different *valences* or *oxidation states*. An example is iron, which can exist in nature as Fe^{2+} (also known as ferrous iron) or Fe^{3+} (also known as ferric iron). In fact, redox state is one of the two master variables, along with pH, that controls the fate and transport of trace metals in water. In the field, redox state can be measured with an Eh (electrical potential) or ORP (oxidation-reduction potential) electrode. Eh and ORP are conceptually similar, and they have similar units (volts or mV) but are not the same number. This is discussed further in Chapter 9.

In projects that involve trace-metal geochemistry, it is useful to show field data on an Eh-pH diagram. Two example Eh-pH diagrams for iron are shown in Figure 8.4. The following are some helpful things to keep in mind when interpreting these diagrams:

- The dashed diagonal lines near the top and bottom of the diagram show the upper and lower stability fields of liquid water at atmospheric pressure. Natural waters should always plot within these two limits.

- Samples that plot closer to the upper stability boundary of water are said to be *oxidized*: elements will tend to exist in their highest (most positive) valence state. Samples closer to the lower stability of water are *reduced*: elements will exist in their lowest (most negative) valence state.

- Shaded fields show regions where iron should precipitate as a given mineral. Fields that are unshaded show regions where iron should dissolve as a given aqueous species. The boundary between precipitated and dissolved iron is a function of the Fe concentration that was used to construct the diagram (in this case, 1 ppm). If the Fe concentration is increased, the solid fields will expand. If the Fe concentration is decreased, the solid fields will contract.

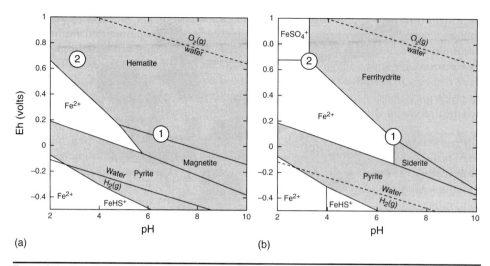

Figure 8.4 Eh-pH diagrams for Fe-S-C-H$_2$O system at 25°C. (a) Showing stability fields of the thermodynamically stable minerals hematite and magnetite; (b) showing stability field of the metastable mineral, ferrihydrite, and the carbonate mineral, siderite. Conditions of the diagram: ΣS = 100 ppm, ΣC = 100 ppm, ΣFe = 1 ppm. Circled numbers are discussed in the text.

- Phase boundaries that are independent of redox plot as vertical lines. An example would be Fe^{2+}/siderite. In this case, both sides of the reaction contain an iron compound that is in the +2 (ferrous) oxidation state. All other phase boundaries are redox-dependent, meaning that the transition from one side to the other involves a change in valence (oxidation state) of iron and/or another element.

When constructing an Eh-pH diagram, one often needs to make choices as to which minerals to plot. For example, Figure 8.4a shows the stability fields of the thermodynamically stable minerals hematite, magnetite, and pyrite. However, experience shows that low-temperature waters are rarely in equilibrium with hematite or magnetite. In oxidized waters, it is more common to find waters in equilibrium with ferrihydrite, also known as ferric hydroxide, Fe(OH)$_3$. This solid is essentially the "rust" that forms wherever dissolved iron is oxidized. Also, in moderately reduced waters at higher pH, it is more likely that a groundwater will be near equilibrium with siderite (FeCO$_3$) as opposed to magnetite. The fields of ferrihydrite and siderite stability are shown in Figure 8.4b.

The greatest utility of an Eh-pH diagram is that it gives the user a "road map" of what kinds of compounds and reactions could occur in a given environmental system. For example, consider a groundwater with pH = 6.7 and Eh = +0.12 V containing 1 ppm dissolved iron. From Figure 8.4b, the most likely form of dissolved iron is Fe^{2+} (ferrous iron), and the water is likely buffered by the coexistence of ferrihydrite + siderite. Furthermore, we can predict that an increase in Eh (e.g., due to exposure to oxygen in air) will cause ferrihydrite (i.e., "rust") to precipitate. This is a common problem when sampling reduced, anoxic groundwater which can be eliminated by adding an acid preservative in the field. At very low Eh, Fe^{2+} may precipitate as pyrite.

Example 8.3 A sample of acidic groundwater seepage from a mine dump in Montana has Eh = +0.68 V and pH = 3.3. Plot this on an Eh-pH diagram for Fe and speculate as to what might be buffering the composition of this water.

The Eh-pH conditions are first plotted (sample #2 of Figure 8.4). As stated above, it is unlikely that hematite or magnetite would be influencing the water chemistry in a low-temperature setting, therefore we will consider Figure 8.4b. The water sample plots near the triple point of dissolved ferrous iron (Fe^{2+}), dissolved ferric iron ($FeSO_4^+$), and ferrihydrite. Therefore, we can infer that the Eh and pH of the water sample are fixed by the coexistence of these three compounds. An increase in pH should precipitate ferrihydrite, whereas an increase in Eh should convert Fe^{2+} to $FeSO_4^+$. By the way, $FeSO_4^+$ is an example of an aqueous *ion pair* formed by a weak bond between Fe^{3+} and SO_4^{2-}.

8.3 Thermodynamics Applied to Water Chemistry

The Equilibrium Constant

Consider a generic chemical reaction:

$$aA + bB = cC + dD \tag{8.1}$$

In this reaction, A and B are reactants, C and D are products, and a, b, c, and d are stoichiometric coefficients that indicate the proportion of each reactant and product in the balanced reaction. The **equilibrium constant**, K_{eq}, for this reaction is written as

$$K_{eq} = \frac{(a_C)^c (a_D)^d}{(a_A)^a (a_B)^b} \tag{8.2}$$

where K_{eq} is the thermodynamic equilibrium constant (a function of temperature) and is the **activity** of each compound. Activities are discussed in more detail in the next section, but for now a_i can be thought of as the "thermodynamic concentration" of species i. In words, the equilibrium constant is "the product of the activities of the product species raised to the power of their respective stoichiometric coefficients divided by the product of the activities of the reactant species raised to their stoichiometric coefficients." The usefulness of the equilibrium constant is that it tells us what the concentrations of reactants and products should be for a given reaction when it is has reached an equilibrium state.

The value of K_{eq} for a given reaction can be determined in an experiment or can be computed from tabulated values of **Gibbs free energy** (G). The value of the equilibrium constant is related to ΔG°_{rxn}, the standard-state Gibbs free energy change of the reaction of interest, as follows:

$$\Delta G^\circ_{rxn} = -RT \ln K_{eq} \tag{8.3}$$

where R is the gas constant (8.3144 J K^{-1} $mole^{-1}$ or 1.9872 cal K^{-1} $mole^{-1}$) and T is temperature in Kelvin. For the generic reaction listed above, ΔG°_{rxn} can be written out as follows:

$$\Delta G^\circ_{rxn} = c \cdot \Delta G^\circ_{f,C} + d \cdot \Delta G^\circ_{f,D} - a \cdot \Delta G^\circ_{f,A} - b \cdot \Delta G^\circ_{f,B} \tag{8.4}$$

where ΔG°_f is the standard-state Gibbs free energy of formation of each compound (a function of temperature). Units of ΔG°_f and ΔG°_{rxn} are kJ/mol (kilo-joules per mole) or kcal/mol (kilo-calories per mole). Values of ΔG_f° are tabulated in various textbooks for tens of thousands of naturally occurring and manmade compounds.

8.4 Activities and Activity Coefficients

By convention, the activity of any solid or liquid (including water) is one provided it is pure. An impure solid or liquid has an activity that is approximately equal to its mole fraction (X_i) in the solid or liquid mixture. For solids, this approximation is reasonable at X_i values > 0.8, but becomes poor for trace impurities. The activity of water can be assumed to be unity in most but not all applications. For example, if a hypothetical solution contains 1.0 molal NaCl, then the mole fraction of water in the solution would be $55.55/(55.55 + 1) = 0.98$, realizing that there are 55.55 moles of water in 1 kg of H_2O. Determination of activities and solute speciation is difficult for highly saline brines, and it is not covered in detail in this chapter.

The activity of a gas is equal to its partial pressure (in bars) multiplied by an **activity coefficient** to account for nonideal behavior. For calculations near atmospheric temperature and pressure, the activity coefficients for gases are near unity and can be ignored. Therefore,

$$a_{i,gas} = P_{i,gas} = X_i \cdot P_{total, gas} \tag{8.5}$$

where $a_{i,gas}$ is the activity of gas i, P_{total} is the total gas pressure (bars), X_i is the mole or volume fraction of gas i in the gas mixture, and $P_{i,gas}$ is the partial pressure of gas i.

The activity of an aqueous species (any compound dissolved in water) is equal to its molal concentration multiplied by an activity coefficient, γ, to account for nonideal behavior:

$$a_{i,aq} = m_i \cdot \gamma_i \tag{8.6}$$

For uncharged aqueous compounds, the value of γ_i is close to one and can be ignored. However, values of γ_i for charged ions can deviate significantly from one, and they need to be computed for accurate thermodynamic calculations. One widely used equation for computing γ_i is the **Debye-Hückel equation** (e.g., Robinson and Stokes 1970):

$$\log \gamma_i = \frac{-A \cdot z_i^2 \cdot \sqrt{I}}{1 + B \cdot \mathring{a} \cdot \sqrt{I}} \tag{8.7}$$

where z is the charge of the solute for which γ_i is being computed ($= 0$ for H_2CO_3, -1 for Cl^-, $+2$ for Ca^{2+}, and so on), å is the hydrated radius ("ion size") of the solute, A and B are temperature-dependent constants, and I is the ionic strength of the solution. **Ionic strength** is calculated from the molal concentrations of all ions in the solution as follows:

$$I = \frac{1}{2} \cdot \sum_i m_i \cdot z_i^2 \tag{8.8}$$

Values of A and B at several temperatures are listed in Table 8.3, along with values of å for several common ions.

T(°C)	A	B	Ion	Ion Size Parameter
0	0.4883	0.3241	NH_4^+	2.5
5	0.4921	0.3249	K^+, Cl^-, OH^-	3.5
10	0.4960	0.3258	Na^+,	4.0
15	0.5000	0.3262	Ca^{2+}, Sr^{2+}, Ba^{2+}, SO_4^{2-}	5.0
20	0.5042	0.3273	HCO_3^-, CO_3^{2-}	5.4
25	0.5085	0.3281	Mg^{2+}	5.5
30	0.5130	0.3290	Fe^{2+}, Mn^{2+}, Li^+	6.0
40	0.5221	0.3305	H^+, Al^{3+}, Fe^{3+}	9.0
50	0.5319	0.3321		
60	0.5425	0.3338		

TABLE 8.3 Values of Input Parameters for the Extended Debye-Hückel Equation [Data adapted from Drever (1997)]

The Debye-Hückel equation has proven to be accurate for solutions with ionic strength less than about 0.1 molal. At higher ionic strengths (up to about 0.7 molal), an alternative to the Debye-Hückel equation is the **Davies equation** (cf. Langmuir 1997):

$$\log \gamma_i = -A \cdot z_i^2 \left[\frac{\sqrt{I}}{1+\sqrt{I}} - 0.3 \cdot I \right] \tag{8.9}$$

Figure 8.5 plots values of γ for several common solutes as a function of ionic strength. The greater the charge (positive or negative) and the smaller the ion size, the more the value of γ deviates from unity. Values of γ_i based on the Debye-Hückel versus Davies equation begin to deviate dramatically above $I = 0.1$. This is because of the add-on term $(0.3I)$ in the Davies equation which causes γ to increase toward unity at very high ionic strength, which has proven to be the case in experimental studies. A simplification of the Davies equation is that it cannot take differences in ion size into account. For this reason, all univalent (+1, −1) ions have the same γ value, all divalent (+2, −2) have the same γ value, and so on.

Example 8.4 Calculate activity coefficients for Na^+, Cl^-, and Ca^{2+} for a solution at 10°C containing $0.1m$ NaCl + $0.2m$ $CaCl_2$. Use both the Debye-Hückel and the Davies equations.

The first step is to calculate the ionic strength. If we assume that each salt dissociates completely when dissolved in water, then it follows that $mNa^+ = 0.1$, $mCa^{2+} = 0.2$, and $mCl^- = 0.5$:

Then, $I = \frac{1}{2}[mNa \cdot 1^2 + mCa \cdot 2^2 + mCl \cdot (-1)^2] = \frac{1}{2}[0.1 \cdot 1^2 + 0.2 \cdot 2^2 + 0.5 \cdot (-1)^2]$

Or, $I = 0.7$ (molal)

Parameters for the Debye-Hückel equation are as follows: $I = 0.7$ (as above); A = 0.4960; B = 0.3258; and ion size parameters (å) for Na^+, Cl^- and Ca^{2+} are 4, 3.5, and 5, respectively. For the Na^+ ion, the

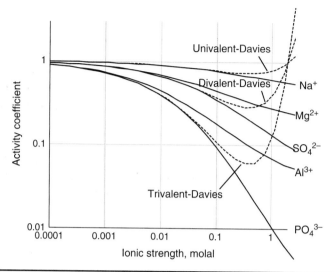

FIGURE 8.5 The dependence of the activity coefficients of cations on ionic strength. Solid curves were calculated with the extended Debye-Hückel equation and ion size parameters in Table 8.3. Dashed curves were calculated using the Davies equation.

equation becomes log $\gamma Na^+ = -0.4960 \cdot (1)^2 \cdot (0.7)^{0.5}/(1 + 4 \times 0.3258 \cdot (0.7)^{0.5})$. Or, log $\gamma Na^+ = -0.198$, and $\gamma Na^+ = 0.63$. The calculations for Cl^- and Ca^{2+} are the same but with their respective values of å and z^2 inserted. The final answers for the Debye-Hückel equation are $\gamma Na^+ = 0.63$, $\gamma Cl^- = 0.61$, and $\gamma Ca^{2+} = 0.20$.

For the Davies equation, we use the same values of I, A, and z, to calculate the following:

$$\gamma Na^+ = \gamma Cl^- = 0.96, \ \gamma Ca^{2+} = 0.20.$$

Temperature Dependence of the Equilibrium Constant

The value of the equilibrium constant for a given reaction changes with temperature. For relatively small temperature changes (e.g., in the range of most surface and groundwaters excluding geothermal waters), the temperature dependence can be quantified using the **Van't Hoff equation**:

$$K_{eq,T} = K_{eq,298} \cdot \exp\left(\frac{-\Delta H^\circ_{rxn}}{R}\left(\frac{1}{T} - \frac{1}{298}\right)\right) \tag{8.10}$$

where $K_{eq,298}$ is the value of the equilibrium constant at 25°C (298.15 K), ΔH°_{rxn} is the standard-state enthalpy change of the reaction (kJ/mol), R is the gas constant (8.3144 mol·J^{-1}·K^{-1}), and T is the temperature of interest (in Kelvin, K). Reactions that have a positive value of ΔH°_{rxn} are said to be **endothermic**, whereas **exothermic** reactions have a negative ΔH°_{rxn}. An easy way to remember this is to think of heat as a reactant or a product in a given reaction. For endothermic reactions, reactants + [heat] = products, whereas for exothermic reactions, reactants = products + [heat]. Equilibrium constants for endothermic reactions always increase with increase in temperature, whereas K_{eq} values for exothermic reactions decrease. Table 8.4 gives ΔH°_{rxn} for many common mineral-dissolution reactions. From this table, it can be seen that the solubility of most carbonate minerals decreases with increase in temperature, since the sign of ΔH°_{rxn} is negative.

Mineral	Solubility-Controlling Reaction	Log K_{eq} 25°C	$\Delta H°_{rxn}$ kJ/mol
Calcite	$CaCO_3(s) + H^+ = Ca^{2+} + HCO_3^-$	1.85	−22.86
Dolomite	$CaMg(CO_3)_2(s) + 2H^+ = Ca^{2+} + Mg^{2+} + 2HCO_3^-$	3.57	−65.98
Siderite	$FeCO_3(s) + H^+ = Fe^{2+} + HCO_3^-$	−0.56	−23.63
Rhodohrosite	$MnCO_3(s) + H^+ = Mn^{2+} + HCO_3^-$	−1.10	−19.23
Strontianite	$SrCO_3(s) + H^+ = Sr^{2+} + HCO_3^-$	1.06	−14.93
Cerussite	$PbCO_3(s) + H^+ = Pb^{2+} + HCO_3^-$	−2.80	7.05
Malachite	$Cu_2(OH)_2CO_3(s) + 3H^+ = 2Cu^{2+} + 2H_2O(l) + HCO_3^-$	5.15	−81.42
Smithsonite	$ZnCO_3(s) + H^+ = Zn^{2+} + HCO_3^-$	0.33	−31.49
Gypsum	$CaSO_4·2H_2O(s) = Ca^{2+} + SO_4^{2-} + 2H_2O(l)$	−4.58	−0.46
Barite	$BaSO_4(s) = Ba^{2+} + SO_4^{2-}$	−9.97	26.60
Celestite	$SrSO_4(s) = Sr^{2+} + SO_4^{2-}$	−6.63	−4.34
Anglesite	$PbSO_4(s) = Pb^{2+} + SO_4^{2-}$	−7.79	9.00
Ferrihydrite	$Fe(OH)_3(s) + 3H^+ = Fe^{3+} + 3H_2O(l)$	4.89	
Gibbsite	$Al(OH)_3(s) + 3H^+ = Al^{3+} + 3H_2O(l)$	8.11	−95.4
Brucite	$Mg(OH)_2(s) + 2H^+ = Mg^{2+} + 2H_2O(l)$	16.84	−113.4
Quartz	$SiO_2(s) + 2H_2O(l) = H_4SiO_4(aq)$	−3.98	25.06
Hematite	$Fe_2O_3(s) + 6H^+ = 2Fe^{3+} + 3H_2O(l)$	−4.01	−129.06

TABLE 8.4 Values of log K_{eq} and $\Delta H°_{rxn}$ for Some Mineral Solubility Reactions [Compiled from Data in Drever (1997)]

Using Equilibrium Constants to Solve Geochemical Problems

Gas Solubility

The solubility of a gas in water can be written in terms of an equilibrium constant as follows: $K_{eq} = a_{i,aq}/a_{i,g}$. Since activity coefficients of gases are near unity, both in the vapor and when dissolved in water, the expression simplifies to $K_{eq} = m_{i,aq}/P_{i,g}$. Equilibrium constants involving gas solubility in water are known as **Henry's law constants**, K_H. Values of K_H between 5°C and 30°C are listed in Table 8.5 for several common gases. All gases become less soluble with increase in temperature, which is why carbonated beverages go flat when they warm up. At 25°C, the order of solubility is $NH_3 > SO_2 > H_2S > CO_2 > O_2 > N_2$. In other words, it is much easier to dissolve NH_3, SO_2, or H_2S in water as opposed to N_2 or O_2. However, since the partial pressures of NH_3, SO_2, and H_2S are extremely low in most settings (with the possible exception of volcanic fumaroles), the concentrations of these dissolved gases in natural waters are usually vanishingly small.

T(°C)	O_2	N_2	CO_2	H_2S	SO_2	NH_3
5	1.91×10^{-3}	9.31×10^{-4}	6.35×10^{-2}	1.77×10^{-1}	2.60	158
10	1.70×10^{-3}	8.30×10^{-4}	5.33×10^{-2}	1.52×10^{-1}	2.12	122
15	1.52×10^{-3}	7.52×10^{-4}	4.55×10^{-2}	1.31×10^{-1}	1.74	95
20	1.38×10^{-3}	6.89×10^{-4}	3.92×10^{-2}	1.15×10^{-1}	1.44	74
25	1.26×10^{-3}	6.40×10^{-4}	3.39×10^{-2}	1.02×10^{-1}	1.20	59
30	1.16×10^{-3}	5.99×10^{-4}	2.97×10^{-2}	9.09×10^{-2}	1.00	47

Units of K_H in this table are mol L^{-1} bar^{-1}.

TABLE 8.5   Values of the Henry's Law Constant $(K_H)^a$ for Various Gases at 1 Bar Total Pressure [Computed from Sander (2015)]

Example 8.5   What is the solubility of $N_2(g)$ in water at 10°C at sea level? Give your answer in molal and mg/L units. From Table 8.5, K_H is 8.30e-4 for $N_2(g)$ at 10°C. Applying Henry's law, $mN_2(aq) =$ 8.30e-4 × P_{N2}. Since air is 78.1% N_2 by volume and since atmospheric pressure at sea level is 1.01325 bar, $P_{N2} = 0.781 \times 1.01325 = 0.791$ bar. Solving through, one gets $mN_2(aq) = 6.57$e-4. Since 1 mole of N_2 weighs 28.02 g, this is equal to 18.4 mg/L of $N_2(aq)$. This result may surprise some readers as it implies that $N_2(aq)$ should be a major solute in surface waters and shallow groundwater. It is! However, since $N_2(aq)$ is chemically inert, it is very rarely quantified.

Mineral Solubility

Table 8.4 lists values of log K_{eq} for the dissolution of some common carbonate, sulfate, and oxide minerals. Problems involving mineral-solubility reactions can be solved using the equilibrium constant approach. If we return to the generic reaction 8.1 and rewrite Equation 8.2 in logarithmic form, we obtain the following relationship:

$$\log K_{eq} = c \cdot (\log aC) + d \cdot (\log aD) - a \cdot (\log aA) - b \cdot (\log aB) \tag{8.11}$$

If log K_{eq} is known and if the activities of three of the compounds involved in the reaction are known, then the activity of the fourth compound can be computed. This is the basic approach used in the following examples.

Example 8.6   A water sample at 25°C contains 520 mg/L of SO_4. What is the predicted concentration of Ba^{2+} in this water if we assume equilibrium with the mineral barite? Calculate your answer assuming $\gamma = 1$ for aqueous species and again assuming an ionic strength of 0.01m with γ values calculated using the Debye-Hückel equation.

From Table 8.4, log $K_{eq} = -9.97$ for the reaction $BaSO_4(s) = Ba^{2+} + SO_4^{2-}$. This can be rearranged: $-9.97 =$ log aBa^{2+} + log aSO_4^{2-} − log $a$barite. As usual, we assume barite is pure and has an activity of 1. For the first part of the answer, we also assume $\gamma = 1$ for Ba^{2+} and SO_4^{2-}, so that $-9.97 = \log mBa^{2+} + \log mSO_4^{2-}$. To compute the molality of sulfate, we divide 530 mg/L SO_4 by the gram formula weight (96 g/mol) to get $SO_4^{2-} = 5.42$e-3 molal. Solving through gives log $mBa^{2+} = -7.70$, and the predicted concentration of dissolved barium is 1.98e-8 molal, or 2.7 μg/L Ba (2.7 ppb).

For the second part of the answer, we compute values of γ for Ba^{2+} and SO_4^{2-} using the Debye-Hückel equation. From Table 8.3, Ba^{2+} and SO_4^{2-} both have the same ion size parameter of 5. Plugging this into Equation 8.5 and using $I = 0.01$ m (specified above) gives $\gamma = 0.67$. The equilibrium constant expression then becomes:

Log K_{eq} = −9.97 = log($mBa^{2+} \cdot \gamma Ba^{2+}$) + log($mSO_4^{2-} \cdot \gamma SO_4^{2-}$) = log(0.67 · mBa^{2+}) + log(0.67 · mSO_4^{2-}). Substituting mSO_4^{2-} = 5.42e-3 (above) and solving, we obtain mBa^{2+} = 4.36e-8, or 6.0 µg/L Ba.

From this example, we see that barite has an extremely low solubility. We also note that the computed solubility more than doubles if we incorporate activity coefficients compared to the same calculation assuming $\gamma = 1$.

Example 8.7   What is the solubility of quartz at 5°C, 25°C, and 100°C? Express your answer in ppm as Si, ppm as SiO_2, and mmol/L.

From Table 8.4, log K_{eq} = −3.98 at 25°C for the reaction SiO_2(s) + $2H_2O$(l) = H_4SiO_4(aq). Putting this into equilibrium constant format, log K_{eq} = −3.98 = log aH_4SiO_4(aq) − log $aSiO_2$(s) −2log aH_2O(l). If we assume the activities of water and quartz are 1 and the γ for H_4SiO_4(aq) is 1 (a good assumption since it is an uncharged aqueous species), then we obtain log mH_4SiO_4(aq) = −3.98, or mH_4SiO_4 = 1.05e-4. This is equivalent to 0.104 mmol/L, 2.94 ppm Si as Si, and 6.29 ppm Si as SiO_2.

To calculate the solubility at 5°C and 100°C, we use the Van't Hoff equation. Plugging in T = 5°C (278 K), $K_{eq,298}$ = $10^{-3.98}$, $\Delta H°_{rxn}$ = 25.06 kJ/mol, and R = 8.3144e-03 kJ K^{-1} $mole^{-1}$ gives

$$K_{eq,278} = 10^{-3.98} \cdot \exp\left(\frac{-25.06}{8.3144e-03}\left(\frac{1}{278} - \frac{1}{298}\right)\right) = 10^{-4.30}$$

Based on the new value of log K_{eq} at 5°C = −4.30, we can compute dissolved silica concentrations of 0.051 mmol/L, 1.4 ppm as Si, and 3.0 ppm as SiO_2. At 100°C (373 K) and following the same approach, we obtain log K_{eq} = −3.09, 22.5 ppm Si as Si, and 48 ppm Si as SiO_2. The solubility of quartz increases with increase in temperature because the solubility-controlling reaction has a (+) enthalpy change (endothermic).

8.5   Geochemical Controls on Water Chemistry

Speciation of Dissolved Inorganic Carbon

Unless water is unusually acidic (e.g., acid mine drainage or a volcanically acidified crater lake), its pH and major ion chemistry will be strongly influenced by the **inorganic carbon system**. Inorganic carbon exists in the atmosphere as CO_2(g) at a concentration of about 4.0e-4 bars, or roughly 410 ppm. The exact CO_2 content of air varies seasonally and geographically, and it is slowly increasing with time due to burning of fossil fuels (NOAA 2018). Carbon dioxide dissolves into water as carbonic acid, H_2CO_3(aq). Carbonic acid is chemically equivalent to CO_2(aq), and both ways of writing dissolved carbon dioxide are used interchangeably. At higher pH, carbonic acid can lose a proton, or **dissociate**, to yield bicarbonate ion (HCO_3^-), which in turn can dissociate to yield carbonate ion (CO_3^{2-}). These reactions can be written as follows:

$$CO_2(g) + H_2O = H_2CO_3(aq) \tag{8.12}$$

$$H_2CO_3(aq) = HCO_3^- + H^+ \tag{8.13}$$

$$HCO_3^- = CO_3^{2-} + H^+ \tag{8.14}$$

Figure 8.6 shows the predominance areas of the aqueous dissolved inorganic carbon (DIC) compounds as a function of pH, and Table 8.6 gives values of the equilibrium constant for reactions (8.12), (8.13), and (8.14) between 5°C and 60°C. The pH values of the H_2CO_3/HCO_3^- and HCO_3^-/CO_3^{2-} crossover points correspond to $pK_{8.13}$ and $pK_{8.14}$,

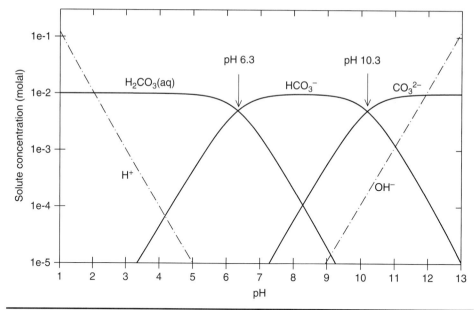

FIGURE 8.6 Speciation of dissolved inorganic carbon as a function of pH at 25°C. The diagram assumes a total DIC concentration of 0.01 molal.

respectively, where the prefix "p" indicates "– log." At 25°C, $pK_{8.13} = 6.35$ and $pK_{8.14} = 10.33$. In other words, $H_2CO_3(aq)$ is dominant when pH < 6.35, and CO_3^{2-} is dominant when pH > 10.33. At intermediate pH values, HCO_3^- is the dominant DIC species. In the pH range of most natural waters (roughly pH 3 to pH 10), the concentrations of one or more DIC compounds are likely to be much greater than the concentrations of either H^+ or OH^- (Figure 8.6) The latter are calculated from $\log K_{eq}$ for the dissociation of water:

$$H_2O(l) = H^+ + OH^- \tag{8.15}$$

T°C	Log $K_{8.12}$	Log $K_{8.13}$	Log $K_{8.14}$	Log $K_{8.15}$	Log $K_{8.16}$
0	−1.11	−6.58	−10.63	−14.94	2.25
5	−1.19	−6.52	−10.55	−14.73	2.16
10	−1.27	−6.46	−10.49	−14.53	2.08
15	−1.34	−6.42	−10.43	−14.35	2.00
20	−1.41	−6.38	−10.38	−14.17	1.93
25	−1.47	−6.35	−10.33	−14.00	1.85
30	−1.52	−6.33	−10.29	−13.83	1.78
45	−1.67	−6.29	−10.20	−13.40	1.58
60	−1.78	−6.29	−10.14	−13.02	1.38

TABLE 8.6 Equilibrium Constants for Reactions in the Dissolved Inorganic Carbon System [Compiled from Data in Drever 1997)]

This helps to explain why DIC plays such a large role in buffering pH of surface water and groundwater.

Chemical Weathering and Cave Formation

Calcite, the most common carbonate mineral, will dissolve in the presence of weak acid to liberate Ca^{2+} and HCO_3^- ions:

$$CaCO_3(\text{calcite}) + H^+ = Ca^{2+} + HCO_3^- \tag{8.16}$$

However, unless the water is strongly acidic, the source of protons for calcite dissolution is usually dissolved carbonic acid, not H^+. In fact, dissolved CO_2 plays a key role in *chemical weathering* of soils and bedrock. Some example reactions include the following:

$$CaCO_3(\text{calcite}) + H_2CO_3(\text{aq}) = Ca^{2+} + 2HCO_3^- \tag{8.17}$$

$$CaMg(CO_3)_2(\text{dolomite}) + 2H_2CO_3(\text{aq}) = Ca^{2+} + Mg^{2+} + 4HCO_3^- \tag{8.18}$$

$$2KAlSi_3O_8(\text{K-feldspar}) + 2H_2CO_3(\text{aq}) + H_2O = 2K^+ + Al_2Si_2O_5(OH)_4(\text{kaolinite})$$
$$+ 4SiO_2(\text{quartz}) + 2HCO_3^- \tag{8.19}$$

An increase in the concentration of carbonic acid pushes each of the above reactions to the right, resulting in dissolution of minerals in the bedrock, formation of secondary minerals in soil (kaolinite, quartz), and delivery of major cations and anions to the hydrosphere. Over geologic time, chemical weathering provides a check on the rate of rise of CO_2 in Earth's atmosphere. Unfortunately, the rates of chemical weathering on a global scale are slow relative to present rates of fossil-fuel combustion.

Although the partial pressure of CO_2 in air is only $10^{-3.8}$ bars, its concentration can be much higher in the subsurface, especially in soils or shallow bedrock, primarily due to the microbial decay of organic carbon compounds. This process, known as **aerobic respiration**, can be written in shorthand form as follows:

$$C_{\text{org}} + O_2 = CO_2 \tag{8.20}$$

Partial pressures of CO_2 in soils can be as high as 10^{-2} or 10^{-1} bars. This leads to high concentrations of carbonic acid and a consequent drop in pH. If calcite is present in the soil, then some of it will dissolve to form HCO_3^- and Ca^{2+}. The higher the value of P_{CO2}, the more calcite will dissolve, and the higher the final concentration of HCO_3^- and Ca^{2+} will be. Other minerals, such as feldspars and micas, will also chemically weather in the presence of dissolved CO_2, but typically at a slower rate compared to calcite.

Reactions between DIC and calcite help to explain certain features in karst settings (Figure 8.7). The formation of caves commonly begins with a buildup of $CO_2(g)$ in a soil zone overlying limestone bedrock. As CO_2-charged soil water drains downward by gravity, it eventually encounters limestone and dissolves calcite to form open space. As long as the P_{CO2} is fixed to a high value, for example, by microbial respiration in the overlying soil, this process will continue, eventually forming a network of open fractures and caves. If the cave network vents somewhere to the surface, mixing of air between the cave and the atmosphere will result in a decrease in P_{CO2} back down to atmospheric levels, with a proportional drop in the concentration of dissolved carbonic

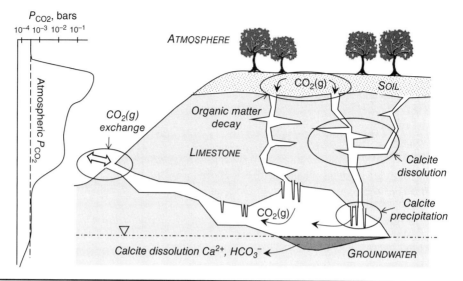

FIGURE 8.7 Schematic diagram summarizing changes in P_{CO2} of water in the subsurface and its influence on cave and stalactite formation.

acid (reaction 8.12). The drop in H_2CO_3 pulls reaction 8.17 back to the left, thus precipitating calcite. Stalactites and stalagmites form in this way where shallow, CO_2-rich water exits from a crack in the ceiling of a cave that is connected to outside air.

Dissolution and Precipitation of Other Minerals

As water migrates through soil and bedrock, it will interact with a host of rock-forming and trace minerals. Each of these minerals will dissolve to a greater or lesser extent into the water, and this is one fundamental control on groundwater chemistry. The longer the residence time of water in the subsurface, the more likely that water will reach chemical saturation with minerals in the rock, and the more variety of minerals that water will be exposed to. The closeness to equilibrium that water has with respect to a given mineral is expressed as the **saturation index** (*SI*):

$$SI = \log Q / K_{eq} \tag{8.21}$$

where Q is the **activity quotient** for the products and reactants involved in a mineral solubility reaction and K_{eq} is the equilibrium constant for that reaction. Q can be thought of as "what actually is dissolved in the water based on chemical analysis," and K_{eq} can be thought of as "what should be dissolved in the water according to thermodynamics." A positive *SI* value means that there are more solutes dissolved in the water than there should be. Such water is said to be **supersaturated** with the mineral in question. Conversely, a negative *SI* means there is not enough solutes and the water is **undersaturated** with the mineral. Values of *SI* near zero suggest the water is **near equilibrium** with the mineral. In practice, any mineral with an *SI* value within ±0.3 units of zero is considered close to equilibrium.

Example 8.8 A filtered groundwater sample contains 750 ppm sulfate and 0.6 ppm Pb. Is this water supersaturated, undersaturated, or close to equilibrium with anglesite (PbSO$_4$)?

The first step is to write the solubility controlling reaction with the mineral of interest on the left:

$$PbSO_4(s) = Pb^{2+} + SO_4^{2-}$$

Table 8.4 lists a log K_{eq} value of -7.79 for this reaction at 25°C. Next we write out the expression for the activity quotient:

$$Q = (aPb^{2+}) \cdot (aSO_4^{2-}) \cdot (aPbSO_4)^{-1}$$

If we assume that anglesite is pure and that the activity coefficients for Pb^{2+} and SO$_4^{2-}$ are close to 1 (not necessarily a good assumption!), then $Q = mPb^{2+} \cdot mSO_4^{2-}$. Dividing the ppm values by the gram formula weights, we obtain $mSO_4^{2-} = 7.8e\text{-}3$, $mPb^{2+} = 2.9e\text{-}6$, and $Q = 2.2e\text{-}8$. SI can now be computed:

$$SI = \log(Q/K_{eq}) = \log(2.2e\text{-}8)/(10^{-7.79}) = +0.14$$

Since the value of SI is within ± 0.3 log units of zero, we conclude that the water is "near equilibrium" (i.e., slightly supersaturated) with anglesite. Anglesite is *probably* present and exerting a solubility control on the concentration of dissolved lead in this water sample.

A rigorous calculation of saturation index is never a simple task, as it requires (1) a complete and accurate chemical analysis, (2) accurate thermodynamic data (log K_{eq} values) for each mineral of interest, and (3) computation of activity coefficients for all solutes involved in the reaction. A number of geochemical modeling programs, such as Visual Minteq, PHREEQC, and Geochemist's Workbench, will quickly compute SI values for dozens to even hundreds of minerals for a particular water sample based on its chemical composition (including pH and temperature). It is up to the user to decide what might be the geologic significance of the derived SI values. As always, thermodynamic-based calculations only tell you what *should* be happening in an environmental system, not what *actually* is happening!

Cation Exchange and Adsorption

Mineral grains can dissolve and precipitate out of aqueous solution, and this obviously has a major control on the chemistry of natural waters. A different but equally important set of reactions involves adsorption of solutes onto a mineral surface. At a microscopic scale, a mineral surface can be thought of as a parking lot. Ions in solution can settle onto a parking spot on the surface, but can also be bumped off by another ion that comes along that has a stronger attraction to the site. The size of the parking lot (cm^2 or m^2) per gram of solid is known as the **specific surface area**. Specific surface areas increase exponentially with decrease in grain size. Submicron-sized colloidal particles can have surface areas greater than 10 m^2 per g. That's a lot of parking spaces!

Three classes of minerals that play an especially important role in adsorption/desorption processes are clays, zeolites, and hydrous metal oxides. Adsorption can also occur onto organic matter, both living and dead. In the following discussion, **adsorption** refers to the attachment of a dissolved solute onto a mineral surface, **desorption** refers to detachment, the **sorbent** is the solid, and the **sorbate** is the solute being adsorbed. Also note the difference between adsorption and absorption. **Absorption** is a biological process in which a solute is assimilated through a cell wall into the interior of the organism. Adsorption is a nonbiological reaction that takes place on the solid surface. Adsorption is important in studies of the fate and transport of harmful solutes in surface or groundwater because freshly formed mineral surfaces can remove many potential toxins out of solution to concentrations that are much lower than any kind of mineral-solubility limit. On the flip side, those same surfaces, once loaded with contaminants, can release the toxins at a later time if the chemistry of the water changes. A highly polluted site may need to be flushed with clean water for decades or even centuries to strip off all of the adsorbed contaminants of concern to safe levels. This is a common problem with groundwater pump-and-treat systems.

Clays and Cation Exchange

Clay minerals have a structure that consists of alternating, two-dimensional sheets of silica layers and Al- or Mg-hydroxide layers. Adjacent sheets are weakly bonded, explaining why all clays are soft minerals. In the tetrahedral layer, a small percentage of Si atoms can be substituted by similar-sized Al atoms. Because Si is a +4 cation and Al is a +3 cation, this process, termed **isomorphous substitution**, results in a (−) charge on the surface of the tetrahedral layer. Dissolved cations that diffuse into the interlayer space of clay minerals are attracted to these (−) sites. Different cations adsorb to different degrees based on physical factors (charge and radius of the ion) and chemical factors (the concentrations in solution and any special chemical attractions between the sorbent and sorbate). At equivalent concentrations, cations with a high value of z/r will adsorb more strongly. Because the bond between the adsorbed cation and the mineral surface is weak, one cation can be replaced by another if the chemistry of the water changes. This process is known as **cation exchange**. Clay minerals that form in marine environments are typically loaded up with Na^+, whereas clays that form on land tend to be richer in Ca^{2+} and Mg^{2+}. Groundwater that interacts with marine shales will see an increase in $(Na + K)/(Ca + Mg)$ ratio, due to cation exchange with Na-rich clays. This can be a problem if the water is used for irrigation, since most crops are sensitive to Na^+.

The **cation exchange capacity** (CEC) of a given clay mineral is a measure of how many meq of cations can be adsorbed per 100 g of solid. Typical values of CEC for clay minerals range from about 10 for kaolinite to higher than 100 for montmorillonite (aka smectite) and vermiculite, the so-called "swelling clays." Good agricultural soil ideally has a significant percentage of clay with high CEC to store and release mineral nutrients mixed with silt and sand to provide porosity and permeability.

> **Example 8.9**   If 1 kg of montmorillonite were allowed to equilibrate with a NaCl-rich water, how much Na by mass would the clay contain? How much Ca would the same clay contain if it equilibrated with a $CaCl_2$-rich water?
>
> If we assume a CEC of 100 meq/100 g solid, then this is equivalent to 1 mole of [+] charge per kg of solid. Since each mole of Na^+ weighs 22.99 g, there would be 22.99 g of Na in 1,000 g of clay (2.299% by mass). For Ca, 100 meq/100 g = 1 mole × 40.078 g/mol/2 = 20.03 g of Ca per kg of clay. You need to divide by 2 in the Ca example since each mol of Ca^{2+} has 2 equivalents of [+] charge.

Adsorption Isotherm and Surface Complexation Models

In an isotherm experiment, a beaker containing a known mass of solid (e.g., fine soil or a pure clay or hydrous oxide mineral) is stirred and a known mass of the solute of interest (e.g., Cu^{2+}) is added to the system in increments. After each addition, the partitioning of Cu^{2+} between solid and aqueous solution is determined. The data are then fit to a mathematical expression. The simplest expression is the **linear distribution coefficient**, or linear K_d model:

$$\{Cu\}_{adsorbed} = K_d \cdot \{Cu\}_{dissolved} \tag{8.22}$$

The brackets {} indicate the concentration of Cu adsorbed (moles/kg or mg/kg) and the concentration of Cu dissolved (moles/L or mg/L). In other words, the higher the concentration of Cu^{2+} adsorbed to the soil, the higher the concentration of Cu^{2+} in the water, and vice versa. The **Freundlich isotherm** (K_f) model follows the same approach, but an adjustable parameter, n, is added to obtain a better fit to the experimental data:

$$\{Cu\}_{adsorbed} = K_f \cdot (\{Cu\}_{dissolved})^n \tag{8.23}$$

A weakness of both the linear K_d and the Freundlich models is that they cannot take into account the fact that there is always a limited number of sorption sites in any solid medium. Once these sites are filled, there can be no more adsorption. The **Langmuir isotherm** (K_L) model takes this into account, as follows (still using Cu as an example sorbate):

$$\{Cu\}_{adsorbed} = \{total\ sites\} \cdot \frac{K_L \cdot \{Cu\}_{dissolved}}{1 + K_L \cdot \{Cu\}_{dissolved}} \tag{8.24}$$

To use a Langmuir isotherm model, the concentration of total adsorption sites (number of moles of site per kg of solid) must be known. This can be determined by fitting a horizontal line to the upper limit of $\{Cu\}_{adsorbed}$, as shown in Figure 8.8.

A complication associated with adsorption isotherm models is that the values of K_d, K_f, and K_L are often strongly dependent on pH, and to a lesser extent, temperature. To completely understand how a given solute adsorbs onto a given solid, parallel experiments would need to be conducted over a range of pHs and temperatures. In general, adsorption of cations is favored by an increase in pH, whereas anions adsorb more strongly at low pH. Figure 8.9 summarizes the relative affinity of different cations

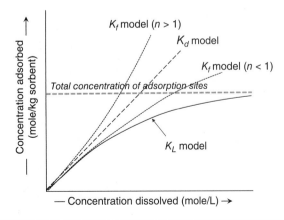

FIGURE 8.8 Schematic diagram showing sorbent-sorbate relationships for different adsorption isotherm models.

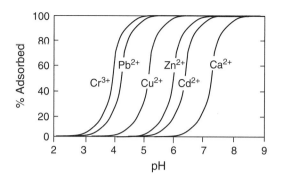

FIGURE 8.9 Extent of adsorption of solutes onto hydrous ferric oxide, $Fe(OH)_3$, as a function of pH. [*After Drever (1997).*]

to adsorb onto ferrihydrite ($Fe(OH)_3$) as a function of pH. As shown in Figure 8.9, each different metal cation has a characteristic pH, or **adsorption edge**, where it transitions from being mostly dissolved to being mostly adsorbed. Similar adsorption edge diagrams can be drawn for anions, although the pH dependence is often more complex. In general, most anions follow a % adsorbed versus pH relationship that is the mirror image of the cation diagram.

The most complex and theoretically robust approach for predicting adsorption is a **surface complexation model** (SCM). SCMs are true thermodynamic models based on hundreds or even thousands of individual experiments, each examining sorption of a particular solute onto a particular solid sorbent. SCMs can take changes in pH and temperature into account, can model competition between different solutes for available surface sites, and can accommodate more than one type of adsorption site. Adsorption site concentrations are either determined experimentally or can be predicted from first principles. Further explanation of the theory behind SCM modeling is beyond the scope of this chapter. Langmuir (1997) gives a good overview of the topic.

Redox Reactions

Chemical reactions that involve a change in valence, or oxidation state, of reactants and products are known as **redox reactions**. Such reactions can be written as a transfer of electrons, as in the reduction of ferric iron to ferrous iron:

$$Fe^{3+} + e^- = Fe^{2+} \tag{8.25}$$

However, free electrons do not physically exist in aqueous solution. In nature, the reduction half-reaction is balanced by a simultaneous oxidation half-reaction so that the overall reaction is conservative with respect to electrons. One way to balance the above reaction is by oxidation of organic carbon:

$$C_{org} + 2H_2O = CO_2 + 4H^+ + 4e^- \tag{8.26}$$

Combining reactions 8.25 and 8.26 gives the following:

$$4Fe^{3+} + C_{org} + 2H_2O = 4Fe^{2+} + CO_2 + 4H^+ \tag{8.27}$$

In the above reaction, iron undergoes a reduction in valence state from +3 to +2, while carbon undergoes an oxidation from 0 to +4. Consequently, each mole of organic carbon can reduce 4 moles of ferric iron. Although the overall reaction is thermodynamically favorable to proceed from left to right, it will do so at a negligible rate in the absence of microbial catalysts. This is true for many redox reactions. Myriad different forms of bacteria and archaea have evolved that facilitate the transfer of electrons in redox reactions. Without microscopic life, the chemistry of soil and groundwater would be much simpler and much less interesting!

In the subsurface, once the supply of atmospheric O_2 is cut off, a series of redox reactions begins. The sequence is predictable from thermodynamics: aerobic respiration occurs first, then nitrate reduction, reduction of Mn- and Fe-oxides, sulfate reduction, and finally reactions that produce methane gas (methanogenesis) (Figure 8.10). Aerobic respiration of organic matter (reaction 8.19) is the most common reason that some groundwaters become anoxic (devoid of measureable DO). The source of organic carbon can be natural, as in a wetland sediment or carbonaceous shale, or it can be anthropogenic, as in a petroleum spill. On an Eh scale, all dissolved oxygen is consumed within 100 mV of the upper stability limit of water.

If nitrate is present, it will be the next compound to be reduced by microbes:

$$4NO_3^- + 5C_{org} + 4H^+ = 2N_2(g) + 2H_2O + 5CO_2 \tag{8.28}$$

This reaction, called **denitrification**, is important because it is one of the only ways that the concentration of nitrate, a toxic compound, can be lowered in the subsurface.

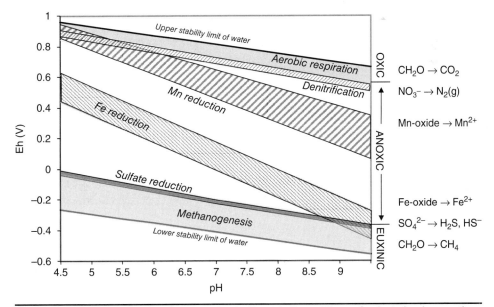

FIGURE 8.10 Diagram showing the approximate Eh and pH conditions of selected redox reactions that take place in natural waters. Thermodynamic data needed to construct this diagram were taken from Drever (1997). Further details are given in Gammons et al. (2009).

Microbes that make denitrification happen require anoxic conditions. The product of denitrification, $N_2(g)$, is chemically and for the most part biologically unreactive.

Once dissolved oxygen and nitrate are consumed, the redox state of most natural waters will drop to the next Eh-buffering reaction in the sequence, **reduction of manganese and/or iron oxides**. Mn-oxides are less common than Fe-oxides, but are almost always present in trace to minor amounts. The reduction reactions for the common minerals pyrolusite (MnO_2) and goethite (FeOOH) can be written as follows:

$$2MnO_2 + C_{org} + 4H^+ = 2Mn^{2+} + CO_2 + 2H_2O \tag{8.29}$$

$$4FeOOH + C_{org} + 8H^+ = 4Fe^{2+} + CO_2 + 6H_2O \tag{8.30}$$

Again, these reactions are catalyzed by microbes. The end result is a groundwater that contains elevated concentrations of dissolved Mn^{2+} and Fe^{2+}. Such waters are sensitive to oxidation and may precipitate a floc of rust-colored ferric hydroxide after pumping to the surface.

At strongly reducing conditions, **bacterial sulfate reduction** (BSR) is the next Eh-buffering reaction in the sequence. Sulfate ion contains S in the +6 valence state, whereas the product of BSR, hydrogen sulfide, contains S in the −2 valence state. The overall reaction can be written as

$$SO_4^{2-} + 2C_{org} + 2H_2O = H_2S + 2HCO_3^- \tag{8.31}$$

Hydrogen sulfide has a strong odor of rotten eggs, and it is toxic to many forms of life, including humans. Luckily for us, most H_2S that is produced by BSR reacts with iron in sediment to form pyrite, FeS_2. If the sediment has a low Fe content, then H_2S can accumulate to levels that are problematic to water quality, although this is uncommon. Besides pyrite, H_2S can react with other metals, such as Cu, Pb, or Zn, to form insoluble metal sulfides. For example, the following reaction explains the formation of sphalerite:

$$Zn^{2+} + H_2S = ZnS(s) + 2H^+ \tag{8.32}$$

In special environments, such as the sea floor of a restricted ocean basin, reactions of this type can sometimes lead to the formation of a mineable metal resource. BSR can also be engineered to help strip dissolved toxic metals out of groundwater or surface water using bioreactors or constructed wetlands that optimize production of H_2S.

In the most strongly reducing environments, microbes exist that can produce methane gas by the breakdown of organic matter or the metabolism of $H_2(g)$. The overall reactions, both of which are different types of **methanogenesis**, can be written as

$$2C_{org} + 2H_2O = CH_4 + CO_2 \tag{8.33}$$

$$CO_2 + 4H_2(g) = CH_4 + 2H_2O \tag{8.34}$$

The first reaction is an example of a **disproportionation reaction** because the organic carbon, which has a valence of zero, is split into a mixture of reduced carbon (−4 valence) in CH_4 and oxidized carbon (+4 valence) in CO_2. The second reaction is common in certain hot springs settings, such as Yellowstone National Park, where geothermal waters with trace amounts of $H_2(g)$ cool to temperatures where microbes can live.

Methane is the main constituent of natural gas, and accumulations of biogenic CH_4 can sometimes be recovered as a fossil fuel. Because methane is insoluble in water, it is easily extracted into the vapor phase by pumping. Examples include the recovery of methane from municipal landfills and pumping CH_4 from buried coal deposits (i.e., coal-bed methane). At high latitudes, methane can be released to shallow groundwater during the destabilization of crystalline gas hydrates, a process that is happening worldwide due to the thawing of permafrost. Most large natural gas reservoirs contain methane that formed by nonbiological breakdown of petroleum and other hydrocarbons at the elevated temperatures and pressures corresponding to deep sedimentary basins.

Nitrogen in Groundwater

Nitrogen, like carbon and sulfur, exists in nature in many different oxidation states, ranging from −3 to +5 (Figure 8.11). The nitrate ion, NO_3^-, is the most stable form of dissolved N in oxidized surface waters and shallow groundwater. Nitrate is an essential nutrient for plants, but it is toxic to humans at high concentrations. The U.S. drinking water standard is 10 mg/L NO_3 as N (44.3 mg/L NO_3 as NO_3) (see "Water-Quality Parameters and Their Significance" online at www.mhprofessional .com/weight3e). The main sources of elevated nitrate in groundwater are human related, being chemical fertilizers and human or livestock waste. However, small amounts of nitrate can enter the hydrosphere via precipitation or by leaching of N-compounds from certain rocks, such as organic-rich shales. Nitrogen gas (N_2) is the dominant constituent in our atmosphere, but it is chemically unreactive, and can be directly utilized only by a small group of N-fixing organisms. Trace amounts of NO_x (nitrogen oxide gases of unspecified composition) exist in air, especially near urban or industrial areas. NO_x, along with $SO_2(g)$, is a contributor to acid precipitation.

The most reduced form of nitrogen is ammonium (NH_4^+), and NH_3 functional groups are the main form of N in organic compounds. When organic material decays, N is released into groundwater or soil water primarily as ammonium. However, in the presence of dissolved oxygen, bacteria quickly oxidize ammonium to nitrate via

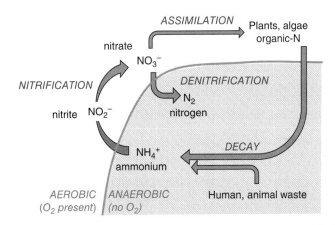

Figure 8.11 Simplified nitrogen cycle showing reactions taking place in the subsurface. The shaded area denotes anaerobic conditions.

nitrification. Nitrification is a complex process involving several different reactive intermediate species, the most common of which is nitrite, NO_2^-. Nitrite is metastable and will oxidize to nitrate if DO is not limiting. Usually if nitrite is detected in a groundwater, it means that the sample is located near an oxidation front where nitrification of ammonium is actively occurring.

Nitrate is a highly *conservative* anion, meaning that it typically does not take part in adsorption or mineral dissolution/precipitation reactions. Once nitrate forms, the only way that its concentration can be diminished is by dilution, assimilation (plant uptake), or denitrification (reduction to N_2 gas). *Assimilation* takes place in the root zone of trees, but at greater depth this pathway becomes negligible. *Denitrification* is the most efficient way to decrease concentrations of dissolved nitrate in groundwater, but the bacteria that perform this function require anaerobic (O_2-absent) conditions. Such conditions can be artificially engineered in a bioreactor, or by injecting O_2-consuming chemicals into the subsurface, a common practice in remediation strategies.

Pyrite Oxidation and Mine Drainage

Oxidation of pyrite is the most common way that a surface water or groundwater sample can become strongly acidic (pH < 4). Pyrite oxidation and generation of acidic drainage is perhaps the biggest environmental challenge to the modern mining industry (Jambor et al. 2003), and many historic mine sites have legacy pollution related to careless disposal of waste rock and tailings. The overall pyrite oxidation reaction can be written as follows:

$$FeS_2(\text{pyrite}) + 15/4O_2 + 7/2H_2O = Fe(OH)_3(s) + 2SO_4^{2-} + 4H^+ \qquad (8.35)$$

Each mole of pyrite oxidized generates 4 moles of protons and 2 moles of sulfate. Unless the pH of the system is very low (<3), most of the iron will precipitate as ferric hydroxide or some other secondary Fe mineral. Streams that are impacted by acid mine drainage are characteristically orange-red stained, and aquatic life may be severely impaired. Although less common, acid rock drainage can also form naturally in headwater streams draining rock that is rich in pyrite and other sulfide minerals (Williams et al. 2015). These naturally acidic streams may have geologic deposits of ferricrete (Fe-oxide-cemented alluvium) along their channels.

Most deposits of pyrite also contain other metal-sulfide minerals, such as sphalerite, galena, chalcopyrite, arsenopyrite, etc. These minerals are also unstable in the presence of atmospheric O_2 and will release their metals to the environment upon weathering. For this reason, water that drains an abandoned mine site may have a complex chemistry with multiple contaminants of concern. Some examples of trace solutes that may be present in such waters are listed in Table 8.7. The solutes are arranged based on whether they dissolve into water as cations or as anions. As discussed above in the section on adsorption, metal cations tend to adsorb more strongly onto mineral particles as pH increases, whereas metal anions show the opposite behavior. For this reason, an acidic stream, lake, or groundwater plume will most likely contain high concentrations of many of the cationic solutes, whereas the anionic solutes, with the exception of sulfate ion, may be scarce or absent. In contrast, a neutral-pH or alkaline water

Cationic Constituents				Anionic Constituents	
Aluminum	Al^{3+}	Manganese	Mn^{2+}	Arsenic	$HAsO_4^{2-}$, H_3AsO_3
Cadmium	Cd^{2+}	Mercury	Hg^{2+}, $Hg(aq)$	Chromium	CrO_4^{2-}
Cobalt	Co^{2+}	Nickel	Ni^{2+}	Molybdenum	MoO_4^{2-}
Copper	Cu^{2+}	Silver	Ag^+	Selenium	SeO_4^{2-}
Iron	Fe^{3+}, Fe^{2+}	Uranium	UO_2^{2+}	Sulfur	SO_4^{2-}
Lead	Pb^{2+}	Zinc	Zn^{2+}	Tungsten	WO_4^{2-}

TABLE 8.7 Solutes of Potential Environmental Significance in Mine Drainage

may pick up problematic amounts of dissolved As (as arsenate), Mo (as molybdate), Se (as selenate), or W (as tungstate). Other contaminants in mine drainage sometimes include nitrate (residue from blasting powder) and cyanide (used in beneficiation of gold ore) and its breakdown products.

The most common method of treating acidic mine water is addition of lime (CaO), a strong base that raises pH and precipitates dissolved metals as a mix of hydroxide and carbonate minerals. This process is expensive, and the fine-grained solids, termed "sludge," are difficult to handle and reclaim. An alternate approach is to engineer a bioreactor in which sulfate-reducing bacteria are established. The H_2S produced by BSR causes toxic metals to precipitate out of solution as metal-sulfides (reaction 8.32), and the overall process generates bicarbonate alkalinity that raises pH (reaction 8.31). Once set up, this approach, sometimes referred to as "passive treatment," involves no addition of chemicals. The microbes do the hard work, putting SO_4^{2-} and dissolved metals back into the ground as sulfide minerals, which is how they were originally present in the rock prior to mining.

8.6 Stable Isotope Hydrogeology

An isotope is a nuclide that has the same number of protons but a different number of neutrons. For example, the element hydrogen has three isotopes, ^1H, ^2H (also known as deuterium, D), and ^3H (also known as tritium, T). All of these isotopes have one proton. Deuterium and tritium have one and two neutrons, respectively. Both ^1H and ^2H are stable, whereas ^3H undergoes spontaneous radioactive decay to ^3He, with a half-life (time for ½ of the ^3H atoms in a sample to decay) of 12.32 years. Oxygen has three isotopes, ^{16}O, ^{17}O, and ^{18}O, all of which contain eight protons, and all of which are stable.

A stable isotope analysis of water involves precise measurement of the $^2H/^1H$ and $^{18}O/^{16}O$ ratios in the H_2O molecule. Historically, such analyses were performed with an isotope ratio mass spectrometer (IRMS). However, with the advent of cavity ring-down technology, low-cost, laser-based isotope analyzers have largely replaced IRMS for routine analysis of water. Regardless of the method, stable isotope compositions are reported in "delta" notation, which compares the isotope ratio of a sample to the isotope ratio of a standard. The isotopic standard for O and H analysis of water is Vienna

Standard Mean Ocean Water (VSMOW). The defining equations can be written out as follows:

$$\delta^2H \ (or \ \delta D) = \left(\frac{\left(\frac{^2H}{^1H} \right)_{sample}}{\left(\frac{^2H}{^1H} \right)_{VSMOW}} - 1 \right) \times 1{,}000 \qquad (8.36)$$

$$\delta^{18}O = \left(\frac{\left(\frac{^{18}O}{^{16}O} \right)_{sample}}{\left(\frac{^{18}O}{^{16}O} \right)_{VSMOW}} - 1 \right) \times 1{,}000 \qquad (8.37)$$

Because the differences in isotope ratio of the sample and VSMOW are very small, the analytical results are multiplied by a factor of 1,000 resulting in units of parts per thousand, or "per mil." A per mil symbol (‰) looks like a percent symbol (%), but with an extra zero in the denominator. If a given water sample has a higher $^2H/^1H$ or $^{18}O/^{16}O$ ratio than VSMOW (average ocean water), then the value of δ^2H or $\delta^{18}O$ will be positive, and the sample is said to be *isotopically heavy* (i.e., enriched in the heavy isotope). Likewise, if the $^2H/^1H$ ratio is less than VSMOW, then the Δ^2H value will be negative, and the sample is *isotopically light*.

Early work by Craig (1961) showed that the $\delta^{18}O$ and δ^2H composition of Earth's meteoric waters (water that was recently recharged from the atmosphere as rain or snowmelt) follows a systematic relationship known as the global meteoric waterline (MWL): $\delta^2H = 8\delta^{18}O + 10$. Although there have been several attempts to refine Craig's MWL on a global scale, it is still widely cited in the literature. Precipitation becomes isotopically more negative (lighter) in both δ^2H and $\delta^{18}O$ with increase in latitude, increase in elevation, and increase in *continentality*, or distance from the ocean. In areas that experience seasonal difference in climate, precipitation will also show seasonal variation, with warm summer rains being heavier than winter snows. Detailed sampling of rain and snow in a given geographic region may show local, relatively minor deviations from the global MWL. In such cases, a local meteoric waterline (local MWL) may be developed and used in subsequent isotope hydrology studies in that area.

Except in highly arid areas, shallow groundwater tends to have an H- and O-isotope composition that is similar to average recharge water. In western Montana where the author lives, most groundwater is recharged by snowmelt and spring rains. Given this seasonal timing, plus the location of Montana at a high elevation, midlatitude, and a long way from either the Pacific Ocean or the Atlantic Ocean, our groundwater is isotopically very light, usually less than −16‰ in $\delta^{18}O$ and less than −120‰ in δ^2H. Groundwater feeds the network of mountain and valley streams that flow with little isotopic change along their course, provided there are no lakes or reservoirs in the watershed. If lakes are present, the isotopic composition of water may be shifted along an evaporation line. The slope of a local evaporation line (LEL) will vary depending on a region's climate, becoming shallower in areas that have low relative humidity. Water samples that have experienced significant evaporation are shifted off of the local MWL to heavier (more positive) $\delta^{18}O$ values. The more evaporation occurs, the further shifted the

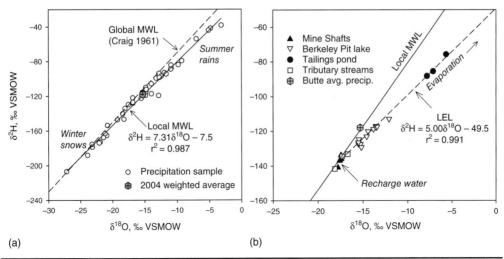

FIGURE 8.12 The isotopic composition of (a) precipitation and (b) surface water and flooded mine shafts in Butte, Montana. [*Data from Gammons et al. (2006).*] MWL = meteoric waterline; LEL = local evaporation line.

isotopic compositions become. Unlike evaporation, transpiration of water by trees and crops does not discriminate between isotopes of H or O. Simultaneous measurement of water chemistry and stable isotopes is a useful way to discriminate "E" and "T" in the "ET" variable.

Figure 8.12 summarizes the above relationships in a δ^2H versus $\delta^{18}O$ plot using Butte, Montana, as an example (Gammons et al. 2006). The global MWL of Craig (1961) and a local MWL established for Butte are shown for comparison, along with the isotopic compositions of individual precipitation events measured in the 2004 calendar year. Also shown are the isotopic compositions of two surface water bodies that experienced evaporation, a tailings pond and a lake that formed by flooding of an open-pit copper mine (the Berkeley Pit). Together, the mine lakes help define the LEL for Butte. The intersection of the LEL and the local MWL defines the isotopic composition of average recharge water for the groundwater system on Butte Hill. Headwater streams and flooded mine shafts have isotopic compositions that are close to that of average recharge waters. An interesting point is that the isotopic composition of groundwater recharge is not equal to the weighted average of all precipitation for a calendar year. This is because recharge is skewed toward snowmelt and spring rains. Most of the rain that falls in midsummer is transpired by plants or is evaporated.

Given the above discussion, it is easy to appreciate how stable isotopes of water could be useful in a hydrogeological study. They can be used to track seasonal precipitation and recharge events through the hydrologic cycle, to distinguish evaporated versus nonevaporated waters, to identify source regions for recharge water based on elevation, to quantify mixing of waters that have different isotopic compositions, and to conduct paleoclimate or future climate change studies. Best of all, a stable isotope analysis of a water sample can now be performed for a cost that is typically much less than a conventional chemical analysis. The interested reader is referred to Clark (2015) or Sharp (2017) for more information about environmental stable isotopes.

8.7   Summary

Water chemistry is an important facet of hydrogeology. This chapter has outlined some of the basic chemical principles applied to natural waters and has reviewed common units and ways to graphically present water-quality data. It is becoming increasingly common to include stable isotope analysis of water in hydrogeological studies, and stable isotopes give a different and sometimes illuminating approach to solving water-balance problems. The more you learn about any topic the more interesting and complex it becomes. This is certainly true of water chemistry, and the reader is encouraged to explore peer-reviewed journals and other textbooks that focus on this topic. Building on the foundation provided here, Chapter 9 discusses practical considerations for conducting a water chemistry study.

8.8   Problems

8.1   The following analysis was obtained from a shallow groundwater sample in the contaminated alluvial aquifer of Butte, Montana. The temperature of this water sample was 10°C, the pH was 5.63, and the specific conductance was 2930 μs/cm.

a)   Fill in the following table. Remember that "meq/L" = "mmol/L" × the charge of the ion.

Species	mg/L	Molar Weight (g)	mol/L	$m_i z_i^2$ (for I)	mmol/L	meq/L	% of Major Cations for Piper Plot	% of Major Anions for Piper Plot
Ca^{2+}	49	40.08	1.22e-3	4.88e-3	1.22	2.44		0
Mg^{2+}	154							
Na^+	81							
K^+	22							
		Σmeq/L common cations for Piper plot						
Fe^{2+}	0.078							
Mn^{2+}	43.6							
Cu^{2+}	12.1							
Zn^{2+}	42.1							
Sr^{2+}	6.5							
		Σmeq/L all cations for charge balance						
HCO_3^-	109							
SO_4^{2-}	1,924							
Cl^-	113							
NO_3^-	29*							
		Σmeq/L common anions for Piper plot						
		Σmeq/L all anions for charge balance						
		$\Sigma m_i z_i^2$ for ionic strength						

*ppm NO_3 as NO_3

b)   Plot this groundwater sample on a Piper diagram (e.g., Figure 8.1) and a Stiff diagram. What type of water is this based on its major cation and anion composition?

c)   Perform a charge balance analysis on these data (see Equation 9.8, next chapter). Does the analysis look good? (An acceptable charge balance would be ±5% or less). Is there an excess of cations or anions?

d)   Calculate the total dissolved solids concentration for this water (TDS, mg/L). To do this, simply add up the mg/L of each solute. It is sometimes said that the TDS of a water sample in mg/L can be estimated as $0.7 \times SC$ (in µs/cm). How well does this approximation hold in the present case?

e)   Calculate the "total hardness" of this water (see Equation 9.9, next chapter). Classify this water as "soft," "moderately hard," "hard," or "very hard":

Hardness range (mg/L CaCO$_3$)	Classification (Hem 1985)
0–60	Soft
61–120	Moderately hard
121–180	Hard
More than 180	Very hard

f)   Calculate the ionic strength (I) of this water.

g)   Use the Davies equation to calculate γ for Ca^{2+} and HCO_3^-.

h)   What is the activity of H^+ in this water?

i)   Calculate the value of Q (activity quotient) and SI (saturation index) for calcite for this water sample (see Table 8.4 for the equilibrium constant you need). Is the water undersaturated, supersaturated, or near equilibrium with calcite?

j)   Calculate saturation indices (SI) for malachite, smithsonite, and gypsum for this water (see Table 8.4 for K_{eq} values). Is the water undersaturated, supersaturated, or near equilibrium with each of these minerals?

k)   Does this water exceed the U.S. drinking water standard of 10 mg/L NO_3 as N?

8.2   Repeat questions 1(a) through 1(j) for selected surface water samples in Table 8.2. Note that the Davies equation is not reliable for the highly evaporated samples of Lake Poopo.

8.3   The Berkeley Pit in Butte, Montana, is one of the worlds' largest accumulations of acid mine drainage. The former open-pit copper mine is now filled with more than 100 billion liters of acidic, metal-rich water. The following is a chemical analysis (concentrations in mg/L) of a sample collected at 220-m depth in 2003 (Pellicori et al. 2005). The pH of the water sample was 2.44 and the temperature was 8.3°C.

Al	278	Fe(2+)	771	Mn	233
As	0.93	Fe(3+)	295	Na	75
Ca	422	K	7	Zn	596
Cu	167	Mg	483	SO_4^{2-}	8,160

As the lake continues to fill, it will eventually need to be pumped and treated to prevent leakage of acidic water into the local stream system. The treatment process will add lime (CaO) to raise pH and precipitate dissolved solutes as gypsum and metal-hydroxide solids. The amount of lime that will need to be added is directly proportional to the "acidity" of the pit water. For the Berkeley Pit, the total acidity can be approximated as follows:

$$\Sigma Acidity \ (mol/L) = mH^+ + 2(mFe^{2+} + mCu^{2+} + mMn^{2+} + mZn^{2+}) + 3(mFe^{3+} + mAl^{3+})$$

Based on the data given, calculate the total acidity of the pit water, and also the percent contribution from each of the solutes in the equation. (Estimate mH^+ from pH.) Which solute is the biggest contributor to acidity? When CaO dissolves in water, it produces two moles of base: $CaO + H_2O \rightarrow Ca^{2+} + 2OH^-$. This being the case, how much lime (g) would need to be added to each liter of water to neutralize the acidity? It is estimated that the pit will need to be pumped and treated at a rate of 10×10^6 L/day to prevent it from rising above the safe water level. How much lime will be consumed (kg or tons) in the treatment plant in 1 year?

8.4   Read the paper by Van Voast (2003) on the geochemical signature of groundwater associated with coal-bed methane in Montana and Wyoming (see reference list). Explain: (1) How are the CBM waters chemically distinct from surrounding groundwater? (2) What chemical reaction(s) are believed to be responsible for this distinct chemistry?

8.5   Read the paper by Plummer et al. (1990) on the geochemistry of groundwater in the Madison limestone aquifer in the northern Rocky Mountains. Explain the processes responsible for regional changes in chemical composition as groundwater flows down gradient.

8.6   Consider the difference between a highly contaminated site that has received pollutants for more than 50 years versus a pristine site where a small chemical spill has just occurred. For which site would a linear distribution coefficient (linear K_d) adsorption model be more valid? To help your answer, explain why the Langmuir adsorption model is a significant improvement over the linear K_d and Freundlich isotherm models.

8.7   The following is an Eh-pH diagram for the element selenium.

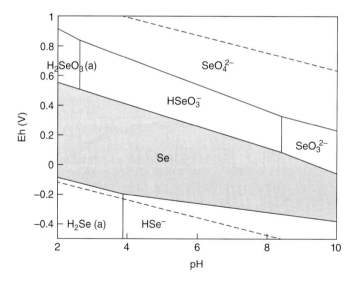

If water has a pH of 6 and an Eh of +0.8 V, which aqueous form of selenium is predicted to be dominant? Research the toxicity and chemical behavior of selenium on the internet. Which forms shown in the diagram are most toxic? What type of chemical reaction is necessary to precipitate dissolved selenium out of solution as a solid?

8.8   The sketch below shows the location of an abandoned open-pit mine that is now filled with water, a waste rock dump, and a pristine tributary stream (example from Gammons et al. 2013). The lake has no obvious surface water inlet or outlet. There are several small groundwater seeps at the toe of the dump. The stream, the dumps, and the lake were sampled for δ^2H and $\delta^{18}O$ in an attempt to see if the groundwater seeps were fed by background recharge water, by water leaking out the bottom of the pit lake, or by a combination of the two.

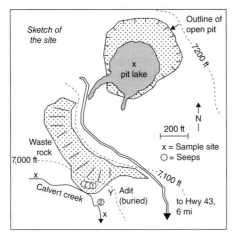

Sample	δ^2H, ‰	$\delta^{18}O$, ‰	Streamflow, $L\,s^{-1}$
Pit lake	−105	−10.8	—
Seepage	−122	−14.6	Not taken
Creek above site	−141	−18.7	7.5
Creek below site	−139	−18.2	Not taken

This example was taken from the study of Gammons et al. (2013).

a) Plot the data on a δ^2H versus $\delta^{18}O$ diagram. Add the equations for the global MWL of Craig (1961), the local MWL for southwest Montana (shown in Figure 8.12a), and the local evaporation line for southwest Montana (shown in Fig. 8.12b). Which waters appear to be least evaporated? Most evaporated?

b) Based on your plot, do you think the lake is leaking water to feed the groundwater seeps near the stream?

c) Stable isotopes are conservative, and therefore it is possible to use isotope data in a water-balance equation. Use the $\delta^{18}O$ data to estimate the total discharge of seepage at the toe of the waste rock pile, based on the following equation:

$$Q_1 * \delta^{18}O_1 + Q_2 * \delta^{18}O_2 = Q_3 * \delta^{18}O_3$$

where Q refers to flow (L/s) and 1, 2, and 3 refer to the stream above the mine, the total flow of the seeps, and the stream below the mine, respectively. [Hint: Note that $Q_3 = Q_1 + Q_2$.]

8.9   References

Allison, J.D., Brown, D.S., and Novo-Gradac, K.J., 1991. MINTEQA2/PRODEFA2, a Geochemical Assessment Model for Environmental Systems. Version 3.0 Users' Manual. U.S. Environmental Protection Agency.

Ayotte, J.D., Gronberg, J.M., and Apodaca, L.E., 2011. Trace Elements and Radon in Groundwater across the United States, 1992–2003. *U.S. Geological Survey, Science Investment. Report 2011-5059*, 115 pp.

Bedient, P.B., Rifai, H.S., and Newell, C.J., 1999. *Ground Water Contamination, 2nd Edition.* Prentice Hall, Upper Saddle River, NJ.

Bethke, C.M., and Yeakel, S., 2015. *Geochemist's Workbench Release 10.0, Reference Manual.* Aqueous Solutions, L.L.C. https://www.gwb.com/pdf/GWB10/GWBreference.pdf

Clark I., 2015. *Groundwater Geochemistry and Isotopes.* CRC Press, Boca Raton, FL.

Cleveland, C.C., and Liptzin, D., 2007. C: N: P Stoichiometry in Soil: Is There a "Redfield Ratio" for the Microbial Biomass? *Biogeochemistry,* Vol. 85, pp. 235–252.

Craig, H., 1961. Isotopic Variations in Meteoric Waters. *Science,* Vol. 133, pp. 1702–1703.

Drever J.I., 1997. *The Geochemistry of Natural Waters: Surface and Groundwater Environments, 3rd Edition.* Prentice Hall, Upper Saddle River, NJ.

Eby, G.N., 2004. *Principles of Environmental Geochemistry.* Brooks/Cole, Pacific Grove, CA.

Fetter, C.W., 1999. *Contaminant Hydrogeology, 2nd Edition.* Prentice Hall, Upper Saddle River, NJ.

Gammons, C.H., Poulson, S.R., Pellicori, D.A., Roesler, A., Reed, P.J., and Petrescu, E.M., 2006. The Hydrogen and Oxygen Isotopic Composition of Precipitation, Evaporated Mine Water, and River Water in Montana, USA. *Journal of Hydrology,* Vol. 328, pp. 319–330.

Gammons, C.H., Duaime, T.E., Parker, S.R., Poulson, S.R., and Kennelly, P., 2010. Geochemistry and Stable Isotope Investigation of Acid Mine Drainage Associated with Abandoned Coal Mines in Central Montana, USA. *Chemical Geology,* Vol. 269, pp. 100–112.

Gammons, C.H., Harris, L.N., Castro, J.M., Cott, P.A., and Hanna, B.W., 2009. *Creating Lakes from Open Pit Mines: Processes and Considerations, with Emphasis on Northern Environments.* Canadian Technical Report of Fisheries and Aquatic Sciences 2826, 106 pp.

Gammons, C. H., Pape, B. L., Parker, S. R., Poulson, S. R., and Blank, C., 2013. Geochemistry, Water Balance, and Stable Isotopes of a "Clean" Pit Lake at an Abandoned Tungsten Mine, Montana, USA. *Applied Geochemistry,* Vol. 36, pp. 57–69.

Gustafsson, J.P., 2013. Visual MINTEQ. Version 3.1. https://vminteq.lwr.kth.se

Guyot, J.L., Roche, M.A., Noriega, L., Calle, H., and Quintanilla, J., 1990. Salinities and Sediment Transport in the Bolivian Highlands. *Journal of Hydrology,* Vol. 113, pp. 147–162.

Hem, J.D., 1985. *Study and Interpretation of the Chemical Characteristics of Natural Water, 3rd Edition.* U.S. Geological Survey. Water-Supply Paper 2254. 263 pp. http://pubs.usgs.gov/wsp/wsp2254/pdf/wsp2254a.pdf

Jambor, J.L., Blowes, D.W., and Ritchie, A.I.M., (eds.), 2009. *Environmental Aspects of Mine Wastes.* Mineralogical Association of Canada, Short Course Series, Vol. 31, 430 pp.

Kehew, A.E., 2001. *Applied Chemical Hydrogeology.* Prentice Hall, Upper Saddle River, NJ.

Langmuir, D., 1997. *Aqueous Environmental Geochemistry.* Prentice Hall, Upper Saddle River, NJ.

NOAA, 2018. *Trends in Atmospheric Carbon Dioxide.* https://www.esrl.noaa.gov/gmd/ccgg/trends/data.html

Parkhurst, D.L., and Appelo, C.A.J., 2013. *Description of Input and Examples for PHREEQC Version 3—A Computer Program for Speciation, Batch-Reaction, One-Dimensional Transport, and Inverse Geochemical Calculations.* U.S. Geological Survey Techniques and Methods, Book 6, Chapter A43, 497 pp.

Pellicori, D.A., Gammons, C.H., and Poulson, S.R., 2005. Geochemistry and Stable Isotope Composition of the Berkeley Pit Lake and Surrounding Mine Waters, Butte, Montana. *Applied Geochemistry,* Vol. 20, pp. 2116–2137.

Piper, A.M., 1944. A Graphic Procedure in the Geochemical Interpretation of Water Analyses. *American Geophysical Union Transactions*, Vol. 25, pp. 914–923.

Plummer, L.N., Busby, J.F., Lee, R.W., and Hanshaw, B.B., 1990. Geochemical Modeling of the Madison Aquifer in Parts of Montana, Wyoming, and South Dakota. *Water Resources Research*, Vol. 26, pp. 1981–2014.

Robinson, R.A., and Stokes, R.H., 1970. *Electrolyte Solutions, 2nd Edition*. Butterworths, London.

Sander, R., 2015. Compilation of Henry's Law Constants (Version 4.0) for Water as Solvent. *Atmospheric Chemistry and Physics*, Vol. 15, pp. 4399–4981.

Schoeller, H., 1962. *Leseaux Souterraines*. Mason and Cie, Paris, 642 pp.

Sharp, Z.D., 2017. *Principles of Stable Isotope Geochemistry, 2nd Edition*. Published online at http://digitalrepository.unm.edu/unm_oer/1

Sprague, J.B., 1971. Measurement of Pollutant Toxicity to Fish—III: Sublethal Effects and "Safe" Concentrations. *Water Research*, Vol. 5, No. 6, pp. 245–266.

Stiff, H.A., Jr., 1951. The Interpretation of Chemical Water Analysis by Means of Patterns. *Journal of Petroleum Technology*, Vol. 3, No. 10, pp. 15–17.

Stumm, W., and Morgan, J.J., 1996. *Aquatic Chemistry: Chemical Equilibria and Rates in Natural Waters, 3rd Edition*. John Wiley & Sons, New York.

Van Voast, W.A., 2003. Geochemical Signature of Formation Waters Associated with Coalbed Methane. *AAPG Bulletin*, Vol. 87, pp. 667–676.

Weiner, E.R., 2015. *Applications of Environmental Aquatic Chemistry—A Practical Guide, 3rd Edition*. CRC Press , Boca Raton, FL.

Welch, A.H., Westjohn, D.B., Helsel, D.R., and Wanty, R.B., 2000. Arsenic in Ground Water of the United States: Occurrence and Geochemistry. *Groundwater*, Vol. 38, pp. 589–604.

Williams G.P., Petteys, K., Gammons, C.H., and Parker, S.R., 2015. An Investigation of Acidic Head-Water Streams in the Judith Mountains, Montana, USA. *Applied Geochemistry*, Vol. 62, pp. 48–60.

Wolfe, M.F., Schwarzbach, S., and Sulaiman, R.A., 1998. Effects of Mercury on Wildlife: A Comprehensive Review. *Environmental Toxicology and Chemistry*, Vol. 17, pp. 146–160.

CHAPTER 9

Water Chemistry Sampling

Chris Gammons

Professor of Geological Engineering, Montana Technological University, Butte, Montana

C hapter 8 of this textbook reviews the basics of water chemistry theory. The purpose of Chapter 9 is to give an overview of some practical considerations regarding how to plan, conduct, and review the results of a water-quality study. The discussion focuses mainly on groundwater but can be applied to surface-water sampling as well. Because this topic is very broad, it is not possible to cover everything in a single chapter. A good overview of groundwater sampling protocols and procedures is the U.S. Geological Survey (USGS) Open-File Report 95-399 (USGS 1995). A more recent reference focusing on field methods is the National Field Manual for the Collection of Water-Quality Data (USGS 2005).

9.1 Have a Plan

The most important step in any water-quality investigation is to understand what question or problem you are trying to address, and to design a Sampling and Analysis Plan (SAP) that will best answer this question within budgetary constraints. With modern instrumentation, it is now possible to analyze any water sample for a bewildering number of analytes and field parameters. But how many of these are critical to the project? Can the project goals be met without collecting *any* samples for laboratory analysis? Although most large-scale projects will require water-quality data, some field problems can be solved by use of simple manual equipment, such as a pH meter, a conductivity meter, or a portable spectrophotometer. In most cases, a rapid field survey using these methods will help screen locations for more detailed field sampling and laboratory work at a later date.

A useful question to ask early in the planning stages is the following: is the question or problem I am trying to address better served by collection of samples from many different locations at one time, or by collection of samples at many different times at a few locations? For example, a study that seeks to characterize the concentrations of nitrate in an alluvial aquifer on a basin-wide scale would most likely start off by sampling as many groundwater wells as possible, distributed more or less evenly across the area of interest. Ideally, the sample set would include several locations that are far removed from any nearby source of nitrate contamination to establish a reference or background concentration against which groundwater that has been "polluted" by human activity can be compared. On the other hand, if monitoring of drinking water is the main issue of concern, then it may make more sense to place a few monitoring wells up gradient of the supply well and sample these for a large number of analytes at regular intervals (e.g., monthly, quarterly, or annually) over a long period of time. Other examples of how time-series data may be useful include tracking the rate of migration of contaminant plumes, evaluating the long-term success of remediation activities, or looking at the effects of climate or seasonal land use practices (e.g., irrigation) on groundwater quality.

Once a method of approach has been defined to meet the project goals, it is highly recommended to formalize these ideas into a SAP project document. Rather than a collection of hastily jotted down notes, the SAP should be a carefully written, complete summary of the overall project objectives and methods. To begin with, the SAP should detail the location and frequency of all samples to be collected. It should specify to what level of concentration each analyte will be quantified (often referred to as the "critical reporting value"), the laboratory that will perform this work, and what the estimated costs will be. Standard procedures should be summarized in appendices and should include information on field safety, equipment operation and maintenance, pre-sampling procedures (such as acid-washing of bottles, preparation of standards, equipment checkout), sampling procedures (calibration protocols, well purging, sample collection), and post-sampling procedures (sample storage, holding times, equipment cleaning and storage). The SAP should also detail any specific quality assurance/quality control (QA/QC) procedures and targets that must be met during the course of the project. The SAP should be reviewed and endorsed by the project supervisor, and ideally will be sent out for independent review prior to collection of any critical data. Once approved, the SAP should travel with field personnel for reference during collection of field data. Any deviations from the SAP should be noted by the field worker and reported to the project leader.

Quality Assurance

The term "quality assurance" is widely used in the environmental field, but its meaning may be vague to a newcomer. Table 9.1 lists six principal quality attributes as defined by the U.S. Environmental Protection Agency (EPA) that should be optimized in any environmental study (EPA 2002). These include *precision, bias, representativeness, comparability, completeness,* and *sensitivity*. The combined error from *imprecision* and *bias* determines the overall *accuracy* of the data that are collected. Accuracy is optimized by making sure that all field equipment is clean and properly serviced, and by developing and following standard operating procedures (SOPs) for all sampling and analytical procedures. *Representativeness* refers to how well a given sample or measurement truly reflects the environmental condition of interest. For example, how representative of the aquifer conditions is a groundwater sample that is obtained by bailing a single grab sample from static water in the well (Chapter 4)? Not very. This is why hydrogeologists have adopted the procedure of purging a well volume at least three times before collecting a sample. Following a standardized procedure will also ensure *comparability* and *completeness*. A common problem in many big projects is that different field personnel wind up collecting data using different pieces of equipment and different procedures. This makes it difficult

Attribute	Definition	Practical Considerations
Precision	The measure of agreement among repeated measurements of the same property under identical or substantially similar conditions	Precision is optimized if the same equipment and SOPs are followed for all measurements. Quantified by field replicates.
Bias	Systematic or persistent distortion of a measurement process that causes error in one direction	Bias is minimized if field instruments are properly calibrated and SOPs are followed. Check lab bias by running spikes and submitting certified standards as unknowns.
Representativeness	The measure of the degree to which data suitably represent a characteristic of a population, parameter variations at a sampling point, a process condition, or an environmental condition	Examples: purge three well volumes before sample collection; avoid sample contamination from external sources; measure pH and alkalinity in the field (before samples degrade).
Comparability	A qualitative expression of the measure of confidence that two or more data sets may contribute to a common analysis	Comparability will be optimized if different field personnel follow the same standard procedures. Use same equipment when possible.
Completeness	A measure of the amount of valid data obtained from a measurement system	Avoid skipping measurements and procedures. Make checklists. Take good notes.
Sensitivity	The capability of a method or instrument to discriminate between measurement responses representing different levels of the variable of interest	Verify that the analytical methods are capable of detecting the analyte of interest to the required reporting value. Take measures to minimize sample contamination.

Table 9.1　Data Quality Attributes for Water Sampling

to compare the data from one field season to another. As well, people should be discouraged from skipping measurements in the interest of saving time, as this can lead to gaps in the final project database. Finally, with respect to *sensitivity*, the SAP should ensure that the method of analysis being employed is capable of detecting the analyte of interest at concentrations below the critical reporting value, and that the sampling SOPs minimize any chance of trace contamination (from filtration equipment, etc.).

　　It is always a good idea to submit your own QA samples to the laboratory, along with each batch of environmental samples. The type and frequency of QA samples to be collected will be spelled out in your project SAP, but typically these samples include field blanks (i.e., laboratory water passed through the same sampling and filtration equipment and preserved in the same manner as the field samples) and field replicates (identical samples collected in different bottles). You may also consider sending a certified multielement standard of known concentration to the lab (being sure not to identify it as such by the sample ID name). A wide variety of certified standards are available from major chemical supply companies. Another good practice is to occasionally send a few replicate samples to different labs.

What Are You Sampling For?

Each project will have its own specific set of data to collect, and it is difficult to make general recommendations. However, as a rule it is always a good idea to collect a complete set of field parameters, including water temperature, specific conductance, pH, dissolved oxygen (DO), Eh or oxidation-reduction potential (ORP), and alkalinity. A routine chemical analysis will also usually include quantification of a full suite of major and minor cations by inductively coupled plasma (ICP-ES or ICP-MS), and a suite of major anions by ion chromatography (IC). Additional data may help to characterize the chemistry of the waters of interest, as shown in Table 9.2. Most of these procedures are discussed in more detail below. It is worth stressing that seemingly random observations about the color or general appearance of the water sample (such as "the sample was effervescing an odorless gas upon pumping to the surface") may become very important later on. So take good notes and pass on any information of this type to your project leader.

In projects that involve regulatory compliance, it is usually important to specify between data collection that is "critical" versus "noncritical." Critical data include any measurements or results that have a major bearing on the overall objectives and outcomes of the project. Such data will typically undergo rigorous QA scrutiny to ensure they are of the highest possible quality. Noncritical data include any supporting measurements that help to characterize the environmental system but do not have a direct bearing on the project outcomes. Although noncritical data may not require QA testing and validation, good field procedures should always be followed.

Make a Checklist

Once your SAP is complete and you are ready to go in the field, make a checklist of supplies and equipment that you will need. There is nothing worse than spending half a day loading and driving to a field site and then realizing you forgot the key to unlock the well back in the office.

Basic/Essential Information	Class	Optional Information	Class
Water temperature	*i*	Stable isotopes ($\delta^{18}O$, δ^2H) of water	*a*
Conductivity (SC, C25)	*i*	Nutrients (nitrate, phosphate, ammonia)	*ts*
pH	*i*	H_2S (very important if present)	*i*
Dissolved oxygen	*i*	Dissolved Fe(II)/Fe(III)	*ts*
ORP or Eh	*i*	Organic compounds	
Alkalinity	*ts*	Dissolved organic carbon	*ts*
Dissolved metals	*a*	Organic contaminants	*ts*
Major anions	*ts*	Pathogens	*ts*
Turbidity	*ts*	Total dissolved solids	
Color/appearance of water	*i*	Age dating	*ts*

i = immediate (collect data *in situ* or immediately after bringing to surface); *ts* = time-sensitive (analyze as soon as practical); *a* = can be archived (shelf life ≥ 6 months).

TABLE 9.2   List of Possible Types of Data to Collect for Groundwater Studies

SAP and SOPs	Rope and reel	Alkalinity kit
Field notebook	Cooler and ice	Turbidity meter
Pencils, pens, sharpies	Water level indicator	Portable spectrophotometer
Sample bottles (assorted)	Datasonde (calibrated)	Reagents for Fe(II)/Fe(III)
GPS	Laptop, calculator	Reagents for H_2S
Keys	Communication cables	Spare batteries
Chain of custody forms	pH-SC-DO-Eh meters	Multi-tool knife, wrench set
Filtration supplies	Calibration standards	Duct tape, packing tape
Peristaltic pump and tubing	Nitric and hydrochloric acid	Kimwipes
Battery (charged)	Automatic pipet and pipet tips	Garbage bags
2-L point sampler	Powder-free gloves	Ziplock bags
Messenger	Safety glasses	10 L deionized (DI) water
Flow cell	5-gal calibrated bucket	DI squirt bottles

TABLE 9.3 *Example Checklist*

Example 9.1 Table 9.3 is an example checklist from a recent project involving collection of groundwater samples in a flooded mine shaft near Butte, Montana. In this project, water samples were collected using a point sampler, a nylon rope, and a hand reel. Downhole water-quality parameters were collected with a datasonde, and a second set of parameters were collected at the surface using handheld probes. Additional measurements at the surface included alkalinity, turbidity, Fe(II)/Fe(III) speciation, and H_2S concentration. Filtration was accomplished using a peristaltic pump and large diameter filters.

9.2 Collecting Water Samples

Sample Retrieval

For collection of groundwater samples, the most common methods are to use a portable submersible pump, a peristaltic pump, a bailer, or a point sampler. The simplest system for relatively shallow (less than about 20 ft deep), small-diameter monitoring wells is to use a peristaltic pump, flexible plastic tubing, and a portable battery. This method is limited to relatively low flow rates (<1 or 2 gal/min), and therefore will not be practical if it is necessary to purge a well that is more than 2 in in diameter. One advantage of a peristaltic pump is that no moving parts contact the water being sampled, making it possible to pump water that is highly corrosive or turbid. For deeper or larger-diameter wells, a submersible pump powered by a generator or DC battery will be needed. A relatively small 1-hp pump similar to what one uses in a hot tub can be carried around by hand and will deliver up to 30 gal/min. Make sure the discharge from the pump is directed as far as possible from the well and does not enter any nearby surface-water body.

A bailer is a hollow cylinder, usually plastic with a check valve at the bottom, that is lowered down a monitoring well to retrieve a water sample. The bailer is lowered until it is completely submerged, and then raised and dropped a few times to fill the cylinder. The bailer is then retrieved and the water is discarded to a 5-gal bucket to monitor the purge volume. This process is repeated until the well is sufficiently flushed, usually after three well volumes, or when all field parameters have stabilized. Although this sounds like hard work, bailing shallow 2-in monitoring wells is often faster than using a peristaltic pump. One problem is that the turbulence caused by raising and lowering the bailer can make the water muddy. To minimize this problem, avoid lowering the bailer all the way to the screen or bottom of the well. Provided the well is properly purged, a representative sample will be obtained simply by bailing from a few feet below the static water level. For some projects with multiple wells it may be more convenient to prevent cross contamination to have a dedicated bailer permanently assigned to each well. Each bailer can be stored inside its designated well above the waterline with a piece of string.

Our group in Butte does a lot of flooded mine shaft sampling, and in this case it is impossible to "purge the well." Instead, we use point samplers, which are hollow cylinders with two articulated caps that are spring-locked in the open position. After the point sampler is lowered to the desired depth, a small weight or "messenger" is sent down the cable which trips the spring and causes the caps to clamp shut, trapping the water sample inside. The sampler is then lifted out of the hole as quickly as possible. Using this method we have been able to get samples of mine shaft water from depths up to 1,000 ft below the static water level. Point samplers are also widely used when sampling lakes (Figure 9.1).

Passive Sampling

Wells work great in sand and gravel, but how does one collect a representative groundwater sample from sediment with a high percentage of silt and mud? Driving a

FIGURE 9.1 Using a 2-L point sampler (white device in upper right) to collect water samples from an acidic lake in Patagonia, Argentina.

monitoring well into the muck at the bottom of a shallow lake or wetland is a dirty job, and it is virtually impossible to pump such a well to obtain a decent water sample. In this case, a better option may be to use a *passive* sampling method, such as a "peeper" or "diffusion sampler." Most peepers are some variation of a vertical column containing several hollow compartments at different depths that are separated from the water-saturated sediment by a semi-permeable synthetic membrane. The sample compartments are typically filled with de-oxygenated, distilled water before deployment. The peeper is then pushed into the subsurface to the desired depth and left for several weeks. During this time solutes diffuse across the membrane until the water chemistry inside each sample compartment is the same as that in the immediately adjacent sediment. The peeper is then removed and each of the compartments is sampled as quickly as possible, sometimes in a N_2- or Ar-glovebox to avoid oxidation of redox-sensitive species. The use of peepers to collect pore-water chemistry profiles in sediment with low hydraulic conductivity soils is still not widely practiced, but has much potential. Another application of diffusion samplers is in the sampling of VOCs (volatile organic carbon compounds). Some recent studies suggest that pumping a monitoring well can change the concentration of VOCs that partition into the gas phase, and therefore a more representative sample is collected by a variant of the diffusion method described above. More information can be found in the excellent review article of ITRC (2006).

Filtration, Preservation, and Labeling

Most SOPs will call for collection of filtered groundwater samples, whereas surface waters may be either filtered or non-filtered. The term "dissolved metal" is operationally defined as any substance that passes through the filter that is used in the field. For groundwater sampling 0.45-μm filters are most commonly used, although filters with smaller diameter are also commercially available, down to about 0.1 μm. The advantage of using 0.45-μm filters is that they will capture almost all of the algae and bacteria and most of the suspended solid in the water, while still being of sufficient pore size to allow rapid filtration of relatively large (>1 L) sample volumes. Speaking of volume, be sure to collect enough sample for all your laboratory needs. When in doubt, collect a larger sample size. It is very frustrating to run out of sample before all of the analytical work is completed.

There is a bewildering array of filter styles and materials on the market, and the choice of what product to use is non-trivial. Many workers use a disposable in-line filter with a plastic housing that attaches to any flexible tubing (Figure 9.2f, g). These are one-use-only filters that can process a large volume of water (depending on the turbidity) but are also somewhat expensive. A slightly different approach is to use a screw-tight plastic filter holder with individual filter sheets (Figure 9.2b, c, d). The filter housing is awkward to carry around and rinsing between samples consumes large amounts of distilled water, but this method turns out to be much less expensive per sample. Finally, in cases where it is impossible to carry a battery and pump, it may be more convenient to use a small hand-operated filter (Figure 9.2a) or plastic syringes (e.g., 60 mL) and syringe filters (Figure 9.2e). These small filters will clog quickly if the water is turbid, and they are also notorious for trace metal contamination, especially zinc. Preliminary QA tests with blank water are always recommended to determine the concentrations of trace metals that are leached out of a given filtration setup. Once a filter size and type is selected, the same filters should be used throughout the life of the project.

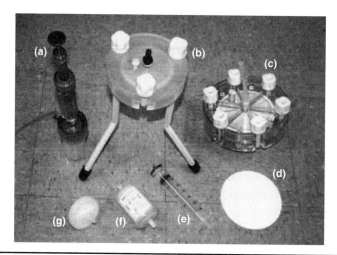

Figure 9.2   Some example filter styles for water sampling: (a) hand-pump portable filter; (b) and (c) plate filters with disposable filter membranes (d); (e) syringe with small disposable filter (~$2 each); (f) high-capacity in-line disposable filter (>$10 each); (g) medium-capacity in-line disposable filter (>$5 each).

Nitric acid (HNO$_3$) is the most common reagent used for preservation of samples for trace metal analysis (ICP-ES or ICP-MS). Sulfuric acid (H$_2$SO$_4$) is used for nitrate and phosphate, and hydrochloric acid (HCl) is usually used to preserve redox-sensitive species, such as Fe^{2+}/Fe^{3+}. Different levels of acid purity may be required, depending on your project. "Trace metal grade" acid is the most common type used for routine sample collection. A small bottle of concentrated acid can be transported into the field (e.g., in the sample cooler), and a pipet can be used to dispense a precise amount of acid into each bottle after the samples are collected. The exact amount of acid to add will be specified in your SAP; we typically acidify groundwater samples to 1% v/v (by volume). Samples collected for dissolved metal analysis must be filtered first prior to addition of the acid. Some workers prefer to use pre-weighed capsules of acid in individual plastic ampoules. Whereas this method avoids having to transport a bottle of concentrated acid into the field, it is somewhat awkward to open the capsules and this can lead to minor spills as well as non-identical amounts of acid going into different sample bottles. Some labs may provide sample bottles with a small amount of acid already in them. Although this may be okay for your application, be aware that concentrated acid can leach metals out of plastic bottles, even pre-acid-washed bottles. All appropriate safety precautions should be made when handling nitric acid, including wearing safety goggles and gloves.

It is worth saying a few things about labeling. Chances are good that the sampler will be using a sharpie or similar waterproof pen to label plastic sample bottles in the field. Essential information that should go on the bottle includes, at a minimum, the date of collection and the sample ID, as well as some indication as to whether or not the sample was filtered and preserved. Our group uses the prefixes "F" or "R" for filtered or raw and "U" or "A" for unacidified or acidified. Thus, a sample that has been filtered and preserved with HNO$_3$ would be labeled "FA-HNO$_3$." Be aware that sharpies are notorious for smudging or wearing off completely. To avoid this, wrap a sheet of transparent packing tape over the label. This will save a lot of headaches, especially if the samples will be transported a long distance. Most SOPs for analysis of water samples call for refrigeration during transport and storage, so bring a cooler with plenty of ice. Be sure to double-bag all samples and pack them so that they will not be "swimming" in water when the ice melts. Frozen ice packs do not keep samples cold enough in the cooler so use ice.

Chain of Custody

Regulatory work requires chain of custody (COC) forms to be filled out by the field sampler and shipped along with the environmental samples to the laboratory. COC forms can be custom-made for each project or can be supplied by the lab. Essential information includes a list of the sample IDs, the dates of collection, the sample volumes, whether the samples were filtered and preserved, the project name, who collected the sample, and a description of what types of laboratory analyses are required for each sample. Always double-check the sample IDs marked on the bottles with the list on the COC form. Anyone who handles the samples on route to the lab must sign and date when they received and relinquished them. The laboratory will also sign and date the COC form when the samples arrive, and will check the condition of the samples, noting any spillage or whether the cooler had warmed above an unacceptable temperature during shipping. The laboratory then checks the sampling dates to determine how quickly the various analyses need to be performed (e.g., anion analysis usually needs to be completed within 2 weeks of sampling). The COC forms are eventually mailed back to the project leader and become part of the permanent record of the project.

Cleaning and Decontamination

Following clean field procedures is as much a valuable skill as taking good field notes or having the stamina to put in a long field day. This is particularly important when analysis of trace metals or organic compounds is going to be performed. The U.S. Geological Survey Water-Quality Field Manual (USGS 2005) has an entire chapter devoted to cleaning field equipment between samples. Although some of these procedures may be overly rigorous for your specific application, there are several reasons why clean sampling methods are important. These include minimizing sample contamination between sites, extending the lifetime of your sampling equipment, and avoiding exposure of yourself or your employees to potentially toxic chemicals. Here are a few pointers:

- The sampler should wear disposable, powder-free gloves and change the gloves frequently, that is, between samples or whenever the gloves are potentially contaminated by contact with uncleaned surfaces or objects.
- Always rinse any sampling or filtration equipment with copious amounts of distilled water before use, and again after use while the materials are still wet. Do not allow any equipment to be stored for an extended period of time without proper cleaning.
- Transport equipment and supplies (including sample bottles) in a sealed storage container or in plastic bags. Avoid contact with dust. This is especially critical when driving on dirt roads.
- Avoid airborne particles getting into the water samples. This is really important on windy days near sites where soil is highly contaminated.

- Use alcohol, alconox, or a dilute bleach solution to rinse equipment when sampling for organic compounds (exception: don't use for dissolved or total organic carbon analyses). In this case, the sampler should wear sturdy plastic gloves that can tolerate organic solvents. Alcohol is flammable, so dispose of waste rinse in a suitable container.

- Avoid generator or vehicle exhaust fumes when sampling for volatile organics

- Be sure all electrodes are properly stored in appropriate storage solutions (usually supplied when the electrode is purchased).

- Make sure any electronic device is turned off and remove batteries from any piece of equipment that will not be used within the next month or two.

9.3   Field Measurements

Prior to collecting a water sample, a number of field observations and measurements are typically recorded. These include the well ID number, depth to static water level, date and time, method of sample collection (e.g., whether a well was bailed or pumped), whether or not the well was purged prior to sampling, and the volume of purge water. Several important water-quality measurements should be taken at the same time as the sample is retrieved, including water temperature, pH, specific conductance, dissolved oxygen, and redox state. Ideally, these parameters will be measured in situ, with a submersible datasonde (Figure 9.3). If this is not possible, then make measurements as soon as possible after the sample is pumped or bailed to the surface. If the well is being pumped, then a good practice is to divert all or a portion of the water through a "flow cell," which has ports for inserting electrodes and thus making measurements without allowing the water to contact air. Stabilization of field parameters after one or more well volumes is an indication of successful purging of a well. The following are additional notes and suggestions for each of the field parameter measurements.

FIGURE 9.3   Using a Hydrolab® Minisonde to record water-quality measurements.

pH and Temperature

Most pH meters have a built-in temperature probe, and routine measurements should be capable of a precision of ±0.1 pH units and ±0.1°C. For best results, pH meters should be freshly (within 6 h) calibrated with at least two pH buffers that span the range of pH of the waters being sampled. Bring your buffers with you for calibration checks in the field. Also, bring an ample supply of distilled or deionized water for rinsing the electrode between measurements, but don't use deionized water to store the electrode. Instead, the electrode tip should be stored in a pH 4 or pH 7 buffer solution, in a KCl buffer solution, or in whatever water with moderately high salinity is available. Do not allow the tip of the pH electrode to dry out.

There are many types of pH electrodes on the market. Glass electrodes are the most sensitive and are well suited to lab work but are breakable and have a port for adding electrolyte solution that often leaks. Many field workers prefer gel electrodes which are unbreakable (or nearly so) and are permanently sealed. However, gel electrodes can be very slow to equilibrate, especially at cold temperatures or for water whose pH is poorly buffered. In such cases, a gel electrode may take as much as 15 min or longer to reach a stable pH value.

The pH of a deep groundwater sample can change after the sample is pumped or bailed to the surface. Examples 9.2 and 9.3 underscore the need to take pH measurements in the field, and to do this in situ (with a submersible probe) if possible.

Example 9.2   Vertical gradients in pH were recently measured in a flooded mine shaft near Butte, Montana. A detailed pH profile was obtained by lowering a submersible datasonde down the shaft. The pH of samples bailed to the surface was also measured using a manual pH meter. As shown in Figure 9.4, the surface pH value for the 100-ft-deep sample was significantly higher as compared to the in situ datasonde pH value. The deeper water effervesced CO_2 gas while bailing to the surface, which is the probable reason for the change in pH.

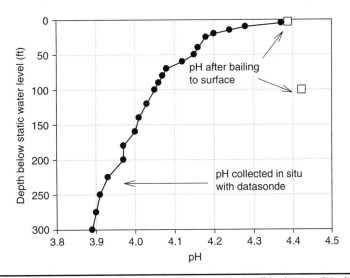

Figure 9.4   Comparison of in situ pH (determined with a submersible datasonde) with the pH of samples bailed to the surface for mine shaft water in Butte, Montana. (*C. Gammons, unpublished data.*)

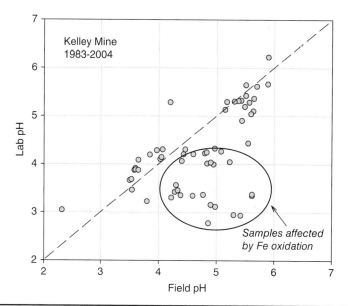

FIGURE 9.5   Comparison of field pH and lab pH for Fe-rich water in a flooded mine shaft. (*C. Gammons, unpublished data.*)

Example 9.3   The same water samples mentioned in the previous example were observed to turn reddish-brown upon storage (even in tightly sealed bottles with no headspace), due to oxidation of dissolved ferrous iron to insoluble ferric hydroxide:

$$Fe^{2+} + 1/4O_2 + 5/2H_2O = Fe(OH)_3(s) + 2H^+ \qquad (9.1)$$

Because reaction (9.1) produces protons, the pH of these mine waters measured in the laboratory were very often much lower than the field pH (Figure 9.5).

Specific Conductance

Specific conductance (SC or C25) is the ability of a water sample to conduct electricity at a reference temperature of 25°C. Salty water conducts electrons much easier than freshwater, and the conductivity of water of a given salt content increases with increase in temperature. Most modern SC meters have a built-in temperature sensor so that the absolute conductivity of the sample at the field temperature is automatically converted to specific conductivity at 25°C. Many older conductivity meters do not have this option, and the measured conductance will need to be adjusted manually to 25°C. SC is most commonly expressed in units of µS/cm (micro-siemens per cm) or mS/cm. A siemen is the same as a mho, and the latter is easy to remember since it is the backwards spelling of "ohm," the unit of electrical resistance. As usual, be careful with units. Some SC meters will automatically shift from µS/cm to mS/cm when the electrode is placed in water with high salinity.

An SC meter is a very useful tool for rapid water-quality assessment, as it can be used as a proxy for total dissolved solids. This can help, for example, to predict how large a dilution factor should be applied prior to laboratory analysis. As well, our group has found SC to be particularly useful in investigations of groundwater/surface-water interaction (Chapter 10). For example, if a given reach of a stream is gaining flow from discharge of shallow groundwater, then it will usually show a change in SC since the influent groundwater almost always has a different (usually higher) SC value than that of the stream. In contrast, if the SC of a stream is invariant with distance over a given reach, it is likely that the river is not receiving influent groundwater or is losing water to the subsurface. Before collecting any stream sample, it is strongly advised to measure SC at a number of locations across the profile to see if the surface water is well mixed. If the water along the near bank has a much different chemistry than the main flow of the stream, a grab sample at this location will not be representative of the entire stream flow.

Example 9.4   Figure 9.6 shows changes in SC and dissolved Zn concentration in a section of High Ore Creek, Montana (Gammons et al. 2007). The steady increase in both SC and Zn was due to discharging shallow groundwater that had been contaminated by interaction with buried mine waste. Because the increase in streamflow through this reach was small (<0.1 cfs), the influent groundwater must have had very high SC and Zn concentration.

Example 9.5   In some cases, field SC measurements can also be used to rapidly estimate stream flow. For example, the following data were obtained at a section of High Ore Creek where a small tributary stream entered: $SC_1 = 595$, $SC_2 = 82$, $SC_3 = 526$ μS/cm, where the subscripts 1, 2, and 3 refer to High Ore Creek above the tributary, the tributary, and High Ore Creek below the tributary, respectively. The flow of High Ore Creek upstream (Q_1) was measured from a Parshall Flume (Chapter 9) and found to be 1.1 cfs (ft³/s). An SC mass balance equation was then used to estimate the flow of the tributary and High Ore Creek downstream:

$$SC_1 \cdot Q_1 + SC_2 \cdot Q_2 = SC_3 \cdot Q_3 \tag{9.2}$$

Since $Q_1 + Q_2 = Q_3$, we can set $Q_2 = x$, and $Q_3 = (1.1 + x)$. Rearrangement of Equation (9.2) gives

$$X = \frac{Q_1 \cdot (SC_1 - SC_3)}{(SC_3 - SC_2)} = \frac{1.1 \cdot (595 - 526)}{(526 - 82)} = 0.17 \tag{9.3}$$

from which $Q_2 = 0.17$ cfs and $Q_3 = 1.27$ cfs were obtained.

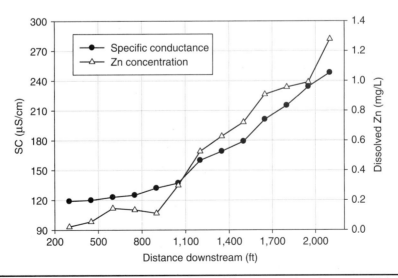

Figure 9.6   Changes in specific conductance (SC) and Zn concentration along upper High Ore Creek, Montana, near the abandoned Comet Mine. [*From Gammons et al. (2007).*]

Dissolved Oxygen and Redox State

The redox state of water is typically characterized by a combination of DO and oxidation-reduction potential (ORP or Eh) measurements. In very broad terms, groundwater can be classified into three categories with respect to its redox state: (1) *oxic* (containing DO well above the practical quantification limit (PQL) of around 1 mg/L), (2) *suboxic* or *transitional* (DO near or below detection but no H_2S), and (3) *anoxic* or *sulfidic* (H_2S present at detectable levels). The fate and transport of trace metals and other contaminants can change dramatically depending on the redox state of the water. Depletion of DO in groundwater may be an indication of O_2-consuming reactions such as decay of organic matter or oxidation of sulfide minerals.

Dissolved oxygen is measured by Winkler titration or using a DO meter with a galvanic or luminescent DO (LDO) electrode. Winkler titrations are cumbersome to do in the field but are a good way to check the accuracy of your DO electrode. Galvanic DO electrodes were once the most common type used on multiparameter instruments, although LDO probes have become the industry standard. Because a galvanic DO electrode will consume O_2 during its operation, the solution must be stirred or flowing to get an accurate measurement. LDO electrodes do not have this problem and are also reputedly more accurate and less prone to instrument drift. Using any of the above methods, it is straightforward to quantify DO in the range of 1.0 to above 10 mg/L. Below about 1 mg/L, DO measurements are problematic, and caution should be used to interpret data near the detection limit. In many cases, the trace amounts of DO that are recorded by the equipment will be a false artifact of atmospheric contamination, chemical interferences, or pressure effects.

For suboxic or anoxic water, an Eh or ORP electrode should be used to quantify redox state. The ORP electrode has a platinum wire tip that should be kept clean and protected from damage. To take an ORP reading, simply attach the ORP electrode to your pH meter, and switch the "mode" button from pH units to mV. Although Eh and ORP both have units of volts or mV, they are different things, and the user needs to be careful. Whereas Eh is defined as the electrical potential of a sample relative to the standard hydrogen electrode (SHE), ORP is the potential of the water sample relative to the particular reference electrode that is used in the field. Most ORP electrodes use an Ag-AgCl reference cell, although there are also some calomel (Hg-$HgCl_2$) electrodes around. The conversion from Ag-AgCl to SHE is typically on the order of 200 to 220 mV (see Column B of Table 9.4), which makes a big difference in the final reported Eh values. This voltage correction is temperature dependent, and it also varies slightly with the concentration of KCl in the electrode filling solution. To check the calibration of an ORP electrode, measure the mV reading in a bottle of fresh ZoBell's solution at a known temperature. The result should be close to what is shown in Column C of Table 9.4. To convert from ORP to true Eh, the following equation can be used:

$$Eh_{sample,\ T} = ORP_{sample,T} + Eh_{Zobells,T} - ORP_{Zobells,T} \tag{9.4}$$

where $ORP_{sample,T}$ and $ORP_{Zobell's,T}$ are the measured ORP values of the water sample and ZoBell's solution at the field temperature, and $Eh_{ZoBells,T}$ is the theoretical Eh of ZoBell's at the field temperature (given in Column A of Table 9.4).

Example 9.6   Our group recently sampled a river in southwestern Spain that was highly acidic due to acid mine drainage. The field ORP of the river water was +420 mV and the temperature was 20.2°C.

Temp °C	Column A Eh (SHE) of ZoBell's Solution	Column B Eh (SHE) of Ag/AgCl (3.5M KCl)	Column C ORP of Zobell's Relative to Ag-AgCl (3.5M KCl)
0	490	221	269
5	478	218	260
10	467	215	252
15	455	212	243
20	443	208	235
25	431	205	226
30	418	201	217

TABLE 9.4 The Temperature Dependence of the Electrical Potential (mV) of Zobell's Solution and the Ag-AgCl Reference Electrode

The ORP of a ZoBell's solution standard placed in the river at the same temperature was 219 mV. The true Eh of the water sample was then calculated using Equation 9.4 and was determined to be $420 + 443 - 219 = 644$ mV.

Alkalinity

Alkalinity is defined as the sum of the concentrations of titratable bases in a water sample. For practical purposes alkalinity can be expressed as

$$\text{Alkalinity (moles/L)} = mHCO_3^- + 2mCO_3^{2-} + mOH^- \, (+ \, mHS^-) \tag{9.5}$$

Although natural waters may contain other basic compounds (such as borate, phosphate, or $Mg(OH)^+$), these are usually very minor contributors to the total alkalinity. Alkalinity is an important measurement because it is the most convenient way to determine the concentrations of dissolved bicarbonate (HCO_3^-) and carbonate (CO_3^{2-}) in an environmental water sample. Although many laboratories will measure alkalinity (for a price), it is always a good idea to perform a quick alkalinity titration in the field soon after the samples are brought to the surface. Even samples that are filtered and stored on ice can "go bad" between the time you collect them and when the lab gets around to doing the measurement (see Example 9.7).

Alkalinity is measured by potentiometric titration. A known volume (usually 100 mL) of filtered, unacidified groundwater is titrated with dilute acid (usually H_2SO_4) to a pH end point of ~4.5. The volume of acid that is needed to lower the pH of the sample to 4.5 is then used to calculate the alkalinity of the sample. For field titrations, the most convenient approach is to use a digital titrator (Figure 9.7) and an appropriate cartridge (usually 0.16 N H_2SO_4 for dilute samples, 1.6 N H_2SO_4 for concentrated samples).[1] You will also need a 250-mL Erlenmeyer flask, a portable magnetic stirrer and stir bar, and a way to measure out 100 mL of sample, such as a pipet, a volumetric flask, or a portable balance. The accuracy of your alkalinity titration will be poor if you simply "eyeball" the 100-mL mark on the Erlenmeyer flask. For rapid measurement

---

[1]0.16N H_2SO_4 = 0.08M H_2SO_4, where N = "normal" and M = "molar."

Figure 9.7   Using a digital titrator to determine alkalinity of a river in Peru.

without the need for a pH meter, a small packet of bromocresol green-methyl red indicator dye can be added to your water sample. As long as pH is above 5, the color will be blue-green. As the pH drops below 5 with continued addition of H_2SO_4, the color will shift to lavender, and then pink at the pH end point of 4.5. Although alkalinity often increases with increase in pH, this is not always the case. However, any water sample with an initial pH less than 4.5 has zero alkalinity (by definition).

Units of alkalinity are particularly troublesome. Common units include mg/L (as $CaCO_3$), mg/L (as HCO_3^-), and meq/L (milli-equivalents per liter of total alkalinity). If you are using a HACH digital titrator, the units corresponding to the digital readout are specified on the side of the cartridge. For example, if a 0.16N H_2SO_4 cartridge is used, the alkalinity (for a 100-mL sample) in units of mg/L as $CaCO_3$ is equal to the digital readout divided by 10. To convert from mg/L as $CaCO_3$ to meq/L, the alkalinity value is divided by 50 (this is the gram formula weight of calcite divided by two, realizing that each mole of $CaCO_3$ has two equivalents of base). To convert from meq/L to mg/L as HCO_3^-, the alkalinity value is multiplied by 61 (the gram formula weight of bicarbonate). Be careful with units!

Example 9.7   Field alkalinity measurements and filtered water samples were taken each hour from a river in western Montana that had very high pH (pH > 9) due to abundant photosynthesis by algae and aquatic macrophytes. The alkalinity of each sample was analyzed in the laboratory several days later. Figure 9.8 summarizes the results. Whereas the field alkalinity data showed a smooth trend over 24 h, the lab data were very scattered, and most of the values were lower than the field measurements. This problem was attributed to precipitation of calcite in the nonacidified bottles after they were sealed. As shown by the following equation, precipitation of calcite would have lowered the bicarbonate alkalinity of the samples:

$$Ca^{2+} + 2HCO_3^- = CaCO_3(s) + CO_2(g) + H_2O \tag{9.6}$$

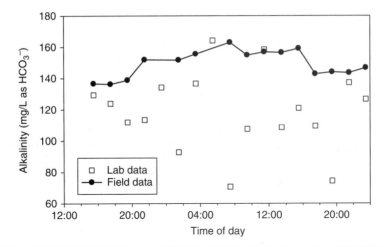

Figure 9.8 A comparison of field and laboratory alkalinity data collected from the Mill-Willow Bypass, a tributary to the upper Clark Fork River, over a 24-h period in August 2005. (*C. Gammons, unpublished data.*)

Thermodynamic modeling later showed that the river simples were strongly supersaturated with calcite owing to the high pH of the water, supporting the hypothesis that calcite precipitation was the cause of the poor lab alkalinity data.

Turbidity and Total Suspended Solids

Turbidity is an important measurement to collect if suspended solids are a concern, and it is a useful measure of how well a given monitoring well has been developed prior to sampling. If the water being pumped is turbid and brown, then chances are the well was never properly developed (Chapter 15) or was completed in fine-grained sediment with no filter packing. Turbidity is quantified using a turbidity meter which measures the amount of light that is scattered at 90° to the incident beam, and it is usually reported in units of nephelometric turbidity units (NTU).

The concentration of total suspended solids (TSS) is determined by filtering a known volume of raw water (usually 1 L) and weighing the residual dried solid. Because groundwater samples are usually filtered prior to laboratory analysis, TSS measurements are often omitted in the SAP. However, with some relatively simple equipment, filtration of a 1-L sample can be done quickly in the field, with the pre-weighed filters stored for later drying and weighing. The stored filter papers can also be used for mineralogical, chemical, or microbiological characterization of the suspended particles. The particles will transform as they dry up and oxidize, so forethought should be given as to how any such data will be used.

9.4 Laboratory Analyses

Dissolved Metals

Groundwater samples collected for metals analysis will typically be filtered and then acidified in the field (Section 9.2) and stored on ice or in the refrigerator prior to analysis. Concentrations of dissolved (filtered) metals in environmental water samples are usually quantified by inductively coupled plasma emission spectroscopy (ICP-ES, also known as ICP-OES or ICP-AES). In cases where a lower detection limit is needed, samples may be analyzed by inductively coupled plasma mass spectroscopy (ICP-MS) or graphite furnace atomic absorption spectroscopy (GFAAS). The advantage of ICP-ES and ICP-MS over GFAAS is that a full suite of 15 to 30 elements can be analyzed at the same time with the former methods, whereas with GFAAS each metal must be analyzed separately. Because of its higher sensitivity, ICP-MS is more expensive than ICP-ES. Beginners should not confuse these two techniques, as they employ quite different instrumentation. Also, one should not refer to their analytical method as simply "ICP," as this does not discriminate between mass versus emission spectroscopy. Most EPA methods for metal analysis specify a maximum 6-month holding time. Many commercial laboratories will supply coolers, sample bottles, acid reagents, and COC forms at little or no additional cost. If this interests you, call the lab ahead of time and arrange for the supplies to be mailed in plenty of time before your field work.

Anions

Samples for anion analysis are normally filtered and unacidified and stored on ice or in the fridge until analysis. Most analytical methods for anions specify a maximum 2-week holding time, so be sure to send your samples to the lab quickly. Ion chromatography (IC) is the method of choice for the geologically common anions including bromide (Br^-), chloride (Cl^-), nitrate (NO_3^-), and sulfate (SO_4^{2-}). Nitrite (NO_2^-) and phosphate (PO_4^{3-}) may be added to the list if concentrations are expected to be above the analytical detection limits. IC cannot quantify bicarbonate or carbonate ions: these solutes are determined by alkalinity titration (Section 9.3). Colorimetric methods have also been developed for most anions, and these are sometimes used when concentrations of species such as nitrate, nitrite, or phosphate are below the detection limit of the IC. HACH Company has developed a large number of tests that can be done on site relatively cheaply with a portable spectrophotometer. These tests should be backed up with plenty of QA checks (including standards, spikes, and duplicate samples) if the data are critical.

Organic Compounds

There are many types of organic substances that can be quantified in a given groundwater sample. These include naturally occurring soluble organic compounds (usually referred to as dissolved organic carbon, or DOC), man-made contaminants, and pathogens (protozoa, bacteria, viruses). Although DOC is often skipped in a sampling plan, most groundwater will contain at least a few mg/L of organic carbon from natural sources, such as humic and fulvic acids, carboxylic acids, and so on (Drever 1997). Some forms of DOC can complex trace metals, and DOC is an essential nutrient for many forms of bacteria. Water samples for DOC measurement are gently filtered (to avoid rupture of suspended bacteria or algal cell walls) and collected in opaque glass bottles prior to analysis using a total carbon analyzer. There is typically a 2-week holding time for DOC measurement, so get your samples to the lab quickly.

The list of man-made compounds that have been detected in groundwater is exceedingly long. Many of these chemicals are toxic, and the U.S. EPA has created a list of 126 priority pollutant organic compounds, as well as a list of standard methods for their analysis (e.g., see Fetter 1999). Most of the target compounds are analyzed by gas chromatography (GC), or a combination of gas chromatography and mass spectroscopy (GC/MS). Many of the priority pollutants are volatile, and some are sparingly soluble in water, leading to the accumulation of "free product," or non-aqueous-phase liquid (NAPL). For their own safety, field workers should take great care handling environmental samples from highly contaminated sites, and also need to be diligent about decontaminating equipment between samples. If you are involved with a project that deals with organic contaminants, chances are good that detailed SOPs regarding safety, sampling methods, and quality control already exist. Follow these procedures.

There has been increased attention to the presence of *pesticides* and *pharmaceutical* compounds in groundwater (Barbash and Resek 1996; Heberer 2002; EPA 2006). Pesticide contamination of groundwater is a problem that has long been recognized in agricultural areas. Pharmaceutically active compounds (PhACs), also referred to as pharmaceuticals and personal care products (PPCPs), are a concern because many of these commercial products are very slow to biodegrade, are mobile in shallow groundwater or surface water, and pose an unknown threat to human health. Although a complete chemical analysis for the full range of pesticides or PPCPs that may be present in a given sample is very expensive, the methods of sample collection and preservation are not. As well, preliminary screening may identify one or more target compounds that can serve as a proxy for the full suite of contaminants likely to be present.

Total Dissolved Solids

Before ~ 30 years ago, a total dissolved solid (TDS) analysis was performed by evaporating a known volume of filtered water to dryness and weighing the mass of residual solids. This mass divided by the original mass of water gives the TDS, usually expressed in mg/L. However, in recent decades it has become a common practice to calculate TDS as the sum of the concentrations of all major and minor dissolved species in the water as determined by ICP-ES, alkalinity, and IC. Interestingly, the values obtained by these two methods are usually different. The reason is that a sample with moderate or high Ca-bicarbonate content will precipitate a fair amount of calcite on evaporation (reaction 9.6), and in the process will liberate $H_2O(g)$ and $CO_2(g)$ that was originally bound in dissolved bicarbonate ion. So, TDS measurements based on evaporation are usually lower than TDS based on chemical analyses. This distinction may be very important if TDS is a parameter that is closely monitored or regulated, as is the case for some high-salinity waters used for irrigation or produced as a by-product of oil and gas extraction (see Van Voast 2003, for a good example). Although TDS is generally proportional to SC, the relationship is semi-quantitative. Therefore, SC should not be substituted for TDS if the latter is a critical measurement.

Other Types of Analyses

Depending on the project, there are a number of additional types of chemical and isotopic analyses that can be performed on a given groundwater sample. For example, our group has found Fe(+2)/Fe(+3) speciation to be particularly helpful in the study of waters impacted by mining (Pellicori et al. 2005). Taking advantage of the fact that dissolved iron compounds are often brightly colored, several different colorimetric tests have been developed to discriminate one Fe oxidation state from the other. Our group uses the FerroZine procedure (Stookey 1970), which is both sensitive and low cost. Regardless of which method you choose, Fe speciation should be performed as quickly as possible after sample collection. If this is not possible, samples may be preserved by filtration (to remove Fe-rich suspended solids and also to remove Fe-oxidizing bacteria), acidification with HCl (not HNO_3) to pH < 2, minimizing contact with air by filling the sample bottle with no headspace, refrigeration, and storage in an opaque container or in the dark to avoid photoreduction of Fe(+3) to Fe(+2).

Some strongly reducing groundwaters will contain dissolved sulfide, and, if so, they will have a strong smell of rotten eggs. HACH offers a simple and highly sensitive field colorimetric method for quantifying sulfide (sum of H_2S and HS^-) in such waters based on the methylene blue reagent (Figure 9.9). Because dissolved sulfide concentrations drop quickly after sample collection (due to the combined effects of oxidation plus loss of gaseous H_2S), any sulfide measurement should be performed on site immediately after sample collection. The presence of even trace (<0.01 mg/L) levels of H_2S in a groundwater will make it undrinkable and will drastically reduce the concentrations of heavy metals, such as Fe, Cu, and Zn, that form insoluble metal sulfides.

If your project involves identifying groundwater sources or recharge rates, you may consider collecting samples for stable isotope analysis or age dating. The stable isotope ratios ($^{18}O/^{16}O$ and $^{2}H/^{1}H$) of water can be very useful to track sources of recharge water, to provide evidence of groundwater mixing, and also to quantify the extent of evaporation (Chapter 8). Many hydrogeologists are surprised when they discover how inexpensive stable isotope analysis can be (often less than $20 for a combined $\delta^{18}O$ and $\delta^{2}H$ analysis) and how easy it is to collect the samples: simply fill a small bottle to the top (no headspace) with filtered water, seal tightly, and send to the lab. No refrigeration is needed. Even if your project has no plans for isotopic work this year, it may be worth archiving a set of filtered samples in case you decide to employ isotopes at a later date. Water samples can be stored indefinitely prior to stable isotope analysis as long as they are tightly sealed. Also, because 10 mL is plenty of sample for O and H isotopic analysis, shelf space should not be a problem.

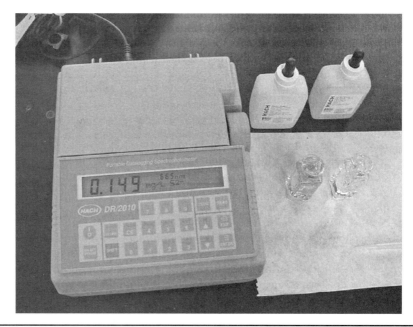

FIGURE 9.9 Using a portable spectrophotometer to determine the concentration of dissolved sulfide in a flooded underground mine in Butte, Montana.

A detailed treatment of different methods to date groundwater would fill a textbook in its own right (e.g., see Kazemi et al. 2006). Cook and Solomon (1997) provided an early overview of different methods that have been developed, including tritium (^3H) analysis, ^3He-^3H analysis, and use of chlorofluorocarbons (CFCs). ^3He-^3H (helium-tritium) is the most precise method for dating relatively young (<50 years) groundwater, but sample collection requires specialized equipment and the cost may be prohibitive for routine monitoring studies. Analysis of tritium by itself is much simpler and cheaper than ^3He-^3H but can only be used to discriminate between water that is older than around 1952 from water that is younger, 1952 being the date when nuclear testing produced a "slug" of tritium in the atmosphere. CFC analyses are also relatively inexpensive and can be used (with caution) to date relatively young groundwater to a precision of a few years, but again require specialized sampling equipment and are prone to contamination from CFC sources other than the atmosphere. More recent references covering methods to date groundwater include Bethke and Johnson (2008), Busenberg and Plummer (2008), IAEA (2013), and Clark (2015).

9.5 Reviewing Lab Results

Data Validation

All commercial laboratories will follow an EPA-approved analytical method that is appropriate for their equipment and the type of analysis being performed. The analytical method will include a large number of internal quality control checks, such as lab blanks, lab duplicates, lab spikes, and continuing standards [often referred to as instrument performance checks (IPCs) or continuing calibration verification (CCV)]. Most laboratories will report the results of the QA checks along with the table of analytical results, although sometimes QA reporting is only supplied upon request at an additional cost. If the lab sends a QA report, do not lose this information. Each user (or his or her supervisor) should verify that all QA checks for elements of critical importance were passed during the analytical session.

Normally a full set of lab QA checks is run every batch of 10 unknown samples, including a lab duplicate, a lab spike, an IPC, and a lab blank. The *lab duplicate* is a reanalysis of the first unknown sample in the preceding set. Typically, lab duplicates should agree within 10%, unless concentrations are very close to the instrument detection limit (IDL). In a *lab spike*, a known amount of spike solution containing a mixture of all of the elements of interest is added to the first unknown sample. Knowing the mass of spike added and its concentration, the nominal increase in concentration for the sample is calculated, and the % spike recovery in the QA sample is given as follows:

$$\% \text{ spike recovery} = [(C_1 - C_0)/C_s] \times 100 \tag{9.7}$$

where C_1 is the concentration of the spiked sample, C_0 is the concentration of the same sample before the spike, and C_s is the expected increase in concentration from the spike. Lab spikes are an excellent way to test for matrix interferences in a given batch of samples and should have recoveries within ±20% of 100%. Poor spike recovery is common when the sample concentration is high relative to the spike contribution, and in such cases does not necessarily indicate a problem with the analysis. An IPC (or CCV) is a multielement standard of known composition that is run every 10 unknowns to check that the instrument calibration is still acceptable. IPCs should agree within

10% of their nominal values. Finally, the laboratory blank is a sample of distilled and deionized water with the same acid matrix as the unknowns. The lab blanks should give concentrations below the IDL for all elements of concern. If not, there may be a problem with acid reagent contamination, or with carryover of metal between samples (e.g., from insufficient flushing).

   If an EPA standard method is being followed, then it is up to the analyst to examine the results of the QA check to be sure they pass all validation criteria. If not, then the data are flagged for the element of concern or the samples reanalyzed. In addition to the laboratory QA checks discussed above, field blanks and field duplicate samples should be collected during each period of sample collection. The results of these QA checks should also be scrutinized for signs of possible contamination (e.g., from sample bottles or filtering equipment) and summarized as an important part of the final report.

Example 9.8   Table 9.5 shows an example ICP-ES laboratory report collected for some river samples. In this analytical session, the % spike recovery and RPD of the IPC sample were both outside of the acceptable ranges for Na. Consequently, the Na data are shaded to show that the QA checks did not pass. Because Na was not a critical analyte for this project, the samples did not need to be reanalyzed.

Sample Name	Ca	Cu	Fe	K	Mg	Mn	Na	Zn
IDL	0.02	0.003	0.001	0.5	0.1	0.0005	0.5	0.0005
IPC-nominal	10	0.5	0.5	0.5	10	0.5	10	0.5
Spike contribution	19.5	0.489	0.5	4.9	4.89	0.5	9.77	0.489
Sample unknowns								
CG-1	12.7	0.122	0.046	0.76	3.42	0.146	1.82	0.040
CG-2	13.2	0.111	0.054	0.67	3.56	0.144	1.95	0.039
CG-3	13.3	0.096	0.031	0.71	3.55	0.141	1.90	0.039
CG-4	13.2	0.101	0.038	0.71	3.48	0.132	2.24	0.038
CG-5	13.1	0.115	0.035	0.73	3.45	0.133	2.08	0.040
QA checks								
CG-1 lab dup	13.1	0.125	0.047	0.77	3.49	0.148	1.87	0.040
RPD of dup	2.4%	2.5%	1.9%	1.1%	2.0%	2.0%	2.8%	1.0%
CG-1 lab spike	29.8	0.586	0.501	5.29	7.97	0.612	13.7	0.493
% Spike recovery	87.6%	94.9%	90.9%	92.3%	93.0%	93.3%	**122%**	92.6%
IPC	9.96	0.499	0.493	5.20	10.2	0.510	13.5	0.499
RPD of IPC	−0.4%	−0.3%	−1.4%	3.9%	2.1%	2.1%	**35.4%**	−0.3%
Lab blank	<.02	<.003	<.001	<.50	<.10	<.0005	<.50	<.0005

IDL = instrument detection limit; IPC = instrument performance check; Lab Dup = laboratory duplicate sample; RPD = reproducibility (RPD = difference/mean).

Table 9.5   Example ICP-ES Laboratory Report. Data That Did Not Pass QA Checks Are Shaded and Underlined. All Data Are in mg/L

Charge Balance

All water samples are electrically neutral, meaning that the sum of the positively charged cations must be exactly equal to the sum of the negatively charged anions. Therefore, a very useful test of the completeness and accuracy of your field and laboratory data is to perform a charge-balance calculation. Charge balance is usually calculated by the following equation:

$$\text{Charge balance (\%)} = \frac{\sum \text{meq cations} - \sum \text{meq anions}}{\sum \text{meq cations} + \sum \text{meq anions} \div 2} \times 100 \qquad (9.8)$$

where "meq" means "milli-equivalents per liter" and is the concentration of each solute in mg/L divided by its gram formula weight multiplied by its charge. (A Periodic Table to determine gram formula weights is found in Appendix C.) For example, 29.2 mg/L Mg^{2+} will contribute $29.2/24.31 \times 2 = 2.4$ meq of (+) charge. Example 9.9 shows a charge-balance calculation for a mine water in Butte. As a general rule, the charge-balance error should be within ±5% of zero using Equation (9.8), although in most cases the error will be <1 or 2% if good field and lab procedures are followed. If the charge balance is outside 5%, it could mean one of several things: (1) problems with field measurements (e.g., alkalinity), (2) problems with the lab analysis (e.g., poor standardization or failure to correct the results for laboratory dilutions), (3) incorrect assignment of the charge for one or more of the major solutes, or (4) the list of compounds that were analyzed was incomplete. The third type of error is a problem for acidic, Fe-rich waters such as the Berkeley Pit lake where it is often uncertain whether the dissolved iron is ferrous (Fe^{2+}) or ferric (Fe^{3+}). (For groundwater with pH > 4, you can assume that all Fe will be Fe^{2+} since the solubility of ferric iron is very low at near-neutral pH.) As an example of the fourth source of error, our group was once puzzled by consistently poor charge balances for surface-water samples from the Big Hole River in southwestern Montana. The problem was eventually attributed to high concentrations of DOC (>10 mg/L as C). Because many forms of DOC carry a negative charge (e.g., the COO^- groups on humic and fulvic acids), failure to include DOC in the charge-balance calculation resulted in an apparent excess of cations.

Example 9.9 The following data (Table 9.6) were collected from a flooded mine shaft sample in Butte, Montana (GWIC ID#142793, collected April 8, 2003). The sum of the meq/L of cations and anions agreed within 0.5%, indicating a very good charge balance. Note that in this calculation, it was necessary to estimate the concentrations of HCO_3^- and CO_3^{2-}. Because the pH of this water was less than 7, the concentration of CO_3^{2-} was vanishingly small, and HCO_3^- could be approximated from alkalinity. These assumptions may be invalid for high pH water samples, as discussed later in the next section.

Speciation of Dissolved Inorganic Carbon

A complete chemical analysis of a given groundwater sample will require information on the concentrations of bicarbonate (HCO_3^-) and carbonate (CO_3^{2-}) ions. For most waters with pH < 8, the concentration of bicarbonate can be computed from the measured alkalinity because at these pH values HCO_3^- is the dominant titratable base in the water. However, as pH rises above about 8.0, there is an increasing contribution to the total alkalinity from CO_3^{2-} and OH^- ions (Table 9.7). In this case, the speciation of alkalinity and inorganic carbon is best done using a geochemical modeling program, such as Visual MINTEQ (a recent, user-friendly variant of the original EPA program described

Cations	mg/L	gfw	meq/L	anions	mg/L	gfw	meq/L
Ca^{2+}	110.0	40.08	5.49	HCO_3^-	218.1	61.03	3.57
Mg^{2+}	29.2	24.31	2.40	CO_3^{2-}	0.0	60.03	0.00
Na^+	24.5	22.99	1.07	SO_4^{2-}	240.1	96.06	5.00
K^+	4.6	39.10	0.12	Cl^-	38.7	35.45	1.09
Mn^{2+}	11.2	54.94	0.41	F^-	0.4	19.00	0.02
Zn^{2+}	1.8	65.37	0.06	Br^-	0.2	79.91	0.00
Fe^{2+}	0.3	55.85	0.02				
		Total	9.55			Total	9.69
Charge balance = (9.55 − 9.69)/(9.55 + 9.69)/2 × 100 = −0.36%							

TABLE 9.6   Charge-Balance Calculation for the Ophir Flooded Mine Shaft

by Allison et al. 1991) or PHREEQC (developed and distributed by the USGS). Both of these programs are widely used and are freely available from the Internet, but it will take a while for a beginner to get familiar with them. The user first creates an input file that specifies the field temperature, pH, alkalinity, and concentrations of solutes as determined by the laboratory. The input file is then run through a speciation program, and an output file is created that lists the concentrations and chemical activities of each solute. Both Visual MINTEQ and PHREEQC can do redox calculations and gas-water reactions and will compute the partial pressure of dissolved $CO_2(g)$ for your water sample. These programs will also perform a "speciated" charge-balance calculation, which is an excellent QA check on the reliability of your field and lab data. More advanced functions such as calculation of mineral saturation indices and adsorption modeling are also built into the software (see Chapter 8 for a review of these concepts).

pH	% as HCO_3^-	% as CO_3^{2-}	% as OH^-
7	99.9	0.1	0.0
7.5	99.7	0.3	0.0
8	99.1	0.8	0.0
8.5	97.3	2.6	0.1
9	91.9	7.7	0.3
9.5	78.1	20.8	1.1
10	52.5	44.0	3.4
10.5	24.3	64.8	10.8

TABLE 9.7   The Distribution of Total Alkalinity between HCO_3^-, CO_3^{2-}, and OH^- at Different pH Values. The Calculations Assume $T = 20°C$, Ionic Strength = 0.001 m, and $PCO_2 = 10^{-3.5}$ bars (Close to Atmospheric Saturation)

Hardness and Sodium Adsorption Ratio

Hardness is related to the concentration of dissolved Ca^{2+} and Mg^{2+} in a water sample. It is easily calculated by the following formula (from Drever 1997):

$$\text{Hardness (in mg/L of } CaCO_{3,eq}) = (\text{mg/L Ca}) \times 2.5 + (\text{mg/L Mg}) \times 4.1 \qquad (9.9)$$

Historically, the term "hard water" had to do with the tendency of a water sample to precipitate calcium carbonate scale upon evaporation, hence the convention of reporting hardness in units of $CaCO_3$ equivalents. However, it is possible for a water to have high Ca and Mg content (and therefore high hardness) but low dissolved carbonate content. Acid mine waters and Ca-Cl-rich brines are two examples.

Sodium adsorption ratio (SAR) is frequently used to quantify the potential for "sodium toxicity" from irrigation water. SAR is calculated as follows:

$$SAR = \frac{[Na]}{\sqrt{\{[Ca] + [Mg]\}}} \qquad (9.10)$$

where the brackets [ ] denote concentrations in millimoles per liter. [A millimole per liter is simply the concentration in mg/L divided by the gram formula weight.] Continuous irrigation with water containing SAR > 10 may be detrimental to crops that are sensitive to sodium (U.S. Salinity Lab 1954). This is another serious issue with respect to discharge waters from coal-bed methane wells (Van Voast 2003).

9.6   Tips on Reporting Data

Sooner or later you will be required to summarize your field and laboratory data for a final project report, thesis, or journal publication. The following are a few tips and suggestions.

First, pay close attention to *units*. This is the most common source of confusion and error in data reporting. Each column in the data table should have units explicitly stated, either at the top of the column or in a footnote. For compounds such as SO_4, SiO_2, HCO_3, PO_4, or NO_3, be clear about whether the value in the table denotes mg/L as the element or as the compound. For example, 20 mg/L NO_3 as NO_3 is equivalent to 4.5 mg/L NO_3 as N. Big difference! It matters little which convention you choose in your own table as long as you are clear which unit convention you have adopted.

Second, pay attention to *significant figures*. Most commercial laboratories will round off their analytical results to an appropriate number of significant figures, which is usually 1, 2, or 3 depending on the method of analysis and the concentration of the solute. However, if you receive lab data in "raw" form, or do manipulations in a separate spreadsheet, the data will likely have an undetermined number of significant figures. All such data should be rounded off to a number of digits that is appropriate to the precision of the analysis. For example, 342.45 mg/L SO_4 is more realistically reported as 340 mg/L SO_4, given the ±5% precision of a typical IC analysis.

Third, be consistent about how you report data that are below the IDL. Many authors will simply list the IDLs in a separate row at the top or bottom of the table, and then specify the low analyte concentrations as "b.d." (below detection). However, if the sample was diluted prior to analysis (which is often the case for ICP-ES or ICP-MS analyses of polluted waters), then the PQL for that sample is actually the IDL multiplied by the dilution factor. To avoid confusion, the best approach is to specify the PQL for all analytes that are below detection within the main body of the table. For example, if a sample diluted 10 times had a Cd concentration below an IDL of 2 µg/L, the result would be reported in the data table as "<20 µg/L." For *statistical analysis*, a common convention is to assign below-detect data a value of ½ the PQL. Whereas this may be allowable for the spreadsheet that serves as the statistics database, it is not a good idea to do this in tables that are published in your final report, as you are essentially making up numbers that may or may not be valid.

In some cases, parameters such as pH, SC, ORP, and alkalinity will be determined both in the field and in the laboratory. The question then becomes, which set of data should be given in the final report? There is no universal recommendation that can be made here. In general, laboratory measurements will have higher precision, accuracy, and comparability than data collected in the field by different workers using different field equipment that may or may not be properly calibrated. However, if field SOPs are followed with diligence, then these data may in fact be more valuable than the lab results because the chemistry of a given sample can change after collection (see Examples 9.2, 9.3, and 9.7 above). A good compromise would be to report *both* the field and the lab measurements in the final table of results.

Finally, it is stressed that the *date* of sample collection should be included in all data tables. This is especially true for surface-water samples, but groundwater samples can also show seasonal or long-term variations in water chemistry. For surface-water samples, it is also recommended to include the *time of day* alongside the date. Our research has shown that the chemistry of rivers and streams can change dramatically on a diurnal (24 h) basis.

Example 9.10   Figure 9.10 is an example data set collected from the Big Hole River, Montana, during midsummer, low-flow conditions. The pH and DO of the water both increased during the day, owing to photosynthesis by algae and aquatic plants. As shown by the following reaction, photosynthesis not only produces DO, but also consumes dissolved CO_2:

$$CO_2(aq) + H_2O = CH_2O \text{ (organic carbon)} + O_2(aq) \tag{9.11}$$

Because CO_2 is a weak acid, a decrease in its concentration typically results in an increase in pH. Although this example focuses on pH and DO, almost any parameter that you can measure, including the concentrations of major solutes, trace metals, and nutrients, will, in some cases, show diurnal fluctuations in a surface-water body (Nimick et al. 2011), making the time of sample collection a critical variable that should be reported in the final table of results.

9.7   Summary

The most important step in any water-quality study is to carefully formulate the question that you are trying to address, and then to write a SAP for how you will carry out the work. Careful thought needs to be given as to where, when, and how often you will sample; what equipment and supplies you will need; which measurements will be taken in the field; what types of samples will need to be collected for laboratory work; which laboratory will do the work; and the estimated total project costs. Depending on the nature of your work, you may also need to specify what QA steps will be taken, and how the final data will be validated. Regardless of whether rigorous QA documentation is mandatory for your project, always take steps to ensure that the environmental

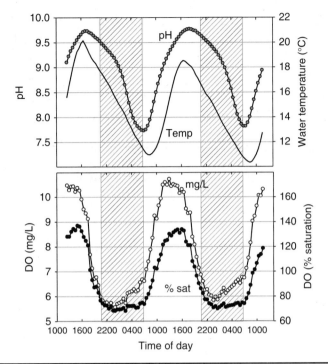

FIGURE 9.10 Diurnal variations in pH, temperature, and dissolved oxygen (DO) in the upper Big Hole River, August 2000. (*C. Gammons, unpublished data.*)

samples and field data that you collect are of the highest quality possible given the amount of time and money you have available.

This chapter has given an overview of most of the common technical procedures that water specialists are likely to encounter when collecting or analyzing samples for chemical parameters. However, it is always a good idea to seek additional information before heading into the field, especially if you are not confident in a given procedure. A lot of material is available on the Internet, but it is sometimes of questionable reliability. The resources mentioned at the beginning of this chapter have all been carefully peer-reviewed, and much of this information is available online. Also, do not hesitate to talk to another human being! This author has spent many hours speaking on the phone (1-800 numbers are very handy) with technical service representatives for supply companies, and sometimes that is the fastest way to solve a question about how to do something.

9.8   Problems

9.1.   Design a sampling plan and a field checklist for a possible study to investigate 24-h changes in the concentrations of dissolved and total metals for a stream or river near your home. A good reference to help get you started is Gammons et al. (2005). Using the lab you are most familiar with, how much would the lab analyses cost (include the cost of any QA samples)? How much will consumables (sample bottles, filters, etc.) cost?

Sample Name	As	Cd	Cu	Fe	Mn	Pb	Zn
IDL	0.2	0.2	1	5	2	0.2	1
IPC-nominal	100	100	100	500	500	100	100
Spike-nominal	20	20	20	100	100	20	20
Samples							
S-1	2.7	0.12	0.05	3.7	3.4	0.14	37.8
S-2	1.5	0.11	0.01	1.5	3.6	0.14	29.4
Field blank	−0.01	0.01	0.03	2.9	1.1	−0.05	32.2
S-3	27.8	0.67	0.95	26.3	3.8	0.13	25.5
S-4	86.6	2.50	17.7	155	33.5	0.55	180
S-5	455	9.53	59.1	12,095	7,850	9.85	555
QA samples							
Lab blank	1.65	0.16	0.85	26.5	5.6	0.08	0.6
IPC	96.8	98.8	98.2	426	98.8	96.5	102
S-1 DUP	2.9	0.11	0.09	6.2	3.5	0.26	37.1
S-1 SPIKE	21.6	20.6	19.9	88.2	101.6	20.6	134

TABLE 9.8   Example of ICP-MS Laboratory Report. All Data Are in µg/L

9.2.   Table 9.8 summarizes results of an ICP-MS session. Comment on the quality of these data based on RPD of IPC, RPD of Dup, % spike recovery, and results for the lab blank and the field blank. Which samples and which analytes should be "flagged"? See Example 9.8 for guidance.

9.9   References

Allison, J.D., Brown, D.S., and Novo-Gradac, K.J., 1991. MINTEQA2/PRODEFA2, a Geochemical Assessment Model for Environmental Systems. Version 3.0 Users' Manual. Environmental Research Laboratory, U.S. Environmental Protection Agency.

Barbash, J.E., and Resek, E.A., 1996. *Pesticides in Ground Water: Distribution, Trends, and Governing Factors*. CRC Press, Boca Raton, FL.

Bethke, C.M., and Johnson, T.M., 2008. Groundwater Age and Groundwater Age Dating. *Annual Reviews Earth Planetary Science*, Vol. 36, pp. 121–152.

Busenberg, E., and Plummer, L.N., 2008. Dating Groundwater with Trifluoromethyl Sulfurpentafluoride (SF5CF3), Sulfur Hexafluoride (SF6), CF3Cl (CFC-13), and CF2Cl2 (CFC-12). *Water Resources Research*, Vol. 44, W02431.

Clark, I., 2015. *Groundwater Geochemistry and Isotopes*. CRC Press, Boca Raton, FL.

Cook, P.G., and Solomon, D.K., 1997. Recent Advances in Dating Young Groundwater: Chlorofluorocarbons, ^{3}H/^{3}He and ^{85}Kr. *Journal of Hydrology*, Vol. 191, pp. 245–265.

Drever, J.I., 1997. *The Geochemistry of Natural Waters, 3rd Edition*. Prentice Hall, Upper Saddle River, NJ.

EPA, 2002. *Guidance for Quality Assurance Project Plans*. U.S. Environmental Protection Agency, EPA QA/G-5, 111 pp. Available online at http://www.epa.gov/QUALITY/qs-docs/g5-final.pdf

EPA, 2006. *Pharmaceuticals and Personal Care Products (PPCPs) as Environmental Pollutants.* U.S. Environmental Protection Agency. Available online at http://www.epa.gov/esd/chemistry/pharma/index.htm

Fetter, C.W., 1999. *Contaminant Hydrogeology, 2rd Edition.* Prentice Hall, Upper Saddle River, NJ.

Gammons, C.H., Milodragovich, L., and Belanger-Woods, J., 2007. Influence of Diurnal Cycles on Monitoring of Metal Concentrations and Loads in Streams Draining Abandoned Mine Lands: An Example from High Ore Creek, Montana. *Environmental Geology*, Vol. 53, pp. 611–622.

Gammons, C.H., Nimick, D.A., Parker, S.R., Cleasby, T.E., and McCleskey, R.B., 2005. Diel Behavior of Fe and Other Heavy Metals in a Mountain Stream with Acidic to Neutral pH: Fisher Creek, Montana, USA. *Geochimica et Cosmochimica Acta*, Vol. 69, pp. 2505–2516.

Heberer, T., 2002. Occurrence, Fate, and Removal of Pharmaceutical Residues in the Aquatic Environment: A Review of Recent Research Data. *Toxicology Letters*, Vol. 131, pp. 5–17.

IAEA, 2013. *Isotope Methods for Dating Old Groundwater.* International Atomic Energy Agency, Vienna. Available online at https://www-pub.iaea.org/MTCD/Publications/PDF/Pub1587_web.pdf

ITRC, 2006. *Technology Overview of Passive Sampler Technologies.* Interstate Technology and Regulatory Council, Washington, DC, 115 pp.

Kazemi, A.A., Lehr, J.L., and Perrochet, P., 2006. *Groundwater Age.* Wiley-Interscience, Hoboken, NJ, 325 pp.

Nimick D.A., Gammons C.H., and Parker S.R., 2011. Diel Biogeochemical Processes and Their Effect on the Aqueous Chemistry of Streams: A Review. *Chemical Geology*, Vol. 283, pp. 3–17.

Pellicori, D.A., Gammons, C.H., and Poulson, S.R., 2005. Geochemistry and Stable Isotope Composition of the Berkeley Pit Lake and Surrounding Mine Waters, Butte, Montana. *Applied Geochemistry*, Vol. 20, pp. 2116–2137.

Stookey, L.L., 1970. Ferrozine—A New Spectrophotometric Reagent for Iron. *Analytical Chemistry*, Vol. 42, pp. 779–781.

USGS, 1995. *Ground-Water Data-Collection Protocols and Procedures for the National Water-Quality Assessment Program: Collection and Documentation of Water-Quality Samples and Related Data.* U.S. Geological Survey Open-File Report 95-399. Available online at http://water.usgs.gov/nawqa/OFR95-399.html

USGS, 2005. *National Field Manual for the Collection of Water-Quality Data.* U.S. Geological Survey Techniques of Water-Resources Investigations, Book 9. Available online at http://water.usgs.gov/owq/FieldManual/

U.S. Salinity Lab, 1954. *Diagnosis and Improvement of Saline and Alkali Soils.* U.S. Department of Agriculture Handbook 60, 160 pp.

Van Voast, W. A., 2003. Geochemical Signature of Formation Waters Associated with Coalbed Methane. *AAPG Bulletin*, Vol. 87, pp. 667–676.

CHAPTER 10

Groundwater/ Surface-Water Interaction

William W. Woessner

Emeritus Regents Professor of Hydrogeology, University of Montana, Missoula, Montana

Willis D. Weight

Professor of Engineering, Carroll College, Helena, Montana

Hydrogeologists must be familiar with how groundwater interacts with rivers, lakes, and wetlands. Researchers in the U.S. Geological Survey (USGS) and other federal and state regulators promote viewing these two systems as a single water resource (e.g., Winter et al. 1998). The exchange of groundwater with streams was briefly addressed in Section 4.4 in Chapter 4 under gaining and losing streams. In this chapter, an extended treatment of this topic including concepts of groundwater exchange at multiple scales and methods used to characterize drivers, locations, and rates are presented.

Successfully implementing water management, resolving contaminated transport and fate, understanding resource health (e.g., Meyer 1997), and planning and executing resource restoration efforts often require qualitatively and quantitatively characterizing groundwater/surface-water exchanges. In addition, applying for or using water rights to develop water requires analyses of how extracting water from one system (e.g., groundwater) will or will not impact the uses and functions of the other system (e.g., surface water). Finally, groundwater exchange with surface-water features often underpins ecological systems of streams, lakes, and wetlands (e.g., Hansen 1975; Grimm and Fisher 1984; Dahm et al. 1998; Hauer and Resh 2006; Hauer and Lamberti 2017). The process of developing appropriate conceptual models of the exchange drivers and the selection and application of methods to evaluate surface-water/groundwater exchange is the focus of this chapter.

10.1   Conceptual Models

When hydrogeologists undertake a project such as a resource assessment, water-supply development, groundwater contamination source identification and cleanup analyses, or a wetland restoration, a well-defined conceptual model of how groundwater and surface-water interface is critical to project success. The conceptual model guides the choices and locations of site instrumentation and interpretation of the field data. A conceptual model of groundwater/surface-water exchange recognizes the nature and control of the geological framework in concert with the physical and geochemical components of the hydrologic system. The following discussion presents generalized conceptual models of the exchange process associated with stream, lake, and wetland features.

The exchange process occurs at multiple scales. Conceptualizing links using cross sections of the surface-water features introduce stage (surface water) and head relationships that drive groundwater exchange. Cross sections are used to represent conditions where a stream, lake, or wetland feature is receiving groundwater discharge (Figure 10.1),

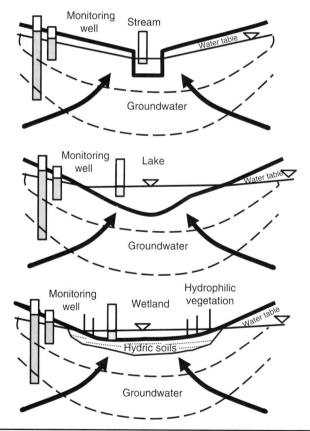

Figure 10.1   Schematic cross sections in the vicinity of a gaining, effluent, or upwelling stream, lake, and wetland where groundwater flow is from the groundwater to the surface-water feature. Dashed lines are equipotential lines, and black arrows show groundwater-flow direction assuming earth material properties are isotropic and homogenous. [*With permission Woessner (2018).*]

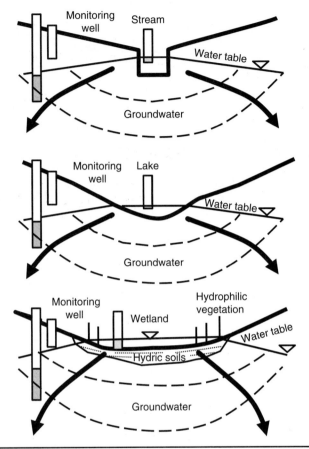

FIGURE 10.2 Schematic cross sections in the vicinity of a losing, influent, or downwelling stream, lake, and wetland where groundwater flow is from the surface-water feature to the groundwater. The surface-water feature and groundwater system are connected. Dashed lines are equipotential lines, and black arrows show groundwater-flow direction assuming earth material properties are isotropic and homogenous. [*With permission Woessner (2018).*]

leaking water to the adjacent and underlying groundwater (Figures 10.2 and 10.3), both receiving groundwater discharge and losing surface water to the adjacent groundwater system (Figure 10.4), and a neutral setting where no exchange occurs (Figure 10.5). The surface-water/groundwater exchange conditions take on different names depending on whether hydrogeologists, hydrologists, or ecologists are describing conditions. A hydrogeologist thinks of the exchange process from the groundwater perspective and would describe conditions at a site where water is leaving the groundwater system as effluent exchange conditions (Figure 10.1). A hydrologist will often view a river, lake, or wetland receiving groundwater discharge as a gaining surface-water feature (e.g., gaining stream) because the flow or volume is increasing with the addition of groundwater. Ecologists studying exchanges in streams have used the term upwelling to describe effluent groundwater discharge settings. This seems a bit confusing; however, it is logical when the context is described. A surface-water feature seeping water into the adjacent and/or underlying

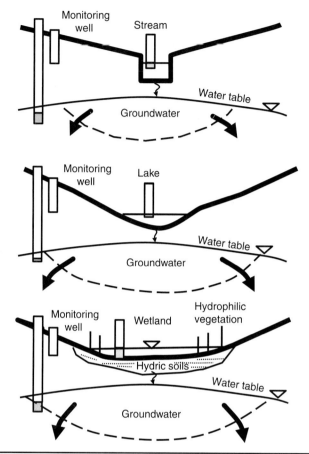

FIGURE 10.3   Schematic cross sections in the vicinity of a losing, influent, or downwelling stream, lake, and wetland where groundwater flow is from the surface-water feature to the groundwater. The perched feature results in surface-water percolation through the unsaturated zone (small back arrows) to the water table. Dashed lines are equipotential lines and thick black arrows show the groundwater-flow direction assuming earth material properties are isotropic and homogenous. [*With permission Woessner (2018).*]

groundwater system is referred to as influent relative to the groundwater system, a losing surface-water body (e.g., losing stream as flows decrease in the down-gradient direction) or a downwelling feature (Figures 10.2 and 10.3). Influent conditions occur with the water table connected to the surface-water feature (surface-water stage represents the water table) or when the surface water is perched above the water table and leakage occurs as infiltration/percolation (the stage does not equal the water table). These influent conditions are also referred to as losing and downwelling. In some settings, the water table is higher than the surface-water stage at one side of the feature and lower at a different location. In these flow-through settings, groundwater enters along one stream bank, lake shoreline, or wetland and the surface water flows into the groundwater at another portion of the channel, lakeshore, or wetland boundary (Figure 10.4). Flow-through conditions (e.g., flow-through lake) are used to describe this setting. With the surface-water

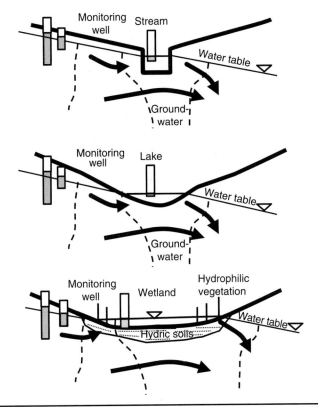

FIGURE 10.4 Schematic cross sections in the vicinity of a flow-through stream, lake, and wetland where groundwater flow is from the groundwater to the surface-water feature on the left-hand side and from the surface-water feature to the groundwater on the right-hand side. Dashed lines are equipotential lines and black arrows show groundwater-flow direction assuming earth material properties are isotropic and homogenous. [*With permission Woessner (2018).*]

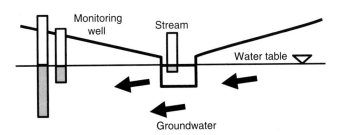

FIGURE 10.5 Schematic cross sections in the vicinity of a surface-water feature where no exchange is taking place. This parallel flow or zero-exchange condition in the vicinity of a stream channel occurs where the water-table and surface-water stages are equal (no equipotential lines shown). Black arrows show a groundwater-flow direction paralleling the surface-water feature (coming out of the page). Earth material properties are isotropic and homogenous. [*With permission Woessner (2018).*]

feature stage connected to the groundwater system, the stage represents the water-table elevation. When equipotential lines are perpendicular to a surface-water feature or portion thereof (e.g., stream channel) and the stage elevation is equal to the water-table elevation, no hydraulic gradient is present, and exchange cannot occur (Figure 10.5). This condition, referred to as zero exchange, is observed in some stream settings and may occur at some portions of lake and wetland boundaries.

It is important to note that characterizing exchange using the described conceptual models represents only a snapshot in time. In some settings, exchange conditions may remain relatively stable; however, as groundwater heads and flow rates, and stream discharges and stages vary over short (e.g., flooding events) or long periods (e.g., periods of increased snowmelt and drought), exchange conditions dominating a location may change. As illustrated in Figure 10.6, exchange for a stream, lake, or wetland system may change as site conditions are impacted by seasonal climatic controls. Investigations should be designed to identify how the exchange process varies over time. Formulating the appropriate conceptual model is a necessary and critical step prior to designing field investigation and selecting appropriate methodologies.

Exchanges at the Regional Scale

Conceptual cross sections are valuable for illustrating basic relationships and principles controlling groundwater/surface-water exchange. Groundwater/surface-water interactions take place at multiple scales. Conceptually, exchange conditions can be set in landscape and smaller hydrologic unit scales (e.g., river reaches and channel bottoms).

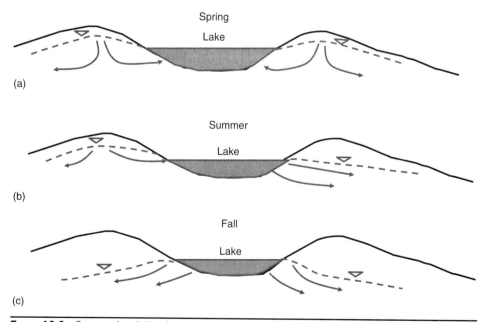

FIGURE 10.6   Seasonal variation in groundwater lake exchange in a hypothetical setting in which spring groundwater recharge drives the lake groundwater system. The dashed line is the water table, and gray arrows represent groundwater flow. (a) Effluent lake. (b) Flow-through lake. (c) Influent lake. [*With permission Woessner (2018).*]

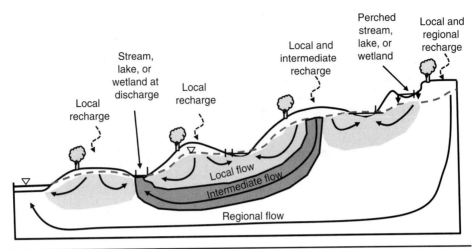

FIGURE 10.7   Schematic cross section of the location of surface-water features (streams, lakes, or wetlands) in a landscape with varying topography and multiple groundwater-flow systems. [*Carter (1996); with permission Woessner (2018).*]

Ground pavewater-flow and surface-water exchange at the landscape scale can be illustrated by viewing the groundwater system as dominated by recharge and discharge areas (Figure 10.7). Local, intermediate, and regional flow systems exchange water between groundwater and surface water (Tóth 1963). Tóth's work suggested that in settings with sufficient net precipitation, anisotropic earth materials and little topographic relief, the flow of groundwater originates (recharged) at topographic highs and discharges at surface-water features found at topographic lows (e.g., sites of rivers, lakes, and wetlands). As the topographic setting becomes more complex, similar groundwater-flow patterns would develop within local topographic features, and in some settings, groundwater flow with longer flow paths form intermediate and regional flow systems (Figure 10.7). These systems with longer flow paths discharge at locations not immediately adjacent to local discharge locations (exchange sites). Exchange relationships of course are highly variable within the broad range of possible geologic, topographic, and climatic settings in which surface-water features are found. For example, an arid basin with limited groundwater recharge may be dominated by a regional flow system and local perched and/or ephemeral surface-water features. Generally, exchange can be described using the conceptual models presented in Figures 10.1 to 10.5. To illustrate how the exchange process varies with the scale of the field investigation, the following section focuses on groundwater exchange with streams principally at the fluvial plain, reach, and channel scale.

Focus on Exchange with Streams

Groundwater exchange with streams at multiple scales has been the focus of researchers attempting to understand how natural and modified streams exchange water, support ecological systems, and whether exchange processes can be reestablished in impacted streams. River systems can be viewed at the watershed/basin scale 20 to 200 mi² (50 to 500 km²), the valley segment scale [colluvial, bedrock, and alluvial valleys,

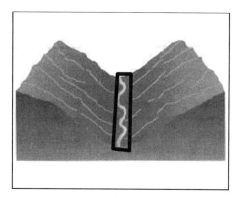
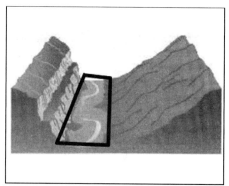

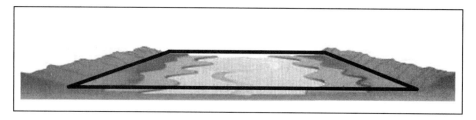

FIGURE 10.8 Examples of river corridors in three geomorphic settings outlined with black lines. The river is shown as a meandering light gray line. (*Modified from https://cfpub.epa.gov/ watertrain/pdf/modules/new_streamcorridor.pdf.*)

1,100 to 110,000 ft² (100 to 10,000 m²)]; channel reaches 108 to 10,800 ft² (10 to 1,000 m²), and channel/habitat units, 11 to 110 ft² (1 to 10 m²) (Bisson et al. 2006).

The focus of this discussion is the valley segment or active stream corridor (Figure 10.8). The stream corridor includes the active and inactive channel, floodplain riparian area, and older unconsolidated river terraces. Woessner (2000) described a stream corridor as a fluvial plain, a fairly planar feature consisting of the active stream channel, floodplain, and associated fluvial sediments, which also includes older sediments associated with previous stream positions (Figure 10.9). In most basins, the fluvial plain is bounded by uplands and would be considered the river corridor based on hydrologic conditions. The fluvial plain system is also referred to as an alluvial valley or a riverine valley (Dahm et al. 1998; Winter et al. 1998) and in a generalized landscape model proposed by Winter (2001) a portion of the lowland down gradient of the valley slope.

Stream corridors can be described geomorphically as constrained and unconstrained, and contain floodplains and river systems that are aggrading or degrading, and meandering or braided. The lateral continuity of sediments within fluvial plains formed from fluvial processes changes rapidly over distances of tens of feet (meters to tens of meters) (e.g., Figure 10.10a). Geological models of these systems vary significantly as the differences between meandering streams with lower gradients and sediment-laden braided-stream systems generate different sequences of sediments (Figures 10.10b and 10.10c) (e.g., Anderson 1989; Mial 1996; Gross and Small 1998; Huggenberger et al. 1998).

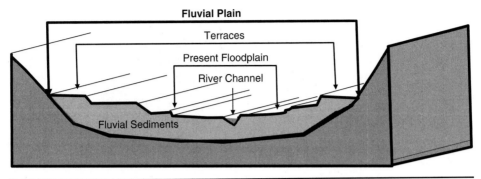

FIGURE 10.9   Schematic of the fluvial plain. [*With permission Woessner (2018).*]

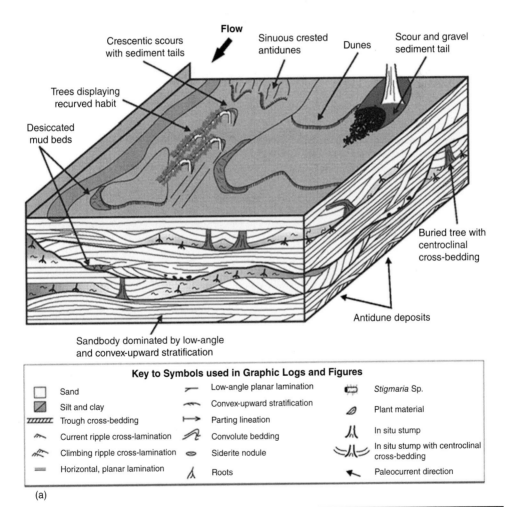

Key to Symbols used in Graphic Logs and Figures

Sand	Low-angle planar lamination		*Stigmaria* Sp.
Silt and clay	Convex-upward stratification		Plant material
Trough cross-bedding	Parting lineation		
Current ripple cross-lamination	Convolute bedding		In situ stump
Climbing ripple cross-lamination	Siderite nodule		In situ stump with centroclinal cross-bedding
Horizontal, planar lamination	Roots		Paleocurrent direction

(a)

FIGURE 10.10   Character of the fluvial plain. (a) Geologic facies model of a fluvial depositional setting. River systems erode and deposit sediment based on differences in sediment input and discharge conditions (after Fielding et al. 2009; Anderson et al. 2015). (b) Braided stream system north of Banff, Canada. (c) Meandering stream in southwestern Montana.

(b)

(c)

Figure 10.10   *(Continued)*

The depositional processes within the fluvial plain result in heterogeneous aquifer-property conditions. Commonly the bulk hydraulic conductivity of the fluvial sediments is greater than the adjacent uplands (Winter et al. 1998; Woessner 1998). Combined with the down-gradient slope of the fluvial plain, groundwater flow is often down plain where it interacts with the stream, floodplain, and riparian corridor (Figure 10.11a) (Woessner 2000).

At the channel reach scale [miles to hundreds feet (kilometers to tens of meters)], groundwater stream exchange can be dominated by effluent, influent, flow-through, or zero-exchange conditions. However, it is not uncommon that multiple exchange conditions occur within a single reach (Figures 10.11b and 10.11c). In some settings a stream reach may be characterized as effluent yet also have sections of influent, flow-through, and zero exchange. However, for the overall defined reach length, a net gain or loss in stream flow may be observed. Groundwater interaction in reaches with single or multiple channels will vary depending on the spatial relationships of water-table elevations and stream stage (e.g., Figures 10.11b and 10.11c).

The complexities of the reach-scale exchange processes are often generalized when standard terms are used to describe streams based on stream flow. For example, stream reaches described as perennial (water is present year-round) are generally dominated by effluent conditions. These streams are typically fed by groundwater systems at rates that maintain stream flow year-round, also known as base flow (Viessman and Lewis 2002). Stream reaches that are ephemeral (flow only part of the year) are usually recharging to a local groundwater system and would be considered influent. Intermittent streams (flow in some sections but not over the entire length) would contain effluent portions where groundwater-flow discharges initiating stream flow and influent sections where stream water seeps into the stream bottom and stream flow ceases.

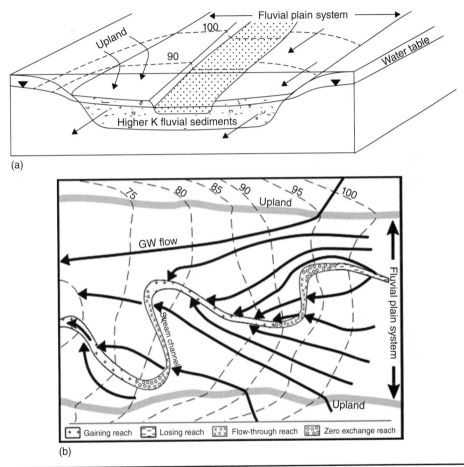

(a)

(b)

Figure 10.11 Groundwater exchange in the fluvial plain of a stream. Black arrows represent groundwater flow assuming isotropic and homogenous conditions and dashed lines are equipotential lines. (a) In this setting, the fluvial sediment has a higher hydraulic conductivity than the surrounding uplands. The fluvial plain is referred to as the lower topography formed by fluvial erosion and deposition. It includes the channel, active floodplain, and older river terraces. Groundwater moves from the uplands to the fluvial plain and continues generally down valley where it interacts with the stream channel (Woessner 2000). (b) Map view of the stream channel and area of fluvial sediments. This schematic illustrates the potential complex interaction of groundwater and the stream. Portions of the channel are gaining (effluent), losing (influent), flow-through, and are not exchanging with the groundwater (zero exchange) (Woessner 2000). (c) A multichannel setting with stream flow from east to west showing groundwater flow, equipotential lines, and groundwater exchange. [*With permission Woessner (2018).*]

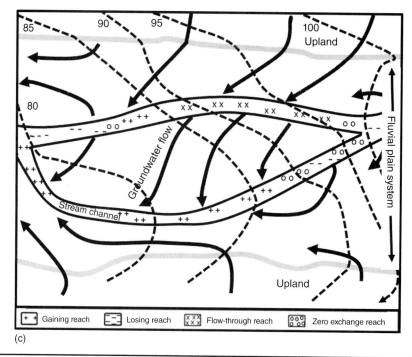

Figure legend: + Gaining reach   Losing reach   Flow-through reach   Zero exchange reach

(c)

Figure 10.11   (*Continued*)

Example 10.1   Clark Fork River, Missoula, Montana
The city of Missoula, Montana's water supply is derived from an unconfined sand, gravel, cobble, and boulder aquifer underlying the mountain valley floor that is approximately 120 ft (36 m) thick. The Clark Fork River is unconstrained and perennial with average flows of about 2,000 cfs (60 cm). An evaluation of how high-yield production wells located near the river in the eastern portion of the valley may impact the river was initiated by a hydrogeologist whose training included the work of Tóth (1963; e.g., Figure 10.7), experiences in temperate climate settings, and the desert Southwest. Observing that the river was a large perennial regional river, the valley was surrounded by mountains, and tens of feet of snow fell in the higher elevations, an initial conceptual model suggested how the valley aquifer interacted with the river. This model proposed the river was a discharge zone for the valley aquifer, thus the Clark Fork River was an effluent or gaining stream. Therefore, pumping wells near the stream would intercept groundwater heading to the stream and possibly at some point directly deplete stream flow by capturing stream water (Figure 10.12). However, after conducting an initial analysis of groundwater levels, stream-stage data, and stream-discharge records, it was determined that the Clark Fork River was not a naturally gaining stream in the eastern portion of the valley. Instead it was perched at an elevation up to 30 ft (10 m) above the water table of the valley's unconfined aquifer. Leakage from the stream was providing significant recharge to the underlying aquifer. Under these conditions pumping would not deplete or influence stream flow, since the valley aquifer water table was not connected to the stream (Figure 10.13). The conceptual model was revised to reflect these revelations.

When the scale of investigation is focused at the channel unit/habitat scale (10 to 110 ft²; 1 to 10s m²), documenting the details of the exchange becomes more complex. This occurs because any changes in bed topography; stream stage and groundwater head distributions; bed and bank hydraulic conductivities, and the distribution of zones

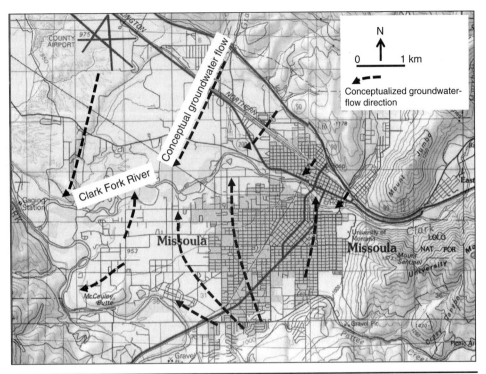

Figure 10.12 Conceptual model of groundwater exchange with the perennial Clark Fork River assuming the river (flowing right to left) is a gaining stream. The aquifer is a sand, gravel, and cobble unconfined system. The city of Missoula, Montana, is surrounded by mountains; elevations are in meters. [*With permission Woessner (2018).*]

of groundwater discharge drive exchange at this scale (Figure 10.14). In some settings, groundwater from local, intermediate, and regional flow systems discharges to the stream channel or is recharged from the channel. The movement of surface water into the adjacent sediments and groundwater system is referred to as hyporheic flow (Figures 10.14 and 10.15). The exchange of river water with the associated saturated sediments occurs within the hyporheic zone, or that portion of the local groundwater system where a mixture of surface water and groundwater can be found (Woessner 2017). Stonedahl (2013) also suggests that hyporheic exchange includes meander-driven (Figure 10.11b), bar-driven (Figure 10.11c), and bedform-driven exchange (Figure 10.14). Works by Boano et al. (2014), Cardenas (2015), and Buss et al. (2009) present detailed literature reviews and descriptions of methodologies used to investigate the hyporheic zone in a wide range of settings.

Hyporheic zones were initially described by river ecologists and included areas in the adjacent saturated stream sediments where distinctive biota were observed (Orghidan 1959). Hyporheic-zone waters, including those found in the stream-channel sediments and circulating in the adjacent floodplain, originate from the stream and return to the stream (Figures 10.14 and 10.15). Both river ecologists and hydrogeologists recognize the importance of hyporheic exchange in streams (e.g., Danielopol 1980; Hynes 1983; Stanford and Ward 1988; Triska et al. 1989; Stanford and Simons 1992; Valett et al. 1993;

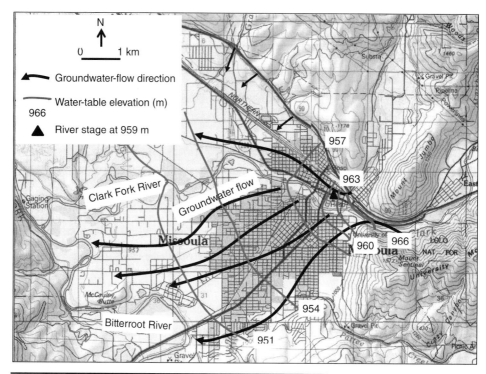

FIGURE 10.13 Groundwater and river conditions in May 2005. The Clark Fork River is perched above the water table as it enters the valley (right). It is a losing stream (influent conditions) until the multichannel section forms near the confluence with the Bitterroot River and it becomes a gaining stream. The aquifer is a sand, gravel, and cobble unconfined system. The city of Missoula, Montana, is surrounded by mountains; elevations are in meters. [*With permission Woessner (2018).*]

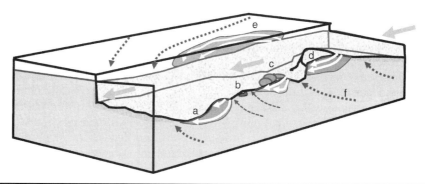

FIGURE 10.14 Schematic of a section of channel receiving both groundwater discharge (dashed arrows) and exchange of hyporheic water (white arrows in gray zones). Stream water (large gray arrows) leaves the stream and mixes with groundwater then reenters the stream at: (a) pool and riffle sequence; (b) bedform; (c) obstruction; (d) mid-channel bar; (e) by flowing into the adjacent floodplain and returning to the stream. (f) Groundwater discharge (dashed black arrows) is shown as focused in portions of the stream bottom and banks (*Woessner 2017*).

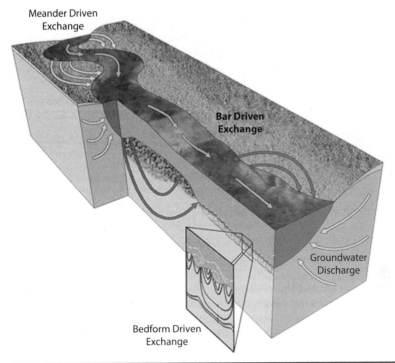

FIGURE 10.15 Stonedahl et al. (2013) describe hyporheic exchange a multiple scales including bedform driven, bar driven, and meander driven.

Findlay 1995; Jones and Holmes 1996; Brunke and Gonser 1997; Morrice et al. 1997; Boulton et al. 1998; Winter et al. 1998; Woessner 2000; Edwards 2001; Malard et al. 2002; Hancock et al. 2005; Dahm et al. 2006; Buss et al. 2009; Ward 2016). The exchange process has been observed to circulate nutrients and support ecosystems including microbes to macroinvertebrates (Wright 1995; Wallace et al. 1996; Brunke and Gosner 1997; Dahm et al. 1998; Hauer and Resh 2006; LaLiberte 2006). In some large gravel-bedded streams in the western United States, hyporheic exchange in adjacent flood-plains can result in the presence of unique macroinvertebrates in the groundwater that are supported by the hyporheic conditions found in the valley aquifers 1.25 to 1.8 mi (2 to 3 km) from the active channel (e.g., Stanford and Ward 1993). Dahm et al. (1998) note that sufficient nutrient retention times can occur within the hyporheic zone and that a complex variety of organisms and biochemical process are supported in some stream settings (Figure 10.16 and Table 10.1) (Stumm and Morgan 1996). These processes may also be complicated by an intermittent vadose zone within gravel bars and the riparian zone that provides the circulating water with additional oxygen.

10.2   Field Methods to Determine Groundwater/Surface-Water Exchange

Conducting field work to characterize groundwater exchange with streams, lakes, and wetlands requires standard hydrologic methods as well as specialized tools. This section addresses the application of these methods at multiple scales.

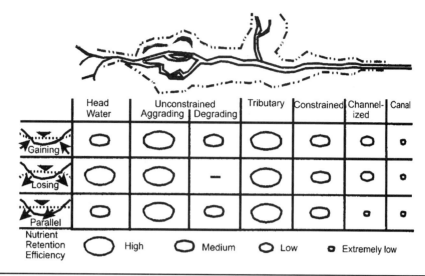

FIGURE 10.16 Hypothesized nutrient retention efficiencies in various groundwater/surface-water stettings and channel configurations. Authors state that losing sections rarely occur in degrading sections of unconstrained streams. [*Adapted from Dahm et al. (1998).*]

Water Budget

Water budgets, hydrologic balances, and/or groundwater budgets are standard hydrologic tools that compartmentalize the inflows, outflows, and changes in storage within a defined area/volume and over a specified period of time. Components of a water budget for a lake represented as the lake-shore boundary and for a designated time period (units in L^3/T) are presented in Figure 10.17. To estimate the groundwater discharge to the lake (GW_{in}) would be expressed in Equation 10.1 as

$$GW_{in} = SW_{out} - SW_{in} - PTT_{in} + Evap_{out} + Trans_{out} + GW_{out} \pm \Delta S \qquad (10.1)$$

Chemoautotrophic Processes	Kcal/Equiv	Electron Accepting Processes	Kcal/Equiv
Nitrification $NH_4^+ \rightarrow NO_3$	−10.3	Sulfate reduction $SO_4^{-2} \rightarrow S^2$	−5.9
		Methanogenesis $CO_2 \rightarrow CH_4$	−5.6
Iron oxidation $Fe^{+2} \rightarrow Fe^{+3}$	−21.0	Iron reduction $Fe^{+3} \rightarrow Fe^{+2}$	−7.2
Sulfide oxidation $S^2 \rightarrow SO_4^2$	−23.8	Denitrification $NO_3 \rightarrow N_2$	−28.4
Methane oxidation $CH_4 \rightarrow CO_2$	−9.1	Aerobic respiration $CH_2O \rightarrow CO_2$	−29.9

TABLE 10.1 Chemoautotrophic and Electron Acceptors in the Fluvial Plain [From Stumm and Morgan (1996)]

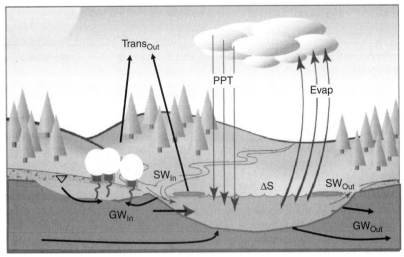

Not to scale

Explanation

Evap	Evaporation
GW_{in}	Groundwater inflow
GW_{Out}	Groundwater outflow
PPT	Precipitation
SW_{in}	Surface-water inflow
SW_{Out}	Surface-water outflow
$Trans_{out}$	Transpiration outflow

ΔS Net change in storage
(lake, soil, and groundwater)

FIGURE 10.17 Flow-through lake-water budget with streams entering and leaving the lake boundary. Phreatophyte (white trees) transpiration causes outflow of some lake water and aquatic vegetation (gray ovals) also transpires water. [*Modified from Robertson et al. (2003).*]

where GW_{in} is discharge into the surface-water domain, GW_{out} is the seepage of lake water into the groundwater, SW_{in} is flow of surface water in, SW_{out} is flow of surface water out, PPT_{in} is precipitation falling within the domain, $Evap_{out}$ is direct evaporation from the water surface, $Trans_{out}$ is flux of water flowing from the lake to the shoreline phreatophytes and the loss of water from aquatic plants in the water body, and $\pm\Delta S$ is the volumetric net change in storage of lake water, groundwater, and soil water over the time duration of the balance.

Quantifying the components of a stream, lake, or wetland water budget requires measuring each parameter and identifying likely errors and uncertainty. Ideally, whenever possible, all components of a water budget should be measured or estimated. When a parameter is solved for as the unknown, errors associated with the other water-budget parameters compound and become part of the solution (Winter 1981; Healy et al. 2007). When formulating a water budget to assess exchanges, care should be given to determining which components dominate the budget and whether other components contribute little to the overall budget. Such an analysis will help identify how available resources should be allocated. Quantifying parameters of water budgets covers a broad range of hydrology and hydrogeology methods. Addressing this wide range of methods is beyond the scope of this text; therefore, it is suggested the reader consult with hydrology texts such as Watson and Burnett (1993), Hornberger et al. (1998), Viessman and Lewis (2002), and Dingman (2014) and hydrogeology texts such as Freeze and Cherry (1979), Domenico and Schwartz (1998), Fetter (2001), and Schwartz and Zhang (2003).

Example 10.2   Mirror Lake Water Budget

Healy et al. (2007) present an example of a water budget computed for Mirror Lake, New Hampshire, the United States. The lake balance was computed using historical records and reported as a yearly budget. Based on Equation 10.1 the Mirror Lake water budget would be stated as

$$\mathrm{PTT_{in} + SW_{in} + GW_{in} = SW_{out} + ET_{out} + GW_{out} \pm \Delta S}$$

where ET is computed as the direct loss of water from the lake surface (energy-budget methods) and $\pm\Delta S$ is the change in lake storage, lake volume change per year.

In their work they examined three methods of defining the water budget and how each method affected the computed values of groundwater inflow and outflow. Groundwater inflows and outflows were first computed using knowledge of the geology, a single aquifer test to estimate hydraulic conductivity, gradients measured from networks of wells and lake stage, and Darcy's law (Table 10.2). A second approach evaluated the use of oxygen isotope mixing models and balances. Results yielded an estimated groundwater inflow of about twice the initially computed value. The third approach used a field-based calibrated basin-scale numerical groundwater model from which the lake-water balance could be extracted. Modeling results suggested groundwater-inflow rates were higher than the initial estimate by about 2.8 times (113,000 m^3/year). Healy et al. (2007) state that the final value of groundwater inflow was agreed upon after reviewing site data and methods as 113,000 m^3/year. This higher inflow rate also required adjusting the outflow rate to reflect the magnitude of the inflow parameter. Clearly, deciding between the approaches required identification of the limitations of each method used and some professional judgment.

	Initial	Final	Estimated Uncertainty
	values in 1,000 m^3/y		
Inflows			
Precipitation	182	182	5%–10%
Surface-water inflow	417	417	5%–10%
Groundwater inflow	47	113	30%–50%
Outflows			
Evapotranspiration	77	77	10%–15%
Surface-water outflow	257	257	5%–10%
Groundwater outflow	281	347	30%–50%
Lake volume change	16	16	Not reported
Imbalance	**15**	**15**	

TABLE **10.2**   Water Budget and Estimated Uncertainty for Mirror Lake, New Hampshire, the United States. Initial values are from field measurements and equations. Final values are based on basin-scale modeling and evaluation of geochemical results [Modified from Healy et al. (1997).]

Other initial inflow and outflow values were considered to be representative of site conditions (Table 10.2, "Final"). Estimates of parameter uncertainty reported in the Mirror Lake study are also presented in Table 10.2. Note that the largest uncertainty is associated with the groundwater terms of the balance (supported by the evaluation of the three methods). It is important to note the budget does not fully balance (off by 15,000 m³/year). When individual components are all measured or estimated independently, an imbalance is expected and, hopefully, relatively small. If the imbalance is large it may indicate some components are poorly known or an important factor was left out of the budget. In the Mirror Lake case Healy et al. (2007) report an overall estimated budget error of 13%.

Surface-Water Flow Characterization

Stream-discharge measurements are key components of most surface-water/groundwater exchange analyses. They are also used to document short-and long-term volumes of water moving through a watershed and to quantify inputs and outputs in water budgets of rivers, lakes, and wetlands at multiple scales. The measurement of stream flow is referred to as stream gauging. Stream discharge, Q (L^3/T), is computed by determining the stream velocity, V (L/T), at a stream cross section, A (L^2):

$$Q = V \times A \tag{10.2}$$

Though this appears to be relatively straightforward, both water velocities and cross-sectional dimensions vary spatially and temporally at stream sites.

Stream gauging is conducted by suspending a measuring tape over the stream (secured to the banks) and measuring the stream depth and velocity at a series of locations (Figure 10.18). This gauging approach is referred to as using the velocity-area method. These individual discharge values are summed to determine the total stream discharge. The USGS recommends the midsection method where the discharge measurement locations are used to represent conditions for a cross-sectional area represented by a rectangle centered at the velocity measuring point (Figure 10.19). Details on making velocity measurements using a number of instruments and methods are described in a useful USGS publication "Discharge Measurements at Gaging Stations, USGS Techniques and Methods Book 3, Chapter A8 by Turnipseed and Sauer (2010). They provide details of recommended methods to measure velocities and compute discharges. They also describe appropriate methods for large and small streams, address error, and safety considerations. The most commonly used methods are briefly described here.

Selecting the Stream Gauging Site

Selection of the location of a stream-discharge measurement would initially seem to be a straightforward exercise, as it would correspond with a site where you would like information on the stream flow. However, not all gauging sites are created equal, and as a result, moving a site slightly may improve the gauging measurement or increase the stability of the stream cross section (a property desirable for more permanently instrumented gauging sites). An ideal site would be a stream stretch that is relatively straight and free from obstructions and vegetation (Sanders 1998). Eddies and large shallow areas near the bank should be avoided. Water depths are addressed using the appropriate equipment and safety standards (Turnipseed and Sauer 2010).

Figure 10.18 Stream gauging Prickly Pear Creek near Helena, Montana. A graduated tape is stretched at a right angle across the stream, and a current meter (Marsh McBirney) is attached to a graduated wading rod. The wading rod is used to measure stream depth and the current meter generates measurements of stream velocity.

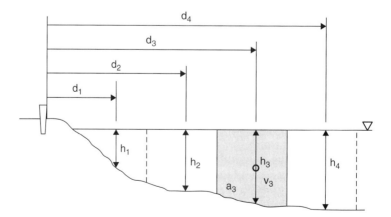

Figure 10.19 Schematic of the current meter midsection method used for computing stream discharge. The discharge (velocity times area) is derived for each subinterval and the values are summed to yield the total discharge. (*Modified from a class handout prepared by Gary Fischer, Carroll College, Montana, and used with permission.*)

When a site is being considered for a more permanent discharge recording station, safe stream access during varied flow conditions and a site with a relatively stable cross section (e.g., constrained by stream conditions or modifications, and/or a naturally occurring stable geomorphic setting) are desirable (Turnipseed and Sauer 2010). At some sites wading will be appropriate all or most of the time; at other sites depths and flow conditions may require access by raft or boat, and/or fixed structures like cableways and bridges (Turnipseed and Sauer 2010). Determining discharge at a gauging site requires careful planning, knowledge of the equipment operation and limitations, and safety. Working in teams, wearing approved personal flotation devices and following all organization safety requirements are critical when working at stream sites.

Measuring Stream Cross-Sectional Areas and Velocity

Discharge measurements require accurate characterization of the stream cross-sectional area at the site and the distribution of stream velocity within that section. Cross-sectional area data are derived by measuring and subdividing stream widths and using sounding methods to determine stream depths at defined locations (Figure 10.19). Methods to characterize cross sections vary depending on the gauging site and methods used to obtain discharge measurements. Stream widths can be determined by stretching a cloth measuring tape or pre-marked cable (tag line) across the stream at right angles to the channel/flow. At permanent gauging stations using cableways or bridge crossings, reference points may be marked on the structures as discharge measurements are made routinely. When cross sections are relatively stable, survey equipment may be used to establish measuring locations and distances (Turnipseed and Sauer 2010). Once the stream width is determined (changes under varying flow conditions) the length of each cross section (velocity measuring point) is determined using the criteria that at least 20 velocity measurement locations should be established and each rectangular subsection should contain less than 10% of the stream flow (Figure 10.19). This requires some estimation of the spacing of measurements at the start of the gauging so that more measuring points may be included in zones with higher discharges and few in areas with low discharges (Turnipseed and Sauer 2010). Water depths at velocity gauging locations collected by wading are obtained by holding a graduated wading rod on the stream bottom that supports a velocity meter and recording the stream depth. It is advised that intervals between readings be uniform; however, obstructions or variable bottom conditions may require that interval spacing be irregular. When gauging from a boat, cableway, or bridge, equipment is lowered through the water to determine the streambed depth. In some cases sonic methods are used for this task.

Once stream depth is established velocity measurements are made. Streams have a parabolic velocity profile that is affected by the bottom and sides of the stream (Figure 10.20). The average velocity is found at approximately the 0.6 depth from the surface. When applying the midsection velocity area method to compute discharge (Figure 10.19) in streams where the subsection depth is 2.5 ft or greater, two measurements should be made, at the 0.2 and 0.8 depths below the surface (zero depth) (Figure 10.20 and Table 10.3). These two velocity measurements are then averaged to produce a value representing the velocity (Buchanan and Sommers 1969; Turnipseed and Sauer 2010). When stream depths are less than 2.5 ft, it is recommended that one measurement be taken at the 0.6 depth below the stream surface.

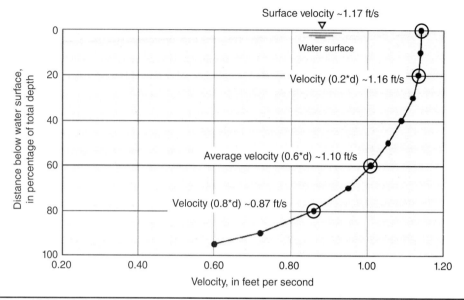

Figure 10.20 Typical velocity profile of a stream plotted as velocity versus percent of depth below the stream surface. The average velocity is the value at 60% of the depth or 0.6 × times the water depth (d), measured from the stream surface. [*Modified from Turnipseed and Sauer (2010).*]

Note that the measurement depths *are not* computed from the bottom of the stream but are representing distances from the stream surface. Measurement errors for a particular instrument and operator are established when total errors of measuring width, depth, and velocity are defined (Sauer and Meyer 1992). The standard error is reported to be about 2% for ideal field and operator conditions and up to 20% for very poor conditions. Average or normal gauging results are expected to have errors between 3% and 6% (Turnipseed and Sauer 2010). In the USGS, hydrographers also qualitatively rate discharge measurements depending on what conditions are present at the discharge measurement location (e.g., measuring section, velocity conditions, equipment, spacing and number of vertical velocity measurements, stability of the river stage, compilations from ice, wind conditions, and any other general

Depth (ft)	Current Meter	Velocity Method
2.5 and greater	Price Type AA	0.2 and 0.8
1.5–2.5	Price Type AA	0.6
0.3–1.5	Price Pygmy	0.6
1.5 and greater	Price Pygmy	0.2 and 0.8
0.3–1.5	ADV	0.6
1.5 and greater	ADV	0.2 and 0.8

Table 10.3 Water Depth Range, Current Meter Type, and Velocity Measurement Method (see text for explanation of current meter types) [Turnipseed and Sauer (2010)]

Mechanical Current Meters	Electromagnetic Current Meters	Acoustic Current Meters

Vertical Axis
Price AA meter
Price AA meter (slow velocity)
Price AA Pygmy meter
Price AA Winter meter
  Horizontal axis
  Contact heads (vertical axis)
	Marsh-McBirney 2000
	Ott electromagnetic
		Acoustic doppler velocimeter (ADV)
		Acoustic digital current meter (ADC)
		Acoustic Doppler current profiler (ADCP)

Other methods	**Direct discharge methods**	**Indirect measurments**
Optical current meter Float	Portable weir-plate Measurement Portable Parshallflume measurements Volumetric measurements	Tracer measurements

TABLE **10.4**   Current Meters and Other Stream Discharge Measuring and Estimating Methods Described in Turnipseed and Sauer (2010)

site or weather conditions that might influence the quality of the measurement). Discharge measurement errors of 2% are applied for a measurement rated as excellent, good (5%), fair (8%), and poor (over 8%) (USGS 1980; Turnipseed and Sauer 2010).

Instruments to measure stream velocity come in a variety of shapes and sizes. To use each meter attention must be paid to operational manuals and descriptions of techniques such as described in Turnipseed and Sauer (2010). Deployment of a velocity measuring device also varies. For small streams where wading is safe, meters are attached to a wading rod. Where conditions are not wadable, meters are suspended from cableways, bridges, rafts, and boats. Examples of commonly used mechanical, electromagnetic, and acoustic Doppler velocity (ADV) meters are briefly described in the following sections (Table 10.4). Other direct and indirect methods to determine stream discharge will also be addressed.

The mechanical Price-type A meter has a horizontal wheel approximately 5 in (12.7 cm) in diameter, with small cups attached. It is a standard mechanical meter used by the USGS and other regulatory agencies (Figures 10.21a and 10.21c). The wheel rotates in the current on a cam attached to the spindle. An electrical contact creates a "click" sound with each rotation when headphones are used. The clicks are counted for 30 or 60 s or sent to a direct readout meter. The number of clicks is compared with a calibration curve to obtain the velocity. Equations and tables to compute velocities from readings are supplied by the manufacturer. A smaller Price Pygmy meter is recommended for stream depths less than 2.5 ft (Turnipseed and Sauer 2010) (Table 10.4; Figure 10.21b). When using either meter it is important they are in good working condition. It is recommended to perform a "spin test" prior to use, and properly maintain this equipment. A spin test is conducted by manually spinning the cups in the air. Properly working meters will spin for about a minute. If they stop within 45 s or so, adjustments should be made, as laid out in the meter manual or possibly repairs are needed. It may be that a drop of lubricant is needed or a nick in the spindle tip may be the problem. Without repair, a meter should not be used if the spin test fails.

The electromagnetic current meter reports velocities directly. It is connected to a wading rod with a digital reading unit that hangs around the neck or shoulder of user (Figure 10.18). These meters create a magnetic field that is altered as the moving water passes the sensing device (Figures 10.21e and 10.21f). Velocity values are correlated with changes in voltage amplitude. An advantage of this type of device is that it can be

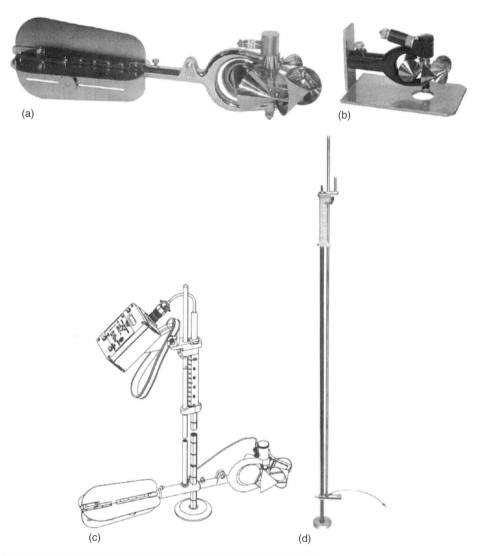

(a)

(b)

(c)

(d)

FIGURE 10.21   Examples of mechanical and electromagnetic current meters and wading rod. (a) Price AA current meter. (b) Price Pygmy meter. (c) Top setting wading rod (graduated) with mounted Price AA meter and current meter digitizer attached (computes velocity). Mechanical current meters can also be outfitted with headphones to count rotations of the meter over time. (d) Top setting graduated wading rod. (e) Mash-McBirney Model 200 electromagnetic flowmeter and display. The sensor is the black knob on the end of the cord. (f) Ott electromagnetic current meter. [*After Turnipseend and Sauer (2010).*]

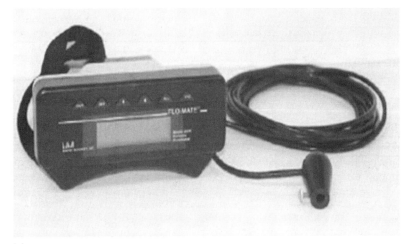

(e)

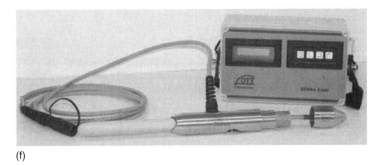

(f)

FIGURE 10.21 *(Continued)*

used in streams or irrigation ditches with weeds or other debris that would normally clog up a spinning wheel (mechanical meters). In comparing the results using both mechanical and electromagnetic current meters during a hydrogeology field camp, velocities measured by both meters were found to be within the error ranges when multiple readings were made using the same device.

Example 10.3 Stream Gauging
Approximately 5 mi west of Butte, Montana, is the Sand Creek drainage basin. A hydrogeologic study was conducted to create a "transitional hydrologic baseline" in the midst of industrial development and water-use changes (Borduin 1999). A significant area of the drainage basin has been zoned heavy industrial. Stream gauging was conducted along Silver Bow Creek and Sand Creek and other tributaries to document gaining and losing stream sections. Later, this information was used to compare basin numerical modeling results with field gauged data. Agreement of potentiometric surface data and stream fluxes was an indication that groundwater model calibration objectives were achieved. Stream reaches less than 5 ft (1.5 m) wide were divided up every 0.5 ft for velocity measurements. Wider stream reaches were divided into enough widths to represent approximately 10% of the total flow. A direct readout current meter was used to obtain point velocities at the 0.6-ft depth. Gauging data collection and discharge results are shown in Table 10.5.

Distance (ft)	Depth (ft)	Velocity (ft/s)	Area (ft²)	Discharge (ft³/s)
Left bank = 1	0.9	0.16	0.9	0.144
3	0.97	0.26	0.97	0.252
4	1.4	0.36	1.4	0.504
5	1.5	0.48	1.5	0.72
6	1.87	0.76	1.87	1.421
7	2.1	1.04	2.1	2.184
8	2.35	1.05	2.35	2.468
9	2.4	1.15	2.4	2.76
10	2.65	1.09	2.65	2.889
11	2.45	1.05	2.45	2.573
12	2.2	0.99	2.2	2.178
13	2.0	0.94	2.0	1.88
14	1.77	0.74	1.77	1.31
15	1.9	0.68	1.9	1.292
16	1.8	0.48	1.8	0.864
17	1.8	0.43	1.8	0.774
18	1.95	0.34	1.95	0.663
19	1.7	0.26	1.7	0.442
20	1.3	0.14	1.3	0.182
21	0.75	0.0	0.75	0.0
			Total	25.5

TABLE 10.5 Example of Stream-Flow Measurements and Calculations Site 1, Silver Bow Creek, August 27, 1998 [From Borduin (1999)]

The mechanical and electromagnetic flow meters are used to measure the water velocity. A third method or application, generally referred to as an acoustic Doppler current meter (ADV) is also widely used. The ADV meter uses sound waves to measure the velocity of particles suspended in the water. The meter has two transducers that generate and receive pulses so the reflection of an instrument derived acoustic signal off suspended particles can be used to compute a phase shift in the water column and a surface-water velocity. The SonTek/YSI Flow Tracker acoustic velocity meter used by the USGS can be mounted on a standard wading rod (Figure 10.22g). This meter measures both velocity and velocity direction making orientation of the meter critical (Turnipseed and Sauer 2010). An extension of the ADV methodology has resulted in the acoustic Doppler current profiler (ADCP) (Figure 10.22). This methodology produces a complete vertical profile (cross section) of the velocity distribution. These instruments compute three-dimensional velocities for the profile. When using ADV meters and the midsection method to compute the mean velocity is generated as the average velocity for the designated section. Profilers are deployed at the surface using rafts or boats (Mueller and Wagoner 2009).

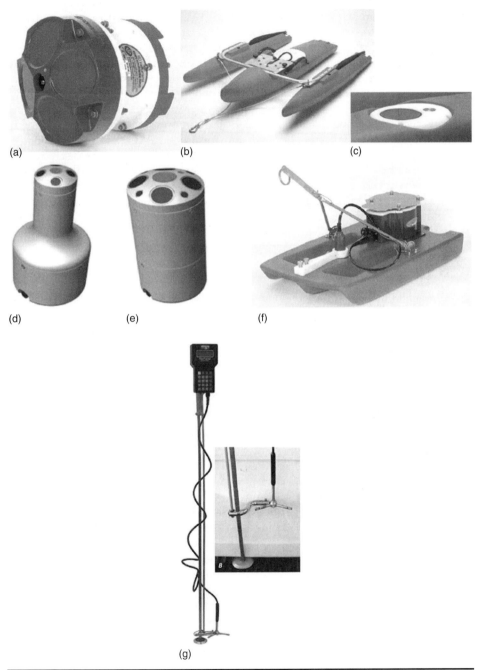

FIGURE 10.22 ACDPs presented by Turnipseed and Sauer (2010). (a) Teledyne RD Instruments 600 kHz RioGrande ADCP. (b) Teledyne RD Instruments 600 kHzRiverRay phased array ADCP. (c) Close-up view of phased array tranducer. (d) SOnTek/YSI RiverSurveyor S5. (e) Sonteck/YSIRIver Surveyor M9. (f) Teledyne RD Instruments Stream Pro ADCP. (g) SonTek/YSI Flow Tracker acoustic Doppler velocity meter.

When the variation in stream flow over time is required at a location, a stream gauging station can be established (e.g., Figure 10.23b). As stated previously, the location selection requires a relatively stable cross section either occurring naturally or constrained by a structure like a culvert or bridge. At the selected location a relationship between the stream stage and discharge measurements are derived by establishing a stage-measuring device and frequent stream gauging. The stage gauge can be as simple as a steel fence post driven in the stream bottom with a hardware store metal ruler attached that is surveyed to a reference, or more sophisticated stage measuring methods such a recording floats, transducers, bubblers, and other devices used at more permanent gauging stations (e.g., Sauer and Turnipseed 2010) (Figures 10.23a and 10.23b). At the gauging station the stage is recorded at each time a stream discharge measurement is made. It is important to gauge the stream under both low-water and high-water flow events. These data are then used to form a stage discharge relationship for the gauging site (Figure 10.23c). Once such a curve is established the discharge can be estimated by simply measuring the stage. The USGS uses this relationship to post real-time discharge information. If the cross section is relatively stable (cross section is not changing), the frequency of field measurements is reduced to a few per year.

Float Method

If no velocity-measuring equipment is available, a quick estimate of flow can be obtained by using the float method. In the float method, a distance of 10 to 20 ft (3 to 6 m) is paced off along the bank of a straight stream. If a bridge or other structure is used the width of the structure is the travel length. A float travel time should be at least 20 s where possible (Turnipseed and Sauer 2010). The cross-sectional area is determined by measuring

(a)

Figure 10.23 An example of stream gauging station and a stage discharge curve. (a) Staff gauge. (b) Example of a USGS stream gauging station. (c) Stage discharge curve. (*From USGS websites.*)

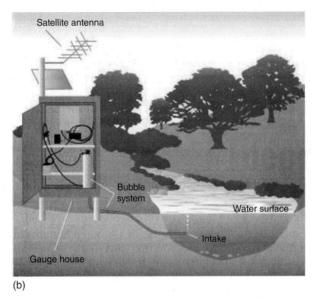

(b)

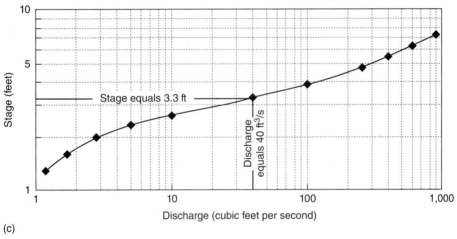

(c)

FIGURE 10.23 *(Continued)*

or estimating stream width and several depths across the stream where the flow "looks" average. The float can be made of almost anything including pinecones, ice pieces, debris, oranges, partially filed bottles, etc. A float is most representative of average flow if it is partially submerged, rather than something floating on the surface, like a small stick. As with the current-meter method, it is recommended that a number of intervals (representative areas) be selected and the float velocities determined. When streams are small or only rough estimates of discharge are needed, a single float may be appropriate. In this case the float is tossed into the middle of the stream above the first marking and timed to determine how long it takes to move from the first to the second marking. This yields an average velocity for the center of the stream. The float velocity is not the average stream velocity as seen in Figure 10.20. An estimate of the mean velocity needed for discharge measurements requires a correction by multiplying by 0.85 to 0.88 (Turnipseed and Sauer 2010). Discharge is then computed by multiplying the estimated velocity and the representative cross-sectional area, and then summing the subsectional discharges. If float methods are used during high-flow events, where the stream depth (cross section) is difficult to determine, high-water bank elevations can be staked and revisited when flows are lower and safer. Multiple float measurements under ideal conditions may have an error of ±10%, where measurements with few float runs can have errors of 25% (Turnipseed and Sauer 2010).

Crest-Stage Gauge Peak Discharge Measurement

At sites without gauging stations, a crest-stage gauge can be used to obtain a single estimate of the maximum stage that occurred during the time period of monitoring (monthly to annually) (Buchanan and Sommers 1968). A crest-stage gauge is a tube [steel or PVC 2 to 3 in (5.1 to 7.6 cm) in diameter] capped on both ends attached to a post or bridge abutment. It contains a measuring rod and a floatent like ground cork or a measuring rod coated with water-soluble ink or paint. The tube is constructed so that water can enter at the bottom and air is allowed to escape at the top. The inside rod will not move once the gauge is closed. The gauge elevation is surveyed as is the adjacent stream cross section. When stream stages rise and fall, the cork will float and adhere to the calibrated stick at the maximum stage, or the ink or paint will wash off that part of the rod that was underwater. This allows preservation of only the highest stage that occurred between observations. The gauge is reset by removing the cork from the rod or recoating it with a water-soluble material. Typically, the Manning equation is used to compute an estimate of the velocity at peak discharge. The Manning equation uses four parameters to estimate the velocity:

$$V = \frac{1.49 R^{2/3} S^{1/2}}{n} \tag{10.3}$$

where V = average estimated velocity (ft/s)
$\quad\quad R$ = hydraulic radius (ft^2/ft and R = area/WP)
$\quad$ Area = cross-sectional area of the stream at the crest stage
$\quad\quad WP$ = wetted perimeter (bank and bed length from the cross-section survey and crest-stage reading) (Figure 10.24)
$\quad\quad S$ = estimated slope of the water surface
$\quad\quad n$ = Manning roughness coefficient

The Manning equation (Equation 10.4), where length is measured in meters and time in seconds is

$$V = \frac{R^{2/3} S^{1/2}}{n} \tag{10.4}$$

The Manning roughness coefficient ranges from 0.025 for streams that are straight, without brush and weeds or riffles and pools to 0.15 for channels containing dense willows (Chow 1959; Phillips and Tadayon 2006). Some of the more common values for streams with grasses, rocks, and pools range from 0.035 to 0.045. See Chow (1959) and USGS publications [e.g., Barnes (1967) and Phillips and Tadayon (2007)] for tables of roughness coefficients and photographs of channel conditions with corresponding coefficients.

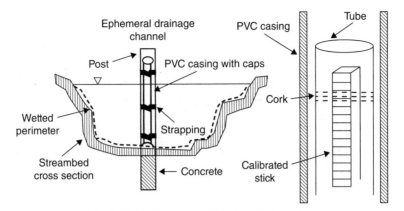

Figure 10.24 Crest-stage gauge attached to a metal fence post with metal strapping. Holes are drilled into the base of the PVC tube and at the top to allow water into the tube and air out. The diagram to the right shows a calibrated measuring rod. In this example, ground cork is placed in an open container or trough near the bottom of the rod. As water fills the tube and floods the cork reservoir, the cork floats. When the water level reseeds a line of cork is left at the high-water mark, producing a high-stage measurement. The wetted perimeter of the stream (dotted line) is defined as the cross-sectional length of the channel cross section below the stream surface. [*From Weight (2008).*]

Example 10.4   Crest-Stage Monitoring

While performing baseline studies for a prospective coal mine in southwestern Wyoming, it was necessary to obtain spring runoff peak discharges for several ephemeral drainages. None of the drainages contained water except in response to spring snowmelt and precipitation events. The crest-stage gauge design presented in Figure 10.24 was used at locations along each drainage to obtain upstream and downstream discharge estimates. Each post was set approximately 3 ft (0.9 m) into the stream channel in concrete. Each crest-stage gauge location was surveyed to obtain a cross-sectional area. Stream discharge values were derived once crest-stage data were obtained, stream cross-sectional areas estimated, and Manning equation velocities computed.

Engineered Structures for Determining Discharge

The two most common structures that can be placed in the stream channel to determine discharge are weirs and flumes (Figure 10.25). These engineered tools force the stream flow through a channel cross section of known dimensions so that only a stage measurement and appropriate equations are needed to determine the discharge. In some case these structures can be installed temporarily in small stream channels. In most cases they are installed to create permanent gauging data (e.g., irrigation ditches or diversions channels).

Weirs   A weir is a notched single plate placed perpendicular to the stream. Its sides and bottom are driven or excavated and sealed into the banks and bed. The stream is backed up behind the weir as water flows through the cutout portion of the plate. The V-notch weir has a notch in the form of a V cut at a known angle (e.g., 90°) (Figure 10.25a). The height of the water above the base of the V as measured in the pool behind the weir (not at the lip of the discharge) and is used to compute stream discharge (Equation 10.5). For a 90° V notch weir (Fetter 2001):

$$Q = 2.5h^{5/2} \tag{10.5}$$

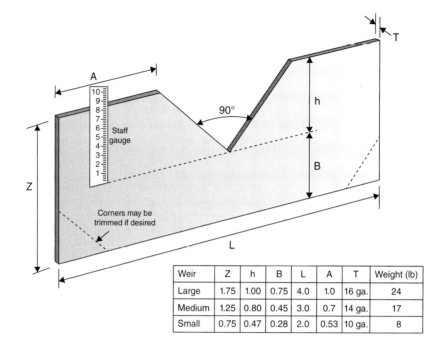

Weir	Z	h	B	L	A	T	Weight (lb)
Large	1.75	1.00	0.75	4.0	1.0	16 ga.	24
Medium	1.25	0.80	0.45	3.0	0.7	14 ga.	17
Small	0.75	0.47	0.28	2.0	0.53	10 ga.	8

(a)

FIGURE 10.25 Engineered stream discharge tools. (a) 90 and 20° V-notch weirs. The pool stage is measured so it does not reflect the hydraulics of the V-notch. (b) Parshall flume with point of stream stage measurement shown as h (https://www.openchannelflow.com/assets/ uploads/ blog images/12-inch-fiberglass-parshall-flume-b.png). Both of the field installation pictures show the need to support the instrument and prevent leakage underneath and around the sides of the instruments. [*Modified from Weight (2008); Turnipseed and Sauer (2010).*]

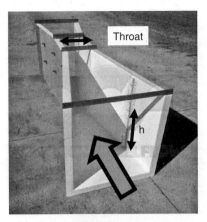

(b)

Figure 10.25   *(Continued)*

where Q is discharge in ft³/s; h is the head of the backwater above the weir base in feet (h in Figure 10.25a). When units are in meters and seconds, use Equation 10.6:

$$Q = 1.379h^{5/2} \tag{10.6}$$

Weirs come in a number of configurations such as rectangular, trapezoidal and broad-crested. Each has a unique equation to compute discharge. In some settings portable weirs constructed of sheet metal or plywood can be installed on small streams when a temporary or singular measurement of discharge is required. If a permanent weir is established it is important that it is designed for flood-stage events as well as low flows, otherwise it may be overtopped under the higher flows (see Example 10.5). Weirs block stream flow and collect sediment in the pool behind the weir. If the pool becomes filled with sediment, the weir will not function properly.

Example 10.5 Monitoring with a V-Notch Weir

A new subdivision was being proposed in an area of limited groundwater recharge in south-central Montana. A legal dispute between the developer and the long-time residents concerned about their groundwater supply ensued. The residents formed a water-users group and documented the existing groundwater levels in wells, and monitored spring and creek discharges over a period of 2 years. Creek discharge was monitored using 90° V notch weirs, with flows ranging between 5 gpm (0.32 L/s) and a peak runoff event of more than 400 gpm (26 L/s) (Figure 10.26). Ultimately, the data and 5 years of effort resulted in a decision to limit additional groundwater development.

(a)

(b)

FIGURE 10.26 90° weir used to measure flow rates between 5 and 405 gpm on Horse Creek in south-central Montana. (a) Low-flow conditions (<5 gpm). (b) High-flow conditions (>400 gpm). (c) Discharge over time graph. [*With permission Weight (2018).*]

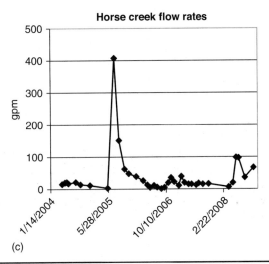

Horse creek flow rates

(c)

Figure 10.26 *(Continued)*

Flumes   Flumes are designed to line a channel and force flow through a box-like structure. A popularly used flume is the Parshall flume. The Parshall flume is essentially a tapered box with a rectangular cross section, flaring both upstream and downstream, and with a slight constriction in the middle (the throat) (Figure 10.25b). The total discharge is computed by knowing the depth of the water (h) and the width of the flume at the throat (University of Wyoming 1994; Turnipseed and Sauer 2010). The flume is designed to pass bedload so it should be free of sediment. Problems with measurements occur when obstructions are present and/or stream water leaks around the flume. Although Parshall flumes are made to rigid specifications, problems can occur with faulty installation or wear and tear. Because the relationship of flume width, water depth, and stream discharge is nonlinear, there is no single equation used to calculate stream discharge. Instead, the hydrogeologist must consult a Parshall flume table for the appropriate conversion (e.g., Johnson 1963; University of Wyoming 1994).

Tracer Injection for Small Discharges

Kilpatrick and Cobb (1985) describe using a continuous tracer injection method to determine discharges of small low-discharge streams. These low-flow streams are typically gaining streams and discharge is too low to measure using standard gauging equipment (depth of flow is too limited). It is important to stress that this method works well for reaches that are gaining and 1 km or less in length. Details on tracers, and their use in surface-water studies and surface-water exchange studies are described in additional detail in this chapter when hyporheic exchanges are addressed later. Equations for estimating the distance downstream of an injection site at which full mixing of a tracer and stream water occurs are addressed in a following section and described in Hubbard et al. (1982).

Tracer methodology, used to estimate stream discharge in small streams, starts by sampling the stream for the background concentration of the tracer selected for use (e.g., KBr, NaCl, specific conductance, fluorescent dyes, etc.) (Kilpatrick 1970) (Chapters 13 and 14). Once the tracer is selected a reservoir is filled with the tracer solution and kept well mixed. It is sampled to provide a value of the tracer concentration. Then a very accurate feed pump (such as a one made by Fluid Metering Inc. (www.fmipump.com)) is installed and operated to pump tracer into the stream at a known discharge (flow rate). Downstream of the tracer introduction site, where the tracer is fully mixed into the flow (see suggestions of Hubbard et al. 1982, and the discussion of tracer use to characterize exchange further along in this section), samples of the tracer concentration in the stream are recorded over time until tracer concentrations stabilize. Once a steady-state concentration is observed downstream, the estimated value of the tracer concentration in the stream water is generated.

Stream flow is calculated using the following simple relationship: Total stream flow downstream × (concentration of tracer measured) = upstream flow entering the tracer site × (background concentration of tracer) + Flow rate of tracer injection × (concentration of tracer). If the background concentration is *not* zero and recognizing that the total stream flow downstream is $(Q_{upstream} + Q_{injection})$, then the concentrations related to flow rates are shown algebraically in Equation 10.7:

$$(Q_{upstream} + Q_{injection}) \times C_{tracer\ downstream} = (Q_{upstream} \times C_{background}) + (Q_{injection} \times C_{tracer}) \qquad (10.7)$$

It follows then that:

$$(Q_{upstream} \times C_{tracer\ downstream}) - (Q_{upstream} \times C_{background})$$
$$= (Q_{injection} \times C_{tracer}) - (Q_{injection} \times C_{tracer\ downstream})$$

For example, if the tracer concentration introduced is 100 mg/L and the injection rate is 1 L/min, and the background concentration of the tracer is 10 mg/L and the downstream total tracer concentration is 25 mg/L then:

$$(Q_{upstream} \times 25\ mg/L) - (Q_{upstream} \times 10\ mg/L) = (1\ L/min \times 100\ mg/L) - (1\ L/min \times 25\ mg/L)$$

$$Q_{upstream} \times 15\ mg/L = 75\ mg/min,\ so\ Q_{upstream} = (75\ mg/min)/(15\ mg/L) = 5.0\ L/min.$$

If a tracer is used where the background concentration *equals* zero then the relationship simplifies to Equation 10.8:

$$(Q_{upstream} + Q_{injection}) \times C_{tracer\ downstream} = (Q_{injection} \times C_{tracer}) - (Q_{injection} \times C_{tracer\ downstream}) \qquad (10.8)$$

It follows then that:

$$(Q_{upstream} \times C_{tracer\ downstream}) - (Q_{injection} \times C_{tracer\ downstream})$$
$$= (Q_{injection} \times C_{injection}) - (Q_{injection} \times C_{tracer\ downstream})$$

Rearranging and dividing through by $C_{tracer\ downstream}$ results in:

$$Q_{upstream} = (Q_{injection} \times C_{injection})/(C_{tracer\ downstream}).$$

This simplifies to Equation 10.9:

$$Flow\,(L/min) = \frac{conc.\,injection\,rate\,(mg\,tracer\,/\,min)}{tracer\,concentration\,in\,downstrem\,sample\,(mg\,/\,L)} \qquad (10.9)$$

For example, if the background tracer concentration of the tracer injected is zero, the injection rate is 1.2 L/min, the tracer concentration is 200 mg/L, and the downstream tracer steady state value is 53 mg/L, the stream flow would be computed as:
Flow (L/min) = (1.2 L/min × 200 mg/L)/ 53 mg/L = 4.5 L/min.

This methodology is very useful when attempting to characterize discharge in streams with very low discharges.

Example 10.6 Stream Flow Measurement Using a Tracer

A study was conducted to determine the source of zinc loading at the reclaimed Comet mine site up High Ore Creek in southwestern Montana (Sudbrink 2007) (Figure 10.27). The stream flow during spring runoff was measurable using gauging equipment (>2 cfs, 56.6 L/s), however, by late summer

(a)

Stream Flow Estimates on High Ore Creek in Southwestern Montana Using the Continuous Injection Tracer Method. [*From Sudbrink (2007)*.]

	July 28, 2005		August 25, 2006	
Station	Tracer cfs	Flow meter	Station	Tracer cfs
HOC 1	0.19	0.13		
HOC 2	0.19			
HOC 3	1.9			
HOC 4	0.26		HOC 4	0.246
HOC 5	0.26		HOC 5	0.250
HOC 6	0.27		HOC 6	0.275
HOC 7	0.27		HOC 7	0.278
HOC 8	0.28		HOC 8	0.283
HOC 9	0.29		HOC 9	0.305
HOC 10	0.30		HOC 10	0.325
HOC 11	0.32		HOC 11	0.344
HOC 12	0.33	0.23	HOC 12	0.379
Gain	0.14	0.10	Gain	0.133

(b)

FIGURE 10.27 Dye tracer test setup and discharge results. (a) Rhodamine WT dye was mixed and stored in a 5-gal jug 150 m above the first station (HOC 1). HOC 1 to HOC 12 represent 12 downstream stations, each spaced at 150 m apart. A continuous injection of the solution of rhodamine WT dye was accomplished using a battery-powered feed pump. The delivery tube is attached to the steel fence post driven into the stream bottom. At a downstream site, the tracer concentration in the stream was measured and discharge computed. (b) Table of results showing computed discharges at varying distances downstream (e.g., HOC 4 to HOC 12 furthest stations downstream). [*Weight (2008)*.]

the flow dropped to a fraction of a cfs (<28.3 L/s), which was estimated using the continuous injection tracer method. Two tracer injection tests were performed (July 28, 2005 and August 25, 2005) to compare flow data at 12 gauging stations under a gaining stream scenario. The injection concentration injection rate ($C_{injection}$) on July 28, 2005 was 12 mg/min and 13.5 mg/min on August 25, 2005. The tracer was injected 150 meters upstream from station HOC 1. It was determined that after 6 h the stream had equilibrated at which time stream concentrations were recorded. Total creek discharge was determined at each of the 12 stations using Equation 10.9 (Figure 10.27).

Synoptic Survey

When investigations of an exchange in a stream segment are required, a seepage run or synoptic survey is often completed (Figure 10.28). The stream is divided into sections and discharge is measured at the beginning and end of each section. All water inputs and outflow are accounted for [gauged ditches and tributaries, direct evaporation (ET), etc.].

The selection of the lengths of the stream reach chosen for analyses may be constrained by field conditions and/or the ability to clearly differentiate measurable changes in stream flow. Generally the length of stream between measurements has to be sufficiently long so that a change in discharge of between ±5% (good field conditions) to ±10% (poorer field conditions) is detected in the measurements. For example, if the upstream measurement was 95 cfs (2.66 m³/s) and the lower reading was 101 cfs (2.82 m³/s), and the measurement error was 5% then the 95 cfs (2.66 m³/s) measurement is really 95 ± 4.9 cfs (2.66 ± 0.14 m³/s) and the downstream measurement is 101 ± 5 cfs (2.82 ± 0.14 m³/s). Because the error ranges overlap, no change in stream flow (groundwater exchange) can be reported. Additional sources of error will occur if surveys are conducted as flows are changing (dynamic conditions) (Roberts and Warren 1999). It is important to identify and gauge all surface-water inputs or outputs. This includes irrigation diversions and irrigation return flows. In the semi-arid western United States, irrigation ditches are extremely common and are often difficult or impossible to locate from a topographic map. Ditches are sometimes easier to spot in the field or from an aerial photo. However, there is no substitute for simply walking the entire survey site and locating significant features with a GPS. In a synoptic study of Big Lake Creek, south of Wisdom, in southwestern Montana, a 10 mile (16 km) downstream traverse uncovered 10 to 12 irrigation diversions with significant flow!

It is evident that the net gain or loss of stream water is determined by subtracting the beginning discharge value from the downstream value plus or minus sources or sinks (inflows or outflows). If the net change is positive, groundwater has discharged to the stream, a gaining stream; if the net change is negative, the stream is losing water (recharging to the groundwater system). The measurement of net change assumes steady-state conditions such that flows are not changing at a given gauging location. Whenever possible, surveys should be completed over a short time period, e.g., a single day. In some settings riparian vegetation limits groundwater flow into the stream or enhances stream water flow into the groundwater (phreatophytes creating ET water losses and locally altering the water-table position) (Weight and Chandler 2010). In colder climates some synoptic surveys are purposely planned to occur after killing frosts have shut-off plant transpiration. Thus, the described net exchange values assume that ET effects are insignificant.

Monitoring Wells and Flow Field Analyses

Monitoring wells and monitoring well networks from which water-level data are collected and interpreted are essential components of any hydrogeologic investigation including those centered on groundwater/surface-water exchange analyses. When

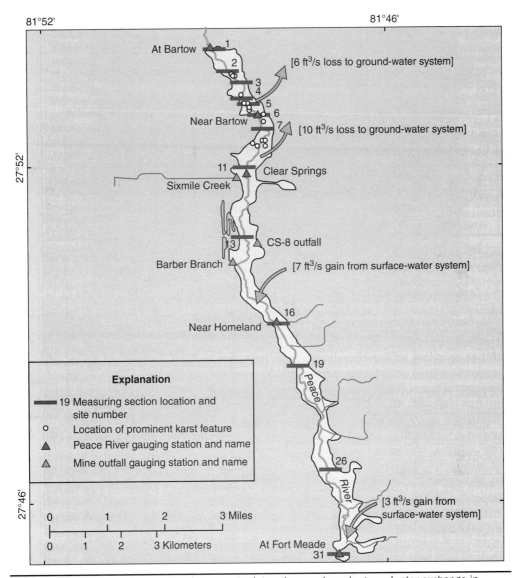

81°52'   81°46'

At Bartow
[6 ft³/s loss to ground-water system]

Near Bartow
[10 ft³/s loss to ground-water system]

Clear Springs
Sixmile Creek

CS-8 outfall
Barber Branch
[7 ft³/s gain from surface-water system]

Near Homeland

Explanation

— 19 Measuring section location and
        site number
  o   Location of prominent karst feature
  ▲   Peace River gauging station and name
  ▲   Mine outfall gauging station and name

Peace

River

[3 ft³/s gain from
surface-water system]

At Fort Meade

0   1   2   3 Miles

0   1   2   3 Kilometers

FIGURE 10.28 Seepage run or synoptic survey to determine reach scale groundwater exchange in Florida, the United States. A complete water budget is needed for each defined reach to correct stream flow measurements for changes in flow not related to groundwater exchange (*Knochenmus 2004*).

shallow unconfined groundwater systems are instrumented with a monitoring well network, comparisons of surface-water feature stages and groundwater heads allow interpretations of groundwater relationships with streams, lakes and wetlands (Figure 10.29). Shallow wells that penetrate the water table are used to interpret horizontal groundwater gradients and groundwater-flow directions (e.g., Freeze and Cherry 1979). Vertical groundwater gradients are obtained when wells adjacent to each other are completed at various depths in the same groundwater system (Section 4.3). The differences in heads are

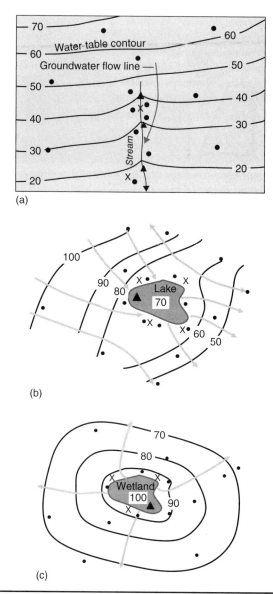

Figure 10.29 Map views of groundwater monitoring networks and interpreted groundwater-flow interaction with surface-water features. Labeled black lines are equipotential lines representing the shallow water table. Black dots are shallow monitoring well locations. Arrows are the interpreted direction of groundwater flow. Xs are locations of nested monitoring wells providing vertical-gradient information. Black triangles represent surface-water stage-measurement locations. (a) Gaining stream where groundwater flows to the stream. Vertical gradients would be upward toward the stream. [*Modified from Healy et al. (2007).*] (b) Flow-through lake where groundwater enters the northwest side of the lake and exits at the southeast shoreline. Vertical gradients would be upward at the northwest shore, downward at the southeast shore, and near neutral at the two remaining sides. [*With permission Woessner (2018).*] (c) A wetland losing water to the groundwater as groundwater flows away from the feature on all sides. Vertical gradients would be downward at all locations. [*With permission Woessner (2018).*]

divided by the vertical distance of the midpoint to midpoint of the perforated intervals of each well. Groundwater flow is upward when the deepest well has a head higher than the shallowest well. When multiple wells are located close together and completed at varying depths, they are often referred to as monitoring well nests or nested piezometers (see also Chapter 15).

When investigating groundwater exchange some of the monitoring wells should be placed adjacent to surface-water features. This allows for direct comparisons of groundwater levels and surface-water stages (Figure 10.29). However; these investigations also benefit from wells installed at greater distances from the feature so that the surface-water feature can be placed within the framework of the larger groundwater system.

Mini-Piezometers

When investigations require the collection of groundwater head data within the stream channel, lake littoral zone, or wetland bottom, a scaled-down monitoring well has been developed. Certainly, in some large-scale investigations, drill rigs may be mounted on barges and wells installed using conventional drilling or auguring techniques; however, this approach is expensive and logistically complex. Researchers studying lakes and streams have developed a small-diameter monitoring tool referred to as a mini-piezometer or piezometer that can be driven into river, lake, and wetland bottoms, and shorelines to gather groundwater head data and surface-water stage information (e.g., Lee and Cherry 1978, Cox et al. 2005, Buss et al. 2009). These instruments provide information about how the groundwater system (e.g., local, intermediate, and regional) interfaces with the surface-water feature and can be used to identify the fine-scale hyporheic exchange of surface water circulating within the beds, banks, and floodplains of surface-water features (e.g., Woessner 2000; Buss et al. 2009).

Mini-piezometers are typically less than 2.0 in (5.1 cm) in diameter (Figure 10.30). They may be 3 to 6 ft (1 to 2 m) in length and are often hand-driven from the water surface a few feet (~0.75 m) into the surface-water feature bed, banks, or shorelines (Figure 10.31a). Selection of the mini-piezometer diameter will depend on its intended use: water-level monitoring (including transducer installation), water-quality sampling and/or preliminary aquifer-property determination, and the degree of difficulty of installation.

Mini-piezometers can be constructed of variety of materials including steel, PVC, galvanized electrical conduit, and flexible clear tubing (Rosenberry et al. 2008). Electrical conductor casing and setups using a steel conductor casing and interior drive rods are often used (Figure 10.30). When driving casing, care must be taken to not bend or crimp the top. Driving with a block of wood placed over the top of the casing can provide some protection. In Figure 10.30a, a special cap was built that fits over both the conductor casing and driving rod so that the driving force does not damage the casing top. When the conductor casing with a solid-center driving-rod method is used, the driving rod is often designed with a top that is rounded and overlaps the conductor casing. The conductor casing center-rod apparatus works well and one initial setup may be used to install tens of mini-piezometers. These installations most often use PVC as the mini-piezometer. When bed sediments are cobble-dominated, hand driving may be ineffective and mechanical assist tools like a gas-powered jackhammer may be needed (Figure 10.31b). When steel conductor casing tubes are installed and need removal, sometime conditions preclude hand pulling. In such situations, a jack and chain or fence post puller

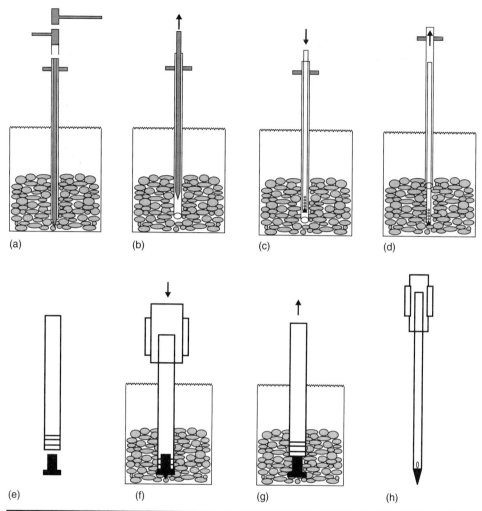

Figure 10.30 Installation of mini-piezometers. (a) Electrical conductor casing with solid center rod driven to depth. (b) Center rod withdrawn. (c) Mini-piezometer inserted. (d) Conductor casing removed. (e) Ridged tube with perforations and loose bolt inserted to protect the end during pounding. (f) Fence post pounder used to install the mini-piezometer. (g) Once the planned depth is reached, the well is pulled back up a few inches so that the bolt is released and the mini-piezometer is connected to the saturated sediments. (h) Ridged steel pipe with a welded steel pointed tip. Pipe is driven to depth and used as the mini-piezometer. [*Modified from Woessner (2017).*]

may be required (Figure 10.31c). Weight (2008) describes what he calls the plate-jack method to extract casing that cannot be hand pulled. A ¼ in (0.63 cm) steel plate about 1 by 2 ft² (0.3 to 0.6 m²) with a slot greater than the diameter of the conductor casing is torch-cut out from one side to center. A handyman jack is bolted to the plate and a length of chain is used to capture the casing and wrap around the jack-lift plate. By working the handyman jack the conductor casing can be extracted.

(a)

(b)

FIGURE 10.31 Mini-piezometer installation and when necessary removal. (a) Hand-driven steel mini-piezometer with welded tip being installed in the Clark Fork River near Missoula, Montana. [*With permission Woessner (2018).*] (b) Power drilling a mini-piezometer in the bed of the cobble-rich Gallatin River, Montana. (*Photo courtesy of Steve Custer*). (c) Plate-jack method to remove piezometers or electrical conductor casing under difficult conditions (*Weight 2008*).

(c)

Figure 10.31 *(Continued)*

Field conditions often include operating in the surface-water environment. An approved safety plan needs to be developed, including a requirement to work in pairs and the use of appropriate personal floatation devices while installing equipment. Researchers should also be aware of current and wave conditions, water depths, changing flow conditions, and the presence debris in the water body.

Piezometers are often finished with the bottom end open to the sediment and/or a short section of drilled hole or saw slot perforations near the bottom (Figure 10.32). Once a piezometer is installed, the bottom sediments are tamped around the casing to prevent surface water from short circuiting along the casing. In addition, it is important of be sure the mini-piezometer is freely connected with the saturated sediments. This is achieved by measuring the water level in the piezometer, extracting a volume of water from or adding a volume of water to the small diameter well, and then remeasuring the water level. After a period of time if the piezometer is well connected to the groundwater, the water level should return to the original static value. In some instances it may useful to pump water from the mini-piezometer or surge the well by raising and lowering a solid rod (or willow branch) in the mini-piezometer to ensure that perforations are not plugged (Figure 10.33).

Mini-piezometers are used to compare the head in the saturated bed, bank, and shoreline sediment with the surface-water stage. These data allow for the determination of the vertical hydraulic gradient (VHG), and exchange directions at the monitoring point. The water level in the mini-piezometer is measured using mechanical and electric tapes, transducers, and/or manometers (Figure 10.34). The difference in the stage and head measurement is used to suggest the direction of exchange at the location (Figure 10.35). If the head is lower than the surface-water stage, surface water is flowing into the sediments. If the stage is lower than the head, groundwater is moving into the stream. VHGs can be quantified by dividing the difference in head (water level in the piezometer minus the surface-water stage), by the distance below the bed to the bottom of the piezometer (open-hole completion) or the distance to the midpoint of the

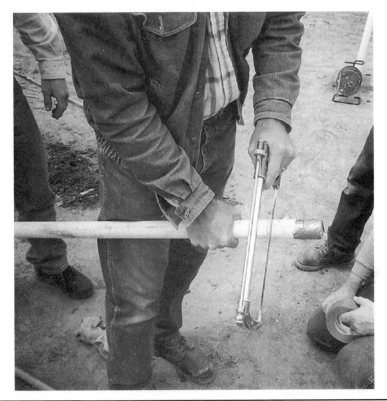

Figure 10.32 Mini-piezometers can be perforated over a short interval using a hacksaw. In this case, the bottom is "capped" with duct tape.

perforated interval (Figure 10.35). A positive VHG indicates effluent conditions and a negative VHG implies influent conditions.

Mini-piezometers can also be used to determine horizontal hydraulic conductivities of the sediments by conducting rising-head or falling-head slug testing (Chapter 7). Ideally, min-piezometers need to be perforated over a short interval (e.g., 0.5 ft or 15 cm) with sufficient perforations so that the water can freely enter and leave the piezometer without it becoming plugged (Figure 10.32). Then a known volume of water is either removed or added to the piezometer and water-level changes in the well over time are monitored. A transducer installed in the mini-piezometer set to record at a rapid-linear logging scale will capture the quick changes in water levels and is a good approach (Section 4.3). Analyses account for open-area design, the rate of change of head and hydraulic conductivity values are often completed using equations by Hvorslev (1951) and Bouwer and Rice (1976) (Chapter 7). To convert horizontal hydraulic conductivities to vertical hydraulic conductivities, horizontal values are often reduced by an order of magnitude. However, horizontal to vertical ratios can vary widely (Anderson et al. 2015).

Example 10.7   Head and Flux Measurements Using Mini-Piezometers
Farinacci (2009) was tasked with establishing the groundwater/surface-water exchange along 11 km of the Clark Fork River, a large gravel-bedded stream (average flow 2,200 cfs, ~60 m^3/s) near Milltown,

Figure 10.33   Photo along the Milk River in northern Montana, where Willis Weight is improvising by using a willow stick as a surge tool to obtain connectivity of the mini-piezometer with the saturated sediments. If the mini-piezometer is filled with water to the point of crowning and does not change in level, a hydraulic connection was not accomplished and surging action is needed to unclog the screen slots.

MT (Figure 10.36). His characterization was used as input for a regional groundwater model developed for the assessment of how the removed of Milltown Dam would impact the shallow valley-wide unconfined groundwater system.

He constructed mini-piezometers from ¾ in (1.9 cm) diameter black steel pipes and established VHGs at 23 sites (Figure 10.36a). Fifteen falling-head tests were conducted at select mini-piezometer locations and horizontal hydraulic conductivities were determined (see Chapter 7 for methods). Vertical hydraulic conductivities were estimated by assuming a 10/1 ratio for K_h/K_v (e.g., Anderson et al. 2015). Temperature monitoring and modeling was used to determine flux rates at 11 of the 23 sites. Flux rates computed from VHGs and hydraulic conductivities ($K_v \times$ VHG) were compared and contrasted with results of temperature modeling (see temperature methods discussed in this chapter). Final values were assigned to streambed segments by splitting the distance between the adjacent measuring-point locations (Figure 10.36b). These data were used as initial input to the groundwater model. Field-measured exchange parameters were allowed to vary within reasonable ranges during flow model calibration.

Seepage Meters

The flux between groundwater and surface water can be measured directly using a seepage meter (Figure 10.37). A seepage meter is constructed by inserting an open-bottomed container into the lake, stream, or wetland sediments and then measuring the time it takes for a volume of water to flow into or out of a thin-walled bag connected to the

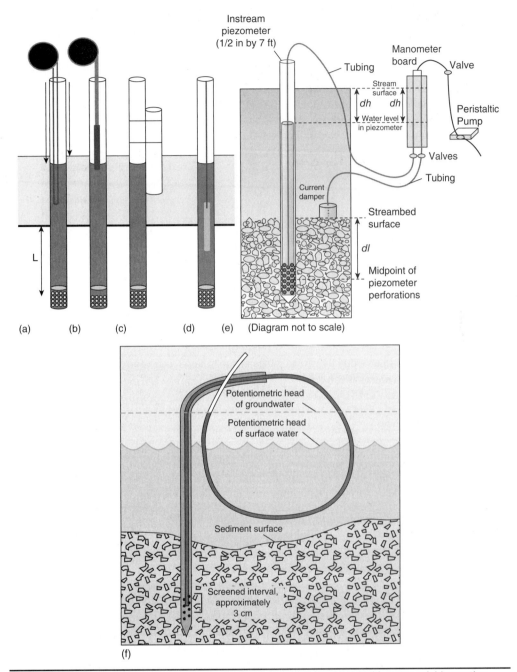

Figure 10.34 Measuring water levels in mini-piezometers. (a) Steel tape with water-soluble coating. (b) Electric water-level tape. (c) Mini-piezometer with a second hollow tube extended below the surface-water stage that forms a stilling well for a more accurate measurement of the stage. (d) Installation of a transducer. (e) Use of a manometer board, an open-ended clear flexible loop of tubing that is submerged both in the stream and in the well. As it extends above the water, it is fitted with a T where suction is applied. The tube is mounted on a board (as shown) with parallel rulers and the difference in water levels is directly read as both water levels are raised equally above the water surfaces (Cox et al. 2005; Woessner, 2017). (f) Mini-piezometer constructed from a commercial root feeder with a coil tube acting as a manometer (*Rosenberry et al. 2007*).

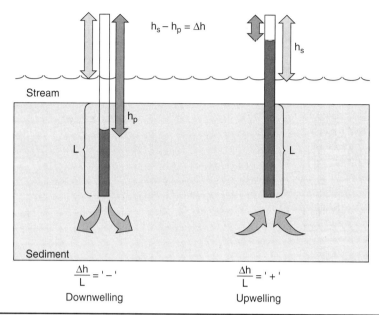

$$h_s - h_p = \Delta h$$

$$h_s$$

Stream

$$h_p$$

L

L

Sediment

$$\frac{\Delta h}{L} = ' - '$$

$$\frac{\Delta h}{L} = ' + '$$

Downwelling

Upwelling

Figure 10.35 Comparison of the stream stage and mini-piezometer head as measured from the top of the mini-piezometer. hs is the stream stage and hp is the mini-piezometer head. The vertical hydraulic gradient is computed by differencing the two measurements as shown and dividing the results by the depth of penetration of the mini-piezometer into the bed (*Woessner 2017*).

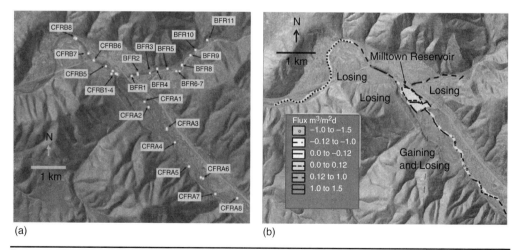

(a)

(b)

Figure 10.36 A research project conducted by Farinacci (2009) where he characterized the exchange of groundwater and Clark Fork River (flowing from the southeast to the northwest) using mini-piezometers, falling-head tests, and temperature studies in the vicinity of the Milltown Reservoir [area of light gray shading in (b)] Superfund site just east of Missoula, Montana. (a) Locations of mini-piezometer sites. (b) The dashed and solid lines in the 11-km study area represent reaches he assigned as losing and gaining. The negative values in his work indicated that groundwater was discharging from the groundwater system creating effluent or gaining conditions. His work was used to calibrate a numerical groundwater model of this site.

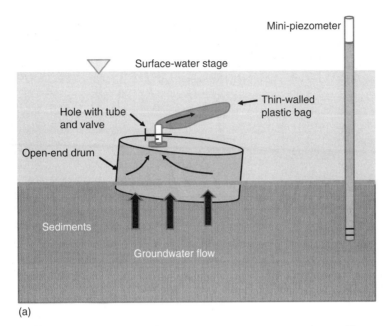

(a)

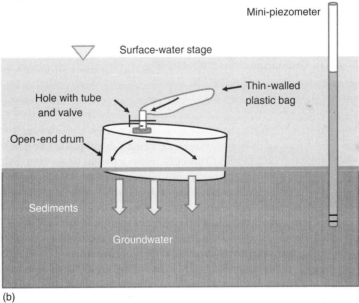

(b)

FIGURE 10.37 Seepage meter design consists of an open-end container (like a 55-gal drum top) that is outfitted with a valve and thin-walled bag. The bag is used to collect seepage or release water when the drum edges are inserted into the bottom sediment. Seepage meters have been used successfully in streams, lakes, wetlands, and oceans. (a) Installation in an area with effluent groundwater. (b) Installation in an area where influent conditions are present. The bag must be prefilled with a known volume of water and once the meter is installed, the change in volume over time is computed. (c) Seepage meter installed in an effluent shallow-water setting. The value and bag are fitted to the side. A hollow tube is installed through the top of the meter to release gases that may collect in the meter from interfering with seepage rates. In this setting, a mini-piezometer is installed nearby each meter to determine flow directions and VHGs. [*With permission (Woessner 2018).*]

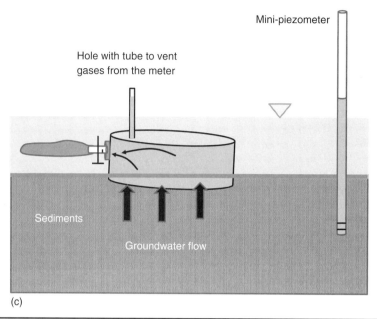

Mini-piezometer

Hole with tube to vent
gases from the meter

Sediments

Groundwater flow

(c)

FIGURE 10.37 *(Continued)*

container (e.g., Israelsen and Reeve 1944; Lee and Cherry 1978). Zamora (2008) provides an extensive list of authors reporting on the use of seepage meters in stream, lake, and wetland settings. Seepage meters were originally used to evaluate exchanges in irrigation ditches and lakes.

A seepage meter is constructed of a container open at the bottom (Figures 10.37 and 10.38). Good results can be obtained by cutting a 5-gal (18.9 L) bucket in half horizontally (Weight 2008) or more traditionally using a 55-gal (0.208 m³) drum with sides cut to approximately 1 ft (30 cm) in length. Generally, the larger the surface area of the container the more representative the seepage rate. A hole is drilled into the container and fitted with a plastic barbed fitting or rubber stopper and tube. Tygon or plastic tubing attached to a valve (or clamp) to provide a site to attach the bag. Seepage meters seem to work best in lakes, wetlands, or surface streams with low current velocities that do not have significant gravel or rocky armoring.

The seepage meter is installed by slowly pushing and twisting the container back and forth into the sediments with the container slightly tilted (Figure 10.37). It is essential that the container cavity be filled with water and no air bubbles. The bag is attached to the plastic tubing with a valve or mechanical clamp. Most commonly the collection bag is prefilled with a measured volume, for example, 100 to 1,000 mL of water, to limit "bag memory" (e.g., Shaw and Prepas 1989). All the air should be removed from the bag. Weight (2008) reports that a deflated mylar balloon works well because it is strong, lightweight, and has a narrow filling access port that is easily clamped to the tubing. Zamora (2008) has also tested bag types and provides recommendations. The tubing diameter has to be large enough so that flow is not limited (Zamora 2008). Bags are often attached to the tubes using rubber bands wrapped around the tube or a hose connector that snaps into the tube above the valve. The bag should be submerged so that the head at the bag is equal to the surface-water head.

Once the bag is installed and the valve is open, the meter is operating. At a later time the valve is closed and the bag is retrieved without allowing water to leak into or from the bag. The volume change in the bag is determined, and the flux rate [vol/(time (meter area)] is calculated by dividing the volume – the prefill volume/time (L^3/T) by the collection area of the meter. Volumes can be measured by weighing or with a graduated cylinder. This generates a flux value that is either positive or negative depending on whether water flowed into the bag or out of it. Water-quality samples can be collected in gaining locations once the original surface water trapped during meter installation has been flushed by the inflowing groundwater. To collect a sample, a deflated bag is placed on the meter. Care must be taken that air is not present in the bag or the sample geochemistry may not be representative.

Rosenberry et al. (2008) discuss in detail how to avoid instrument issues (some previously discussed) when using seepage meters (Table 10.6). Rosenberry and Menheer (2006) also discuss a system they developed to calibrate seepage meters. When seepage rates are low, multiple meters can be interconnected to a single bag to better measure seepage rates (Figure 10.38c). Rosenberry (2005) describes this approach as ganged meter setups. Seepage meters use in larger rivers and in some tidal situations are affected by the hydraulics of water flowing over the bag and the potential for water short circulating under the edge of the meter. Current effects on bags have been addressed by some researchers who suggest methods of shielding the bag (e.g., Libelo and MacIntyer 1994; Zamora 2008) (Figure 10.38b). Zamora (2006) also notes operational short circuiting, including changes in surface-flow conditions that result in meter scour, and sediment burial. If a long tube is used to set up a bag out of a stream current (e.g., in a more protected bank area), it should be at the same stream stage as the meter. In lake settings, Lee and Cherry (1978) suggest that a flexible perforated tubing can be installed into the bedding materials to act as a mini-piezometer. This can be operated like a seepage meter by attaching a prefilled bag to the flexible tube. The bag should be attached so that it remains below the water surface. By observing the change in bag

---

Sources of Error

Sources of error when using seepage meters include:

1. Incomplete seal between seepage-meter chamber and sediments, unstable cylinders;
2. Insufficient time between meter installation and first measurement;
3. Improper bag-attachment procedures, bag resistance, and moving water;
4. Leaks;
5. Measurement error;
6. Flexible seepage-meter chamber;
7. Insufficient or excessive bag-attachment time;
8. Accumulation of trapped gas;
9. Incorrect coefficient to relate measured flux to actual flux across the sediment-water interface; and
10. Insufficient characterization of spatial heterogeneity in seepage through sediments.

TABLE 10.6   Rosenberry et al. (2007) List of Sources of Error When Using Seepage Meters. They Discuss Each Source of Error in Their Publication

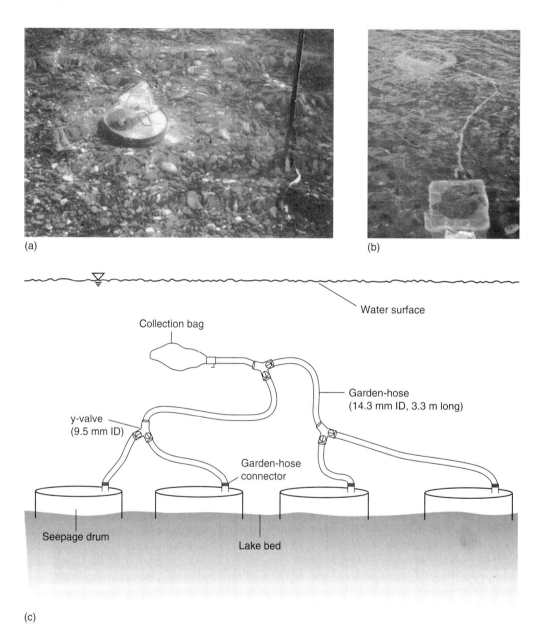

(a)

(b)

(c)

FIGURE 10.38 Seepage meters. (a) 55-gal drum seepage meter deployed in Flathead Lake, Montana. [*With permission Woessner (2018); Weight (2008).*] (b) Seepage meter deployed in Lake Lacawac, Pennsylvania, with a plastic box protecting the collection bag (Heaney et al. 2007). (c) Ganged seepage meters used to collect seepage from multiple meters (*Rosenberry 2005*).

volume over time and estimating the seepage area as the bottom open area of the tube, a flux rate is estimated. However, in this case only a very small sampling area is being used and it will likely not be very representative of seepage rates over a wider area.

When the seepage meter is paired with a mini-piezometer and a VHG is determined, the vertical hydraulic conductivity of the bed sediment can be computed using Darcy's law. The hydraulic gradient [VHG = I, change in head per depth below the bottom (l/l)] and the seepage-meter flux rate q yields:

$$K_v = q/I \tag{10.10}$$

where q = flux, (L³/L²T), and
K_v = the vertical hydraulic conductivity (L/t)

The average linear vertical seepage velocity (V_s) is determined by multiplying K_v by the VHG (I) and dividing by the effective porosity, n_e, of the unconsolidated sediments (e.g., 25% to 35% for sand rich settings) (Chapter 3).

$$V_s = \frac{K_v}{n_e} I \tag{10.11}$$

Example 10.8   Seepage Meters and Mini-Piezometers
A graduate class studying surface-water/groundwater exchange investigated seepage along the shoreline of Frenchtown Pond State Park near Frenchtown, Montana (Figure 10.39a). The pond was excavated in sand and gravel deposits to supply gravel for construction of the adjacent interstate. The pond surface represents the local water table. Seepage meters were constructed from 1 ft (30 cm) diameter steel drums cut so the tops and bottoms could be used. Meter sides were about 30 cm in length. A rubber stopper and hard plastic tube were sealed into the meter top. A thin-walled food storage bag was prefilled with 100 mL of water, then a 4 in (10 cm) long piece of tygon tubing was inserted into the bag corner and a rubber band used to seal the bag to the tube. Air was forced out of the bag and a hose clamp was used to pinch the tube closed. Next, meters were installed by pushing and turning them into the sediments about 3 to 4 in (7 to 10 cm) (Figure 10.39b). Final installation produced a slight tilt to the meter (Figure 10.37). Bags were installed, unclamped, and left in place for about 1 h. Care was taken so bags remained completely submerged (sometimes a small rock was placed on the corner of the bag on top of the meter to keep it submerged). After a measured time the clamp was installed on the tygon tubing and the tubing and bag were removed for measurement. The volume of water in the bag was measured and the prefilled volume subtracted from the total. Seepage rates were computed as volume/(meter area (sampling time)) (Figure 10.39c). During the seepage meter operation 5 ft (1.5 m) long mini-piezometers constructed of 0.5 in (1.27 cm) diameter electrical conduit and fitted with a loosely fitting carriage bolt were driven into the pond bottom near the meters to about 1 to 1.5 ft (30 to 50 cm) using a fence post driver (Figures 10.39b and 10.31a). Each piezometer was twisted and pulled back about 2 in (5 cm). The sediment around the piezometer was tamped down using the wading boot shoe.

Water-level measurements were taken with either an electric tape or a small-width carpenter's tape marked with water-soluble marker. Then the mini-piezometers were pumped (developed) for 10 to 15 s and a second water-level measurement was taken. Measurements were repeated until levels stabilized to ensure that the piezometers were well connected to the sediments and water levels represented groundwater heads. VHGs were computed by determining the head difference between the pond level (top of the pipe to the pond-water level) and the groundwater head (top of pipe to inside water level) and then dividing this value by the length of the piezometer penetrating the pond-bottom sediments (Figure 10.35). After computing both VHGs and the seepage meter rates, q = groundwater flux (L³/(L²T), vertical hydraulic conductivities were computed as q/(VHG). Results of field work form March 27, 1999 are reported in Figure 10.39c.

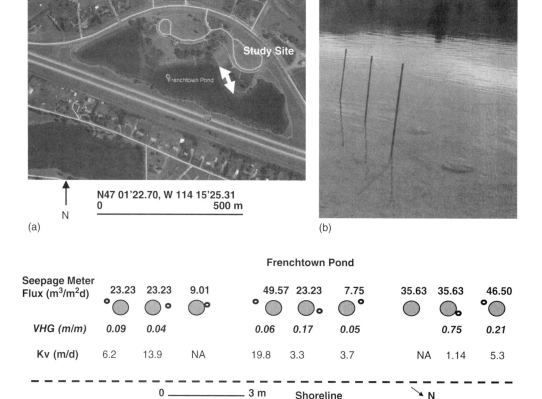

Figure 10.39 A March 27, 1999 seepage meter and mini-piezometer study of Frenchtown Pond State Park, Frenchtown, Montana. (a) A site map showing the location the shoreline study site (double white arrow). (b) Seepage meter and mini-piezometer installation. (c) A shoreline map showing the approximate location of seepage meters (gray circles) and mini-piezometers (open black dots). Groundwater was discharging into the pond and all seepage values are positive. The study results are presented as seepage rates above the meter symbols; VHGs are located below the meter symbols and the computed vertical hydraulic conductivities, Kv, are presented. Note the seepage, VHG, and Kv values are quite variable reflecting the heterogeneous nature of sand and gravel deposits and pond-bed sedimentation. [*With permission Woessner (2018)*.]

Modifications of the traditional seepage-meter methodologies have included the use of a flow-through chamber to record flow rates directly as water enters or leaves the meter (Figure 10.40). Rosenberry and Morin (2004) tested a commercially available electromagnetic-flow meter (EFM) placed in the tubing between the seepage meter and the traditional plastic bag (Figure 10.40a). They found that the EFM was able to measure seepage-rate changes over three orders of magnitude, from 30 mL/min to 30 L/min (Rosenberry and Morin 2004). The USGS has used electromagnetic seepage meters to characterize flux in both shallow and deepwater settings (e.g., at the sea

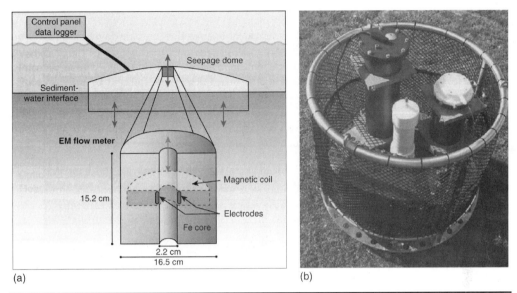

Figure 10.40 Seepage meters with water flow measuring devices. (a) Electromagnetic flow meter used by the USGS that is tethered to a power source and data logger. (b) Ultra Seep meter designed by the U.S. Navy that uses sonic technology to record seepage exchange, record geochemical parameters, and sample seepage water. The meter is retrieved to obtain data.

floor; https://coastal.er.usgs.gov/capabilities/shipboard/water/em.html). These types of meters allow data collection in deep-water locations but require that they are attached to a power and data recording source (Figure 10.40a). The US Navy developed the Ultra Seep instrument designed to be deployed and retrieved in deep-water settings. It included an ultrasonic flow meter, on board geochemical sensors and water-sampling capabilities (https://clu-in.org/programs/21m2/navytools/gsw/) (Figure 10.40b). Weighted seepage meters deployed by boats in deep settings like the Great Lakes are described by Cherkauer and McBride (1988) and Boyle (1994). Meter designs use long meter tubing to allow bag systems to be accessible within a meter or so of the surface. Some more standard meters have also been placed at depths where scuba diving is required for installation and servicing.

Proper use of seepage meters requires careful planning and review of the literature (e.g., Zamora 2008, Table 10.6). Seepage rates in streams, wetlands, and lakes are expected to vary spatially and temporally. Multiple instruments are often needed to generalize seepage conditions (e.g., Example 10.8). McBride and Pfannkuch (1975) suggested that in some lakes, seepage rates at lake shorelines decrease with distance and as water depth increases away from the shore. Meter use in streams can be problematic. Depending on the setting, seepage rates may represent true groundwater exchange (local, intermediate, and regional systems) or near-bed hyporheic exchange (Figures 10.14 and 10.15). Constantz et al. (1994) reported that the seepage rates in losing streams varied with the overall surface-water temperature, a condition attributed to the dependence of vertical hydraulic conductivity values on fluid temperature (Freeze and Cherry 1979; Fetter 2001). Daily and seasonal surface-water temperature change effects on seepage rates should at least be evaluated to determine if significant variations are likely (Zamora 2008).

Water-Temperature Studies

In some settings, the heat budget of a stream, lake, or wetland can be developed to evaluate locations and rates of exchange with groundwater (Figure 10.41). Isolating a change in surface-water temperature caused by an effluent groundwater system requires accounting for dominant components of the heat budget, the defined boundary conditions, and time frame, and when possible, elimination of components that are considered minor contributors to the budget (e.g., Bartolino and Niswonger 1999; Anderson 2005; Schnidt et al. 2007; Rosenberry and LaBaugh 2008; Webb et al. 2008; Buss et al. 2009; Boana et al. 2014). For example, in an effluent reach of stream the overall stream temperature could be impacted by the influx of groundwater with a contrasting temperature (Figure 10.42). However, in settings where the reach is in influent or groundwater fluxes and temperature contrasts are small relative to surface water, the influence of groundwater impacting stream temperatures could easily be undetectable (Figure 10.42). A complete heat balance is needed when attempting to identify and assess groundwater-specific temperature changes in surface-water bodies. Anderson (2005) produced an excellent review paper that discusses the theory of using heat as a groundwater tracer. In addition, an extensive review paper by Boana et al. (2014) cites methods using temperature to assess hyporheic exchange. Glose et al. (2017) presents an example of using modeling tools to simulate a heat budget and fitting the unknown groundwater exchange term.

Temperature monitoring of streams, lakes, and wetlands at various scales can reveal information on the location of groundwater exchange when groundwater-inflow temperatures contrast with surface, bed, and/or sediment temperatures. Groundwater/surface-water exchanges can be diffuse and distributed over a broad area or focused at specific discharge zones. When groundwater fluxes occur at banks or shorelines and are

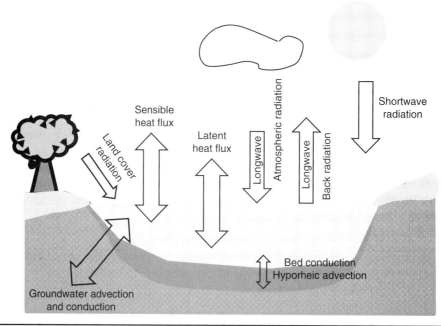

FIGURE 10.41   Components of an energy budget for a surface-water feature interacting with bed hyporheic water and underlying groundwater. [*With permission Woessner (2018).*]

(a) Effluent or Gaining Stream

Q1*TSW1

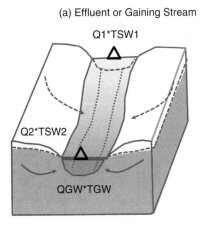

Q2*TSW2

QGW*TGW

(b) Influent or Losing Stream

Q1*TSW1

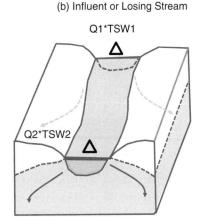

Q2*TSW2

Q1*TSW1 +QGW*TGW = Q2*TSW2

FIGURE 10.42 Changes in stream temperature as a function of groundwater exchange. Q1 and Q2 are stream flows, QGW is groundwater seepage to the stream, TSW1 and TSW2 are stream temperatures, and TGW is the groundwater temperature. Triangles represent surface-water monitoring points. (a) Effluent conditions in which a measurable change in stream temperature may be observable and attributed solely to an influx of groundwater. A complete heat budget would be needed to examine other influences on stream temperature between the monitoring points. (b) Influent conditions, no change in the surface-water temperature attributed to groundwater would occur as no groundwater discharge to the stream occurs. QGW inflow is zero. [*With permission Woessner (2018).*]

sufficiently different from the surface-water temperatures, their locations can often be identified. To enhance detection of exchange locations, temperature surveys are often performed when temperature contrasts are maximized (e.g., during winter when surface-water temperatures may be lower than groundwater temperatures, or in summer when surface-water temperatures may be higher than groundwater temperatures). Determining exchange sites using temperature requires measurements that are sufficiently accurate and precise that small differences can be observed and mapped. Handheld thermistors set in metal probes can be used to survey surface water and bed-sediment temperatures. A survey of the water- and bed-sediment temperatures in a spring brook near Ronan, Montana, revealed evidence of cooler groundwater discharging to the creek along its right bank (Figure 10.43).

Example 10.9 Streambed Temperatures and Groundwater Exchange
Conant (2004) investigated the fate of a dissolved PCE (tetrachloroethene) plume discharging in the Pine River, in Ontario, Canada. He measured subsurface temperatures in summer and winter along 1-m transects, spaced across the stream at 2-m intervals over a 60-m reach of the Pine River in Ontario, Canada (Figure 10.44). The surficial streambed geology consisted of fine sand with zones of gravel and cobbles. The surficial fluvial sediments were underlain by semi-confining silts, clays, and peats. Bed hydraulic gradients (VHG) were upward with mini-piezometer water levels extending 0.5 to 1.5 m above the stage of the river.

Summer and winter streambed groundwater temperatures were used to determine locations of concentrated (focused) groundwater seepage into the stream. Temperatures were measured by inserting a temperature sensor 0.2 m into the sediments. Readings were taken after about 30 s when the instrument had stabilized. Data were mapped by contouring (Figures 10.44a and 10.44b). At one location, vertical temperature profiles were measured at 0.1-m intervals to a depth of 0.5 to 0.6 m

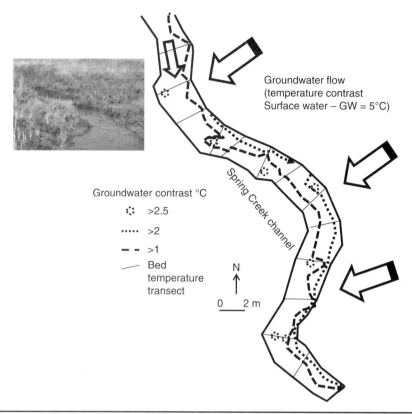

Groundwater flow
(temperature contrast
Surface water – GW = 5°C)

Spring Creek channel

Groundwater contrast °C

٭ >2.5
..... >2
– – >1
— Bed
temperature
transect

N
↑

0   2 m

FIGURE 10.43 Spring Creek temperature bed sediment survey, Ronan, Montana (picture). Spring Creek is flowing south. Groundwater monitoring data show groundwater flow is dominantly from the east. A metal probe thermistor was inserted 3 to 6 cm into the streambed along transects and temperatures were recorded. Background groundwater temperatures were 5°C lower than the creek temperature. The lower bed temperatures along the right bank and at the stream bottom show groundwater is discharging to the stream. Bed and flow system heterogeneity are indicated by focused locations of colder groundwater (dashed circles) and irregularity of the bed temperature zones. [*Modified from Loustaunau (2003).*]

(Figure 10.44c). Results showed that August diurnal surface-water temperature changes impacted characterization in the bed sediments to a depth of 0.3 m. Conant Jr. recognized a penetration depth of 0.2 m reflected some temperature error in his mapping. The summer and winter temperature distribution implied that concentrated groundwater discharge areas (and dissolved PCE) were occurring at three locations and other more isolated areas.

Conant (2004) also used slug testing and VHG measurements at selected piezometers to calculate groundwater flux (Figures 10.45a and 10.45b). He then paired flux values with temperatures measured at these locations, generating a relationship between recorded temperatures and measured flux rates of summer and winter conditions (Figure 10.45c). These relationships allowed estimates of flux for the temperate mapped area, including locations where no flux values were directly obtained. It should be noted that these relationships are influenced by site conditions (geologic and hydrologic setting) and are time specific, therefore the temperature flux relationships shown in Figure 10.45c are unique to this site.

It is also valuable to use heat as a tracer and then determine not only locations of exchange but also rates of exchange (e.g., Stonestrom and Constantz 2003). The collection

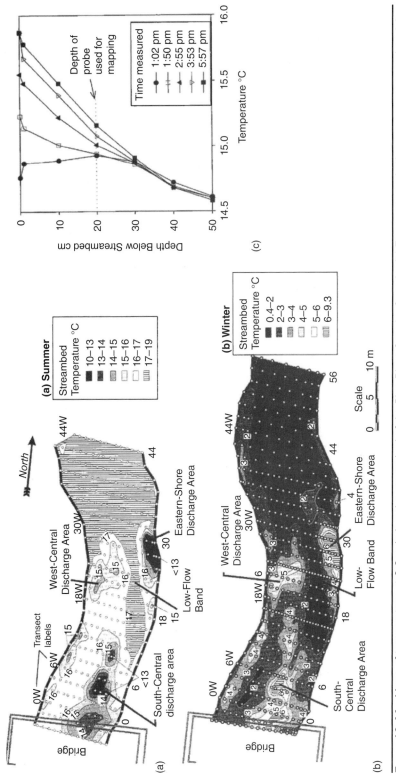

Figure 10.44 Maps of temperature at 0.2 m below the streambed of the Pine River, Ontario, Canada. (a) Temperatures in the summer measured on July 28 and 29, 1998. (b) Temperatures in the winter measured on February 18 and 20, 1999. (c) Vertical profile of bed temperatures measured in sand at five different times on August 14, 1997, to examine the influence of depth on temperature measurements (Conant 2002).

489

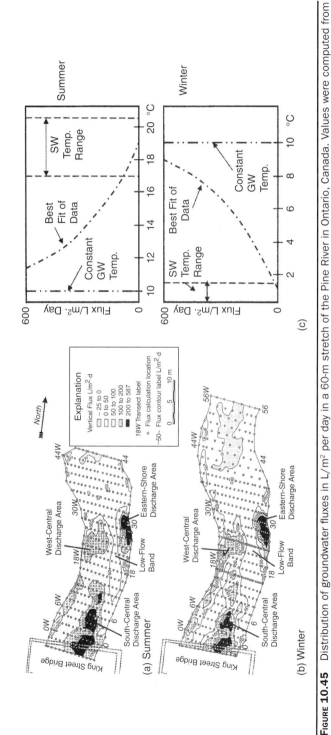

Figure 10.45 Distribution of groundwater fluxes in L/m² per day in a 60-m stretch of the Pine River in Ontario, Canada. Values were computed from slug-test-derived hydraulic conductivity values and the computed relationship between temperature and flux. (a) Summer distribution. (b) Winter distribution. (c) Flux discharge in L/m² per day versus temperature in °C (x-axis) plotted for the summer and winter conditions. [*Modified from Conant (2004).*]

of time-dependent temperature data in the surface-water body and in the underlying sediments at single or multiple depths provides an opportunity to compute exchange rates at these locations using heat-transport equations (e.g., Constantz and Thomas 1997; Constantz 1998; Constantz et al. 2002a, 2002b; Goto et al. 2005; Hatch et al. 2006; Keery et al. 2007; Constantz 2008; Swanson and Cardenas 2011; Bhaskar et al. 2012; Gordon et al. 2012; Boano et al. 2104). Constantz et al. (2008) summarized stream and groundwater interactions at temperature-instrumented bed locations. When the temperature of surface-water fluctuates with the air temperatures (e.g., summer daytime temperatures are often higher than during the nighttime) a surface-water diurnal temperature signal is generated (Figure 10.46). Groundwater temperatures tend to remain more constant (Conant 2004). Therefore, if temperatures at a bed-sediment location remain fairly uniform over time, one can interpret that groundwater discharge (upwelling) is having a strong influence on streambed temperatures (Figure 10.46a). Conversely, if stream-sediment temperatures fluctuate and are similar to surface-water temperatures, it is likely that groundwater is being recharged by downwelling stream water (e.g., Figures 10.46b, 10.46c, and 10.46d). These observed trends can be used separately or in concert with mini-piezometer and seepage-meter data sets to locate exchange sites and directions of water movement, and to compute flux rates.

Flux rates can be computed from time-dependent surface-water and groundwater-temperature data and application of models based on heat-transport theory. Basically, as water moves heat (measured as temperature) is transported with the water, it acts as

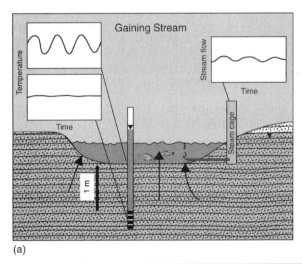

(a)

Figure 10.46 Schematic of surface water temperature, bed temperature, and stream flow change over time. Temperature monitors are place in the stream (or lake or wetland; top temperature record on the left hand side) and at the bottom a mini-piezometer inserted into the saturated sediment below the bed (bottom temperature record on the left hand side). These monitors are often isolated from the complete water column by placing a packer above the thermistor. (a) Response of water temperatures where groundwater is discharging to the stream. (b) Response of water temperatures where surface water is discharging to the groundwater. (c) Response of water temperatures to infiltration of water from a disconnected or perched losing stream. (d) Response of water temperature to an ephemeral stream (losing) with no flow until an event in the middle of the hydrograph. [*Modified from Constantz (2005).*]

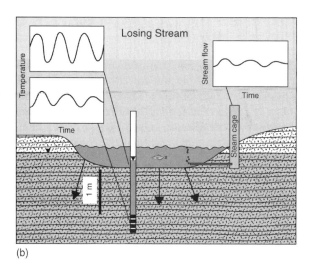

(b)

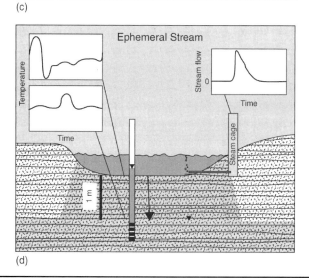

(c)

(d)

FIGURE 10.46 *(Continued)*

492

a tracer of the water source. The process of transporting heat with the flow of the groundwater is referred to as advection or convection (Stonestrom and Constantz 2003). Heat also migrates from areas of higher temperatures to lower temperatures independently of flow by conduction (analogous to molecular diffusion in a contaminant transport setting, Anderson 2005). In addition, as heat is transported in the groundwater water it also interacts with the solid porous media (heating or cooling), thus this process is also incorporated into heat-flow tracer methods. Based on these principles, a steady-state one-dimensional saturated heat-transport equation can be used to estimate exchange.

Modeling assumes that thermal properties of the sediments vary within a limited range and that observed temperature variations are controlled by basic heat-flow principles. The process requires estimates of sediment porosity and measurements of VHGs and the fitting of flux and groundwater velocities (Figure 10.47). Anderson (2005), Stonestrom and Constantz (2003), and Constantz (2008) provide tables of thermal properties of sediments useful in heat-flow modeling.

Both analytical solutions and numerical models are used to examine exchange rates (e.g., Stallman 1965; Goto et al. 2005; Hatch et al. 2006; Keery et al. 2007;

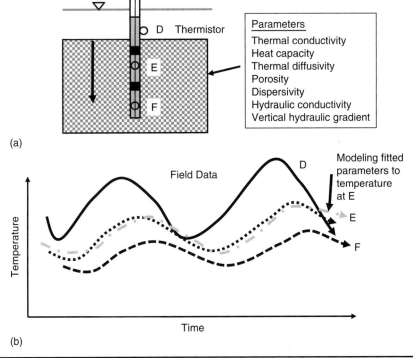

FIGURE 10.47 Temperate data collected in a surface water feature and within the bed sediments. (a) A mini-piezometer with recording temperature instruments located in the surface water (D) and in the underlying sediments (E) and (F). Thermistors are separated inside the piezometer with friction closed-cell packers taped to the outside of a small diameter center rod. (b) Thermographs of the recording thermistors. Note the surface water is losing water to the sediment. The peak temperature signals recorded in the bed sediments are lagged and the amplitude decreases with depth. The dash-dotted line represents the results of modeling the heat transport. Once an acceptable fit is achieved flux rates are computed. [*With permission Woessner (2018).*]

Schmidt et al. 2007). Software codes developed by the USGS, for example, VS2DH (Healy and Ronan 1996), are used to compute exchange rates. Models assume that streambeds are near saturation and therefore use a linear thermal-conductivity function. The software is free and can be downloaded along with the documentation at http://water.usgs/nrp/gwsoftware. Swanson and Cardenas (2011) developed a program Exstream coded in MATLAB to analyze field thermographs. The code VFLUX (Gordon et al. 2012) relates stream temperatures to travel time in bed-sediment temperature profiles. The software is available at http://hydrology.syr .edu/lautz_group/vflux.html. Each representation requires water moving vertically into and out of the boundary sediments. Some authors have cautioned that two- and three-dimensional components of exchange may not be ignored in some cases (e.g., Cuthbert and Mackay 2013).

Characterizing temperature distributions in surface water and groundwater can be accomplished using a wide range of instruments, from a standard handheld thermometer to arial/satellite survey tools. For the collection of point measurements (single time), handheld thermometers or thermistor-tipped probes can be used (e.g., Figures 10.43 and 10.44). They need to be of sufficient construction so they will not easily break. They should respond quickly so temperature readings stabilize and be of sufficient accuracy and precision to ensure that contrasts in temperatures can be differentiated (Example 10.9). Some electric measuring-tape manufactures have developed electric water-level probes that measure water level and temperature and are of value when both data sets are desired (e.g., https://www.solinst.com).

In settings where surface-water temperatures are monitored to detect the effects of discharging groundwater, instruments that sense differences in water-surface temperatures are useful. A standard thermistor inserted into the surface water or temperature readings from a near-surface submerged transducer will provide point temperature measurements. Tools used to map or view contrasts in water-surface temperatures include thermal digital cameras operated from the land surface or from a drone or aircraft. The USGS has overview information on the use of handheld thermal imaging cameras for groundwater/surface-water interaction studies at https://water.usgs .gov/ogw/bgas/thermal-cam. This tool records the differences in surface temperatures and requires calibration by including field-based temperature measurements of the range of anticipated temperatures (Figure 10.48). Surveys mapping larger areas are often completed using aircraft and FLIR (Forward Looking InFrared) imaging sensors to compile temperature distributions (e.g., Cox et al. 2005).

Mapping stream, lake, or wetland-bottom sediment-interface temperatures has been accomplished using fiber-optic distributed temperature sensing (FO-DTS) methodologies where fiber-optic cable sensors are laid on the surface-water body bottom and temperature readings recorded at fixed intervals (e.g., Selker et al. 2006). The USGS has a good informational overview: FO-DTS technology demonstration and evaluation project (https://water.usgs.gov/ogw/bgas/fiber-optics). The technology can be used to obtain measurements to 0.01°C over miles (several kilometers) and at a resolution from seconds to hours (e.g., Mwakanyamale et al. 2012) (Figure 10.49).

Temperature devices that record temperature at a selected time interval are used for field investigations, especially when a single-point measurement or a one-time temperature survey is inadequate to address project goals. In addition to the properties mentioned above, instruments must have a useful recording interval and battery life. Some recording devices may require hard wiring to a recording base on shore. However, water and site conditions usually limit the use of these instruments. As a result, independent wireless

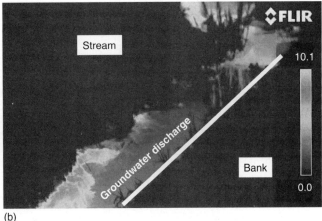

(a)

(b)

Figure 10.48 Application of thermal camera to detect groundwater discharge. (a) An example of a thermal camera used by the USGS. (b) Thermal image (about 6 m) of a stream and stream bank (dark shades) where warmer groundwater (light shades) is discharging to stream in late fall at Tidmarsh Farms, Massachusetts. The dark shades represent cooler bank and stream temperatures and the light shades are the stream temperatures where warmer groundwater is discharging to the stream. (*Modified from https://water.usgs.gov/ogw/bgas/thermal-cam.*)

instruments are preferred. When selecting tools for a field investigation, the physical size of the instrument and the installation conditions must be considered (Figure 10.50). Small-diameter transducers that record changes in head over time also provide temperature readings as transducer readings are temperature dependent. The use of a transducer provides two sets of data and is commonly used when budgets are adequate (see also Chapter 4). Another alternative is a stand-alone, less expensive sensor that only records temperature and time. Manufacturers have designed a variety of sensor equipment at increasingly reduced costs. Some of the more commonly used sensors are pictured and described in Stonestrom and Constantz (2003) (Figure 10.50). Johnson et al. (2005) performed an evaluation of wireless/stand-alone temperature loggers. Many of the devices previously described in the references above were compared with the Thermochron iButton manufactured by Dallas semiconductor or Dallas, Texas at (http://www.ibutton.com).

Example 10.10   iButton Application

The iButton temperature logger was originally developed to monitor temperature-sensitive cargo during transit (e.g., refrigerated products). Different models cover a temperature range from −40°C to 120°C and have accuracies of 1.5°C to 0.625°C. Johnson et al. (2005) pointed out four principal advantages of the iButton logger over other wired and other standalone temperature loggers.

1. Being wireless, the instrument does not have to be located in a control-recording system.
2. Its small size (just over 17 mm) allows for deployment in small-diameter, hand-driven, or direct-push monitoring wells.
3. Multiple loggers are easily suspended within a fully screened monitoring well to obtain a high-resolution (within 0.5°C) vertical temperature profile.
4. The relatively low cost (ratio of 10 units to 1) allows for detailed spatial and temporal studies (costs have increased since this article as have recording and resolutions options; check the website for current prices).

Before using iButtons in exchange studies, they should be tested to ensure they are sufficiently water sealed. Waterproof containers for the iButtons are also available from the company. It is useful

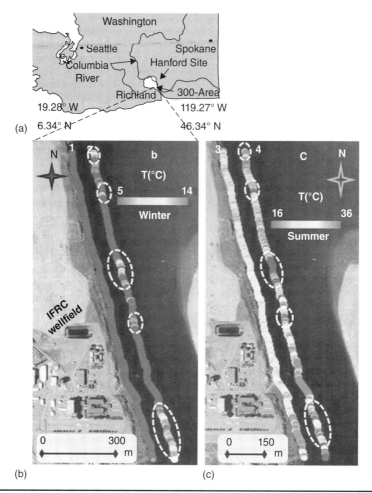

FIGURE 10.49 Results of a 1.6-km FO-DST bottom survey of the Columbia River at the Hanford Site in Richland, Washington. The cable was placed 2 m from shore. Results (1 to 4) are shown with an 80-m offset but all readings were taken at the same location. (a) Location map. (b) Results of winter bottom temperatures during high-river stage, February 28, 2009 (line 1) and low-river stage, March 15, 2009 (line 2). (c) Summer temperature distribution at high stage, August 20, 2009 (line 3) and low-river stage, August 3, 2009 (line 4). Lighter shades highlighted by dashed white lines represent focused groundwater exchange that is most easily observed in low stage records (*Mwakanyamale et al. 2012*).

to check factory resolution and operation by placing instruments in a constant-temperature bath that is adjusted to progress through the anticipated field temperature conditions. In addition to checking for instrument failure, a set of calibration data is also generated. These data should be used to adjust instrument-reported values so that readings of all instruments are comparable. In some field settings, two iButtons have been installed side by side (backup) to ensure no temperature data are lost because of a failed recorder. Each iButton logger has a unique identification code; data are downloaded using a reader. An example of an installation of four iButtons in a mini-piezometer installed in the sediments of a losing reach of the Umatilla River, Oregon, is shown in Figure 10.51. Temperature data were used to identify the site as losing and the temperature versus time data to compute flux rates at the site.

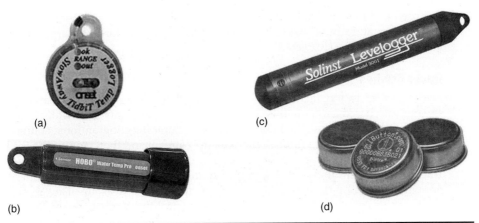

FIGURE 10.50 Examples of stand-alone temperature measuring devices. (a) Onset Tidbit (http://www.onsetcomp.com). (b) Hobo Water Temp Pro (http://www.onsetcomp.com). (c) Solinst transducer Levelogger (with temperature; http://www.solinst.com). (d) iButton Maxim-Dallas Semiconductor. (http://www.Thermoncron.com.)

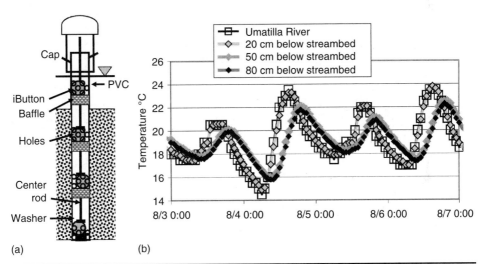

FIGURE 10.51 iButton data characterizing the surface water and hyporheic temperatures in the gravel-bedded Umatilla River in Oregon, the United States. (a) iButton installation (gray circle). A steel mini-piezometer was perforated at four locations and iButtons were installed using a center rod and closed foam baffles (to isolate water/temperature circulation). (b) The symbols on the temperature versus time plot (8/3 0:00 represents August 3 at midnight) represent individual iButton readings. Note that this section of the river was losing. Surface-water temperatures and the iButton readings set at 20 cm suggest surface water is freely circulating to a depth of ~20 cm, as temperatures are very similar to river temperatures. However, iButton readings obtained at 50 and 80 cm are degraded (lower) and peaks are lagged [*Modified from Johnson et al. (2005).*]

In settings with water conditions and currents that limit the use of shallow-water tools, multiple sensor tools such as the Trident probe and some deep-water seepage meters (Figure 10.40b), have been developed to sample head, water quality, and temperature (https://clu-in.org/programs/21m2/navytools/gsw/#trident). This boat-deployed tool was used in the Columbia River, Washington, Hanford site, to characterize conditions of pore water at salmon spawning sites (https://pdw.hanford.gov/arpir/index.cfm/viewDoc?accession=1107060717).

Tracer Studies

Tracer studies in streams have been used to determine stream discharge in small streams (discussed earlier in this chapter) and the "time of travel" of water in the channel (Wilson 1968; Kilpatrick 1970). Methods include injecting an ionic solution (e.g., chloride or bromide) or a fluorescent dye into a stream and then monitoring the breakthrough curve of the tracer at some distance downstream (Smart and Laidlaw 1977). Tracer studies conducted to determine aquifer velocities and aquifer properties are presented in Chapters 13 and 14. In some settings, tracer tests are also valuable tools for characterizing reach-scale exchanges between the surface water, in this case streams, and its associated adjacent groundwater system (e.g., Bencala and Walters 1983; Castro and Hornberger 1991).

In most cases tracer tests are used in streams with small discharges, as those with large discharges require adding large quantities of tracers which is expensive and logistically difficult. Generally, two types of tracer addition tests are used. The tracer can be mixed in a volume of water and then the volume of water is added to the stream all at once as a slug or pulse source. The second approach is to add a tracer solution into a stream at a constant rate, a constant source. In both settings, the change in concentration over time is monitored at a downstream location. A mass balance is also part of a tracer analyses allowing the determination of what portion of the tracer has passed the observation point or entered temporary storage.

Once the concentration breakthrough curve is derived, it is analyzed to determine if results fit theoretical sets of transport conditions (Figure 10.52). If tracer transport is uninterrupted by exchange or in-channel delays, temporary storage, or the influx of groundwater, the theoretical breakthrough curve will only be a function of the natural spreading and dilution occurring during transport, and the breakthrough curve will account for all the mass added to the stream (Figure 10.52, curve D). Under these conditions, when a test is performed in a losing reach and stream containing tracer water seeps into the underlying groundwater system, the mass balance computed at the observation point will not account for the initial tracer mass though the BTC will look like D. If groundwater is discharged to the stream between injection and observation points, the shape of the curve will look the same but the maximum tracer concentration will be lower. It is more likely that breakthrough data will reflect temporary storage of some of the tracer resulting from channel storage (e.g., dead zones, backwaters, eddies) and/or hyporheic circulation (Figure 10.52 curves E and F). Analyses of these BTC can be used to suggest the degree of exchange occurring in the stream reach.

Analyses of breakthrough field data require the formulation of a model that reflects some or all of the factors impacting the behavior of the breakthrough curves. Boana et al. (2014) provide a very good review paper that describes multiple conceptual models of the transport process. Their paper includes a description of the appropriate governing equations: advection-dispersion and fractional advection-dispersion equation, the space-time fractional advection-dispersion equation, fractional spatial-derivative advection-dispersion equation, and the fractional temporal-derivative advection-dispersion equation. Early transport modeling included a one-dimensional transient storage model with a finite-sized storage zone (e.g., Hays 1966; Thackston and

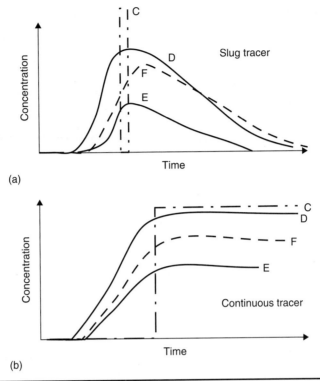

FIGURE 10.52 Examples of conservative tracer breakthrough curves (BTCs) from stream sampling. (a) Slug input of a tracer. (b) Continuous introduction of a tracer. C. The BTC if no spreading or temporary loss of tracer occurs. The BTC would reflect a plug flow sharp tracer front. This is usually never seen as the stream flow process transports the tracer at multiple velocities spreading the mass of the tracer as seen in D. The BTC representing transport without storage or hyporheic exchange. E. Data represents the same behavior as D but mass is unaccounted for possibly lost to temporary storage in streambed and/or floodplain waters. F. A redistribution of tracer mass and a delay in peak concentrations. Such an observation may represent channel storage and/or hyporheic exchange processes. [*With permission Woessner (2018)*.]

Schnelle 1970; Valentine and Wood 1979; Bencala and Walters 1983; Jackman et al. 1984; Kim et al. 1992; Wörman 1998; Bencala et al. 2011). Parameter calibration is achieved by model fitting to the field data. Modeling was often performed using the One-Dimensional Transport with Inflow and Storage (OTIS) model that was developed by the USGS (e.g., Runkel and Chapra 1993; Runkel, 1998; software is available at http://water.usgs.gov/software/OTIS). However, it was noted that the use of a small number of parameters to represent what are most likely a larger number of complex interactions influencing the observed tracer response is a limitation of this approach (Bencala et al. 2011).

Efforts to improve identification of factors controlling tracer exchanges require field investigations of both physical site conditions and stream-exchange and storage within the tracer test area. This additional information allows the development of models that include additional parameters accounting for observed breakthrough curves (e.g., Haggerty et al. 2000; O'Connor et al. 2010). Haggerty and Reeves (2002) expanded BTC model methods to include a multi-rate, mass-transfer using the model STAMMTL (available at http://science .oregonstate.edu/~haggertr/STAMMLT). Additional modifications of this approach include averaging advective storage paths allowing exchange by purely advective transport (Worman et al. 2002). Other model techniques using random-walk transport representations (e.g., Boano et al. 2007); the separation of storage locations and timing (stream and hyporheic) are executed in the code the Solute Transport in Rivers Model (software available at http://www.wetengineering.com/en/downloads/62-stir-solute-transport-in-rivers.html) (Marion et al. 2008). More detailed discussions of these analytical methods and models are beyond the scope of this chapter and the reader is referred to Boano et al. (2014) and the references cited by them.

Methods

Conducting stream tracer tests requires the input of a tracer and monitoring its change in concentration over time. Tracers need to be inexpensive, their detection methods low cost, conservative, and in low to zero background concentrations in the surface water and groundwater. Since they are added to streams they also must be nontoxic to aquatic life or other water users. Permits may be required by local, state, or federal agencies and should be obtained prior to conducting tracer experiments. In designing a tracer experiment, an evaluation of the type of tracer and amount needed to obtain a detectable concentration at the observation location is necessary. Often salt tracers (e.g., NaCl, KCl, Na Br, etc.) are considered as they can be analyzed for one of the ions or, if they act conservatively, as changes in specific conductance. An alternative to salt tracers is the use of fluorescent dyes (Chapter 14). Their advantage is that they are not commonly found in background concentrations and dyes can be detected at low concentrations (using a fluorimeter set to the appropriate wave length). A detailed comparison of eight fluorescent dyes used as tracers was presented by Smart and Laidlaw (1977). Their evaluation yielded a recommendation of three: rhodamine WT (pink), lissamine FF (green), and amino G (blue). Fluorescein (green) is sometimes used for stream studies. However, it is not recommended as it will photo-degrade and is more likely to sorb onto sediments and some components of the stream water. Kilpatrick (1970) found that rhodamine WT could be used at a lower concentration than rhodamine BA (20% solution compared to 40% solution) and at a lower cost. One should consult Chapter 14 for the latest info on fluorescent dyes and their properties.

Kilpatrick (1970) developed an equation to estimate likely dye concentrations that can be used to trace surface-water flow in a stream using a slug input of dye. He provides a guide for determining the volume of dye that can be added to the stream to obtain a desired observable concentration at some distance downstream.

$$V_d = 3.4 \times 10^{-4} \left(\frac{Q_m L}{V} \right)^{0.93} C_p \tag{10.12}$$

where V_d = Volume of 20% rhodamine WT dye in liters (Figure 10.53)
Q_m = Discharge in the reach in ft^3/s
L = Length of reach in miles
V = Mean stream velocity in ft/s
C_p = Peak concentration and micrograms per liter (µg/L) desired at the lower end of the reach

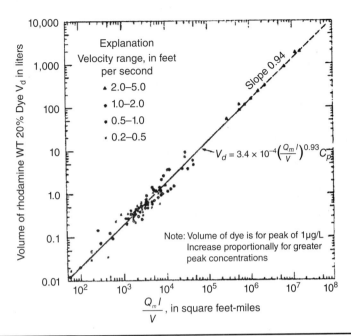

FIGURE 10.53 Quantity of rhodamine WT 20% dye required for a slug injection that will produce a peak concentration of 1 µg/L at a distance downstream, L, at a mean velocity, V, and a maximum discharge, Q_m for the reach. [*From Kilpatrick (1970).*]

Equation 10.12 represents the case where $C_p = 1$ µg/L at the lower end of the reach and for stream velocities that range between 0.2 and 5.0 ft/s (0.06 to 1.52 m/s). To increase the peak concentration at the lower end of the reach (C_p), a proportionately greater volume of dye is needed. The USGS does not recommend exceeding 10 µg/L as a peak concentration (Kilpatrick 1970; Kilpatrick and Cobb 1985). Equation 10.12 is presented graphically in Figure 10.53 to evaluate dye volumes directly. Dye tracing work by Hubbard et al. (1982) presents methodologies in metric units focusing on time-of-travel and dispersion in streams. Most tracer tests are initiated as a point source by introducing a volume of tracer at a known concentration. To characterize a downstream breakthrough curve generated from field sampling data, samples must be taken at a sufficiently distant location so that complete mixing of the tracer in the stream flow has occurred. Complete lateral mixing from a point-source at midstream is required. The distance downstream (L_m) (Hubbard et al. 1982) where complete mixing occurs is estimated by Equation 10.13:

$$L_m = 0.1 \, (V \times W^2)/E_z \qquad (10.13)$$

where E_z is the transverse mixing coefficient (m²/s) and is equal to, $E_z = 0.2dv_s$, and
d is mean depth (m), v_s is shear velocity (m/s), defined as: $v_s = gds$, where g = the acceleration of gravity (9.81 m/s²), d = mean depth (m), s = water surface slope (m/m).
V is mean stream velocity (m/s), and
W is average channel width (m).

For example, if dye is introduced at mid-channel and the stream is 13 ft (4 m) wide, flowing at 7 ft/s (2 m/s), has an average depth of 3.28 ft (1 m), and a surface-water slope of 0.01 m/m, complete dye mixing would occur at a channel length (L_m) of

$$E_z = 0.2 \times (1\ \text{m}) \times (9.8\ \text{m/s}^2) \times (0.01\ \text{m/m}) = 0.02\ \text{m}^2/\text{s}$$

$$L_m = 0.1 \times ((2\ \text{m/s}) \times (4)^2)/0.02\ \text{m}^2/\text{s} = 320\ \text{m} \ (1{,}050\ \text{ft})$$

Slightly different equations are used if the dye is injected at a shoreline. More rapid mixing is accomplished if the dye is introduced over a large portion of the stream surface. Once the dye is introduced, a sampling event should occur at two locations below the complete mixing length and at a site further downstream to establish transport conditions (Hubbard et al. 1982).

Data analysis techniques have been outlined in the previous section. It is recommended that slug breakthrough curve analyses be addressed initially using simple models such as OTIS and then more complex approaches depending on whether additional field observations or measurements of the storage and transport mechanisms were made.

Example 10.11 OTIS Model

The USGS OTIS model was developed to analyze slug or constant tracer source input and stream tracer breakthrough curve data sets (Runkel 1998; https://water.usgs.gov/software/OTIS/doc/fs.pdf). The model addresses advection and dispersion of the tracer in the stream, the effects of lateral inflow and transient storage and chemical transformations (first-order decay and kinetic sorption). Model parameters include the channel cross-sectional area, dispersion coefficient, storage-zone cross-sectional area, lateral inflow rates and solute concentrations, and main-channel and storage-zone decay coefficients and sorption-distribution coefficients.

Batchelor and Gu (2014) applied OTIS to assessing the role of hyporheic exchange and nutrient uptake in a forested and urban stream in the southern Appalachians of North Carolina. They examined how storage impacted the transport of a bromide (conservative) and nitrate (reactive) tracer injected at a constant rate on what they described as urbanized Boone Creek (mean annual discharge 3.5 cfs, 0.1 m³/s) and the forested Winkler Creek (mean annual discharge of 7 cfs, 0.2 m³/s). They assumed total tracer mixing at 20 stream widths. A continuous tracer addition was designed to raise background concentrations of bromide 1 to 2 mg/L above background (non-detect) and nitrate concentrations 2 to 3 mg/L above background (1.5 to 7 mg/L). Prior to the addition of the bromide and nitrate tracers, a preliminary test using sodium chloride was conducted to adjust input rates and examine stream mixing. The main tracers were added using a Mariotte bottle at a rate of about 30 L/h. Sampling was conducted at 165 ft (50 m) and 330 ft (100 m) downstream until the recession tail of the curve was captured after tracer injection stopped (Figure 10.54). OTIS modeling was used to determine the likely role hyporheic transient storage played in the behavior of the observed bromide and nitrate breakthrough curves. Model fit is shown in Figure 10.54.

Batchelor and Gu (2014) reported that the absolute and relative cross-sectional areas of the storage zone in the Winkler Creek reach are larger than in Boone Creek. However, the time spent in transient storage over a 660-ft (200-m) reach was only 0.35% and 1.05% of the reach average median travel time at base flow for Boone and Winkler creeks, respectively. This suggested the overall hyporheic exchange in both creeks is small. Nutrient uptake was higher in the forested Winkler Creek. The investigators state that urbanization appears to reduce hyporheic exchange and water and solute retention capacity.

Tracers are also helpful in determining the movement of groundwater in streambeds and near-channel sediments. Both qualitative and quantitative analyses can be

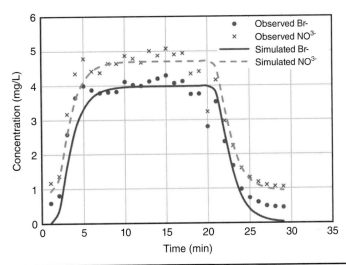

Figure 10.54 Observed bromide and nitrate concentrations at 50 m downstream from the Winkler Creek injection site. Solid and dashed lines are the result of OTIS simulations (*Batchelor and Gu 2014*).

used to establish connections between river water and shallow groundwater, and hyporheic exchange. For example, in shallow stream settings tracers can be injected into the bed sediments at the head of a riffle; hyporheic-discharge locations and concentrations can be observed and sampled downstream. Stream ecologists have also recognized the value of sampling bed and floodplain groundwater during stream tracer experiments (Stanford and Ward 1988, 1993; Meyer 1997; Dahm et al. 1998; Dahm et al. 2006).

Example 10.12   Stream Tracer Hyporheic Flow
A stream tracer study was conducted in the North Fork of Dry Run Creek (gradient 0.057) in Shenandoah National Park in Page County, Virginia, by Castro and Hornberger (1991). They designed the test to evaluate groundwater/surface-water interaction with fluvial sediments located within and adjacent to the stream. The 225-m reach contained pools and riffles. The floodplain width of the alluvial materials though which the stream flowed varied between 22 and 77 m (Figure 10.55).

A KBr tracer was mixed with stream water in a 100-gal (378-L) polyethylene tank. On the day before the test, the solution was mixed via a small pump for 8 h. During the test, the tank was stirred manually twice a day during the injection period. The tracer concentration in the tank was 48.6 g/L. The tracer was introduced into the stream at a riffle zone at a rate of 50 mL/min over the 93-h injection period (~0.05% to 0.1% of the total discharge). The injection rate was checked using a bucket and stopwatch several times during the injection period.

Six surface-water sampling sites (SH1 through SH6) and a series of transects were established to evaluate the distribution of tracer in the sediments (Figure 10.55). Groundwater samples were obtained from hand-dug wells along each transect near the active channel and from an old abandoned channel within the alluvial plain. Additionally, a few cased wells were drilled with completion depths ranging from 0.75 to 1.75 m. These were up to 24 m away from the active stream (Figure 10.55).

Samples were collected from the streambed and gravel bed every 0.5 to 1.0 h until the arrival of the leading edge of the tracer. Subsequently, samples were collected every 4 h while concentrations were relatively constant during injection of the tracer. Following tracer injection, samples were collected at 0.5 to 1, 4, 8, and 24 h, and ultimately weekly (Castro and Hornberger 1991).

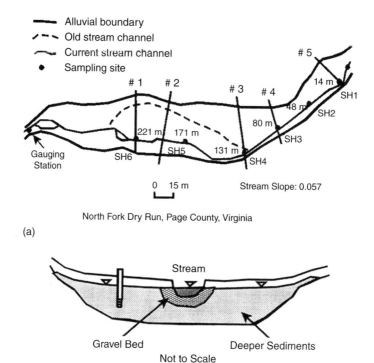

FIGURE 10.55 Study site on the North Fork of Dry Run Creek. (a) Plan view of study area in Shenandoah National Park, Virginia. The study area represents a 225-m reach of the North Fork of Dry Run Creek, Page County, Virginia. (b) Schematic cross section showing mixing zones in the stream, the gravel bed, and deeper sediments. [*Adapted from Castro and Hornberger (1991) with permission from the American Geophysical Union (1991).*]

A specific ion electrode (accurate to 0.25 mg/L) was used to measure bromide concentrations (peak concentrations were near 30 mg/L). This was used in conjunction with a pH/conductivity/temperature meter. Because of the low ionic strength of the stream water prior to analysis, all samples and standards were mixed with an ionic strength adjuster (ISA), $2M$ potassium nitrate, to the ratio of 1:10 (ISA:sample) (Castro and Hornberger 1991). Standards were made with upstream water and compared with standards made with deionized water with no differences found.

Significant tracer concentrations were found in wells at distances of 8 and 16 m from the active stream channel after concentrations within the stream had returned to background. The presence of tracer supported the concept that stream water can enter and remain in adjacent sediments for an extended period. Concentration tails (abrupt concentration arrivals followed by gradual decreasing trends; see Chapter 13) at surface-water sites waned after 4 to 6 days, while similar breakthrough curve tails at wells lingered approximately 20 days before returning to background levels. This study showed how mixing within the hyporheic zone was active at significant distances away from the main channel and that stream-water movement away from and back into the channel through the bottom sediments occurred.

Geochemical Methods

Rates, locations, and timing of exchanges are characterized by tracing the path of the water through water budgets and the application of site-specific measurement methods. Exchanges of surface water and groundwater are not solely characterized by the physical processes as described in the preceding sections. As was presented in the discussion of temperature methods, the water source also is represented by its geochemistry. The geochemistry of groundwater and surface water is always a reflection of the landscape, water source, and geologic materials from which it originates (Chapter 8). Often groundwater and surface water have different histories, therefore exchanges can be inferred from characterizing the geochemical components of each resource and examining locations of contrast and developing mixing models. Ecologists examining the influence of exchange on water bodies and near-surface groundwater suggest exchanges cycle key elements including nutrients, supply dissolved oxygen, and sustain conditions supporting microbes, macroinvertebrates, and higher-level organisms (e.g., fish) (e.g., Brunke and Gonser 1997; Hauer and Lamberti 2017).

The chemistry of a surface water and groundwater is described by factors such as gross ionic components, temperature, pH, DO (dissolved oxygen), TDS (total dissolved solids and/or specific conductance), and stable and unstable isotopes. One approach when contrasting groundwater and surface-water chemistries is to generate mass balances of selected parameters, measuring concentrations of chemical components and discharge rates. For example, if one looks at the original water-budget model as a vehicle for accounting for flows and then includes concentrations of a water-quality constituent, a mass balance model is produced (Equation 10.14):

$$\text{Net GW exchange} = (\text{GW}_{in} \times (C_{GWin}) - \text{GW}_{out} \times (C_{GWout}))$$
$$= ((\text{SW}_{out} \times (C_{SWout}) - \text{SW}_{in} \times (C_{SWin})) - (\text{PPT}_{in} \times (C_{PTTin})$$
$$+ (\text{Evap}_{out} \times (C_{Eout})) + (\text{Trans}_{out} \times (C_{Transout})) \pm \Delta S(C_{storage})))$$

$$(10.14)$$

where GW_{in} is groundwater discharge to the surface-water domain, GW_{out} is the flow of surface water into the groundwater, SW_{in} is the flow of surface water into the domain, SW_{out} is the flow of surface-water out of the domain, PPT_{in} is precipitation falling within the domain, E_{out} is direct evaporation from the water surface, Trans_{out} is the loss of water from transportation, and $\pm\Delta S$ is the volumetric net change in storage of stream, lake water, groundwater, and soil water over the time duration of the balance. Concentrations of a selected constituent are represented by the concentration terms [e.g., (C_{SWout}) concentration in mg/volume of the surface water flowing out of the domain]. Healy et al. (2007) and Winter (1981) caution that if some components of mass balances are poorly defined, large errors are likely.

An initial simplified mixing model can be evaluated to examine if contrasts in flows and chemistries are sufficiently large to allow for measurable changes in geochemical constituents. Healy et al. (2007) describe how natural isotopic signals, historical components added by human activity in the past, and the application of tracers can be used to represent water sources (Figure 10.56). Usually useful tracers need to behave conservatively or have known degradation rates (e.g., Sacks et al. 1998). For example, a mass balance for a stream reach may be formulated as:

$$(Q_{SW,in} \times WQP) + (Q_{GW,in} \times WQP) = (Q_{SW,out} \times WQP) \qquad (10.15)$$

Examples of tracers used in water-budget studies.

Use	Naturally occurring in the environment	Historical — Added to the environment from human activity in the past	Applied — Added to the environment in the present	Example study
Ground-water age — Time since recharge water became isolated from the atmosphere	^{35}S, ^{14}C, ^{3}H/^{3}He, ^{39}Ar, ^{36}Cl, ^{32}Si	^{3}H, ^{36}Cl, ^{85}K, chlorofluorocarbons, herbicides, caffeine, pharmaceuticals		Plummer and others (2001)
Temperature of recharge	N^{2}/Ar solubility			Plummer (1993)
Tracing ground-water flow paths	^{18}O, ^{2}H, ^{13}C, ^{87}Sr	Chlorofluorocarbons, herbicides, caffeine, pharmaceuticals	Cl, Br, dyes	Renken and others (2005)
Exchange of surface water and ground water	^{18}O, ^{2}H, ^{3}H, ^{14}C, ^{222}Rn		Cl, Br, dyes	Katz and others (1997)
Surface-water discharge and traveltime			Cl, Br, dyes	Kimball and others (2004)

FIGURE 10.56   Table of geochemical components of surface water and groundwater that can be used to develop mass balances and mixing models. Healy et al. (2007) refer to these parameters as tracers of water types.

where $Q_{SW,in}$ = stream flow into the area, in L^3/T
   WQP = water quality parameter of choice (consistent with flow term mg/L^3)
   $Q_{GW,in}$ = groundwater flow into the area, in L^3/T
   $Q_{SW,out}$ = stream flow out of the area, in L^3/T

The objective is to use the stream flow and water-quality data to evaluate the groundwater exchange, in this case $Q_{GW,in}$. This approach is applicable to defined volumes over specified time domains. It is applicable to lakes and wetlands if the contribution of groundwater to the water body is detectable using geochemical constituents (e.g., Dinçer 1968; Turner et al. 1984; Stauffer 1985; Krabbenhoft et al. 1990; Pollman et al. 1991; Krabbenhoft and Webster 1995). Sacks et al. (1998) provide a number of examples of using mass balances to examine the contribution of groundwater to lakes in Florida and discuss applications of ionic and isotope components (Figure 10.57).

Example 10.13   Groundwater Exchange Using Geochemistry
In a study conducted by English (1999) near Silver Gate, Montana, near Yellowstone National Park, a mass-balance approach was used to estimate the influx of groundwater into Soda Butte Creek. Stream-gauging stations 7, 11, and 20 were used in the analysis along with their respective water chemistries, along with local shallow wells (<65 ft below ground surface) (Figure 10.58). Station 7 is on Soda Butte Creek just downstream from the confluence of a tributary (Sheep Creek). Station 11 is on Wyoming Creek, a tributary in the middle of the study area, just before it merges with Soda Butte Creek. The water-quality and gauging data reflect this contribution. Station 20 is on Soda Butte Creek at the exit point of the study area. The respective water qualities and gauging data are shown in Table 10.7. A mass-balance equation was formulated (Equation 10.15) using the measured values of an individual constituent in each water-budget component and ($Q_{gw,in}$) was solved for as the unknown (Table 10.8).

Radon-222 is often used to examine the rates of leakage of surface water into adjacent groundwater systems. Streams, lakes, and wetlands usually have very low

Lake	Computed ground-water inflow					
	Using chloride		Using sodium		Using δD		Using δ¹⁸O
	median gw inflow conc.	minus - plus 25 % of median conc.	median gw inflow conc.	minus - plus 25 % of median conc.	assuming δ_a in equilibrium with rainwater	δ_a backcalculated from Starr Cl results	δ_a backcalculated from Starr Cl results
Annie	180	[1]28 - n a	293	[1]30 - n a	[1]90	[1]78	[1]100
George	[1]8	[1]4 - 51	[1]19	[1]9 - n a	54	[2]31	[2]33
Grassy	15	9 - 49	8	7 - 11	30	15	19
Hollingsworth	90	n a - 45	123	n a - 52	n a	n a	n a
Isis	n a	n a - 61	8	2 - n a	101	90	95
Olivia	57	24 - n a	33	18 - 166	33	22	37
Round	48	10 - n a	8	6 - 11	6	n a	n a
Saddle Blanket	21	14 - 42	42	19 - n a	n a	n a	n a
Starr	11	9 - 13	10	9 - 11	23	[3]11	[3]11
Swim	29	16 - 122	29	17 - 116	183	155	160

[1]Negative ground-water outflow computed, which cannot be explained by uncertainty in 1996 surface-water ouflow estimate.
[2]Negative ground-water outflow computed, but within uncertainty of 1996 surface-water outflow estimate.
[3]Not independently computed.

[Units in inches per year; gw, ground water; conc., concentration; %, percent; δD, delta deuterium, δ¹⁸O, delta oxygen-18; δ_a, isotopic composition of atmospheric moisture; n/a, numerically invalid solution; Cl, chloride]

Figure 10.57 Groundwater inflow to a number of lakes in Florida using mass balances of chloride and sodium, and of deuterium and oxygen 18. Groundwater inflows are reported in units of inches per year and computed using median concentrations. Sensitivity of groundwater inflows calculations is reported as ranges when ionic mass-balance concentrations are allowed to vary by 25%. The isotopic groundwater concentrations used to compute groundwater inflows used two approaches to setting groundwater concentrations: assuming atmospheric equilibrium and using data from Starr Lake measurements of Cl and the isotopes (Sacks et al. 1998).

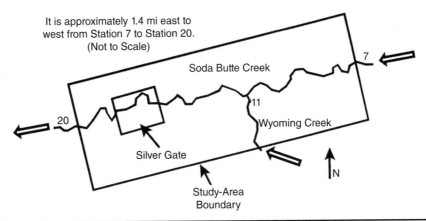

Figure 10.58 Schematic of Soda Butte Creek near Silver Gate, Montana, showing surface-water gauging stations and water-quality sites used in a mass-balance analysis of groundwater influx.

Well	Depth (ft bgs*)	SC (lmhos/cm)	Ca (mg/L)	Mg (mg/L)	Na (mg/L)	Cl (mg/L)
5-well ave	35–65	270	32.68	7.35	4.23	0.66
SW Station	Flow (L/s)	SC μmhos/cm	Ca (mg/L)	Mg (mg/L)	Na (mg/L)	Cl (mg/L)
7	192.86	189	22.1	5.84	3.93	0.52
11	15.86	81.1	5.37	1.96	8.1	0
20	277.82	218	26.5	6.02	3.87	0.6

*Below ground surface

TABLE 10.7   Water-Quality and Stream-Gauging Data near Silver Gate, Montana (from English 1999)

Specific cond.	Calcium	Magnesium	Sodium	Chloride	Average
82.1	93.4	70.8	45.3	99.1	78.1

TABLE 10.8   Summary of Estimated Groundwater Influx (L/s) Based on Groundwater Quality Data (from English 1999)

concentrations of radon because the water is open to the atmosphere. When surface water with low radon concentrations infiltrates into the bottom, bank, or perimeter sediments and flows away from the surface-water source, natural radon being generated in the sediments becomes incorporated into the new groundwater. So, the longer the new surface-water recharge is in contact with the saturated sediments the more radon it incorporates until equilibrium is reached. When radon samples are taken along a flow path of infiltrating surface water, flow rates can therefore be estimated (e.g., Sacks et al. 1998; Baskaran et al. 2009).

10.3   Summary

Effluent, influent, flow-though, and zero-exchange conditions describing groundwater exchange with streams, lakes, and wetlands occur at multiple scales and within varied time frames. Exchange magnitudes and locations can be fairly uniform or heterogeneous. A large body of work has concentrated on groundwater interactions with streams at river corridor, fluvial plain, and channel scales. Methods used to characterize exchange rates include water-budget and mass-balance analyses, stream-gauging techniques, synoptic surveys, mini-piezometer studies, use of seepage meters, temperature contrast analyses, tracer tests, and contrasts in geochemical parameters. The investigation of exchanges allows hydrologists, hydrogeologists, and ecologists to both qualitatively and quantitatively describe hydrologic systems in additional detail. The ecological health of surface-water systems requires some degree of exchange. Understanding the exchange processes can also be used to assess if hydrologic systems are impacted. The methods presented in this chapter provide exchange analyses at multiple levels.

10.4   Problems

10.1   Conceptual models of groundwater/surface-water exchange are presented in cross sections and map views.

A.   Cross sections representing a gaining stream segment that include flow lines and equipotential lines often are presented with flow lines crossing equipotential lines at right angles (e.g., Figure 10.1). What hydrogeologic conditions are represented by such a presentation?

B.   Examine Figure 10.11 and think about what the groundwater head distribution would look like if a vertical line were drawn across the center of the figure. Sketch a cross section along this line.

C.   If a colleague said that to study exchanges at rivers you need to construct transects of wells at right angles to the river channel. Would you agree with your colleague? Why or why not?

10.2   Do this problem prior to problem 10. In many data sets it is necessary to interpolate values from known points in a table. For example, using a Price AA flow meter, you measure a rotational rate on the meter of 43 revolutions in 55 s. Looking up the standard rating table for the meter, you find the following information:

Revolutions	Seconds	Water Velocity (ft/s)
30	55	1.20
40	55	1.59
50	55	1.98

Use Equation 10.16 interpolate the appropriate water velocity:

$$y = y_o + (x - x_o) \times \frac{y_1 - y_o}{x_1 - x_o} \tag{10.16}$$

y = outcome one is looking for; $y_1 = 1.98$; $y_0 = 1.59$; $x = 43$; $x_1 = 50$; $x_o = 40$.

10.3.   You and another teammate gathered the following data (in table). It is up to you to fill in the rest of the columns and determine the total discharge in CFS. Stations (column 1) represent where a 100-ft cloth tape was stretched across the stream [e.g., Station 14 is where the cloth tape coincides with the edge of water EOW and 4 ft later (Station 18) the water depth was 0.5 ft]. To determine the velocity (V) use the following relationships: V = 2.14N + 003 when N < 1.00 revolutions/sec and V = 2.19N + 0.01 when N is greater than or equal to 1.00 revolutions/sec. V is in ft/sec and N is computed by dividing the number of revolutions by the total seconds.

STREAM FLOW DATA

Date: _____   Transect: ___1___ of: ___1_____

Stream Name: ___Pine Creek_____   Begin Time: ___11:00 AM_____

Location: ___Sasquatch Co., MT____   End Time: ___11:50 AM_____

Client/Project: _____   Flow Meter ID: ___Price AA_____

Station	Depth (ft)	Rev/S	Velocity (fps)	Width (ft)	Area (ft²)	Flow (cfs)
14.0	EOW					
18.0	0.5	15/41				
22.0	0.9	30/40				
26.0	1.4	35/45				
28.0	1.5	40/46				
30.0	1.5	45/44				
32.0	1.5	40/42				
34.0	1.5	40/40				
37.0	1.3	45/42				
40.0	1.1	50/45				
43.0	0.9	50/47				
47.0	0.9	40/43				
51.0	0.8	40/41				
55.0	0.8	40/43				
59.0	0.9	30/46				
63.0	1.2	30/46				
67.0	1.0	30/41				
71.0	0.8	25/50				
75.0	EOW					
				Answer:		**cfs**

Survey Team (Your Name).

10.4   A hydrologist was preparing a monthly (30 days) water budget for a circular lake covering an area of 1 km². A stream with a daily discharge of 25 m³/day enters the lake but there is no surface-water discharge from the lake. The lake is set in a 10-m-thick gravel outwash plain that has an average hydraulic conductivity of 100 m/day and is underlain by till. There is a shallow water table and on one side groundwater flows toward the lake. Over a month period rainfall is 30 cm. Direct evaporation computed from an energy budget is 1.2 cm/day. After installing monitoring wells, it appears the groundwater is discharging to the lake along 1.75 km of shoreline through a saturated thickness of 5 m under a gradient of 0.004. On the other side of the lake, there is 1.75 km of shoreline where the water table is slightly lower than the lake level, the saturated thickness is 6 m and the gradient from the lake to the groundwater is 0.005. Compute the groundwater outflow based on the information given. Using a water-budget estimate the groundwater outflow by setting it as the unknown. Explain why the values are different and how would you choose a value to represent GW_{out}. What uncertainties would you assign to each parameter used in your calculations?

10.5   A crest-stage gauge is installed on a small stream. The stream has a sandy bed and few bushes are present on the bank. The gauge records a stage of 2.56 m. A survey of the channel cross section and slope yield the flood cross-sectional area of 25 m², a wetted perimeter of 16 m, and a channel slope of 0.03. Compute the discharge at the peak stage. Justify your selection of a value for any undefined parameters used in your calculation.

10.6 A number of seepage meter and mini-piezometer investigations were performed at Frenchtown Pond near Frenchtown, Montana (Figure 10.39). For this field work, seepage meters that are 60 cm in diameter were installed in a line at right angles from shore: M1 at 3 m, M2 at 4 m, and M3 at 5 m from the shoreline. One-liter lightweight prefilled bags with 100 mg/L of pond water were attached and allowed to operate for 1 h; 2.54-cm-diameter PVC mini-piezometers (MP1-MP3) about 2.2 m long with 5 cm of perforations at the bottom were installed to a depth below the pond bed of about 0.5 m. They were located about 50 cm from each seepage meter. They were constructed by driving a conductor casing with a solid central rod to a depth of 0.5 m, removing the center rod, installing the PVC mini-piezometer, and withdrawing the conductor casing. Mini-piezometers were developed by adding about 50 mL of pond water and measuring water levels to ensure that original static water levels returned once the added water left the bottom of the piezometer. The following data were collected at the site.

Seepage Meter	Vol after 1 h	Mini Piezometer	Head in the Piezometer Measured from the Top of MP	Pond Stage Measured from the Top of MP
M1	507 mL	MP1	65 cm	70 cm
M2	342 mL	MP2	33 cm	40 cm
M3	156 mL	MP3	9 cm	20 cm

A. Compute the seepage rate as mL/(m²h).
B. Compute the VHG at each mini-piezometer (MP).
C. Compute the vertical hydraulic conductivity in m/day.
D. Some literature suggests that seepage rates decrease logarithmically from the lake shoreline (e.g., McBride and Pfannkuch 1975). After reviewing your results, do they fit this conceptual model? Why or why not?
E. In the pond area sampled by the three meters what parameter seems to be controlling the seepage rates?

10.7 Suppose you wish to perform a tracer test using 20% rhodamine WT dye. Select a stream of your choice with a mean annual stream discharge between 1.0 and 5.0 cfs. Assume the conditions surrounding Equation 10.12 are valid for the stream you chose. However, the peak concentration at the downstream end is between 1.0 and 1.25 μg/L.
A. Estimate the volume of 20% rhodamine WT dye you will need for your study.
B. Use Equation 10.13 to estimate the distance downstream where complete mixing occurs. Justify your work.

10.8 Examine the temperature data for a losing section of the Umatilla River in Oregon (Figure 10.51). Choose a one-dimensional heat-flow model that could be used to interpret the transient temperature signals of the river and the groundwater at 50 cm and compute exchange rates [e.g., VS2DI, Healy and Ronan (1996), http://water.usgs/nrp/gwsoftware; Ex-stream, Swanson and Cardenas (2011), http://www.academia.edu/6228766/Ex-Stream_A_MATLAB_program _for_calculating_fluid_flux_through_sediment_water_interfaces_based_on_steady_and _transient_temperature_profiles; VFLUX, Gordon et al. (2012), http://hydrology.syr.edu/lautz _group/vflux.html]. Review one of the models and make a list of the required input parameters.

10.9 Read the paper by Batchelor and Gu (2014) that discusses interpreting tracer test in Boone and Winkler creeks of North Carolina. The results of the continuous injection are presented in Figure 10.54. They used the USGS model OTIS to match observations (e.g., OTIS, Runkel and Chapra 1993; Runkel 1998, http://water.usgs.gov/software/OTIS). Review the model documentation and input requirements. Make a list of the needed input data. How is the model adjusted to match field observation data when used by Batchelor and Gu (2014)?

10.10 A surface-water and groundwater mass-balance model was used to estimate groundwater discharge into a section of Soda Butte Creek near Silver Gate, Montana (Figure 10.58). Using the data presented in Table 10.7, solve for the total discharge of groundwater between Station 7 and Station 20 using calcium as the chemical constituent.

A. Write out the mass-balance equation in symbols and define the terms.
B. Compute the discharge in L/s of groundwater flow into the creek.
C. Discuss likely factors that you think would add uncertainty to the value.

10.5   References

Anderson, M.G., 1989. Physical Hydrology. *Progress in Physical Geograph: Earth and Environment*, Vol. 13, No. 1, pp. 93–102.

Anderson, M.P., 2005. Heat as a Ground Water Tracer. *Ground Water*, Vol. 43, No. 6, pp. 951–968.

Anderson, M.P., Woessner, W.W., and Hunt, R.J., 2015. *Applied Groundwater Modeling: Simulation of Flow and Advective Transport*. Academic Press-Elsevier, London, 564 pp.

Bartolino, J.R., and Niswonger, R.G., 1999. *Numerical Simulation of Vertical Ground Water Flux of the Rio Grande from Ground-Water Temperature Profiles, Central New Mexico*. U.S. Geological Survey Water-Resources Investigations Report 99-4212, 34 pp.

Baskaran, S., Ransley, T., Brodie, R.S., and Baker P., 2009. Investigating Groundwater–River Interactions Using Environmental Tracers. *Australian Journal of Earth Sciences*, Vol. 56, No. 1, pp. 13–19. doi:10.1080/08120090802541887

Batchelor, C., and Gu, C., 2014. Hyporheic Exchange and Nutrient Uptake in a Forested and Urban Stream in the Southern Appalachians. *Environmental and Natural Resources Research*, Vol. 4, No. 3, pp. 56–66.

Bencala, K.E., Gooseff, M.N., and Kimball, B.A., 2011. Rethinking Hyporheic Flow and Transient Storage to Advance Understanding of Stream Catchment Connections. *Water Resources Research*, Vol. 47, WH00H03. doi:10.1029/2010WR010066

Bencala, K.E., and Walters, R.A., 1983. Simulation of Solute Transport in a Mountain Pool-and-Riffle Stream: A Transient Storage Model. *Water Resources Research*, Vol. 19, No. 3, pp. 718–724.

Bhaskar, A.S., Harvey, J.W., and Henry, E.J., 2012. Resolving Hyporheic and Groundwater Components of Streambed Water Flux Using Heat as a Tracer. *Water Resources Research*, Vol. 48, W08524. doi:10.1029/2011WR011784

Bisson, P.A., Buffington, J.M., and Montgomery, D.R., 2006. Valley Segments, Stream Reaches and Channel Units. In Hauer, F.R., and Lamberti, G.A., eds., *Methods in Stream Ecology, 2nd Edition*, Academic Press, Amsterdam, pp. 23–49.

Boano, F., Harvey, J.W., Marion, A., Packman, A.I., Revelli, R., Ridolfi, L., and Worman, A., 2014. Hyporheic Flow and Transport Processes: Mechanisms, Models, and Biogeochemical Implications. *Reviews of Geophysics*, Vol. 52, pp. 603–679. doi:10.1002/2012RG000417

Boano, F., Packman, A.I., Cortis, A., Revelli, R., and Ridolfi, L., 2007. A Continuous Time Random Walk Approach to the Stream Transport of Solutes. *Water Resources Research*, Vol. 33, W10425. doi:10.1029/2007WR006062

Borduin, M.B., 1999. *Geology and Hydrogeology of the Sand Creek Drainage Basin, Southwest of Butte, Montana*. Master's Thesis, Montana Tech of the University of Montana, 103 pp.

Boulton, A.J., Findlay, S., Marmonier, P., Stanley, E.H., and Valett, H.M., 1998. The functional significance of the hyporheic zone in streams and rivers. *Annual Review of Ecology and Systematics*, Vol. 29, pp. 59–81.

Bouwer, H., and Rice, R.C., 1976. A Slug Test Method for Determining Hydraulic Conductivity of Unconfined Aquifers with Completely or Partially Penetrating Wells. *Water Resources Research*, Vol. 12, No. 3, pp. 423–428.

Boyle, D.R., 1994. Design of a Seepage Meter for Measuring Groundwater Fluxes in the Nonlitoral Zones of Lakes-Evaluation in a Boreal Forest Lake. *Limnology and Oceanography*, Vol. 39, No. 3, pp. 670–681.

Brunke, M., and Gonser, T., 1997. The Ecological Significance of Exchange Processes between Rivers and Groundwater. *Freshwater Biology*, Vol. 37, No. 1, pp. 1–31.

Buchanan, T.L., and Sommers, W.T., 1968. *Stage Measurement at Gauging Stations*. Techniques of Water-Resources Investigations of the U.S. Geological Survey, Book 3, Chapter A7. U.S. Geological Survey, Washington, DC.

Buchanan, T.L., and Sommers, W.T., 1969. *Discharge Measurements at Gauging Stations*. Techniques of Water-Resources Investigations of the U.S. Geological Survey, Book 3, Chapter A8. U.S. Geological Survey, Washington, DC.

Buss, S., Cal, Z., Cardenas, B., Fieckenstein, J., Hannah, D., Heppell, K., Hulme, P., Ibrahim, T., Kaeser, D., Krause, S., Lawier, D., Lerner, D., Mant, J., Malcolm, I., Old, G., Parkin, G., Pickup, R., Pinay, G., Porter, J., Rhodes, G., Richie, A., Riley, J., Robertson, A., Sear, D., Shields, B., Smith, J., Tellam, J., and Wood, P., 2009. *The Hyporheic Handbook: A Handbook on the Groundwater-Surface Water Interface and Hyportheic Zone for Environment Managers*. Integrated Catchment Science Programme. Science Report: SC050070. Environment Agency, Bristol, 264 pp.

Cardenas, M.B., 2015. Hyporheic Zone Hydrologic Science: A Historical Account of Its Emergence and a Prospectus. *Water Resources Research,* Vol. 51, pp. 3601–3616. doi:10.1002/2015WR017028

Carter, V., 1996. *Technical Aspects of Wetlands: Wetland Hydrology, Water Quality, and Associated Functions*. National Water Summary on Wetland Resources, U.S. Geological Survey Water Supply Paper 2425, 35–48 pp.

Castro, N.M., and Hornberger, G.M., 1991. Surface-Subsurface Water Interaction in an Alluviated Mountain Stream Channel. *Water Resources Research*, Vol. 27, No. 7, pp. 1613–1621.

Cherkauer, D.S., and McBride, J.M., 1988. A Remotely Operated Seepage Meter for Use in Large Lakes and Rivers. *Groundwater*, Vol. 26, pp. 165–171.

Chow, V.T., 1959. *Open Channel Hydraulics*. McGraw-Hill, New York.

Conant, B., Jr., 2004. Delineating and Quantifying Ground Water Discharge Zones Using Streambed Temperature. *Ground Water*, Vol. 42, No. 2, pp. 243–257.

Constantz, J., 1998. Interaction between Stream Temperature, Streamflow, and Groundwater Exchanges in Alpine Streams. *Water Resources Research*, Vol. 34, No. 7, pp. 1609–1616.

Constantz, J., 2008. Heat as a Tracer to Determine Streambed Water Exchanges. *Water Resources Research*, Vol. 44, W00D10. doi:10.1029/2008WR006996

Constantz, J., Jasperse, J., Seymour, D., and Su, G., 2002a. Use of Temperature to Estimate Streambed Conductance, Russian River, California. *American Water Resources Association*, pp. 595–600.

Constantz, J., Niswonger, R., and Stewart, A., 2008. Analysis of Sediment-Temperature Gradients to Determine Stream Exchanges with Ground Water, in Rosenberry, D., and Labaugh, J., eds., *Field Techniques for Estimating Water Fluxes between Surface Water and Ground Water Plea*, U.S. Geological Survey Techniques and Methods Report 4-D, pp. 128.

Constantz, J., Stewart, A.E., Niswonger, R.G., and Sarma, L., 2002b. Analysis of Temperature Profiles for Investigating Stream Losses Beneath Ephemeral Channels. *Water Resources Research*, Vol. 38, No. 12, pp. 52–1 to 52–13.

Constantz, J., and Thomas, C.L., 1997. Streambed Temperatures Profiles as Indicators of Percolation Characteristics beneath Arroyos in the Middle Rio Grande Basin, USA. *Hydrologic Processes*, Vol. 11, No. 12, pp. 1621–1634.

Constantz, J., Thomas, C.L., and Zellweger, G., 1994. Influence of Diurnal Variations in Stream Temperature on Streamflow Loss and Groundwater Recharge. *Water Resources Research*, Vol. 30, pp. 3253–3264.

Cox, S.E., Simonds, F.W., Doremus, L., Huffman, R.L., and Defawe, R.M., 2005. *Groundwater/Surface Water Interactions and Quality of Discharging Groundwater in Streams of the Lower Nooksack River Basin, Whatcom County, Washington*. U.S. Geological Survey Scientific Investigations Report, 2005–5255, 46 pp.

Cuthbert, M.O., and Mackay, R., 2013. Impacts of Non-Uniform Flow on Estimates of Vertical Streambed Flux. *Water Resources Research*, Vol. 49, No. 1, pp. 19–28. doi:10.1029/2011WR011587

Dahm, C.N., Grimm, N.B., Mamonier, P., Valett, H.M., and Vervier, P., 1998. Nutrient Dynamics at the Interface between Surface Waters and Groundwaters. *Freshwater Biology*, Vol. 40, pp. 427–451.

Dahm, C.N., Valett, H.M., Baxter, C.V., and Woessner, W.W., 2006. Hyporheic Zones. In Hauer, F.R., and Lamberti, G.A., eds., *Methods in Stream Ecology, 2nd Edition*, Academic Press, Elsevier, Burlington, MA, pp. 119–142.

Danielopol, D.L., 1980. The Role of the Limnologist in Groundwater Studies. *Internationale Revue der gesamten Hydrobiologie*, Vol. 65, pp. 777–791.

Dingman, S.L., 2014. *Physical Hydrology*. Waveland Press, Long Grove, IL, 643 pp.

Dinçer, T., 1968. The Use of Oxygen 18 and Deuterium Concentrations in the Water Balance of Lakes. *Water Resources Research*, Vol. 4, No. 6, pp. 1289–1306. doi.org/10.1029/WR004i006p01289

Domenico, P.A., and Schwartz, F.W., 2000. *Physical and Chemical Hydrogeology*, Wiley, New York.

Edwards, R.T., 2001. The Hyporheic Zone. In Naiman, R.J., and Bilby, R.E., eds., *River Ecology and Management*, Springer, New York, pp. 399–429.

English, A., 1999. *Hydrogeology and Hydrogeochemistry of the Silver Gate Area, Park County, Montana*. Master's Thesis, Montana Tech of the University of Montana, 131 pp.

Farinacci, A.J., 2009. Surface Water and Groundwater Exchanges in Fine and Coarse Grained River Bed Systems and Responses to Initial Stages of Dam Removal, Milltown, Montana. Graduate Student MS Theses, Dissertations, & Professional Papers. 10828. https://scholarworks.umt.edu/etd/10828, 222 pp.

Fetter, C.W., 2001. *Applied Hydrogeology, 4th Edition*. Prentice Hall, Upper Saddle River, NJ, 598 pp.

Fielding, C.R., Allen J.P., Alexander, J., and Gibling, M.R., 2009. Racies Model for Fluvial Systems in the Seasonal Tropics and Subtropics. *Geology*, Vol. 37, pp. 623–626.

Findlay, S., 1995. Importance of Surface-Subsurface Exchange in Stream Ecosystems—The Hyporheic Zone. *Limnology and Oceanography*, Vol. 40, pp. 159–164.

Freeze, R.A., and Cherry, J.A., 1979. *Groundwater*. Prentice Hall, Englewood Cliffs, NJ, 604 pp.

Glose, A.M., Lautz, L.K., and Baker, E.A., 2017. Stream Heat Budget Modeling with HFLUX: Model Development, Evaluation, and Application across and Contrasting Sites and Seasons. *Environmental Modelling and Software*, Vol. 92, pp. 217–228. doi:10.1016/j.envsoft.2017.02.021

Goto, S., Yamano, M., and Kinoshita, M., 2005. Thermal Response of Sediment with Vertical Fluid Flow to Periodic Temperature Variation at the Surface. *Journal of Geophysical Research*, Vol. 110, B01106. doi:10.1029/2004JB003419

Gordon, R.P., Lautz, L.K., Briggs, M.A., and McKenzie, J.M., 2012. Automated Calculation of Vertical Pore-Water Flux from Field Temperature Time Series Using the VFLUX Method and Computer Program. *Journal of Hydrology*, Vol. 420, pp. 142–158.

Grimm, N.B., and Fisher, S.G., 1984. Exchange between Interstitial and Surface Water: Implications for Stream Metabolism and Nutrient Cycling. *Hydrobiologia*, Vol. 3, pp. 219–228.

Gross, L.J., and Small, M.J., 1998. River and Floodplain Process Simulation for Subsurface Characterization. *Water Resources Research*, Vol. 34, No. 9, pp. 2365–2376.

Haggerty, R., McKenna, S.A., and Meigs, L.C., 2000. On the Late-Time Behavior of Tracer Test Breakthrough Curves. *Water Resources Research*, Vol. 36, No. 12, pp. 3467–3479. doi:10.1029/2000WR900214

Haggerty, R., and Reeves, P.C., 2002. STAMMT-L 1.0: *Formulation and User's Guide*, Technical Report, ERMS #520308, Sandia National Laboratory, Albuquerque, NM.

Hansen, E.A., 1975. Some Effects of Groundwater on Brown Trout Redds. *Transactions of the American Fisheries Society*, Vol. 104, No. 1, pp. 100–110.

Hancock, P.J., Boulton, A.J., and Humphreys, W.F., 2005. Aquifers and Hyporheic Zones: Towards an Ecological Understanding of Groundwater. *Hydrogeology Journal*, Vol. 13, pp. 98–111.

Hatch, C.E., Fisher, A.T., Revenaugh, J.S., Constantz, J., and Reuhl, C., 2006. Quantifying Surface Water-Groundwater Interactions Using Time-Series Analysis Using Thermal Streambed Records: Methods Development. *Water Resources Research*, Vol. 42, W10410. doi:10.1029/2005WR004787

Hauer, F.R., and Lamberti, G.A., eds., 2017. *Methods in Stream Ecology, 3rd Edition*. Vols. 1 and 2, Academic Press, Elsevier. Burlington, MA, 886 pp.

Hauer, F.R., and Resh, V.H., 2006. Macroinvertebrates. In Hauer, F.R., and Lamberti, G.A., eds., *Methods in Stream Ecology, 2nd Edition*. Academic Press, Elsevier, Burlington, MA, pp. 435–463.

Hays, J.R., 1966. *Mass Transport Phenomena in Open Channel Flow*. PhD Dissertation, Department of Chemical Engineering, Vanderbilt University, Nashville, TN.

Healy, R.W., and Ronan, A.D., 1996. *Documentation of the Computer Program VS2DH for Simulation of Energy Transport in Variably Saturated Porous Media-Modification of the U.S. Geological Survey's Computer Program VS2DT*. U.S. Geological Survey Water-Resources Investigation Report 96-4230, 36 pp.

Healy, R.W., Winter, T.C., LaBaugh, J.W., and Franke, O.L., 2007. *Water Budgets: Foundations for Effective Water-Resources and Environmental Management*. U.S. Geological Survey Circular 1308, 90 pp.

Heaney, M.J., Nyquist, J.E., and Toran, L., 2007. Marine Resistivity as a Tool for Characterizing Zones of Seepage at Lake Lacawac, PA. *Conference Symposium on the Application of Geophysics to Engineering and Environmental Problems.* doi:10.4133/1.2924642

Hornberger, G.M., Raffensperger, J.P., and Eshleman, K.N., 1998. *Elements of Physical Hydrology.* Johns Hopkins University Press, Baltimore, MD, 312 pp.

Hubbard, E.F., Kilpatrick, F.A., Martens, L.A., and Wilson, J.F., Jr., 1982. *Measurement of Time of Travel and Dispersion in Streams by Dye Tracing.* U.S. Geological Survey Techniques of Water-Resources Investigations 03-A9, 44 pp.

Huggenberger, P., Hoehn, E., Beschta, R., and Woessner W., 1998. Abiotic Aspects of Channels and Floodplains in Riparian Ecology. *Freshwater Biology,* Vol. 40, pp. 407–425.

Hvorslev, M.J., 1951. Time Lag and Soil Permeability in Ground Water Observations. *Bulletin 36,* U.S. Army Corps of Engineers, Waterways Experimentation Station, Vicksburg, MS.

Hynes, H.B.N., 1983. Groundwater and Stream Ecology. *Hydrobiologia,* Vol. 100, pp. 93–99.

Israelsen, O.W., and Reeve, R.C., 1944. *Canal Lining Experiments in the Delta Area.* Utah Technical Bulletin 313, Utah Agricultural Experiment Station, Logan, UT.

Jackman, A.P., Walters, R.A., and Kennedy, V.C., 1984. Transport and Concentration Controls for Chloride, Strontium, Potassium and Lead in Uvas Creek, a Small Cobble-Bed Stream in Santa Clara County, California, U.S.A. Part 2: Mathematical Modeling. *Journal of Hydrology,* Vol. 75, pp. 111–141.

Johnson, A.I., 1963. *Modified Parshall Flume.* U.S. Geological Survey Open File Report, Hydrologic Laboratory, Denver, CO, 6 pp.

Johnson, A.N., Boer, B.R., Woessner, W.W., Stanford, J.A., Poole, G.C., Thomas, S.A., and O'Daniel, S.J., 2005. Evaluation of an Inexpensive Small-Diameter Temperature Logger for Documenting Ground Water–River Interactions. *Groundwater and Remediation,* Vol. 25, No. 4, pp. 101–105.

Jones, J.B., and Holmes, R.M., 1996. Surface-Subsurface Interactions in Stream Ecosystems. *Trends in Ecology and Evolution,* Vol. 11, pp. 239–242.

Keery, J., Binley, A., Crook, N., and Smith, J.W.N., 2007. Temporal and Spatial Variability of Groundwater-Surface Water Fluxes: Development and Application of an Analytical Method Using Temperature Time Series. *Journal of Hydrology,* Vol. 336, No. 1–2, pp. 1–16.

Kilpatrick, F.A., 1970. Dosage Requirements for Slug Injections of Rhodamine BA and WT Dyes. In *Geological Survey Research,* U.S. Geological Professional Paper 700-B, pp. B250–B253.

Kilpatrick, F.A., and Cobb, E.D., 1985. *Measurement of Discharge Using Tracers.* Techniques of Water-Resources Investigations of the U.S. Geological Survey, Book 3, Chapter A16.

Kim, B.K.A., Jackman, A.P., and Triska, F.J., 1992. Modeling Biotic Uptake by Periphyton and Transient Hyporheic Storage of Nitrate in a Natural Stream. *Water Resources Research,* Vol. 28, pp. 2743–2752. doi:10.1029/92WR01229

Knochenmus, L.A., 2004. *Streamflow Losses through Karst Features in the Upper Peace River Hydrologic Area, Polk County, Florida, May 2002 to May 2003.* U.S. Geological Survey Fact Sheet 102-03, 4 pp.

Krabbenhoft, D.P., Bowser, C.J., Anderson, M.P., and Valley, J.W., 1990. Estimating Groundwater Exchange with Lakes 1. The Stable Isotope Mass Balance Method. *Water Resources Research,* Vol. 26, pp. 2445–2453.

Krabbenhoft, D.P., and Webster, K.E., 1995. Transient Hydrogeological Controls on the Chemistry of a Seepage Lake. *Water Resources Research,* Vol. 31, pp. 2295–2305.

Lee, D.R., and Cherry, J.A., 1978. A Field Exercise on Groundwater Flow Using Seepage Meters and Mini-Piezometers. *Journal of Geological Education*, Vol. 27, pp. 6–10.

Libelo, E.L., and MacIntyre, W.G., 1994. Effects of Surface-Water Movement on Seepage-Meter Measurements of Flow through the Sediment-Water Interface. *Applied Hydrogeology*, Vol. 2, No. 4, pp. 49–54.

Loustaunau, K.P., 2003. *Transport and Fate of Methyl Tertiary Butyl Ether (MTBE) in a Floodplain Aquifer and a Stream Interface, Ronan, Montana*. M.S. Thesis, Department of Geosciences, University of Montana, Missoula, 86 pp.

Lowe, R.L., and LaLiberte, G.D., 2006. Benthic Stream Algae: Distribution and Structure. In Hauer, F.R., and Lamberti, G.A., eds., *Methods in Stream Ecology, 2nd Edition*. Academic Press, Elsevier, Burlington, MA, pp. 327–339.

Malard, F., Tockner, K., Dole-Olivier, M.J., and Ward, J.V., 2002. A Landscape Perspective of Surface-Subsurface Hydrological Exchanges in River Corridors. *Freshwater Biology*, Vol. 47, pp. 621–640.

Marion, A., Zaramella, M., and Dottacin-Busolin, A., 2008. Solute Transport in Rivers with Multiple Storage Zones: The STIR Model. *Water Resources Research*, Vol. 44, W10406. doi:10.1029/2008/WR007037

McBride, M.S., and Pfannkuch, H.O., 1975. The Distribution of Seepage within Lakebeds. Journal of Research of the U.S. Geological Survey, Vol. 3, No. 5, pp. 505–512.

Meyer, J.L., 1997. Stream Health: Incorporating the Human Dimension to Advance Stream Ecology. *Journal of the North American Benthological Society*, Vol. 16, No. 2, pp. 439–447.

Mial, A.D., 1996. *The Geology of Fluvial Deposits: Sedimentary Facies, Basin Analysis, and Petroleum Geology*. Springer-Verlag, New York, 582 pp.

Morrice, J.A., Valett, H.M., Dahm, C.N., and Campana, M.E., 1997. Alluvial Characteristics, Groundwater-Surface Water Exchange and Hydrological Retention in Headwater Streams. *Hydrological Processes*, Vol. 11, pp. 253–267.

Mueller, D.S., and Wagner, C.R., 2009. *Measuring Discharge with Acoustic Doppler Current Profilers from a Moving Boat*. U.S. Geological Survey Techniques and Methods 3A–22, 72 pp. http://pubs.water.usgs.gov/tm3a22

Mwakanyamale, K., Slater, L., Day-Lewis, F., Elwaseif, M., and Johnson, C., 2012. Spatially Variable Stage-Driven Groundwater-Surface Water Interactions Inferred from Time-Frequency Analysis of Distributed Temperature Sensing Data. *Geophysical Research Letters*, Vol. 39, L06401. doi:10.10292011GL050824

O'Connor, B.L., Hondzo, M., and Harvey, J.W., 2010. Predictive Modeling of Transient Storage and Nutrient Uptake. *Journal of Hydraulic Engineering*, Vol. 136, No. 12.

Orghidan, T., 1959. Ein neuer Lebensraum des unterirdischen Wassers: Der hyporheische Biotop. *Archiv für Hydrobilogie*, Vol. 55, pp. 392–414.

Phillips, J.V., and Tadayon, S., 2006. *Selection of Manning's Roughness Coefficient for Natural and Constructed Vegetated and Non-Vegetated Channels, and Vegetation Maintenance Plan Guidelines for Vegetated Channels in Central Arizona*. U.S. Geological Survey Scientific Investigations Report 2006–5108, 41 pp.

Pollman, C.P., Lee, T.M., Andrews, W.J., Sacks, L.A., Gherini, S.A., and Munson, R.K., 1991. Preliminary Analysis of the Hydrologic and Geochemical Controls on Acid-Neutralizing Capacity of Two Acidic Seepage Lakes in Florida. *Water Resources Research*, Vol. 27, pp. 2321–2335.

Roberts, M., and Warren, K., 1999. *North Fork Blackfoot River Hydrologic Analysis*. Montana Department of Natural Resources and Conservation, Helena, MT, 36 pp.

Robertson, D.M., Rose, W.J., and Garn, H.S., 2003. *Water Quality and the Effects of Changes in Phosphorous Loading, Red Cedar Lakes, Barron and Washburn Counties, Wisconsin.* U.S. Geological Survey Water-Resources Investigations Report 03-4238, 42 pp.

Rosenberry, D.O., 2005. Integrating Seepage Heterogeneity with the Use of Ganged Seepage Meters. *Limnology and Oceanography Methods*, Vol. 3, pp. 131–142.

Rosenberry, D.O., and LaBaugh, J.W., 2008. *Field Techniques for Estimating Water Fluxes Between Surface Water And Ground Water.* U.S. Geological Survey Techniques and Methods 4–D2, 128 pp.

Rosenberry, D.O., LaBaugh, J.W., and Hunt, R.J. 2008. Use of Monitoring Wells, Portable Piezometers, and Seepage Meters to Quantify Flow between Surface Water and Ground Water. In Rosenberry, D.O., and LaBaugh, J.W., eds., *Field Techniques for Estimating Water Fluxes Between Surface Water and Ground Water*, U.S. Geological Survey Techniques and Methods 4-D2, pp. 39–70.

Rosenberry, D.O., and Menheer, M.A., 2006. *A System for Calibrating Seepage Meters used to Measure Flow between Ground Water and Surface Water.* U.S. Geological Survey Scientific Investigations Report 2006–5053, 21 pp.

Rosenberry, D.O., and Morin, R.H., 2004. Use of an Electromagnetic Seepage Meter to Investigate Temporal Variability in Lake Seepage. *Groundwater*, Vol. 42, pp. 68–77.

Runkel, R.L., 1998. *One-Dimensional Transport with Inflow and Storage (OTIS): A Solute Transport Model for Streams and Rivers.* U.S. Geological Survey Water Resources Investations Report 98-4018, 73 pp.

Runkel, R.L., and Chapra, S.C., 1993. An Efficient Numerical Solution of the Transient Storage Equations for Solute Transport in Small Streams. *Water Resources Research*, Vol. 29, pp. 211–215. doi:10.1029/92WR02217

Runkel, R.L., McKnight, D.M., and Andrews, E.D., 1998. Analysis of Transient Storage Subject to Unsteady Flow: Diel Flow Variation in an Antarctic Stream. *Journal of the North American Benthological Society*, Vol. 17, pp. 143–154.

Sacks, L.A., Swancar, A., and Lee, T.M., 1998. *Estimating Ground-Water Exchange with Lakes Using Water-Budget and Chemical Mass-Balance Approaches of Ten Lakes in Ridge Areas of Pol and Highlands Counties, Florida.* U.S. Geological Survey Water-Resources Investigations Report 98-4133, 52 pp.

Sanders, L.L., 1998. *A Manual of Field Hydrogeology.* Prentice Hall, Upper Saddle River, NJ, 381 pp.

Sauer, V.B., and Meyer, R.W., 1992. *Determiniation of Error in Individual Discharge Measurements.* U.S. Geological Survey Open File Report 92-144, 21 pp.

Sauer, V.B., and Turnipseed, D.P., 2010. *Stage Measurement at Gaging Stations.* U.S. Geological Survey Techniques and Methods Book 3, Chap. A7, 45 pp. http://pubs.usgs.gov/tm/tm3-a7.

Schmidt, C., Conant, B., Jr., Bayer-Raich, M., and Schirmer, M., 2007. Evaluation and Field-Scale Application of an Analytical Method to Quantify Groundwater Discharge Using Mapped Streambed Temperatures. *Journal of Hydrology*, Vol. 341, Nos. 3–4, pp. 292–307.

Schwartz, F.W., and Zhang, H., 2003. *Fundamentals of Ground Water.* Wiley, New York, 592 pp.

Selker, J.S., Thévenaz, L., Huwald, H., Mallet, A., Luxemburg, W., van de Giesen, N., Stejskal, M., Zeman, J., Westhoff, M., and Parlange, M.B., 2006. Distributed Fiber-Optic Temperature Sensing for Hydrologic Systems. *Water Resources Research*, Vol. 42. W12202. doi:10.1029/2006WR005326

Shaw, R.D., and Prepas, E.E., 1989. Anomalous, Short-Term Influx of Water into Seepage Meters, *Limnology and Oceanography*, Vol. 34, No. 7, pp. 1343–1351.

Smart, P.L., and Laidlaw, I.M.S., 1977. An Evaluation of Some Fluorescent Dyes for Water Tracing. Water *Resources Research*, Vol. 13, No. 1, pp. 15–33.

Stallman, R.W., 1965. Steady One-Dimensional Fluid Flow in a Semi-Infinite Porous Medium with Sinusoidal Surface Temperature. *Journal of Geophysical Research*, Vol. 70, No. 12, pp. 2821–2827.

Stanford, J.A., and Simons, J.J., eds., 1992. *Proceedings of the First International Conference on Ground Water Ecology*, American Water Resources Association, Bethesda, MD.

Stanford, J.A., and Ward, J.V., 1988. The Hyporheic Habitat of River Ecosystems. *Nature*, Vol. 335, pp. 64–66.

Stanford, J.A., and Ward, J.V., 1993. An Ecosystem Perspective of Alluvial Rivers: Connectivity and the Hyporheic Corridor. *Journal of the North American Benthological Society*, Vol. 12, No. 1, pp. 48–60.

Stauffer, R.E., 1985. Use of Solute Tracers Released by Weathering to Estimate Groundwater Inflow to Seepage Lakes. *Environmental Science and Technology*, Vol. 19, pp. 405–411.

Stonedahl, S.H., Harvey, J.W., and Packman, A.I., 2013. Interactions between Hyporheic Flow Produced by Stream Meanders, Bars, and Dunes. *Water Resources Research*, Vol. 9, pp. 5450–5461. doi:10.1002/wrcr.20400

Stonestrom, D.A., and Constantz, J., 2003. Heat as a Tracer of Water Movement Near Streams. In Stonestrom, D.A., and Constanz, J., eds., *Heat as a Tool for Studying the Movement of Ground Water Near Streams*, USGS Circular 1260. USGS, Reston, VA, pp. 1–6.

Stumm, W., and Morgan, J.J., 1996. *Aquatic Chemistry, 3rd Edition*. John Wiley & Sons, New York, 1022 pp.

Sudbrink, A., 2007. Investigation of Contaminated Groundwater within the Comet Mine Reclamation Site, High Ore Creek, Drainage Basin, Southwestern MT. MS Thesis Montana Tech of the University of Montana, Butte, MT.

Swanson, T.E., and Cardenas, M.B., 2011. Ex-Stream: A MATLAB Program for Calculating Fluid Flux through Sediment-Water Interfaces Based on Steady and Transient Temperature Profiles. *Computers and Geosciences*, Vol. 37, No. 10, pp. 1664–1669.

Thackston, E.L., and Schnelle K.B., 1970. Predicting Effects of Dead Zones on Stream Mixing. *Journal of the Sanitary Engineering Division*, Vol. 96, pp. 319–331.

Tóth, J., 1963. A Theoretical Analysis of Groundwater Flow in Small Drainage Basins. *Journal of Geophysical Research*, Vol. 68, pp. 4795–4812. doi:10.1029/JZ068i016p04795

Triska, F.J., Kennedy, V.C., Avanzino, R.J., Zellweger, G.W., and Bencala, K.E., 1989. Retention and Transport of Nutrients in a Third-Order Stream in Northwestern California: Hyporheic Processes. *Ecology*, Vol. 70, No. 6, pp. 1893–1905.

Turner, J.V., Allison, G.B., and Holmes, J.W., 1984. The Water Balance of a Small Lake Using Stable Isotopes and Tritium. *Journal of Hydrology*, Vol. 70, pp. 199–220.

Turnipseed, D.P., and Sauer, V.B., 2010. *Discharge Measurements at Gaging Stations*. U.S. Geological Survey Techniques and Methods Book 3, Chap. A8, 87 pp. http://pubs .usgs.gov/tm/tm3-a8

University of Wyoming, College of Agriculture, 1994. *Irrigation Water Measurement, Irrigation Ditches and Pipelines*. Cooperative Extension Service Bulletin 583 R, University of Wyoming Laramie, WY, 70 pp.

USGS (U.S. Geological Survey), 1980. *National Handbook of Recommended Methods for Water-Data Acquisition (Updates)*. Office of Water Data Coordination, U.S. Geological Survey, U.S. Department of Interior, Reston, VA.

Valentine, E., and Wood, I., 1979. Experiments in Longitudinal Dispersion with Dead Zones. *Journal of the Hydraulics Division*, Vol. 105, pp. 999–1016.

Valett, H.M., Hakenkamp, C.C., and Boulton, A.J., 1993. Perspectives on the Hyporheic Zone: Integrating Hydrology and Biology: Introduction. *Journal of the North American Benthological Society*, Vol. 12, pp. 40–43.

Viessman, W, Jr., and Lewis, G.L., 2002. *Introduction to Hydorlogy, 5th Edition*. Pearson, New York, 612 pp.

Wallace, J.B., Grubaugh, J.W., and Whiles, M.R., 1996. Biotic Indices and Stream Ecosystem Process: Results from an Experimental Study. *Ecological Applications*, Vol. 6, pp. 140–151.

Ward, A.S., 2016. The Evolution and State of Interdisciplinary Hyporheic Research. *WIREs Water*, Vol. 3, pp. 83–103. doi:10.1002/wat2.1120

Watson, I., and Burnett, A.D., 1993. *Hydrology: An Environmental Approach (Theory and Applications of Ground Water and Surface Water for Engineers and Geologists)*. Buchanan Books, Cambridge, 702 pp.

Webb, B.W., Hannah, D.M., Moore, R.D., Brown, L.E., and Nobilis, R., 2008. Recent Advances in Stream and River Temperature Research. *Hydrological Processes*, Vol. 22, pp. 901–918. doi:10.1002/hyp.6994

Weight, W.D., 2008. *Hydrogeology Field Manual, 2nd Edition*. McGraw-Hill, New York, 751 pp.

Weight, W.D., and Chandler, K.M., 2010. Hydraulic Properties of Rocky Mountain First-Order Alluvial Systems and Diurnal Water-Level Fluctuations in Riparian Vegetation. *Journal of Environmental Science and Engineering*, Vol. 4, No. 9, pp. 12–23.

Wilson, J.F., 1968. *An Empirical Formula for Determining the Amount of Dye Needed for Time-of-Travel Measurements in Geological Survey Research*. U.S. Geological Survey Professional Paper 600-D, pp. D54–D56.

Winter, T.C., 1981. Uncertainties in Estimating the Water Balance of Lakes. *Water Resources Bulletin*, Vol. 17, pp. 82–115.

Winter, T.C., 2001. The Concept of Hydrologic Landscapes. *Journal of the American Water Resources Association*, Vol. 37, No. 2, pp. 335–349.

Winter, T.C., Harvey, J.W., Franke, O.L., and Alley, W.M., 1998. Ground Water and Surface Water: A Single Resource. *USGS Circular*, 1139, 79 pp.

Woessner, W.W., 1998. Changing Views of Stream-Groundwater Interaction. In Van Brahana, J., Eckstein Y., Ongley L.W., Schneider R., and Moore, J.E., eds., *Proceedings of the Joint Meeting of the XXVIII Congress of the International Association of Hydrogeologists and the Annual Meeting of the American Institute of Hydrology*, American Institute of Hydrology, St. Paul, MN, pp. 1–6.

Woessner, W.W., 2000. Stream and Fluvial Plain Ground-Water Interactions: Re-Scaling Hydrogeologic Thought. *Ground Water*, Vol. 38, No. 3, pp. 423–429.

Woessner, W.W., 2017. Hyporheic Zones. In Hauer, F.R., and Lamberi, G.A., eds., *Methods in Stream Ecology, 3rd Edition*. vol. 1, Academic Press-Elsevier, Cambridge, MA, pp. 129–157. doi:10.1016/B978-0-12-416558-8.00008-1

Woessner, W.W., 2018. Permission for reproduction of original figures and tables, Missoula, MT.

Wörman, A., 1998. Analytical Solution and Timescale for Transport of Reactive Solutes in Rivers and Streams. *Water Resources Research*, Vol. 34, pp. 2703–2716. doi:10.1029/98WR01338

Wörman, A., Packman, A.I., Johansson, H., and Jonsson, K., 2002. Effect of Flow-Induced Exchange in Hyporheic Zones on Longitudinal Transport of Solutes in Stream and Rivers. *Water Resources Research*, Vol. 38, No. 1, pp. doi:1001, 10.1029/2001WR00769

Wright, J.F., 1995. Development and Use of a System for Predicting the Macroinvertebrate Fauna in Flowing Waters. *Australian Journal of Biology*, Vol. 20, pp. 181–197.

Zamora, C., 2006. Estimating Rates of Exchange across the Sediment/Water Interface in the Lower Merced River, CA. Master's Thesis, California State University, Sacramento, CA, 110 pp.

Zamora, C., 2008. *Estimating Water Fluxes across the Sediment–Water Interface in the Lower Merced River, California.* U.S. Geological Survey Scientific Investigations Report 2007–5216, 47 pp. http://pubs.usgs.gov/sir/2007/5216

CHAPTER 11

Vadose Zone Hydrogeology: Basic Principles, Characterization, and Monitoring

Daniel B. Stephens

Principal, Daniel B. Stephens & Associates, Inc.

Todd Umstot

Senior Hydrogeologist, Daniel B. Stephens & Associates, Inc.

The vadose zone is generally defined as the geologic media between land surface and the regional water table (Figure 11.1). The upper part of the vadose zone commonly includes the plant root zone and weathered soil horizons. Within the vadose zone, soils and bedrock are usually unsaturated, that is, their pores are only partially filled with water.

In places, however, the vadose zone may become water saturated; consequently, the vadose zone is not synonymous with the unsaturated zone. One obvious location where this occurs is just above the regional water table where capillary rise causes water to fill the pore spaces. Here, the pores are saturated but the water is held under tension, that is, the fluid pressure is less than atmospheric pressure. Saturated regions with positive pressures may also be found in the vadose zone; such as above a low-permeable layer where perched conditions may develop (Chapter 3). A perched aquifer occurs within and is part of the vadose zone, and it is separated from the regional water table below by an unsaturated zone. These unsaturated zones above and below the perched aquifer are critical to the definition and field identification of a perched aquifer. In the vadose zone, saturated conditions also may develop locally beneath surface impoundments

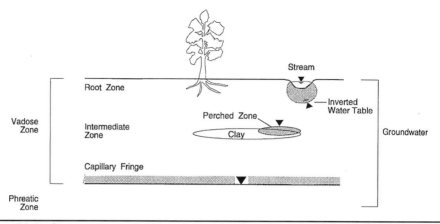

FIGURE 11.1 Conceptual model of the vadose zone. [*From Stephens (1996). With permission of Taylor & Francis Group LLC.*]

and drainages as well as over more extensive areas near land surface during infiltration following precipitation events or flood irrigation.

To many hydrogeologists, groundwater occurs at a depth where water first enters a well. This is not uncommon for definitions set for various groundwater-related regulations. But Figure 11.1 shows that, by our preferred definition, the groundwater system includes the vadose zone as well. The groundwater system also includes the geologic medium below the regional water table, the phreatic zone, which will be an aquifer, aquiclude, or aquitard.

11.1 Energy Status of Pore Water

What causes water to flow in the vadose zone? Just as in aquifers, the vadose zone water flows from areas of high-potential energy to areas of low-potential energy. To determine the direction of the driving force acting on water in the vadose zone, one needs to quantify the total potential energy of the soil water. The gradient of this total soil-water potential (the change in potential energy with direction in space) determines the magnitude of the driving force acting on the fluid. Mathematically, the total soil-water potential, ϕ_T, is expressed as

$$\phi_T = \phi_{sw} + \phi_G \tag{11.1}$$

Notice that there are two primary components to the total soil-water potential, just as there are for the total hydraulic head in an aquifer (Chapter 4). The first component is the soil-water potential, ϕ_{sw}, and the second is simply the gravitational potential, ϕ_G.

The soil-water potential accounts for soil-water movement due to gradients in capillarity, pressure, temperature, chemical, and electrical potential. These components of the soil-water potential have been derived from fundamental thermodynamic principles. We will avoid these details here and summarize the main components of the soil-water potential:

$$\phi_{sw} = \phi_{matric} + \phi_{pressure} + \phi_{osmotic} \tag{11.2}$$

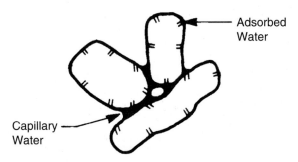

Adsorbed
Water

Capillary
Water

Figure 11.2 Matric potential is due to water held in soil by capillary and adsorptive effects. [*From Stephens (1996). With permission of Taylor & Francis Group LLC.*]

The matric component, ϕ_{matric}, recognizes that the energy state of water, relative to the reference state of an open body of pure water at 20°C, is influenced by the mutual attraction between the liquid, gas, and solid phases of the soil. In the definition of matric potential, we assume that the other components, such as air pressure and chemistry of the pore fluids, are the same as those in the reference reservoir. Matric potential is attributed to capillary forces which predominate in wet soils and adsorptive forces which are more important in dry soils. Water held by capillarity occurs in the main pore bodies, whereas adsorbed water occurs where the polar water molecules coat the charged surfaces of the solid soil particles (Figure 11.2). In clayey soils, water can also be bound to soil particles in the electrostatic double layers. Where the soil is fully saturated and no air-water interface is present, $\phi_{matric} = 0$.

In the relatively wet range, which is generally at matric potentials greater than about −1 bar, the matric component is expressed by the capillarity equation:

$$\phi_{matric} = \frac{2\sigma}{r} \tag{11.3}$$

where σ is the interfacial tension of the liquid (MT^{-2}), and r is the radius of curvature on the air-water meniscus. Ideally, the surface tension of water is 72 dyn/cm. If the meniscus is concave toward the air, r is negative by convention. Therefore, the matric potential is negative as referenced to a free-water surface at zero potential. Equation 11.3 can be written in terms of the radius of a capillary tube, r_t, by recognizing that $r = r_t / \cos \alpha_c$:

$$\phi_{matric} = \frac{2\sigma \cos \alpha_c}{r_t} \tag{11.4}$$

where α_c is the contact angle made by the liquid-gas interface where it contacts the solid.

The osmotic component, $\phi_{osmotic}$, takes account of the chemical concentration differences that may influence the energy state of the water. Increasing the chemical concentration of the pore fluid will lower the potential energy of the liquid relative to the pure-water reference state.

After considering all the components of total soil-water potential in the vast majority of practical vadose zone problems, matric potential remains as the most important to characterize, inasmuch as gradients of the other components of the soil-water potential are usually small relative to the soil-water potential gradient and gravitational gradient.

For computations, it is most convenient, however, to express soil-water potential as the potential energy per unit weight, so that the gravitational potential can be expressed in units of length and calculated simply as the elevation of the measurement point above some datum (Chapter 4). When the soil-water potential is expressed as a potential energy per unit weight, the soil-water potential is called the pressure head, ψ. Many soil scientists and engineers use different units of pressure to express soil-water potential, so it is useful to be familiar with some unit conversions:

$$1 \text{ bar} \approx 0.99 \text{ atm} \approx 101 \text{ kPa} \approx 1{,}017 \text{ cm H}_2\text{O} \qquad (11.5)$$

To convert the volume-based soil-water potential, ϕ_{sw}, to pressure head, ψ, the following equation is used:

$$\psi = \frac{\phi_{sw}}{\rho g} \qquad (11.6)$$

where ρ is the fluid density and g is the gravitational constant. When the potential energy of water in soil is expressed on a per unit weight basis, the total potential energy Equation 11.1 becomes

$$H = \psi + z \qquad (11.7)$$

where H is the total hydraulic head and z is the elevation head relative to an arbitrary datum. This is exactly the same equation routinely used by hydrogeologists to compute hydraulic head and hydraulic gradients in aquifers, except that this form is more general in that, with our definition of ψ, water movement can occur under unsaturated as well as saturated conditions.

One of the key characteristics of the vadose zone is that the soil-water potential in partially saturated soil pores is less than zero, relative to a free-water surface. The soil-water potential, including pressure head, would be positive within perched zones and below the water table, whereas at the water table, the soil-water potential would be essentially zero (where zero represents atmospheric pressure). Elsewhere, the soil-water potential is negative. There is a continuum of pressure in the vadose zone and this continuum extends into the underlying phreatic zone as well (Chapter 4).

11.2 Water Content

The water content of a soil is probably the easiest property of the vadose zone to understand. However, different disciplines deal with this property in different ways. Three common expressions for water content are the following:

$$\text{Volumetric water content: } \theta = \frac{\text{volume water}}{\text{bulk volume soil}} \ (\text{cm}^3 \text{ cm}^{-3}) \qquad (11.8)$$

$$\text{Gravimetric water content: } \theta_g = \frac{\text{mass water}}{\text{mass dry soil}} \ (g \ g^{-1}) \qquad (11.9)$$

$$\text{Saturation percentage: } S = \frac{\theta}{n} \times 100 \ (\text{cm}^3 \text{ cm}^{-3}) \qquad (11.10)$$

Most hydrologists and soil scientists prefer the volumetric water content, while geotechnical engineers generally use gravimetric water content, and petroleum engineers often work with saturation percentage. The volumetric water content is typically 40% to 60% greater than the gravimetric water content because of the following relationship:

$$\theta = \frac{\rho_b}{\rho_w} \theta_g \qquad (11.11)$$

where the dry bulk density, ρ_b, is typically 1.4 to 1.6 g/cm³, and the density of water, ρ_w, is 1.0 g/cm³.

11.3 Soil-Water Retention Curves

There is a very important relationship that exists between the soil-water potential and the water content that is illustrated in the so-called soil-water retention curve or **soil-water characteristic curve** (Figure 11.3).

One way to illustrate the relationship between tension (soil-water potential) and water content is to represent the variable pore sizes of a field soil as a bundle of vertical capillary tubes having different diameters. If this bundle of tubes is inserted into the water table, water will be drawn up into the tubes until a static equilibrium condition is achieved (Figure 11.4b). The equation to compute the height of water rise in a capillary tube (Figure 11.4a) reveals that the smaller the tube radius, the greater the height of capillary rise. Up to a certain distance above the water table, all the tubes are filled with water. This height above the water table defines the upper limit of the capillary fringe. For clayey soils, the height of the capillary fringe may exceed 2 m in clay, and in gravel, it may only rise 10 cm. At greater distances above the water table, notice that fewer tubes are filled with water. By looking at the proportion of the tubes which are water-filled at progressively greater elevations, it is evident that the mean

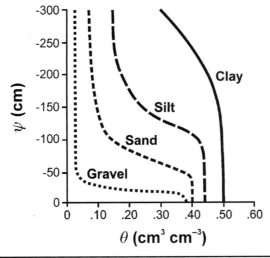

FIGURE 11.3 Effect of texture on the soil-water characteristic curve. [*From Stephens (1996). With permission of Taylor & Francis Group LLC.*]

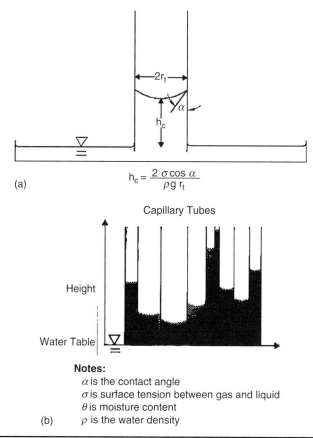

$$h_c = \frac{2\,\sigma\cos\alpha}{\rho g\, r_t}$$

(a)

Capillary Tubes

Height

Water Table

Notes:
α is the contact angle
σ is surface tension between gas and liquid
θ is moisture content
(b) ρ is the water density

FIGURE 11.4   (a) Height of water rise in a capillary tube and (b) water rise in a bundle of capillary tubes representing a range of pore sizes. [*From Stephens (1996). With permission of Taylor & Francis Group LLC.*]

water content within the bundle of tubes decreases with increasing height above the water table (Figure 11.4b).

Thus, by analogy to the capillary tube bundle, the amount of water held in the soil at a prescribed pressure head or tension is dependent on soil texture, as shown in Figure 11.3. If a soil is saturated and begins to drain (e.g., water table falls), coarse soils drain faster than fine-textured soils.

There are several features of a soil-water characteristic curve that are commonly referenced or measured (Figure 11.5). During drainage, the pressure head at which the air first begins to enter the previously saturated soil is called the air-entry value. Sand and gravel lose a large portion of their water just after the pressure head decreases below the air-entry value, whereas clay tends to lose much less water with decreasing pressure head.

Another term associated with soil-water characteristic curves is the field capacity. After 2 or 3 days of drainage following a thorough rain or irrigation, the rate of gravity drainage slows considerably, and the water content at this stage of drainage is called the

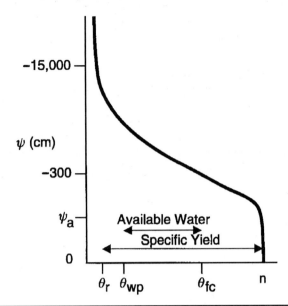

FIGURE 11.5 Indices describing the soil-water characteristic curves: θ_r is residual water content, θ_{wp} is permanent wilting point, θ_{tc} is field capacity, n is porosity, and Ψ_a is air-entry value. [*From Stephens (1996). With permission of Taylor & Francis Group LLC.*]

field capacity. Although there is no good alternative for measuring field capacity other than in the field, field capacity often is arbitrarily quantified in laboratory tests as the water content corresponding to –100 cm or –300 cm pressure head (about –1/10 to –1/3 bars; Figure 11.5).

With additional slow drainage or evapotranspiration (ET) to a very dry condition, the water content may become so low that many plants wilt. By convention, the water content corresponding to –15,000 cm pressure head (about –15 bars) is called the permanent wilting point, although the soil-water potential near many desert plants is even less, in the range of –40 to –80 bars or less. The water-content difference between field capacity and permanent wilting (or –1/3 bar minus –15 bar water contents) is called the available soil water (to plants) (Figure 11.5).

As the soil dries below the permanent wilting point, the water content approaches an asymptote, so that with decreasing pressure head no further water drains from the soil. This is called the residual water content.

11.4   Darcy's Equation and Unsaturated Flow Parameters

Perhaps the most widely recognized equation among soil scientists, hydrologists, and petroleum engineers is Darcy's equation. Buckingham (1907), a soil scientist, demonstrated that Darcy's equation could be extended to unsaturated conditions as well.

If only water is the fluid of interest, then Darcy's equation is written as

$$q_i = -K(\theta)_{ij}\left(\frac{\partial \psi}{\partial x_i} + \frac{\partial z}{\partial x_i}\right) \tag{11.12a}$$

where i and j are cartesian coordinates [e.g., horizontal (x,y) and vertical direction (z)], and z is positive upward. Where the soil is homogenous and isotropic and flow is vertical, then Darcy's equation becomes

$$q_z = - K(\theta) \left(\frac{\partial \psi}{\partial z} + \frac{\partial z}{\partial z} \right) = - K(\theta) \left(\frac{\partial \psi}{\partial z} + 1 \right) \qquad (11.12b)$$

Darcy's equation simply states that fluid flow is a function of the driving force called hydraulic gradient (pressure and gravity terms in brackets) and a constant of proportionality called the hydraulic conductivity, K (Chapter 3). The hydraulic conductivity accounts for the viscous flow and frictional losses that occur as a fluid moves through the porous medium.

Hydraulic Gradient

The hydraulic gradient in the vadose zone exhibits interesting characteristics that contrasts markedly with those hydrogeologists are accustomed to in aquifers. In aquifer systems, flow is primarily horizontal, and the regional hydraulic gradient is often in the range of 10–4 to 10–3; it is rare that the hydraulic gradient ever exceeds 0.01. There are a few exceptions such as at seepages faces, where groundwater flows across faults and aquitards, and in close proximity to pumped wells. But in the vadose zone, hydraulic head gradients near 1.0 (unit gradient) are common. Near unit hydraulic gradients may occur in deep vadose zones with uniform texture where the soil-water content is constant with depth. A unit hydraulic gradient indicates that the soil water is flowing vertically downward. When the gradient is unity, the magnitude of the flux, q, equals the hydraulic conductivity, K(θ).

Although the hydraulic gradient is often near unity, the hydraulic gradient can be many orders of magnitude larger near sharp wetting fronts in dry soils. On the other hand, the hydraulic gradient may also be much less than unity and, in fact, is zero where no flow occurs.

Unsaturated Hydraulic Conductivity

The following equation further explains how the hydraulic conductivity is a function of the fluid properties, the media properties, and the water content, θ:

$$K(\theta) = \left(\frac{k \rho g}{\mu} \right) k_r (\theta) \qquad (11.13)$$

where k = intrinsic permeability of the medium (L^2), ρ = fluid density (ML^{-3}), g = gravitational constant (LT^{-2}), μ = dynamic viscosity of fluid ($MT^{-1} L^{-1}$), and $k_r(\theta)$ = relative permeability (dimensionless, ranges from 0 to 1). In Equation 11.13, the quantity in brackets represents the familiar saturated hydraulic conductivity for isotropic conditions. The relative permeability, sometimes called relative hydraulic conductivity, is a dimensionless parameter that accounts for the dependence of the hydraulic conductivity on pressure head or water content, as shown in Figure 11.6.

The relative hydraulic conductivity decreases rapidly with decreasing water content. As drainage progresses, smaller and smaller pores are left holding water. As the water content decreases, the path of water flow becomes more tortuous and the cross-sectional area of water in the pores decreases. In the dry range, the relative

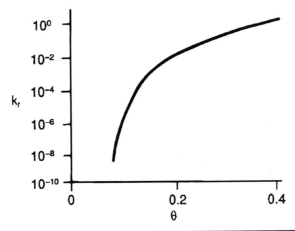

Figure 11.6   Relative hydraulic conductivity, K_r, versus water content, θ. Porosity is 0.4 cm³/cm⁻³. [*From Stephens (1996). With permission of Taylor & Francis Group LLC.*]

hydraulic conductivity becomes very small, so at low-water content, the hydraulic conductivity may be perhaps more than a million-fold smaller than the saturated hydraulic conductivity.

The hydraulic conductivity of variably saturated media is highly dependent upon soil texture (Figure 11.7). Hydrogeologists and engineers are well aware of the nature of spatial variability in saturated hydraulic conductivity that is attributed to variability in the intrinsic permeability (Equation 11.13) of the geologic material (Chapters 2 and 3). For instance, well-sorted sand typically has a saturated hydraulic conductivity of about 10^{-2} cm/s, whereas clay may have a saturated hydraulic conductivity of about 10^{-6} cm/s. But over the range of water contents likely to be encountered in the vadose zone, the unsaturated hydraulic conductivity of a single soil sample may change by one-million- or one-billion-fold or more.

It is especially important to recognize that at low-pressure head or water content, the unsaturated hydraulic conductivity of a fine-textured soil may be greater than that of a coarse soil. For most hydrogeologists and engineers, this is a paradox, in that the soil with the highest intrinsic permeability (Equation 11.13) can have the lowest hydraulic conductivity. However, this fact can be very important in forming conceptual models about vadose zone processes of flow and transport, particularly in heterogeneous or layered media.

11.5   Soil-Water Budget

The general equation for the soil-water budget is derived by considering the mechanisms by which water can enter, exit, or be stored in a predefined region of the vadose zone. For many problems, the inflow across the upper boundary of the vadose zone is infiltration, while outflow from the upper boundary is evaporation and transpiration, and outflow from the lower boundary is groundwater recharge. Net inflow (inflow minus outflow) must equal the change in soil-water stored in the vadose zone:

$$I - E - T - R = \Delta S \qquad (11.14)$$

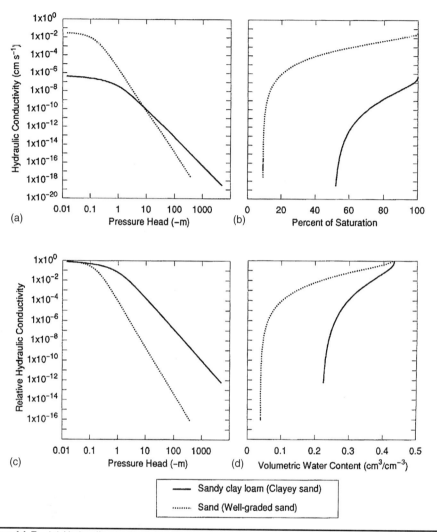

FIGURE 11.7   (a) Hydraulic conductivity, K, versus pressure head, Ψ, for sand and sandy clay loam; (b) hydraulic conductivity versus water content; (c) relative hydraulic conductivity versus pressure head; and (d) relative hydraulic conductivity versus percent saturation. [*From Stephens (1996). With permission of Taylor & Francis Group LLC.*]

where I is cumulative infiltration, E is evaporation, T is transpiration, R is deep percolation or recharge (all four variables in units of L^3T^{-1}), and ΔS represents the change in water storage over the time interval Δt. Now we will briefly describe each of the principal soil-water budget components, and present some of the methods to obtain them.

Infiltration

Perhaps the most widely studied subsurface hydrological process has been infiltration. Infiltration is the process whereby water enters the soil from surficial sources, such as rainfall, snowmelt, flooding, irrigation, liquid waste spills, etc. Infiltration is usually regarded as a process that occurs predominantly in the vertical direction when water is applied over extensive areas of the land surface. The infiltration process in any geometry is controlled by many factors, such as the precipitation characteristics, rate of water application, antecedent moisture in the soil, soil hydraulic properties, topography, and others.

First consider a thick, uniform-textured, permeable soil that is subject to a gentle but constant rainfall rate over a flat-lying terrain (Figure 11.8a). In this case, we are describing a one-dimensional infiltration process in which the rainfall intensity of the rain is less than the magnitude of the saturated hydraulic conductivity of the soil. All the water supplied is free to infiltrate, so none is available for runoff. Because the capacity of the soil to conduct water is greater than the rate of supply, the soil will be unsaturated. And, if the rain continues for a prolonged period at this rate, the unsaturated hydraulic conductivity of the soil will approach the rain intensity.

Next consider the case in which a constant rainfall rate in excess of the saturated hydraulic conductivity is applied to initially dry soil. The soil initially takes up the rainwater as rapidly as it falls (Figure 11.8b), owing to the strong capillary forces in the dry soil. As the soil pores fill with water, the influence of the capillary forces on infiltration diminish gradually. Therefore, as the soil becomes wetter and as the depth to the wetting front increases, the hydraulic head gradient decreases. Consequently, the infiltration rate decreases over time. If the rain persists, ponding on the soil surface will occur at time, t_p, when the soil cannot transmit water at a rate greater than the rainfall rate. When ponding occurs, the soil surface layer is saturated, except perhaps for entrapped air. After ponding, the infiltration rate continues to diminish and approaches the field-saturated hydraulic conductivity of the soil. Finally, for the case where storm intensities that are high relative to the saturated hydraulic conductivity, it is probable that ponding will begin almost instantly (Figure 11.8c). Figure 11.8 also illustrates

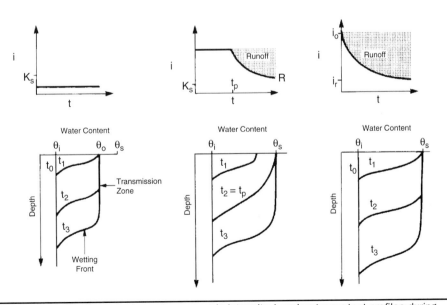

FIGURE 11.8 Infiltration rate dependence on rain intensity, i, and water-content profiles during (a) low, (b) moderate, and (c) intense precipitation. [*From Stephens (1996). With permission of Taylor & Francis Group LLC.*]

the water-content profiles associated with the low- and moderate-intensity rain infil-tration as well as ponded infiltration into a uniform soil.

Infiltration is usually quantified in one of three ways: (1) by the residual from a surface-water budget analysis, (2) by field measurements, or (3) by calculation based on soil hydrau-lic properties. In a surface-water budget, infiltration may be computed by measurements of precipitation, applied irrigation, and surface run-on, and subtracting from this sum the surface runoff, interception of water on plant canopies, direct evaporation, and increases in surface-water storage. Methods to quantify each of these components of a surface-water budget can be found in standard engineering hydrology texts and reference books.

In the field, infiltration from precipitation as rainfall can be measured by weighing and totaling in precipitation gages, estimated from surrounding recording stations [e.g., Daymet (Thorton et al. 2018) and PRISM (Daly et al. 1994)], or it is simulated by using a rainfall simulator to quantify infiltration over an area of roughly 1 to 10 m² for controlled duration and intensity storms (e.g., Zegelin and White 1982). Under ponded-conditions a ring infiltrometer, a short cylinder approximately 0.1 to 1 m² in cross section driven into the soil, can be used to determine the infiltration rate by measuring flow through the cylinder under equivalent ponded-head conditions. Beneath impound-ments, lakes, or reservoirs, seepage meters (Chapter 10) may provide reliable infiltra-tion measurements (e.g., Stephens et al. 1985). Infiltration from stream channels, where evaporation can be neglected, is most readily obtained by subtracting discharges measured at stream gauging stations along the reach of the channel. Infiltration can also be calculated from hydraulic properties of the soil using mathematical expressions which describe the infiltration process, such as Darcy's equation.

To describe the one-dimensional vertical infiltration process, Green and Ampt (1911) developed the following simple but very useful equation:

$$i = K_o \left[\frac{(H_p + L) - \psi_f}{L} \right]$$
(11.15)

where K_o is the hydraulic conductivity behind the wetting front, H_p is depth of ponding, L is depth to the wetting front, and $\psi_f < 0$. This equation is virtually identical to Darcy's equation, with the term in brackets representing the hydraulic gradient. The cumula-tive infiltration is given by $L \times (Qs-Qi)$ where Qs is the saturated water content and Qi is the initial water content. Unlike the usual application of Darcy's equation to flow in soil columns or aquifers, L is time dependent in the Green-Ampt equation. Inspection of Equation 11.15 shows that for a constant depth of ponding, as the depth to the wet-ting front increases, the gradient eventually approaches one regardless of the magni-tude of the ponding depth. Consequently, the late-time, steady infiltration rate is independent of the ponding depth.

Undoubtedly, the most widely referenced mathematical equation on infiltration is that by Philip (1957). Where the depth of ponding is very thin, the simplified form of Philip's transient infiltration equation is

$$i = \frac{1}{2} St^{-1/2} + K_o$$
(11.16)

where S is called the sorptivity of the soil, t is time since infiltration began, and K_o is the hydraulic conductivity of the soil's upper layer. Sorptivity is a soil hydraulic property that embodies several factors describing the capacity of the soil to imbibe water at early time, such as the initial dryness of the soil (e.g., Brutsaert 1977). Note that as $t \to \infty$, the late-time infiltration rate approaches K_o, just as we discussed previously for the Green-Ampt equation. Solutions to Philip's equation can be used to predict not only the transient infiltration rate, but also water-content profiles.

Evaporation and Transpiration

Evaporation refers to the water lost from the vadose zone by vapor phase transport from the soil directly to the atmosphere. Transpiration is the water depleted from the vadose zone by plant root uptake. For most practical problems it is both difficult and unnecessary to separate these two processes, so the two are combined and called **evapotranspiration**. There are two different approaches to determine ET: by measurement and by estimation.

Soil lysimeters provide the most accurate method to measure ET, and in fact most estimation methods for ET have been verified by comparing the estimates to measured lysimeter data. Unfortunately, soil lysimeters are usually cumbersome, in some cases expensive to construct, and require rather long periods of data collection. There are three types of lysimeters used for water-balance analysis: weighing, non-weighing, and floating. All soil lysimeters share the same concept. A small monolith of soil with vegetation is placed in a container and is returned to its original position in the landscape. Instrumentation is emplaced to allow measurements of precipitation, soil-water storage, and deep drainage. From these components, one can compute ET by simple arithmetic using the water-balance equation (Equation 11.14). In a weighing lysimeter, the lysimeter is placed on a scale (Figure 11.9). In a non-weighing lysimeter, ET from the soil monolith is

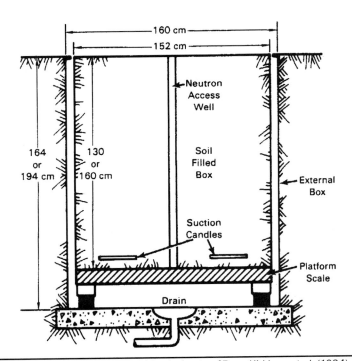

Figure 11.9 Cross-sectional view of weighing lysimeter. [*From Kirkham et al. (1984). With permission of ASA, CSSA, and SSSA.*]

determined by measuring the rate of water supply to the monolith container that is necessary to maintain a constant depth to water in the base of the container. And in a floating lysimeter, the soil monolith is placed on a liquid-filled pillow so that water gains and losses can be obtained by measuring fluid pressure through a manometer tube.

Another approach to measure ET is to place a canopy over the plant and measure airflow rate and water content of the inflowing and outflowing air (Sebenik and Thomas 1967).

Methods for making micrometeorological measurements above a vegetated surface have been developed over the past 50 years or so to determine the actual ET (Rosenberg et al. 1983). Two such methods are the Bowen ratio method (e.g., Tanner 1960) and the eddy-correlation method (Swinbank 1951). The Bowen ratio method is based upon a simplified energy-budget equation (e.g., Marshall and Holmes 1992).

Instrumentation requirements include a net radiometer, soil heat flux, and temperature and vapor pressure at two elevations above the surface. The eddy-correlation method is based on the principle that water vapor flux across the land surface can be measured by correlating the vertical variations of wind speed, w, with variations of vapor density, q_v (Tanner et al. 1985). The instrumentation requirements include an anemometer, hygrometer, and thermocouples that are connected to a data logger.

The so-called eddy-correlation/energy-budget method (e.g., Czarnecki 1990) is used to determine the actual evaporative flux when field instrumentation accounts for net radiation and heat conduction into the ground, as in the Bowen ratio method, and when sensible heat flux is determined by the eddy-correlation technique.

Because of the obvious logistical difficulties associated with measuring ET over extensive areas, estimates of ET are usually preferred. There are two broad approaches to estimate this component of the soil-water budget: climatological and micrometeorological methods. Among the climatological methods, some are based on air temperature (e.g., Thornthwaite 1948; Blaney and Criddle 1950); others are derived from solar radiation measurements (e.g., Jensen and Haise 1963), and others incorporate both energy supply data and turbulent transfer of water vapor away from the surface (e.g., Penman 1948). A study by Jensen et al. (1990) concluded that these combination energy balance/aerodynamic methods provide the most accurate estimates.

While actual ET is the quantity we seek, it is important to recognize that the above climatological methods calculate the potential evapotranspiration (PET); that is, the amount of ET that would occur from a short green crop that fully shades the ground exerts negligible resistance to the flow, and is always well supplied with water.

To compute the ET from PET when the water supply is limited requires an additional calculation based upon plant type, water availability, and vegetation coverage on the landscape:

$$ET = K_c \, (PET) \tag{11.17}$$

where K_c is a crop coefficient. Class A type evaporation pans are often used to determine PET:

$$PET = Cp \times Epan \tag{11.18}$$

where Cp is a pan coefficient (typically about 0.7) and Epan is the pan evaporation. The crop coefficient is usually obtained by establishing an experimental relationship

Vegetation	K_{co}							
	Nov to Mar	Apr	May	Jun	Jul	Aug	Sep	Oct
Sagebrush-grass	0.50	0.60	0.80	0.80	0.80	0.71	0.53	0.50
Pinyon-juniper	0.65	0.70	0.80	0.80	0.80	0.80	0.69	0.65
Mixed mountain shrub	0.60	0.67	0.81	0.85	0.82	0.74	0.65	0.60
Coniferous forest	0.70	0.71	0.80	0.80	0.80	0.79	0.75	0.71
Aspen forest	0.60	0.67	0.85	0.90	0.86	0.75	0.65	0.60
Rockland and miscellaneous	0.50	0.60	0.65	0.65	0.65	0.60	0.50	0.50
Phreatophytes	1.00	1.00	1.00	1.00	1.00	1.00	1.00	1.00

From McWhorter and Sunada, 1977. With permission.

TABLE 11.1 Estimated Plant-Water-Use Coefficients K_{co} for Native Vegetation

between ET (measured with lysimeters) and PET (calculated by a specific method) for some brief period. Crop coefficients for agricultural crops are summarized by Doorenbos and Pruitt (1975) and values estimated for selected native vegetation are presented in Table 11.1 from McWhorter and Sunada (1977). The dependence of the crop coefficient upon available water (AW) is described by a wide variety of formulations (Moridis and McFarland 1982). Perhaps currently the most widely used method is that published as Food and Agriculture Organization (F.A.O.) Irrigation and Drainage Paper 56 (Allen et al. 1998). Allen et al. (1998) also describe the application of remote sensing techniques to obtain crop coefficients.

11.6 Water Storage and Deep Percolation

To quantify water-storage changes requires repeated measurements of water content within the soil-water budget volume. Over a year or several years, the water-content change is usually small, and for some sites and climatic conditions, the long-term change is negligible.

Neutron probe logging affords a means to nondestructively measure water-content changes at the same depths without introducing the sampling bias accompanying destructive soil sampling techniques. From the discrete water-content measurements, θ, the volume of water in storage is calculated as

$$\Delta S = \frac{1}{\Delta t} \left[\int_0^D \theta \, dz \right] \cdot (\text{area}) \tag{11.19}$$

The rate of change in water storage is calculated by subtracting the water storage to depth D at two different time periods and dividing this by the time between monitoring events. In weighing soil lysimeters, the change in water storage within the monolith can be simply obtained from the change in mass divided by the water density.

Deep percolation is the water that moves downward below the root zone. Often the deep percolation below the root zone will become groundwater recharge when it reaches the water table. Groundwater recharge is water that enters either a perched aquifer or the phreatic zone. One of several ways water can enter aquifers is by migration of deep percolation through the vadose zone. Deep percolation beneath the root zone has been observed to follow preferential flow paths that occupy only a small portion of the vadose zone (Kung 1990). In the following discussion, we summarize methods to calculate deep percolation and recharge to an aquifer from the vadose zone by physical and chemical methods. For a detailed review of methods to compute groundwater recharge, see, for example, Scanlon et al. (2002).

11.7　Physical Methods

Soil Lysimeters

Soil lysimeters for the purpose of collecting deep drainage and estimating recharge are constructed by excavating soil to the desired depth, installing casing with a sealed base, and repacking the casing with soil to the in situ bulk density (Figure 11.10). Water percolating under tension can be collected at the bottom of the lysimeter by extracting it from porous ceramic cups or tubes that are subject to a vacuum that exceeds the soil-water potential. Alternatively, the water accumulated at the base of

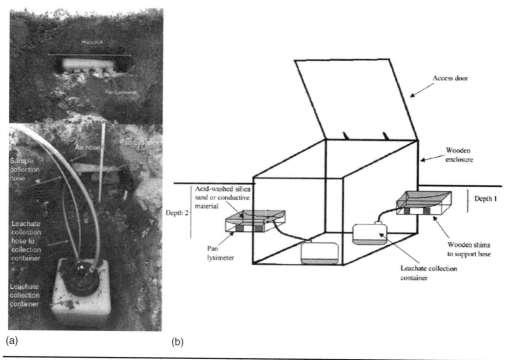

(a)　　　　　　　　　　　　(b)

Figure 11.10　(a) Field design of a pan lysimeter and (b) and installation sketch of the pan lysimeter. [*Used by permission of Singh et al. (2018).*]

the lysimeter can be obtained by measurements of water-content change with a neutron probe or by piezometers to measure the depth of saturation.

Water Balance

One of the most widely adopted approaches to determine deep percolation or groundwater recharge is the water-balance method. In this approach, recharge is calculated as the residual in the water-budget equation (Equation 11.14).

Darcy Flux in the Vadose Zone

Darcy flux calculations also comprise a physical means to calculate recharge from the vadose zone. The components in Darcy's equation for unsaturated vertical flow include the hydraulic conductivity at the field water content (or potential) and the hydraulic gradient. In soils below the root zone where the matric potential is nearly constant with depth, it is a good assumption that the hydraulic gradient is approximately unity, and that flow is downward. Consequently, where this assumption is reasonable and where vapor phase transport downward is negligible, the recharge rate is approximately equal to the in situ vertical, unsaturated hydraulic conductivity.

Numerical Models of Soil-Water Flow

There are two types of numerical models relevant to calculating deep percolation and recharge: water-balance models and the models based on Richards' equation. Richards' equation for unsaturated flow can be written as

$$\nabla \cdot K(\psi)\,\nabla H = C(\psi)\,\frac{\partial \psi}{\partial t} \tag{11.20}$$

where C is the specific water capacity, which is the slope of the soil-water retention curve.

The water-balance models include codes such as HELP (Schroeder et al. 1984), GLEAMS (Leonard et al. 1989), PRZM-2 (Mullins et al. 1993), PRMS (Markstrom et al. 2015), and INFIL (USGS 2008).

All these vadose zone water-balance models partition precipitation into runoff and infiltration. Infiltration is further separated into components such as ET, lateral drainage or interflow, soil-water storage, and deep percolation by applying deterministic and empirical equations that describe each of the processes. Actual ET is computed from climatic data (e.g., precipitation, temperature, solar radiation), input from on-site measurements or from default daily historic data for the nearest location stored in the program library. Other factors such as the vegetation cover and rooting characteristics also enter into the ET analysis. Water, which cannot be held in storage or extracted by the plants, becomes available for deep percolation. Some models such as PRZM-2 and INFIL3 (Hevesi et al. 2003) take the deep percolation output from the water balance in the root zone and also route this through the deeper vadose zone using Darcy's equation for one-dimensional, unsaturated flow.

One of the algorithms often used to compute recharge in the soil-water budget models is based on the concept of field capacity. Percolation below any soil layer is allowed in the models only if the water content exceeds the field capacity.

There are a large number of numerical models for simulating soil-water processes, including finite difference and finite element forms, based on one-, two-, or three-dimensional forms of Richards' equation (e.g., HYDRUS, VS2D, and TOUGH2).

To account for infiltration and ET in these codes, in lieu of detailed meteorological information, the upper boundary of the model and/or the root zone is usually specified as either a constant or a time-varying flux or pressure head. In contrast to some of the water-balance models, the numerical models allow the user to more realistically represent the physical properties of the porous medium, including complex geology with spatially varying hydraulic conductivity and water retention characteristics. When the lower boundary of the model is specified as the water table, the water flux out the base of the model represents the groundwater recharge.

Chemical Methods in the Vadose Zone

Among the chemical methods for calculating recharge, there are stable and radioactive isotopes that serve as a means to compute recharge, including tritium, chlorine-36, chloride, and oxygen-18 and deuterium. One of the advantages of some of these methods is that the analysis may represent an integration of hydrologic events over decades or even tens of thousands of years. Another advantage is that the data are derived from in situ sampling, without need for field instrumentation for monitoring.

Tritium

Tritium, a radioactive isotope of hydrogen with a half-life of about 12.4 years, is well suited as a hydrologic tracer because it is part of the water molecule. During the atmospheric nuclear testing beginning in the 1950s, tritium in the atmosphere increased substantially over a relatively short time, culminating in the period 1963–1964 (Phillips et al. 1988). The record of tritium has been preserved in atmospheric water that infiltrated the soil profile. Recharge, or more precisely, net infiltration, is obtained from the depth to the center of mass of the tritium pulse, L, with the following equation:

$$R = \theta \frac{L}{\Delta t} \tag{11.21}$$

where θ is the mean water content through depth L.

Chlorine-36

Another tracer of soil-water flux or recharge is chlorine-36. This is a radioactive isotope with a half-life of about 300,000 years, produced as a by-product of thermonuclear testing near the oceans in the 1950s (Bentley et al. 1982). Chloride is very stable in the environment and enters the hydrologic cycle as the chloride ion dissolved in water and as a component of dust fallout. Because it is soluble and non-volatile, chloride is an excellent tracer for liquid-phase transport. Recharge can be determined from the depth of the chlorine-36 peak, in the same manner as described above for tritium.

Chloride Mass Balance

The chloride mass-balance method relies upon the slow accumulation in the soil profile of natural chloride that dissolves in precipitation and infiltrates. The concentration of chloride typically decreases with increasing inland distance from the coasts. The expected chloride pattern in the soil profile, at least in areas of modest precipitation, is that chloride is concentrated with increasing soil depth, as water is extracted by the plant roots; and below the root zone the chloride concentration is expected to be constant where the deep percolation migrates toward the water table.

To interpret chloride distribution in the vadose zone, three fundamental assumptions are made: (1) all chloride in the vadose zone originates from atmospheric deposition, (2) the only long-term sink for chloride is downward advection or dispersion (e.g., no runoff), and (3) chloride behaves conservatively during soil-water transport. If dispersion and macropore flow are neglected, a simple piston-displacement model is derived in which the chloride concentration increases in proportion to the ratio of precipitation to recharge (Allison and Hughes 1978):

$$R = P \frac{Cl_p}{Cl_s} \tag{11.22}$$

where P is the average precipitation rate, Cl_p is chloride concentration in bulk precipitation (wet and dry fallout), and Cl_s is the average soil chloride concentration in pore water below the root zone. If runoff is present, the rate of recharge estimated using Equation 11.22 may be overestimated.

Chloride patterns that depart from this model may produce a bulge in the chloride concentration at some depth in the profile, and below this proturbance the concentration decreases to approach a near constant value. Phillips (1994) discussed possible explanations for this behavior in desert soils such as preferential or bypass flow in macropores but suggested that the low concentration of chloride at depth most likely reflects greater recharge during a wetter paleoclimatic period when the indigenous plants were less efficient at capturing the soil moisture.

11.8 Characterizing Hydraulic Properties

In this section, we present some of the field and laboratory techniques that are available for obtaining parameters that can be used in quantitative analyses and model simulations of vadose zone processes. For additional details, refer to Stephens (1996).

Soil-Water Characteristic Curve

To obtain the moisture retention or soil-water characteristic curve, laboratory methods are most often chosen. There are essentially two types of laboratory methods for measuring soil-water characteristic curves. The first is the hanging water-column method in which the soil-water in the sample is subjected to a tension, and the second is the family of pressure plate techniques. In both techniques, the hanging water-column apparatus uses a Büchner funnel (available from scientific glassware suppliers), a glassware that contains a fritted glass, porous plate (Figure 11.11). The pore size of the plate is so fine that after the plate is saturated, the water tension must decrease to −0.1 or −0.3 bars before the air will displace water from the pores of the plate.

The soil sample is placed in the Büchner funnel on the upper side of the porous plate. A plastic tubing and burette assembly control the tension on the underside of the Büchner funnel which causes water to drain from the soil. After equilibrium is achieved, the amount of drainage is recorded in the burette, and then the burette is lowered again and the process continues to a mean negative pressure head equal to the distance between the center of the sample and the water level in the burette. Additionally, the distance between the center of the soil sample and water level in the burette is recorded as a measure of the pressure head in equilibrium with the water still held in the soil. Upon completion, the final water content of the sample is determined. The water

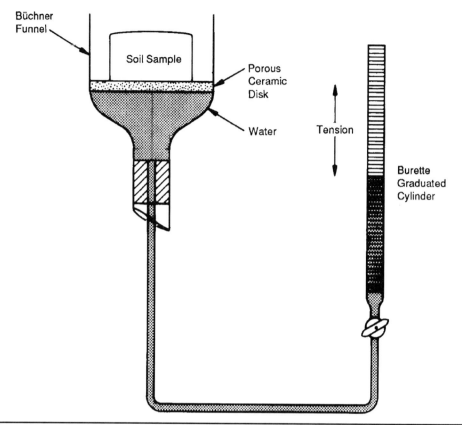

FIGURE 11.11 Hanging water-column (Büchner funnel) apparatus. [*From Stephens (1996). With permission of Taylor & Francis Group LLC.*]

content associated with all other previous "tensions" achieved in the test is calculated by adding to the final water content the incremental volumes of water drained or imbibed from each step. In this way, one obtains pairs of water-content and pressure head data.

The hanging water-column method is practical for relatively wet conditions to pressures as low as about –0.3 bars, depending upon the air-entry value of the porous plate in the Büchner funnel.

The pressure cell methods include Tempe cells (Figure 11.12a), pressure plate apparatus, and pressure membrane apparatus (Figure 11.12b). Tempe cells have ceramic plates that remain saturated to approximately –2 bars; the pressure plate apparatus uses ceramic plates with a range to about –15 bars, and pressure membranes of cellulose acetate are useful to about –150 bars.

The basic assembly contains a pressure chamber comprised of a rigid cylinder fitted tightly on the top by a lid or removable plate. Attached to the bottom of the pressure chamber is another removable or fixed end plate. Above the bottom plate is a water-saturated porous disc (ceramic plate, cellulose acetate, or visking membrane) that is seated tightly against the wall of the cylinder by a rubber gasket or O-ring. A wet-soil

FIGURE 11.12 Water retention apparatus: (a) Tempe cells and (b) pressure membrane apparatus. (*Courtesy of Soilmoisture Equipment Corp.*)

sample in a sample ring is placed on the porous plate and the top and bottom plates are tightly attached. A nitrogen gas is applied at constant pressure to the sample chamber through a fitting in the cylinder wall. Water is forced out of the sample through the porous disc and into a collection tube, until water outflow ceases. The moisture content still retained in the soil sample is presumed to be in equilibrium with the applied pressure.

For the very dry range, psychrometric and centrifuge methods are also useful.

Expressions for Soil-Water Characteristic Curves

Expressions for soil-water characteristic curves provide a mathematical relation between pressure and water content, which are commonly needed for numerical models of soil-water flow. The expressions assume that the soil follows a particular model or relationship between pressure and moisture content. Two commonly used expressions for water retention characteristics are the Brooks and Corey (1964) and the van

Genuchten (1978, 1980) equations. The Brooks and Corey (1964) equation for water retention characteristics is as follows:

$$S_e = \left(\frac{\psi_{cr}}{\psi}\right)^{\lambda} \quad \text{for } \psi < \psi_{cr} \tag{11.23a}$$

$$S_e = 1 \quad \text{for } \psi \geq \psi_{cr} \tag{11.23b}$$

where ψ_{cr} is the critical pressure or bubbling pressure and λ is the pore size distribution index. The van Genuchten (1978; 1980) equation for water retention characteristics is as follows:

$$S_e = [1 + |\alpha_v \psi|^N]^{-m} \tag{11.24}$$

where α_v (1/L) and N (dimensionless) are fitting parameters, m is $1 - 1/N$, and S_e is effective saturation $(\theta - \theta_r) \div (\theta_s - \theta_r)$. Parameters for the water retention curves are obtained by fitting the equations to observed moisture retention (θ versus ψ) data or by using pedotransfer functions. Fitting the equations to observe data is typically performed using a nonlinear least-squares computer routine [e.g., RETC (van Genuchten et al. 1991)] (Table 11.2).

Saturated Hydraulic Conductivity

Laboratory Methods

Because saturated hydraulic conductivity is one of the most widely characterized hydraulic properties of soils, representative values have been published for a wide range of soils (Figure 11.13). Laboratory tests involve removing a small sample of soil and testing it in the laboratory under controlled conditions. Because of concern for problems of sample disturbance and representativeness of the relatively small sample, field methods are often preferred. Although field methods may overcome the limitations of the laboratory methods, they often suffer from lack of control on the experimental conditions and other logistical problems.

There are in general two types of laboratory tests for saturated hydraulic conductivity: constant-head and falling-head permeameters. In a constant-head permeameter, water is introduced into the sample by maintaining inflow and outflow reservoirs at constant positions relative to the sample (Figure 11.14). The steady-flow rate, sample length and cross-sectional area, difference in reservoir elevations, and water temperature are used to calculate hydraulic conductivity according to Darcy's equation (Equation 4.1; Chapter 4). In a falling-head permeameter, water is introduced to a water-saturated sample by gravity drainage from a standpipe, while the head on the downstream end of the sample remains constant (Figure 11.15). There are standards published by the American Society for Testing and Materials (ASTM) for laboratory tests on granular and fine-textured soils (ASTM D-2434-68 and D-5084-90). For compressible soils in particular, such as clay liner material, it is also important to conduct the tests under confined conditions that reproduce the overburden pressures. Soil engineers use fixed wall and flexible wall cells (e.g., triaxial cells) to accomplish this (e.g., Daniel 1989, 1993). For additional guidance refer to appropriate ASTM standards or reference materials such as Klute and Dirksen (1986).

Soil Type	Catalog Number	van Genuchten Curve Parameters		Residual Water Content
		α (1/cm)	N	
Silt "Columbia"	2,001	0.015511	1.7676	0.1369
Silt Mont Cenis (limon Silteaux)	2,002	0.013647	1.3234	0.0000
Silt of Nave-Yaar	2,003	0.072010	2.1969	0.3979
Rideau clay loam	3,101	0.069118	2.0604	0.2863
Yolo light clay	3,102	0.027000	1.6000	0.1800
Caribou silt loam	3,301	0.047125	1.6981	0.2956
Grenville silt loam	3,302	0.030702	1.2878	0.0326
Ida silt loam(>15 cm)	3,305	0.040000	1.2700	0.0000
Ida silt loam (0–15 cm)	3,306	0.089975	1.1768	0.0000
Touched silt loam	3,308	0.027302	3.5385	0.0993
Silt loam G.E. 3	3,310	0.004233	2.0594	0.1313
Gilat loam	3,402	0.017000	2.3000	0.0846
Guelph loam	3,407	0.073566	1.7844	0.2193
Rubicon sandy loam	3,501	0.052321	1.8570	0.1388
Loamy Sand-Hamra Sharon	4,004	0.018695	5.1537	0.1997
Plainfield sand (210–250 μ)	4,101	0.045177	3.9979	0.0102
Plainfield sand (177–210 μ)	4,102	0.038611	4.0409	0.0099
Plainfield sand (149–177 μ)	4,103	0.032170	4.0570	0.0069
Plainfield sand (125–149 μ)	4,104	0.024903	5.8327	0.0283
Plainfield sand (104–125 μ)	4,105	0.022127	4.4446	0.0148
Sand	4,106	0.094490	2.0422	0.0000
Sand	4,107	0.060000	2.6400	0.0400
Del Norte fine sand	4,108	0.016254	4.3600	0.0505
Oakley sand	4,112	0.095194	2.0136	0.0255
G.E. 3 sand	4,115	0.035965	4.4892	0.0409
Crab Creek sand	4,117	0.118896	2.4506	0.0000
Sinai sand	4,122	0.023803	5.3076	0.0326
Sand (50–500 μ)	4,124	0.019116	4.6747	0.0693
Gravelly sand G.E. 9	4,135	0.015048	2.8391	0.0793
Fine sand G.E. 2	4,136	0.007192	3.8937	0.0608
Plainfield sand (0–25 cm)	4,146	0.033730	3.8518	0.1133
Plainfield sand (25–60 cm)	4,147	0.031813	4.1948	0.0724
Aggregated glass bead	5,003	0.039748	6.4676	0.0983
Monodispersed glass bead	5,004	0.036049	7.6171	0.0363

TABLE 11.2 Unsaturated Flow Parameters from van Genuchten's (1978) Three-Parameter Solution of Mualem's (1976) Model Using Imbibition Data (Mualem and Dagan 1976)

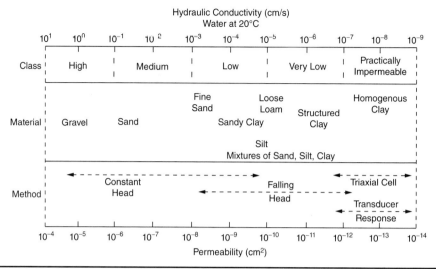

FIGURE 11.13 Hydraulic conductivity of various materials at saturation. [*From Klute and Dirksen (1986). With permission of ASA, CSSA, and SSSA.*]

Field Methods

Here we discuss only the air-entry permeameter and borehole permeameter, although the disc permeameter and sealed double-ring infiltrometer are other viable approaches, as discussed in Stephens (1996).

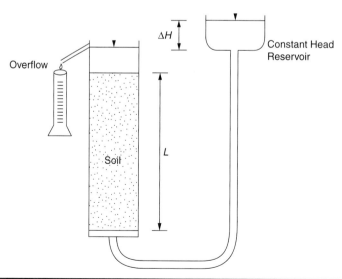

FIGURE 11.14 Constant-head permeameter. [*From Stephens (1996). With permission of Taylor & Francis Group LLC.*]

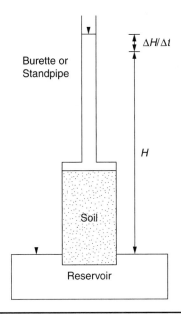

Burette or
Standpipe

$\Delta H/\Delta t$

H

Soil

Reservoir

FIGURE 11.15 Falling-head permeameter. [*From Stephens (1996). With permission of Taylor & Francis Group LLC.*]

Air-Entry Permeameter

The air-entry permeameter (Figure 11.16) was first proposed by Bouwer (1966) as a means to determine the vertical, field-saturated hydraulic conductivity of geologic materials above the water table. The air-entry permeameter consists of a single ring, typically 30 cm in diameter and 25 cm deep, that is driven vertically into the material to be tested. A water-supply reservoir is mounted on top of an adjustable standpipe. To begin the test, the air-entry permeameter reservoir is filled and water is allowed to infiltrate into the ring. When the water displaces the air between the soil and top plate, the air-escape valve is closed. The rate of decline of head applied at the soil surface is then recorded, either through the use of a transducer set within the ring at the soil surface or by observing the decline of water in the graduated supply reservoir (ASTM D-5126). Infiltration continues until the wetting front approaches the bottom of the ring, as indicated usually by a change in soil tension monitored by a tensiometer or the change in electrical conductivity at buried resistance probes, or as inferred by a mass-balance approach using an assumed fillable porosity of the soil. The water-supply valve is closed, and the soil water is allowed to redistribute. The negative pressure created by the redistributing soil water is monitored by a vacuum gauge or pressure transducer mounted at the top of the ring. When air begins to enter into the saturated soil from below the wetting front, the gauge will have recorded a minimum pressure. The minimum pressure achieved during this portion of the test is used to calculate the air-entry pressure of the soil, ψ_a. The field saturated hydraulic conductivity, K_{fs}, is calculated using the following equation:

$$K_{fs} = \frac{(R_r/r)^2\, L\, (dH/dt)}{(H_p + L - \psi_a)} \tag{11.25}$$

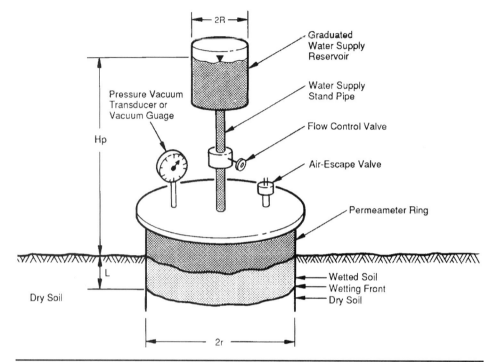

FIGURE 11.16 Air-entry permeameter. [*From Havlena and Stephens (1991). With permission of ASTM.*]

where R is the radius of the reservoir, r is the radius of the cylinder, L is the depth to the wetting front when dH/dt is measured, dH/dt is the rate of decline in head in the reservoir, and H_p is the depth of ponding when dH/dt is measured.

Borehole Permeameters

The vadose zone borehole permeameter is a method to determine in situ saturated hydraulic conductivity, K_s, of unfissured, homogenous, isotropic soil and rock above the water table. The borehole is constructed utilizing an auger (or drilling) to the desired depth, and, if the soil contains fines, the sides of the borehole must be scraped to remove any smear zone. In caving formations, well screen and a coarse sand or gravel pack must be placed in the borehole. In a typical borehole permeameter, a constant head of water is maintained in the borehole until the infiltration rate is steady. The borehole testing equipment typically consists of a water-supply reservoir and a means of controlling the head of water within a 5- to 15-cm-diameter borehole.

There are two types of equations to compute hydraulic conductivity: deep and shallow water-table conditions, as described by the U.S. Bureau of Reclamation (1977, described below):

$$K_s = Q_s \left[\frac{\sinh^{-1}\left(\dfrac{H}{r}\right) - 1}{2\pi\left(\dfrac{H}{r}\right)} \right]$$

(11.26)

In general, deep water-table conditions exist when the distance between the water level in the borehole and the water table is greater than three times the depth of water in the borehole (U.S. Bureau of Reclamation 1974). Most of the steady-state solutions are valid where $H/r > 10$. Analytical solutions for the deep water-table case were derived by Nasberg (1951), Glover (1953), Terletskaya (1954), Cecen (1967), and Reynolds et al. (1983). An equation for K_s under shallow water table conditions was developed by Zanger (1960). All these formulae neglected capillarity. The equation of Glover (1953) is

$$K_s = \left(\frac{Q_s}{C_u rH} \right) \tag{11.27}$$

where

$$C_u = \frac{2\pi(H/r)}{\sin h^{-1}(H/r) - 1} \tag{11.28}$$

and Q_s is the steady infiltration rate, H is the water depth, and r is the borehole radius.

Philip (1985) and Reynolds et al. (1985) developed approximate analytical solutions to compute K, which take capillarity into account. Stephens et al. (1987) developed an empirical equation for deep water-table conditions based on the van Genuchten soil-water retention model (Equation 11.31) parameters α_v (cm^{-1}) and N, water depth ($H(m)$), and borehole radius $r(m)$:

$$\log_{10}(C_u) = 0.653 \cdot \log_{10}(H/r) - 0.257 \cdot \log_{10}(\alpha_v) - 0.633 \cdot \log_{10}(H) + 0.021(H/r)^{0.5}$$
$$- 0.313N^{-0.5} + 1.456r + 0.453 \tag{11.29}$$

where $C_u = Q/rHKs$.

The van Genuchten parameters can be calculated from laboratory measurements of the soil-water retention curve, they can be estimated from soil texture as described in the section on pedotransfer functions, or they can be estimated from tables such as in Stephens 1987 (Table 2). For additional discussion on borehole permeameters refer to ASTM Standard D-5126, and Elrick and Reynolds (1992).

Unsaturated Hydraulic Conductivity

The purpose of this section is to review methods to measure unsaturated hydraulic conductivity in the field and laboratory. For additional detail on many of the methods discussed here, refer to Dirksen (1991) and Stephens (1993).

Laboratory Methods

Laboratory methods have been grouped into two general categories: (1) steady-state flow methods and (2) transient flow methods. In most cases, the transient techniques offer comparable accuracy with considerably less time to complete testing.

Steady, unsaturated flow can be introduced into vertical laboratory columns under constant-head or constant flux conditions. By whatever plumbing apparatus this is achieved, the unsaturated hydraulic conductivity is calculated from Darcy's equation as simply the ratio of steady flow rate per unit cross-sectional area to the hydraulic gradient (Equation 11.12). The hydraulic conductivity is associated with the mean water content and/or pressure head established in the column. Tensiometers (see the section "Tensiometers") are often installed through the wall of the soil column at two different positions to measure soil-water pressure head for computing both the hydraulic gradient and the mean pressure head at steady state. A series of steady-state tests are run to obtain a sequence of conductivity measurements, beginning with a nearly saturated column and ending with a low-water-content column.

Transient laboratory techniques to measure unsaturated hydraulic conductivity include instantaneous profile method, Bruce-Klute method, pressure-plate method, one-step outflow method, and ultracentrifuge method. Refer to Stephens (1996) for additional details.

Field Methods

Field methods include the instantaneous profile method, flux control methods, flow-net method, borehole point-source method, and air-permeameter method as described in Stephens (1996). The instantaneous profile method is probably the most widely used of the field techniques.

In the instantaneous profile method, a square plot, approximately 3 to 10 m on a side, is prepared at a level site and a berm is made on the perimeter. Tensiometer nests and a neutron probe access tube are emplaced inside the plot, near the center where the flow field is likely to be unaffected by lateral flow beyond the perimeter of an unbounded plot. The "bermed" area is filled with water until the profile is saturated to the depth of interest. Usually this depth is within 2 to 3 m of the surface, but for clay soils the practical depth of testing may be much less. During infiltration, the tensiometers and neutron probe measurements show when the soil has reached maximum saturation. Either the infiltration rate at constant ponding when the soil is saturated, or the rate of decline in ponded depth upon cessation of water application may be used to estimate the field saturated hydraulic conductivity if lateral seepage is negligible.

To obtain the unsaturated hydraulic conductivity by the instantaneous profile method, the water supply is shut off, the plot is covered to prevent evaporation, and pressure head and water content are measured as the profile drains. The transient data are used to calculate the unsaturated hydraulic conductivity at some depth below the top of the soil profile, L, according to the following equation:

$$K(\bar{\theta}) = \int_0^L \frac{\partial \theta}{\partial t}\, dz \div (dH/dz) \tag{11.30}$$

At discrete depths, simultaneously, the hydraulic gradient is calculated from tensiometric data, and the rate of change in moisture content is calculated from the slope of the moisture content versus time plot.

Estimating Unsaturated Hydraulic Conductivity

Field and laboratory methods to characterize unsaturated hydraulic conductivity are either tedious, time consuming, or have other logistical difficulties. It is more convenient to estimate unsaturated hydraulic conductivity in lieu of measuring it in the laboratory or in the field. Calculating hydraulic conductivity ($K - \psi$) from moisture retention ($\theta - \psi$) curves is perhaps the most popular of all means in current use, in part because it is based on measured hydraulic properties (porosity, saturated hydraulic conductivity, and moisture retention). The hydraulic properties can also be estimated using pedotransfer functions.

The most commonly used expressions for calculating unsaturated hydraulic conductivity are the Brooks and Corey (1964) and the Mualem (1976) models derived using the van Genuchten (1980) retention function (the van Genuchten-Mualem model). The Brooks and Corey equation for unsaturated hydraulic conductivity as a function of water content is

$$K_r(\theta) = \left(\frac{\theta - \theta_r}{\theta_s - \theta_r}\right)^\eta \tag{11.31}$$

where

$$\eta = \frac{2}{\lambda} + 3 \tag{11.32}$$

and, as a function of pressure,

$$K_r = (\psi/\psi_{cr})^{-\lambda}, \quad \text{for } \psi > \psi_{cr} \tag{11.33a}$$

$$K_r = 1, \quad \text{for } \psi > \psi_{cr} \tag{11.33b}$$

where ψ_{cr} is the critical pressure or bubbling pressure and λ is the pore size distribution index. The van Genuchten-Mualem model as a function of water content is

$$K_r(\theta) = S_e^l \left\{1 - \left[1 - S_e^{1/m}\right]^m\right\}^2 \tag{11.34}$$

The pore-connectivity parameter (l) was estimated by Mualem (1976) to average around 0.5. The van Genuchten-Mualem model as function of capillary pressure is

$$K_r(\psi) = \frac{\{1 - |\alpha_v\,\psi|^{N-1}\,[1 + |\alpha_v\psi|^N]^{-m}\}^2}{[1 + |\alpha_v\,\psi|^N]^{m/2}} \tag{11.35}$$

where α_v (1/L) and N (dimensionless), as indicated in Equation 11.28, are fitting parameters m is $1 - 1/N$, and S_e is effective saturation $(\theta - \theta_r) \div (\theta_s - \theta_r)$.

Correction of Hydraulic Parameters for Large Particles

Natural soils in the vadose zone often contain large particles within a matrix of finer-grained (<2 mm) soil such as in some alluvial fan deposits. Commonly, laboratory methods for measuring water content and saturated hydraulic conductivity are for the fine fraction (<2 mm or <4.75 mm) of the soil after removing the large particles. Therefore, the laboratory measurements must be corrected to obtain the "bulk" measurement for the natural soil. The laboratory measurements are converted to bulk parameters representative of the field by measuring the mass fraction of large particles and then adjusting the laboratory measurements based on the mass fraction. Collection of samples from the field must be careful to sample all of the representative particle sizes. In the field, large particles are commonly contained within a matrix of fine materials. When the large particles are contained within a matrix of finer-grained soil, the large particles reduce the hydraulic conductivity and volumetric water content of the bulk sample of the vadose by removing volume from the vadose zone where water can flow or be stored. Methods for correcting the hydraulic properties for the presence of large particles (e.g., stones) are presented in Bouwer and Rice (1984) and Brakensiek and Rawls (1994). Stony soils are defined as having a particle size greater than 2 mm

(USDA Soil Classification) or greater than 4.75 mm (USCS Soil Classification). Using the mass-fraction of rock fragments greater than 2 mm, the matrix hydraulic conductivity is converted to a bulk hydraulic conductivity as follows (Brakensiek and Rawls 1994):

$$K_b = K_m (1 - Z1)$$ (11.36)

where K_b is the bulk hydraulic conductivity, K_m is the hydraulic conductivity of the matrix, and Z1 is the mass fraction of rock fragments. Volumetric matrix-water content is converted to bulk volumetric water content as follows (Brakensiek and Rawls 1994):

$$\theta_b = \theta_m (1 - Z1)$$ (11.37)

where θ_b is the bulk volumetric water content, θ_m is the volumetric water content of the matrix and Z1 is the mass fraction of rock fragments greater than 2 mm in diameter.

Pedotransfer Functions

Parameters fitted to the soil-water retention curves (e.g., the van Genuchten parameters in Equation 11.28) can be estimated from observations or measurements of soil texture using what are termed pedotransfer functions. Typically, the soil texture expressed as the relative percentages of sand, silt, and clay is used to predict the retention function parameters but some pedotransfer functions improve the parameter estimates by including bulk density, organic matter, or field-capacity measurements. When only the soil texture class is known, the van Genuchten parameters are commonly estimated using the parameters published by Carsel and Parrish (1988). If the relative percentages of sand, silt, and clay are known, the Rawls and Brakensiek (1985) pedotransfer function based on multiple linear regression provides estimates of both the saturated hydraulic conductivity and the van Genuchten soil-water retention curve parameters using the general regression model:

$$\begin{aligned} f(S, C, \theta_s) = {} & b_0 + b_1 S + b_2 C + b_3 \theta_s + b_{11} S^2 + b_{22} C^2 \\ & + b_{33} \theta_s^2 + b_{12} SC + b_{13} S\theta_s + b_{23} C\theta_s \\ & + b_{112} S^2 C + b_{223} C^2 \theta_s + b_{113} S^2 \theta_s + b_{112} SC^2 \\ & + b_{233} C\theta_s^2 + b_{1133} S^2 \theta_s^2 + b_{2233} C^2 \theta_s^2 \end{aligned}$$ (11.38)

Table 11.3 provides the parameter values for the Rawls and Brakensiek pedotransfer function (Carsel and Parrish 1988; Rawls et al. 1992; Lee 2005).

If the percentage of sand is greater than 70%, the Rawls and Brakensiek pedotransfer estimate of saturated hydraulic conductivity should be modified as follows (Cronican and Gribb 2004):

$$K_s = \exp(99.49815337 - 2.808839211 \times S + 0.017873264$$
$$\times S^2 - 0.019881556 \times SC + 0.000268919 \times S^2 C)$$ (11.39)

The Brooks-Corey parameters can be estimated from the relative percentages of sand, silt, and clay, as shown in Figure 11.17 from McCuen et al. (1981).

Term	Natural Log Saturated Hydraulic Conductivity (KS) ln[cm/hr]	Residual Water Content (θr) [cm³/cm³]	Natural Log Empirical Constant Alpha ($1/\alpha$) ln[cm]	Natural Log Empirical Constant N − 1 ln[− −]
(Constant)	−8.96847	−0.0182482	5.3396738	−0.7842831
S	0	0.00087269	0	0.0177544
C	−0.028212	0.00513488	0.1845038	0
θs	19.52348	0.02939286	−2.48394546	−1.062498
S^2	0.00018107	0	0	−5.30E-05
C^2	−0.0094125	−0.00015395	−0.00213853	−0.00273493
θs^2	−8.395215	0	0	1.11134946
SC	0	0	0	0
$S\theta s$	0.077718	−0.0010827	−0.0435649	−0.03088295
$C\theta s$	0	0	−0.61745089	0
$S^2 C$	0.0000173	0	−1.282E-05	−2.35E-06
$C^2\theta s$	0.02733	0.00030703	0.00895359	0.00798746
$S^2\theta s$	0.001434	0	−7.2472E-04	0
SC^2	−0.0000035	0	5.40E-06	0
$C\theta s^2$	0	−0.0023584	0.5002806	−0.00674491
$S^2\theta s^2$	−0.00298	0	0.00143598	2.6587E-04
$C^2\theta s^2$	−0.019492	−0.00018233	−0.00855375	−0.00610522

Notes: Coefficients for residual water content corrected based on Rawls et al. (1992, p. 334) coefficients for alpha corrected based on Lee (2005).

TABLE 11.3 Rawls and Brakensiek (1985) Multiple Regression Coefficients [From Carsel and Parrish (1989)]

The ROSETTA software by Shaap et al. (2001) uses a hierarchical approach depending on the set of input data. If only soil textural class is known, the software uses a lookup table. If sand, silt, and clay percentages are known, the software uses a neural network analysis. The neural network analysis can be improved in the software with the addition of bulk density and water-content measurements at 1/3 bar and 15 bar (Schaap et al. 2001).

11.9   Vadose Zone Monitoring

Vadose zone monitoring includes methods to measure pressure head and moisture content as well as methods to sample pore liquids and soil gas. Some of these have been in use for many decades, primarily in applications to agricultural problems. For additional details, refer to Stephens (1996).

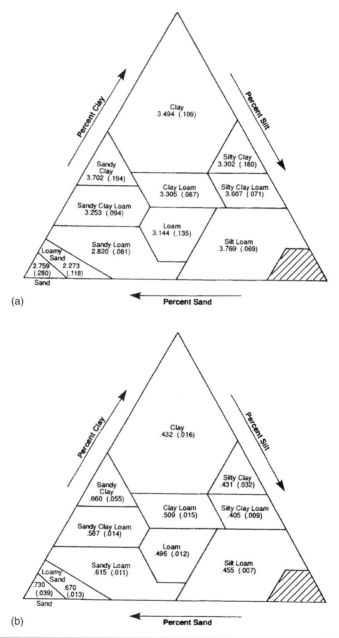

(a)

(b)

Figure 11.17   Mean and (standard error) of Brooks and Corey parameters: (a) in Ψ_b and (b) $\lambda^{1/2}$. [*From McCuen et al. (1981). With permission of American Geophysical Union.*]

Pressure Head Soil-Water Potential

Pressure head is an important measure of the energy status of soil water. In the vadose zone, spatial differences in the potential determine the direction of soil-water movement.

Tensiometers

A tensiometer consists of a porous cup, usually composed of a moderately permeable ceramic, that remains saturated under significant tensions (Figure 11.18). Tensiometers are commercially available from Soilmoisture Equipment Corp. in Santa Barbara, California, for example. The cup is attached to a small diameter water-filled pipe that is sealed on the upper end with a removable cap. The purpose of the removable cap is to allow for filling of the tensiometer with water and for purging of air accumulations. A manometer installed through the top part of the water-filled pipe measures the pressure of the water in the tensiometer. When the tensiometer is inserted into the soil, the soil imbibes water from the tensiometer, and as this occurs, the water tension in the tensiometer increases until the tensiometer fluid pressure is in equilibrium with soil water outside the cup. The principal differences among tensiometers are attributable to the types of manometers used to measure pressure (Figure 11.18).

All tensiometers share a variety of problems. Care must be taken to use de-aired water in the tensiometer and maintain tight fittings to minimize air accumulation in the system. A small amount of air is tolerable, but as the air accumulates, the response time of the system decreases. Consequently, air must periodically be removed from the system.

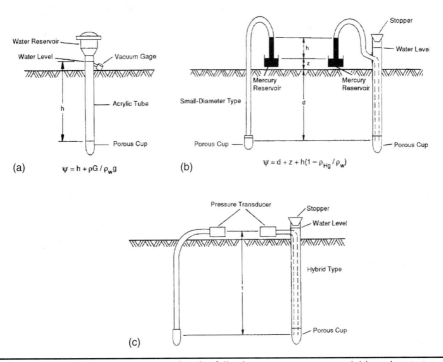

FIGURE 11.18   Tensiometer systems using the following pressure sensors: (a) bourdon gage, (b) mercury manometer, and (c) pressure transducer. [*Modified from Stannard (1990). USGS.*]

The most significant limitation is that tensiometers cannot measure soil-water potentials less than about −0.8 bars, because near −1 bar at sea level, the water in the tensiometer would vaporize. For most irrigated soils as well as for uncultivated coarse, uniform-textured soils this is not a severe limitation. In fact, tensiometers have functioned quite well even in semi-arid climates in sand dunes where the moisture content is only about 5% (Stephens and Knowlton 1986). Freezing conditions can create severe problems for some systems unless antifreeze is used as the solution in the tensiometer (McKim 1976).

Psychrometers

A soil psychrometer measures the relative humidity within the soil atmosphere from the difference between the wet bulb and dry bulb temperature. The lower the relative humidity, the faster will be the rate of evaporation, and the lower will be the temperature of an evaporating liquid relative to the dry bulb temperature. In contrast to tensiometers, the pressure head determined from psychrometers includes both the matric (capillary plus adsorbed water) and the osmotic (solute) potential components, as presented in Equation 11.2. The contribution of osmotic potential to the soil-water potential is usually small and is generally ignored. Figure 11.19 illustrates the general features of a Peltier psychrometer, often referred to as a Spanner (1951) psychrometer, which is the common design used in field applications. Psychrometer calibration is accomplished by using the sensor to determine the output voltage (temperature depression) from a series of sodium or potassium chloride solutions of known different concentrations. These aqueous solutions have known osmotic potentials likely to be encountered in the field (e.g., Lang 1967). This calibration curve is used to compute the in situ soil-water potential from the measured field output voltage. One of the potentially serious problems with all psychrometers is that the calibration can change over time due to microbial growth on the thermocouple wires (Merrill and Rawlins 1972) or corrosion (Daniel et al. 1981). Psychrometers are used only for dry conditions, usually where soil-water potential is in the range of about −2 to −70 bars or even lower.

Indirect Methods

Indirect methods include electrical resistance blocks, heat-dissipation sensors, and the filter-paper method. All indirect methods share the same general operational principles. When the sensor is placed in firm contact with the soil, water flows into or out of the sensor due to the hydraulic gradient between the potential in the sensor and the soil-water potential, until there is equilibrium. The water content of the sensor depends on the pore-size distribution of the sensor as well as the soil-water potential. To relate the measured water content of the sensor to the soil-water potential requires calibration in a pressure-plate apparatus or an independent laboratory determination of the soil-water characteristic curve for the porous sensor.

Electrical Resistance Blocks

Electrical resistance blocks consist of porous gypsum, nylon, or fiberglass within which is embedded an electrode with leads connected to a Wheatstone Bridge to measure resistance (Figure 11.20). As the water content of the resistance block decreases, the electrical conductivity of the block decreases. The solubility of the gypsum blocks limits their longevity to perhaps a couple years in wet soils. Daniel et al. (1992) found that wet gypsum blocks disintegrated when repeatedly installed and retrieved. Although the

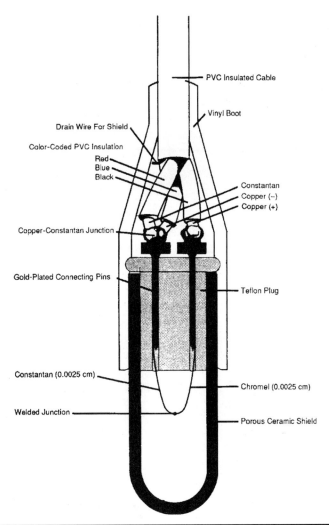

PVC Insulated Cable

Vinyl Boot

Drain Wire For Shield

Color-Coded PVC Insulation

Red

Blue

Black

Constantan

Copper (–)

Copper (+)

Copper-Constantan Junction

Gold-Plated Connecting Pins

Teflon Plug

Constantan (0.0025 cm)

Chromel (0.0025 cm)

Welded Junction

Porous Ceramic Shield

FIGURE 11.19 Peltier psychrometer/hydrometer with porous ceramic thermocouple shield. [*From Briscoe (1986). With permission.*]

nylon and fiberglass blocks are not readily soluble, Campbell and Gee (1986) report that in some cases these units may fail in 2 to 3 years because of corrosion of the welds. According to Campbell and Gee (1986), gypsum blocks are most sensitive in the dry range at potentials below about –0.3 bars, whereas nylon and fiberglass blocks are most sensitive in the range of 0 to –1 bars.

Heat-Dissipation Sensors

Heat-dissipation sensors are comprised of a porous material (ceramic in the commercially available ones) that surrounds a heating element and thermal detector (Phene et al. 1971) (Figure 11.21). The principle is based on the dependence of the thermal diffusivity (thermal conductivity + heat capacity) on moisture content of the porous material in the

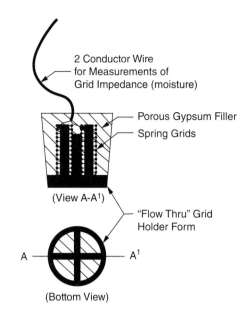

2 Conductor Wire
for Measurements of
Grid Impedance (moisture)

Porous Gypsum Filler

Spring Grids

(View A-A¹)

"Flow Thru" Grid
Holder Form

A ———— A¹

(Bottom View)

(a)   Soilmoisture Blocks

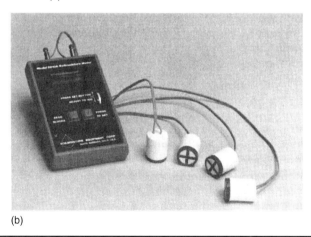

(b)

FIGURE 11.20   (a) Schematic cross-sectional view and bottom view of a gypsum block. (b) Gypsum block. (*Courtesy of Soilmoisture Equipment Corp.*)

sensor. Current is passed into a coil of copper wire in the center of the porous material, and after the heating phase, the rate of temperature decline is determined with a diode circuit located inside the heating element. As the moisture content of the porous material increases, the thermal diffusivity increases, and the voltage output from the diode circuit decreases (Campbell and Gee 1986). The output voltage is related to the potential of the soil by a laboratory calibration curve. The heat-dissipation sensor is insensitive to changes in soil salinity. Table 11.4 compares the range of sensitivity of the heat-dissipation sensor with electrical resistance blocks and the portable salinity probe.

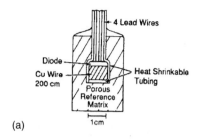

(a)

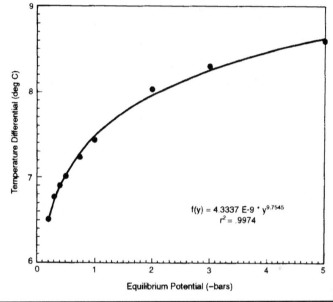

(b)

FIGURE 11.21   (a) Cross-sectional sketch of the Phene heat-dissipation sensor [*modified from Phene et al. (1971), with permission of ASA, CSSA, and SSSA*] and (b) heat-dissipation probe calibration.

Type of Monitoring Probe	Range Over Which Probe Exhibited Maximum Sensitivity of Response	
	Volume Water Content (%)	Soil Suction (Bars)
Gypsum block	20–50	0–15
Fiberglass res. device	40–55	0–4
Heat-dissipation sensor	30–50	0–10
Resistivity probe	20–35	5–15

From Daniel et al. (1992). With permission of ASTM.

TABLE 11.4   Comparison of Calibration Tests Conducted on Four Vadose-Zone Monitoring Probes in a Kaolin Clay

Moisture Content

Gravimetric Method

The gravimetric method derives its name from the fact that moisture content is determined by weighing. There are two approaches that are commonly in use to express water content. The first is volumetric water content and the second is the mass-based or gravimetric water content. The volumetric moisture content represents the water volume in a known volume of soil, whereas the gravimetric water content represents the mass of water in a known dry mass of soil. For typical soils, the volumetric water content (% cm^3/cm^3) will be roughly 20% to 50% larger than the gravimetric water content (% g/g).

To determine moisture content by either approach, a sample of soil is collected from a particular depth by hand auger, power auger, or drilling rig. If the sample is undisturbed and collected so that the bulk volume of soil can be quantified, such as in a known volume of a ring sample (100-cm^3 and 250-cm^3 sizes are common), then the volumetric water content can be determined. Undisturbed samples are usually obtained by carefully pushing a sampling ring in to the soil by hand, hydraulic press, or weight of the drill rig. In the laboratory, the testing process is almost identical for gravimetric or volumetric water-content determinations. After removing the field wrapping, the samples are placed in a tare dish and weighed. The samples are then dried at a temperature of about 105°C in either a conventional convection oven (for 24 h), a forced draft oven (for 10 h), or a household microwave oven (for 6 to 20 min) (Gardner 1986). The samples are removed from the oven, and after cooling in a desiccator, they are reweighed to determine the loss of water. This loss on a mass basis, or on a volume basis if converted by dividing by the water density, is directly used to compute the gravimetric or volumetric water content, respectively.

Subsurface Geophysical Methods

For many applications, it is important to monitor moisture content profiles over time. This necessitates a nondestructive measurement that can be made repeatedly. Geophysical techniques are well suited to this purpose, both subsurface and surface techniques.

Time Domain Reflectometry (TDR)

TDR is based on measuring the dielectric constant of soil. Topp et al. (1980), who first proposed the method for soil-water investigations, showed that the dielectric constant of soil is dependent primarily upon the water content through a nearly universal calibration equation that is very insensitive to soil type. Most soil minerals have a dielectric constant of less than 5, whereas water has a dielectric constant of about 78. Thus, water-content variations should be easily detectable from measured variations in dielectric constant.

To measure the dielectric constant in situ using TDR requires determining the rate of travel of electromagnetic energy through the soil along two parallel metallic conductor rods pushed into the soil (Figure 11.22). The TDR source introduces a voltage pulse, and an oscilloscope maps the travel time for the wave to propagate along the transmission lines and reflect back to the source. From this velocity, one can calculate the apparent dielectric constant, K_{ad}, and then use the generic calibration equation (Topp et al. 1980) to compute the water content:

$$\theta = -0.053 + 0.029\, K_{ad} - (5.5 \times 10^{-4})\, K_{ad}^2 + (4.3 \times 10^{-6})\, K_{ad}^3 \qquad (11.40)$$

This calibration equation reportedly has an accuracy of ±0.02 cm^3/cm^3.

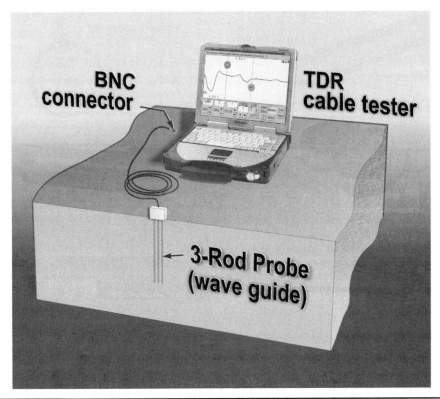

Figure 11.22 Time-domain reflectometry (TDR) equipment setup. (*Adapted from Greenlee Textron, Inc., with permission.*)

Neutron Probe

The neutron probe is a geophysical logging tool that is lowered into a cased borehole on a cable that connects the neutron source with electronic readout equipment on the surface (Figure 11.23). The principle is based upon the neutron thermalization process, in which a radioactive source (e.g., americium–241) emits high-energy neutrons into the soil where, through collisions primarily with hydrogen atoms, the energy of the neutrons is reduced to lower (thermal) energy levels (Gardner 1987). The neutrons are emitted more or less radially and form a sphere around the source within which the fast neutrons are attenuated. For dry soils the diameter of the sphere is approximately 70 cm, and at saturation the diameter is only about 16 cm; this diameter is unaffected by the strength of the radioactive source (Gardner 1987). The detector of thermal neutrons commonly consists of a boron trifluoride gas, which, when encountering a thermal neutron, emits an alpha particle as it decomposes to lithium-7 (Goodspeed 1981); this causes an electrical pulse to be released within the detector. In other words, the number of electrical pulses counted by the detector is proportional to the number of thermal neutrons it encountered.

If hydrogen is the only variable affecting the density of slow or thermal neutrons, then a calibration curve can be developed to relate count rate to water content. The calibration of a neutron probe can be done in the field or in the laboratory.

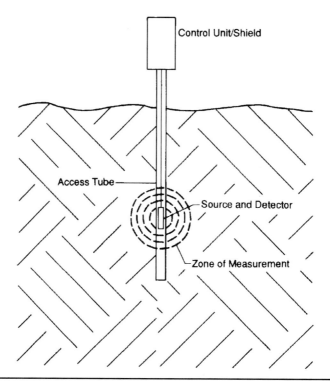

FIGURE 11.23   Schematic diagram of water-content measurement by the neutron depth probe method. (*From ASTM D-5220-92, 1992. With permission of ASTM.*)

The preferred borehole construction for shallow vadose zone investigations calls for 5-cm-diameter (2-in-diameter), thin-walled aluminum tubing within which the slightly smaller-diameter neutron probe can be easily raised and lowered. The tubing fits tightly in a smooth-walled borehole to minimize air pockets outside the casing. Other casing materials, such as polyvinyl chloride and steel, tend to absorb the slow neutrons, but it is possible to prepare a separate calibration curve for each different casing material, with some loss of sensitivity.

Temperature Monitoring

Fiber Optic Distributed Temperature Sensing

Fiber Optic Distributed temperature sensing (FO-DTS) is an emerging technology that provides large scale (meter to kilometer scales) measurements of temperature at high spatial and temporal resolution that can monitor soil hydrologic data in situ, including soil moisture and temperature. Temperature is sensed with FO-DTS by pulsing a laser and timing the return signal. Photons from the laser are scattered due to the inelasticity of the molecules (Raman and Brillouin scattering) produce frequencies higher (anti-Stokes) and lower (Stokes) than the transmitted light. The intensity of Stokes backscatter is largely independent of temperature while the anti-Stokes backscatter is strongly dependent on temperature. The ratio of Stokes and anti-Stokes scattering can be interpreted to obtain a measurement of temperature. The measurements are obtained over a given length of the fiber optic cable (e.g., 1 to 3 m) and a given length of time where longer time and lengths give more accurate temperature measurements (Tyler et al. 2009). For details on FO-DTS, refer to Selker et al. (2006) and Tyler et al. (2009).

FO-DTS measurement of soil moisture relies on either passive or active methods. Passive methods rely on ambient temperature fluctuations (e.g., Steel-Dunne et al. 2010) while active methods combine the FO-DTS with the application of heat to the soil (Sayde et al. 2010). In the active method, a heat pulse is generated by applying current to stainless steel winding along the length of the fiber optic cable. Temperature measurements from the FO-DTS are obtained prior and during the application of the heat pulse and the temperature signal is interpreted to obtain an estimate of soil moisture (Sayde et al. 2010). Care must be taken not to overheat and dry the soil. Passive methods are more economical and don't require a large energy source allowing for installations in remote locations. In the passive method, temperature is sensed at different depths and the difference in the responses at depth to diurnal temperature fluctuations at the surface is interpreted using numerical models to estimate the soil moisture (Steele-Dunne et al. 2010). FO-DTS experimental designs are not without complications and the performance of FO-DTS strongly depends on the design of the experiment, operating conditions, the choice of fiber optic cable and connectors, and the experience of the operator (Tyler et al. 2009).

Soil Gas Sampling and Monitoring

Soil gas monitoring is a technique that is applicable to highly volatile organic chemicals that partition into the air-filled soil voids. There are both passive and active types of soil gas monitoring techniques. The passive devices are tubes containing absorbent material, usually activated carbon, that are placed in the soil for a period of time and later retrieved for chemical analysis (e.g., Bisque 1984; Stutman 1993). The absorbing material picks up volatile and semi-volatile chemicals as they diffuse from the soil and are adsorbed. The passive absorbers must be buried for approximately several days to 2 weeks before retrieval. The active type of soil gas sampler is still by far the most popular. It uses vacuum to pump soil vapor from an in situ sampling probe directly into a chemical analyzer for a more rapid turnaround compared to the passive systems. There are several different installations for collecting soil gas in active sampling systems, including permanent soil gas monitor wells, a semi-permanent monitoring system, soil gas surveys, and soil gas profiles. Depth profiling for soil gas is accomplished either by the permanent or semi-permanent installations described above or by driving probes and sequentially sampling as the drive point is advanced to greater depths. For most unconsolidated soils a cone penetrometer rig (Chapter 15) can collect soil gas samples as the cone is advanced by the hydraulic jack system. An advantage of the cone penetrometer method is that other information on soil texture can be collected continuously with depth at the same time and location as the soil gas sample is collected. This approach can lead to improved interpretations and more rapid decisions in the field program that may follow.

Pore Liquid Sampling

Pore liquid sampling includes extracting liquids from the matrix of soil cores and in situ samplers to collect liquids from the vadose zone. Samples of soil can be collected from the soil by hand tools or mechanical drilling equipment. Conventional drilling equipment such as the hollow-stem flight auger with core sampling generally produces excellent results, except in very loose dry soils, and cemented or stony soils. In most solid or hazardous waste site sampling projects, soil cores are sent to a laboratory for chemical analysis, under strict preservation and quality assurance/ quality control procedures. The main point to recognize in considering soil cores for pore liquid sampling is that this is a destructive technique. For time-series monitoring, repeated coring to collect samples from the same general location potentially introduces significant bias that may preclude distinguishing chemical changes attributable to contaminant migration from changes due to spatial variability in fluid chemistry.

Porous cup samplers, also referred to as suction lysimeters or porous suction samplers, consist of a porous tip attached with an airtight seal to the lower end of a thin plastic casing. The upper end is also tightly capped, so that a vacuum can be applied and maintained to the casing through connections in the upper cap. The sampler cup is designed with pores sufficiently fine that they remain saturated with water while the vacuum is applied to the casing section. The principle of operation is to place the sampler in a boring so that there is good communication with the formation. Usually a slurry of soil cuttings or 200-mesh silica flour is placed around the sampler (in a thin pancake batter consistency), and bentonite seals are set above this interval to prevent channeling of runoff or perched fluids. A vacuum is applied to the sampler with the intent that it will establish a hydraulic gradient between the formation and the sampler. This system may fail for a variety of reasons, for example, if the applied vacuum exceeds the air-entry value for the porous cup or if the soil tension exceeds the air-entry pressure of the cup. In all these instances, consequently, the porous cup will dewater partially, the air in the sampler casing will be in communication with the soil air, and vacuum on the liquid phase will be lost.

Porous cup samplers differ in two general ways: by the material used to construct the cup and by the design of the component. Most of the cups consist of porous ceramic or Teflon, although Alundum and nylon have been used (Creasey and Dreiss 1985). In the simplest design of the porous cup sampler, the cup is located at the bottom of the sampler tube. A two-hole rubber stopper caps the upper end of the casing so that a short tube inserted through the stopper can be used to apply a vacuum to collect the liquid in the sampler. Then a vacuum is applied to the long discharge tube which is set to the base of sampler to bring the liquid to the surface (Figure 11.24). This design, referred to as the vacuum sampler, is usually used at depths of less than 2 m but potentially could be installed at depths of about 8 m, the maximum practical suction lift. For deeper sampling, a pressure-vacuum sampler (Parizek and Lane 1970) is required. When vacuum is applied a check valve opens and the sample is drawn from the cup to an upper chamber within the sampler. Then positive pressure is applied as the check valve closes to prevent liquid from the upper chamber from falling back into the cup and the sample is forced to the surface (Figure 11.25). A type of porous cup sampler called the BAT system (Torstensson 1984) is a unique design to allow sampling of volatile chemicals and constituents involved in redox reactions (Figure 11.26). Perhaps the greatest problem with porous cup samplers is that they simply do not function if the soil-water potential is less than about 1 bar. In humid climates, this may not be a problem, except at shallow depths after prolonged dry conditions. But in semi-arid and Mediterranean climates, many fine-textured soils are too dry for these samplers to function.

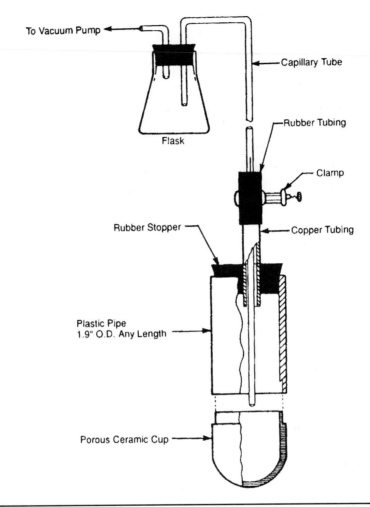

To Vacuum Pump

Capillary Tube

Flask

Rubber Tubing

Clamp

Rubber Stopper

Copper Tubing

Plastic Pipe
1.9" O.D. Any Length

Porous Ceramic Cup

FIGURE 11.24 Pore liquid sampling with porous cup vacuum sampler (lysimeter). *Note:* Capillary tube is only inserted to recover the water sample. [*From Parizek and Lane (1970). With permission from Elsevier.*]

11.10 Summary

The vadose zone is an integral part of the hydrogeologic system consisting of the porous or fractured media above the regional water table. Understanding the processes governing flow through the vadose zone are highly relevant to hydrogeologic applications including quantifying natural and artificial recharge, assessing impacts from chemical spills and leaching from waste sources such as landfills for instance. Characterizing the hydraulic properties of the vadose zone is more complex than characterizing aquifer properties, because in the vadose zone properties such as unsaturated hydraulic conductivity not only varies spatially but they exhibit large variations with water content.

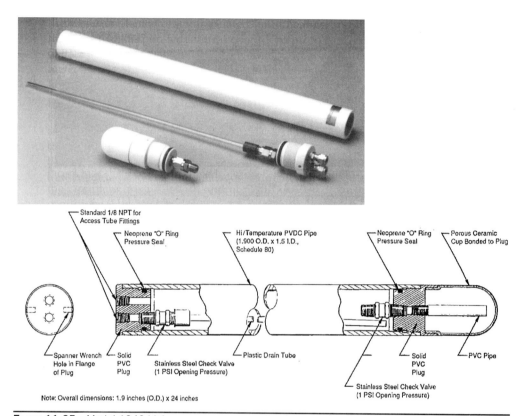

Figure 11.25   Model 1940 high-pressure-vacuum soil-water sampler. (*Courtesy of Soilmoisture Equipment Corp.*)

Fortunately, there is a wide array of field and laboratory methods to characterize vadose zone hydraulic properties and to monitor water and chemical movement, most of which have been in wide use for many decades.

11.11   Problems

11.1   Direction and Rate of Soil-Water Flow   Determine the direction and rate of soil-water flow from in situ measurements of pressure head and hydraulic conductivity in a soil having a uniform texture. Figure 11.27 shows the location of two tensiometers (discussed in Section 11.7) for measuring pressure head. Table 11.5 indicates the pressure head measurements at the two depths. It has already been determined from laboratory analyses of cores that the saturated hydraulic conductivity is 1 cm/day. Assume that the unsaturated hydraulic conductivity fits the exponential model:

$$K(\psi) = K_s \exp(\alpha\psi) \tag{11.41}$$

with $\alpha = 0.02$ cm^{-1} for this soil. Compute the direction and the magnitude of the total hydraulic head gradient and compute the unsaturated hydraulic conductivity.

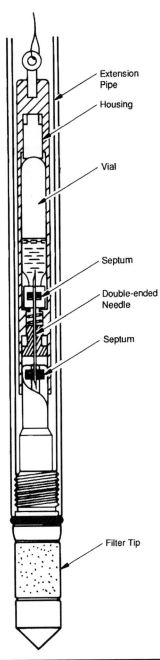

Extension
Pipe

Housing

Vial

Septum

Double-ended
Needle

Septum

Filter Tip

FIGURE 11.26 BAT system. [*From Torstensson (1984). With permission of the National Water Well Association.*]

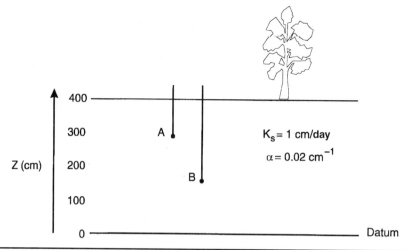

FIGURE 11.27   Example to calculate hydraulic gradient, flow direction, and flow rate. [*From Stephens (1996). With permission of Taylor & Francis Group LLC.*]

	Measured Pressure Head ψ (cm)	Elevation Head Z (cm)	Total Head H (cm)
A	−100	300	200
B	−90	200	110

TABLE 11.5   Pressure Head and Total Head Measurements at Two Depths

11.2   Estimate the Saturated Hydraulic Conductivity from a Borehole Permeameter   Estimate the saturated hydraulic conductivity from a borehole permeameter installed in a loam soil under deep water-table conditions. A cased and screened borehole, 5 cm in diameter was set at a depth of about 110 cm below ground surface and the hole was filled with water to 76-cm depth. This depth was maintained constant with a carburetor valve and styrofoam floats. Prior to water delivery, carbon dioxide gas was pumped into the borehole to minimize entrapped air. Figure 11.28 illustrates the infiltration rate. For simplicity, ignore capillarity.

11.3   Calculate the Unsaturated Hydraulic Conductivity from an Instantaneous Profile Test   Figure 11.29 shows an excellent example of an instantaneous profile test conducted on fine sandy loam at Sandia National Laboratory in Albuquerque, New Mexico (Bayliss et al. 1996; Goering et al. 1996). The 15.5-m × 15.5-m bermed area was flooded with water to a constant depth of about 5 cm until the infiltration rate was constant (Figure 11.30) and the soil profile was thoroughly wetted, as indicated by periodic measurements with neutron probe measurements to 3 m and tensiometer measurements to 1.8 m, as well as time domain reflectometry probes and frequency domain reflectometry probes. After about 1.5 days, the profile was saturated to the 2-m depth. Water-content profiles show the rapid initial rate of drainage and much slower drainage until the test was terminated after 400 days (Figure 11.31). The hydraulic head data as shown in Figure 11.32. Calculate the unsaturated hydraulic conductivity (Table 11.6).

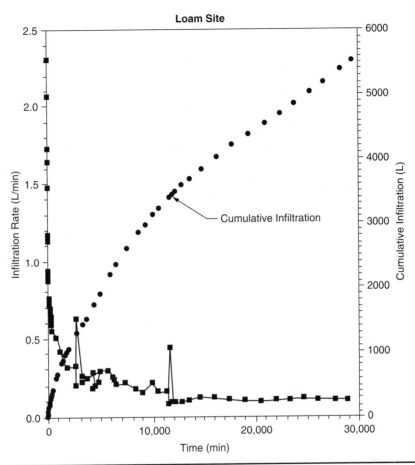

FIGURE 11.28 Infiltration rate and cumulative infiltration at the M-mountain loam site. [*From van Genuchten, Leij, and Lund (1989).*]

11.4 Calculate Leakage and Time for Leakage to Reach Water Table A concrete-lined canal has a saturated hydraulic conductivity (Ks_{liner}) of 10^{-6} cm/s. The soil beneath the canal concrete liner is a loam with an initial volumetric water content (θi) of 0.10 and the hydraulic properties are saturated hydraulic conductivity of 25 cm/day, van Genuchten curve parameters of $\alpha = 0.036$ 1/cm and $n = 1.56$, saturated volumetric water content (θs) of 0.43, and residual volumetric water content (θr) of 0.078. Wastewater discharge is released into the canal and maintains a head of 5 cm of water in the canal. The water table is at a depth (d) of 20 m. What is the leakage from the canal and how long will it take for the initial leakage to reach the water table?

11.5 Pedotransfer with Rock Correction A soil survey reports that a bulk soil sample has a coarse mass fraction of 20% gravel and that the fine mass fraction separate from the bulk soil sample is composed of 62% sand and 23% silt. What is the saturated hydraulic conductivity of the bulk soil sample?

11.6 Chloride Mass Balance The average chloride dissolved in precipitation (Clp) is 0.2 mg/L. Dry deposition (D) is 2.7 mg/day/m². Measurements of chloride in soil pore water (Cls) beneath

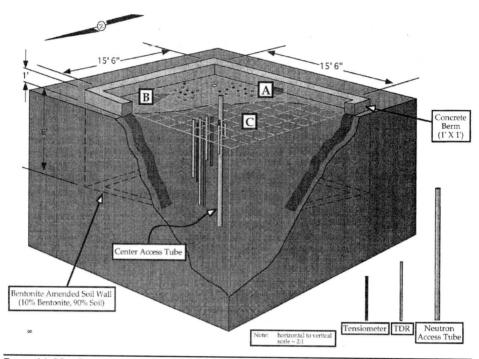

FIGURE 11.29 Cross section of IP plot showing instrument Cluster C. [*From Bayliss et al. (1996), Sandia National Laboratories.*]

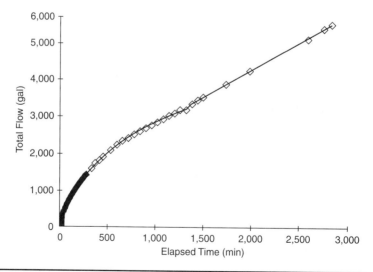

FIGURE 11.30 Cumulative flow applied to the IP test plot. [*From Goering et al. (1996), Sandia National Laboratories.*]

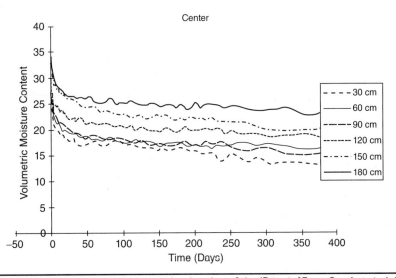

Figure 11.31 Moisture content data over the duration of the IP test. [*From Goering et al. (1996), Sandia National Laboratories.*]

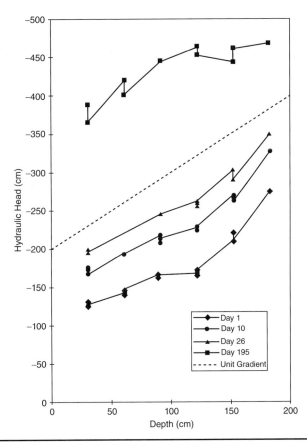

Figure 11.32 Hydraulic head profile changes with time. [*From Goering et al. (1996), Sandia National Laboratories.*]

Depth	Volumetric water content			
(cm)	Day 1	Day 10	Day 26	Day 195
30	24.6	19.8	17.8	15.9
60	29.4	21.7	19.3	17.0
90	33.9	21.5	20.1	17.2
120	31.3	24.4	22.0	19.9
150	28.3	27.7	26.0	22.5
180	26.2	28.2	26.6	24.1

TABLE 11.6   Volumetric Water Content over the Duration of the IP Test

the root zone of natural vegetation averaged 50 mg/L. Precipitation (P) averages 15 in per year. What is the mean annual rate of recharge (R)?

11.7   Initial Conditions Problem   The initial conditions for a one-dimensional vadose zone model need to be determined. Plot the vertical distribution of head for a one-dimensional infiltration problem where the infiltration rate at the top of the model is steady at 100 cm/day and the water table is at a depth of 10 m. The vadose zone has two layers with sand in the top 5 m and loam in the bottom 5 m. The hydraulic properties of the loam are saturated hydraulic conductivity of 25 cm/day, van Genuchten curve parameters of $\alpha = 0.036$ 1/cm and $n = 1.56$, saturated water content of 0.43, and residual water content of 0.078. The hydraulic properties of the sand are saturated hydraulic conductivity of 700 cm/day, van Genuchten curve parameters of $\alpha = 0.145$ 1/cm and $n = 2.68$, saturated water content of 0.43 and residual water content of 0.045.

11.12   References

Allen, R.G., Pereira, L.S., Raes, D., and Smith, M., 1998. *Crop Evapotranspiration, Guidelines for Computing Crop Water Requirements*. FAO Irrigation and Drainage Paper 56, Food and Agriculture Organization of the United Nations, Rome.

Allison, G.B., and Hughes, M.W., 1978. The Use of Environmental Chloride and Tritium to Estimate Total Local Recharge to an Unconfined Aquifer. *Australian Journal of Soil Research*, Vol. 16, pp. 181.

American Standards for Testing and Materials, 1992. D-5220-92 Method for Water Content of Soil and Rock In-Place by the Neutron Depth Probe Method. ASTM International, West Conshohocken, PA.

Bayliss, S.C., Goering, T.J., McVey, M.D., Strong, W.R., and Peace, J.L., 1996. *Preliminary Data from an Instantaneous Profile Test Conducted Near the Mixed Waste Landfill, Technical Area 3, Sandia National Laboratories, NM*, SAND96-0813.

Bentley, H.W., Phillips, F.M., Davis, S.N., Gifford, S., Elmore, D., Tubbs, L.E., and Gove, H.E., 1982. Thermonuclear C1-36 Pulse in Natural Water. *Nature*, Vol. 30, No. 23/30, pp. 737–740.

Bisque, R.E., 1984. Migration Rates of Volatiles from Buried Hydrocarbon Sources through Soil Media. In *Proceedings of the National Conference on Petroleum Hydrocarbons and Organic Chemicals in Groundwater—Prevention, Detection, and Restoration*. National Well Water Association, Worthington, OH.

Blaney, H.F., and Criddle, W.D., 1950. *Determining Water Requirements in Irrigated Areas from Climatological and Irrigation Data.* Technical Paper No. 96, U.S. Department of Agriculture Soil Conservation Service, Washington, DC.

Bouwer, H., 1966. Rapid Field Measurement of Air-Entry Value and Hydraulic Conductivity of Soil as Significant Parameters in Flow System Analysis. *Water Resources Research*, Vol. 2, pp. 729.

Bouwer, H., and Rice, R.C., 1984. Hydraulic Properties of Stony Vadose Zones. *Ground Water*, Vol. 22, No. 6.

Brakensiek, D.L., and Rawls W.J., 1994. Soil Containing Rock Fragments: Effects on Infiltration. *Catena*, Vol. 23, pp. 99–110.

Briscoe, R.D., 1986. Thermocouple Psychrometers for Water Potential Measurements. In Gensler, W.G., ed., *Advanced Agricultural Instrumentation: Design and Use, NATO ASI Series*. Martinus Nijhoff, Dordrecht, the Netherlands.

Brooks, R.H., and Corey, A.T., 1964. Properties of Porous Media Affecting Fluid Flow. *ASAE Journal of Irrigation Drainage Division*, Vol. IR2, pp. 61.

Brutsaert, W., 1977. Vertical Infiltration in Dry Soil. *Water Resources Research*, Vol. 13, pp. 363.

Buckingham, E., 1907. *Studies on the Movement of Soil Moisture.* U.S. Department of Agriculture Bureau of Soils, 38 pp.

Campbell, G.S., and Gee, G.W., 1986. Water Potential: Miscellaneous Methods. In Klute, A., ed., *Methods of Soil Analysis, Part 1, 2nd Edition.* Agronomy Monograph No. 9, ASA, CSSA, and SSSA, Madison, WI, pp. 619.

Carsel, R.F., and Parrish, R.S., 1988. Developing Joint Probability Distributions of Soil Water Characteristics. *Water Resources Research*, Vol. 24, pp. 755–769.

Cecen, K., 1967. *The Investigation of the Coefficient of Permeability in Connection with Construction Engineering Soil Investigations.* Verlag Wasser and Boden, Hamburg, Germany.

Creasey, C.L., and Dreiss, S.J., 1985. Soil Water Samplers: Do They Significantly Bias Concentrations in Water Samples? In *Proceedings of the Conference on Characterization and Monitoring of the Vadose (Unsaturated) Zone.* National Well Water Association, Dublin, OH, 173 pp.

Cronican, A.E., and Gribb, M.M., 2004. Hydraulic Conductivity Prediction for Sandy Soils. *Groundwater*, Vol. 42, No. 3, pp. 459–464.

Czarnecki, J.B., 1990. *Geohydrology and Evapotranspiration at Franklin Lake Playa, Inyo County, California.* Open-File Reptort 90-356, U.S. Geological Survey, Denver, CO.

Daly, C., Neilson, R.P., and Phillips, D.L., 1994. A Statistical-Topographic Model for Mapping Climatological Precipitation over Mountain Terrain. *Journal of Applied Meteorology*, Vol. 33, No. 2, pp. 140–158.

Ple Daniel, D.E., Burton, P.M., and Hwang, S.D., 1992. Evaluation of Four Vadose Zone Probes Used for Leak Detection and Monitoring. In Nielsen, D.M., and Sara, M.N., eds., *Current Practices in Ground Water and Vadose Zone Investigations, ASTM STP 1118.* American Society for Testing and Materials, Philadelphia, PA.

Daniel, D.E., 1989. In Situ Hydraulic Conductivity Tests for Compacted Clay. *ASCE Journal of Geotechnical Engineering*, Vol. 115, pp. 1205.

Daniel, D., Hamilton, J., and Olson, R., 1981. Suitability of Thermocouple Psychrometers for Studying Moisture Movement in Unsaturated Soils. In Zimmie, T.F., and Riggs, C.O., eds., *Permeability and Groundwater Contaminant Transport, ASTM STP 746.* American Society for Testing and Materials, Philadelphia, PA, pp. 84.

Daniel, D.E., 1993. State-of-the-Art: Laboratory Hydraulic Conductivity Tests for Saturated Soils. In Daniel, D.E., and Trautwein, S.J., eds., *Hydraulic Conductivity and Waste Contaminant Transport in Soils, ASTM STP 1142*. American Society for Testing and Materials, Philadelphia, PA.

Dirksen, C., 1991. Unsaturated Hydraulic Conductivity. In Smith, K.A., and Mullins, C.E., eds., *Soil Analysis: Physical Methods*. Marcel Dekker, New York, pp. 209.

Doorenbos, J., and Pruitt, W.O., 1975. Crop Water Requirements. In *Irrigation and Drainage Paper*. FAO, Rome.

Elrick, D.E., and Reynolds, W.D., 1992. Infiltration from Constant-Head Well Permeameters and Infiltrometers. In Topp, G.C., Reynolds, W.D., and Green, R.E., eds., *Advances in Measurement of Soil Physical Properties: Bringing Theory into Practice*. SSSA Special Publication No. 30, ASA, CSSA, and SSSA, Madison, WI.

Gardner, W.H., 1986. Water Content. In Klute, A., ed., *Methods of Soil Analysis, Part I, 2nd Edition*. Agronomy Monograph No. 9, ASA, CSSA, and SSSA, Madison, WI, pp. 493.

Gardner, W.R., 1987. Water Content: An Overview. In *International Conference on Measurement of Soil and Plant Water Status, Vol. 1, Soils*. Utah State University, Logan, UT, 7.

Glover, R.E., 1953. Flow for a Test Hole Located above Groundwater Level. In Zangar, C.N., ed., *Theory and Problems of Water Percolation*. Engineering Monograph No. 8. U.S. Bureau of Reclamation, Washington, DC, pp. 69.

Goering, T.J, McVey, M.D., Strong, W.R., and Peace, J.L., 1996. *Analysis of Instantaneous Profile Test Data from Soils Near the Mixed Waste Landfill, Technical Area 3, Sandia National Laboratories, NM*. SAND95-1637.

Goodspeed, M.J., 1981. Neutron Moisture Meter Theory. In Greacen, E.L., ed., *Soil Water Assessment by the Neutron Method*. CSIRO, East Melbourne, Australia, pp. 17.

Green, W.H., and Ampt, G.A., 1911. Studies in Soil Physics, I. The Flow of Air and Water through Soils. *Journal of Agricultural Science*, Vol. 4, pp. 1–24.

Havlena, J.A., and Stephens, D.B., 1991. Vadose Zone Characterization Using Field Permeameters and Instrumentation. In Nielsen, D.M., and Sara, M.N., eds., *Current Practice in Ground Water and Vadose Zone Investigations, ASTM STP 1118*. ASTM, Philadelphia, PA.

Hevesi, J.A., Flint, A.L., and Flint, L.E., 2003. *Simulation of Net Infiltration and Potential Recharge Using a Distributed-Parameter Watershed Model of the Death Valley Region, Nevada and California*. U.S. Geological Survey Water-Resources Investigations Report 03–4090.

Jensen, M.E., Burman, R.D., and Allen, R.G., 1990. *Evapotranspiration and Irrigation Water Requirements*. Engineering Practice Manual No. 70. American Society of Civil Engineers, New York, 332 pp.

Jensen, M.E., and Haise, R., 1963. Estimating Evapotranspiration from Solar Radiation. *ASCE Journal of Irrigation and Drainage Division*, Vol. 89, pp. 15.

Kirkham, R.R., Gee, G.W., and Jones, T.L., 1984. Weighing Lysimeters for Long-Term Water Balance Investigations at Remote Sites. *Soil Science Society of America Journal*, Vol. 48, No. 5, pp. 1203–1205.

Klute, A., and Dirksen, C., 1986. Hydraulic Conductivity and Diffusivity: Laboratory Methods. In Klute, A., ed., *Methods of Soil Analysis, Part I, 2nd Edition*. Agronomy Monograph No. 9, ASA, CSSA, and SSSA, Madison, WI, pp. 687.

Kung, K-J.S., 1990. Preferential Flow in a Sandy Vadose Zone: 1. Field Observation. *Geoderma*, Vol. 46, pp. 51–58.

Lang, A.R.G., 1967. Psychrometric Measurement of Soil Water Potential In Situ under Cotton Plants. *Soil Science*, Vol. 106, pp. 460–464.

Lee, Do-Hun, 2005. Comparing the Inverse Parameter Estimation Approach with Pedo-Transfer Function Method for Estimating Soil Hydraulic Conductivity. *Geosciences Journal*, Vol. 9, No. 3, pp. 269–276.

Leonard, R.A., Davis, F.M., and Knisel, W.G., 1989. *Groundwater Loading Effects of Agricultural Management Systems (GLEAMS): A Tool to Assess Soil-Climate-Management-Pesticide Interactions*. U.S. Department of Agriculture, Tifton, GA.

Markstrom, S.L., Regan, R.S., Hay, L.E., Viger, R.J., Webb, R.M.T., Payn, R.A., and LaFontaine, J.H., 2015. PRMS-IV, the Precipitation-Runoff Modeling System, version 4. U.S. Geological Survey Techniques and Methods, Book 6, Chap. B7, 158 pp., https://dx.doi.org/10.3133/tm6B7.

Marshall, T.J., and Holmes, J.W., 1992. *Soil Physics*. Cambridge University Press, London.

McCuen, R.H., Rawls, W.H., and Brakensisk, D.L., 1981. Statistical Analysis of the Brooks-Corey and the Green-Ampt Parameters across Soil Textures. *Water Resources Research*, Vol. 17, No. 4, pp. 1005–1013.

McKim, H.L., Berg, R.L., McGaw, R.W., Atkins, R.T., and Ingersoll, J., 1976. Development of a Remote Reading Tensiometer/Transducer System for Use in Subfreezing Temperatures. In *Proceedings of the Second Conference on Soil-Water Problems in Cold Regions*. AGU, Edmonton, Alberta, Canada.

McWhorter, D.B., and Sunada, D.K., 1977. *Ground-Water Hydrology and Hydraulics*. Water Resources Publication, Fort Collins, CO.

Merrill, S.D., and Rawlins, S.L., 1972. Field Measurement of Soil Water Potential with Thermocouple Psychrometers. *Soil Science*, Vol. 113, pp. 102.

Moridis, G.J., and McFarland, M.J., 1982. Modeling Soil Water Extraction from Grain Sorghum. In *Proceedings Winter Meeting*. American Society of Civil Engineers, St. Joseph, MI.

Mualem, Y., 1976. A New Model for Predicting the Hydraulic Conductivity of Unsaturated Porous Media. *Water Resources Research*, Vol. 12, pp. 513–522.

Mualem, Y., and Dagan, G., 1976. *A Catalog of the Hydraulic Properties of Unsaturated Soils*. Technical Report. Technion Israel Institute of Technology, Haifa, Israel, 100 pp.

Mullins, J.A., Carsel, R.F., Sarbrough, J.E., and Ivery, A.M., 1993. *PRZM-2, A Model for Predicting Pesticide Fate in the Crop Root and Unsaturated Soil Zones, User's Manual for Release Version 2.0*. U.S. Environmental Protection Agency, Environmental Research Laboratory, Athens, GA.

Nasberg, V.M., 1951. *The Problem of Flow in an Unsaturated Soil or Injection under Pressure*. Izvestja Akademia Nauk, SSSR odt tekh Nauk, no. 9, translated by Mr. Reliant, 1973, B.R.G.M., France.

Parizek, R.R., and Lane, B.E., 1970. Soil-Water Sampling Using Pan and Deep Pressure-Vacuum Lysimeters. *Journal of Hydrology*, Vol. 11, pp. 1.

Penman, H.L., 1948. Natural Evapotranspiration from Open Water, Bare Soil and Grass. *Proceedings of the Royal Society London*, Vol. 193A, pp. 120.

Phene, C.J., Hoffman, J., and Rawlins, S.L., 1971. Measuring Soil Matric Potential In Situ by Sensing Heat Dissipation within a Porous Body: I. Theory and Sensor Construction. *Soil Science Society of America Proceedings*, Vol. 35, pp. 27.

Philip, J.R., 1957. The Theory of Infiltration. 1. The Infiltration Equation and Its Solution. *Soil Science*, Vol. 83, pp. 345.

Philip, J.R., 1985. Approximate Analysis of the Borehole Permeameter in Unsaturated Soil. *Water Resources Research*, Vol. 21, pp. 1025.

Phillips, F.M., Mattick, J.L., Duval, T.A., Elmore, D., and Kubik, P.W., 1988. Chlorine 36 and Tritium from Nuclear Weapons Fallout as Tracers for Long-Term Liquid and Vapor Movement in Desert Soils. *Water Resources Research*, Vol. 24, pp. 1877.

Phillips, F.M., 1994. Environmental Tracers for Water Movement in Desert Soils: A Regional Assessment for the American Southwest. *Soil Science Society of America Journal*, Vol. 58, No. 1, pp. 15–24.

Rawls, W.J., Ahuja, L.R., and Brakensiek, D.L., 1992. Estimating Soil Hydraulic Properties from Soils Data. In van Genuchten, M.T., et al., eds., *Proceedings of the International Workshop on Indirect Methods for Estimating the Hydraulic Properties of Unsaturated Soils.* USDA and University of California, Riverside, pp. 329–340.

Rawls, W.J., and Brakensiek, D.L., 1985. Prediction of Soil Water Properties for Hydrologic Modeling. In *Proceedings of Symposium on Watershed Management.* American Society of Civil Engineers, New York, 293–299 pp.

Reynolds, W.D., Elrick, D.E., and Topp, G.C., 1983. A Reexamination of the Constant Head Well Permeameter Method for Measuring Saturated Hydraulics Conductivity above the Water Table. *Soil Science*, Vol. 136, pp. 250.

Reynolds, W.D., Elrick, D.E., and Clothier, B.E., 1985. The Constant Head Well Permeameter: Effect of Unsaturated Flow. *Soil Science*, Vol. 139, pp. 172.

Rosenberg, N.J., Blad, B.L., and Verma, S.B., 1983. *Microclimate: The Biological Environment, 2nd Edition.* Wiley-Interscience, John Wiley & Sons, New York.

Sayde, C., Gregory, C., Gil-Rodriguez, M., Tufillaro, N., Tyler, S., van de Giesen, N., English, M., Cuenca, R., and Selker, J.S., 2010. Feasibility of Soil Moisture Monitoring with Heated Fiber Optics. *Water Resources Research*, Vol. 46, pp. W06201. doi:10.1029/2009WR007846

Scanlon, B.R., Healy, R.W., Cook, P.G., 2002. Choosing Appropriate Techniques for Quantifying Groundwater Recharge. *Hydrogeology Journal*, Vol. 10, pp. 18–39.

Schaap, M.G., Leij, F.J., and van Genuchten, M.T., 2001. Rosetta: A Computer Program for Estimating Soil Hydraulic Parameters with Hierarchical Pedotransfer Functions. *Journal of Hydrology*, Vol. 251, pp. 163–176.

Schroeder, P.R., Gibson, A.C., and Smolen, M.G., 1984. *The Hydrologic Evaluation of Landfill Performance (HELP) Model.* USEPA Document No. EPA/530-SW-84-010. U.S. Environmental Protection Agency, Office of Solid Waste and Emergency Response, Washington, DC.

Sebenik, P.G., and Thomas, J.L., 1967. Water Consumption by Phreatophytes, Progessive Agriculture in Arizona, 19, 10.

Selker, J.S., The venaz, L., Huwald, H., Mallet, A., Luxemburg, W., van de Giesen, N., Stejskal, M., Zeman, J., Westhoff, M., and Parlange, M.B., 2006. Distributed Fiber-Optic Temperature Sensing for Hydrologic Systems. *Water Resources Research.* Vol. 42, pp. W12202. doi:10.1029/2006WR005326

Shaap, M.G., Lei, F.J., and van Genuchten, M.T., 2001. ROSETTA: A Computer Program for Estimating Soil Hydraulic Parameters with Hierarchical Pedotransfer Functions. *Journal of Hydrology*, Vol. 251, pp. 163–176.

Singh, G., Kaur, G., Williard, K., Schoonover, J., and Kang, J., 2018. Monitoring of Water and Solute Transport in the Vadose Zone: A Review. *Vadose Zone Journal*, Vol. 17, pp. 160058. doi:10.2136/vzj2016.07.0058

Spanner, D.C., 1951. The Peltier Effect and Its Use in the Measurement of Suction Pressure. *Journal of Experimental Botany*, Vol. 11, pp. 145.

Stannard, D.I., 1990. *Use of a Hemispherical Chamber for Measurement of Evapotranspiration.* Open-File Report 88-452. U.S. Geological Survey, Denver, CO.

Steele-Dunne, S.C., Rutten M.M., Krzeminska D.M., Hausner M., Tyler S.W., Selker J.S., Bogaard T.A., and van de Giesen N.C., 2010. Feasibility of Soil Moisture Estimation Using Passive Distributed Temperature Sensing. *Water Resources Research*, Vol. 46, pp. W03534. doi:10.1029/2009WR008272

Stephens, D.B., 1993. Hydraulic Conductivity Assessment of Unsaturated Soils. In Daniel, D.E., and Trautwein, S.J., eds., *Hydraulic Conductivity and Waste Contaminant Transport in Soils*. ASTM STP 1142. American Society for Testing and Materials, Philadelphia, PA.

Stephens, D.B., 1996. *Vadose Zone Hydrology*. CRC Press, Boca Raton, FL.

Stephens, D.B., and Knowlton, R., Jr., 1986. Soil Water Movement and Recharge through Sand at a Semi-Arid Site in New Mexico. *Water Resources Research*, Vol. 22, pp. 881.

Stephens, D.B., Knowlton, R.G., Jr., Stanfill, M., and Hirtz, E.M., 1985. *Field Study to Quantify Seepage from a Fluid Impoundment*. In *Proceedings of the Conference on Characterization and Monitoring of the Vadose (Unsaturated) Zone*. National Water Works Association, Dublin, OH, 283 pp.

Stephens, D.B., Lambert, K., and Watson, D., 1987. Regression Models for Hydraulic Conductivity and Field Test of the Borehole Permeameter. *Water Resources Research*, Vol. 23, No. 12, pp. 2207–2214.

Stutman, M., 1993. A Novel Passive Sorbent Collection Apparatus for Site Screening of Semivolatile Compounds. In *Proceedings of the Third International Conference on Field Screening Methods for Hazardous Waste and Toxic Chemicals*, Las Vegas, NV. February 24–26, 1993.

Swinbank, W.C., 1951. A Measurement of Vertical Transfer of Heat and Water Vapour and Momentum in the Lower Atmosphere with Some Results. *Journal of Meteorology*, Vol. 8, pp. 135.

Tanner, B.D., Tanner, M.S., Dugas, W.A., Campbell, E.C., and Bland, B.L., 1985. Evaluation of an Operational Eddy Correlation System for Evapotranspiration Measurements, in Advances in Evapotranspiration. In *Proceedings of the National Conference on Advances in Evapotranspiration*, December 16–17, 1985, Chicago. American Society of Agricultural Engineers, St. Joseph, MI, 1985, pp. 87–99.

Tanner, C.B., 1960. Energy Balance Approach to Evapotranspiration from Crops. *Soil Science Society of America Proceedings*, Vol. 24, pp. 1–9.

Terletskaya, N.M., 1954. Determination of Permeability in Dry Soils. Hydroelectric Waterworks No. 2, February, Moscow.

Thornthwaite, C.W., 1948. An Approach toward a Rational Classification of Climate. *Geographical Review*, Vol. 38, No. 1, pp. 55–94.

Thornton, P.E., Thornton, M.M., Mayer, B.W., Wei, Y., Devarakonda, R., Vose, R.S., and Cook, R.B., 2018. *Plea Daymet: Daily Surface Weather Data on a 1-km Grid for North America, Version 3*. ORNL DAAC, Oak Ridge, Tennessee, USA. https://doi.org/10.3334/ORNLDAAC/1328

Topp, G.C., Davis, J.L., Annan, A.P., 1980. Electromagnetic Determination of Soil Water Content: Measurements in Coaxial Transmission Lines. *Water Resources Research*, Vol. 16, pp. 574–582.

Torstensson, B.A., 1984. A New System for Groundwater Monitoring. *Ground Water Monitoring Review*, Vol. 4, pp. 131.

Tyler, S.W., Selker, J.S., Hausner, M.B., Hatch, C.E., Torgersen, T., Thodal, C.E., and Schladow, S.G., 2009. Environmental Temperature Sensing Using Raman Spectra DTS Fiber-Optic Methods. *Water Resources Research*, Vol. 45, pp. W00D23. doi:10.1029/2008WR007052

U.S. Bureau of Reclamation, 1974. *Earth Manual. 2nd Edition*. U.S. Government Printing Office, Washington, DC.

U.S. Bureau of Reclamation. 1977. *Ground Water Manual. 1st Edition*. U.S. Government Printing Office, Washington, DC.

U.S. Geological Survey, 2008. *Documentation of Computer Program INFIL3.0—A Distributed-Parameter Watershed Model to Estimate Net Infiltration below the Root Zone*. U.S. Geological Survey Scientific Investigations Report 2008–5006, 98 pp.

van Genuchten, M.T., 1978. *Calculating the Unsaturated Hydraulic Conductivity with a New Closed-Form Analytical Model*. Research Report 78-WR-08. Princeton University, Princeton, NJ.

van Genuchten, M.T., 1980. A Closed-Form Equation for Predicting the Hydraulic Conductivity of Unsaturated Soils. *Soil Science Society of America Journal*, Vol. 44, pp. 892.

van Genuchten, M.T., Leij, F.J., and Lund, L.J., 1989. Indirect Methods for Estimating the Hydraulic Properties of Unsaturated Soils. In *Proceedings of the First International Workshop for Estimating the Hydraulic Properties of Unsaturated Soil*, Riverside, CA, October 11–13, 1989. U.S. Salinity Laboratory, Agricultural Research Service, USDA, University of California, Riverside.

van Genuchten, M.T., Leij, F.J., and Yates, S.R., 1991. *The RETC Code for Quantifying the Hydraulic Functions of Unsaturated Soils*. Prepared by U.S. Salinity Laboratory, U.S. Department of Agriculture, Agricultural Research Service. Riverside, California. Prepared for Robert S. Kerr Environmental Research Laboratory, Office of Research and Development, U. S. Environmental Protection Agency, Ada, OK. 9250EPA/600/2-91/065. December 1991.

Zanger, C.N., 1953. *Theory and Problems of Water Percolation*. Engineering Monograph No. 8. U.S. Bureau of Reclamations, Denver, CO.

Zegelin, S.J., and White, I., 1982. Design for a Field Sprinkler Infiltrometer. *Soil Science Society of America Journal*, Vol. 46, pp. 1129.

CHAPTER 12
Karst Hydrogeology

David M. Bednar, Jr.

Fort Smith, Arkansas

12.1   Introduction

Karst is an area of land, predominantly underlain by carbonate rocks, where unique surface and subsurface landforms and hydrology are created by the reaction of water with the bedrock. Carbonate rocks consist of limestone and dolostone. Calcite ($CaCO_3$) is the dominant mineral in limestone and dolomite ($CaMg(CO_3)_2$) is the dominant mineral in dolostone. Water that falls as precipitation to the earth contains dissolved carbon dioxide (CO_2) that produces a weak carbonic acid (H_2CO_3). The weak acidic water dissolves the calcite and dolomite in the rock and transports the mineral matter into and through the groundwater-flow system (Figure 12.1). Water that moves through the soil increases in acidity by further absorption of CO_2 from microorganisms living in the soil and decayed plant matter. Once groundwater reaches the bedrock, it passes through bedding planes, joints, and fractures in the rock and openings become wider through time.

Distinctive surface landforms that form in karst include sinkholes, sinking streams, and many types of sculpted bedrock surfaces. Distinctive subsurface features include caves and the unique formations that develop in them. Some karst terrains are a rough and jumbled land of deep depressions, isolated towers, and pointed hills while others may be gently rolling plains with only the slightest number of depressions to label them as karst (White 1988). For example, some karst areas such as the Mitchell Sinkhole Plain of Indiana and the Pennyroyal Sinkhole Plain in Kentucky show obvious signs of karst while other areas are not so obvious.

Karst landscapes are found in many areas from tropical to colder alpine settings. About 25% of the United States is karst with 40% located east of the Mississippi River (Figure 12.2). About 25% of the world's population depends on freshwater from karst aquifers (Ford and Williams 1989). Some of the karst areas in the United States include the Great Valley of the Appalachian Mountains, the Ozark Plateau of southern Missouri and northern Arkansas, the Pennyroyal Plain of western Kentucky, the Mitchell Plain of Indiana, the Edwards Plateau in southwestern Texas, the Nashville and Lexington Plains in Tennessee and Kentucky, the Black Hills of western South Dakota, and Florida. Karst areas are found in other countries some of which include Slovenia, Russia;

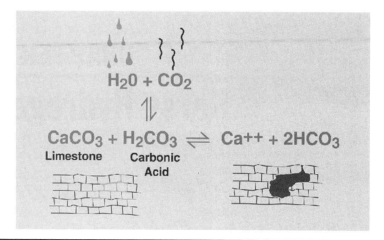

FIGURE 12.1 The generalized chemical reaction that occurs when carbonic acid flows over limestone in areas of karst. (*Used with permission, North Carolina Geological Survey.*)

the southern and western Alps of Italy; the Grand Causses and Jura Mountains in France; the Franconian Alps in Germany; the Padis Karstic Plateau in the Bihor Mountains of the western Carpathians in Rumania; the Carboniferous Limestone, Peak District, Yorkshire Dales, North Wales, Northern Pennies, and the Mendip Hills of Great Britain; the Cockpit Plateaus of Jamaica; Lunan in Yunnan, China; the Nullarbor Plain in Australia; and the Bonito karst region in midwestern Brazil (Figure 12.3).

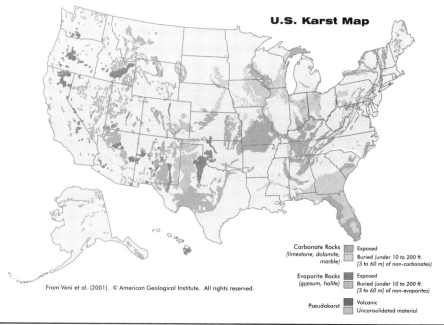

FIGURE 12.2 General representation of karst and pseudokarst areas in the United States. (*Used with permission, American Geological Institute.*)

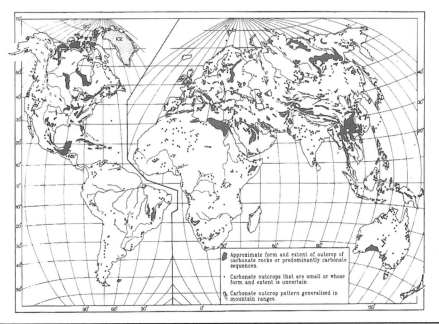

FIGURE 12.3 Global distribution of major outcrops of carbonate rocks. (*Used with permission of Springer Science and Business Media.*)

Karst is the Germanized word for Kras. In the Slovene language, Kras is a plateau region underlain by limestone, now a part of Slovenia and Croatia, along the northernmost part of the Adriatic Sea. The plateau stretches in a northwest-southeast orientation and is 25 mi long and up to 8 mi wide, covering about 170 mi^2 (Kranjc 1997). The earliest origin of the word Kras means "stony ground." The land was left stony and barren because deforestation and overgrazing led to the loss of most soil into sinkholes and caves (Gams 1994). It is not yet completely documented how the name Kras became the general term for land with karstic characteristics and who was the first or the most important to make the transition (Kranjc 1998). In the 1800s, geologists and geographers entered the Slovene term "Kras" into the geologic and geographic literature. Jovan Cvijic's 1893 composition of Das Karstphaenomen is still considered a landmark investigation on the geomorphology (study of landforms) of karst areas. Some considered Cvijic to be the father of karst geomorphology.

12.2 Karst Types

Not all karst landscapes look identical. The degree to which karst forms is dependent on several factors, some of which include climate type, the amount of rainfall, the depth of groundwater circulation, the amount of soil cover, and the structure of the rock and its solubility. Some karst areas are easily recognizable based on surface landforms while other areas may show no surface expression of karst development but contain solution caves beneath the surface.

Covered and Exposed Karst

Geoscientists have distinguished karst landscapes from one another by the degree to which sediment covers the karst and by the karst feature that dominates the landscape. Quinlan (1967, 1978) developed a classification scheme for karst landscapes according to what covers the karst and placed them into two major types: covered karst and exposed karst. Covered karst represents areas where the soluble rock is covered by some material and not exposed at the surface. Five subcategories within this type include subsoil karst, mantled karst, buried karst, interstratal karst, and subaqueous karst.

Subsoil karst occurs when the karst surface is covered with either soil transported or its own sediment (White 1988). A karst area that is covered by a thick deposit of unconsolidated sediments is known as mantled karst. Buried karst, also known as paleokarst, occurs when a karst surface is completely filled within sediments and buried with younger rocks. Interstratal karst refers to soluble rocks buried by less soluble rocks prior to karstification. Even though these rocks are not exposed at the surface, soluble rocks can still develop karst features if located within the groundwater-flow system. Subaqueous karst forms beneath bodies of water such as rivers and lakes, or within tidal zones.

Exposed karst refers to areas of soluble rock that are exposed at the surface. Four subcategories occur within this type and include naked karst, denuded karst, exhumed karst, and relict karst. Naked karst occurs primarily in alpine regions where soils are poorly developed (White 1988). Denuded karst occurs when an insoluble rock located above the soluble rock has been removed by erosion and the soluble rock is exposed at the surface. Exhumed karst refers to karst that was once buried and later exposed by erosion. Relict karst is soluble rock that has been removed by erosion without any trace of topographic expression.

Cone and Tower Karst

The requirements for cone and tower karst (Figure 12.4) seem to be a thick massive limestone, high relief, and a well-developed fracture system (White 1988). Cone and tower karst, sometimes called tropical karst, develops in tropical regions where the greater amount of rainfall aids in the dissolution of the vertical fracture system. Cone and tower karst is found in Central America and the South Pacific (White 1988). The classic tower karst can consist of isolated or nested towers that are often relatively narrow with vertical sides while cone karst refers to more rounded or conical hills. The Yangshuo region of Guangxi, China provides an excellent example of cone and tower karst.

Cockpit Karst

The type area for well-developed cockpit karst lies in Jamaica (Figure 12.5). Additionally, known as polygonal karst, the depressions are a series of touching polygons. The topography of the cockpit karst resembles an egg carton consisting of large star-shaped closed depressions or sinkholes surrounded by cone-shaped hills. The cockpits are steep sided and surrounded by ridges, and many cockpits are connected to each other (Day and Chenoweth 2004).

Pavement Karst

Pavement karsts (Figure 12.6) occur in high-latitude or alpine terrains where soils are thin or stripped (White 1988). Sometimes called bare karst, the lack of soils and vegetation exposes vertical or nearly vertical fractures in the rock and dissolution creates wide crevices in the rock known as kluftkarren. Sweeting (1973) preferred the term glaciokarst. A fine example of pavement karst is found in the Burren Plateau in Ireland.

FIGURE 12.4   Cone and tower karst of China. (*Photo courtesy of Andrej Kranjc.*)

FIGURE 12.5   Cockpit karst of Jamaica. (*Photo courtesy of Mick Day.*)

FIGURE 12.6   Limestone pavement, Pyrenees Mountains, Spain. (*Photo courtesy of John Mylroie.*)

Coastal Karst

Many islands that form along coastal areas, located in tropical and subtropical climates, can be found where limestone deposition is still in process. Coastal karst can be found along New Zealand, whereas the Bahamas represents the typical island karst type. Carbonate islands and coastlines consist of many caves and karst features. Their differences result from (1) freshwater mixing with seawater, (2) sea-level changes, and (3) the constant changing of the coastline from the erosional forces of the sea.

Fluviokarst

A karst landscape that shows evidence of a past drainage system is called fluviokarst. Many karst features occur in fluvial karst such as swallow holes, sinkholes, blind valleys, and caves. Limestone surfaces in fluviokarstic areas tend to be more soil covered due to the presence of soil washing and debris caused by fluvial processes (Sweeting 1973). Much of the karst in the eastern United States is fluviokarst; some of the best examples occur along the Cumberland Plateau in Tennessee and northern Alabama (White 1988). Other examples of fluviokarst are located in the Mendip Hills and the Peak District in Britain and the Causses of Quercy in central France.

Sinkhole Karst

The dominant surface feature in this type of karst is the sinkhole. Bedrock underlying sinkhole karst is relatively flat. Examples of sinkhole karst can be found in northern Florida, south-central Kentucky, and southern Indiana (Figure 12.7).

Karst in the Literature

The geology, hydrology, and geomorphology of karst landscapes have been studied for over 100 years and resulted in an extensive amount of literature including textbooks, encyclopedias, conference proceedings volumes, and original papers. Textbooks on karst written within the last 35 years include those by Herak and Stringfield (1972),

Figure 12.7 Sinkhole plain, Indiana, USA. (*Photo courtesy of Samuel S. Frushour.*)

Sweeting (1973), Jakucs (1977), Bogli (1980), Milanovic (1981, 2004), Jennings (1985), Dreybrodt (1988), White (1988), Bosak et al. (1989), Ford and Williams (1989), Drew and Hotzl (1999), Klimchouk et al. (2000), Waltham et al. (2005), Klimchouk (2007), Palmer (2007), Palmer and Palmer (2009), Stafford et al. (2009), Mylroie and Lace (2013), Kresic (2013), Cooper and Mylroie (2015), Stevanovic, (2015), White (2015), Benson and Yuhr (2016), Feinberg et al. (2016), Hobbs et al. (2017), Klimchouk et al. (2017), and White et al. (2018).The *Encyclopedia of Cave and Karst Science* (Gunn 2004) and the *Encyclopedia of Caves* (Culver and White 2005, 2012) provide some of the most recent knowledge on karst landscapes. Ford and Williams (2007) is the second edition prepared by the authors that presents a systems-oriented approach and integration of geomorphology and hydrology.

Probably the most recognized conferences held within the last 25 years on sinkholes and the engineering and environmental impacts of karst have been those sponsored by P.E. LaMoreaux and Associates, Inc. and recently by the Geo Institute of the American Society of Civil Engineers. In 2011, The National Cave and Karst Research Institute (NCKRI) assumed a leadership role the in the organization, hosting the sinkhole conference held at the NCKRI headquarters in Carlsbad, New Mexico (Stephenson, 2015). Other publications by the NCKRI include Chavez and Reehling (2016), Klimchouk et al. (2014), Klimchouk (2016). Additionally, original karst research has been published by several mainstream professional journals some of which include the *Journal of Hydrology*, *Ground Water*, and *Environmental Geology*.

Other books published recently on karst include Cooper (2015), Feinberg et al. (2016), and Upchurch S et al. (2018), LaMoreaux, J.W. (2012, 2013).

12.3 Carbonate Rocks

Carbonate rocks are those rocks that contain more than 50% by weight of the minerals calcite, aragonite, or dolomite. To be considered pure, carbonate rocks must have at least 90% of such minerals (Jennings 1985). A carbonate rock that predominantly contains the minerals calcite is called limestone. There are over 100 uses for carbonate rocks in industry; some of these include filler material for asphalt, fertilizer, cement, building stones, road stone, and the manufacture of glass and paper (Siegel 1967).

The color of a carbonate rock is due to such factors as the environment of deposition, variations in the size of grains within the rock, mineral content, and the amount of organic material. Carbonate sediments buried in shallow water in an environment with low amounts of oxygen with organic material would likely produce a gray or black color. Sediments buried with traces of iron and exposed to the atmosphere and weathering are usually a buff or faded yellow or cream color. Carbonates formed in deeper water may be red, pink, or purple through preservation of iron and manganese oxide pigments (Wilson 1975).

Carbonate rocks commonly form in shallow marine areas known as carbonate platforms. Carbonate platforms are known to occur attached to landmasses such as in South Florida and the southern coast of the Persian Gulf (Blatt 1982), but also can occur in deep ocean basins, nonmarine environments in lake depressions, and as isolated accumulations of carbonate-producing organisms that rise from greater depths like the Pacific atolls. Currently, carbonate sediments are forming in shallow waters in the Bahama-Florida platform, the Gulf of Batabanno off southern Cuba, Campeche Bank off the Yucatan Peninsula, the continental shelf of British Honduras, the Trucial coast of the Persian Gulf, and the coasts of Western Australia and Queensland (Bathurst 1975).

The Bahama Platform represents a well-studied and excellent example of a modern isolated carbonate depositional area (Blatt 1982). The northwestern Bahama Islands consist of isolated masses that rise above sea level from two carbonate platforms: Little Bahama Bank and Great Bahama Bank (Onac et al. 2001). This platform consists of 20 major islands with a surface area of 434 mi (698 km) by 186 mi (299 km) (Tucker and Wright 1990). It is generally submerged to a depth of less than about 30 ft (9.1 m), the greater part of the seafloor being covered by less than about 21 ft (6.4 m) of water (Bathurst 1975). Drilling on the islands has encountered about 15,000 ft (4,572 m) of shallow carbonate and associated rocks.

Much has been written about the origin and occurrence of carbonate depositional environments and rocks that comprise them. Some of these works include Chilingar et al. (1967), Lippman (1973), Bathurst (1975), Wilson (1975), Cook and Enos (1977), Tucker (1981), Blatt (1982), Scholle et al. (1983), Boggs (1987, 2001), Tucker and Wright (1990), Blatt and Tracey (1999), Moore (2001), Braithwaite (2005), and James and Jones (2016).

Limestone

Limestone ($CaCO_3$) commonly forms in shallow marine environments within 30° latitude of the equator. These rocks are composed of calcite grains along with visible or microscopic shells from marine organisms. To completely understand the formation of limestone one must have knowledge of the biochemical interactions between marine organisms and the waters in which they live (Blatt 1982).

Several classification schemes have been developed over the years to describe the differences in particles or grains that appear in limestones. The classification schemes developed by Folk (1959, 1962) and Dunham (1962) are the most widely used by geologists who study carbonate sediments. Folk's scheme has been widely used, but Dunham's classification is recommended for practical use (Braithwaite 2005).

Folk Classification

Folk (1959, 1962) developed a classification that was based on the size and shape of grains within the rock and the crystals that hold or cement the grains together. Folk defined these particles as allochems and orthochems (Figure 12.8). Allochems consist of the particles in the rock and are subdivided into four categories: fossils, peloids, ooids, and intraclasts. All four constituents listed above may be mixed in a wide range or proportions to form limestone beds (Folk 1962).

Intraclasts are fragments of existing limestone that were disturbed and redeposited within the area of deposition. Intraclasts may be of any shape or size. Ooids are small spherical grains coated with calcium carbonate that have formed around small shell fragments or quartz grains. The carbonate coating is chemically precipitated from agitated water that is evident of being transported by strong currents (Blatt 1982) in high-energy environments such as tidal channels. Larger ooids are called pisoids.

Pellets are muddy carbonate grains that have no apparent internal structure (Braithwaite 2005). They are thought to be of fecal origin excreted by marine organisms (Selley 2000) such as gastropods and worms. Some pellets harden over time while others weather and fall apart. Pellets are smaller than ooids and range in size from silt to fine sand and are characteristic of low-energy environments such as lagoons.

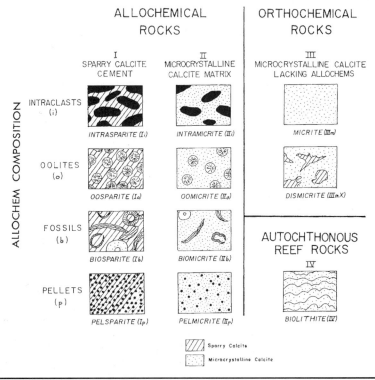

FIGURE 12.8 Folk classification for limestones (Folk 1962). (*Courtesy of the American Association of Petroleum Geologists.*)

Fossil skeletal particles, also known as bioclasts, consist of broken fragments of or whole marine organisms such as pelecypods, brachiopods, and ostracods. Skeletal particles are by far the most common allochem in carbonate rocks and are so abundant that in some limestones they make up the entire rock (Boggs 1987).

Orthochems are the smaller grains that hold or act as a cement to hold the larger grains together. Microcrystalline calcite or micrite are smaller particles that are deposited at the same time as the allochems. Micrite is composed of very small carbonate crystals deposited under quiet-water conditions. It can be present as matrix material among larger carbonate grains or it may make up most or all of the limestone (Boggs 1987). Carbonate sediments are accumulating in many modern environments from tidal flats and shallow lagoons to the deep-sea floor. Spar or sparry calcite cement infiltrates the deposits of allochems after deposition and cements the allochems in place. Sparry calcite cement fills the pore spaces within the carbonate particles.

A limestone can exhibit various amounts of allochems and orthochems. To name the rock, Folk developed a prefix for each allochem and a suffix for each orthochem. Fossil was replaced by the prefix "bio," "pel" replaced peloid, "oo" replaced ooid, "intra" replaced intraclast, "mic" replaced micrite, and "spar" replaced sparite. For example, a limestone composed of ooids within a micrite matrix is identified as an oomicrite.

Dunham Classification Scheme

The classification developed by Dunham was also based on the depositional texture of the rock with an additional subdivision based on the abundance of grains (Figure 12.9). Grains are carbonate particles found in the rock. The grains can be composed of skeletal particles or nonskeletal particles. A limestone with less than 10% of the rock composed of grains with a mud matrix are named mudstones, mudstones with greater than 10% grains but still with a mud matrix are wackestones. Grain-supported muddy carbonate rocks are termed packstone. Those rocks that are mud free and grain supported are grainstones.

Dolomite

Dolomite $CaMg(CO_3)_2$ is a rock that is composed of greater than 50% of the mineral dolomite. Some researchers prefer to use the term dolostone to refer to the rock composed of dolomite. However, both terms have been used to identify the rock.

Dolomite has been known to be precipitated directly from water such as some lagoons and lake settings. However, this is a rare event in the geologic record. Predominantly, dolomite forms when existing limestone rocks are chemically altered after

DEPOSITIONAL TEXTURE RECOGNIZABLE					DEPOSITIONAL TEXTURE NOT RECOGNIZABLE
Original Components Not Bound Together During Deposition				Original components were bound together during deposition... as shown by intergrown skeletal matter, lamination contrary to gravity, or sediment-floored cavities that are roofed over by organic or questionably organic matter and are too large to be interstices.	Crystalline Carbonate
Contains mud (particles of clay and fine silt size)			Lacks mud and is grain-supported		
Mud-supported		Grain-supported			(Subdivide according to classifications designed to bear on physical texture or diagenesis.)
Less than 10 percent grains	More than 10 percent grains				
Mudstone	Wackestone	Packstone	Grainstone	Boundstone	

FIGURE 12.9   Dunham classification for limestone (Dunham 1962). (*Courtesy of the American Association of Petroleum Geologists.*)

deposition by magnesium-rich water that infiltrates the rock. Reasons to suggest that dolomite forms after deposition is that it has irregular boundaries within the rock strata and lacks even bedding (Selley 2000).

Dolomites form at low latitudes in shallow marine water just as limestones. Recent dolomite deposits are known to form in arid hypersaline coasts termed sabkha—Arabic for salt marsh. Dolomite crusts occur on tidal flats of Andros Island in the Bahamas; Sugarloaf Key, Florida; Bonaire Island of the Venezuela mainland, Netherlands Antilles; in the high intertidal sediments of the Trucial Coast and within coastal evaporitic lakes of the Coorong, South Australia; in saline lakes of Victoria, Australia; and in hypersaline lagoons of Baffin Island, Texas, and Kuwait (Tucker and Wright 1990).

Dolomite is rare in modern carbonate environments and more abundant in older rocks. Investigators have called this "the dolomite problem." Dolomite has been intensely studied throughout the century and remains one of the most thoroughly researched, but poorly understood problems in sedimentary geology (Arvidson and MacKenzie 1997; Boggs 2001). Although authors agree that the source of magnesium in the rock must come from seawater, the mechanism required to precipitate dolomite is still up for debate. This inability of scientists to agree on a single mechanism for dolomite formation has resulted in the development of several models of deposition and there are critics and supporters of each. All current models of dolomite formation are primarily physical descriptions of how to move seawater, or modify seawater, through calcium carbonate sediments (Burns et al. 2000).

12.4   Recharge Areas

The karst landscape and its subsurface are sculpted by flowing water. Over time, water slowly dissolves the rock away as it works its way through joints and bedding planes. Eventually, surface and groundwater flow function as one continuous unit. In a true karst landscape, no water exists on the surface, all water is circulating through fissures underground (Sweeting 1973). This section discusses the various types of features that commonly develop on karst landscapes.

Surface water that infiltrates the karst surface through openings in the bedrock is known as recharge. Several terms have been used in the karst literature to describe recharge water that enters the karst landscape. The four terms that are commonly used included allogenic, autogenic, diffuse, and concentrated recharge. Allogenic recharge refers to surface water that flows from adjacent non-karst areas onto the karst landscape and sinks into the subsurface. Allogenic recharge enters the subsurface through concentrated areas of the bedrock that have been dissolved out or solutionally enlarged. This commonly occurs along stream segments and through the bottoms of sinkholes. Autogenic recharge is defined as infiltration supplied by precipitation that falls directly on the karst landscape and enters the subsurface through smaller openings in the bedrock, spread out over the entire area of karst landscape. Diffuse recharge refers to the water that enters the subsurface through small openings in the bedrock at less volume. Concentrated recharge enters the subsurface through larger openings and has greater flow volume. Concentrated recharge and diffuse recharge represent the maximum ends of a continuum and most karst landscapes exhibit both types of recharge.

Gunn (1985) developed a conceptual model for flow in karst aquifers and introduced four additional terms to describe recharge: diffuse autogenic, concentrated autogenic, diffuse allogenic, and concentrated allogenic. Diffuse autogenic recharge falls directly on the karst landscape and enters the subsurface through small openings in the bedrock. Concentrated autogenic recharge is recharge that falls on the karst landscape and flows toward closed depressions such as sinkholes. Diffuse allogenic recharge falls onto non-permeable rocks that overlie the karst. Concentrated allogenic recharge falls on non-karst lands and flows to the karst where it enters the subsurface through sinking streams not far from the non-karst/karst boundary.

Ewers (1992) conducted a study with data loggers to closely monitor the physical and chemical reactions of karst springs during storm events in Kentucky. Two additional terms were used to describe recharge. Quickflow recharge described relatively rapid movement of water into the subsurface, and seepage recharge was used to identify the movement of water that flowed slower.

Recharge water can move into the subsurface at various speeds, depending on the size of the opening in the bedrock. Eight terms have been introduced in this section to describe water flow into the karst subsurface. More terms have been developed in the literature to describe the difference between slow and faster recharge into the karst subsurface. The beginner to karst investigations needs to be aware of the differences in these terms, use caution, and not confuse terms. All terms can be appropriate depending on the objective of the study.

12.5   Surface Karst Features

Surface landforms come in a variety of sizes and shapes ranging from small dissolutional grooves on exposed rock surfaces, shallow to deep circular depressions to large isolated towers of rock that can be seen at great distances. In some areas, karst landforms are overly abundant at the surface and easily recognizable while in other areas they are not so easy to find.

Sinkholes

One of the most recognizable surface karst landforms is the sinkhole. Sinkholes are circular or enclosed depressions of various sizes. The Serbian word doline meaning a little dole or valley, is commonly used as the equivalent of sinkhole in karst areas (Stringfield et al. 1979), especially outside of North America. In some karst areas like the Mitchell Plain in Indiana and Pennyroyal Plain in Kentucky, sinkholes are very obvious with the use of aerial photography and topographic maps and from viewpoints on the surface that are much higher in elevation that surround the valley. Sinkholes can be broad, shallow, and barely noticeable to tens or hundreds of feet in diameter. Depths of sinkholes can range from a few feet to several hundreds of feet.

Sinkholes have been classified into different types several times in the literature (Sweeting 1973; Bogli 1980; Jennings 1985; Beck and Sinclair 1986; White 1988; Ford and Williams 1989; Lowe and Waltham 2002; Williams 2004; Waltham et al. 2005). Beck (2005) provides a process-oriented definition of a sinkhole: sinkholes are the surface and near-surface expressions of the internal drainage and erosion process in karst terrane, usually characterized by depressions in the land surface. Beck (2005) classifies sinkholes into five types including solution sinkholes, cover-collapse sinkholes, bedrock or cave collapse sinkholes, cover-subsidence sinkholes, and buried sinkholes. These five sinkhole types are the result of two different processes: the transport of surface material downward along solutionally enlarged channels or collapse of the rock roof over large bedrock cavities (Beck 2005). Dissolution of the limestone occurs most rapidly at the surface. The complexity of natural processes often results in more than one mechanism being involved in the formation of sinkholes such that they can be considered as having a polygenetic origin (Williams 2004).

Solution sinkholes develop when the rock is exposed at the surface or very close to the surface. Water drains downhill and enters the rock where the water can easily enter the subsurface and begins to enlarge vertical pathways. Over time, soil slowly drains into the enlarged openings in the underlying limestone bedrock. Eventually, these sinkholes provide a direct connection to the underlying groundwater-flow system. A number of these sinkholes may in time coalesce, or join, by lateral enlargement and form into a uvala (Stringfield et al. 1979).

Cover-subsidence sinkholes (Figure 12.10) develop in areas where unconsolidated materials such as alluvium, glacial moraine, or sand overly the karst bedrock (Williams 2004) and is called a mantled karst (Beck 2005). Soils slowly wash down through the vertical drain within the limestone. Soil piping is the term used to describe the slow upward erosion of soils within the sinkhole. European investigators split subsidence sinkholes into two types based on the amount of time it takes for the sinkhole to form and includes the terms suffusion and dropout sinkholes. Suffusion sinkholes form from subsidence that occurs over a period of months or years while dropout sinkholes are the result from an instantaneous failure (Waltham et al. 2005).

Buried sinkholes are sinkholes that have been filled in over time by natural processes and are no longer visible at the surface. Although no longer visible at the surface, it still functions as a drain for water to enter the groundwater-flow system (Beck 2005).

Cover-collapse sinkholes cause the most amount of damage whether they occur in residential or urbanized areas (Figure 12.11). Cover-collapse sinkholes can develop in a relatively short period of time and are typically the result of the collapse of an underlying

FIGURE 12.10   Sinkholes on Greenbrier County, West Virginia. (*Photo courtesy of John Mylroie.*)

FIGURE 12.11   Collapse sinkhole, Trigg County, Kentucky. (*Photo courtesy of John Mylroie.*)

soil arch above the limestone drain or the collapse of an underlying cavity roof within the bedrock. A cover-collapse sinkhole is a subtype of subsidence sinkholes.

Cenotes (a Spanish word meaning well) (Figure 12.12) are collapse sinkholes that form from the collapse of flooded underlying caves in carbonate islands and are especially found in the Yucatan. Blue holes are similar collapse features that form on the carbonate islands in the Bahamas and represent steep vertical openings in the carbonate rock (Mylroie 2013). The largest and best-known blue holes are found on North Andros, South Andros, and Grand Bahama Islands in the northwestern Bahamas (Mylroie 2004).

FIGURE 12.12   Cenote taken near Mayan ruins in Yucatan, Mexico. (*Photo courtesy of John Mylroie.*)

FIGURE 12.13   Dean's Blue Hole on Long Island, Bahamas, the world's deepest blue hole at over 200 m deep. (*Photo courtesy of John Mylroie.*)

Blue holes (Figure 12.13) are subsurface voids that are developed in carbonate banks and islands; are open to the earth's surface; contain tidally influenced waters of fresh, marine, or mixed chemistry; and extend below sea level for a majority of their depth (Mylroie et al. 1995). Blue holes differ from cenotes in that the cave water, the blue holes are connected to, is tidally influenced groundwater and creates strong current within the caves. Mylroie and Lace (2013) provided a comprehensive reference on karst developed in coastal settings and islands.

Bedrock or cave collapse sinkholes occur in rocks that overly cavities in limestone. The roof of a cave continuously or suddenly collapses and extends upward into the overlying non-karst rock until ground failure eventually occurs at the surface to develop a sinkhole. An example of this type of sinkhole is found in the Namurian Sandstone in South Wales where the overlying conglomeritic sandstone has collapsed into the underlying limestone caves (Williams 2004).

Sinking Streams

Stream valleys underlain by less permeable non-karst rocks like shale maintain their flow across the landscape. In karst areas, this is not always the case and internal drainage dominates. In places, usually at or near the contact of the karst and non-karst rocks, all the water is lost or sinks through the streambed or along the stream bank into the subsurface through dissolved openings in the underlying limestone and are called sinking, losing, or disappearing streams. Large volumes of concentrated recharge from losing and sinking rivers are central to the evolution of most of the world's largest and most significant caves and springs in karst (Ray 2005).

Swallow holes, additionally known as swallets and ponors, represent the openings in the bedrock where the surface water enters the subsurface in large volume and vary in

FIGURE 12.14a   Lost River, Hardy County, West Virginia photographed during high-flow conditions.

shape and size. Some swallow holes are pits and open cave entrances (White 1988). Downstream from the swallow hole, the streambed is mostly dry and this segment of the stream is typically known as a lost stream, disappearing stream, or sinking stream. Eventually, these waters reappear at the surface farther downstream at springs or a series of springs. One example is the Lost River in Hardy County, West Virginia, USA (Figures 12.14a and 12.14b).

FIGURE 12.14b   Lost River, Hardy County, West Virginia photographed during low-flow conditions.

The Lost River loses surface water continuously to the groundwater-flow system at a portion in the river known as the "sinks." In this stretch, the river is underlain by the limestone of the Helderberg Formation. During the dry season, river flow decreases and the entire discharge of the river sinks underground and reappears at a series of springs on the bank 2 mi (3.2 km) downstream. Throughout the summer, the Lost River stream channel is dry except for some springs that drain the surrounding ridges. However, as precipitation increases later in the year, the groundwater-flow system can no longer handle the entire discharge of the river and flow continues downstream. Sinking streams represent one mode of recharge to the underlying karst aquifers.

Karst Valleys

Three distinctive types of valleys can occur in karst: dry valleys, blind valleys, and poljes. In dry valleys, the entire discharge from the river basin is lost to the subsurface. Overland flow will only occur during major flood events. A blind valley occurs when water is able to flow across the karst for some distance and then terminates into a cliff of soluble rock. The River Rak entering Tkalca Cave in Slovenia represents a good example.

Poljes are another form of a karst valley (Figure 12.15). Meaning "field" in Slovene, poljes are large closed depressions with a well-developed underground drainage system bounded by steep-sided uplands. The best occurrence of poljes is in the Adriatic karst. Runoff from the surrounding uplands discharges at springs in the polje to form alluvial streams. These streams sink into caves or swallow holes on the opposite side during periods of low to moderate flow and discharge from them when the groundwater system is at full capacity during periods of high flow.

Karren

Karren refers to the variety of small to larger-scale grooves, pits, or channels carved into the rock through controlled flowing water on the karst rock. The name karren was

Figure 12.15 Cerknisko polje seen from Mt. Slivica in Slovenia. (*Photo courtesy of Martin Knez.*)

originally used to describe solution channels or runnels cut into limestone, but it is used for the whole complex of microforms that occur on outcrops of karst limestones (Sweeting 1973). Karren develops best in massive, thick-bedded pure limestone (White 1988). Factors that affect the formation of karren include the amount of precipitation, slope of the bedrock, nature of the bedding, texture of the rock, and the presence or lack of vegetative cover. Sweeting (1973), White (1988), and Ford and Williams (1989) provide lengthy discussions on the variety of karren types.

Rillenkarren is the smallest karren type and forms on steep to nearly vertical slopes and consists of small, rounded troughs with sharp ridges produced by direct rainfall, and it was the first type of karren to be described in the Alps and Dianaric Karst (Sweeting 1973). Rinnenkarren is similar to rillenkarren but at a larger scale. They are found on limestone surfaces of all slopes and on bare as well as soil-covered surfaces (White 1988). The crests between the troughs on rinnekarren are sharp and the troughs are generally rounded and can contain pits and scallops. When the crests are peaked they are called spitzkarren (Sweeting 1973). On shallower slopes the solution channels take on the appearance of a meandering river channel and are called meanderkarren.

Kluftkarren develops along near-vertical-to-vertical joints in the exposed bedrock surface and represents one of the largest forms of karren in karst. Through time the rock dissolves away and the joints widened by the solution channel can become wider and deeper depending on the structure of the rock. As water flows down the vertical joint it can come into contact with a horizontal bedding plane in the rock at depth. The water can then flow along the fracture and dissolve the rock adjacent to the bedding plane. Examples of kluftkarren are found in the limestone areas in Clare and Galway Counties in Ireland (Sweeting 1973). Trittkarren, also known as stepped karren, forms on near-flat surfaces and takes on the appearance of heel prints. As water flows down the rock in repeated waves, a stepped surface is formed. Once started, the steps stabilize and enlarge into a step-like form with a nearly flat tread and a nearly vertical riser (White 1988).

Epikarst

The epikarst can be generally defined as the uppermost portion of the weathered bedrock surface or the interface between the soil and bedrock in karst landscapes and is typically about 30 ft (9.1 m) deep (Williams 2003) (Figure 12.16). Williams (1972, 1983) and Mangin (1975) are commonly referred to in the literature as calling attention to this occurrence of karst bedrock. Mangin (1975) referred to the karst zone near the surface as the epikarst while Williams (2004) called it the subcutaneous zone. The development of the epikarst is controlled by several factors which include the structure of the bedrock, lithology, climate, time since the last glaciation, solubility of the rock, vegetal cover, and depth of groundwater circulation (Aley 1997). As a result, the nature of the epikarst is highly variable with location (Williams 2003).

The top portion of the epikarst is dominated by vertical infiltration of water from the karst surface. With depth, the vertical movement of water decreases as openings in the rock get narrower and become less abundant. The path of water begins to move laterally, then becomes increasingly concentrated to form shafts (Klimchouk 1996). A limited number of vertical drains cause a backup in the flow of water through the groundwater-flow system and cause the water to be stored there temporarily. Results from hydrochemical and isotopic studies from various regions demonstrate that such a delay can last from several days to a few months (Gunn 1983; Williams 1983). However, as the number of vertical shafts and their ability to transmit water increase with time,

Figure 12.16   An epikarst exposure of the Boone Formation in north-central Arkansas along U.S. 65. (*Photo courtesy of Debbie Burris Bednar.*)

the ability of the epikarst to retain water diminishes despite an increasing porosity of the epikarstic zone (Klimchouk 2000).

Aley (1997) recognized three types of epikarstic zones in unconfined carbonate aquifers based on results from about 30,000 quantitative dye analysis samples and about 1,000 positive groundwater traces. These include rapidly draining, seasonally saturated, and perennially saturated epikarsts. These epikarst types were categorized based on the capacity of the epikarst to retain water.

Rapidly draining epikarsts typically develop in areas of high topographic relief in bedrock with high solubility, negligible sediment infiltration, little water storage, and are saturated with water for short periods of time, especially after storm events. Seasonally saturated epikarsts develop in areas of moderate relief where the solubility of the bedrock has resulted in the development of soil and residuum thickness at elevations greater than local perennial streams. Water is typically stored seasonally and after major storm events lasting periods of weeks to months. Perennially saturated epikarsts occur in areas of low to moderate relief along perennial streams and are mostly saturated with water.

In 2003, the Karst Waters Institute sponsored a symposium about the epikarst that was attended by hydrologists, biologists, and geoscientists. The purpose of this interdisciplinary group of professionals was to gain a better understanding of the physical, chemical, and biological processes of the epikarst and to compose a better definition. As the conference proceeded, it was apparent that the epikarst was difficult to characterize and its properties remain unpredictable (Jones et al. 2004).

Karst Springs

The location where groundwater appears at the surface is called a spring. Karst springs are some of the largest springs in the world and provide substantial quantities of water for human consumption and commercial enterprises (Figure 12.17). Big springs rather

Figure 12.17   Blanchard Springs located at Blanchard Springs Caverns in Arkansas.

than small springs are the general rule in mature karst regions (LeGrand 1983). The largest reported karst springs are located in Papua New Guinea where several large rivers flow directly from caves (White 2005). Figeh Spring, in Syria, which is the third largest spring in the world, on average discharges 63,200 gal/min (3,990 L/s) and supplies the entire city of Damascus with water (Veni et al. 2001). Several spring types have been recognized by geologists and the name given for the spring is dependent on where the spring appears at the surface, and the characteristics of the flow rate and chemistry of the water. The following paragraphs provide a few examples of studies conducted to learn more about karst springs and the general conclusions about them Kresic and Stevanovic (2010).

White (2005) recognized two broad types of karst springs: conduit flow springs and diffuse flow springs which were named after the work done by Shuster and White (1971) from a study of 14 springs that discharged from the Ordovician dolomites and limestones in central Pennsylvania. They suggested that flow type within the aquifer controlled the chemical character of the springs and used the statistical variation in water hardness (the amount of dissolved carbonate in the water) to distinguish between the two types. The first group of springs that discharged from the fractured dolomites had a constant hardness with a variation of less than 5%, displayed a nearly constant chemical character regardless of season or storm event and were recharged by diffuse recharge and White (2005b) called them diffuse flow springs. Diffuse flow springs discharge from many fractures and areas of high permeability and are considered to be non-flashy and groundwater flow moves slower and can take months to travel short distances.

The second group of springs discharged from karstic limestones with concentrated recharge water from sinkholes. Hardness was found to be very variable throughout the year with a variation of greater than 10% and these are called conduit flow springs. Jacobson and Langmuir (1974) produced similar results regarding chemical variation at carbonate springs. Springs that discharge from conduits are known to be flashy based on the high ratios between maximum discharge and base flow discharge (Chapter 10). Base flow is the lowest discharge from the spring. Flow is turbulent while water hardness is low but highly variable. The variation in hardness ranges from 10% to 25% or more (Quinlan and Ewers 1985; Quinlan 1989). Groundwater-flow velocities are commonly rapid and can travel up to a mile per day (1.6 km/day).

The use of statistical variations in water hardness to distinguish between spring flow types appeared to work well for the study conducted by Shuster and White (1971) for small drainage basins in temperate climates. However, larger drainage basins showed less variation when springs were fed by conduits (White 1988). Subsequent studies found that the binary classification developed by Shuster and White (1971) was not applicable for all karst settings.

Aley (1978) applied different terms for spring types in Missouri. He used the terms high transit springs and high storage springs. High transit springs derive water from areas of discrete recharge and are similar to conduit flow springs. High storage springs are similar to diffuse flow springs where water is derived from areas of diffuse recharge.

Scanlon and Thrailkill (1987) conducted a study in the relatively flat lying limestones and shales of the Inner Bluegrass karst region of central Kentucky. The physical and chemical characteristics of springs from the Inner Bluegrass were compared with those from the study conducted by Shuster and White (1971). Results from their study showed that seasonal water chemistry variation in major springs did not correspond with the physical characteristics of the springs over time. Chemical similarities were attributed to the mode of recharge. Bedrock type and structure between the two studies were considered to be controlling factors regarding the relationships between physical and chemical attributes of the springs.

Worthington et al. (1992) performed a statistical analysis using data from 39 springs in six countries with temperate climates taken from the literature. Results from their study demonstrated no evidence that hardness variation was an indicator of flow conditions within an aquifer as suggested by Shuster and White (1971) and that greater than 75% of hardness variation was explained by recharge type.

Smart and Worthington (2004) stress that recharge and the conduit network within the karst aquifer result in the water quality and quantity variations we experience at karst springs and named two broad spring types. Resurgences are springs that have variable flow rates, high variations in water quality, and receive most of their flow from allogenic recharge. Springs that receive most of their discharge from autogenic recharge are called exsurgences and show less variability in water chemistry and flow rate.

12.6 Caves

A cave can be defined as a natural opening formed in the rocks below the surface of the ground large enough for a human to enter (Field 1999). The scientific study of caves is known as speleology, derived from the Greek word spelaion, meaning cave, and logos, meaning study. The study of caves to determine how and why they originate and develop is known as speleogenesis. Of the several different types of natural caves that have been identified, those that form in soluble rocks are known as solution caves. Klimchouk et al. (2017) provided a comprehensive source on the current science established on the formation of caves. Two major genetic types of caves have been identified in the upper part of the earth's crust and are known as epigene caves and hypogene caves (Klimchouk et al. 2017). The two distinct types of cave formation are based on the hydrodynamic properties of the source of the groundwater flow.

Caves in Epigenetic Karst Systems

Caves formed from water from the vadose zone and are within the epigenic karst systems and are local systems predominantly formed by recharge water derived from within the vadose zone (Stafford et al. 2009). The most cavernous states in the United States, ranked by the number of known caves, are Tennessee, Missouri, Virginia, Alabama, Kentucky, West Virginia, Arkansas, and Indiana (Palmer 2005). Gulden (2006) provided a list of the longest and deepest caves in the United States and the world and the longest and deepest caves in Canada. Krubera (Voronja) Cave is the deepest cave in the world. It is located in the alpine glaciokarst of the Arabika limestone massif in the western Caucasus Mountains in Abkhazia, a republic within Georgia with a depth of 7021 ft (2,140 m). An open shaft was first documented by Georgian researchers in the early 1960s who named it after Alexander Kruber, a founder of karst science in Russia (Klimchouk 2005).

Mammoth Cave is the longest cave in the world and is located about 100 mi (160 km) south of Louisville, Kentucky near Park City and Cave City, Kentucky. All passages combined total about 367 mi (591 km). As of 2016, over 405 miles (650 km) of passages have been mapped (Hobbs et al. 2017). Most of the cave lies within Mammoth Cave National Park and is developed within Mississippian limestone rocks in an area known as the central Kentucky karst (Brucker 2005). Mammoth Cave is a great tourist attraction and scientists have used the cave for many decades to understand how caves are formed. The cave formed within the Girkin Formation, St. Genevieve Limestone, and the Upper St. Louis Formation. The Girkin Formation is primarily composed of limestone, the St. Genevieve of limestone and dolomite with chert beds, and the St. Louis of limestone, dolomite, chert, and gypsum. The stratigraphy and structure of the bedrock have influenced the regional groundwater-flow patterns while bedding planes appear to have greater influence on passage orientation than do variations in rock type (Deike 1989).

Caves in Hypogenic Karst Systems

Hypogenic caves are formed from recharge of regional or intermediate flow dominated by upward flow by sulfuric acid-rich waters. Production of hydrogen sulfide is common where sulfate rocks are exposed to hydrocarbons (Palmer 2016). Also known as sulfuric acid speleogenesis (Palmer 2016), the dominant characteristic is that hypogenic caves lack any genetic relationship to groundwater recharge with the overlying or adjacent rock strata. In 2007, Klimchouk defined hypogene speleogenesis as the formation of solution-enlarged permeability structures (void-conduit systems) by fluids that recharge the cavernous zone from below, driven by hydrostatic pressure or other sources of energy, independent of direct recharge from the overlying or immediately adjacent surface (Veni 2016). However, there were two views of thought among researchers as to the approach to define hypogene speleogenesis. In the first view, Palmer (2000) indicated a geochemical origin near the surface with respect to acidic sources. The second approach (Klimchouk 2007, 2015) was based on a hydrogeological point of view. Klimchouk believed that the preferential routes formed by fluid flows and subsequent dissolution of the bedrock to create cave-forming zones are from deeper sources. As the sulfuric acid-rich water ascended from great depth, the water moves toward outlets such as fractures or zones of high permeability that determine the location of sulfuric acid caves (Palmer 2016). As the result, the dominant factor influencing hypogene speleogensis is thought to be driven by hydrostatic pressure from below; therefore, an updated definition was developed. To include a hydrogeologic approach to the process of hypogene karst, the definition was redefined as the formation of solution-enlarged permeability structure (void-conduit) systems by upwelling fluids that recharge the cavernous zone from hydrostratigraphically lower units, whereas fluids, originate from distant estranged (by lower permeability strata) or deep sources, independent of recharge from the overlying or immediately adjacent surface (Klimchouk 2015).

However, metamorphic petrologists have recognized that permeability in the middle/deep crust of the earth responds to tectonic stress, dewatering and fluid production, and other geochemical reactions that are compatible with permeability in karst (Ingbriten and Manning 2010). It has been recognized that deep-seated volatiles, including carbon dioxide and other noble gases, can rise into the upper crust (Crossey et al. 2016) and that karst regions are located in these areas. The production of sulfuric acid is most common where sulfate rocks are exposed to hydrocarbons, and sulfuric acid is the most soluble gas (Palmer 2016).

Hypogenic caves have developed in the United States, the United Kingdom, France, Spain, Austria, Italy, the Middle East, central Asia, South America, Africa, and Australia (Klimchouk et al. 2017).

The deepest limestone cave in the United States and fifth longest in the world is Lechuguilla Cave (233 km long, 489 m deep). Lechuguilla Cave is located within the Guadalupe Mountains in Carlsbad Caverns National Park in southeastern New Mexico. Lechuguilla Cave represents one of the world's best documented examples of a cave formed by the dissolution of carbonate rocks by sulfuric acid and is being used for important studies in geology, cave microbiology, microclimatology, geomicrobiology, and geochemistry (Kambesis 2005). Over 300 caves have been documented in the Guadalupe Mountains, the most widely known of these caves are Carlsbad Caverns and Lechuguilla Cave.

The origin of the sulfuric acid that dissolved the carbonate rock that formed Lechuguilla Cave was hydrogen sulfide gas. Evidence to support the generation and migration of gas is from water analysis of oil wells within the nearby Delaware Basin. The hydrogen sulfide ascended through the rock strata into the overlying carbonate rocks of the Guadalupe Mountains. Hydrogen sulfide mixed with the descending groundwater at the water table to form sulfuric acid that dissolved the carbonate rocks. Other evidence to support dissolution by sulfuric acid is the presence of gypsum blocks and native sulfur not commonly found in caves developed by carbonic acid dissolution and isotopic evidence from cave deposits similar to that of petroleum reservoirs. Hypogene karst development in the Delaware Basin is consistent with traditionally defined hypogene karst, but variations in the dominance of controlling mechanisms occur throughout the region creating a complex and diverse speleogenetic history for the gypsum plain and evaporite karst is far more complicated that previously recognized (Stafford 2017). Because speleogenesis events in the Guadalupe Mountains have stopped, there is speculation concerning the origin of water flow paths and the H_2S

source, therefore there are differing models on cave development (DuChene et al. 2017). Four models were developed and include the Hill model, the Kirkland model, the Duchene Cunningham model, and the Queen model. Details about each model are discussed in Chapter 31 of Klimchouk et al. (2017).

Kane Caves in the Bighorn Basin of the Rocky Mountains represent an active hypogenic cave formed by sulfuric acid (Hill et al. 1976). The cave smells like hydrogen sulfide, are warm and contain gypsum crusts (Palmer et al. 2017). Kane Cave has an upper inactive cave and a lower active cave. Hydrogen sulfide gas in the cave is at a level that can make cavers sick. Seeps and balls of hydrocarbons observed in the upper and lower caves provide support for a petroleum-related source of hydrogen sulfide. The gypsum deposits on the cave walls are hypothesized to result from the degassing of hydrogen sulfide, where the springs form sulfuric acid that dissolves the limestone of the walls of the cave to form a crust of gypsum. NOVA, a public broadcast station, provides about a 15-min video of footage from within Kane Cave and explains the extreme conditions of the chemistry and addition knowledge about the cave (NOVA 2008).

The cave portion of the NOVA shows can be viewed at https://aetn.pbslearning-media.org/resource/ess05.sci.ess.earthsys.kanecave/cave-formation-kane-cave/?#.Ww1Q4vZFwdU.

In the Appalachian plateaus of New York, Palmer et al. (2017) demonstrate a hypogenic system at Saratoga Springs based on high carbon dioxide water rising along faults in the Ordovician limestone. Tritium values measured in 2004 at various sulfide-rich springs in the vicinity indicate a residence time of about 10 years and a water chemistry that supports values of high concentrations of sulfide. Additionally, work conducted by Claypool et al. (1980) and Terrell (2008) suggests no chemical alternation had taken place such that sulfate was not derived from oxidation from hydrogen sulfide. Nine sulfide-rich caves are actively developing in the state of Tabasco, Mexico due to active production in petroleum fields (Hose and Lagarde 2017) near the Villa Luz Park near the town of Tapijulapa. These are the longest and largest caves. Cueva de Vill Luz (CVL) is the most studied cave and contains a toxic sulfide-rich atmosphere. The cave is developing in middle Cretaceous limestone along a northwest-plugging anticline. The cave atmosphere is consistently high in hydrogen sulfide that is related to the cave floor springs that emit hydrogen sulfide gas with a pH of 3.

Twentieth-Century Theories on Cave Formation

European and American geoscientists have studied caves for over 100 years. By 1900, caves were thought to have formed in the vadose zone, shallow phreatic zone, or deep phreatic zone. European geomorphologists were concerned with active drainage systems and saw caves only as conduits through which the water moved while most American geomorphologists studied dry caves (White 1988). Americans focused on whether caves formed above or below the water table. A summary of some of the theories and models proposed on the formation of caves in the twentieth century will be presented by the author in a forthcoming beginners guide to karst.

The Ford and Ewers (1978) model forms the common ground of most current international speleogenetic thinking (Lowe 2004). Although no model is perfect, the Ford and Ewers (1978) model was the first model to provide an explanation that caves could form above, at, and below the water table. In 1978, their paper, *The Development of Limestone Cave Systems in the Dimensions of Length and Depth*, presented a model of speleogenesis derived from detailed field studies in southwest England, the Rocky and Mackenzie Mountains of Canada, and other general observations elsewhere.

Controlling Factors in Cave Formation

Several factors influence or control the development of cave passageways. These controls include the interaction of chemically aggressive groundwater within the openings in the bedrock, the structure of the bedrock, and the elevation or base level of nearby rivers. Within individual caves, the fundamental passage pattern is controlled mainly by the nature of the groundwater recharge to the karst aquifer (Palmer 2000). Early in the dissolution process, recharge enters the subsurface through many flow paths in the rock and the widest openings in the rock throughout their lengths are the ones that enlarge into caves (Palmer 1987). Cave development by diffuse recharge generally takes longer due to the large surface area which takes the water a longer time to dissolve the rock. However, over time, the water manages to find the best route through the rock and dissolve larger openings with depth which increases the water's ability to dissolve the rock.

Concentrated recharge is the most effective recharge type to develop large conduits in the subsurface. At least 60% of all caves of explorable size are fed by small point sources (sinkholes) or concentrated recharge (Palmer 2000). Through time, conduits grow larger where the rock is at or close to the surface, and sinkholes will develop to provide a direct connection to the subsurface. As additional sinkholes form in the recharge area, more water is added to the groundwater-flow system to increase the cave development process.

The structure of the bedrock has an influence on passage development. In bedded strata, groundwater in the vadose zone follows pathways in the most down-gradient direction by following the openings in the rock along vertical joints and along the horizontal bedding-plane partings. Once the groundwater reaches the water table or phreatic zone, the groundwater will generally follow the strike of the geologic formation (Chapter 2) and form a tubular passage. Regardless of the inclination or dip of the rock, the vadose dip orientation and phreatic strike orientation are most probable in well-bedded rocks regardless of dip (Palmer 2000). Additionally, Palmer (1987) recognized that beds resistant to dissolution, such as shale or chert, can control cave development by perching or confining groundwater flow.

The elevation or base level of nearby rivers exerts an influence on cave development. The longer the base level remains static, the longer in time caves have the opportunity to become more developed. As base level falls, cave passages that were saturated will soon become abandoned and dry as the water moves to the lower base level.

Cave Passageways

Cave passageways reflect the local hydrogeologic setting (Palmer 1991). Palmer and Audra (2004) identified several distinctive cave patterns: branchwork, anastomotic mazes, network mazes, spongework mazes, and ramiform patterns (Figure 12.18). Branchwork caves are similar to dendritic surface water drainages. Concentrated recharge from sinkholes contribute water to form these cave passageways. Early in cave formation, passages are isolated and not hydraulically connected. However, over time, these passages are forced to converge due to the intersection of structural features in the bedrock. Several cave passageway geometries are encountered in branchwork caves.

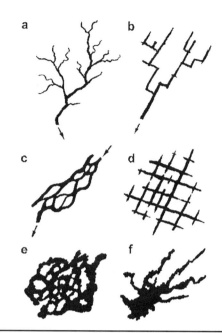

FIGURE 12.18   Types of cave passageways. (*Used with permission from Arthur N. Palmer.*)

Passages influenced by bedding-plane partings are sinuous and curvilinear (Palmer 1991). Single conduit caves can exhibit linear, angulate, or sinuous passageways. Linear passages are straight without any bends. Angulate passages have sharp bends and straight segments. Sinuous passages meander like surface streams. Examples of branch-work caves include Mammoth Cave and the caves in the Mendip Hills of England (Palmer and Audra 2004).

Palmer (1975) performed an exhaustive study of maze caves and identified two types. Network maze caves can form where permeable insoluble rock such as sandstone overlies limestone. The network pattern arises when all joints, regardless of their original permeability, are dissolved at a uniform rate, so that all possible mechanical openings enlarge into the cave (White 1988). Clifty Cave in Indiana and Anvil Cave in Alabama are examples of this type of cave.

Anastomotic maze passages look like a braided stream on the surface. Spongework passages are random, interconnected passageways that vary in size and consist of a three-dimensional pattern. Parts of Carlsbad Caverns exhibit a spongework maze. Ramiform caves resemble ink plots in plan view and passageways grade from spongework to network caves (Palmer 1991). An example of a ramiform cave is Carlsbad Caverns in New Mexico.

Sometimes a cave will exhibit several layers or tiers of nearly horizontal cave passageways. Tiered caves are developed as the regional groundwater table lowers over time. As the water level in adjacent streams erodes to lower elevations, the water flowing through the cave will seek that same elevation and eventually abandon the higher-level cave. The Mammoth Cave-Flint Ridge Cave system is an exceptionally large and complex tiered cave (White 1988).

Speleothems

Speleothems represent secondary mineral deposits that form along the roof, walls, and floors of caves as carbon dioxide–rich waters enter the cave atmosphere. The cave environment has influenced the mineral's deposition (Hill and Forti 1997). Over 200 cave minerals have been documented. Calcite is the most important and, together with aragonite, constitutes perhaps 95% of all cave minerals (Gillieson 1996). Speleothems are rarely composed of dolomite. Hill and Forti (1997) provided a comprehensive book on the cave minerals of the world and recognize 38 types, subtypes, and varieties of speleothems.

Stalactites and stalagmites are the most known types of speleothems. Stalactites grow as water drips from the cave ceiling. They commonly develop in lines along joints or bedding planes through which the water percolates (Jennings 1985). Calcite crystals deposit around the circumference of the water drop to form a hollow ring of calcite. Through time, successive rings of calcite are deposited. Usually young stalagmites are called soda straws. Drops of water that reach the floor of the cave accumulate and form calcite crystals. These crystal deposits grow vertically and are called stalagmites. A column is formed when a stalactite and stalagmite meet and form a single vertical speleothem. Figure 12.19 shows an example of a stalactite, stalagmite, and a column formed in Mystic Caverns in central Arkansas.

Figure 12.19 Speleothems at Mystic Caverns in central Arkansas.

12.7   The Karst Aquifer

During the last 40 years, several researchers proposed generalized conceptual models of groundwater flow through karst aquifers. Investigative methods primarily used to collect data to support models included results from groundwater dye tracing studies (Chapter 14), spring hydrograph and chemograph analysis on karst springs, water budgets, cave mapping, and quantitative mathematical analysis. As more data became available from research conducted throughout the world, conceptual models were revised and improved. The following paragraphs provide summaries of some of the studies and critical thinking conducted over the years to develop a better understanding of how groundwater flows through karst aquifers. The models proposed by various authors to characterize karst aquifers are based on their own interpretation and have been collectively used to further the knowledge of groundwater flow through karst. Karst aquifers have been documented in limestone, dolomite, marble, chalk, anhydrate, gypsum, and halite, and classified as unconfined, semi-confined, or confined types (Stevanovic 2015).

Because the entire groundwater-flow system cannot be observed physically, emphasis is placed on understanding the differences in the water as it enters and exits the subsurface. Chemical and flow characteristics of recharge water and spring water are examined in detail. The storage characteristics or volume of water within the karst aquifer is of importance. Water that moves quickly through the aquifer by large passageways, called conduits, are considered to have low storage while water that moves slower through small openings expresses a higher storage capacity or volume. Conduits are a feature of karst aquifers that makes them distinctive from other aquifers because they act as networks of pipes carrying water rapidly through the aquifer (White 2005).

White (1969) developed conceptual models for flow in carbonate aquifers. His models were based on hydrogeology and included diffuse flow aquifers, free flow aquifers, and confined flow aquifers; where each flow-type included subtypes. Diffuse flow and conduit flow represented end members along a flow continuum and most carbonate aquifers contain characteristics of both while being more toward the conduit flow type. Diffuse flow exhibits laminar to slightly turbulent flow through a system of widened joints or bedding planes that are being dissolved very slowly. Diffuse flow was thought to commonly occur in less mature or developed karst areas.

Conduit flow refers to water that enters the groundwater system and is transported quicker through a network of interconnected passageways or conduits to discharge at big springs or spring systems. Conduit flow is more turbulent in larger passageways where the velocity is greater. Pollutants can travel rapidly and can have catastrophic effects on water quality more than 10 mi away (16 km) in just a week during base flow (Vandike 1982) and even sooner during flood flow. White (1977) revised his conceptual model from 1969 to consider what effect relief, geologic structure, and the aerial extent of the aquifer had on its development.

LeGrand and Stringfield (1971) recognized three types of karst aquifers based on the uneven distribution of permeability. Fine-textured karst aquifers were defined by closely spaced openings in the bedrock and were of young geologic age. Coarse-textured karst aquifers were characterized by larger openings spaced farther apart in the bedrock as displayed in areas of mature karst. Reactivated karst aquifers are the oldest of the three aquifers that have undergone submergence by an advancing sea and buried by younger sediments.

Atkinson and Drew (1974) performed a 7-year detailed hydrological study using water budgets and measurements from several limestone springs in the Mendip Hills of England to determine the nature of a limestone aquifer. Results indicated that conduits provide a direct connect from swallets to springs and that recharge to underlying conduits from swallets accounted for less than 30% of discharge to springs. Recharge from percolation water accounted for greater than 70% of the discharge to springs and that percolation water ultimately reaches water-filled conduits. Atkinson (1977) studied the Cheddar Spring catchment area within the Carboniferous Limestone of the Mendip Hills in Great Britain. His study was conducted to determine the proportions of conduit flow and diffuse flow within an aquifer. Atkinson used the terms quick flow for conduit flow and base flow for diffuse flow. Quick flow includes water derived from sinking streams and percolation water while base flow obtains water from slow percolation from groundwater leakage and areas not drained by closed depressions. Results from the study indicated that limestone aquifers are a two-component system where the majority of water in storage comes from narrow fissures (base flow) while most of the water is transported through the solutionally enlarged conduits (Atkinson 1977).

Gunn (1983) recognized that closed depressions were an important source of concentrated recharge to limestone aquifers dominated by conduit flow. He proposed a six-component depression hydrology model within the polygonal karst of the Oligocene Te Kuiti Group limestones in New Zealand. Storage was derived from the overlying soil and surface deposits, the epikarst, phreatic conduits, and the saturated rock mass. Additionally, Gunn's (1983) model provided for 10 types of distinctive flow regimes to the aquifer that consisted of overland flow, infiltration, throughflow, percolation, epikarstic flow, shaft flow, vadose flow, vadose seepage, diffuse flow, and conduit flow. Gunn (1983) believed that storage in the soil and epikarst zone and autogenic concentrated recharge was not sufficiently recognized while diffuse flow and storage in the bulk rock mass was overemphasized.

Worthington (1991) conducted a study of the karst springs at Crowsnest Pass in the Canadian Rocky Mountains. His analysis of the discharge and hydrochemical variations of the springs resulted in the development of a new conceptual model regarding the development and function of karstic aquifers. The model enables predictions to be made of sink to resurgence flow velocities, of conduit depth below the water table, of the ratio of beds and joints used by conduits, of the spacing between cave tiers, and of the depth of vauclusian springs (Worthington 1991). Additionally, he provided the catalyst toward a new view on flow in carbonate aquifers. Worthington (1991) proposed, based on documented discharge rates of springs, that all karst springs discharge from conduits and that the idea of a diffuse karst aquifer is a contradiction in terms (Worthington 1991).

Quinlan et al. (1992) developed a conceptual model that used recharge, flow, and storage as a measure of the aquifers susceptibility to groundwater contamination and proposed four dominant aquifer types: hypersensitive aquifers, very sensitive aquifers, moderately sensitive aquifers, and slightly sensitive non-karst aquifers. Hypersensitive aquifers are the most karstic, thus most susceptible to contamination, and are characterized by discrete recharge, conduit flow, and low storage. Any contaminants that enter this type of aquifer will most likely pass through the aquifer relatively quickly to discharge at springs.

Very sensitive aquifers are characterized by discrete recharge, conduit flow, and low to moderate storage. Moderately sensitive aquifers would have recharge somewhere between discrete and dispersed through epikarstic drains, intermediate flow through dissolutionally enlarged fissure networks, and moderate to high storage. Slightly sensitive aquifers are not karst aquifers, but some carbonate aquifers are similar in character to clastic aquifers in which flow is via intergranular pores or fractures developed during unloading or exposure in the near-surface environment (Quinlan et al. 1992).

Worthington et al. (2000, 2001) provided support for a triple-porosity/triple-permeability model for unconfined carbonate aquifers. Four contrasting carbonate aquifers were studied in Ontario, Canada; Kentucky; USA, Great Britain; and the Yucatan Peninsula in Mexico to approximate the porosity, permeability, storage, and proportions of flow within the rock matrix, fractures, and channels. Matrix permeability refers to the pore spaces in the bedrock, fracture permeability refers to all joints and bedding-plane partings that have not been enlarged by dissolution and channels refer to all joints, faults, and beddings planes that have been enlarged by dissolution (Chapter 2). Channel and conduit permeability are synonymous terms used in the literature to refer to pathways enlarged by dissolution within the karst aquifer. Aquifers were chosen to provide a mixed range of rock type (limestone and dolostone), recharge, (allogenic and autogenic), and age (Paleozoic, Mesozoic, and Cenozoic) (Worthington et al. 2000).

Investigative methods included results from analysis of bedrock core samples, packer tests, pumping tests, videos of fractures in boreholes, dye trace tests, cave passage geometries, geochemical studies, and numerical models and demonstrated that at least 96% of the storage or volume of groundwater in all four karst aquifers was in the matrix portion of the rock (Worthington et al. 2001). In all four cases, it was shown that the enhancement of permeability caused by dissolution created efficient dendritic networks of interconnecting channels that are able to convey 94% or more of the flow in the aquifer (Worthington et al. 2001).

Currently it is still accepted by karst practitioners that carbonate aquifers should be classified as "triple porosity" or "triple permeability" aquifers (Aley 2018, Kuniansky 2018, Mylroie 2018, Spangler 2018, Worthington 2018), where water flow exists within the rock matrix, fractures, and dissolutional channels. However, the great degree of variation in deposition of carbonate and evaporites and the great variation in post-depositional diagenesis result in extreme variations in properties of karst aquifers throughout the world and even in the United States (Kuniansky 2018). It appears to be that one type does not fit all when it comes to karst aquifers. Some karst aquifers can have characteristics of dual porosity while others can exhibit more triple-porosity characteristics and subsets of both. It is still valid to think in terms of triple porosity, with matrix, fracture, and channel components, though it is harder to collect the information for three porosity elements than for two. It is more common to combine fracture and channel elements and think of aquifers as dual porosity (Worthington 2018). Additionally, Green (2018) states that for practical application karst aquifers function as dual porosity in carbonate systems in Minnesota, where they are well-developed conduit systems that include cave passages. Spangler (2018) states that in the Floridan and other geologically young aquifers, primary porosity (matrix) is much higher, so along with fractures and conduits, the aquifer definitely has all three types of porosity. Aquifers developed in most Paleozoic-age carbonate rocks, however, the matrix porosity is very low and the fractures and conduits dominate, so effectively, they are more of dual-porosity aquifers. However, in the Basin and Range province of the western United

States, matrix porosity is low, as well as conduit development and fractures dominate. Chalk is dominated by matrix porosity and conduit development is often limited.

12.8   Land Use Problems in Karst

Approximately 20% of the earth's land surface is located in karst areas (White 1988). Karst represents areas of land that are highly vulnerable to groundwater contamination due to the ease at which surface waters can move into the subsurface and through the groundwater-flow system. Improper land use planning on karst areas can result in the contamination of private and public water supplies and have potential impact to sensitive cave species. Additionally, development can and has resulted in substantial damage to private property and local infrastructure. Unfortunately, development on karst will continue, but with proper land use planning, strategies, and construction practices, impacts to the vulnerable landscape can be minimized. The following paragraphs provide a few examples of problems that have occurred in karst areas.

Human Impact on Karst Groundwater Resources

Many people depend on karst groundwater resources for their source of water. It is estimated that 25% of the world's population, including many large cities and extensive rural areas, depend on karst water supplies to sustain their daily activities (Ford and Williams 1989). Springs serve as natural outlets from karst aquifers and are used as water supplies. Groundwater can travel through large open conduits in the karst aquifer quicker than non-karst aquifers with extreme velocities of 7,500 ft/h (2,286 m/h) while a range of 30 to 1,500 ft/h (10 to 460 m/h) is typical (Quinlan and Ewers 1985). As a result, any type of contamination that enters recharge waters that feed springs can impact water supplies within days or weeks (Examples 12.1 and 12.2).

> **Example 12.1**   Rubin and Privitera (1997) reported on the construction of a 107-acre (43.3 ha) industrial park site located in New York east of the Catskill Mountains and slightly west of the Hudson River. The park was designed to be located along the top of a narrow forested ridge underlain by carbonate rocks. The karst aquifer beneath the park is within the dolostone and limestone of the Upper Silurian Rondout Formation and the limestones of the Lower Devonian Group. A storm-water drainage system was designed to direct runoff into an unroofed deep vertical mine. Once in the mine the storm water would flow through a mound of newly blasted rock and through a short horizontal adit where it would drain into a sinkhole (Rubin and Priniterra 1997). This project consisted of a major federal action that by law required an environmental impact analysis. Consulting engineers and geologists, hired by the development company, denied the presence of sinkholes, caves, and sinking streams. No site-specific karst investigations were conducted (Rubin and Priniterra 1997).

> **Example 12.2**   In Indiana, two sinkholes were used as a municipal dump, known as the Lemon Lane Landfill, from 1933 to 1964 on the west-central edge of the city of Bloomington, Indiana. Bloomington is located in the Mitchell Plain, a low plateau underlain by carbonates of Mississippian age on which a characteristic karst terrane has developed (McCann and Krothe 1992). A large number of electrical capacitors containing polychorinated biphenyls (PCBs) were dumped at the Lemon Lane Landfill. Within one mile of the landfill there are approximately 90 homes that obtain drinking water from private wells and it was shown that several drinking water wells were found to be contaminated with PCBs (EPA 2006). Several local springs have been contaminated with PCBs as a result of contaminant migration from the landfill. Currently, the landfill is listed as a Superfund site.

George et al. (1999) provided the results of a multiphase characterization study at an industrial site in central Alabama. The site is underlain by the Ketona Dolomite and preliminary investigative results showed releases of coal tar and light oils into the groundwater system. It was concluded that slow groundwater-flow rates keep the contaminants on site.

Berry et al. (2001) report on the accidental release of 900 gal of gasoline at a convenience store in Bulverde, Texas. The site is underlain by the karst Trinity aquifer located 300 ft beneath the surface. A few days later, residents who lived downslope of the spill detected a gasoline odor in their drinking water. Results for groundwater sampling indicated unacceptable levels of MTBE and BTEX, typical chemicals associated with gasoline.

Improper agricultural and industrial activities have impacted groundwater in karst areas. Huntoon (1992) reported on deforestation of the uplands above the stone forest aquifers within the south China karst belt from 1958 to the mid-1970s. The karst uplands were left barren, resulting in an alternation of the hydrologic conditions of the area. Before deforestation, the uplands temporarily stored the groundwater which was slowly released to the stone forest aquifers in the valley during the dry season. After deforestation, the uplands could no longer retain the water that resulted in less discharge to springs and decreased water levels during the dry season.

12.9   Investigative Methods in Karst

Groundwater Dye Tracing

Groundwater dye tracing has been used in karst areas to determine groundwater-flow routes and velocities from sinking points in recharge areas to springs. Springs serve as excellent sampling locations to monitor for dye that has been introduced into the karst groundwater system. A successfully designed dye trace study should determine all possible discharge points to be sampled and also some unlikely ones. A few examples are given below; however, an expanded discussion of this is found in Chapter 14.

Dye tracing is used on a wide variety of projects, some of which include proposed highway corridor studies, delineation of springhead-protection areas, recharge areas and subsurface basins, leakage in sewage disposal systems, identification of sources of potential pollution from hazardous waste sites, and detection of leakage from dam sites. Nontoxic fluorescent dyes have been used in karst since the late 1800s. The first reported use of a fluorescent dye as a tracer was conducted in southwestern Germany in 1877 where dye was injected at infiltration points in the bed of the Danube River and reappeared 2 days and 7.5 mi (12 km) away at the Aach Spring (Jones 2005). The location of dye-introduction points, the manner in which the dyes are introduced into the subsurface, and the sampling strategy and analytical approach used must be tailored to the hydrogeologic setting, the issues of concern, and the quality and credibility of the data needed for the study (Bednar and Aley 2001).

Commonly used dyes include fluorescein, eosine, rhodamine WT, and sulfluorescein. Fluorescein dye has the most resistance to adsorption to soil particles and can be used for longer dye trace studies. Fluorescein dye has more resistance to adsorption to sediments in cave streams, therefore better to use for those types of project than rhodamine WT. Rhodamine WT dye should be used in karst aquifers with lower flow rate and has less resistance to adsorption to soil particles. A good source for learning more about dyes can be found in the *Groundwater Tracing Handbook* (Aley 2002).

Example 12.3 Groundwater dye tracing is an effective tool to use during the highway development process to avoid or minimize impacts to karst groundwater resources. Detailed hydrologic studies must be employed when planning highway projects through karst because underlying karst aquifers are highly vulnerable to contaminants. Contaminants can enter the groundwater-flow system and reach important springs used for human consumption or recreation activities relatively quickly.

Spangler (2002) conducted a study of the Dewitt Spring near Logan, Utah. Dye trace studies determined travel times ranging from 22 to 31 days from losing stream segments located approximately 7 mi from the spring. Results indicated that groundwater velocities ranged from 530 to 1,740 ft/day. In the southern Missouri karst, travel times for dye arrival from surface sinks to springs took 16 days to travel 39.5 mi (Aley 1997).

Example 12.4 Bednar and Aley (2001) provided the results of studies conducted along highway corridors in Arkansas, Tennessee, and West Virginia. In West Virginia, three dye traces were conducted along the preferred alignment for the Corridor H Highway project near Greenland Gap in Grant County. Greenland Gap was formed from the down-cutting of New Creek Mountain by Patterson Creek. Patterson Creek was considered a high-quality trout stream that received discharge from a karst spring upstream from the gap. The preferred alignment crossed the mountain upslope of many sinkholes. Sinkholes that contribute water to the groundwater-flow system could adversely impact the spring. Results demonstrated that water derived along the preferred alignment would discharge at two springs along Patterson Creek within 7 days.

Example 12.5 Duwelius et al. (1995) described a dye trace study that was conducted by the Indiana Department of Transportation during the State Route 37 Bedford to Mitchell highway improvement project. The project area was located in the Mitchell Plain of Indiana and was characterized by numerous sinkholes. Semiquantitative dye tracing involved injection of fluorescent dyes into 15 sinkholes along the highway right-of-way as part of an underground drainage assessment. Results from the study demonstrated that all sinkholes along the project alignment were hydrologically interconnected.

Example 12.6 Dye tracing was conducted at Sulphur Springs Pool in Hillsborough County, Florida, in response to unsafe levels of coliform bacteria in the water (Wallace 1993). The pool had been used as a swimming and recreational area since the 1920s. Dye testing was conducted on several occasions since 1958 to determine the source of the problem. Results from the dye tracing showed that the source of contamination was from a sinkhole that was located within a storm-water basin adjacent to Interstate 275 that provided a direct connection to Sulphur Springs Pool.

Hydrographs and Chemographs

Spring hydrograph and chemograph analysis have been widely used in concert with dye tracing to gain insight about groundwater flow through karst. A hydrograph shows a plot of spring discharge versus time in the shape of a curve. Discharge is plotted on the vertical axis and time on the horizontal axis. The hydrograph is used to gain information as how the spring responses to storm events. A spring hydrograph is divided into three segments for analytical purposes and include lag time, the rising limb, and the falling limb (Figure 12.20). The lag time portions of the curve show the spring discharge before the recharge event reaches the spring. The rising limb portion shows the first arrival and increase of discharge at the spring. The peak of the graph indicates the maximum amount of spring discharge from the storm event. The peak discharge reflects older water that is already within the aquifer in the most downstream direction and not the recharge water that enters the subsurface from the storm event (Huntoon 2000). The falling limb, more commonly called the recession curve, shows the spring discharge as it returns to pre-storm conditions.

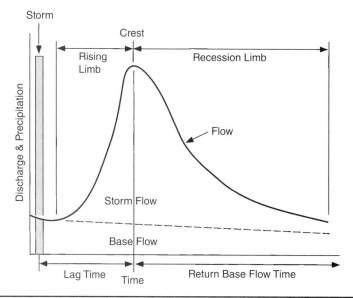

FIGURE 12.20 A spring hydrograph. (*Used with permission, Karst Waters Institute.*)

A chemograph (Figure 12.21) can show several characteristics of the water chemistry, sediment content, and discharge rates of spring water over time. Although many parameters can be used to characterize the chemistry at karst springs, common parameters include turbidity, dissolved oxygen, hardness, stable isotopes, specific conductance, and temperature (Ford and Williams 1989).

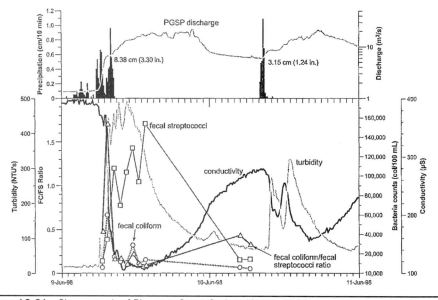

FIGURE 12.21 Chemograph of Pleasant Grove Spring. (*Currens 2005.*)

Geophysical Techniques

Geophysical methods measure the physical, electrical, and chemical properties of soil, rock, and pore fluids (ASTM 1999; Benson 2000). The bedrock-soil interface in karst areas is commonly composed of an irregular surface consisting of high and low points from the dissolving action of the groundwater. As a result, geophysical surveys have been used to better define the bedrock soil boundary and aid in the interpretation of subsurface karst features, cavities, and potential groundwater-flow routes. Common geophysical surveys used in karst include seismic, electrical resistivity, electromagnetic, ground penetrating radar, and gravity.

Electrical Survey

The electric resistivity survey measures the resistance to flow of electricity in subsurface materials (EPA 1993). Electrodes are inserted into the soil and the electrical resistance between them is measured (White 1988). As the resistance to current flow varies vertically and horizontally, changes in the electrical potential reflect these differences (Nowak 1999).

Frequency-Domain Electromagnetic Survey

Frequency electromagnetic (EM) surveys measure the electrical conductivity or the ability of the subsurface materials to react to an induced electromagnetic field produced by a transmitter coil. A receiver coil then intercepts the electromagnetic fields created by the transmitter coil. Measurements are made as an operator walks along the ground carrying an instrument and results are typically presented in map form (Hoover 2003).

Ground Penetrating Radar

The ground penetrating radar (GPR) geophysical method is a rapid, high-resolution tool for noninvasive investigation (Cardimona 2002). This method involves the transmission of ultra-high-frequency radio waves into the subsurface by a transducing antenna. When the subsurface signal encounters a boundary between the media with different electrical properties a portion of the signal is reflected back to the surface, detected by the antenna, and recorded on a graphical recorder (Smith 2002). Results of the survey are shown as a profile of the subsurface.

Gravity Survey

Microgravity surveys measure differences in the density of the underlying soil and rock and can be used to characterize geologic conditions. It is a powerful tool for establishing the presence of karst features that often give rise to measurable negative anomalies (McDonald et al. 1999). Identification of underlying material of high and low density can aid the investigators to determine the characteristics of the subsurface. Characteristics include the depth to bedrock, extent and shape of a collapse area below the surface, location of the crevice or crevices into which regolith (all earth material that includes soil and rock fragments above the bedrock) is collapsing, and locations of additional regolith voids in the vicinity of a collapse (Crawford et al. 1999). Low-density zones can identify the location of cavities and buried sinkholes.

Seismic Survey

Seismic surveys are regularly used in karst terrain to answer geotechnical questions about the soil and bedrock (Smith 2005). There are two types of seismic surveys: seismic reflection and seismic refraction. Seismic reflection measures the amount of time that a sound wave takes to bounce off rocks in the subsurface. Seismic refraction measures the velocity of the energy that is transmitted through the subsurface. Seismic methods were developed out of the pioneering earthquake studies in the mid to late nineteenth century (Reynolds 1997). Since the early twentieth century, seismic methods have been used often for hydrocarbon exploration. Seismic methods have been used for engineering site investigations, groundwater exploration, location of faults and fractures zones and rock types, geologic studies of aquifers, and shallow coal exploration. Seismic methods can also be used to determine subsurface cavities.

> **Example 12.7** In May 2000, the U.S. Geological Survey (USGS) conducted a survey at the Gregg chemical waste landfill site in northwestern Arkansas, an area underlain by the karst limestone of the Boone Formation, to locate filled pits where glass containers of laboratory chemicals were buried (Stanton and Schrader 2001). Electromagnetic conductivity, magnetometer, and electrical resistivity surveys were used to locate the glass containers. The use of these geophysical methods in concert helped to determine the location of five potential areas of concern where hazardous materials may be found.

> **Example 12.8** Styles and Thomas (2001) conducted a microgravity survey to detect cavities beneath the surface in an area under construction for a container Port in Freeport, Grand Bahama, Bahamas. The survey resulted in the delineation of a large cave system beneath the site. Jansen et al. (1993) used electrical resistivity and seismic refraction surveys to determine the location of bedrock pinnacles for the expansion of a landfill site in Monroe County, Wisconsin. Results from the survey demonstrated that the geophysical methods employed were generally successful in detecting bedrock pinnacles but an extensive boring program was necessary to confirm the geophysical results.

> **Example 12.9** Seismic methods were used at the Oak Ridge Reservation in Tennessee to assess karst features (Doll et al. 2005). Seismic refraction and seismic reflection surveys were used to determine the depth to bedrock and karst features that could direct contaminants to the site. Daniel et al. (1999) reported on this use of gravity, electromagnetic, and electrical resistivity surveys to evaluate karst features for roadway design in association with a maximum-security prison in Lee County, Virginia. Gravity, electrical resistivity, ground penetrating radar, and seismic refraction surveys were used to detect subsurface cavities in a karst area along a proposed highway alignment in Logan County, Ohio (Nowak 1999). A geological investigation, including the use of electrical resistivity surveys, was conducted as part of a comprehensive engineering design study for a planned highway widening project south of Front Royal, Virginia (Rehwoldt and Reed 2001). A concern of the Virginia Department of Transportation was to protect Skyline Caverns from the effects of roadway construction.

12.10   Summary

Karst landscapes are unique geologic environments formed from the dissolving of the rock by natural waters. Sinkholes and sinking streams serve to transmit the water into the subsurface at greater velocities. As openings in the rock grow larger and larger, so does the amount of water that can travel into and through the groundwater-flow system to eventually discharge at springs. Karst springs used for public water supplies or recreational activities are highly vulnerable to contamination due to the relatively quick speeds at which water can move through the groundwater-flow system as demonstrated in this chapter.

Proper land use planning must be used when residential, commercial, and transportation development occurs in karst areas. Investigative techniques such as groundwater dye tracing and geophysical surveys are essential tools for use to help avoid or minimize impacts to karst environments. The more we learn about karst the more we can reduce pollution of its natural waters. See "Sources of Cave and Karst Information" online at www.mhprofessional.com/weight3e for a list of agencies and organizations where the beginner can learn about cave and karst science.

12.11 Problems

12.1 The project area is located in the Frederick Valley that lies near the western edge of the Piedmont Physiographic. The Frederick Valley contains a karst terrain. Eosine dye was injected into two Epikarstic Dye Introduction Points (EDIPs) into borings that terminated into the approximate location of the epikarst zone (about 20 to 50 ft below grade). Approximately 2,500 gal of water was used to flush the dye into the epikarst zone. The straight-line distance from the EDIPs to the spring where the dye was recovered was 1,425 ft away. The dye was recovered 19 days after dye injection. Calculate the velocity of the groundwater from the EDIP to the spring. What might the actual distance to the recovery sight have been?

12.2 Rhodamine WT dye was injected into a sinking stream at the bank. Based on visual observation, no dye was observed downstream, therefore, all the dye was lost to the underground at the injection site. A week later, Rhodamine WT dye was found in groundwater samples. The sample point was approximately 1 mile away. Approximately 3,024,000 gal of water reaches the sampling station in 7 days. What was the approximate groundwater-flow rate where the dye was injected?

12.3 Name the four different types of groundwater recharges and articulate the differences between them.

12.4 Name the four commonly used dyes from groundwater dye tracing and what each dye is used commonly for. (See also Chapter 14 for additional information.)

12.5 The author recommends the following to any newcomer to the field of karst studies:

a. Read, everything you can about karst written by American and non-American professors/workers in karst and recognize the confusing terminology and beware of that issue. For example, there are two ways for rainwater to enter the underling bedrock: slowly and quickly. Then why are there three terms to describe the same process associated with the way surface water enters the subsurface?

b. Be aware of the masters who have worked in karst and read everything they have written: Palmer, Klimchouk, Aley, Ewers, Quinlan, Green, John Mylrorie, John Gunn, etc.

c. Get a job with one of the masters of karst. Don't think you can learn everything from just reading.

d. Visit as many caves as you can in different geologic settings. Use your observational skills and write things down.

12.12 References

Aley, T., 1978. Ozark Hydrology: A Predictive Model. *Journal of the Missouri Speleological Survey*, Vol. 18, pp. 185.

Aley, T., 1997. Groundwater Tracing in the Epikarst. In Beck and Stephenson, eds., *The Engineering Geology and Hydrogeology of Karst Terranes*. Balkema, Rotterdam, pp. 207–211.

Aley, T., 2002. *The Ozark Underground Laboratory's Groundwater Tracing Handbook*. Ozark Underground Laboratory, 44 pp. http://www.ozarkundergroundlab.com/assets/groundwater-tracing-handbook-2016.pdf

Arvidson, R.S., and MacKenzie, F.T., 1997. Tentative Kinetic Model for Dolomite Precipitation Rate and Its Application to Dolomite Distribution. *Aquatic Geochemistry*, Vol. 2, pp. 273–298.

ASTM. 1999. *Standard Guide for Selecting Surface Geophysical Methods*, D6429. ASTM International, West Conshohocken, PA, 11 pp.

Atkinson, T.C., 1977. Diffuse Flow and Conduit Flow in Limestone Terrain in Mendip Hills, Somerset (Great Britain). *Journal of Hydrology*, Vol. 35, pp. 93–110.

Atkinson, T.C., and Drew, D.P., 1974. Underground Drainage of Limestone Catchments in the Mendip Hills. *Fluvial Processes in Instrumental Watersheds*, Special Publication 6. Institute of British Geographers, 87–106 pp.

Bathurst, R.G.C., 1975. Carbonate Sediments and Their Diagenesis. In *Developments in Sedimentology 12*. Elsevier, Amsterdam, 658 pp.

Beck, B.F., 2005. Soil Piping and Sinkhole Failures. In Culver, D.C., and White, W.B., eds., *Encyclopedia of Caves*. Elsevier, Amsterdam, pp. 521–526.

Beck, B.F., and Sinclair, W.C., 1986. Sinkholes in Florida: An Introduction. *Florida Sinkhole Research Institute Report 85-86-4*, 16 pp.

Bednar, D., and Aley, T., 2001. Groundwater Dye Tracing: An Effective Tool to Use during the Highway Development Process to Avoid or Minimize Impacts to Karst Groundwater Resources. In Beck, B., and Herring, J.G., eds., *Geotechnical and Environmental Applications of Karst Geology and Hydrology*. Balkema, Rotterdam, pp. 201–207.

Benson, R.C., 2000. Case Histories of Six Roadway Investigations in Karst Combining Geophysical and Conventional Methods. In *51st Annual Highway Geology Symposium Proceedings*. Washington State Department of Transportation, Olympia, pp. 3–10.

Benson, R.C., and Yuhr, L.B., 2016. Site Characterization in Karst and Pseudokarst, Practical Strategies and Technology for Practicing Engineers, Hydrologists, and Geologists. Springer Science + Business Media Dordrecht, 421 pp.

Berry, R., Long, R., and Baer, T.M., 2001. Operation and Maintenance of Vadose Zone and Groundwater Remediation Systems in a Karst Aquifer. In Beck, B., and Herring, J.G., eds., *Geotechnical and Environmental Applications of Karst Geology and Hydrology*. Balkema, Rotterdam, pp. 287–291.

Blatt, H., 1982. *Sedimentary Petrology*. W.H. Freeman and Company, New York, 564 pp.

Blatt, H., and Tracey, R.J., 1999. *Petrology, Igneous, Sedimentary, and Metamorphic*. W.H. Freeman and Company, New York, 529 pp.

Boggs, S., 1987. *Principles of Sedimentology and Stratigraphy*. Merrill Publishing Company, New York, 784 pp.

Boggs, S., 2001. *Principles of Sedimentology and Stratigraphy*. Prentice Hall, New York.

Bogli, A., 1980. *Karst Hydrology and Physical Speleology*. Springer-Verlag, Berlin, 284 pp.

Bosak, P., Ford, D.C., Glazek, J., and Haracek, I., 1989. *Paleokarst: A Systematic and Regional Review*. Elsevier, Amsterdam, 725 pp.

Braithwaite, C., 2005. *Carbonate Sediments and Rocks: A Manual for Earth Scientists and Engineers*. Whittles Publishing, Edinburgh, 164 pp.

Brucker, R.W., 2005. Mammoth Cave System. In Culver and White, W.B., eds., *Encyclopedia of Caves*. Elsevier, New York, pp. 351–355.

Burns, S.J., McKenzie, J.A., and Vasconcelos, C., 2000. Dolomite Formation and Biogeochemical Cycles in the Phaserozoic. *Sedimentology*, Vol. 47, No. 1, pp. 49–61.

Cardimona, S., 2002. Electrical Resistivity Techniques for Subsurface Investigation. In *Geophysics 2002, the 2nd Annual Conference on the Application of Geophysical and NDT Methodologies to Transportation Facilities and Infrastructure*. Federal Highway Administration, FHWA-WRC-02-001.

Chavez, T., and Reehling, P., eds. 2016. *Proceedings of Deep Karst 2016, Origins, Resources, and Management of Hypogene Karst*. MCKRI Symposium 6, National Cave and Karst Research Institute, Carlsbad, NM.

Chilingar, G.V., Bissell, H.J., and Fairbridge, R.W., 1967. *Carbonate Rocks, Physical and Chemical Aspects, Developments in Sedimentology 9B*. Elsevier, Amsterdam, 413 pp.

Claypool, G.E., Holser, W.T., Kaplan, I.T., Sakai, H., and Zak, I., 1980. The Age Curves of Sulfur and Oxygen Isotopes in Marine Sulfate and their Mutual Interpretation. *Chemical Geology*, Vol. 28, pp. 199–260.

Cook, H.E., and Enos, P., 1977. *Deep-Water Carbonate Environments*. Society of Economic Paleontologists and Mineralogists. Special Publication, Vol. 25, pp. 336.

Cooper, M.P., and Mylroie, J.E., 2015. Glaciation and Speleogenesis—Interpretations from the Northeastern United States. Springer, 142 pp.

Crawford, N.C., Lewis, M.A., Winter, S.A., and Webster, J.A., 1999. Microgravity Techniques for Subsurface Investigations of Sinkhole Collapses and for Detection of Groundwater Flow Paths through Karst Aquifers. In Beck, B.F., Pettit, A.J., and Herring J.G., eds., *Hydrogeology and Engineering Geology of Sinkholes and Karst*. Balkema, Rotterdam, pp. 203–218.

Crossey, L.J., Karlstrom, K.E., Schmandt, B., Crow, R.R., Colman, D.R., Cron, B., Takacs-Vesbach, Dahm, C.N., Northrup, D.E., Hilton, D.R., Ricketts, J.W., and Lowry, A.R., 2016. Continental Smokers Couple Mangle Degassing and Distinctive Microbiology within Continents. *Earth and Planetary Science Letters*, Vol. 435, pp. 22–30.

Culver, D.C., and White, W.B., 2005. *Encyclopedia of Caves*. Elsevier, New York, 654 pp.

Currens, J.C., 2005. *Changes in Groundwater Quality in a Conduit Flow Dominated Karst Aquifer as a Result of Best Management Practices*, Report of Investigations 11. Kentucky Geological Survey, 72 pp.

Daniel, J.C., Benson, R.C., and Samford, A.M., 1999. Use of Geophysical Techniques to Evaluate Karst Features for Roadway Design. In *50th Annual Highway Geology Symposium Proceedings*. Radford University, pp. 32–41.

Day, M., and Chenoweth, S., 2004. Cockpit County Cone Karst, Jamaica. In Gunn, J., ed., *Encyclopedia of Caves and Karst Science*. Fitzroy Dearborn, New York, pp. 233–235.

Deike, G.H., 1989. Fracture Controls in Conduit Development. In White, W.B., and White, E.L., eds., *Karst Hydrology: Concepts from the Mammoth Cave Area*. Van Nostrand, New York, pp. 259–293.

Doctor, D.H., Land, L., and Stephenson, J.B., eds. 2015. Sinkholes and the Engineering and Environmental Impacts of Karst. *Proceedings of the Fourteenth Multidisciplinary Conference,* October 5–9, Rochester, MN. NCKRI Symposium 5. National Cave and Karst Research Institute, Carlsbad, NM.

Doll, W.E., Carr, B.J., Sheehan, J.R., and Mandell, W.A., 2005. Overview of Karst Effects and Karst Detection in Seismic Data from the Oak Ridge Reservation, Tennessee. In Kuniansky, E.L., ed., *U.S. Geological Survey Karst Interest Group Proceedings*. Rapid City, SD, September 12–15, 2005, Scientific Investigations Report 2005–5160, pp. 20–28.

Drew, D., and Hotzl, H., eds., 1999. *Karst Hydrogeology and Human Activities, Impacts, Consequences, and Implications*. Balkema, Rotterdam, pp. 322.

Dreybrodt, W., 1988. *Processes in Karst Systems*. Springer-Verlag, Berlin, 288 pp.

Duchene, H.R., Palmer, A.N., Paler, M.V., Queen, J.M., Polyak, V.J., Decker, D.D., Hill, C.A., Spilde, M., Burger, P.A., Kirkland, D.W., and Boston, P., 2017. Hypogene Speleogenesis in the Guadalupe Mountains, New Mexico and Texas, USA. In Klimchouk, A., et al., eds., *Hypogene Karst Regions and Caves of the World*. Springer International Publishing, New York, pp. 511–530.

Dunham, R.J., 1962. Classification of Carbonate Rocks. In Ham, W.E., ed., *Classification of Carbonate Rocks, a Symposium*. American Association of Petroleum Geologists, Memoir 1, pp. 108–121.

Duwelius, J., Bassett, J.L., and Keith, J.H., 1995. Application of Fluorescent Dye Tracing Techniques for Delineating Sinkhole Drainage Routes, Highway 37 Improvement Project Lawrence County, Indiana. In Beck, B., ed., *Karst Geohazards*. Balkema, Rotterdam, pp. 227–233.

EPA (Environmental Protection Agency). 1993. *Use of Airborne, Surface, and Borehole Geophysical Techniques at Contaminated Sites. A Reference Guide*. EPA/625/R-92/007, 304 pp.

Feinberg, J.M., Gao, Y., and Alexander, E.C., eds., 2016. *Caves of Karst across Time*. Special Paper 516. Geological Society of America, pp. 300.

Feinberg, J.M., Yongli G., and Alexandar Jr. E.C. (eds), 2016. *Caves and Karst Across Time*. Geological Society of American, Special Paper 516, 300 pp.

Field, M.S., 1999. *The QTRACER Program for Tracer-Breakthrough Curve Analysis for Karst and Fractured-Rock Aquifers*. U.S. Environmental Protection Agency, 137 pp.

Folk, R.L., 1959. Practical Petrographic Classification of Limestones. *Bulletin of the American Association of Petroleum Geologists*, Vol. 43, pp. 1–38.

Folk, R.L., 1962. Spectral Subdivisions of Limestone Types. In Ham, W.E., ed., *Classification of Carbonate Rocks*. American Association of Petroleum Geologists, Memoir 1, pp. 62–84.

Folk, R.L., 1962. Spectral Subdivisions of Limestone Types. In Ham, W.E., ed., *Classification of Carbonate Rocks*. American Association of Petroleum Geologists, Memoir 1, pp. 108–121.

Ford, D.C., and Ewers, R.O., 1978. The Development of Limestone Cave Systems in the Dimensions of Length and Depth. *International Journal of Speleology*, Vol. 10, pp. 213–244.

Ford, D., and Williams, P., 1989. *Karst Geomorphology and Hydrology*. Chapman and Hall, Cambridge, 701 pp.

Gams, I., 1994. Types of Polges in Slovenia, Their Inundation and Land Use. *Acta Carsologica*, Vol. 23, pp. 285–302.

George, S., Aley, T., and Lange, A.L., 1999. Karst System Characterization Utilizing Surface Geophysics, Borehole Geophysics, and Dye Tracing Techniques. In Beck, B.F., Pettit, A.J., and Herring, J.G., eds., *Hydrogeology and Engineering Geology of Sinkholes and Karst*. Balkema, Rotterdam, pp. 225–242.

Gillieson, D., 1996. *Caves: Processes, Development and Management*. Blackwell Publishers, Cambridge, MA, 324 pp.

Gunn, J.A., 1983. Point Recharge of Limestone Aquifers: A Model from New Zealand Karst. *Journal of Hydrology*, Vol. 61, pp. 19–29.

Gunn, J.A., 1985. *A Conceptual Model for Conduit Flow Dominated Karst Aquifers*. Karst Water Resources, No. 161. International Association of Hydrological Sciences, Vol. 161, pp. 587–596.

Gunn, J.A., ed., 2004. *Encyclopedia of Caves and Karst Science*. Fitzroy Dearborn. New York, 902 pp.

Gulden, B., 2006. Available online at http://www.cavediggers.com/Main_Page/LongandDeep/longanddeep.html

Herak, M., and Stringfield, V.T., 1972. *Karst, Important Karst Regions of the Northern Hemisphere*. Elsevier, Amsterdam, 551 pp.

Hill, C.A., and Forti, P., 1997. *Cave Minerals of the World, 2nd Edition*. National Speleological Society, Hunts ville, AL, 463 pp.

Hill, C., Sutherland, W., and Tierney, L., 1976. *Caves of Wyoming*. Geological Survey of Wyoming. Bulletin 59, 1976.

Hobbs, H.H., Olson, R.A., Winker, E.G., and Culver, D.C., 2017. *Mammoth Cave*. Springer International Publishing, New York, 275 pp.

Hoover, R., 2003. Geophysical Choices for Karst Investigations. In Beck, B.F., ed., *Sinkholes and the Engineering and Environmental Impacts of Karst*. Geotechnical Special Publication No. 122, American Society of Civil Engineers, pp. 529–538.

Hose, L.D., and Lagarde, L.R., 2017. Sulfur-Rich Caves of Southern Tabasco, Mexico, In Klimchouk, A., Palmer, A.N., DeWaele, J., Auler, A.A., and Audra, P. , eds., *Hypogene Karst Regions and Caves of the World*. Springer International Publishing, New York, pp. 803–814.

Huntoon, P., 1992. Chairman Mao's Great Leap Forward and the Deforestation Ecological Disaster in the South China Karst Belt. In *Proceedings of the Third Conference on Hydrogeology, Ecology, Monitoring, and Management of Ground Water in Karst Terranes*. Water Well Journal Publishing Company, pp. 149–159.

Huntoon, P.W., 2000. Variability of Karstic Permeability between Unconfined and Confined Aquifers, Grand Canyon Region, Arizona. *Environmental and Engineering Geoscience*, Vol. 6, No. 2, pp. 155–170.

Ingbriten, S.E., and Manning, C.E., 2010. Permeability of the Continental Crust: Dynamic Variations Inferred from Seismicity and Metamorphism. *Geofluids*, Vol. 10, Nos. 1–2, pp. 193–205.

Integrating Science and Engineering to Solve Karst Problems. In Carbonates and Evaporites. Part 1, Vol. 27, No. 2, 2012, Part II Vol. 28. Nos. 1–2. Spring-Verlag Berlin.

Jacobson, R.L., and Langmuir, D.L., 1974. Controls on the Quality Variations of Some Carbonate Spring Waters. *Journal of Hydrology*, Vol. 23, pp. 247–265.

Jakucs, L. 1977. *Morphogenetics of Karst Regions*. Halsted Press, New York, 268 pp.

James, N.P., and Jones, B., 2016. *Origin of Carbonate Sedimentary Rocks*. Wiley, New York, 446 pp.

Jansen, J., Ankam, J., Goowin, C., and Roof, A., 1993. Electromagnetic Induction and Seismic Refraction Surveys to Detect Bedrock Pinnacles. In Beck, B.F., ed., *Applied Karst Geology*. Balkema, Rotterdam, pp. 115–130.

Jennings, J.N. 1985. *Karst Geomorphology*. Blackwell Publishers, New York, 293.

Jones, W.K., 2005. Karst Water Tracing. In Culver, D.C., and White, W.B., eds., *Encyclopedia of Caves*. Elsevier, Amsterdam, pp. 321–329.

Jones, W.K., Culver, D.C., and Herman, J.S., 2004. Introduction. In *Epikarst*. Special Publication 9, Karst Waters Institute, 1–2 pp.

Kambesis, P., 2005. Lechuguilla Cave, New Mexico. In Culver, D.C., and White, W.B., eds., *Encyclopedia of Caves*. Elsevier, Amsterdam, pp. 339–346.

Klimchouk, A.B., Sauro, U., and Lazzarotto, M., 1996. "Hidden" Shafts at the Base of the Epikarstic Zone: A Case Study from the Sette Communi Plateau. Venetian Pre-Alps, Italy. *Cave and Karst Science, Transactions of the British Cave Research Association*, Vol. 23, No. 3, pp. 101–107.

Klimchouk, A. 2000. The Formation of the Epikarst and its Role in Vadose Speleogenesis. In Klimchouk, A., et al., eds., *Speleogenesis, Evolution of Karst Aquifers*. National Speleological Society, Huntsville, AL, pp. 91–99.

Klimchouk, A., 2005. Krubera (Voronja) Cave. In Culver, D.C., and White, W.B., eds., *Encyclopedia of Caves*. Elsevier, Amsterdam, 335–338.

Klimchouk, A., 2007. P *Hypogene Speleogenesis: Hydrogeological and Morphogenetic Perspective*. Special Paper no. 1. National Cave and Karst Research Institute. Carlsbad, NM, 106 pp.

Klimchouk, A., Sasowsky, I., Mylroie, J., and Engel, S.A., eds., 2014. *Hypogene Cave Morphologies*. Selected Papers and Abstracts of the Symposium held February 2 through 7, 2014, San Salvador Island, Bahamas. Karst Waters Institute Special Publication 18, Karst Waters Institute, Leesburg, VA, 112 pp.

Klimchouk, A., 2015. The Karst Paradigm: Changes, Trends, and Perspectives. *Acta Carsologica*, Vol. 44, No. 3, pp. 289–313.

Klimchouk, A., Palmer, A.N., DeWaele, J., Auler, A.A., and Audra, P., 2017. *Hypogene Karst Regions and Caves of the World*. Springer, New York, 911 pp.

Klimchouk, A., Ford, D.C., Palmer, A.N., and Dreybrodt, W., eds., 2000. *Speleogenesis: Evolution of Karst Aquifers*. National Speleological Society, Huntsville, AL, 527 pp.

Klimchouk, A., 2016. Types of Hypogene Karst. In *Proceedings of Deep Karst: Origins, Resources, and Management of Hypogene Karst*. NCKRI Symposium 6, National Cave and Karst Research Institute, Carlsbad, NM, 7–15 pp.

Klimchouck, A., 2017. Types and Settings in Hypogene Caves. In Klimchouk, A., Palmer, A.N., DeWaele, J., Auler, A.A., and Audra, P., eds., *Hypogene Karst Regions and Caves of the World*. Springer, New York, pp. 1–39.

Kranjc, A., 1998. Kras (The Classical Kras) and the Development of Karst Science. *Acta Carsologica*, Vol. 27, pp. 151–164.

Kresic, N., 2013. McGraw-Hill, New York, 708 pp.

Kresic, N., and Stevanovic, Z., 2010. *Groundwater Hydrology of Springs*. Elsevier, Amsterdam, 573 pp.

Kuniansky, E., 2018. Personal communication.

Land, L., Lueth, V., Raatz, W., Boston, P., and Love, D., 2016. *Caves and Karst of Southeastern New Mexico*. New Mexico Geological Society, Inc., 344 pp.

LeGrand, H.E., 1983. Perspective on Karst Hydrology. *Journal of Hydrology*, Vol. 61, pp. 343–355.

LeGrand, H.E., and Stringfield, V.T., 1971. Development and Distribution of Permeability in Carbonate Aquifers. *Water Resources*, Vol. 7, No. 5, pp. 1284–1294.

Lippman, F., 1973. *Sedimentary Carbonate Minerals*. Springer-Verlag, New York, 228 pp.

Louise, D.H., and Lagarde, L.R., 2017. Sulfur-Rich Caves in Southern Tabasco, Mexico. In Klimchouk, A., Palmer, A.N., DeWaele, J., Auler, A.A., and Audra, P., eds., *Hypogene Karst Regions and Caves of the World*. Springer, New York, pp. 1–39.

Lowe, D., and Waltham, T., 2002. *Dictionary of Karst and Caves*, British Cave Research Association, 40 pp.

Lowe, D.J., 2004. Speleogenesis Theories: Post-1890. In *Encyclopedia of Caves and Karst Science*. Fitzroy Dearborn, New York, 670–674 pp.

McCann, M.R., and Krothe, N.C., 1992. Development of a Monitoring Program at a Superfund Site in a Karst Terrane Near Bloomington, Indiana. In *Proceedings of the Third Conference on Hydrogeology, Ecology, Monitoring, and Management of Groundwater in Karst Terranes*. Water Well Journal Publishing Company, pp. 349–372.

McDonald, R., Russill, N., and Davies, R., 1999. Integrated Geophysical Surveys Applied to Karstic Studies. In Beck, B.F., Pettit, A.J., and Herring, J.G., eds., *Hydrogeology and Engineering Geology of Sinkholes and Karst*. Balkema, Rotterdam, pp. 243–246.

Milanovic, P., 1981. *Karst Hydrogeology*. Water Resources Publications, 434 pp.

Milanovic, P., 2004. *Water Resources Engineering in Karst*. CRC Press, Boca Raton, FL, 312 pp.

Moore, C. H., 2001. *Carbonate Reservoirs: Porosity Evolution and Diagenesis in a Sequence Stratigraphic Framework*. Elsevier, Amsterdam, 444 pp.

Mylroie, J., 2004. Blues Holes of the Bahamas. In: *Encyclopedia of Caves and Karst Science*. Fitzroy Dearborn, New York, 155–156 pp.

Mylroie, J.E., 2013. Coastal Karst Development in Carbonate Rocks. In Lace, M.J., and Mylroie, J.E., eds., *Coastal Karst Landforms, Coastal Research Library 5*. Springer Science and Business Media, Dordrecht, pp. 77–109.

Mylroie, J., Carew, J.L., and Moore, A.I., 1995. Blue Holes: Definition and Genesis. *Carbonates and Evaporites*, Vol. 10, No. 2, pp. 225–233.

Mylroie, J.E., and Lace, M.J., 2013. *Coastal Karst Landforms*. Springer Science and Business Media, Dordrecht, 429 pp.

Mylroie, J.E., 2018. Personal communication.

NOVA. 2008. *Mysterious Life of Caves*. Public Broadcasting Service.

Nowak, E., 1999. Near Surface Cavity Detection—Logan County, Ohio. In *50th Annual Highway Geology Symposium Proceedings*. Radford University, pp. 68–84.

Onac, B.P., Mylroie, J.E., and White, W.B., 2001. Mineralogy of Cave Deposits on San Salvador Island, Bahama. *Carbonates and Evaporites*, Vol. 16, No. 1, pp. 8–16.

Palmer, A.N., 1975. The Origin of Maze Caves. *National Speleological Society*, Vol. 37, pp. 56–76.

Palmer, A.N., 1987. Cave Levels and Their Interpretation. *National Speleological Bulletin*, Vol. 49, No. 2, pp. 50–66.

Palmer, A.N., 1991. Origin and Morphology of Limestone Caves. *Geological Society of America Bulletin*, Vol. 103, No. 1, pp. 1–21.

Palmer, A.N., 2000. Hydrogeologic Control of Cave Patterns. In Klimchouk, A., et al., eds., *Speleogenesis: Evolution of Karst Aquifers*. National Speleological Society, Huntsville, AL, pp. 77–90.

Palmer, A.N., 2005. Solution Caves in Regions of Moderate Relief. In Culver, D.C., and White, W.B., eds., *Encyclopedia of Caves*. pp. 527–535.

Palmer, A.N., 2007. *Cave Geology*. Cave Books, Dayton, OH, 454 pp.

Palmer, A.N., 2016. Sulfuric Acid vs. Epigenic Carbonic Acid Cave Origin and Morphology. In *Proceedings of Deep Karst: Origins, Resources, and Management of Hypogene Karst*. NCKRI Symposium 6, National Cave and Karst Research Institute, Carlsbad, NM, 7–15 pp.

Palmer, A.N., and Audra, P., 2004. Patterns of Caves. In Gunn, J.A., ed., *Encyclopedia of Caves*. Elsevier, Amsterdam, pp. 573–575.

Palmer, A.N., and Palmer, W.V., eds., 2009. *Caves and Karst in the ESA*. National Speleological Society, Huntsville, AL, 446 pp.

Palmer, N., Taylor, P.N., and Terrell, L.A., 2017. Hypogene Karst Springs along the Northeastern Border of the Appalachian Plateaus, New Your State. In Klimchouk, A. et al., eds., *Hypogene Karst Regions and Caves of the World*. Springer International Publishing, New York, pp. 709–719.

Quinlan, J. 1967. Classification of Karst Types: A Review and Synthesis Emphasizing the North American Literature, 1941–1966. *National Speleological Society Bulletin*, Vol. 29, pp. 107–109.

Quinlan, J. 1978. *Types of Karst, with Emphasis on Cover Beds in Their Classification and Development*. PhD Thesis, University of Texas, Austin.

Quinlan, J.F., and Ewers, R.O., 1985. Ground Water Flow in Limestone Terranes: Strategy Rationale and Procedure for Reliable, Efficient Monitoring of Ground Water Quality in Karst Areas. In *Proceedings of the Fifth National Symposium and Exposition on Aquifer Restoration and Groundwater Monitoring*, May 21–24, 1985, pp. 197–234.

Quinlan, J.F., and Ralph, O.E., 1985. Ground Water Flow in Limestone Terranes: Strategy Rationale and Procedure for Reliable, Efficient Monitoring of Ground Water Quality in Karst Areas. In *Proceedings of the National Symposium and Exposition on Aquifer Restoration and Groundwater Monitoring*. National Groundwater Association, pp. 197–234.

Quinlan, J.F., 1989. *Ground-Water Monitoring in Karst Terranes: Recommended Protocols and Implicit Assumptions*. U.S. Environmental Protection Agency, 79 pp.

Quinlan, J.F., Smart, P.L., Schindel, G.M., Alexander E.C., Jr., Edwards, A.J., and Smith, A.R., 1992. Recommended Administrative/Regulatory Definition of Karst Aquifer, Principles for Classification of Carbonate Aquifers, Practical Evaluation of Vulnerability of Karst Aquifers, and Determination of Optimal Sampling Frequency at Springs. In Quinlan, J., and Stanley, A., eds., *Conference on Hydrogeology, Ecology, Monitoring, and Management of Groundwater in Karst Terranes Proceedings*. National Groundwater Association, pp. 573–635.

Ray, J.A., 2005. Sinking Streams and Losing Streams. In Culver, D.C., and White, W.B., eds., *Encyclopedia of Caves*. Elsevier, Amsterdam, pp. 509–514.

Rehwoldt, E.B., and Reed, R.E., 2001. Characterization of Karst Systems for a Highway Widening Study in Warren County, Virginia. In *The 52nd Annual Highway Geology Symposium*, Maryland State Highway Administration, pp. 50–59.

Reynolds, J.M., 1997. *An Introduction to Applied and Environmental Geophysics*. Wiley, New York, 796 pp.

Rubin, P.A., and Privitera, J.J., 1997. Engineered and Unregulated Degradation of Karst Aquifers: Two Case Studies in New York, USA. In Beck, B.F., and Brad Stephenson, J., eds., *The Engineering Geology and Hydrogeology of Karst Terranes*. Balkema, Rotterdam, pp. 467–478.

Scanlon, B.R., and Thrailkill, T., 1987. Chemical Similarities among Physically Distinct Spring Types in a Karst Terrain. *Journal of Hydrology*, Vol. 89, pp. 259–279.

Scholle, P.A., Bedout, D.G., and Moore, C.H., eds., 1983. *Carbonate Depositional Environments*, American Association of Petroleum Geologists, AAPG Memoir 33, 708 pp.

Selley, R.C., 2000. *Applied Sedimentology, 2nd Edition*. Academic Press, New York, 523 pp.

Shuster, E.T., and White, W., 1971. Seasonal Fluctuations in the Chemistry of Limestone Springs: A Possible Means for Characterizing Carbonate Springs. *Journal of Hydrology*, Vol. 14, pp. 93–128.

Siegel, F.R., 1967. Properties and Uses of the Carbonates. In Chilingar, G.V., Bissell, H.J., and Fairbridge, R.W., eds., *Carbonate Rocks: Physical and Chemical Aspects*. Series No. 9B, Chapter 9, Elsevier, Amsterdam, 471 pp.

Smart, C., and Worthington, S.R.H., 2004. Springs. In *Encyclopedia of Cave and Karst Science*, New York, pp. 699–703.

Smith, D.V., 2005. The State of the Art of Geophysics and Karst: A General Literature Review. In Kuniansky, E.L., ed., *U.S. Geological Survey Karst Interest Group Proceedings*. Rapid City, SD, September 12–15, 2005, pp. 10–16.

Smith, R., 2002. GPR and EM Surveys to Detect and Map Peat and Permafrost: A Pilot Study. In *Geophysics 2002, the 2nd Annual Conference on the Application of Geophysical and NDT Methodologies to Transportation Facilities and Infrastructure*, Federal Highway Administration, FHWA-WRC-02-001.

Spangler, L.E., 2002. Use of Dye Tracing to Determine Conduit Flow Paths within Source Protection Areas of a Karst Spring and Wells in the Bear River Range, Northern Utah. In Kuniansky, E.L., ed., *U.S. Geological Survey Karst Interest Group Proceedings*. Shepherdstown, WV, Water Resources Investigations Report 02-4174, pp. 75–80.

Spangler, L., 2018. Personal communication.

Stafford, K.W., Land, L., and Veni, G., eds., 2009. *Advances in Hypogene Karst Studies: NCKRI Symposium 1*. National Cave and Karst Research Institute, Carlsbad, NM, 182 pp.

Stafford, K.W., 2017. Hypogene Evaporate Karst of the Greater Delaware Basin. In Klimchouk, A., Palmer, A.N., DeWaele, J., Auler, A.A., and Audra, P., eds., *Hypogene Karst Regions and Caves of the World*. Springer, New York, pp. 531–542.

Stanton, G.P., and Schrader, T.P., 2001. Surface Geophysical Investigation of a Chemical Waste Landfill in Northwestern Arkansas. In Kuniansky, E., ed., *U.S. Geological Survey Karst Interest Group Proceedings*. St. Petersburg, FL, February 13–16, 2001, U.S. Geological Survey, pp. 107–115.

Stevanovic, Z., ed., 2015. *Karst Aquifers-Characterization and Engineering*. Springer, New York, 692 pp.

Stevanovic, Z., 2015. Characterization of Karst Aquifer. In Stevanovic, Z., ed., *Karst Aquifers-Characterization and Engineering*. Springer, New York, 692 pp.

Stringfield, V.T., Rapp, J.R., and Anders, R.B., 1979. Effects of Karst and Geologic Structure on the Circulation of Water and Permeability in Carbonate Aquifers. *Journal of Hydrology*, Vol. 43, pp. 313–332.

Styles, P., and Thomas, E., 2001. The Use of Microgravity for the Characterization of Karstic Cavities on Grand Bahama, Bahamas. In Beck, B., and Herring, J.G., eds., *Geotechnical and Environmental Applications of Karst Geology and Hydrology*. Balkema, Rotterdam, pp. 389–396.

Sweeting, M.M., 1973. *Karst Landforms*. Columbia University Press, New York, 362 pp.

Terrel, L., 2008. *Chemistry of Sulfide-Rich Karst Springs in Otsego and Schoharie Counties, New York*. MA Thesis, State University of New York, Oneonta.

Tucker, M.E., 1981. Sedimentary Petrology: An Introduction. In *Geoscience Texts Volume III*. Wiley, Oxford, 252 pp.

Tucker, M.E., and Wright, V.P., 1990. *Carbonate Sedimentology*. Blackwell Science, London, 482 pp.

Upchurch, S., et al., 2019. The Karst Systems of Florida, Understanding Karst in a Geologically Young Terrain. Springer International Publishing, 450 pp.

Vandike, J.E., 1982. The Effects of the November 1981 Liquid Fertilizer Pipeline Break on Groundwater in Phelps County, Missouri. Unpublished Report on File with Missouri Department of Natural Resources, Division of Geology and Land Survey, 27 pp.

Veni, G., Duchene, H., Crawford, N.C., Groves, C.G., Huppert, G.N., Kastning, E.H., Olson, R., and Wheeler, B.J., 2001. *Living with Karst: A Fragile Foundation*. American Geological Institute, 64 pp.

Veni, G., 2016. A Re-Evaluation of Hypogenic Speleogenesis: Definition and Characteristics. In: *Proceedings of Deep Karst: Origins, Resources, and Management of Hypogene Karst*. NCKRI Symposium 6, National Cave and Karst Research Institute, Carlsbad, NM, pp. 17–19.

Veress, M., 2016. *Covered Karsts*. Springer Science + Business Media, New York, 536 pp.

Wallace, R.E., 1993. Dye Trace and Bacteriological Testing of Sinkholes Tributary to Sulphur Springs, Tampa, Florida. In Beck, B.F., ed., *Applied Karst Geology*. Balkema, Rotterdam, pp. 89–98.

Waltham, T., Bell, F., and Culshaw, M., 2005. *Sinkholes and Subsidence*. Springer/Praxis Berlin, 382 pp.

White, W., 1969. Conceptual Models for Carbonate Aquifers. *Groundwater*, Vol. 7, No. 3.

White, W., 1988. *Geomorphology and Hydrology of Karst Terrains*. Oxford University Press, New York, 464 pp.

White, W.B., 2005a. Hydrogeology of Karst Aquifers. In Culver, D.C., and White, W.B., eds., *Encyclopedia of Caves*. Elsevier, New York, pp. 293–300.

White, W.B., 2005b. Springs. In Culver, D.C., and White, W.B., eds., *Encyclopedia of Caves*. Elsevier, Amsterdam, pp. 565–570.

White, W.B eds. 2015. The Caves of Burnsville Cove, Virginia. Fifty Years of Exploration and Science. Springer International Publishing. 477 pp.

White, W.B., and Culver, D.C., eds., 2012. *Encyclopedia of Caves, 2nd Edition*. Elsevier, New York, 945 pp.

White, W.B., Herman, E.K., Rutigliano, M., Herman, J.S., Vesper, D.J., and Engel, S.A., 2018. Karst Groundwater Contamination and Public Health: Abstracts and Field Trip Guidebook for the Symposium Held January 27–30, 2016, San Juan, PR. Karst Waters Institute Special Publication 19, Karst Waters Institute, Leesburg, VA, 82 pp.

Williams, P.W., 1972. Morphometric Analysis of Polygonal Karst in New Guinea. *Geological Society of America Bulletin*, Vol. 83, pp. 761–796.

Williams, P.W., 1983. The Role of the Subcutaneous Zone in Karst Hydrology. *Journal of Hydrology*, Vol. 61, pp. 45–67.

Williams, P.W., 2003. The Epikarst: Evolution of Understanding. In Jones, W.K., et al., eds., Epikarst, Special Publication 9, Karst Waters Institute, Huntsville, AL, pp. 8–15.

Williams, P.W., 2004. Dolines. In Gunn, J.A., ed., *Encyclopedia of Cave and Karst Science*. Fitzroy Dearborn, New York, pp. 304–310.

Williams, P. 2004. The Epikarst: Evolution of Understanding. In: Jones, W.K., Culver, D.C., and Herman, J.S. Karst Waters Institute, Special Publications 9. 160 pp.

Wilson, J.L., 1975. *Carbonate Facies in Geologic History*. Springer-Verlag, New York, 471 pp.

Worthington, S., 1991. *Karst Hydrology of the Canadian Rocky Mountains*. Unpublished Ph.D. Dissertation, Department of Geography, McMaster University. Hamilton, ON, Canada, 227 pp.

Worthington, S.R.H., 2017. Personal communication.

Worthington, S.R.H., Davies, G.J., and Quinlan, J.F., 1992. Geochemistry of Springs in Temperate Carbonate Aquifers: Recharge Type Explains Most of the Variation. In *Colloque d'Hydrologie en pays Calcaire et en Millieu Fissure (5th Neuchatel Switzerland)*, *Proceedings*, Annales Scientifiques de l'Universitie de Besancon, Geologie—Memoires Hors Series, No. 11, pp. 341–347.

Worthington, S.R.H., Ford, D.C., and Beddows, P.A., 2000. Porosity and Permeability Enhancement in Unconfined Carbonate Aquifers as a Result of Solution. In Klimchouk, A., et al., eds., *Speleogenesis: Evolution of Karst Aquifers*. National Speleological Society, Carlsbad, NM, pp. 463–471.

Worthington, S.R.H, Ford, D.C., and Beddows, P.A., 2001. Characteristics of Porosity and Permeability Enhancement in Unconfined Carbonate Aquifers Due to the Development of Dissolutional Channel Systems. In Gunay, G., Johnson, K., Ford, D., and Johnson, A.I., eds., *Present State and Future Trends of Karst Studies Proceedings*. Technical Documents in Hydrology, 1, 49, UNESCO, pp. 13–29.

CHAPTER **13**

Tracer-Test Techniques

Marek H. Zaluski

Senior Hydrogeologist, Water and Environmental Technologies, Butte, Montana

A tracer test is a field method used in hydrogeology to quantify selected hydraulic or hydrochemical parameters. In its most common form, a tracer test is conducted by emplacing a defined quantity of traceable substance or a heat source within the aquifer and tracking it down hydraulic gradient, where at certain points of space and time its quantity is measured. This form of a tracer test can be viewed as a positive-displacement slug test or an injection test in which stress to the aquifer is delivered in the form of chemical substance or a heat source, rather than by injecting a known volume of water. Dissipation of such a stress with respect to time and the aquifer space provides a data set from which many hydrogeologic conditions may be concluded. Quantification of the groundwater flow direction and rate, aquifer porosity, anisotropy, dispersivity, retardation factor, and other physicochemical characteristics are the most common outcomes of the tracer-test interpretation.

Other types of tracing techniques deal with quantifying the changes in concentration of quite a few elements and their isotopes that naturally occur in groundwater to trace aspects of the water cycle, and for groundwater dating. (Some information on those elements is found in Chapters 10 and 11.) This chapter includes additional information on applicability of tritium in combination with its daughter product, a stable isotope helium, for dating very young groundwater, thus indirectly determining a source of groundwater at the location of interest. Injections of certain substances that react with hydrocarbon contaminants, allowing determination of the contaminant concentration, as well as a large family of surface-water tracer tests, are not discussed in this chapter.

Tracer tests, their objectives, theory, performance, and interpretation have been addressed relatively broadly in the literature. Information provided in this chapter is based on these sources as well as actual field experience of the author; it does not, however, aspire to be a handbook for all or any selected method or application. It summarizes practical guidelines for planning and performing tracer tests for common applications. For details on specific application and a comprehensive review of the subject, the reader is referred to the references included at the end of this chapter.

A particularly important reference is a 581-page book entitled *Tracing Technique in Geohydrology* by W. Käss (1998). This book seems to be the most comprehensive source of information on the tracer technology for groundwater and surface water and includes approximately 1,000 references on tracer tests and related topics. Application of isotopes as tracers is broadly addressed by Clark and Fritz (1997) and Clark (2015).

13.1   Tracer-Test Objectives

In many cases, a tracer test is part of site-characterization efforts. The appropriate tracer-test procedure is selected depending on questions remaining after all other information has been collected. For example, if the dispersivity value is all that is needed, then a single-well tracer test may be performed. However, if the isotropic conditions of the aquifer are questioned, a more time-intensive tracer test, which may involve numerous wells, would need to be considered. Therefore, the kind of the tracer test that needs to be conducted depends on its objectives as specified in Table 13.1.

For all methods, the retardation coefficient may be obtained if adsorptive and conservative tracers are injected simultaneously and the results compared. The only exception is a recirculating test, with the pumped water recirculated to the point source, discussed later on.

13.2   Tracer Material

There are quite a variety of substances that may be used as tracers for groundwater investigations. Detailed information regarding these tracers and the application methods can be found in Davis (1985), Käss (1998), or other publications. Table 13.2 summarizes information on various tracers but *should not* be used exclusively for the final selection of the trace material. Readers are encouraged to look for case studies that would include information on the use of the particular tracer in the given hydrogeologic setting. The statement "one size and kind fits all" is not applicable for the tracer-test technology, as a tracer that might be excellent for certain geological conditions might be unusable for other subsurface environments as shown in the following example.

The rhodamine WT dye was developed by DUPONT (Käss 1998) especially for water, as signified by the WT (water tracer) abbreviation. This dye, which has a reputation for being one of the most conservative (minimal adsorption) tracers with a very low detection limit, was successfully used in limestone–karstic environments and for some nonconsolidated porous deposits. However, its application for sediments deposited between lava flows in the Snake River Plain in Idaho is an example of a mismatch of the tracer material and subsurface environment. After an unsuccessful series of tracer tests, batch tests were conducted to determine sorption and desorption isotherms for rhodamine WT, eosine, and fluorescein to clastic sediments present between lava flows. The result of the investigation demonstrated that a Freundlich adsorption isotherm for rhodamine WT could be expressed by the following equation:

$$C_{ad} = 19.083 \times C_{aq}^{0.915} \tag{13.1}$$

where C_{ad} is rhodamine WT adsorbed by sediment ($\mu g/kg$) and C_{aq} is rhodamine WT present in solution ($\mu g/L$). The very high coefficient value, of 19.083, indicates that rhodamine WT was very strongly adsorbed to sediments, and thus could not be

Test	Subcategory	Source Duration	Information Obtained	Hydraulic Stress	Comments
Single well	Borehole dilution	Instantaneous	Flow direction, seepage velocity	Natural flow conditions	Special instrumentation needs to be installed in the well, e.g., thermistors
	Injection/pumping	Instantaneous	Seepage velocity dispersion coefficient	Injection period precedes pumping period	
Point source/ one sampling well	Natural flow	Instantaneous or continuous	Seepage velocity dispersion coefficient	Natural flow conditions	Wells must be located at the same flow line
	Diverging test	Instantaneous	Porosity, dispersion coefficient	Injection in the point-source; sampling well is not pumped	
	Converging test	Instantaneous	Porosity, dispersion coefficient	Pumping from the sampling well; point-source well is not stressed	Pumping at discrete intervals enables vertical differentiation
	Recirculating test	Continuous	Porosity, dispersion coefficient	Both wells are stressed. Injecting well receives pumped water or "clean" water from other source	Tests a considerable larger portion of the aquifer, than a converging test
Point-source/two sampling wells	Recirculating tests	Continuous	Porosity, dispersion coefficient, anisotropy	All wells are stressed. Injecting well receives pumped water or "clean" water from another source	
Point-source/ multiple sampling wells		Instantaneous	Flow direction, seepage velocity, anisotropy*	Natural flow conditions	

*Only if contours of hydraulic head are known.

TABLE **13.1** Most Common Tracer Tests for Groundwater Investigations

629

Tracer	Sub Category	Permeable Medium	Detection Field Method	Detection Limit	Advantages	Disadvantages	Comments
Ions	Br⁻	Any kind	Bromide electrode, Hach kit	1 ppm	(1) Low [<1 ppm] background concentration in groundwater (2) Very low toxicity (3) No MCL*	At high concentrations tend to sink At low pH may be adsorbed by clay	Anions are more conservative than cations (1) May be the most common tracer used (2) Common source: NaBr
	Cl⁻	Any kind	Chloride electrode, Hach kit	1 ppm	Availability	High background concentration drinking water standard** = 250 ppm	Very conservative
	I⁻	Any kind	Iodine electrode, Hach kit	1 ppm	Very low [<0.01 ppm] background concentration in groundwater	May be affected by microbiological activity	Radioactive isotope of iodine was used as a radioactive tracer
Dyes		Perform best in fracture rock and karst					(1) Maybe the oldest tracer used (2) Less conservative than anions
	Fluorescein	Karst rock, clean sand, sandstone, basalt Works better in porous medium than rhodamine WT and eosine	Fluorimeter	0.3 ppb over background	(1) Less expensive than other dyes (2) Visually detected if >40 ppm	(1) High natural background fluorescence (2) Adsorbs strongly to organic matter (3) Loses color at pH <6	Green opalescent color
	Rhodamine WT	Fracture and karst rocks	Fluorimeter	0.013 ppb over background	Low detection limit	(1) Biologically not as safe as fluorescein (2) May strongly adsorb to certain sediments	Red color

							Superior tracers
Radio-nuclides	^{131}I, ^{82}Br,	Any kind	Radiation meters	Extremely low	(1) Short half life [8 & 1.5 days] in comparison to other radionuclides (2) Very mobile	Despite the demonstrated safety, public perception, and regulations often prohibit the field usage of these tracers	
	Deuterated water (^2H_2O)	Any aquifer; problematic for clay aquitards	None: requires isotopic-hydrogen analysis using isotope-ratio mass spectrometer	0.1 mg deuterium/L over background deuterium ~ 0.02%	Very conservative. No buoyancy effect within injection concentration <500 detection limit (Becker and Copler 2001)		
Heat		Any kind	Thermistors	0.05°C	Direct measure of groundwater flow direction		(1) Used in a single well (2) Feasible for multiwell tracer test, if a point or line sink of warm water is active for long time
Bacteria	Variety	High permeable media	None, requires laboratory procedure			(1) Regulatory concerns (2) Strain used must be mobile	Bacteria must be nonpathogenic
Virus	Variety	High permeable media	None, requires laboratory procedure			(1) Regulatory concerns (2) May move faster than advection (preferential path due to large size)	Virus must be nonpathogenic

*MCL = maximum contaminant level.
**National secondary drinking water regulations.

TABLE **13.2** Most Common Tracers Used for Groundwater

631

considered a viable tracer. For comparison, the Freundlich adsorption isotherm for fluorescein was expressed as

$$C_{ad} = 0.535 \times C_{aq}^{0.890}$$ (13.2)

A coefficient 0.535 in Equation 13.2 indicates that fluorescein was a much better tracer than rhodamine WT, as its retardation coefficient ranged from 3.65 to 1.64 for concentrations of 1.92 μg/L and 750 μg/L, respectively.

This example illustrates the importance of an appropriate match of a tracer with the given hydrogeological condition and/or tracer-test objectives and the danger of making inadequately researched assumptions.

13.3   Design and Completion of Tracer Test

There are several components of a tracer-test procedure that, if considered, planned, and carefully implemented, will usually lead to a successful completion of the project. These are the following:

- Conceptual design
- Selection of the initial mass of the tracer or its concentration
- Point-source infrastructure
- Observation wells
- Sampling schedule
- Monitoring
- Equipment used

Conceptual Design

This preliminary phase includes the following:

- Review of site hydrogeologic information available
- Setting the objectives for the tracer test
- Selection of the kind of the tracer test to be conducted
- Selection of the tracer material and its cost evaluation
- Preliminary costing of a necessary infrastructure
- Assessment of operational costs

If the predicted expenses are within the available budget, the tracer concentration may be assessed in detail. Otherwise, it may be necessary to revise (limit) the objectives of the tracer test and choose a less expensive method as presented in Example 13.1.

Example 13.1

Initial Conceptual Design

Hydrogeologic information   The contaminated site in question was located between two perennial creeks. An unconfined aquifer beneath the site consists of alluvial deposits. A 2-in-diameter (5.1-cm) monitoring well was located in the center of the reported incidental gasoline spill and two 2-in-diameter (5.1-cm) monitoring wells were located adjacent to each creek. There was no information regarding the site location with respect to any groundwater divide.

Objectives of the tracer test:

    Determination of the flow direction at the spill location

    Determination of seepage velocity

    Determination of anisotropy ratio

Tracer test initially selected:

    Point source (an existing central well) and multiple (six) sampling wells located in a hexagonal pattern around the point source well.

Tracer material selected:

One kilogram of bromide provided as NaBr	
NaBr is available in 50 lb bags @ $100 a bag	$100

Necessary infrastructure:

Six 2-in-diameter observation wells @ $900	$5,400

Operational costs:

Technician 12 h @ $80/h	$960
Measuring equipment (bromide electrode and a meter)	$1,500
Miscellaneous expenses (containers, glassware & incidentals)	$500
Total expenses	$8,460
Available budget	$5,000

The deficit of $3,460 could be almost offset by decreasing the number of observation wells from six to two. Such a reduction, however, would jeopardize the achievement of three of the objectives for the tracer test. However, dropping the objective of *determination of anisotropy-ratio* and changing the multi-well tracer test to a borehole-dilution test would allow for achieving two of the remaining objectives within the budgetary requirements. Moreover, it would be possible to determine these parameters for three separate locations as described below.

Revised Conceptual Design

Objectives of the tracer test:

    Determination of the flow direction at the spill location

    Determination of seepage velocity

Kind of tracer test:

    Single-well borehole-dilution test

Tracer material selected:

Heat stress delivered by a down-hole instrument with a central heat electrode and six thermistors	
Rental fee: @ $600 a field-day (shipment cost and time covered by vendor)	$600

Necessary infrastructure:

Three 4-in-diameter (10.2-cm) test wells @ $1,100/well	$3,300

Operational costs:

Technician 9 h (3 h per well) @ $80	$720
Measuring equipment, included in the rental fee	
Miscellaneous expenses (incidental)	$400
Total expenses	$5,020

Selection of the Initial Mass of the Tracer or Its Concentration

As stated in Table 13.1, a tracer can be introduced into the hydrologic system either as an instantaneous (slug) or in a continuous manner. Depending on the type of the tracer test and the method of interpretation, either an initial mass of the tracer or its initial concentration needs to be considered to secure appreciable concentration of the tracer at the receptor point. For continuous injection of a tracer, its initial concentration must be evaluated. For an instantaneous placement, depending on the tracer-test method, either the tracer mass or its concentration needs to be considered. An inherent contradiction of the successful tracer test is that it is necessary to assume the results of the tracer test before it is actually conducted. To do so, the effect of the tracer dilution, dispersion, adsorption, and reactivity need to be pre-evaluated to predict the expected peak concentration at the receptor and to compare it with the detection limit and operational range of available instruments.

The peak concentration at the receptor point can be estimated by using an appropriate closed-form equation, which can be found, for instance, in Bear (1979), Freeze and Cherry (1979), Sauty (1980), Javandel et al. (1984), and Domenico and Schwartz (1990), or using a public domain program SOLUTE (a proprietary version compiled by Paul K.M. van der Heijde is distributed by Integrated GroundWater Modeling Center) or any other solute transport model. In this process, an assumed mass or an initial concentration is plugged into an appropriate method, and the tracer concentration, at its maximum, at the receptor can be calculated together with the time elapsed since the emplacement of the tracer.

At this point, however, it is necessary to emphasize that spending extensive time for pre-modeling of a tracer test is usually not a good time investment. This is because there are too many unknown variables at that stage of the hydrogeologic investigation. Had those variables been known, there would not have been any need to conduct the tracer test in the first place. Nevertheless, a good match of the hydrogeologic conditions and the kind of the tracer test with the stipulations set for applicability of the given formula is of prime importance. For example, for a simple case of a two-well tracer test conducted within a natural flow regime with an instantaneous release of the tracer, the following formulas may be used.

For a three-dimensional flow regime (Freeze and Cherry 1979):

$$C_{(x,y,z,t)} = \frac{M}{8(\pi t)^{3/2} \sqrt{D_x D_y D_z}} \exp\left(-\frac{X^2}{4D_x t} - \frac{Y^2}{4D_y t} - \frac{Z^2}{4D_z t}\right) \tag{13.3}$$

For a two-dimensional flow regime (Bear 1979):

$$C_{(x,y,t)} = \frac{M}{4\pi t \sqrt{D_x D_y}} \exp\left(-\frac{X^2}{4D_x t} - \frac{Y^2}{4D_y t}\right) \tag{13.4}$$

For a one-dimensional flow regime (Bear 1979):

$$C_{(x,t)} = \frac{M}{2\sqrt{\pi D_x t}} \exp\left(-\frac{X^2}{4D_x t}\right) \tag{13.5}$$

In Equations 13.3, 13.4, and 13.5,

C = contaminant concentration at location x, y, z at time t ($M/L^3, M/L^2, M/L$) for three-, two-, and one-dimensional flow, respectively, where L is the linear dimension

x, y, z = coordinates (L) of the receptor with respect to the point source

M = mass (M) of contaminant introduced at the point source—it can be expressed as the product of initial concentration C_0 and initial volume V_0

t = Time elapsed (t) from the instantaneous injection of the tracer

D_x, D_y, D_z = dispersion coefficient (L^2/t) in longitudinal, transversal, and vertical (transversal in the vertical plane) direction, respectively

X, Y, Z = distance (L) to the receptor from the center of gravity of the contaminant mass

For the specific case where the point source and the receptor are at the same flow line, the position of the center of gravity of the contaminant mass at time t will lie along the flow path in the x direction at coordinates x_t, y_t, z_t (L). Consequently $y_t = z_t = 0$, but $x_t = vt$ where $v(L/t)$ is the average velocity (seepage velocity) and n (dimensionless) is the effective porosity. Thus, in Equations 13.3, 13.4, and 13.5, $X = x - vt$ and Y, Z = zero.

Formulas 13.3, 13.4, and 13.5 are valid for calculation of the tracer concentration until it reaches its maximum (the rising limb of the bell-shaped curve). Thus, they may be used for prediction of the time for first arrival of the tracer at the receptor point. Certain modifications of the definition for the time term (t) is needed to have these formulas valid for times longer than this, which correspond to the maximum concentration.

For the prediction of the maximum concentration (C_{max}) at the receptor point, that is, when $x = vt$ and consequently $X = 0$, these formulas simplify to contain only their first member. Such formulas reflect a condition that for a nonreactive transport the maximum concentration of tracer coincides with the time t, which equals an advective travel time from the injection source to the receptor point. For a two-dimensional flow, the maximum concentration at the receptor point is expressed as

$$C_{(max)} = \frac{M}{4\pi t \sqrt{D_x D_y}} \tag{13.6}$$

An example using a spreadsheet program for prediction of the maximum concentration of a tracer at the receptor point is presented in Example 13.2. For the tracer-test configuration as defined in the spreadsheet, the maximum concentration of 116.7 mg/L will be recorded after 1.93 days in the monitoring well located 12 ft from the point source. Using the same spreadsheet, one can calculate that the maximum concentration in a well located twice as far, that is, at the distance of 24 ft, would be only 29 mg/L and would be recorded after 3.86 days.

Example 13.2 Spreadsheet software (e.g., MS Excel) can be used to estimate the maximum solute concentration at the receptor assuming a two-dimensional, advective-dispersive transport equation, and an instantaneous release of a tracer to an aquifer with known natural hydraulic gradient.

The following information is entered in the spreadsheet cells, represented by the table included in this example:

Hydraulic conductivity [ft/day]

d_{20} [mm] if used to estimate hydraulic conductivity

Effective porosity [dimensionless fraction]

Distance [ft] between the injection and receptor points

Hydraulic gradient [dimensionless fraction]

Hydraulic head (H) difference [ft] at the injection and receptor locations if used for hydraulic gradient calculation

Tracer mass [mg]

Tracer initial concentration [mg/L] and volume of water injected with tracer[L] if used for calculating mass of the tracer

Assumptions:

Injection and receptor points are on the same flow line.

Lateral extent of the aquifer does not limit transversal spread of the tracer.

Longitudinal dispersivity is 0.1 of the distance between the injection and receptor points.

Transversal dispersivity is 0.01 of the distance between the injection and receptor points.

Hydraulic conductivity (K) may be estimated based on sieve analysis using the U.S. Department of Agriculture formula:

$$K = 0.36 \times d_{20}^{2.3} \text{ cm/s} \tag{13.7}$$

where d_{20} = particle diameter (mm) that together with smaller particles constitutes 20% of weight of the sample

Parameters used for Example 13.2

Hydraulic conductivity value	91.23 ft/day
d_{20}	0.35 mm
K	0.03 cm/s 91.23 ft/day
Effective porosity	0.25
Distance from injection point	12 ft
Hydraulic gradient	0.02
Del H	0.20 ft
Calc. hydraulic gradient	0.02
Tracer mass, M	10,000 mg
Co	5.000 mg/L
Vo	2.00 L
M	10,000 mg

Formula:

$$C_{max} = M/[4 * \text{PI} * t * (DxDy)^{\wedge}0.5 * 9.29] \tag{13.8}$$

where C_{max} = maximum concentration at the receptor point [mg/dec², an equivalent of mg/L for 3D flow]

Co = tracer initial concentration [mg/L]

Vo = volume of injected water with the tracer [L]

t = travel time for advective flow from the injection point to the receptor point [d]

Dx = coefficient of hydrodynamic dispersion in direction X [ft²/day]

Dy = coefficient of hydrodynamic dispersion in direction Y [ft²/day]

9.29 = conversion factor from ft² to dec²

Calculation:

$$v = 6.20 \text{ ft/day}$$
$$t = 1.93 \text{ days}$$
$$Dx = 1.20 \text{ ft}^2/\text{day}$$
$$Dy = 0.12 \text{ ft}^2/\text{day}$$
$$C_{max} = 117 \text{ mg/dec}^2$$

After the initial mass of the tracer or its concentration is calculated, it needs to be reevaluated with respect to uncertainty of the values used for the calculations. It is common to use a safety factor and increase the calculated initial concentration or its mass by one order of magnitude for tracer tests that will involve hydraulic stress at the point source and observation wells. Depending on hydrogeologic conditions, a safety factor of 1,000 may be used for tracer tests that are conducted under natural hydraulic-gradient of groundwater. Conversely, a recirculating tracer test is the least vulnerable to miscalculations of the initial concentration.

There are two main reasons for using a safety factor:

- Intensive and costly efforts may be undertaken for tracer-test preparation and its monitoring. Thus, it is hardly possible to imagine a more embarrassing scenario than the case with a *no-show* of the tracer at the receptor point.

- It is practically impossible to repeat the tracer test after an apparent failure of an earlier attempt and still have confidence with regards to the origin of detected concentrations of tracer at the receptor point. Either a long time (relatively to the seepage velocity) has to separate two attempts or the second tracer needs to be different.

Point-Source Infrastructure

Injection Wells

In most cases, a tracer is introduced to the hydrogeologic system using an injection well. Construction requirements for such a well depend on the duration of source injection.

For a continuous injection of tracer, a well should have characteristics similar to the pumping well, that is:

- The well needs to have an appropriate transmitting capacity to accommodate the designed injection rate without excessive well losses (Chapter 5).

- Its screen should not straddle the water table, otherwise water will be injected also to the vadose zone.

Conversely, if a tracer is to be injected between two packers that are placed within the screened interval, the well must not be gravel packed, and it should be poorly developed to disallow a shortcut of water through the gravel pack back to the well.

One also needs to realize that for an unconfined aquifer, even if the screen does not straddle the water table, mounding of groundwater will occur. Therefore, the injection rate should be slow enough to accommodate the mass injected. Groundwater mounding might affect the dynamics of the tracer test, especially at the beginning when the transient phase groundwater "invades" the vadose zone whose hydraulic conductivity is always lower until it reaches full saturation (Chapter 11). This is one of the reasons why, for a recirculating test with a continuous tracer injection, the tracer should not be introduced to the system until steady-state flow is approached.

For most tracer tests with an instantaneous release of the tracer, the injection well should be designed so that:

- Its screened interval and its diameter are large enough to allow for placement of the tracer solution as one slug. Otherwise, the requirements of the instantaneous release will not be satisfied.

- Its screened interval and its diameter are small enough to disallow or minimize the effect of dilution of the tracer in the well before it enters the aquifer.

- Its construction allows for placement of the tracer with no or minimal changes in the natural flow regime in the aquifer. Otherwise the requirement of point-source release will not be satisfied, and the calculated values will include additional errors. For instance, values of transversal dispersivity in horizontal and vertical planes will be exaggerated. Needless to say, this requirement is not important for a single well–injection/extraction test or the diverging test for which a radial flow pattern is a designed feature.

- Its screened interval coincides with the section of the aquifer to be tested by the tracer test.

- The screen has high transmitting capacity, and the well is well developed (Chapter 15).

A design that satisfies the above requirements would allow for placement in the well of a tightly fitting and closed container with the entire volume of the tracer. After the natural flow regime has recovered, i.e., the water table in the well has returned to its initial level, the bottom valve is opened and the container removed from the well exposing the tracer to a screened portion of the well. This method of releasing a tracer to a hydrogeologic system is especially recommended for a test that is conducted at natural-flow conditions.

For such a case, a common recommendation of "pushing" the tracer by injecting a three-well-volume slug of clean water is in error. If a well is screened over the tested interval, the tracer does not need to be "pushed" out of the well, because it will leave the well itself being "pushed" by the flowing groundwater that actually converges in the well. The convergence is caused by the principle of streamline-refraction (tangent law, Chapter 4) at the interface of the aquifer material and a circular object (the well) of infinite hydraulic conductivity. Specifically, twice as much of the groundwater will flow horizontally through a circular cross section of the well than through an equally wide section of the aquifer.

Sometimes, however, in a real field situation, some of the above recommendations are difficult to implement and the tracer test is conducted with several simplifications, as described in Example 13.3. While analyzing the results of such a test, it is necessary to limit the conclusions to only those whose correctness was not jeopardized by the deficiencies of the field procedure.

Example 13.3 A tracer test was conducted as an experiment to see whether a forced gradient setting (the hydraulic stress) could mobilize a NaBr tracer approximately 15 ft from an injection point toward a pumping well. The pumping well was pumped at a constant rate of 5 gpm (18.9 L/min) for 2.5 h prior to the introduction of a NaBr solution. The solution was mixed in a 13-gal (49 L) container and pumped via a peristaltic pump over a period of 1 h and 45 min (Figure 13.1). The solution was released at the screen depth and was supposed to be observed at the pumping well at some time in the future.

The concentration of the solution at the pumping well was desired to be approximately 100 mg/L. An estimate of the starting concentration was devised through a simplified version of a pie approach (Domenico and Robbins 1985). The starting concentration at the pumping well was assumed to be 0 mg/L, and the starting concentration at the injection point was the unknown. The pumping rate (Q) was 5 gpm, and the proportion of the contributing Q for the pie shaped area near the injection point was estimated by assuming that it would take 1 h to inject the 13-gal (49 L) solution. The mixing equation was used to estimate the unknown initial concentration to be used in the field test. Schematically this is illustrated in Figure 13.2.

Because of unavailability of a bromide electrode in the field, the pH electrode was used instead. Standards were made up in the lab in mg/L and brought out into the field (Figure 13.3). Table 13.3 lists the values measured in the field.

The injection solution was estimated to be near 2,300 mg/L. The effluent discharge remained at pH 7.05 for 2 1/2 h, and it occurred to the participants that it could take many hours before the plume moved 14.9 ft. Thus, a piezometer was hand-drilled along the pathway at a distance of 6.9 ft from the injection point, and a reading of pH = 8.36 was obtained at 3 h. The gradient was 4/15 ft/ft, the effective porosity was assumed to be 0.15, and the movement of 7 ft in 1/8 of a day yielded a seepage velocity of 56 ft/day. The hydraulic conductivity was calculated to be 31 ft/day. This correlated well with estimates from constant discharge pumping tests, which yielded an estimate between 25 and 40 ft/day.

Figure 13.1 Delivery of bromide trace via a peristaltic pump.

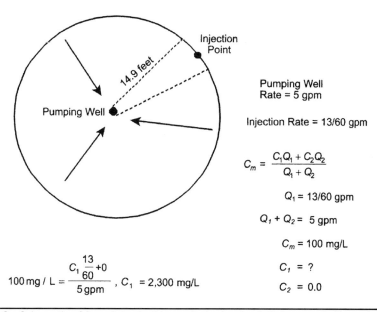

Pumping Well
Rate = 5 gpm

Injection Rate = 13/60 gpm

$$C_m = \frac{C_1 Q_1 + C_2 Q_2}{Q_1 + Q_2}$$

$Q_1 = 13/60$ gpm

$Q_1 + Q_2 = 5$ gpm

$C_m = 100$ mg/L

$C_1 = ?$

$C_2 = 0.0$

$$100\,\text{mg / L} = \frac{C_1 \dfrac{13}{60} + 0}{5\,\text{gpm}}\,, \quad C_1 = 2{,}300\ \text{mg/L}$$

FIGURE 13.2   Schematic of forced gradient tracer test, showing initial concentration calculations.

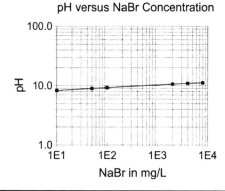

FIGURE 13.3   Plotted field measurements of concentration of sodium bromide versus pH units from Table 13.3.

Concentration NaBr in mg/L	pH units (10^{-n})
Distilled water	7.09
10	8.27
50	8.97
100	9.26
Solution	10.64
4,000	10.93
8,000	11.18

TABLE 13.3   Field Measurements for Example 13.3

Non-Well Injections

In special cases like investigating surface water/groundwater relationship or groundwater recharge conditions, it is necessary to introduce a tracer at the location where the recharge is assumed to take place. In many cases, especially in areas with a well-developed morphology, it would be a body of surface water that recharges a groundwater system. An example of such a case may be a stream, flowing through a karst area that may be losing a portion of its water through a sinkhole located in the stream bottom. Another case might be an artificial pond created either for landscaping or industrial purposes. In such situations, there is hardly a standard method for emplacement of the tracer, leaving investigators to their own inventiveness. This is illustrated in Example 13.4.

Example 13.4   The photograph in Figure 13.4 depicts the emplacement of the bromide tracer in a flow-through settling pond located on the top of a waste rock pile accumulated in front of an adit of the abandoned mine. The pond contained relatively good quality water. The water balance (Chapter 1) of that pond indicated that approximately 10% or 2 gpm of the pond water leaks through the pond bottom and presumably reappears at the toe of a rock pile heavily contaminated (enriched) with heavy metals. To confirm and quantify this stipulation, a tracer test was conducted using bromide as the tracer. The following procedure was used to emplace the bromide.

The required initial concentration of bromide (Br) in the pond was determined to be 2,000 mg/L. The bottom of the pond was surveyed, and volume of water in the pond was estimated to be 46,550 L (12,300 gal). Thus, 93 kg (205 lb) of bromide or 120 kg (264 lb) of dry sodium bromide (NaBr) was needed for the tracer test. Mine waters usually flowing into the pond were diverted, and the outflow of the pond was dammed to create a standing body of water in the pond.

All of the dry NaBr, 120 kg (264 lb), was dissolved within a plastic container to make 465 L (123 gal) of solution at a concentration of 258 g/L of NaBr, an equivalent of 200 g/L of Br. This solution had a specific gravity of 1.193. A battery-powered, boat-trolling motor was used to mix dry NaBr in the container with water.

A 50-ft length of 1.5-in-diameter hose was attached to the bottom of the NaBr solution container. The other end of the hose was extended to a small pontoon floating on the pond (Figure 13.4). A trolling motor was attached to the pontoon to be used for mixing NaBr solution in the pond water.

Figure 13.4   Placing a bromide tracer in a flow-through settling pond mixed with a discharge line and a trolling motor.

Ropes were attached to the pontoon, so that two people standing on the shore could pull the pontoon across and along the pond.

The third person boarded the pontoon and directed the outlet of the hose with gravity-flowing bromide solution to the propeller of the trolling motor, while the pontoon was dragged back and forth across the pond. This procedure provided good 1:100 mixing of the bromide solution with water in the pond, preventing stratification of the heavy bromide solution in the bottom. On mixing, samples of pond water were collected, and the concentration of bromide was checked using a bromide electrode.

An approximately 2,000 mg/L bromide solution was left in the pond to leak through its bottom. After 4 days, approximately 90% of the pond water had leaked through its bottom, and the natural flow condition through the pond was restored to maintain a constant hydraulic head over the moving tracer.

Observation Wells

For a recirculating tracer test, the sampling well should have similar characteristics as a pumping well to accommodate required discharge. For all other tracer tests, the diameter of a sampling well should be small to minimize the influence of the *dead water* residing in the well above the sampling interval. More important, however, is to screen the observation well over the entire investigated interval and have no blank casing (sump) below the screen (Chapter 15).

Sampling Schedule

The sampling schedule is usually a by-product of the preliminary calculation of the tracer's initial concentration. Theoretically, first sampling should be scheduled for the calculated time of arrival of the front portion of the dispersed tracer. Then, the sampling and measurements of the tracer concentration need to be continued until the tracer test is terminated. For a continuously injected tracer, the test may be terminated when there is no change of the tracer concentration as measured in real time at the receptor point. For an instantaneously injected tracer (slug injection), which should yield a bell-shaped curve in the diagram of the concentration versus time, it is best to terminate the test when the measurements are again close to the detection limit.

Prediction of the time for an earliest sampling may be done using Equations 13.3, 13.4, 13.5, or other formulas whose applicability match the configuration of the given test. A spreadsheet for determination of the tracer concentration versus time for a two-dimensional flow at natural hydraulic conditions is given in Example 13.5. In this case, because of an assumption that the point source and the receptor point are placed along the same flow-path, Equation 13.4 simplifies to

$$C_{(x,y,t)} = \frac{M}{4\pi t \sqrt{D_x D_y}} \exp\left(-\frac{x^2}{4D_x t}\right) \qquad (13.9)$$

This simplification is possible because $Y = y = 0$, as explained earlier in this chapter, and there is no movement of the tracer with respect to the y coordinate, except for the transversal dispersion. Example 13.5 uses the same tracer-test configuration as assumed for Example 13.2. The final result of the calculation using Equation 13.9 is shown in Example 13.5. It provides a predicted time for the first measurable record of the tracer at the receptor point for the case that the instrument detection limit is 0.005 mg/L. This time of arrival was 0.21 day. A maximum concentration of 116.7 mg/L was expected to be recorded at 1.93 days.

Example 13.5 Spreadsheet software can be used to predict the tracer concentration versus time assuming a two-dimensional, advective-dispersive transport, and instantaneous release of a tracer to an aquifer with a known natural hydraulic gradient. The following information was entered into the spreadsheet cells, represented by the table shown in this example:

Limitation: Values are calculated for a valid time (t) not greater than the time (t_{max}) corresponding to the maximum concentration.

Time (t) [day] for the calculated concentration of the tracer,

Hydraulic conductivity [ft/day],

d_{20} [mm] if used to estimate hydraulic conductivity,

Effective porosity (dimensionless fraction),

Distance (x) [ft] between the injection and receptor points,

Hydraulic gradient (dimensionless fraction),

Hydraulic head (H) difference [ft] at the injection and receptor locations if used for the hydraulic gradient calculation,

Tracer mass [mg],

Tracer initial concentration [mg/L] and volume of water injected with tracer [L] if used for calculating mass of the tracer.

Assumptions:

Injection and receptor points are on the same flow-path (Equation 13.9).

Lateral extent of the aquifer may not impact transversal spread of the tracer.

Longitudinal dispersivity is 0.1 of the distance between the injection and receptor points.

Transversal dispersivity is 0.01 of the distance between the injection and receptor points.

Instrument detection limit is 0.005 mg/L.

Hydraulic conductivity (K) may be estimated based on sieve analysis using the U.S. Department of Agriculture formula (Equation 13.7) repeated here:

$$K = 0.36 d_{20}{}^{2.3} \text{ cm/s} \qquad (13.10)$$

where d_{20} = particle diameter (mm) that together with smaller particles constitutes 20% of weight of the sample.

Parameters used in the spreadsheet of Example 13.5.

time (t)	0.21 day
Hydraulic conductivity value	91.23 ft/day
d_{20}	0.35 mm
K	0.03 cm/s 91.23 ft/day
Effective porosity	0.25
Distance from injection point	12 ft
Hydraulic gradient	0.02
Del H	0.20 ft
Calc. hydraulic gradient	0.02
Tracer mass, M	10,000 mg
Co	5,000 mg/L
Vo	200 L
M	10,000 mg

Cell formula:

$$Ct = M/[4 * \text{PI} * t * (Dx * Dy)^{0.5} \times 9.29] \exp\left(-X^2/(4 * Dx * t * 9.29)\right) \qquad (13.11)$$

where Ct = concentration of the tracer at the receptor point at time (t) [mg/dec², an equivalent of mg/L for 3D flow]

$\quad Co$ = tracer initial concentration [mg/L]

$\quad Vo$ = volume of injected water with the tracer [L]

$\quad\quad t$ = time for calculated concentration of the tracer at the receptor point [d]

$\quad Dx$ = coefficient of hydrodynamic dispersion in direction X[ft²/day]

$\quad Dy$ = coefficient of hydrodynamic dispersion in direction Y[ft²/day]

$\quad 9.29$ = conversion factor from ft² to dec²

$\quad\quad X = x - vt$

Calculations:

Seepage velocity $\qquad\qquad v = 6.20$ ft/day

$\qquad\qquad\qquad\qquad\qquad X = 10.70$ ft

Max. concentration time $\quad t_{max} = 1.93$ days

$\qquad\qquad\qquad\qquad\quad Dx = 1.20$ ft²/day

$\qquad\qquad\qquad\qquad\quad Dy = 0.12$ ft²/day

$\qquad\qquad\qquad\qquad\quad Ct = 0.005$ mg/dec², for $t = 0.21$ day

The time of arrival calculated in Example 13.5, however, needed to be adjusted (shortened) proportionally to the uncertainty of dispersivity values used for calculations. The frequency of measurements may also be determined through preliminary calculations to ensure that a reasonably smooth curve can be obtained.

Monitoring

Results of the tracer test must be monitored at real time as soon as a sample is collected. The monitoring equipment used depends on the kind of tracer. For instance, for dye tracers, a fluorimeter is required; however, for radionuclide tests a water sampling kit, usually provided by a laboratory, needs to be used and the samples shipped to the laboratory. If bacteria are used as a tracer, it is necessary to provide an appropriate analytical field kit. Unfortunately, field kits for bacteria are notorious for unreliable performance.

For most popular tracers like bromide or iodine, a single ion electrode is commonly used. Such electrodes have a certain operational range, often starting from 0.1 part per million (ppm), and a very specific calibration procedure. If several samples a day are collected, an electrode must be calibrated at least twice a day, otherwise an electrode drift may remain unnoticed (Chapter 8). It is a common mistake to save *a little time* for a field calibration of the measurement instrument, only to end up spending *excessive time* later during the data-reduction process for removing a "guesstimate" drift portion of the measured field values.

Maintaining detailed notes of the recorded noise (i.e., measurements below the detection limit) is also strongly recommended especially for tracer tests with multiple observation wells and a large vertical component of the hydraulic gradient. It happened once to the author that the flow direction was traced only by recording consistent noise at one of many observation wells.

Equipment

Equipment used for a tracer test depends of the procedure used. The least equipment-intensive tracer tests are those conducted at natural flow conditions. Conversely, recirculating tracer tests usually include a lot of equipment that amplify proportionally to the rate of recirculated fluid. Figure 13.5 depicts a setup for a recirculating tracer test conducted with two sampling/extraction wells and an injection well operated at 7-gpm injection rate over a 4-day period. Two 1,000-gal tanks, electrical mixers, a battery of flowmeters, submersible and centrifugal pumps, and the like, were used for this project, which included needs for anisotropy ratio determination. The list of equipment may include the following.

General Equipment

- First-aid kit
- Safety gloves
- Safety goggles
- Cell phone or communication capabilities
- Laptop computer
- Good supply of paper, notebooks, and pencils
- Paper towels
- String, duct tape, and insulation tape
- Measuring tape
- A kit with pH, specific conductance and temperature meters and probes
- Shelter for rainy or hot weather (e.g., canopy or a pickup with a topper)

Figure 13.5   Setting up equipment for a recirculating tracer test.

Tracer-Test Implementation Equipment

- Appropriate amount of a tracer
- Material safety data sheet (MSDS) of the tracer used
- Tarp or waterproof boxes to store supply of a tracer (if continuous injection)
- Weighing scale to weigh bulk tracer while preparing the tracer solution
- Containers for preparing and storing the tracer solution
- Stir or mechanical mixer for tracer solution preparation
- Calibrated bucket for discharge measurements, if applicable
- Flowmeters (for a recirculating tracer test)
- Submersible pumps (for a recirculating tracer test)
- Centrifugal pumps for transfer of tracer solution (for a recirculating tracer test)
- Hoses and valves
- Generator, if applicable
- Data logger with pressure transducers (for recirculating tracer test)
- Water-level indicator E-tape
- Stopwatch (dedicated cell phone or otherwise)
- Good source of light for night sampling
- A tool box with tools appropriate for the infrastructure used

Tracer-Test Monitoring Equipment

- Measuring instrument(s) for monitoring of tracer concentration
- Instruments' instruction manuals
- Weighing scale for preparation of a set of calibration solutions
- Calibrated cylinders and pipettes for preparing calibration solutions
- Tissues for blotting electrodes, and so on.
- Deionized water
- An icebox with ice for cooling calibration solutions if temperature sensitive
- A propane/butane stove for heating water to warm calibration solutions if temperature sensitive
- Bottles (containers) to store calibration solutions
- Bottles (containers) to store and keep electrodes hydrated between measurements
- Boxes with lids to store measuring equipment and calibration solutions between measurements
- Forms for keeping records of instruments calibrating
- Forms for recording measurements
- A laptop computer with a spreadsheet allowing real-time plotting of the concentration curve

13.4 Common Errors

A tracer test justifies the expression "those who do nothing make no errors." A variety of tracer-test procedures, tracer-test materials, and diversity of hydrogeologic conditions make each tracer test a unique task for which no standard operating procedure exists. It is thus inevitable that mistakes will be made despite all efforts undertaken to avoid them. The key issue is to minimize the number of mistakes. Some common field errors are listed below:

- The receptor was located too far from the injection point.
- Monitoring of the tracer test was terminated prematurely because:
  - Assumed effective porosity was too low.
  - Assumed hydraulic conductivity was too high.
- Monitoring of the tracer started too late because:
  - Assumed effective porosity was too high.
  - Assumed hydraulic conductivity was too low.
- Measured concentrations were too low to be used for quantitative interpretation because:
  - Dispersivity values were underestimated.
  - Tracer was less conservative than expected.
  - Ratio of anisotropy and/or principal directions of anisotropy were "misevaluated."
  - Location of the receptor was incompatible with the actual flow pattern.
  - Vertical component of hydraulic gradient was neglected or "misevaluated."
  - Instrument detection level was too high.
- A continuous-injection tracer test was initiated before steady-state conditions for flow had been established.
- Emplacement of the tracer significantly altered natural flow conditions while the conditions were supposed to be undisturbed.
- Calibration check of the instruments used for measuring tracer concentrations at the receptor point were not frequent enough.

Errors committed during the test design or its results interpretation are usually related to inappropriate match of the formula or the method used with the hydrogeologic setting or the tracer test conducted. An example of such an error would be using a formula developed for a three-dimensional flow to design a tracer test in a shallow flow system with a long distance between injection and receptor points.

13.5 Natural Tracers

Recent advancements in laboratory techniques to identify and determine the concentration of element stable isotopes and radioisotopes have created new opportunities for the hydrogeologist to use them for determining groundwater age, origin, and sometimes flow path. In general, stable isotopes are used for analysis of the water cycle while radioisotopes are used for groundwater dating owning to their known decay (half-life) rate and often associated with the gain of daughter products (Chapter 8). Several radioisotopes are known and used for groundwater age dating with their applicability depending on the target age. However, only ^{35}S, ^{3}H with association of ^{3}He, and ^{85}Kr can be considered to be used for similar purposes and time scale as tracer tests. Though these radioisotopes can be used to address groundwater flow occurring within a time scale of a few months to few years, only tritium ^{3}H (aka T) is associated with its daughter product, that is, stable isotope of helium, tritigenic helium, ^{3}He, can be routinely used. The applications of ^{35}S and ^{85}Kr, are still in the experimental phase (Clark 2015).

The $T - {}^{3}$He method is similar to a tracer test conducted at a natural flow regime with a tracer emplaced continuously from a point source (Table 13.1), but for the $T - {}^{3}$He method the emplacement of T is areal. Though a disadvantage of the $T - {}^{3}$He method is that the results can yield only a seepage velocity, its advantage is that T has already been emplaced in groundwater by precipitation and has undergone radioactive decay along its flow path. Therefore, there is no need to wait for its arrival at the location of interest.

Tritium is a hydrogen (H) unstable isotope with the half-life ($T_{1/2}$) of 13.43 years. In addition to hydrogen's one proton ^{3}H includes two neutrons. Tritium decays to a stable isotope of helium (^{3}He) by the release of β particles (β ray). Tritium is a part of a natural water molecule with an abundance of 10^{-18} (that is, the molar ratio of tritium atoms to stable hydrogen atoms). This ratio is used as a tritium unit (TU)[1] to measure its concentration when T is used as groundwater tracer for groundwater dating.

Tritium is produced naturally by cosmic radiation and its concentration in precipitation is a function of geomagnetic latitude, with greater production at higher latitudes.[2] There is also a seasonal variation of T in atmospheric precipitation, with higher values occurring in spring. Though much greater production of T accompanied the atmospheric testing of the thermonuclear bombs between 1951 and 1980, by the 1990s most of the "bomb" tritium had been washed from the atmosphere. Thus, the tritium level in global precipitation is now close to the natural levels (Clark and Fritz 1997).

For dating groundwater (thus also determining its seepage velocity between two locations) by T, the decay equation (13.12) is used:

$$T_t = T_0 e^{-\lambda t} \tag{13.12}$$

where T_t = the measured tritium concentration after some time, t,
  T_0 = the initial concentration of tritium,
  λ = a decay constant in units of time^{-1} and is expressed as Equation 13.13:

$$\lambda = \ln 2 / T_{1/2} \tag{13.13}$$

Thus, knowing T_t concentration in the aquifer at the given time and back extrapolating to T_0, that is, to the known T concentration in the recharge (precipitation) water, its age at the point of water sample collection could be calculated. However, for the precipitation entering recharge areas decades after the detonation of thermonuclear bombs, this

---

[1] 1 TU = 10^{-9} ppb and in terms of activity = 0.119 Becquerls per liter (Bq/L).
[2] E.g., 2 TU at Barbados 13°N in comparison to 11 TU in Ottawa 45°N in 1992.

would lead to ambiguous results since the initial T concentration is poorly known because of the following reasons:

- Groundwater that was recharged after thermonuclear period have T decaying at the same rate as its source.

- In most cases and especially for an unconfined aquifer the recharge occurs over an area rather than at one distant point.

- Reduction of tritium concentration in the water sample collected at the given location has been affected by dispersive transport in the aquifer.

- Seasonal variation of T concentration in atmospheric water is another factor that makes T_0 for groundwater difficult to establish.

However, knowledge of T_0 is not required if the accumulation of its daughter product, ^{3}He, can be measured in the groundwater sample. As the loss (decay) of tritium is expressed by an exponential function (13.12), so is a gain of stable isotope of helium expressed as ^{3}He$_t$ in Equation 13.14:

$$ {}^3\text{He}_t = T_0 \, (1 - e^{-\lambda t}) \tag{13.14} $$

By combining Equations 13.12 and 13.14 the input concentration T_0 is removed, and t can be calculated using Equation 13.15:

$$ {}^3\text{He}_t = T_t \, (e^{-\lambda t} - 1) \tag{13.15} $$

This is done by rearranging Equation 13.15, substituting λ with Equation 13.13, using the half-life for tritium $(T_{1/2})$ of 12.43, and formulating Equation 13.16 for groundwater age (t):

$$ t = 12.43/\ln 2 * \ln(1 + [{}^3\text{He}_t]/[T_t]) \tag{13.16} $$

Square brackets in Equation 13.16 indicate that both concentrations are expressed in TU. However, the value of ^{3}He$_t$ measured in the groundwater sample needs to be corrected for atmospheric ^{3}He that accompanies groundwater during recharge. This is done by measuring either the total helium (He) or neon (Ne) concentration in the water sample. The advantage of making the correction using Ne is that its solubility is less sensitive to temperature. Details on the procedure how to make the correction in question are given in Clark (2015).

The $T - {}^3$He method is particularly useful for determining seepage velocity between two distant locations. In such a case, this method has the following advantages over a tracer test:

- Eliminates long, often unrealistic, waiting time between the tracer emplacement and its arrival

- Does not require injection of any tracer

- Does not require equipment for the tracer emplacement

- Provides an average large-scale seepage velocity value that accounts for inhomogeneity of the medium along the flow path

However, this method requires special water-sampling containers and a collection process that is specific for the isotopic laboratory used for water analyses. Current (2018) analytical cost is approximately $600 for one water sampling set collected at one location and analyzed for all the required analytes, that is, total He, total Ne, ^{3}H, and ^{3}He. The number of sampling locations will be project specific.

13.6   Summary

For hydrogeologic investigations, a tracer test is an attractive field method to quantify selected parameters of an aquifer like groundwater flow direction and rate, aquifer porosity, anisotropy, dispersivity, retardation factor, and other physicochemical characteristics. In its most common form, a tracer test is conducted by emplacing a defined quantity of traceable substance within the aquifer and tracking it down hydraulic gradient: either a natural or a purposely imposed gradient during emplacement and/or extraction of the tracer. The substances may be ions, radionuclides, dyes, bacteria, organic compounds, or isotopes, to list those most commonly used. The emplacement of these substances may be instantaneous or may last for a prescribed time interval. All these variables must be carefully considered while designing a successful tracer test for the given hydrogeologic setting, the test objectives, available budget, and given time period. Therefore, a statement of "one size and kind fits all" is not applicable for tracer-test technology. Tracer materials, their use and emplacement that might be excellent for certain geological conditions might not be usable for other subsurface environments and/or objectives. Readers are encouraged to look for case studies that include information on the applicability of a particular tracer-test technology for a given hydrogeologic setting.

13.7   Problems

Recent excavation on a college campus to install a new water supply line to a new building was completed to a depth of approximately 1.5 m and left open during a season of heavy precipitation in northern latitudes. Sometime later, but still during the precipitation season the grounds personnel noticed a seepage area at the corner of the football field located about 70 m away from the excavation activities at an elevation approximately 4 m lower than the land surface at the place of the excavation. The city water line and other utility lines were checked with no loses confirmed. The excavated material was colluvial sand and gravel with an unusually high content of silt. At a depth of 2 m below ground surface the unconsolidated sediments were underlain by a fractured and weathered (in its top portion) volcanic andesite known in this region to form an unconfined aquifer of local extent. Hand-auger borings installed through the bottom of the excavation were advanced to approximately 1.5 m before they met refusal, showing very moist cuttings retrieved starting 1 m below the bottom of the excavation.

13.1   Having a budget of $15,000 you are asked to design a tracer test to confirm or deny that the excavation left open for the precipitation season caused damage to the football field. The other alternative could be the rise of water level in the unconfined aquifer due to the unusually high precipitation. Use Tables 13.1 and 13.2 to come up with a basic design. Which method would you choose and which tracer medium? You may formulate your answer based on the time elapsed between completion of the excavation and the time when the seepage was first observed. You also have the luxury of choosing the time for conducting the tracer test at any time during the year. In addition, you also have time to perform limited field investigations to gather additional information. State your assumptions and hypothesize on the expected outcome. While doing so, always remember that a tracer test is a valuable tool; however, if performed in a "vacuum" with respect to other hydrogeologic information its usefulness may be questionable.

13.8 References

Bear, J., 1979. *Hydraulics of Groundwater*, McGraw-Hill, New York.

Becken, M.W., and Coplen, T.B., 2001. Technical Note: Use of Deuterated Water as a Conservative Artificial Groundwater Tracer. *Hydrogeology Journal*, Vol. 9, pp. 512–516.

Clark, I.D., 2015. *Groundwater Geochemistry and Isotopes*, CRC Press/Taylor & Francis Group, Boca Raton, FL.

Clark, I.D., and Fritz, P., 1997. *Environmental Isotopes in Hydrogeology*, Lewis Publishers, Boca Raton, FL.

Davis, S.N., Cambell, D.J., Bentley, T.J, Flynn, T.J., 1985. *Ground Water Tracers*. National Water Well Association. National Ground Water Association, Westerville, OH.

Domenico, P.A., and Robbins, G.A., 1985. A New Method of Contaminant Plume Analysis. *Ground Water*, Vol. 23, No. 4, pp. 476–485.

Domenico, P.A., and Schwartz, F.W., 1990. *Physical and Chemical Hydrogeology*, John Wiley & Sons, New York.

Freeze, R.A., and Cherry, J.A., 1979. *Groundwater*. Prentice Hall, Upper Saddle River, NJ.

Javandel, I., Dughty, C., Tsang, C.F., 1984. *Groundwater Transport: Handbook of Mathematical Models*. American Geophysical Union. Water Resources Monograph Series 10. American Geophysical Union, Washington, DC.

Käss, W., 1998. *Tracing Technique in Geohydrology*, A.A. Balkema, Rotterdam.

Sauty, J.P., 1980. An Analysis of Hydrodispersive Transport in Aquifers. *Water Resources Research*, Vol. 16, No. 1, pp. 145–158.

van der Heijde, P.K.M., 1997. *Solute, Analytical Models for Solute Transport in Ground Water*. IGWMC-BAS 15, International Ground Water Modeling Center, Colorado School of Mines, Golden, CO.

CHAPTER **14**

Tracer Tests—Dye

Thomas Aley

President and Senior Hydrogeologist, Ozark Underground Laboratory, Protem, Missouri

14.1 Utility of Groundwater Tracing

Groundwater tracing is a fundamental tool for hydrogeology work in karst areas and in many fractured rock settings. Tracing agents, particularly fluorescent tracer dyes, have a well-established track record for demonstrating flow directions and travel times from both surface and subsurface points to wells, springs, and surface streams fed by groundwater (Hotzl and Werner 1992; Kranze 1977). Fluorescent dyes have also been used to trace water through talus and landslide debris, glacial deposits, through permeable zones in earth-fill dams, and into and through both active and abandoned mines. Dyes are also used to trace inflow into pipes (infiltration) and leakage out of pipes (exfiltration).

At waste sites tracer dyes have demonstrated connections and travel times between contaminant source areas and monitoring and recovery wells (Figure 14.1). They have been introduced into aquifers immediately prior to the addition of groundwater remediation compounds to delineate the portion of the aquifer that will be affected and to characterize travel times. At sites where hazardous waste leaks have been discovered and repaired, the tracer dyes have been used to demonstrate that the repaired leaks can account for all monitoring points where contaminants of concern have been detected. They are also used to verify or refine groundwater modeling.

Groundwater tracing with fluorescent tracer dyes has been used to delineate groundwater protection zones for municipal wells and to help delineate Zones of Influence for quarries pumping appreciable quantities of groundwater. They have also been used to delineate the recharge areas for caves, springs, and wetlands that provide habitat for threatened and endangered aquatic species and to help assess the potential impacts of proposed road construction and timber harvest on groundwater systems supplying water to salmon and trout streams.

With modern analytical equipment and protocols the detection limits for the more useful tracer dyes are in the parts per trillion range. Successful groundwater traces can be conducted with even smaller concentrations of dye by using activated carbon samplers that adsorb and accumulate tracer dyes. These samplers will be discussed under Section 14.5 "Sampling and Analysis for Dyes."

FIGURE 14.1   Dye introduction into a shallow trench dug to bedrock. Solvents had previously been dumped on the ground in this area. Dye infiltrating through the trench was subsequently detected in 13 monitoring wells and piezometers and at two off-site groundwater monitoring stations about 4,000 ft (1,219 m) from the dye introduction trench.

A wide range of tracing agents in addition to fluorescent dyes has been used. These have included soluble salts, radioactive isotopes, sulfur hexafluoride, and minute biological agents including bacteriophage and *Lycopodium* spores. Kass (1998) provides an excellent compendium of information on almost every tracing agent ever used in groundwater tracing. However, this chapter places primary emphasis on tracer dyes since the author estimates that about 90% of all professionally directed groundwater traces in the United States have used these dyes and the use of these tracers is simple, practical, and routinely effective. Tracer dyes in the amounts used for professionally directed groundwater tracing work do not pose health or environmental risks that should limit or preclude their use (Smart 1984; Field et al. 1995).

Three basic questions commonly encountered in groundwater hydrology are the following:

- Where does the water go?
- How long does it take to get there?
- What happens along the way?

Tracing agents are very good for answering the first two questions and in some cases they can address the third question. As an example, fluorescein dye experiences less adsorption onto soil particles than the commonly used herbicides alachlor and atrazine, but these herbicides experience more adsorption onto soils than does rhodamine WT dye (Sabatini 1989). One could thus use these dyes to bracket the performance of these herbicides in an aquifer.

Much hydrogeologic work is based on equations and calculations that presume homogenous and isotropic groundwater conditions. Tracers are particularly useful where these conditions are not met or may not be met. In karst and fractured rock aquifers modeling premised on presumed homogenous and isotropic conditions can yield estimates of travel times that are orders of magnitude in error, and these errors can have serious implications as indicated in Example 14.1.

Example 14.1   Walkersville, Maryland, is underlain by a limestone aquifer. A well field with three municipal wells supplies public water. As part of a wellhead protection study, the aquifer was modeled by the United States Environmental Protection Agency (USEPA) and lines were drawn on maps indicating 5- and 10-year "time of travel" borders. City officials were concerned that the delineations might be substantially in error and funded a groundwater tracing investigation to assess the credibility of the time of travel modeling. One dye introduction made almost on the calculated 5-year time of travel line reached town wells in less than 17.5 h. In the absence of the tracer study the modeling created a false sense of security. A few years after the groundwater tracing work a construction accident resulted in a release into the groundwater system of 900,000 gal of raw municipal sewage (Figure 14.2). Based on the previous mathematical modeling, the calculated travel time to wells from the spill area would have been approximately 8 years. Bacterial sampling and tracer dye introduced at the spill site showed that the actual travel time for first arrival of bacteria and dye at the wells was between 4 and 5 days. Fortunately, because of the earlier dye tracer study, the city responded immediately to the spill and took effective and comprehensive actions to obtain emergency water supplies and protect their residents. There were no resulting cases of waterborne illness. There are two lessons for hydrogeologists. First, the assumptions of uniformity used in the modeling were inappropriate for a soluble rock aquifer and resulted in travel rate estimates that were at least 500 times too slow. Second, dye tracing can be an effective method for validating and refining groundwater models.

Figure 14.2   Site of the Walkersville, Maryland, sewage spill. Fluorescein dye was introduced into the pit in the foreground. It was first detected between 4 and 5 days later in the Walkersville municipal well field about 5,300 ft (1,615 m) from the spill site.

Fractured rock aquifers commonly provide preferential flow routes where travel rates are substantially greater than estimated by groundwater modeling. A tracer test (hereafter referred to as trace) conducted by the author in fractured basalt over a distance of several hundred feet from an injection well to a recovery well resulted in the first arrival of dye within 20 min of introduction. Modeling had predicted first arrival times of 20 to 28 days, a travel rate more than 1,400 times too slow. While this and the Walkersville cases are extreme examples they illustrate the utility and necessity for tracer studies in aquifers where preferential flow routes are likely to exist.

14.2   Five Common Misconceptions about Tracer Dyes

One common misconception is that dyes may be harmful or that large amounts of dye must be used and that this will visibly color water throughout an area. There is extensive technical literature (such as Smart 1984; Field et al. 1995) demonstrating that the dyes discussed in this chapter present no health or environmental problems at concentrations five orders of magnitude or more above the detection limits of modern analytical protocols. Dye tracing, especially when primary sampling reliance uses activated carbon samplers, does not require large quantities of dyes, and the dyes are not harmful in the amounts needed for groundwater tracing.

Most adequately designed and conducted tracer studies will not result in visibly colored off-site water unless this condition is inherent in the design of the tracer study. There are cases where dyes at visible concentrations need to be detectable in the field. This is best done by working in a dark room and shining a flashlight at a right angle to the viewer into a transparent glass jar containing the test water. If fluorescent dye is visually detectable it will be seen in the light beam. This test should obviously not be conducted by a person who is color-blind. All dyes discussed in this chapter have their maximum fluorescence in the visible wavelength range. *The idea that the best way to detect the discussed dyes is with an ultraviolet light is a myth.*

A second common misconception is that dye tracing works only in well-developed karst areas of limited size where one is tracing from features such as sinkholes to springs. While tracing is often needed for investigating water-related issues in such settings, successful groundwater tracing can routinely be conducted in almost any soluble bedrock aquifer as well as in many other hydrogeologic settings. Fluorescein dye has been used to trace from a sinking surface stream through the groundwater system for 39.5 mi to Big Spring, Missouri (Aley 1978). The flow rate at the dye introduction point was 0.5 cubic feet per second (cfs) and the flow rate of Big Spring at the time of dye introduction was 324 cfs. The amount of dye equivalent in the dye mixture used for this trace was 5.0 lb. Fluorescent dyes have also been successfully used in many traces in high-yield limestone aquifers including the Barton Springs portion of the Edwards Aquifer in Texas and the Floridan Aquifer. Hauwert et al. (2002) discuss an extensive tracing program in the Austin, Texas, area of the Edwards Aquifer.

A third common misconception is that tracer dyes will not work in acidic waters such as those commonly encountered in metal or coal mines. While not all dyes are suitable for such conditions, many successful traces into and/or through both active and inactive mines have been conducted with fluorescent dyes. Mine waters present unique challenges and bench tests lasting a few weeks with mine water and/or representative rock units are sometimes prudent for selecting the most appropriate type and quantity of dye for the planned trace. Groundwater traces of up to 1.6 mi with mean travel rates of greater than

100 ft/day have been conducted into mines using fluorescein and eosine introduced along normal faults in sedimentary rocks including sandstone, shale, and coal.

A fourth common misconception is that dyes are rapidly destroyed by sunlight and that this precludes their use for tracing water into or out of surface streams. Rhodamine WT and sulfo rhodamine B are not rapidly destroyed by sunlight. Dyes introduced into a surface stream late in the day usually have adequate time to enter the groundwater system prior to being exposed to appreciable sunlight. Dyes discharged from groundwater into a surface stream at night can be detected with activated carbon samplers or in water samples collected with an automatic sampler programmed to collect samples during the hours of darkness.

A fifth common misconception is that groundwater tracing is impractical because it requires substantial experience, the purchase of hard to find materials, and analytical work necessitating instruments not normally found in water testing laboratories. There are firms that routinely provide these services and assist others in designing and conducting groundwater traces. Since there is no EPA or American Society for Testing and Materials (ASTM) standard protocol for tracer dye analysis, each of these firms has its own protocols and they are not all of equal quality. Desirable analytical protocols are discussed later in this chapter.

14.3   Dye Nomenclature and Properties

There are a number of fluorescent dyes that have been used in groundwater tracing with varying degrees of success. Color Index names and numbers are the technical identifiers for the dyes and should be included in reports in addition to the more common names (such as eosine, fluorescein, etc.).

All of the dyes are sold as mixtures and the dye equivalent is the percent of dye in the mixture. The rest of the mixture is a diluent. Diluents are used to standardize the product and facilitate mixing with water. Dye mixtures that are liquids routinely contain lower dye equivalents than powder mixtures. Liquid mixtures are usually easier to use than powders that must be mixed with water. Unfortunately, dye equivalent values are seldom shown on labels and most retail suppliers of dyes do not know the dye equivalent percent in the mixtures they sell, regardless of whether the mixtures are liquids or powders. Many tracing efforts have failed because mixtures used were assumed to be 100% dye when in reality the dye equivalent was only a few percent. Firms that provide analytical services for dye tracing studies routinely supply dye mixtures of known dye equivalent percentages as part of their quality assurance programs. The dye equivalent percent of dye mixtures used should be included in reports in addition to the amount of the dye mixture introduced.

Eosine is also known as eosin, eosine OJ, and Drug and Cosmetic (D&C) Red 22. Eosine is Acid Red 87 and its Color Index number is 45380. Its CAS number is 17372-87-1. The approximate percent of dye in "as sold" mixtures ranges from about 2% to 75%. Eosine is pink to green depending upon concentration and whether it is viewed in sunlight or shade. When mixed in water this dye is the least visually noticeable of the five dyes discussed in this chapter.

Fluorescein is also known as uranine, uranine C, sodium fluorescein, fluorescein LT, D&C Yellow 8, and fluorescent yellow/green. The fluorescein used in dye tracing is a sodium salt of fluorescein and one should always ensure that the fluorescein mixture being used is the sodium salt of fluorescein. Fluorescein is Acid Yellow 73 and its Color

Index number is 45350. Its CAS number is 518-47-8. The approximate percent of dye in "as sold" mixtures ranges from 2% to 80%. Fluorescein has a green color and is the distinctive coloring agent in most automotive coolant solutions. Fluorescein from vehicle coolants can sometimes be detected in wells, springs, and streams downstream of large parking areas and busy roads due to leakage of coolants from vehicles. Fluorescein is also common in landfill leachates and has been used as an indicator of such waters. Fluorescein is present in some household products and thus is routinely present at small concentrations in municipal sewage. Groundwater heavily impacted by municipal sewage discharge commonly has detectable concentrations of fluorescein since this dye is not removed by sewage treatment plants except to the extent that this occurs by photodegradation. Fluorescein performs well in seawater and was the dye used for the tracing work shown in Figure 14.3.

Pyranine is also known as Solvent Green 7 (SG 7). Its Color Index name is D&C Green 8 and its Color Index number is 59040. Its CAS number is 6358-69-6. The approximate percent of dye in "as sold" mixtures undoubtedly varies among suppliers but is commonly about 77% when sold as D&C Green 8. Pyranine is present in some household products but to a lesser extent than is fluorescein. A major limitation of pyranine is that its emission peak fluorescence and fluorescence intensity vary with pH values in the range normally encountered in groundwater. While pyranine can be used effectively in some settings, its use and analysis requires more technical sophistication than is required for the other four dyes discussed, and one of the other dyes is likely to be more suitable for the tracing work.

Rhodamine WT is sometimes sold as fluorescent red, but this name has also been applied to rhodamine B (Basic Violet 10) which has potential carcinogenic properties, a dramatically different chemical structure, and is not a suitable dye for groundwater tracing. The Color Index name for rhodamine WT is Acid Red 388; no Color Index number has been assigned. The CAS number is 37299-86-8. The approximate percent of dye

FIGURE 14.3 View of a National Park Service visitor center in Sitka, Alaska. A tracer dye was introduced into the on-site sewage treatment system at the visitor center to determine if it contributed to springs (dark areas in the foreground) within the intertidal zone. No dye was detected in the springs.

in "as sold" mixtures ranges from about 3% to 20%; this dye is almost always sold in a liquid form. Rhodamine WT is a dark red dye. Emission fluorescence peak wavelengths for rhodamine WT and sulfo rhodamine B are very close together and these two dyes should generally not be used concurrently in situations where both might appear at a particular sampling station. Dyes with fluorescent characteristics similar to rhodamine WT and sulfo rhodamine B are used as the coloring agents in some hydraulic fluids and thus may be present at some industrial sites.

Sulfo rhodamine B is also known as sulfo rhodamine B, pontacyl brilliant Pink B, Lissamine red 4B, kiton rhodamine B, acid rhodamine B, amido rhodamine B, and fluoro brilliant pink. This dye is sometimes sold simply as "red fluorescent dye" and may be confused with rhodamine WT if only common names are used. The color index name for sulfo rhodamine B is Acid Red 52, and its Color Index number is 45100. The CAS number is 3520-42-1. The approximate percent of dye in "as sold" mixtures ranges from about 3% to 75%. Sulfo rhodamine B is a dark red dye.

Table 14.1 summarizes data on dye degradation in sunlight as measured by fluorescence magnitude. The values are for pans containing 0.5 in of water that were allowed to sit in bright sunlight for periods of 1, 3, and 5 h. As should be expected larger dye concentrations last longer than smaller concentrations. Dyes will experience slower photodegradation in deeper streams, streams that are well shaded, and streams where turbidity limits sunlight penetration. Fluorescein and eosine dyes can be used to trace groundwater into most surface streams if sampling stations are placed at intervals where the mean travel time between stations is 6 h or less. Such sampling stations are able to detect dyes discharging into the stream at night and when the sun is at low angles to the earth. Placing sampling stations on surface streams at about 0.5- to 1.0-mi intervals works well in many landscapes, and in many urban and suburban areas road access makes this general interval possible. When sampling a large stream where dilution is a major concern the intervals

Dye	Initial dye Concentration (ppb)	1 h (%)	3 h (%)	5 h (%)
Eosine	1,000	2	<1	—
Eosine	100	1	<1	—
Fluorescein	1,000	19	<1	—
Fluorescein	100	7	<1	—
Pyranine	1,000	68	3	—
Pyranine	100	55	1	—
Rhodamine WT	1,000	100	100	83
Rhodamine WT	100	97	49	32
Sulfo rhodamine B	1,000	100	97	95
Sulfo rhodamine B	100	81	70	60

Table from Aley (2002); courtesy Ozark Underground Laboratory, Inc.
Note: Water depth 0.5 in. Percentages reflect amount of initial dye remaining.

TABLE 14.1 Dye Degradation in Sunlight as Measured by Fluorescence Magnitude

Dye	Fluorescence Substantially Decreased at pH Less Than	Fluorescence Mostly Eliminated at pH Less Than
Eosine	4.0	2.5
Fluorescein	6.5	5.5
Pyranine	9.5	6.5
Rhodamine WT	5.0	2.5
Sulfo rhodamine B	3.5	2.0

Table from Aley (2002); courtesy Ozark Underground Laboratory, Inc.

TABLE 14.2   Influence of pH on Fluorescence Intensity

should be shorter. In many cases sampling near the bank of a large stream and on the same side as the dye introduction point will also be more likely to detect the tracer dyes than will sampling further out into the stream.

Table 14.2 summarizes the influence of pH on the fluorescence magnitude of the five tracer dyes. Despite the values in Table 14.2, fluorescein dye is still an effective tracing agent in many acidic waters. Low pH values alter the fluorescein from a fluorescent to a nonfluorescent molecule. If the pH of water samples containing fluorescein are pH adjusted to basic values fluorescein will again fluoresce. However, the nonfluorescent molecules of fluorescein appear to have a greater tendency to be adsorbed onto earth materials than is the case for fluorescein at neutral to basic pH values. Activated carbon samplers (discussed in Section 14.5) effectively adsorb fluorescein from both acidic and basic waters. Based on field studies, Aley (2002) ranked the general suitability of the five tracer dyes for use in tracing acidic mine waters as follows: Fl > Eos > SRB > RWT. Pyranine is not suitable for this use.

Table 14.3 compares tracer dye adsorption onto various mineral and organic materials. Note that all dyes are subject to some adsorption and that the amount of adsorptive loss increases as the amount of the adsorbing material increases. Adsorptive losses can be overcome in groundwater tracing work by increasing the amount of dye used. There are no data presented for eosine since this dye was not studied by Smart and Laidlaw (1977). Based upon field experience eosine typically experiences similar, but somewhat greater, sorptive losses than does fluorescein.

Based upon laboratory tests (which did not include eosine), Smart and Laidlaw (1977) ranked the resistance of the dyes to adsorption onto inorganic material as follows: Py > FL > RWT > SRB. They ranked the resistance of the dyes to adsorption onto organic material as follows: Py > SRB > FL = RWT.

Behrens (1986) ranked the resistance of a number of tracer dyes to adsorption. His data were based upon field work with water samples. His rankings were Py = FL > Eos > RWT > SRB. Aley (2002) reported similar findings for sampling based on activated carbon samplers although he noted that his data on pyranine were limited.

Table 14.4 shows typical detection limits for the five tracer dyes. Instrument analysis values are derived from method detection limits currently used in analysis work at the Ozark Underground Laboratory and values for other laboratories may be somewhat different.

Material	Sediment Concentration (g/L)	FL	PY	RWT	SRB
Mineral					
Kaolinite	2	98	95	89	88
Kaolinite	20	93	95	67	51
Bentonite	2	98	100	92	98
Bentonite	20	87	98	79	—
Limestone	2	98	96	93	97
Limestone	20	94	85	66	76
Orthoquartzite	2	98	100	98	—
Orthoquartzite	20	98	87	90	—
Organic					
Sawdust	2	86	70	81	92
Sawdust	20	11	30	42	—
Humus	2	83	76	82	92
Humus	20	17	31	11	63
Heather	2	41	74	81	—
Heather	20	0	18	18	—

Table from Aley (2002); courtesy Ozark Underground Laboratory, Inc. Adapted from Smart and Laidlaw (1977).

Note: Values are percent of dye remaining in solution from a 100-ppb initial solution. FL = fluorescein; PY = pyranine; RWT = rhodamine WT; SRB = sulfo rhodamine B.

TABLE 14.3 Comparison of Tracer Dye Adsorption onto Mineral and Organic Materials

Method	EO	FL	PY	RWT	SRB
Dye in water, instrument analysis	0.015	0.002	0.015*	0.015	0.008
Dye in charcoal elutant, instrument analysis	0.050	0.015	0.010	0.170	0.080
Dye in water; field conditions, experienced person	135	7	175*	125	50
Dye in water; field conditions, general public	13,500	140	3,500*	2,500	1,000
Dye in water; dark room with flashlight, experienced person	10	2	3*	50	5

Table slightly modified from Aley (2002); courtesy Ozark Underground Laboratory, Inc.

*pH adjusted water with pH of 9.5 or greater.

Note: Concentrations are in parts per billion (ppb) and are based upon the as-sold weights of the dye mixtures. Dye equivalent values for the mixtures were 75% for eosine, fluorescein, and sulfo rhodamine B; 77% for pyranine; and 20% for rhodamine WT. EO = eosine; FL = fluorescein; PY = pyranine; RWT = rhodamine WT; SRB = sulfo rhodamine B.

TABLE 14.4 Detection Limits for Five Tracer Dyes under Different Conditions

14.4   Designing a Dye Tracing Study

The following are key issues that should be addressed in designing a dye trace:

- What questions do you wish to answer with the tracing work and how precisely do you need to have the answers? The latter question will, in part, determine how frequently samples must be collected and analyzed.

- What is the hydrogeologic setting? At a minimum, what geologic units are involved? To what extent must preferential flow pathways be expected?

- Where will the dye or dyes be introduced? It is often possible to conduct two or three concurrent traces for only a slightly greater cost than a single trace because samples can be analyzed for multiple dyes at little or no additional analytical cost.

- How much water will be introduced with the dye? This is specific to each trace. Before introducing dye, add sufficient water to moisten all surfaces if they are not already wet. As a general rule the more water used to flush the dye into and through the groundwater system the better the tracing effort will be. However, in some cases one does not wish to distort water elevations in the aquifer. In groundwater sampling it is common to extract a volume of water equal to three to five well volumes within the saturated zone prior to collecting a water sample. For dye introductions one can moisten well surfaces, introduce the dye as a pulse, and then add three to five well volumes to move most of the dye out of the well bore and into the aquifer.

- Can chlorinated tap water be used for the dye introduction? In most cases the answer is yes and it is generally more desirable to use tap water than water from ponds and streams. In some cases it may be desirable to dechlorinate the tap water before introducing it; activated carbon is one way of doing this. Historically much groundwater tracing has used stormwater runoff or waters naturally sinking into the groundwater system in a localized area. Chlorinated tap water will not destroy much of the introduced dye because of the high dye concentrations at the dye introduction point.

- How will the dye be introduced? In general, it is easier and better for tracing purposes to introduce the dye as a short-duration pulse rather than adding it with introduced water at a constant rate for a prescribed period of time.

- How much background sampling will be done prior to dye introduction? Background sampling determines if one or more dyes or compounds with fluorescence characteristics similar to dyes may already be present at some of the sampling stations. Even in remote areas one should usually have at least one set of background samples collected and analyzed from most or all sampling stations prior to any dye introduction. In areas where previous tracing has been done or in urban or industrial areas two or three sets of background samples are desirable. White et al. (2015) report a karst site where 10 pounds of 20% dye equivalent rhodamine WT dye was introduced into a sinkhole in 1995 and not detected in later sampling that year. This dye was subsequently detected with carbon samplers during sampling 18 years later at nine sampling stations located up to 5,000 ft from the dye introduction point. Such long dye residence time in the aquifer illustrates the importance of adequate background sampling.

- What dye or dyes will be used and in what quantities? Different dyes have different properties, and a pound of one dye is not equivalent to a pound of another dye. The proper selection of dye type and quantity is very important to successful groundwater tracing. A study plan should include a preliminary determination of the dye type(s) and quantity(s) to be used, but the final determination of both type and quantity must be made after at least one set of background samples. Laboratories that routinely assist in tracer studies can usually analyze background samples and identify potential dye interferences within a week of the time the background samples are collected.

- Where are all of the potential sampling points? In most cases do not limit sampling to the points where you expect to detect dyes. It is often important to identify points where the dye is not detectable. Control stations to demonstrate that dyes from other sources are not moving into your study area are desirable, especially in urban settings. Maintain flexibility in the study plan to permit adding or modifying sample locations based upon conditions encountered in the field. As an example, major precipitation or snow melt runoff during a trace may create additional sampling points that should be tested.

- Are there any data or calculations indicating likely travel times for the trace? If such data exist they should be used to the extent reasonable. A common approach is to design the study to continue for 1.5 times the estimated travel time to the most distant sampling station where dyes may be detected.

- What is the pH of waters that will be encountered in the trace? If the pH is less than about 6.8 it may affect the type and/or quantity of dye that should be used.

- Develop a good site map showing hydrogeologic features that may be of significance plus all dye introduction and dye sampling locations. Developing a good site map should not be deferred until the report phase of the study.

- Develop a rationale for the conditions under which the study can be ended. A common approach is to identify tentative study duration but then state that sampling will not be ended until certain conditions have been met. Common criteria for ending sampling include (1) no new dye detection sites in the last three sampling events unless the new detection sites are in areas near previous dye detection points, (2) dye concentrations are decreasing at most (and preferably all) sampling stations where dye has been detected, and (3) no other factor(s) warrant extending the sampling period. Sampling can be discontinued earlier than the tentative study duration at some sampling stations where dye has been detected in multiple samples or it can be discontinued if there have been at least two positive dye detections at each of the positive dye sampling stations and it is reasonable to conclude that no additional sampling stations are likely to receive any tracer dye. Bear in mind that the credibility and utility of many dye traces have been compromised because sampling was ended prematurely; this is bad science and a false economy.

- Gain concurrence from all relevant parties for the study plan prior to beginning the study. Be aware that many regulatory agencies are staffed by people with limited experience and that their experience may not include tracer studies. Providing copies of toxicity papers such as Field et al. (1995) and Smart (1984) can be helpful to reviewers.

- Determine if there are any other site-specific conditions that should be addressed in the study plan.

- Few consulting firms or universities have the analytical equipment needed for conducting modern dye traces. An old filter fluorometer in a storage room will not be adequate for most tracing work. Additionally, few entities have personnel with expertise in designing and conducting dye traces. Most consulting firms and universities offset the lack of equipment and experience by contracting with a firm that specializes in the analysis of tracer dyes and has expertise in the design of tracer studies. The quality and level of expertise of the firms varies. If this approach is to be used, then, before starting a trace, get the outside analytical service on board with the project. In addition to analyzing samples, a quality firm can assist in the design of the study, provide activated carbon samplers and certified dye-free sample containers, and provide dye mixtures of known dye equivalent percentages.

- Specify the details of sample collection, shipping, and analysis. Identify the laboratory that will do the analysis work and provide the details on how that work will be done. Firms that routinely assist in tracer studies usually have detailed procedures and criteria documents that can be incorporated as an appendix to the study plan if their services are involved.

- Identify what reports will be prepared and when they will be submitted.

- Determine if the tracing work is to be conducted in a state or area where a short-term authorization, permit, or some other type of official notification or authorization may be required. Where such permitting or notification is required it is usually simple. A little time on the phone can find the person responsible for dealing with any required permits and that person can help you determine exactly what information is needed and what procedure should be followed. Even when a permit or notification may not be required it is a good protocol to inform agencies responsible for water quality protection about the planned dye tracing work. Extending that informing to entities that might receive a phone call from the public about colored water (if that is a possibility) is also a prudent action. If you are not required to obtain a permit from a particular agency then do not ask for their permission; instead, simply inform them.

14.5 Sampling and Analysis for Dyes

Grab samples of water and activated carbon samplers can both be used to detect the tracer dyes. In addition, there are some fluorometers that can be used in the field to provide continuous data of fluorescence intensity at a sampling station. While these instruments are useful for some purposes, they are most useful for time of travel studies in surface waters where there is no question as to where the dyed water will appear.

Grab samples of water can either be collected by hand or with a programmable automatic sampler. The automatic samplers fill bottles at selected time intervals or can collect composite samples. The analytical laboratory should be queried as to the quantity of sample they require; the quantities are typically 40 mL or less, although for some tracing in the vadose zone samples as small as 2.5 mL have been used. For most dye tracing involving monitoring wells it is not necessary to pump and purge the wells prior to sampling since the dyes are not volatile compounds. This greatly simplifies sampling, field time, and wastewater disposal costs.

Analysis of water samples provides a known dye concentration at a particular point in time. Since a dye pulse reaching a sampling station may last only a relatively short period of time, sampling that places primary reliance upon water samples routinely requires a large number of samples. Automatic samplers can be advantageous if there are only a few sampling stations where this equipment must be used. The typical automatic sampler holds 24 bottles. The automatic samplers cost several thousands of dollars each and this routinely limits their use. Also, the samplers must be set on the surface of the ground and can only lift water from relatively shallow monitoring wells. As a result of these conditions, traces that rely primarily or exclusively on grab samples of water are expensive in terms of field time and analytical costs. There are some monitoring wells that are equipped so that they can only be sampled with grab samples of water. In such cases if these wells are to be monitored it must be with grab samples of water.

Activated carbon samplers are fiberglass screen packets partially filled with a laboratory grade of coconut shell activated carbon (Aley 2016). Firms providing tracer dye analysis and assistance routinely provide the activated carbon samplers. The typical samplers contain about 4.25 g of activated carbon, but the amount varies among the firms with some entities not using a standard amount of carbon. Appropriate activated carbon is capable of adsorbing and retaining all five of the tracer dyes discussed in this chapter. Activated carbon samplers can be left in place in the water being sampled for periods ranging from a day to a few weeks. A common sampling interval is about 1 week. Longer sampling intervals may be used for tracing in remote areas, especially if the water being sampled is of high quality. At waste sites sampling intervals should generally not be longer than about a week since the longer the sampler is in place the fewer available activation sites remain on the carbon. The author has found that sampling intervals of 1 week have worked well even in wells with free product petroleum or high concentrations of solvents. Longer sampling intervals can be used after the first 8 to 12 weeks of sampling for longer duration tracer studies.

Using a large database, Aley (2017) compared dye analysis results for sampling periods when both carbon samplers and water samples were analyzed. There were 1,002 sampling periods for springs; dye was detected in carbon samplers for 99% of the sampling periods but in only 44% of the periods for water samples. A similar comparison was made for 939 sampling periods at monitoring wells. Dye was detected in carbon samplers for 96% of the sampling periods but in only 81% of the periods for water samples. The percentage values were similar for eosine, fluorescein, rhodamine WT, and sulfo rhodamine B except that the percent of detections was smallest for sulfo rhodamine B in water samples from springs. The data were from actual studies at numerous sites with widely different conditions. The data clearly demonstrate that carbon samplers are superior to water samples in detecting the presence of eosine, fluorescein, rhodamine WT, and sulfo rhodamine B dyes.

Aley (2017) evaluated dye concentrations from carbon samplers for periods when dye was not detectable in water samples for the same period. Based on 378 sampling periods at springs and a mean sampling duration of 16 days, the mean dye concentration in carbon samplers was 656 times greater than the minimum dye detection limit in water samples. Based on 110 sampling periods at monitoring wells and a mean sampling duration of 27 days, the mean dye concentration in carbon samplers was 517 times greater than the minimum dye detection limit in water samples. Unlike water samples, carbon samplers accumulate the dyes, sample continuously, and do not miss short-duration pulses of dyes. These characteristics are essential for credible groundwater tracing results, especially when relatively small amounts of dye must be detected in large volumes of water and detection sites must not be missed due to unwarranted reliance on water samples.

Example 14.2 White et al. (2015) reported on dye analysis results from two tracer studies conducted 18 years apart at a waste site in karst in the Frederick Valley of Maryland. The 1995 study had 139 sampling stations. Eosine was detected in carbon samplers from 13 stations but was detectable in associated water samples from only 8 of the stations. Fluorescein was detected in carbon samplers from 8 of the stations but was detectable in associated water samples from only 5 of the stations. If sampling in 1995 had relied only upon water samples 38% of the dye detection sites would have been misidentified as non-detection sites.

In the 2013 study there were 127 sampling stations. Eosine was detected in carbon samplers from 17 stations but was detectable in associated water samples from only 7 of the stations. Fluorescein was detected in carbon samplers from 7 of the stations but was detectable in associated water samples from only 6 of the stations. In addition, rhodamine WT introduced in the earlier study was detected in carbon samplers from 9 stations but was not detectable in any water samples. If sampling in 2013 had relied only upon water samples 61% of the dye detection sites would have been misidentified as non-detection sites.

Activated carbon samplers in monitoring wells are often attached to the top of dedicated disposable bailers and suspended in the middle of the screened interval or in the middle of the saturated zone. In some studies it is desirable to learn what zones in the well are contributing most of the dyed water. In such cases a weighted line can be used with an activated carbon sampler attached at each zone of interest.

Example 14.3 In a tracer test to determine the point where contaminants from a former coal gasification plant were entering a city storm sewer, a cord was blown with high-pressure air through the sewer for a city block with carbon samplers placed every 50 ft. Dye was introduced into the contaminated source water. Analysis results identified the area where contaminants were entering the sewer, and dye concentrations indicated that there was a single problem area and thus minimized the amount of street in a high value historic district of Arlington, Virginia, that needed to be excavated to repair the pipe.

Activated carbon samplers placed in springs or surface streams are anchored to stakes, rocks, or bricks and placed where there is visible current. If the bottom of a spring or stream is loose sediment, the samplers should be placed so they remain above the sediment. At streams and springs a minimum of two independently anchored samplers should be placed in the event one is lost and to permit analysis of some duplicate samplers for statistical analysis of the precision of analytical results. The activated carbon is *adsorbing the dye, not filtering it* out of the water, so the velocity of the water makes relatively little difference in the amount of dye that will be adsorbed in the sampler so long as the water in contact with the activated carbon particles is not appreciably depleted of dye due to adsorption.

If pumping wells are sampled, a continuous stream of water equal to about 1 gal/min should pass through the sampler. This can often be achieved by running water continuously into a bucket and having the sampler suspended in the bucket.

When activated carbon samplers are collected they are placed in sterile labeled plastic bags. Most firms that do dye analysis work supply such bags and have done quality assurance work to ensure that the bags are free of any tracer dyes. The bags are labeled as to station number, name, and date and time of collection. The labeling must be on the outside of the bags and with an indelible black felt marker. Do not use red felt markers since many of these contain eosine or other dyes as the coloring agents.

Even if primary sampling reliance is placed on activated carbon samplers it is a good protocol to also collect a grab sample of water each time a sampling station is visited. If dye is detected in the carbon sampler for the station the water sample can also be analyzed to determine the dye concentration at that time. Custody issues are important, especially if the tracer study may be challenged in court. Recognize that you have not maintained continuous custody for activated carbon samplers "in place" at sampling stations such as springs and surface streams. Collecting and analyzing grab samples of water from these stations adds credibility to sampling results and verification that the results from the carbon samplers are reasonable.

Table 14.5 summarizes advantages and disadvantages of designing traces where sampling reliance is placed on water samples, on activated carbon samplers, or with

---

Sampling based on water samples only. *Advantages:* Generally best suited to research projects. Accurate determinations of dye concentrations at known times. Data suitable for developing a mass balance (i.e., accurately knowing how much of the introduced dye was detected at particular sampling stations). Maintains chain of custody of all samples. *Disadvantages:* Requires more sampling time and more analytical costs than is the case with activated carbon samplers. Because samples must be collected frequently, this method is not well suited to traces with a large number of sampling stations. Likely to miss detecting dye at stations reached by only small concentrations of dye; this can be especially important for monitoring wells at hazardous waste sites. Requires the use of substantially more dye than if sampling utilizes activated carbon samplers. Poses an enhanced risk of visually colored water at receiving locations.

Sampling based on activated carbon samplers only. *Advantages:* Minimizes amount of dye needed for the tracing since activated carbon samplers accumulate dyes. Permits continuous sampling of all sampling sites. Excellent approach for identifying all sampling stations reached by the tracer dye. Excellent approach for traces with a large number of sampling stations and stations in remote locations. Minimizes field time since sampling can typically be done once per week or sometimes at longer intervals. Minimizes analytical costs. Suitable for sampling at multiple levels in the same well. Ideal for detecting dyes that may discharge into surface streams during nighttime when they are not destroyed by sunlight. *Disadvantages:* Time of first dye arrival and time of peak dye arrival at sampling stations is known only within a sampling period that is likely to be several days long; precision in this information is often not critical. No quantitative dye concentration data for water at sampling stations. Good chain of custody only for stations (such as wells) that are locked.

Sampling based on primary reliance on activated carbon samplers with secondary reliance on water samples. *Advantages:* Generally best suited to problem-solving projects. Same advantages as sampling with activated carbon samplers plus quantitative dye concentration data for water at sampling stations where dyes are detected. Provides good chain of custody records for water samples. Good method for enhancing the credibility of tracing results. *Disadvantages:* Increases analytical costs over use of activated carbon samplers alone by 15% to 30%; insignificant increase in field expenses.

TABLE 14.5 Advantages and Disadvantages of Sampling Reliance Based on Water Samples Only, Activated Carbon Samplers Only, or Primary Reliance on Activated Carbon Samplers with Secondary Reliance on Water Samples

primary reliance on activated carbon samplers with secondary reliance upon water samples. The table indicates that sampling that places primary reliance on activated carbon samplers with secondary reliance upon water samples is commonly the best strategy for a problem-solving project that must be conducted for reasonable budgets. It is also the best strategy for minimizing the amount of dye needed.

Collected samples, both as grab samples of water and activated carbon, should be kept in the dark (to prevent photodecomposition) and in a cooler after collection. While the tracer dyes are reasonably stable in the activated carbon samplers they can degrade or be destroyed in water samples that are not refrigerated. It is a good protocol to ship samples by overnight courier to an analytical laboratory in coolers packed with refrigerant packs with trade names including Blu Ice, Freez Pak, Koolit Refrigerant, and Chillers. The author once tested a refrigerant called Green Ice and found it to contain fluorescein; it should not be used. Ice should not be used since it melts during shipment and could cross-contaminate samples. There is no established "hold time" for dye samples, but it is a good protocol to have them analyzed as soon as possible.

In the laboratory activated carbon samplers are washed in strong jets of unchlorinated and dye-free water to remove sediment and debris that can interfere with the analysis. The samplers are then eluted in a solution consisting of a strong base, alcohol, and water. There is no standard ASTM or EPA protocol for dye analysis and different laboratories use somewhat different eluting solutions. The different solutions are typically designed to best elute the mixture of dyes that the particular laboratory routinely uses, and all of the eluting solutions work reasonably well. The activated carbon is typically left in the eluting solution for a standard time period; 1 h is common.

The quality of the reagent water used for washing samplers is critical and should be considered in selecting a laboratory for use in a study. A few laboratories use tap water without any treatment. This presents two quality control problems. First, at times public water supplies (and especially those using surface waters) may contain small concentrations of tracer dyes or similar fluorescent compounds that have the potential to contaminate samples. Secondly, public water supplies are chlorinated and the chlorine can destroy some of the adsorbed dye. *This problem is most severe for activated carbon samplers that contain only a small amount of dye.*

After the activated carbon sampler has been eluted for a standard time period, the eluting solution is poured off and a sample of it is then available for analysis. Water samples can be analyzed directly. If fluorescein or eosine is present the water must be pH adjusted prior to analysis by placing open sample vials in a high-ammonia environment until a pH of 9.5 or greater is achieved. Various fluorometric instruments can be used to conduct the analysis but the most sensitive and the ones best suited to discriminating between dyes and other fluorescent materials are spectrofluorophotometers that can synchronously scan excitation and emission wavelengths and print out a graph of the emission fluorescence. Figure 14.4 shows an example of such a graph for an activated carbon sampler that contained fluorescein, eosine, and rhodamine WT dyes. The emission wavelengths of the fluorescence peaks are a function of the particular dye, its matrix, and various instrument settings. The emission wavelength ranges of the dyes are sufficiently separated that traces can often be conducted concurrently with two or three dyes. These dyes are typically fluorescein, eosine, and rhodamine WT. Rhodamine WT and sulfo rhodamine B should generally not be used concurrently if there is any chance that both of the dyes might appear at the same sampling station.

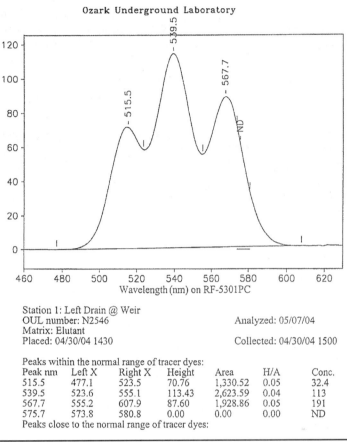

Ozark Underground Laboratory

Station 1: Left Drain @ Weir
OUL number: N2546
Matrix: Elutant   Analyzed: 05/07/04
Placed: 04/30/04 1430
                                                   Collected: 04/30/04 1500

Peaks within the normal range of tracer dyes:

Peak nm	Left X	Right X	Height	Area	H/A	Conc.
515.5	477.1	523.5	70.76	1,330.52	0.05	32.4
539.5	523.6	555.1	113.43	2,623.59	0.04	113
567.7	555.2	607.9	87.60	1,928.86	0.05	191
575.7	573.8	580.8	0.00	0.00	0.00	ND

Peaks close to the normal range of tracer dyes:

Figure 14.4 Emission fluorescence graph from a synchronous scan of the elutant from an activated carbon sampler that contained fluorescein (peak on the left, concentration 32.4 ppb), eosine (peak in the middle, concentration 113 ppb), and rhodamine WT (peak on the right, concentration 191 ppb). Traces were through an earth-fill dam from three different points on the upstream side of the dam.

Table 14.6 summarizes some of the fluorescence characteristics of the five dyes. The maximum excitation and emission wavelengths of each of the dyes are from Ford and Williams (1989). The normal acceptable wavelength ranges for the dyes in water and in the standard elutant used by the Ozark Underground Laboratory are also shown to illustrate how the values differ among the dyes and between water and elutant samples. The normal acceptable wavelength range is derived from a database of positive dye detections from actual traces. The acceptable range equals the mean wavelength plus and minus two standard deviations. All synchronous scan values for eosine, fluorescein, rhodamine WT, and sulfo rhodamine B are from synchronous scans with a bandwidth separation of 17 nanometers (nm) and standard excitation and emission slit settings. The pyranine scan values are from synchronous scans with a bandwidth separation of 35 nm.

Dye	Maximum Excitation	Maximum Emission	Normal Acceptable Wavelength Range for Water Samples	Normal Acceptable Wavelength Range for Elutant Samples
Eosine	515	535	529.6–538.4	533.0–539.6
Fluorescein	490	520	505.6–510.5	510.7–515.0
Pyranine	455	515	501.2–505.2*	500.4–504.6
Rhodamine WT	555	580	569.4–574.8	561.7–568.9
Sulfo rhodamine B	565	590	576.2–579.7	567.5–577.5

*Adjusted to a pH of 9.5 or greater.
Note: See text for explanations. All values in nanometers (nm).

TABLE 14.6   Fluorescence Wavelength Characteristics of Five Dyes

14.6   Detection Limits for Tracer Dyes

Detection limits for tracer dyes are commonly expressed as the concentration of the dye mixture in the water or elutant sample rather than as the dye equivalent. The wider the excitation and emission slit settings in a spectrofluorophotometer the lower the detection limits, but the less discrimination there is among fluorescent compounds. *Never* assume that the best analytical protocol is the one with the lowest reported detection limit. Detection limits should be based on the dye mixture used, not the assumed dye equivalent in the mixture. Spiked laboratory samples provide lower detection limits than spiked field samples; the latter are the more appropriate since they represent "real world" conditions. Table 14.4 (presented earlier) indicates detection limits for the five tracer dyes in samples analyzed at the Ozark Underground Laboratory; the concentrations are based upon typical as-sold dye mixtures rather than dye equivalents. The dye mixtures were 75% dye equivalent for fluorescein, eosine, and sulfo rhodamine B; 20% dye equivalent for rhodamine WT; and 77% dye equivalent for pyranine.

The reason that the detection limits are smaller for water samples than for carbon elutant samples under the Ozark Underground Laboratory protocol is that the slit widths under which carbon elutant samples are analyzed are narrower than the slit widths for water samples. Activated carbon can adsorb many compounds and the narrower slits provide more discrimination for carbon elutant samples than is deemed necessary for water samples.

Quantification of the concentrations of each of the dyes present in a sample is obviously information that any good quality dye analysis laboratory will provide. This applies to both water samples and elutants from activated carbon samplers. Detection limits vary among the dyes because of variations in dye equivalents in the dye mixtures and also because fluorescence intensity varies among the dyes. Quantitative analysis with a spectrofluorophotometer operated under a synchronous scan protocol is more accurate than with a fluorometer.

14.7   Determining Dye Quantities

Experience in groundwater tracing with fluorescent dyes is the best method for estimating the quantity of dye needed for particular traces. Determining the quantity of dye needed for a particular trace cannot be accomplished by using any simple equation. Appreciable amounts of dye can be lost to various processes including adsorption and decomposition. Substantial dilution occurs in many aquifers including some of those with significant preferential flow routes. While there are over 20 equations that have been developed for calculating dye quantities, none is recommended for routine use.

The epikarst (Chapter 12) is the weathered upper portion of the bedrock in carbonate rock landscapes and is commonly about 33 ft thick (Ford and Williams 1989), but there is a large range in thickness. Based on data from about 1,000 traces, Aley (1997) reported that dye detection percentages for traces conducted in permanently saturated epikarstic zones were from 0.1% to 1.0% of the amount introduced into the groundwater system. The percentages were typically 1% to 10% for the seasonally saturated epikarstic zone. Aley (2017) summarized mass balance data for 15 traces in six different U.S. states and found that the median value for percent of introduced dye detected for traces in karst was 4.9%. Hauwert et al. (2004) presented dye detection percentages for 20 traces in the Barton Springs portion of the Edward Aquifer in Texas and found that the median value was about 4.2%. In the author's experience dye recovery rates from traces in fractured rock aquifers are commonly in the range of 1% but are also highly variable.

The amount of dye needed for successful tracing is a function of a number of factors including the sampling and analytical approaches being used. Traces where primary sampling reliance is placed on grab samples of water and dye detections are likely to occur at multiple sampling stations are likely to require at least an order of magnitude more dye than would be necessary if primary sampling reliance were based on activated carbon samplers. When lesser amounts of dye are used there is appreciable risk that some detection sites will not be identified. When larger amounts of dye are used the risk of producing colored water at an undesirable location is increased.

Example 14.4   Benson and Yuhr (2016) discuss a tracing study by the United States Geological Survey in a municipal well-field for Miami-Dade County, Florida. Approximately 50 kg (110 lbs) of rhodamine WT dye mixture (presumably with a dye equivalent of 20%) was introduced about 100 m (328 ft) up gradient of a municipal well in the highly productive Biacayne Limestone Aquifer. Brightly colored water discharged from the production well and some entered the distribution system before the well was shut off. The incident yielded concern in newspapers about discolored water in the water supply and the potential (unlikely) for pink underwear in washed laundry. At least part of the mistake in the obviously flawed study design was that actual groundwater travel rates were about 12 times faster than suggested by modeling. In the author's experience travel rates determined by tracer tests are commonly faster than suggested by modeling or as calculated based upon measured aquifer parameters. An order of magnitude difference in the travel rates between modeling and tracer test results is not unusual. In fact, an important objective of many tracer tests is verification and refinement of groundwater models at important sites.

Filter fluorometers are sometimes used for dye analysis work. These instruments are not capable of separating fluorescence intensity measurements into background fluorescence and fluorescence due to a tracer dye. Background fluorescence can vary by an order of magnitude over the course of a week so dye pulses at a sampling station can readily be masked by background fluorescence. This problem does not affect spectrofluorophotometers operated under a synchronous scan protocol. As a result, if dye analysis is to be dependent upon a filter fluorometer then the amount of dye needed for a trace will be greater than the amount needed with a spectrofluorophotometer operated under a synchronous scan protocol. Additionally, while groundwater traces using two or three dyes concurrently can readily be done using a spectrofluorophotometer operated under a synchronous scan protocol, traces dependent upon filter fluorometers for analysis should use only one dye at a time.

The amount of dye needed for a trace is also a function of the type of dye. In a setting where all of the five tracer dyes are potentially useful a trace that could be done with about 1 pound of fluorescein would typically require about 1.5 lb of eosine for similar results or about 4 lb of rhodamine WT or about 5 lb of sulfo rhodamine B. If all of these dyes could be used pyranine would not be a reasonable selection. If it were selected, the trace would require about 3 lb of pyranine. The above quantity estimates presume that all of the dye mixtures contain about 75% dye equivalents except for rhodamine WT which is assumed to have a dye equivalent of 20%.

Based on the author's experience and fluorescein mixture with a 75% dye equivalent many groundwater traces can be done with 5 lb or less of the fluorescein dye mixture. This presumes that sampling places primary reliance on activated carbon samplers and analysis with a spectrofluorophotometer operated under a synchronous scan protocol. Traces in fractured rock aquifers typically require at least three times more dye than traces in karst aquifers. Traces in the epikarstic zone of a karst aquifer require three to five times more dye than traces below this zone, and this is especially true for perennially saturated epikarstic zones. Traces through alluvial and glacial deposit aquifers typically require about three times more dye than traces in karst aquifers with finer-textured materials requiring more dye than coarser materials.

The best approach for making a final selection of dye type and quantity is to seek the advice of a person with extensive dye tracing experience. Their advice is typically based upon personal experience with similar tracing projects. Such persons are often associated with a firm capable of providing design assistance and analytical services for tracer tests.

14.8   Summary

Groundwater tracing with fluorescent dyes is a cost-effective method for determining flow paths and travel rates through aquifers. Tracing with the fluorescent dyes is particularly useful in karst and fractured rock aquifers but can be used effectively in other types of aquifers.

Successful groundwater traces require the selection of dyes appropriate for the site; all dyes are not equal in performance and detectability. Sampling for the dyes can be done with water or activated carbon samplers. For most problem-solving tracing work primary sampling reliance should be based upon activated carbon samplers and secondary reliance upon water samples. This minimizes costs, the quantity of tracer dyes needed, and the risk that sites receiving only small amounts of dye or short duration pulses of dye will not be identified as detection locations.

For most tracing work the analysis should rely on spectrofluorophotometers operated under a synchronous scan protocol. These instruments are vastly superior to filter fluorometers for tracing work. There are several commercial firms that routinely provide analytical services, supplies, and assistance in the design of tracer studies.

14.9   Problems

14.1   Groundwater inflow rates into a limestone quarry have recently substantially increased. The quarry operator suspects that much of the increased flow is derived from a nearby surface stream that flows adjacent to the quarry for 2,000 ft and has a mean flow rate of about 50 cfs. Water is entering the quarry in four separate and highly localized areas. Design a dye tracing study that will answer relevant questions and be as beneficial as possible to the quarry operation in planning corrective actions to reduce water inflow rates to the quarry. Explain rationales supporting your study plan design.

14.2   A home has a basement excavated into fractured sandstone. Water is seeping through the basement wall and high fecal coliform numbers suggest that the water is largely sewage. There is a sump pump in the basement that discharges a relatively constant 100 gal of water per day. A municipal sewer line is within 100 ft of the house and at a higher elevation than the portion of the basement wall where seepage occurs. The city engineer believes the source of the problem is leakage from the homeowner's sewer line. The homeowner believes the problem is due to leakage from the municipal sewer line. Design a tracer test that will address this dispute and explain rationales supporting your study plan design.

14.3   You are designing a groundwater tracing study for a facility located adjacent to a stream with a typical flow rate of about 3 cfs. Both the facility and the stream are underlain by alluvium. The facility manager is concerned that dye from the tracing work will color the stream and create undesirable publicity. List at least three actions you can take in the study design to minimize the likelihood of the stream being visibly colored.

14.4   Well-designed tracer tests commonly sample with both activated carbon samplers and water samples. Water samples are typically collected, each time carbon samplers are collected, and new samplers are placed. List three reasons that primary sampling reliance is usually based on data from activated carbon samplers rather than from water samples. List two reasons that analysis of selected water samples is also valuable.

14.10   References

Aley, T., 1978. Ozark Hydrology: A Predictive Model. *Missouri Speleology*, Vol. 18, pp. 185.

Aley, T., 1997. Groundwater Tracing in the Epikarst. In *Proceedings of the 6th Multidisciplinary Conference on Sinkholes and the Engineering and Environmental Impacts of Karst*. Balkema, Rotterdam, pp. 207–211.

Aley, T., 2002. *The Ozark Underground Laboratory's Groundwater Tracing Handbook*. Ozark Underground Laboratory, Protem, MO. 35 pp. Available online at www.ozarkundergroundlab.com.

Aley, T., 2016. Using Activated Carbon Samplers to Improve Detections of Fluorescent Tracer Dyes in Groundwater Remediation Studies. In *Tenth International Conference on Remediation of Chlorinated and Recalcitrant Compounds*, Palm Springs, CA. ISBN 978-0-9964071-1-3, Battelle Memorial Institute, Paper A-056, 10 pp.

Aley, T., 2017. Improving the Detection of Fluorescent Tracer Dyes in Groundwater Investigations. *Remediation: The Journal of Environmental Cleanup Costs, Technologies and Techniques*, Vol. 27, pp. 39–46.

Behrens, H., 1986. Water Tracer Chemistry: A Factor Determining Performance and Analytics of Tracers. In *Proceedings of the 5th International Symposium on Water Tracing*. Institute of Geology and Mineral Exploration, Athens, pp. 121–133.

Benson, R.C., and Yuhr, L.B., 2016. *Site Characterization in Karst and Pseudokarst Terraines*. Springer, Dordrecht, pp. 295–306.

Field, M.S., Wilhelm, R.G., Quinlan, J.F., and Aley, T.J., 1995. An Assessment of the Potential Adverse Properties of Fluorescent Tracer Dyes Used for Groundwater Tracing. In *Environmental Monitoring and Assessment*, Vol. 38. Kluwer Academic Publishers, Dordrecht, pp. 75–96.

Ford, D.C., and Williams, P.W., 1989. *Karst Geomorphology and Hydrology*. Unwin Hyman, London, 601 pp.

Hauwert, N.M., Johns, D.A., Sansome, J.W., and Aley, T.J., 2002. Groundwater Tracing of the Barton Springs Edwards Aquifer, Travis and Hays Counties, Texas. *Gulf Coast Association of Geological Societies Transactions*, Vol. 52, pp. 377–384.

Hauwert, N.M., Sansom, J.W., Johns, D.A., and Aley, T.J., 2004. *Groundwater Tracing Study of the Barton Springs Segment of the Edwards Aquifer, Southern Travis and Northern Hays Counties, Texas*. Barton Springs/Edwards Aquifer Conservation District and City of Austin Watershed Protection and Development Review Department. 112 pp. + appendix materials.

Hotzl, H., and Werner, A., eds., 1992. Tracer Hydrology. In *Proceedings of the 6th International Symposium on Water Tracing*, Karlsruhe, Germany. Balkema, Rotterdam, pp. 464.

Kass, W., 1998. *Tracing Technique in Geohydrology* (translated from German). Balkema. Rotterdam, 581 pp.

Kranjc, A., ed., 1997. Tracer Hydrology 97. In *Proceedings of the 7th International Symposium on Water Tracing*, Portoroz, Slovenia. Balkema, Rotterdam, pp. 450.

Sabatini, D.A., 1989. *Sorption and Transport of Atrazine, Alachlor, and Fluorescent Dyes in Alluvial Aquifer Sands*. PhD Dissertation, Iowa State University, Ames, IA, 216 pp.

Smart, P.L., 1984. A Review of the Toxicity of Twelve Fluorescent Dyes Used for Water Tracing. *National Speleological Society Bulletin*, Vol. 46, pp. 21–33.

Smart, P.L., and Laidlaw, I.M.S., 1977. An Evaluation of Some Fluorescent Dyes for Water Tracing. *Water Resources Research*, Vol. 13, No. 1, pp. 15–33.

White, K.A., Aley, T., Cobb, M.K., Weikel, E.O., and Beeman, S.L., 2015. Tracer Studies Conducted Nearly Two Decades Apart Elucidate Groundwater Movement through a Karst Aquifer in the Frederick Valley of Maryland. In *Proceeding of Sinkholes and Engineering and Environmental Impacts of Karst; 14th Multidisciplinary Conference*, Rochester, MN. National Cave and Karst Research Institute Symposium 5, pp. 101–112.

CHAPTER 15

Drilling and
Well Completion

P erhaps one of the more common tasks a hydrogeologist will be involved with is obtaining subsurface information, which can be found through a variety of methods, including geophysical methods, hand tools, and drilling methods. This chapter presents the common drilling methods and explains how monitoring and production wells are installed and completed. A hydrogeologist does not need to know how to run a drill rig but should be familiar with drilling methods, terms, and well-completion strategies so that meaningful subsurface information can be obtained. Also, the hydrogeologist, with some guidance, will be better informed to make decisions in the field.

Hydrogeologists need to be able to work safely with the drill crews in obtaining subsurface information. It is helpful to know the basics of drilling methodologies, how to describe drill cuttings, and what is involved in well completion. The approach taken here is to familiarize entry-level hydrogeologists or professionals unfamiliar with drilling operations with what they should know to safely proceed.

Understanding the geology of an area (Chapter 2) is helpful in knowing which drilling methodologies would be most productive and result in obtaining the best subsurface information. Examples of drilling in different geologic settings are presented throughout to show applications of the appropriate drilling method in a given geologic setting.

15.1   Getting Along with Drillers

Drillers are an interesting breed and can make your life highly productive and successful or exceedingly miserable. The biggest factor, generally, is with you and your attitude. A friendly, helpful, congenial, useful attitude (Boothman 2000) on your part can go a long way to getting the best information possible. During any given day, things can and will go wrong, but it is up to you to decide how you will let your circumstances govern your actions.

> **Example 15.1** If you have little experience working with drillers the following experience, related by Tom Aley (Chapter 14), is a good illustration on the proper attitude to have. This is what I recall Tom had to say:
>
>   "At my first experience working with drillers on a drilling project, I figured I didn't know what I was doing and needed their help, so I bought a dozen donuts and a pot of coffee and brought these

to the site. Then I said, "Guys I don't know much about drilling, maybe you can help me get the job done. We got along great and I had a good experience."

This illustration shows how having a useful attitude can go a long way to a win-win situation.

Drillers are generally very professional and knowledgeable about what they do. They know when "first" water has been found or what the conditions are like at depth (Figure 15.1). Nothing is more irritating to them than to have a young inexperienced person tell them how to do their job. If you get on their bad side, watch out. It has been the author's experience from being involved with a variety of drilling conditions in various parts of the world that disgruntled drillers do unpleasant things to smart-aleck geologists or hydrogeologists. For example, while you are away, your briefcase may be rearranged, your vehicle may be sabotaged, or your person may be adorned with lubrication grease. It is better to be polite, act interested, and ask questions: for example, Did we just go through a gravel layer? Are we making a bit more water? Was that last drill rod 120 ft or 140 ft?

There is a fine line between the drillers making footage and your obtaining complete information. It is your responsibility to obtain good subsurface information.

FIGURE 15.1 Discussing drilling conditions during monitoring well installation. (*Photo courtesy of O'Keefe Drilling.*)

If the drillers are going too fast, you need to politely ask them to slow down and probably have another morning meeting with donuts and coffee. How the job is bid is an important factor. Is payment by the job, by the foot, or by the hour? Being cognizant that a job needs to be done and that drillers need to make a living, too, leads to a cooperative relationship. The bottom line, however, is that you, the hydrogeologist, are responsible for the subsurface information.

Other factors that help or hinder a working relationship are the little things. Do you collect your drill cuttings and then sit in the vehicle until the next drill rod is ready to go in? Or do you always sit in your vehicle while the drill rods are being coupled together and going into the hole (**"tripping in"**) or being pulled out of the ground (**"tripping out"**)? Simple gestures like helping shovel cuttings out of the way, helping guide a **sand line**, or retrieving a tool helps the process run smoother and contributes to a positive working relationship (Figure 15.2). While the driller helper has gone for water do you assist or stand around? This is something each professional has to decide for him- or herself but can make a big difference in how a job gets done. Company policies may address these issues but being helpful is a good way to be.

Sometimes breakdowns occur, the equipment needs attention, or something may be inadvertently dropped down the hole and a "fishing" expedition is under way (see Example 15.12; Section 15.6). In some cases, there is nothing you can do but stay out of the way. Being patient and understanding that "things happen" is more helpful than yelling at your help.

FIGURE 15.2 Driller and geologist in conference. Initial morning conferences with donuts and coffee set a congenial working relationship. (*Photo courtesy of O'Keefe Drilling.*)

15.2   Rig Safety

In the above discussion, no information is required at your peril. Always think safety. There are few pieces of equipment with more moving parts than a drill rig. If you are within the mast length of the drill rig, then you need to be checking and looking around. You need to look up as often as you check your rearview mirror while driving a vehicle (every few seconds) because a hard hat cannot save you from falling drill pipe. Equipment can break loose, cables can break, and conditions can change in a heartbeat. Always have a sense of caution and safety in mind.

The minimum appropriate attire at the drill site is steel-toed shoes, a hard hat, gloves, and safety glasses (Figure 15.3). If you are working at a hazardous waste site, other protective clothing will be required. Loose clothing is susceptible to becoming snagged by a rotating machine, pulling you in or getting torn up.

The safest place to stand near a drill rig is on the driller's side (Figure 15.4). The driller is in charge of the controls and is the most knowledgeable person about what is going on. He or she is the person you need to have an ongoing dialog with and will be the most helpful in answering your questions. The "helper" side is generally the direction where more things fall and more moving items are located. The helper's job is to help make connections, place pipe into position, or place "slips," wrenches, or other equipment in place.

If you want to help, observe what the helper does and then ask at the appropriate time. Never put your hands, fingers, or other items where you can be pinched, trapped, or will compromise safety. No drillhole is worth an injury. If you don't feel comfortable helping, then stand back and get out of the way. Another obvious, but important point is, there is a fair amount of welding and grinding at times. Do not stare at the bright

FIGURE 15.3   Minimum safety equipment: hard hat, steel-toed boots, safety glasses, and gloves. (*Photo courtesy of O'Keefe Drilling.*)

FIGURE 15.4   The driller's side of the rig is the safest place to stand. (*Photo courtesy of O'Keefe Drilling.*)

lights; it can damage your eyes (Figure 15.5). Grinding wheels can throw hot, sharp metals pieces, so please stand back.

If you need to write something down that will take longer than the few seconds needed to keep from looking up, move well away to the driller's side or walk over to your vehicle. If you have to write something down that will take some time and it is critical to the project, tell the drillers and have them wait. You may miss something important if they continue. Work methodically, conscientiously, and safely.

FIGURE 15.5   Welding on another section of casing. Never stare at the bright lights or this will damage your eyes.

FIGURE 15.6   Drill site with spray-painted utility lines marked on the surface (arrows pointing). You must call 48 h in advance, so plan ahead.

Remember, your vehicle should have been parked farther away than the length of the drill mast. The best thing to do is ask the driller where to park because he or she may have some maneuvering plans for the water truck or other equipment.

Before you drill, it is a good idea to call a toll-free number to have the utility lines located. This has to be arranged 48 h in advance, so plan ahead. A person will be dispatched to locate water (marked blue), power (marked red), gas (marked yellow), phone, or TV cable (marked green or black) in spray paint in the ground surface (Figure 15.6). Contacting the local courthouse or city government can be helpful in locating sewer or other utility lines. Installing monitoring wells and making a spark while encountering a gas line during drilling can make for an unsafe interesting situation.

Example 15.2   An elementary school in Butte, Montana, was having basement flooding problems year after year, each spring. School district maintenance officials were interested in knowing why they had a problem and not just how to solve the problem. The biggest problem appeared to be that a 72-in (1.8-m) culvert was directing snowmelt and storm runoff water from a drainage originating at the Continental Divide under an interstate highway toward the school.

Another significant problem was that the school was located on fill materials in a floodplain (Figure 15.7). Surface water and groundwater would move toward the school and back up against the fill materials and basement wall as the water table rose each spring. By summer time, the groundwater levels would drop more than 10 ft (3 m) below the basement slab. A dewatering system was designed to alleviate the problem until the various parties of county and state could work out disputes on how to correct the surface drainage issues.

As part of the investigation, a series of monitoring wells were installed. Before drilling, the toll-free telephone number was called and the utility lines were located (Figure 15.6). While drilling one hole near the front of the school, a light-colored powder came up with the cuttings at a depth of 7 ft (2.1 m). Smelling the power revealed that we were drilling into concrete. A decision was made to abandon the hole, as it was thought that it might be an unmarked sewer line. Later on, it was discovered from an old-timer that it was more likely that drilling was boring into a concrete block that was part of the fill material. This illustrates the need to use all your senses, be cautious, and make good use of the available resources in finding out information about a site.

FIGURE 15.7 Elementary school site (4-pod building at bottom center) in Butte, Montana, built on fill placed in the middle of a drainage. See Example 15.2 on basement flooding. (*Photo courtesy of Hugh Dresser.*)

Drill rigs make good lightning rods. If stormy conditions are possible, keep a wary eye open. Often, you can watch clouds approaching from a particular direction. The sound of thunder is a sure sign that lightning is not far away. A simple conversation with the driller will result in a mutual waiting period for the storm to pass over. Drill rig masts can also extend up into power lines. In conversations with other drillers, it is the author's understanding that a drill rig should be more than 15 to 30 ft (4.6 to 9.1 m) away from power lines or arcing may occur. A phone call to a power utility may result in it placing insulated covers over the power lines so that drilling can proceed safely. Ask your driller what he or she would suggest.

Drilling operations attract people. Drilling in remote areas often draws curious onlookers (Figure 15.8). Flag off your safety area so that onlookers are kept back. People getting too close can slow down working operations and increase the potential for hazards. A smile and firm demeanor usually gets the message across. One of the biggest problems drillers encounter is that clients often do not follow the safety rules. As a hydrogeologist on site, make sure the clients are wearing the minimum safety equipment or politely tell them to back over to a safer place.

Summary of Safety Points

- Have a useful attitude.
- Always look up, as often as you would check a rearview mirror.
- Park your vehicle a full drill mast length away from the rig or ask the driller where to park, as he or she may have a particular method of maneuvering equipment.
- Stand on the driller's side of the rig.

Figure 15.8   Curious children from the People's Republic of China. Flag or rope off your safety zone to keep curious onlookers safely back.

- If you have to write a long time (more than approximately 3 to 5 seconds), move away from the rig to the driller's side or ask the drillers to wait a moment while you write things down at your vehicle.
- Don't stare at the welding process.
- Never drill without locating utility lines or identifying other hazards, such as power lines.
- Thunder usually means lightning, so stop what you are doing until this danger has passed over.
- Use caution when drilling near power lines and make sure the mast is far enough away.
- Rope or flag off your area to keep curious onlookers back.
- Prior to donning personal protective equipment (PPE) or respirators, work out a communication protocol, so that signals are clear.
- Never, never, never compromise safety, it can be fatal.

Other Considerations

Common sense is an important part of your tool box. Most of the time it is the hydrogeologist's responsibility to tell the driller where to drill. The direction of the wind is a significant factor. If you have a serious wind blowing from the front of the rig toward the back, you are going to eat the dust and cuttings swirling around you. This is especially unpleasant when drilling with air and a forward rotary rig is drilling in a coal bed.

Have plenty of layers of clothing to put on or take off. Nothing is more miserable than being too cold or too hot while you are trying to collect data. Your hands do not write well when the fingers are stiff and your lips are blue. At the same time, dripping perspiration on your log book can damage what you have written down. Layers allow

FIGURE 15.9   Performing field tasks dressed out in level "A" personal protective equipment (PPE). Field tasks with PPE can affect your performance and pose a risk of heat stress.

flexibility in staying comfortable. A person too cold or too hot is more inclined to make mistakes. This affects safety and the quality of information collected.

Drilling near hazardous waste sites and other conditions require an extra dose of common sense. Dressing out in level "A" PPE can affect your performance (Figure 15.9). Heat stroke is often as dangerous as the hazards you are being protected from. Communication and frequent breaks will often be necessary. A particular communication protocol needs to be worked out in advance with all parties involved, along with step-by-step procedures before drilling begins. Sometimes methane or other gases may be present in certain situations. In this case, keeping a detection device operating to identify these gases will be necessary. Safety should never be inappropriately compromised.

Example 15.3   In eastern Montana, a number of coal mines are operating in the Fort Union Formation (Chapter 2). Drilling is part of the ongoing exploration and production process. During a very cold day −10°F (−23°C), a propane torch was being used to thaw out some of the fluid lines on the rig. The methane gas detector was inadvertently shut off because of the use of propane. This was a coal well, and methane gas was emitting from the well. The torch ignited the rising methane and the drill rig burned for 3 days before it could be pulled off the drill site!

It is not the intention of this section to provide a treatise on safety; however, it is hoped that pointing out several of the more common hazards found near drill rigs may prove helpful. To obtain more information and discussion about the principles of safety, the reader may consult Spellman and Whiting (1999) and https://www.keystoneenergytools.com/5-rig-safety-precautions-you-need-to-follow.

15.3   Drilling Methods

A variety of drilling methods have been developed over time to account for the many geologic conditions. Some formations are very hard, such as granite, while others are soft, like shale, or unconsolidated, such as sand and gravel found in an alluvial setting.

Drilling and subsequent well completion should be considered together when deciding which drilling method is most appropriate. Some methods may be faster than others but may result in disturbing the aquifer to the point of reducing yield. For example, if the production zone is completed in unconsolidated sediments, high-pressure drilling fluids may actually cause damage to the aquifer (disturb existing natural packing by suspending and mixing grains). In this case, production wells need to be drilled using a slower, less-disruptive method so that production-zone disturbance is minimized. This chapter explores the most common truck-mounted drilling methods. For a more complete discussion of drilling applications in the water-well industry, the reader is referred to Driscoll (1986) and https://denr.sd.gov/des/gw/Spills/Handbook/SOP9.pdf.

Monitoring wells provide a point of access to the aquifer. They need to be constructed so that the hydraulic head and water-quality information collected is representative of the aquifer under investigation. Again, the drilling method will depend on the geologic conditions. The well-completion materials need to be appropriate for the water quality of the site, which may include such factors as pH, presence of organics, water temperature, or depth.

Cable-Tool Method

Perhaps one of the oldest drilling methods used is a percussion approach known as the cable-tool method. These technologies were developed by the Chinese some 4,000 years ago (Driscoll 1986; WelldrillingSchool.com). In this drilling method, a heavy drill bit is repeatedly lifted and dropped by a walking beam or spudding beam attached to a spooled cable via a pitman arm (Figure 15.10). The motion of the bit is up and down in

Figure 15.10 Cable-tool drill rig. The spud beam or rocker arm is angled slightly upward toward the back of the drill rig (uppermost beam).

a stroke-like fashion, so that the bit strikes the bottom of the hole each time it is dropped. The spudding beam is also the reason they are known as "spudder" rigs, although the author also knows of the spudding beam being referred to as a rocker arm (from the up-and-down rocking motion).

The lifting and dropping of the drill tool results in the breaking up and crushing of geologic materials. Drilling continues until the drill tool advances a few feet (1 m) beyond the end of the casing. It is the author's experience that the casing exhibits a "bouncing" action while geologic materials are still inside the casings. Once the bit reaches the end of the casing, the bouncing stops and the drilling sound becomes more like a thud. The drilling action is usually followed by driving casing down the hole as the drilling proceeds.

The casing is driven by means of a "pounding plate" or drive clamp is bolted to the drill tool (Figure 15.11). The stroking action of the rocker arm lifts and drops the drill tool causing the "pounding plate" to strike the casing with a loud clang! The clanging continues until the casing has been driven a few feet (1 m) *below* where drilling ceased. During this part of the operation, it is advised to have hearing protection. It is an effective drilling method in unconsolidated and softer materials, but most effective in brittle formations, such as shales and limestone. It can also be used in hard formations but may be *very* slow. The well casing associated with this method usually ranges from 4 in (10.1 cm) to 24 in (0.61 m), with depth capacities up to 2,000 ft (609 m) (NGWA 1998).

After the materials are broken up or crushed, they are bailed to the surface via a separate cable and spool. If the hole is dry, the bailing operation is done by first pouring a couple of 5-gal (18.9 L) buckets of water down the hole or running a hose attached to a water truck to facilitate mixing with the cuttings. A heavy bailer equipped with a closing valve is lowered to the bottom of the hole through an auxiliary cable. One common name for an auxiliary cable is the **sand line**. The bailer is a

FIGURE 15.11 A drive or "pounding" plate which is bolted to the drilling tool used to drive the casing. Laying beside it (left) is the wrench used to tighten or loosen the plate.

Figure 15.12   Dart valve bailer next to the 5-gallon buckets at a cable-tool drill rig. The driller helper is arranging the buckets to be filled with water and dumped down the hole to mix with drill cuttings. The mixture is retrieved with the bailer.

long tube with an opening/closing valve (Figure 15.12), a **dart valve** bailer is an example. When the bailer is lowered to the bottom of the hole, the dart valve is compressed and the water/cuttings mixture gushes into the bailer. When the bailer is lifted the valve spring closes, thus trapping the mixture. The sand line retrieves the bailer where it is brought to the land surface. This can be lowered into a discharge chamber. When this is done, the dart valve once again compresses, and the water/cuttings mixture is released out through the discharge chamber (Figure 15.13). The cuttings can be directed away from the working area by digging a shallow trench.

Once the hole has been drilled a few feet (a meter or so), casing is pounded or driven to the bottom of the hole to keep the hole from sloughing (as previously described). Depending on the casing and tool-handling capabilities of the rig, the length of casing driven or added varies. The spudding wheel turns proportionally to the drilling rate. One can observe whether the formations are hard or soft by observing the spudding wheel. It has been the author's experience for smaller diameter wells (6 to 8 in, 15 to 20 cm) that 10 ft (3 m) of casing is welded on each time. This results in an overall drilling rate of approximately 10 ft (3 m) per hour. Rates may vary significantly, but this rate is given as a comparison with other methods.

The cable-tool method is ideal when time is not of the essence, the geologic materials are relatively soft or brittle, and the formation disturbance needs to be minimized. Attempting to drill in hard materials (granitic or metamorphic rocks; Chapter 2) may result in drilling rates too slow to be cost effective, unless the rig is capable of handing a much larger and heavier bit. Cable-tool rigs are also useful in drilling shallow, larger-diameter wells, 18 to 24 in (0.46 to 0.61 m) (Figure 15.14), or for larger-diameter holes used for construction supports, such as caissons.

The author has found this drilling method is especially useful in drilling unconsolidated materials that may contain **heaving sands**. Heaving sands are loosely

FIGURE **15.13**   Discharge chamber and shallow trench to direct cuttings away from the drilling.

consolidated sand zones with relatively high pore-water pressures (Chapter 4), found just beneath a confining layer (Chapter 3). When the drillhole breaches into one of these zones, the sand and water gushes into the casing. The remedy is to drill out the intruding sand and attempt to proceed with the hole. This is very disruptive to the formation and is exacerbated using the forward-rotary drilling method.

FIGURE **15.14**   Cable-tool drilling and the installation of 24-in diameter casing. Used by permission from Holt Services Inc. (www.holtservicesinc.com)

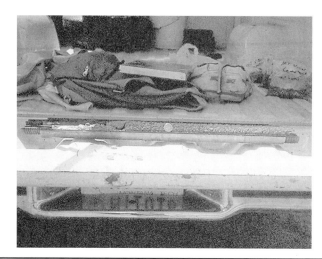

Figure 15.15   Split-spoon core sediments from Example 15.4. Weathered granite cuttings about 2 to 5 mm (U.S. quarter for scale).

Example 15.4   An irrigation well capable of 250 gpm was desired for irrigating a cemetery in Butte, Montana. The sediments consisted of unconsolidated and weathered-granitic materials, alternating beds of fine gravel, sand, and clay (Figure 15.15). The clay beds confined or semiconfined some of the sandy zones (Chapter 6), thus resulting in elevated pore-water pressure zones. Recharge to the sandy zones were from elevations sufficiently high to produce heaving sands or sands with exceedingly high confining pore pressures. These were usually found immediately under a clay horizon. The cable-tool method was used to drill an 8-in (20 cm) cased well. Heaving sand events occurred on several occasions while drilling the 170-ft (52-m) well. The sands would intrude approximately 15 to 20 ft (4.6 to 6.1 m) into the casing each time, requiring that the intruded sands be redrilled out of the casing before continuing. The fine-gravel zones were stable enough to complete a production zone in the well.

Forward (Direct) Rotary Method

The forward rotary or direct rotary method is perhaps the most versatile drilling method for smaller-diameter wells (<12-in diameter wells). Greater drilling depths can be achieved using this method over other methods, and it is well suited for drilling in harder geologic materials. Part of the versatility comes from the variety of drill bits and drilling additives that enhance the drilling environment.

In the forward rotary method, a bit and drill-pipe string are rotated by means of a power-drive system. The power drive can be at the top of the drill string (a top-head drive) (Figure 15.16) or near the working level of the drillers, a table-drive system.

In a table-drive system, at the top of the drill string is a **Kelly** bar, which is turned inside the table. The Kelly slides through drive bushings in the rotary table. When these are in their proper place, the table turns the Kelly bar and drill-string bit configuration and feeds it down hole (Driscoll 1986). The shape of the Kelly bar is often square or hexagonal and approximately 3 ft (1 m) longer than the drill pipe (Driscoll 1986). Connected to the bottom of the Kelly bar is a short (<1 m) removable section used to connect the first drill pipe to the Kelly. This removable section, often referred to as a "sub," is there to save on wear and tear of the threads at the end of the Kelly bar. A similar "sub" is used to connect the bit to the lowermost drill pipe. The drill string is

Figure 15.16 Forward rotary rig with top-head drive. Pictured is the power-drive system located at the top of the drill string.

lifted up the mast via a heavy cable with the top section laterally connected to a heavy chain collectively known as the **drawworks**. The driller can raise and lower the drill string via the drawworks while the Kelly bar slides past the Kelly bushings in the table. The chain moves in a circular fashion within the mast. The hydrogeologist usually observes the rate of drilling by observing chalk markings made on the chain by the driller or numbered increments made down the mast (Figure 15.17).

The diameter of the drill pipe is smaller than the diameter of the borehole being cut by the bit. The drill pipe is thick-walled but hollow enough to allow the passage of drilling fluids. Drilling fluids are injected via a hose connected to the top of the drill string from a pumping system. The fluids exit through ports in the bit that help lubricate it and keep it cool. The cuttings are forced up the annulus between the borehole wall and the drill pipe to the land surface, a concept known as **circulation** (Figure 15.18).

An air compressor on the rig is designed to produce sufficient airflow [in cubic feet per minute (cfm)] to lift the cuttings to the surface. When hole diameters exceed 12 in (0.3 m), most air compressors are not able to supply sufficient air volume to lift the cuttings all the way to the surface. This can be overcome by using an auxiliary compressor. Sometimes two drill rigs are "plumbed" side by side to use the combined strength of both compressors to maintain circulation or to complete 10- or 12-in-diameter (25 or 30 cm) wells. Wells with even larger diameters become impractical using this drilling method unless the drill rigs are very large; even then, maximums are reached at diameters of approximately 22 in (0.56 m) (Driscoll 1986).

During mud drilling, a pit is excavated [or a portable pit with baffles is used when cuttings must be removed from the site (Example 15.5)]. The pump intake hose is placed

FIGURE 15.17 Depth footages marked with chalk on steel casing to keep track of drilling depths. Markings on the mast of a forward rotary drill rig are another way to keep track.

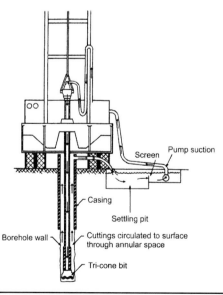

FIGURE 15.18 Direct or forward rotary circulation system. A pump draws drilling fluids from a pit and directs it down the drill pipe. The fluids cool the bit and are lifted back to the surface and discharged to the settling side of the pit. [*From NWWAA (1984).*]

on one side of the baffles where drilling additives can be easily mixed. The returning fluid and cuttings are discharged back into the opposite or "settling" side of the pit (Figure 15.18). The baffles filter or separate the cuttings from the intake side. When there is a continuous movement of fluids through the drill pipe to the bit and a return of fluids and cuttings up the annulus of the borehole to the land surface, circulation is "being maintained." The drilling fluids added during forward rotary drilling are designed to keep the circulation process going.

There is a hierarchy of drilling fluids used in order to increase hole stability. Stability can be defined as the ability of a drillhole to maintain circulation and lift cuttings to the surface during the drilling process. Once the cuttings fail to circulate and reach the surface other drilling fluids are added. The function of the drilling fluids is to create a **"wall cake"** (a lining of the borehole wall with fluids and some of the cuttings to keep it from caving or sloughing or to prevent drilling fluids from disappearing into the formation). Thus, the annular space between the drill pipe and borehole wall is kept open and cuttings can be lifted to the surface. The hierarchy of drilling fluids is as follows (with air being the least stable):

- Air
- Water
- Foam (a detergent)
- Stiff foam (foam with polymer added)
- Mud

Once a certain drilling fluid is used, there is no returning to a previously used fluid. For example, once a stiff foam is required to maintain borehole stability, the driller can't go back to using just water, or else the water would wash away or cut through any wall cake created by the stiff foam. The same thing can be said for each of the above-mentioned drilling fluids. Mud is the last effort to maintain hole stability. Once mud has to be used to achieve circulation, water or other fluids would cut through the wall cake and circulation would be lost once again. Drilling with mud should also be avoided to keep from sealing off production zones.

Mud drilling can be very successful but requires a constant monitoring of the viscosity (fluid thickness) of the fluid is performed using a Marsh funnel test (MF test). The MF test is where a quart of mixed drilling mud is poured through a screen into a 2:1 shaped 1-quart Marsh funnel, with the index finger covering the bottom drain hole (Figure 15.19). The finger is removed and the time it takes to drain the fluid *in seconds* is a measure of the fluid viscosity (Ofite.com, 1992).

Example 15.5 The Kalispell Montana Regional Hospital in northwestern Montana was desirous of saving electrical costs via a geothermal heat exchange system. Heat from the server room of their computer system and from the surgery operating rooms required a significant amount of electrical energy to keep the air-conditioning running. If they could reduce the temperature by 10°F (5.5°C) from these rooms in the hospital and heat-exchange it with the natural 50°F (10°C) water from the deep Kalispell aquifer it would be economically viable. This required the drilling of two production wells and two injection wells and installing a heat-exchanger system. Since the geological materials ranged from clayey sediments to unconsolidated boulder-sized gravels, it was decided to drill with mud. A large pit was excavated and lined to capture the cuttings while using the forward rotary drilling method (Figure 15.20). The viscosity of the drilling mud had to be monitored using an MF test to know whether the mud was of the correct consistency to keep the borehole wall from washing out

FIGURE 15.19 Drillers performing a Marsh funnel test to measure the viscosity of the drilling mud. See Example 15.5.

FIGURE 15.20 A large pit was excavated and lined to capture the cuttings while using the forward rotary drilling method described in Example 15.5.

Formation Materials	Size (mm)	Marsh Funnel Viscosity (Seconds/Quart)
Swelling clays	<0.004	35–40
Clays, silts, and fine sands	0.004–0.25	40–45
Medium sands	0.25–1.0	45–55
Coarse sands to fine gravels	1.0–4.0	55–65
Gravels	4–25	65–75
Very coarse gravels	>25	75–85

TABLE **15.1** Marsh Funnel Mud-Thickness Test Used in Example 15.5. A 1-Quart Marsh Funnel Is Filled with Mixed Drilling Mud. The Timing in Seconds It Takes to Drain Represents the Viscosity of the Fluid. Coarser-Grained Materials Require a Thicker Mix [Modified after Ofite.com (1992)]

(Figure 15.19). Table 15.1 was used as a guide to maintain the proper mud fluid thickness, depending on the materials being drilled through. When drilling in gravels, an MF test timing of 65 to 70 seconds proved adequate to maintain proper control of the borehole wall. This was a successful venture and all boreholes were successfully drilled and developed.

The hierarchy of drilling fluids is also directly correlated with the speed of drill-penetration rates. The author has observed holes drilled with air in softer layered sediments achieve drilling penetration rates of 100 ft/h (30 m/h). However, in unconsolidated sediments and in moisture-laden clays drilling with air is not viable. Unconsolidated sediments will collapse against the drill pipe and clays "ball up" and do not reach the surface. By adding water or mist, the circulation may once again be restored.

Another versatile thing about forward rotary drilling is the variety of bit types available for use. Two of the most common bits are the tri-cone and the drag bit. A tri-cone bit has bearings that turn as the bit is rotated. Each cone has patterned cutting teeth or is studded with bumps or buttons that are designed for the hardness of the geologic materials (Figures 15.21a and 15.21b). Generally, the shorter the teeth or buttons, the harder the materials. Short buttons may be studded with diamond or tungsten carbide while larger teeth may be constructed of hard-surfaced alloy steel or other materials resistant to abrasion. Drag bits are fashioned with nonrotating metal cutters that are used for rapid penetration, such as through soft shales. Ask your driller about bits and their function.

Forward rotary drilling is ideal for harder geologic materials and faster penetration scenarios. Although forward rotary drilling methods can drill through most geologic materials, there are at least four general conditions when this method is *not* recommended:

1. Rock materials that do not allow circulation, such as highly fractured basalts or karstic limestone (Chapters 2 and 12).

2. Soft unconsolidated materials with high confining pressures.

3. Zones that alternate from being soft to hard and then back to soft once again.

4. Large-diameter wells (>22 in or 0.56 m).

Figure 15.21a A series of drag bits used in soft formations. Drag bits are fixed and turn with the drill string.

Figure 15.21b In front are a variety of tri-cone bits that rotate on bearings and cut into the formation. The size of the teeth are inversely proportional to the hardness of the formation. The larger the teeth, the softer the formation.

In highly fractured rocks or limestones with large void spaces, drilling fluids do not remain in the borehole because they are "lost" out into the formation, even with mud. If there are heaving sands or other unstable conditions, forcing drilling fluids up the borehole may actually result in disturbing the formation enough to decrease the hydraulic conductivity of the intended aquifer (Chapter 3). Another problem occurs when the geologic zones alternate between hard and soft horizons. This is illustrated by Example 15.6.

Example 15.6 A family in southwestern Montana was interested in building a home in a topographic saddle between two hills. They called in a drilling company to get their well drilled. The drillers were hopeful they could complete the well before the upcoming Labor Day holiday weekend. A forward-rotary drilling method was employed. The drillers went through some fairly hard zones followed by soft zones. Worried that they might get the bit stuck in the hole, the drillers advised the property owners that if they continued drilling they would have to pay for the bit. Were they prepared to pay if it became stuck (approximately U.S. $5,000 at the time)? The owners said no. The drillers pulled off the well and left with no explanation.

Shortly thereafter the author received a call from the property owners seeking advice on what to do next. After receiving an explanation of the problem and being advised of the location of the property, the geology of the site was researched and drawn on a topographic map (Figure 15.22). The well site, chosen by the property owners, was near one of the hills to protect their home from the prevailing north winter winds. As it turned out, the geology of the hill near the well site consists of Cambrian sedimentary rocks and the saddle area between the two hills consists of basalt (Chapter 2). The author reasoned that the well was drilled into a "chill" zone near the contact between the older sedimentary and the younger igneous rocks. The significance of the chill-zone interpretation is that at the time of the lava flow the cooling rates near the sedimentary rocks was likely uneven resulting in basalt layers that alternated between soft, altered, and easily weathered zones and hard

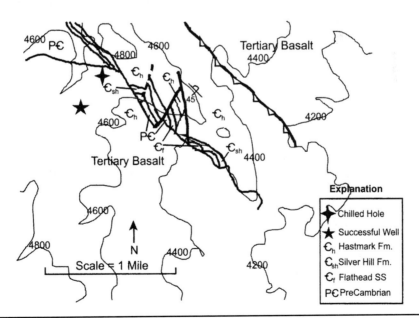

Figure 15.22 Surface geology and elevation contours from Example 15.6. Elevation contours have thinner font and thicker font represents geologic contacts of Cambrian age formations (Chapter 2). The chilled well and successful well are marked with star patterns.

Figure 15.23   Rock layering in igneous rocks; this case, layering is in granite from exfoliation.

unweathered zones. This was confirmed in the field by looking at the drill cuttings and other field evidence. (Figure 15.23 illustrates the type of layering that may occur in igneous rocks.) Under the hard ledges, the softer materials can accumulate and "crowd around" the drill pipe, closing off the diameter of the hole. When the drill string is pulled upward, the bit may then become wedged under one of the hard ledges.

The author suggested that if they could drill away from the "chill" zone, perhaps the rock would be brittle or hard all the way down and the hole diameter would remain uniform. Since an E-tape was not available (Chapter 4), the depth of the water table was estimated by dropping a small rock and timing when it hit water (Equation 15.1). This was repeated three times with consistent values (it took about 2.5 seconds for a rock to reach the water). The depth was estimated to be approximately 100 ft (30 m):

$$D = (v_o \times t) + (0.5 \times a \times t^2) \tag{15.1}$$

where v_o = is the starting velocity of the rock or zero (first term goes to zero)
   a = acceleration due to gravity (32.2 ft/s^2 or 9.81 m/s^2)
   t = time between the release of the rock and the time it reaches the water

With this information, we drove to the top of the hill to examine the property area. It was noticed that a linear drainage extended from the top of the saddle area and seemed to correspond with a fault in the geologic map (Chapter 2). By drilling near the fault, it was hoped that increased fracture permeability would be encountered. A site was recommended, and the author later sought more information on the thickness of the basalt. Fortunately, an oil well had been drilled within the property indicating that the basalt was approximately 400-ft thick (120 m) and underlain by lacustrine shales (Chapter 2). The author reasoned that recharge waters from precipitation would "pile up" on top of the shales and that a drillhole with a depth of 250 ft (76 m) or so would result in a thick enough production zone.

The property owners called another driller to perform the task. The new driller wouldn't drill at that high of an elevation unless he consulted with a water witch (Section 15.7). Interestingly enough, the "witcher" picked the same place that was recommended by the author (the site was flagged) and a successful well was drilled at a depth of 200 ft (61 m).

Reverse Circulation Drilling

Unlike forward rotary methods, where continuous circulation is required before drilling can continue, reverse circulation drilling avoids the problems of losing cuttings and drilling fluids into the formation by taking up all cuttings and drilling fluids up through the drill pipe. This is a reverse drilling process from forward rotary methods. Drilling fluids are introduced between the borehole wall and the drill pipe and then drawn up at the bit. The discharge hose is located at the top of the Kelly bar where water and cuttings may be diverted out through a cyclone and dropped to the ground or are directed back into a mud pit (Figure 15.24).

Perhaps the most common application of reverse circulation methods is for drilling larger-diameter wells from 24 to 72 in (0.61 to 1.8 m) (NGWA 1998). The cost for drilling larger-diameter wells is not much greater than for smaller-diameter ones in unconsolidated materials and the well-completion applications are very straightforward. A summary of useful applications for reverse circulation drilling of larger-diameter wells in unconsolidated or soft formations is given by Driscoll (1986) and WelldrillingSchool .com:

- This method causes very little disturbance of the natural porosity and permeability of formation materials, an important consideration for production zones when other methods may disrupt the aquifer characteristics.
- Large-diameter wells can be drilled quickly and economically.
- No casing is required during the drilling operation.
- Well screens are easily installed as part of the well-completion process.
- Borehole wall erosion is less likely because down-hole fluid velocities are low.

FIGURE 15.24 Author collecting cuttings from the bottom of a cyclone. Application is described in Example 15.16.

There are also some conditions that must be met in order for this method to be effective (Driscoll 1986; WelldrillingSchool.com):

- Formations should be soft sedimentary rocks with no large boulders. (Drawing boulders up through the center of the drill pipe is a problem.)

- The static water level (SWL) should be more than 10 ft (3 m) below the land surface to be able to maintain a sufficiently high hydrostatic pressure along the outside of the borehole. Problems also arise if the SWL is high and adequate water for drilling is lacking.

- Drilling using this method requires a high volume of available water with requirements ranging up to hundreds of gallons per minute (tens of liters per second). If this cannot be obtained, this method will not work well.

- Drill rigs, equipment, and components are very large and expensive and require more room to work.

Another reverse circulation drilling method that has very successful applications is known as the dual-wall reverse circulation method. Instead of fluids moving along the outside of the drilling pipe, there are two fluid passageways within the drill pipe (Figure 15.25). The drilling fluids pass down the outer passageway, and the cuttings and fluids are drawn up through the center passageway. The inner sleeve is sealed by an "O" ring. During this method, the only place the drilling fluids contact the formation is near the bit. This allows a continuous sampling of the formation with minimal disturbance and reduces the chances of any cross contamination.

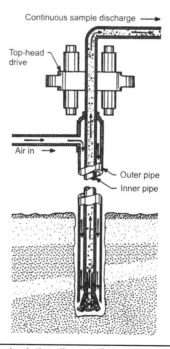

FIGURE 15.25 Dual-wall reverse circulation diagram. [*From Driscoll (1986).*]

The author first became aware of dual-wall reverse circulation drilling in placer gold exploration applications in the 1980s. Any cuttings within 3 in (7.6 cm) of the bit are drawn up through the pipe. Drilling recovery rates using this drilling method are in excess of 95%. Heavy minerals, like gold, would not likely make it to the surface using some other drilling method. This is the method of choice for exploration test drilling at drilling locations with heavy metals, although sonic drilling (discussed later on) has gained in popularity. Representative samples can be obtained using the dual-wall reverse circulation method. Although drill-pipe outer diameters (OD) range from 3½ to 9-5/8 in (8.9 to 24.4 cm), the most common drill pipe size is 4½ OD with a 2-1/5 ID (11.4 cm OD and 6.4 cm ID) (Driscoll 1986), which works well in exploration holes.

Casing Advancement Drilling Methods

In cable-tool drilling, casing advancement is an integral part of the drilling process. Two other casing advancement methods are discussed in this section: top-head forward rotary and wire-line drilling. Casing advancement methods are helpful when the borehole conditions are unstable or there may be a danger of contamination of materials from up-hole to down-hole. In top-head forward rotary methods, the casing driver is usually at the top of the mast. The drill pipe and casing are suspended together below. At the bottom of the first piece of casing is a thick forged drive shoe that is welded to the bottom to keep the casing from crimping as it is being driven. The drill bit and casing advance simultaneously or the drill bit may advance a few feet (a meter or so) ahead of the casing before it is driven.

The casing driver is a piston percussion-driven system activated by air pressure. The bit grinds up the cuttings, as in forward rotary drilling, and the cuttings are blown up through the annular space between the casing and the drill pipe to a discharge tube near the bottom of the casing driver (Figure 15.26). When the bit advances ahead of the casing, the bit can be retracted inside the casing before driving the casing. There are many advantages to this method over traditional forward rotary methods in spite of the additional cost of the casing hammer and the noise of operation. Most drillers who use this

Figure 15.26   Students collecting cuttings from the discharge tube. The drilling fluid is foam.

FIGURE 15.27   Example of hearing protection on hard hats; in this image the driller is measuring flow rate by the time it takes to fill a 5-gal bucket.

method wear hearing protection attached to their hard hats (Figure 15.27). It is also advisable for the hydrogeologist logging the hole to wear hearing protection.

Several advantages of using the top-head casing-driver method have been summarized by Driscoll (1986):

- Unstable formations that would result in too much disturbance using forward-rotary methods can now be attempted with a good rate of success.
- The borehole is stable through the entire drilling operation.
- Penetration rates are rapid in most drilling environments.
- Loss of circulation is no longer a problem since cuttings are lifted by air up through the casing.
- Specific zones can be tested for water production and sampling without "up-hole" contamination.
- This method can be used in all weather conditions.

In wire-line drilling, the casing also serves as the drill string (NGWA 1998). The bit is attached to the bottom of the first piece of casing and advances along with the drilling process. As the hole advances, new casing is added to the drill string (twisted on). All tools can be withdrawn upon completion of the hole. Wire-line drilling is ideal for continuous coring applications at depths where tripping-in or tripping-out can be cumbersome (>160 ft, 50 m). The core barrel (located inside) can be retrieved via a sand line by slinging a brass messenger down the sand line to the top of the core catcher. One engaged, the sand line can then be used to pull up the inner-sleeve core barrel to the surface. A new core barrel can subsequently be sent back to the bottom of the hole via the sand line and reattached to the core-catcher and bit already in place. In this way coring can occur continuously.

Figure 15.28 Wire-line drilling in eastern China. Canyons behind the onlookers reveal silty loess deposits 80 to 100 ft (25 to 30 m) thick. Runoff waters easily cut and erode these sediments, which provide large sediment loading to rivers.

Example 15.7 The author spent a couple of months in the People's Republic of China performing wire-line coring on a bituminous coal property a couple of hundred kilometers west of Beijing. The drilling was unstable because of having to drill through approximately 120 to 150 ft (35 to 45 m) of soft overlying formations before encountering stable layered sediments. The stable layered sediments, Pennsylvanian in age (Chapter 2), contained coal beds with thicknesses in excess of 50 ft (15 m). The geologic materials from the surface down to a depth of 80 to 100 ft (25 to 30 m) consisted of silty loess (wind-blown deposits; Figure 15.28). (Extensive loess deposits in eastern China easily erode and contribute sediments and abundant turbidity to the rivers, resulting in names like the Yellow River.) This was underlain by a few tens of feet (10 m) of unconsolidated fine gravel before reaching the first stable Pennsylvanian shale units. A surface casing was set to isolate the soft, unstable overlying materials before attempting any continuous wire-line coring. Coring efforts often extended downward in excess of 300 ft (100 m). The wire-line coring method was a very convenient way to obtain continuous core at fairly deep depths, below very unstable drilling conditions.

Auger Drilling

Two methods of auger drilling are presented, solid-stem and hollow-stem auger (HSA) drilling. The applications are more common for monitoring-well completion than for production wells, although shallow (approximately 100-ft or 30-m) production wells can be completed in soft formations using this method (NGWA 1998). The chief advantage of auger drilling is that it requires no drilling fluids when drilling the hole. This is especially significant for contamination sites that require monitoring wells. Drilling depths can reach 200 ft or more depending on the geologic formation and capabilities of the drill rig.

Solid-stem auger drilling is often used when the well can be completed "open hole." This means that all of the auger flights can be retrieved from the hole before installing the well casing. Solid-stem auger drilling can usually achieve greater drilling depths than HSAs because the shafts on auger flights are narrower; however, the hole diameters are also usually smaller. The auger flights are rotated via a rotary head drive with a hydraulic-feed mechanism that allows either downward pushing or upward pulling

(Driscoll 1986). The Kelly bar is hexagonal and is attached to the first flight with two rod bolts that are often turned "finger tight." As additional flights are added, they are usually tightened with a wrench or air-powered drill. As one driller put it "anything below ground gets wrenched."

As the augers turn, the geologic materials spiral upward to the surface. It is confusing at first to tell at which depth the cuttings came from, so it is a good idea to communicate with your driller. Cuttings depths can be correlated between the sound and "feel" of the drilling and the associated materials observed. An experienced driller will be able to tell you what you are in. For example, "We hit that gravel about a foot ago," even though the gravel may not be observed at the surface for another 10 ft (3 m) of drilling. Sometimes cuttings do not make it to the surface and cannot be sampled until the auger flights are pulled.

Many auger flights are 5 ft (1.5 m) long except for the first flight that has a cutter head attached, extending the length an additional foot or so (Figure 15.29). This, of course, depends on the diameter of hole being drilled and size of the drill rig, etc. The author's experience is primarily with smaller-diameter monitoring wells (<4 in or 0.1 m). Additional discussion on larger-diameter wells can be found in Driscoll (1986). Completing monitoring wells is discussed in Section 15.5.

Figure 15.29 First 5-ft auger flight with cutter head attached. Application is described in Example 15.2.

FIGURE 15.30 Truck-mounted direct-push drill rig. The setting of the application is described in Chapter 4, Example 4.9.

Direct-Push Methods

Another method of obtaining subsurface information with or without completing a well is through direct-push methods. The technology and expansion of direct-push subsurface tools has grown rapidly since the first edition of this book. There are a wider variety of drill-rig sizes, coring-length capabilities, well-installation diameters, and aquifer testing and sampling tools. Direct-push drill rigs are truck or van mounted (Figure 15.30; Example 4.9) or manufactured as tracking units with track widths as narrow as 12 in (0.3 m), which allows for easy maneuverability in smaller areas (Figure 15.31).

All units are equipped with a hydraulic system to push a 4-ft (1.2-m), 5-ft (1.5-m), or 6-ft (1.8-m) core barrel into the ground (Figure 15.31). A stainless-steel core tube captures the core [typically 1 and 7/8 in (4.8 cm)], inside a plastic tube. The tube can be laid on a core tray or back of a truck, where it can be sliced in half with a knife for inspection (Figure 15.32; Example 4.9) or may be sealed for laboratory soil-gas work. Coring systems can be readily altered to an auguring system in the field within minutes. This allows the flexibility of continuous coring followed by installing a 3/4-in, 1-in, 2-in, or 3-in (1.9-cm, 2.5-cm, 5.1-cm, or 7.6-cm) or larger diameter monitoring wells (Figure 15.33).

The strength in using these devices is in the ability to perform soil-gas surveys or delineating plumes with volatile organic carbon sources or to define bedrock in relatively shallow aquifer systems. There is a capability to extract soil gases directly into a gas chromatograph on the rig or pull a groundwater sample via a peristaltic pump, suction pump, or through the casing with a retractable tip (Figure 15.34). The upper tool

FIGURE 15.31   Track-mounted direct-push drill rig with track width varying between 12 and 24 in for easy maneuverability.

in Figure 15.34 shows the tip refracted to expose the stainless-steel screen for water sampling. A check-ball valve enhances the ability of lifting fluids from deeper depths. It is a relatively quick way to obtain several point samples in a short period of time with minimal surface impact, since no well completions are needed. This is ideal for investigative and site-evaluation work.

FIGURE 15.32   Plastic tubes from direct-push cores, cut in half for inspection. In this case one notes the 20-in sheen of diesel in the core. The setting of the application is described in Chapter 4, Example 4.9.

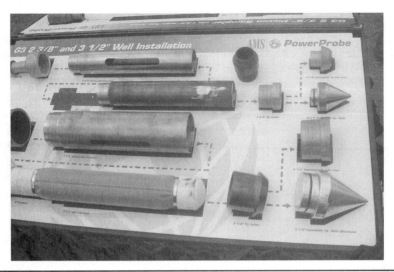

FIGURE 15.33 Array of well-completion and sampling tools for 2 3/8-in (6-cm) to 3½-in (8.9-cm) monitoring wells. See providers for options.

Direct-push methods are only effective in unconsolidated materials or softer sediments. Depth capacities are usually in the 20- to 50-ft (6.1- to 15.2-m) range, although drillers exceeding depths of 100 ft (30 m) are documented by Geoprobe Systems' 100 club (Geoprobe 1999).

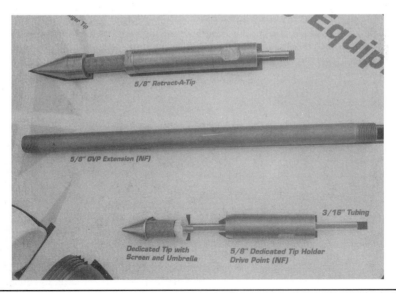

FIGURE 15.34 A series of direct-push down-hole tools. The upper tool is a retractable tip that exposes steel screen for water sampling. In the middle is an extension piece and the lower tool is a dedicated screened zone near the driving tip.

Sonic or Rotosonic Drilling

Sonic drilling is a relatively newer drilling method that is expanding its applications in the obtaining of subsurface information. Sonic drilling is especially known for being able to take continuous cores in unconsolidated sediments, including glacial till and alluvial deposits, and frozen and solid materials, because of its unique process of coring without drilling fluids.

The heart of the sonic system is in the rotation of two imbalanced counter rotating weights or rollers, which are driven by high-speed motors. During rotation, a high-frequency vibration (157 Hz) resonance takes place that is delivered directly to the drilling rods (SonicSampDrill.com). The synchronization of the rollers ensures that the entire length of the drill pipe becomes a penetrating tool with little friction between the drill rod and borehole interface (Midwest Drilling 2004; Precision Sampling 2006). While resonant vibration is occurring, the drill rods are also being rotated to improve the cutting effect. This explains why the name Rotosonic is also applied to this drill rig application. Coring depths in excess of 400 ft (122 m) are achieved with these drill rigs.

Mini-Sonic

A mini-sonic drill rig is approximately a 1/3-size version of the larger rigs (Figure 15.35a). The rig in Figure 15.35a has 18-in (0.46-m) wide tracks and also comes in 24-in (0.61-m) wide track models. Drilling can be vertical or angular and achieve depths of approximately 150 ft (46 m). The imbalanced counter weights are rotated at approximately 5,000 revolutions per minute (rpm), while the drill string can be rotated up to 100 rpm.

Drill rods are laid in a receiving unit and coupled mechanically (a safety measure), which keeps the helper away from the drill string (Figure 15.35b). In monitoring-well

Figure 15.35a   Track-mounted mini-sonic drill rig. It comes in 18-in (450 mm) wide and 24-in (600 mm) wide tracks (www.boartlongyear.com).

Figure 15.35b Coupling of additional drill rod on the mini-sonic drill rig. Coupling is done mechanically to increase safety during drilling.

installation applications a smaller-diameter drillhole can be followed with a larger-diameter casing to seal off the upper contaminated zone. Drilling inside the larger casing allows a "telescoped" drilling and well-completion method that can then proceed without any danger of carrying near-surface contamination downward.

Cone Penetrometer Testing

Another relatively newer application of obtaining subsurface information using an older technology is cone penetrometer testing or CPT. The CPT system is a truck-mounted system using a sensitive cone-shaped probe-head that is hydraulically pushed into unconsolidated sediments. The equipment includes the CPT probe, down-hole tools, and a computer data logger for data collection. The probe head is equipped with a low-pressure transducer, an inclinometer, a friction sleeve (containing strain-gauge load cells), and a pore-pressure transducer (Precision Sampling 2006). The standard cone has a surface area of 10 cm^2 and a tapered angle of 60°. The CPT system can use electrical and acoustical energy to transmit signals to a receiving unit at the surface. The resistance to penetration is continuously measured with depth and correlated with grain size to characterize soil types and identify zones of increased permeability for water sampling or production. An example of a CPT log is shown in Figure 15.36 (see Example 4.6).

Horizontal Drilling

Another older technology that has expanded rapidly since the 1990s is horizontal drilling. Horizontal drilling has been traditionally used in placing utility or small-diameter pipelines but new applications have been discovered in the environmental field.

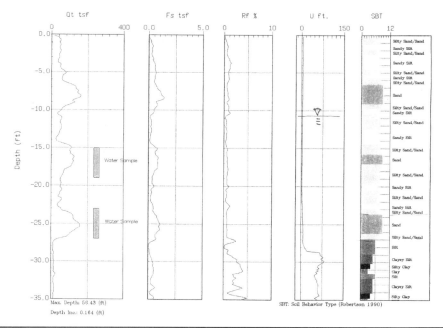

FIGURE 15.36   Information derived from a CTP log, showing relative resistance to cone penetration (water sampling intervals on the left) and analog lithologies and descriptions on the right.

Horizontal drilling has been found to work favorably for drilling and completing wells under permanent structures such as landing strips, airport fueling areas, roadways, tank farms, and other areas that would be problematic for remediation if there were a host of vertical wells.

Drilling is accomplished by rotating a disklike bit that is directed at a low angle to the surface. Control is maintained by rotating the disk in a continuous motion. Changes in the direction are made via the orientation of the bit. Often a "foot" person with a surface-detecting device works with the driller to identify exact positioning of the bit, particularly if one is drilling along existing utility lines (Figure 15.37). There are an increasing number of drilling companies that perform horizontal drilling services and would be happy to provide additional information and most have a web site. One printed source of information is the annual drill rigs issue of *Water Well Journal* (July issue of any given year; NGWA, 1999).

15.4   How to Log a Drillhole

Before newcomers begin logging subsurface information, it is helpful to be familiar with the different drilling methodologies described above and the section on rig safety before attempting to log your first hole. You need to know how the cuttings arrive at the surface and the objectives of the drilling project. In addition to logging a hole, you are also responsible to see that the drilling project is completed in a timely fashion. There needs to be a balance between making good footage for the driller and also obtaining the needed subsurface information. The hydrogeologist is responsible for the data that are collected.

FIGURE 15.37   Horizontal drilling rig performing utility line work under asphalt roadways.

It is typically your responsibility to locate where to drill. How will you decide? Will the locations be pre-staked? Were they located by someone else, which means that you have to be able to read a map and locate yourself? Do you know where the site is and where to begin? Is it necessary to call a utility line locator? Remember that you need 48 h advanced notice (Section 15.2). Are there any issues regarding personal protective clothing? These may sound like obvious or bewildering questions, but for someone new they are a significant source of stress and emphasizes the need for a predrilling meeting with the driller and other involved parties (Section 15.1).

In addition to knowing the drill-site locations (or enough locations to get started), there are a number of tasks that one can prepare before going into the field. Does your company have a standard form they use for drill logs? Maybe you are using a field book from which you will be transferring information to a standard form. Consult Chapter 1, Section 1.6, on field note taking. Will you be taking samples that will be stored? Will you collect them every 5 ft (1.5 m) or at a major lithology change? Will these be chip samples, core samples, or something else (Figure 15.38)? Did you come prepared with whatever containers, bags, or boxes are needed so the drillers don't have to wait for you?

Make a list and be sure you are prepared to go into the field. Some common items needed are the following:

- Site location map or Google Earth image
- A topographic map, if it is not included on the site map
- Geologic map, if possible, or at least have well logs from other wells in the vicinity
- Field book, well-log forms
- Hand lens, grain-size chart, acid bottle, or hardness tester

Figure 15.38 Chip samples organized into compartmentalized container by depth. Example context in Chapter, Example 2.8.

- Containers for samples, or materials needed for samples, core boxes, etc.
- Chain-of-custody information if needed for soil or water samples
- Writing utensils, with extra backup writing capability
- If using a laptop, extra batteries or a power source
- Appropriate clothing, hard hat, rain gear, etc.
- Drinking water and food
- Reading material for breakdowns or delays
- Cell phone or other communication device for emergencies

This is not an exhaustive list but serves to get one thinking. It will be essential to have a cursory idea of the site geology, even if you have never been to the site before. Which drilling method is most appropriate for the conditions? Is forward rotary better or auger drilling?

You will need to know how to tell how deep you are in the ground. When the driller sets up at the hole and raises the mast, he or she will level the rig. When the bit rests at the ground surface, this is ground surface or zero drilling depth on your log. If there are preexisting numbers painted down the mast, you need to note where the zero marking coincides. Another way to keep track of footage is to mark the drawworks chain, every 5 ft (1.5 m) or so with a bright-colored grease marker (the driller will do this for you). With casing advancement methods another way is to mark directly on the drill casing with chalk (Figure 15.16). It is best to ask the driller the most appropriate way to keep track of drilling footage.

Example 15.8 illustrates the dynamics and demands of logging a drillhole while using the forward rotary drilling method. Initially drilling is with air and then switched to water and then to foam.

Figure 15.39 Students describing drill cuttings during a field-camp exercise.

Example 15.8 The driller asks you if you are ready and the deafening sound from the air compressor is directed down through the Kelly into the drill pipe. Air begins gushing out the ports at the end of the bit and dust begins to blow. The driller engages the power-drive system and the rotary table begins to turn. The cuttings begin to blow around at the base of the drill rig and pile up. The driller helper begins to shovel the cuttings to channel the fluids and cuttings away from the back of the rig.

You have prefilled out most of the header of your logging form and stand near the driller with your sieve ready to catch cuttings. You do so frequently and look up few seconds to make sure you are safe and aware of what is going on. In your log book you write down descriptions for every 5 ft (1.5 m) or at major lithology changes. Generally, lithology changes more frequent than 3 ft (1 m) will require too much writing to keep up with the drilling penetration rates. You start piling representative cuttings sequentially on the ground to be able to refer back to them or place them in pre-marked bags (Figure 15.39). After 50 ft or so you notice the samples are getting moist and you ask, did we hit water yet?

The driller nods and then starts pumping water along with the air to maintain circulation. It will be necessary to fill a 5-gal bucket (19-L) from the back of the water truck to rinse off your cuttings now. What you observe in your sieve may represent a mixture of cuttings from the full length of the hole. What are we in? You communicate with the driller and get back on track. The hole seems to be making an increase in water so you ask the driller to stop at the end of the next joint (typically 20 ft each) and blow on it (only inject air) to see how much water the hole is making (Figure 15.27).

Another 50 ft (100 ft now) and the cuttings are not coming to the surface like before. The driller adds detergent to a mixing tank and begins to pump foam down the hole for added stability. It gets pretty messy around the base of the rig. The sieve fills up mostly with foam. You vigorously rinse the cuttings, and the hole-logging operation is a bit more challenging.

Example 15.8 presents some issues worth discussing in more detail. How do you know what you are drilling in when the cuttings represent a mixture from the total length of the hole? As you drill deeper it takes longer for cuttings to reach the surface. How does one tell what the conditions of the subsurface are like?

Describing the Cuttings

When drilling with air and at relatively shallow depths, cuttings arrive almost instantaneously at the surface. As drilling proceeds, cuttings from near the bit become mixed in with cuttings that slough off from "up-hole" as they are continuously blown to the surface. This yields a mixture of cuttings in your sieve. Generally, about half of what you see is from the current drilling depth. This needs to be evaluated along with the sound of drilling, drilling rate, and drilling depth. Shales and other softer materials will drill more quickly and quietly, while sandstone and limestone may be much louder and slower. Once again, consult with the driller in determining where a formation change took place or when you first drilled through fractured rock.

In describing cuttings, it is helpful to find a representative large piece in the sieve and break it in half to obtain a fresh surface to look at with your hand lens (Figure 15.40). Figure 15.40 shows a mixture of cutting materials from over 300 ft (100 m) of drillhole. In this example, what is mostly being observed are quartzite cuttings from gravels and cementing materials (silica cement and sand) between the quartzite clasts (Chapter 2). (Harder cuttings like these will not break easily in half.)

As you proceed a quick decision must be made when assessing the materials before writing something down. This can be stressful and requires immediate recognition of the lithology. However, fine-grained lithologies can be difficult to discriminate between. For example, it may become necessary to distinguish differences among mudstones, claystones, and siltstones. The author recommends a personally used field test, described in Table 15.2. Table 15.2 basically provides a quick way to decide what the relative percentage is of silt in fine-grained formation cuttings. Of course, don't try this if there is any danger of contaminated soils. It was primarily an exploration well-logging technique but can be very helpful.

Figure15.40 Example of cuttings taken from over 300 ft (100 m) of drilling. Observed are mostly a mixture of quartzite fragments from coarse gravels. The context for this example is described in Example 15.5.

% Silt	Lithology	Taste Test (Rub Against Teeth)
0–33% silt	Claystone or clay shale	Cuttings smear on teeth with little or no grit
33–67% silt	Mudstone or mud shale	Very gritty on teeth, grit observable
67–100% silt	Siltstone	Mostly grit observed

TABLE 15.2   Field Methodology for Distinguishing Finer-Grained Sediments Based upon Visual or Response on Rubbing a Chip Sample on the Teeth

Lag Time

The time it takes cuttings to reach the surface from the drill bit is known as the **lag time**. When drilling with air or foam, the fluids and air from the compressor must be able to lift all cuttings to the surface to maintain circulation. There is turbulence during this process and cuttings may take quite a while to reach the surface after exceeding depths of 300 ft (>100 m) or so. The best way to gauge the lag time is to notice a distinctive formation and glance at your watch when you first recognize it. Note the time and continue to log cuttings as they arrive at the surface. Coal or some other marker bed is especially helpful. When the cuttings you noted appear at the surface, look at your watch again; this is the lag time. At depths greater than 100 m, the sound and chatter of the rig will be as helpful as the actual observed cuttings. By being able to have a marker bed or two every so often, the drill log can be adjusted to match the lithology footage depths better. Adjustments can also be made on lithologic logs if geophysical methods are also available.

> **Example 15.9**   While performing coal exploration in Wyoming, each coal bed had a particular "signature." It was easy to tell when you were in coal because of the chatter of the drill rig and rapid penetration rate. When drilling at depths of 500 ft or so, the lag times ran from between 15 and 20 min. The time depended on the drillhole diameter, volume of air being injected in cubic feet per minute (cfm) by the compressor, and fluids being added.
>
> As another example, lag times for cuttings mixed with mud from a 1,000-ft hole at the Big Sky Montana ski resort area required 30 min or more to reach the surface.

How Much Water Is Being Made and Where Did It Come From?

One of the objectives of drilling is to know when you first hit water and which zones the water was likely coming from. It may be that there is a significant difference between "first" water and where the SWL is. Confining units may be keeping the water from rising to its unrestrained levels. You can learn a lot about the hydrogeology of an area by observing which zones produce water and where the SWL is each morning before you continue drilling.

You should make it a practice to take a level measurement (Chapter 4) in the borehole at the beginning of each day and after periods of breakdown, where the hole has sat unperturbed for a significant length of time. This along with your notations of where water-producing zones are all tell part of the story. If you do not have a water-level measuring device you can get an idea of where the water level is by reflecting a sunbeam down the well bore. On a clear day you can see down more than 80 ft (24 m) (Figure 15.41).

FIGURE 15.41 Driller observing depth to water using reflected light off a mirror. On a clear sunny day one can see downward more than 80 ft (24 m).

If you suspect that the hole is producing water, ask the driller to stop at the end of the next joint or rod (usually 20 ft in a truck mounted drill rig) and inject air only. After 5 to 10 min, the water being produced from the hole will *only* be coming from the formation. Realize that this may represent minimum production, since not all of the water makes it out of the hole. Some may be going out into the formation. It is the author's experience that a hole will produce approximately 25% to 30% more than what is being observed at the surface. This can have important implications for water production depending on whether water is being produced from the length of the hole or from a specific zone. This is significant because it is most likely the upper zones have been sealed off by the casing. Table 15.3 gives a rough idea of the difference between volumes blowing from the hole and what could be produced with a pump.

Table 15.3 may be important because if drilling has proceeded for a couple of hundred feet (66 m) and total production is on the order of 2 to 3 gpm (0.13 to 0.19 L/s) one must evaluate the risk associated with obtaining additional water production from deeper zones compared with the added drilling costs. In the United States, the federal housing authority (FHA) requires that a well produce 5 gpm (18.9 L/min) for at least 4 h (1,200 gallons) to be considered a viable water supply before granting approval for funding for a home mortgage (hudgov.prod.parature.com).

Production Observed (gpm)	Possible from Pump Production (gpm)
3–5	5+
5	7
7–8	10
10–12	12–15

TABLE 15.3 Production Estimates Based on Water Exiting the Hole. The Borehole Will Make More Water Using a Pump Than Lifting Water Out Blowing with Air

FIGURE 15.42 A driller helper guiding a bailer to a 5-gal (18.9-L) bucket to estimate well yield.

Estimating the volume takes practice and can be solved a variety of ways. If the volume of water can all be channeled through a single place, the chances are greater of getting an accurate estimate. Using a bucket and timer, a weir, or flume are common methods (Figure 15.27) (see also Chapter 10). Most experienced drillers can estimate flow rates just by looking. If flow rate is an important issue, attempts to measure the volumes accurately are warranted. Figure 15.42 shows drillers filling 5-gal (19 L) buckets in sequence after waiting 1 min to estimate yield. The production at the time was estimated to be 15 gpm (57 L/min).

15.5 Monitoring-Well Construction

Monitoring-well construction is ideally performed to provide a window into the general aquifer properties at a given completion depth. The goal is to complete the well in such a way that you can have confidence in the hydraulic-head information (Chapter 4)

and water-quality samples (Chapter 9. Monitoring wells are usually not designed to evaluate aquifer hydraulics, although they are often used to do so. They are usable as observation wells, but not designed as production wells. Production wells (Section 15.6) are designed to be pumped at a level that actually stresses the aquifer.

Objectives of a Groundwater Monitoring Program

Monitoring-well programs require significant thought and planning to be able to accomplish their intended task. Table 15.4 lists some of the objectives associated with monitoring-well designs.

From Table 15.4 it is apparent that the objectives must be well defined prior to achieving a successful program (Crumbling 2004). Merely placing wells haphazardly is not an effective use of budget funds (Chapter 5). Some of the objectives in Table 15.4 require that wells be placed with a certain component of randomness or the assumptions inherent in the statistical analysis methods to be used will be violated. It is also important to have a knowledge of some of the following aquifer properties:

- Are there multiple aquifers?
- Are there perched zones?
- Is there contamination in the vadose zone?
- Is there more than one flow system? Local and regional? (Chapter 10)
- Where are the recharge and discharge areas?
- What is the nature of a contaminant? What is its solubility or density? Will the contaminant be a floater or sinker? Will it dissolve, spread widely and easily, or be somewhat contained?

Objective	Discussion
Reconnaissance	This may be a regional study to see what is there. It may consist of a few sparsely placed wells. Shallow/deep pairs may be needed for head, vertical flow conditions, and water-quality data.
Fixed station	Long-term monitoring. Each station may have greater expense, such as a more secure access or a protective shelter.
Research	Temporary monitoring design that may have materials that need to be retrieved later on after research is completed.
Cause and effect	This is where monitoring is established to evaluate whether a process or procedure is effective. Usually there is a control area where the standard approaches or status quo is occurring, compared to a site where a treatment is being applied.
Compliance	These may be wells placed to make sure compliance requirements are being met according to some minimum or maximum standard.
Quality control	Monitoring may be occurring to check whether a process or remediation strategy is working properly.
Trend analysis	A monitoring system may be designed to determine whether a decreasing or increasing trend may be occurring.

TABLE 15.4 Monitoring System Types Based upon Objectives of the Analysis [After Weight (1989)]

All of these issues complicate the monitoring design, since it may be necessary for multiple sampling ports at various depths. Are the wells being placed in the groundwater-flow path and at the appropriate elevations to capture what is being monitored? Another consideration that seems to take precedence is what is the budget and how soon does the monitoring program need to be in place? Is the study being conducted "in-house" or is it under public scrutiny? All these issues tend to complicate the design process. In a public setting, there may be a negotiation process among regulators, the public, and the company, before the monitoring program can be implemented.

Since the first edition of this book, an abundance of good information regarding site characterization and monitoring well design has come forth (Neilsen et al. 2006; Sara 2006). In particular are discussions and techniques for multilevel monitoring-well completion using some newer materials that result in a significantly improved understanding of the three-dimensional nature and heterogeneity of natural systems (Einarson 2006; Neilsen and Schalla 2006). It is beyond the scope of this chapter to address all of these topics, but the intent is to give the newcomer tasked with monitoring-well installation a sense of the basics, before going into the field, and having a reference that is handy.

The method used to actually install monitoring wells depends on the geologic setting, the depth to be completed, diameter of borehole project goals, and so on. A common method for drilling and completing monitoring wells in unconsolidated materials is with HSAs, and a step-by-step procedure is presented next.

Installing a Groundwater Monitoring Well

This example will be for a 2-in (5-cm) monitoring well to be used for hydraulic-head measurements and water-quality samples (Figure 15.43):

1. The casing materials and supplies for building the well have been laid out near the well site or are easily accessible from a vehicle nearby.

2. The bottom of the bit is fitted with a knockout plate or a removable plug that fits flush with the bit. The knockout plate or plug functions to keep cuttings from passing up through the hollow stem and getting bound into the hole. This allows the cuttings to spiral upward along the outside of the augers, similar to a solid-stem auger system. Once the desired depth has been reached, the knockout plate can be dislodged so that the well can be constructed (Step 4). The materials for the knockout plate vary, but the author has seen from wood to polytetrafluoroethylene (PFTE, Teflon) used.

   A very important point will be made here. Before dislodging the knockout plate it is important to **load the hole**. There may be a difference between the water level inside the borehole and inside the hollow stem. "Loading the hole" means that water is added to the inside of the hollow stem to equalize the two levels. If the hole is not "loaded," then muddy water from the borehole gushes up into the hollow stem chamber. This may produce a "skin" along the inside of the borehole that is difficult to remove during well development (Chapter 5). This is such a small step but can make a great difference in well development. The author has seen wells become unusable by forgetting to "load the hole."

3. The drillhole penetrates through the first 5 ft (1.5 m). The helper loosens the two rod bolts, and the driller hoists up the Kelly and drive sleeve up the mast.

FIGURE 15.43 Students lifting 2-in well monitoring-well casing and screen into an HSA borehole.

The helper slings over the next auger flight, fitting it on top of the one in the ground (Figure 15.44). Two bolts are pneumatically wrenched into place to secure the flights together. The driller lowers the Kelly and drive sleeve over the new auger flight and the rod bolts are twisted on finger tight once again. The next 5 ft of drilling continues. As a hydrogeologist, you carefully describe the cuttings that spiral upward along the hollow stem. You record the information in the borehole log so that you know what all the critical depths are when constructing the well. To meet the objectives of this study, you decide to penetrate the aquifer an additional 10 ft (3 m) below the water table. Once there, you prepare to construct the well.

Figure 15.44 In HSA auger drilling, each 5-ft augur is attached with two-bolts. Shown is a driller helper (student) guiding the next additional auger flight to connect with the ones already in the ground.

4. It is now time to remove the knockout plate. This is done by lifting a heavy metal rod with a sand line and allowing it to "free-fall" inside the hollow stem to the bottom after loading the hole. This procedure is repeated several times. The driller checks the depth by lowering a cloth tape weighted with a heavy piece of metal, so that it will fall freely to the bottom. The driller can also "feel" if the knockout plate has been dislodged by distinguishing the difference between the bottom and the hard plate. In the case of the removable plug, it must be jarred loose and pulled out so that the well can be constructed.

5. It is time to start building the well (Figure 15.45). A bottom cap is secured onto the bottom of the well screen. It is important to make sure that when the well is

FIGURE 15.45   Building the well. Shown is the construction of a PVC monitoring well bottom cap (addition of PVC glue), short sump, and a 40-slot well screen.

constructed the screen is placed exactly where desired. Therefore, if the well penetrated into a confining layer, a sump consisting of blank casing and a bottom cap will need to be added to the bottom so that the screen is positioned within the aquifer. Blank casing is added (twisted on) to the top of the well screen until it extends a couple of feet (0.6 m) above the top of the hole, unless it is located in a street or pathway where one could trip on the casing. In this case, one should complete the well with a flush mounting (Chapter 4; Figure 4.20). A top cap is placed to keep the inside clean during the rest of the construction operation. Preferably this is all done with threaded pipe. Screens and pipe come manufactured in a variety of materials and lengths. This will be discussed later on.

6. Now that the casing is in the hole, it is time to "complete" the well. Typically, the next step is the placement of a filter pack around the screen. The packing material is poured down through the annulus between the hollow stem pipe and the casing. The filter packing materials should be larger than the screen slot size (<10% passing through the slots). Sieved silica sand is a good choice and these come in 50-lb and 100-lb (22.7- and 44.3-kg) bags (Figure 15.46). The 50-lb (22.7-kg) bags are much easier to handle if used directly, but the 100-lb (44.3-kg) bags may be more cost effective. The sand size should not be confused with the slot size. Sand sizes represent sieved intervals, for example, 10 to 20 sand was sized through a number 10 to 20 sieve (2 to 0.83 mm, very coarse sand). And a 20-slot screen size means that the openings are 20 thousandths of an inch (0.51 mm). This is a commonly used scenario. Another example would be to use 6 to 9 sieved (3.35 to 2.16 mm, very fine gravel) sand with a 40-slot (1.02-mm) screen (Figure 15.45).

7. One effective way to add sand to the hole is to pour it from a 5-gal (18.9-L) bucket with a notched flap cut in the bottom and side. A 1-in (2.5-cm) cut is

FIGURE 15.46 Monitoring-well sand-packing materials. Sieved sand is rated by sieve size not slot size. Observed in the bucket is 6- to 9-mesh sieved sand (Figure 2.2) used for a 40-slot screen monitoring well.

made on the side of the bucket, and then a 1-in (2.5-cm) cut is made to the bottom to make an L-shaped flap. This piece can be flipped up into place while sand is being added to the bucket and flipped down to allow the sand to pour in a controlled manner in the annulus between the hollow stem and the casing (Figure 15.47). This is a good way to keep sand or filter pack materials from spilling all over the place, and it allows the worker to use the more cost-effective 100-lb (44.3-kg) bags. While the sand is being added, the auger flights are being lifted upward to avoid creating a sand lock between the casing and the auger. This is checked by the driller using the weighted cloth tape in a tamping motion to "feel" the level and placement of the sand and make sure there is a space between the bottom of the auger and the top of the sand. This procedure is continued until the sand is at least a foot or two above the top of the well screen (this is different for a production well).

8. Once the sand is in place, depending on the stability of the hole, the rest of the auger flights may be pulled out for easier well completion. As each flight is raised up to where you can see the bolts, the helper shoves in a horseshoe-shaped plate to suspend the drill string below ground surface, while the connecting bolts are loosened with an air wrench. Once the bolts are loosened, the auger flight is pulled upward and the helper guides the auger flight away from the rig and lays it on the ground for cleaning (similar to Figure 15.44). This process is repeated until all auger flights are out of the ground. If the hole is not stable, then the bentonite sealing materials are added in a similar way to the previously described for the sand or gravel pack (Step 9), being poured through the annulus of the hollow stem and casing as the auger string is being incrementally raised.

Figure 15.47 Driller adding sand-packing materials using a bucket with a notch cut in the bottom. This allows adding packing sand without the mess of pouring from the bags they come in.

9. A bentonite seal is placed on top of the sand-packing materials to isolate the zone that the monitoring well is sampled from and to protect this zone from contamination from above. There are a variety of methods to accomplish this. One method is to gradually pour 3/8-in (1-cm) bentonite chips around the casing. Care must be taken not to pour too quickly or a "bridge" can form that creates a void in the completion zone. Bentonite chips will fall through a water column if poured slowly (Figure 15.48). Deeper well completions may require the use of a tremie pipe. A tremie pipe is a smaller-diameter pipe that passes down the hollow stem to the desired depth, thus providing a means of directing the placement of sand or bentonite as a well is being completed. Bentonite also comes in pellet form. Both chips and pellets partially hydrate when they come in contact with water. The pellets can be placed with greater success if water is added to them in a bucket to make a slurry and then poured down the hole. If larger volumes are required, then grouting materials may need to be pumped into place with the drill rig. The length of the seal depends on regulations on a state-by-state basis. An expansive grouting material can be made by mixing cuttings with bentonite chips and then backfilled within approximately 2 ft (0.6 m) from the land surface.

10. A top "plug" is made with concrete with framed sides to form a pad with sides sloping away from the casing to protect from surface contamination. A metal locking security device can be shoved into the concrete to hold it tightly into place. This way the well could have a slip cap over the casing, with a locking painted metal security cover.

Figure 15.48 Pouring bentonite chips to grout a monitoring well.

11. A drain hole should be drilled into the outer security casing to allow for condensation drainage (Fetter 1994). It is also a good idea to drill a hole below the cap of the monitoring well casing to allow for pressure changes to equilibrate easier, unless the site is for monitoring gases. This can also be very important for widely fluctuating water levels (Chapter 4).

Well-Completion Materials

Monitoring wells are supposed to provide hydraulic head and water-quality information representative of the aquifer. One of the most important criteria for selection of construction materials is water quality. The construction materials can actually influence the water-quality (Barcelona et al. 1983; Reynolds and Gillham 1985; Parker et al. 1990). For example, there is a tendency for chlorinated solvents to sorb onto polyvinyl chloride

(PVC) and PFTE (Teflon) from aqueous solutions (Reynolds and Gillham 1985). Stainless steel 304 and 316 may sorb heavy metals from aqueous concentrations as low as 50 to 100 μg/L (Parker et al. 1990). PVC glue, used for bell couplings, may also be leached into the monitoring well (Barcelona et al. 1983). This is one reason threaded couplings are desired over fitted couplings that require an adhesive (Figure 15.45).

Ranney and Parker (1997) performed tests on six different casing materials to see how they would react to 28 different pure "free product" organic chemicals and some acids and bases. PFTE and fluorinated ethylene propylene (FEP) casing showed good resistance to all chemicals, and stainless steel showed good resistance to the organic compounds. Stainless steel is less affected by organic compounds and is suitable for most geologic settings, except for acid rock drainage (ARD) sites where heavy metals may be present or where contamination with strong acids and bases has occurred. A reasonable casing material at an ARD site is PVC.

Some materials that may be the very best for chemical resistance may also be the costliest. The various pros and cons of different construction materials involve chemical resistance, weight, strength, and cost. A compromise may be required depending on the budget and site requirements. Table 15.5 lists some of the more common monitoring-well construction materials and lists respective advantages and disadvantages from Driscoll (1986) and Fetter (1993).

Well Development

One of the problems with monitoring wells, or production wells for that matter, is that they are often not properly developed. If the well-construction process goes as it should, and the water level recovers quickly after bailing the well a couple of times, these wells are often deemed "completed and developed," and the crew moves on to the next location. This may be adequate for establishing a connection with the aquifer

Type	Advantage	Disadvantage	Relative Cost
PVC	Lightweight, great for ARD acids, alkalis, alcohols, and oils	May adsorb VOC organics, react with ketones and esters	1.0
Stainless steel 304, 316 with varying amounts of nickel	Least reactive to organic materials, high strength, and temperature ranges	May corrode in heavy metal or ARD waters, higher cost than plastic, heavy to use	6.0, 11.2
Teflon	Resistant to many chemicals, lightweight	Lower wear resistance and tensile strength, expensive	20.7
Mild steel	Inexpensive, strong, lower temperature sensitivity	Heavy to use, not as chemically resistant as stainless steel	1.1
Polypropylene	Lightweight, resistant to acids, alkalis, alcohols, and oils	Weak, not very rigid, temperature sensitive, hard to make slots for screen	2.1

TABLE 15.5   Well Construction Materials Comparison [Modified from Driscoll (1986) and Fetter (1993)]

to verify hydraulic-head data but may not be adequate for water-quality sampling (Chapter 9). The sediments from the drilling operation have not been properly mobilized from the borehole and may affect water-quality results. This can be kind of a dilemma because there is a tendency to limit well development at contaminated sites because of the need to handle and dispose of the development water. There may also be a concern about disturbing the configuration of the plume. Decisions made about well development must be weighed with the objectives of the study or project.

If the monitoring well is to be used for slug testing (Chapter 7) or the evaluation of aquifer properties (Chapters 5 and 6), then proper well development is a must. This is done by pumping and surging the well to get the finer sediments to move back and forth through the screened interval. This may take an hour or more, but greatly improves the confidence in obtaining reliable water-level and water-quality data.

There are various techniques that can be used for developing monitoring wells. Many wells are simply overpumped until the water runs clear. Using low-volume water-quality sampling pumps is a poor choice for well development because the impellers may become clogged quickly and may even seize up the pump (Figure 5.16). It is better to use some form of surging action to mobilize the fines prior to using a pump designed to handle high sediment content. One good choice for development yields in the range of 10 to 15 gpm (0.63 to 0.95 L/s) is a converted chainsaw engine pump produced by Homelite. The pump needs to be primed but does a great job of producing dirty water (Figure 15.49). Generally, it is *not* a good idea to add external water to the well, particularly if it is to be used for water-quality sampling. An example of a nonconventional well-development method is given in Example 15.10.

Example 15.10   At one site a garden hose was within reach of a 2-in (5-cm) monitoring well. A successful well-development method was devised by using the garden hose with duct tape wrapped around it approximately 2 ft (0.6 m) below the nozzle to form a surging tool. The hose contributed clean water to the well as the surging tool was raised and lowered. The dirty water was lifted out of the well during the process for a successful completion.

Figure 15.49   Using a converted Homelite chainsaw engine as a well-development pump. It is capable of discharging very sediment laden water.

Example 15.10, once again, was not given as an example of standard practices, but as an illustration of the principles of surging and overpumping and an example of creative thinking.

Completion of Multilevel Monitoring Wells

The goal at many monitoring sites is to gain a detailed understanding of the vertical and horizontal distribution of water-quality parameters and hydraulic head (see also Chapter 10). One approach to this is by completing monitoring wells capable of being sampled at discrete vertical intervals. In shallow systems (<100 ft, 30 m) this can be a matter of drilling a larger-diameter hole and placing several small-diameter wells completed at different vertical depths within the borehole (Figure 15.50). This is known as a **nested well**. The deepest well is placed at the bottom of the borehole, with packing materials added to a foot or so above the screened zone followed by adding bentonite sealing materials to the next desired depth. The process is repeated until several vertical zones have been established. Wells completed at different depths, but in close proximity to one another are known as **cluster** wells. Nested wells may have the problem of achieving a good seal between desired intervals (vertical communication along the casing). Improvement in proper spacing

FIGURE 15.50   Nested wells in a constructed wetlands setting. Nested wells were used in the evaluation of the vertical distributions of dilute bromide in a tracer test.

between individual wells in a nest can be accomplished by using centralizers; however, the U.S. EPA, and the California Department of Water Resources discourage the use of nested wells (Einarson 2006).

Another approach to installing dedicated multilevel groundwater monitoring wells is by using the services or materials produced by commercially available systems. Four prominent ones are the Westbay MP, Solinst Waterloo, Solinst CMT, and Water FLUTe systems. These systems allow the vertical sampling of water quality and hydraulic head at 10 or more discrete positions in a single well. Ports from individual vertical positions need to be established prior to placing the materials down hole. A very good discussion of the peculiarities of installing multilevel monitoring wells along with additional references can be found in Einarson (2006).

15.6 Production-Well Completion

Production-well completion could easily comprise another chapter but is included here as a comparison and contrast to monitoring-well completion. Proper production-well completion permits well yields to be maximized while providing favorable well efficiency (Chapter 5). This also leads to confidence in the results from aquifer testing (Chapters 5 and 6).

A thought-provoking question regarding well completion is: What we testing at the well, the aquifer or the packing materials? An aquifer capable of producing a fair amount of water may actually be hindered by drilling practices, choice of well-completion materials, or well-development techniques. Another type of problem arises when well-completion materials are preselected before there is a proper knowledge of the aquifer properties. An improper slot size (use of what's in stock) may be chosen for the well screen. Or the packing material to be used around the well screen may be assumed to be appropriate as with other wells in the area. You really don't know what is suitable until drilling has taken place. Each hole in the ground is different, and aquifer conditions can change dramatically both laterally and vertically. Another fallacy is just because a production well may be designed to yield a particular quantity of water, the aquifer may not be capable of doing so.

Before drilling a larger-diameter production well, the author suggests drilling a small-diameter pilot or test hole to observe directly what the aquifer is like. Larger-diameter production wells can then be drilled with a greater level of confidence of where the production zones are. If the cable-tool drilling method is used and the target aquifer is known to be productive, production diameter casing may be used at the onset. This approach can also be used for "drill and drive" rotary methods. Samples from the production zone can be collected and analyzed to select the appropriate screen- and gravel-packing materials. With the casing in the ground the production zone can later be exposed by jacking up the casing to "expose" the screen (described later on).

Williams (1981) suggests that well-design criteria should include approach velocity and turbulence concepts rather than simply the entrance velocity principle alone. This process begins with selecting the appropriate packing materials around the screen and picking the right screen slot size. This is usually done by performing a sieve analysis on the formation materials. Well development around screens can be for "naturally" developed or artificial materials. Naturally developed wells are less expensive than gravel-packed wells because of the additional steps involved in well completion. If the best production zone has a significant quantity of stratified fines, then gravel packing allows for greater flexibility (Williams 1981).

Sieve Analysis

Assuming the well has been drilled to a desired depth and is cased to the bottom, the hydrogeologist can take the formation cuttings to a lab for analysis or perform a sieve analysis. If the cuttings or formation materials have been bagged every 5 ft (1.5 m), a sieve analysis should be performed over every potential zone to be included within the screened interval. A way to evaluate the grain-size distribution is to plot them on a graph (Chapter 3). Some graph forms are prepared for this purpose and can be provided by any screen manufacturer (Figure 15.51). A general evaluation of grain-distribution or sorting is through the uniformity coefficient (C_u), Equation 15.2 (see also Equation 3.10):

$$C_u = d_{60}/d_{10} \qquad (15.2)$$

where d_{60} = diameter where 60% of the grain sizes are finer
  d_{10} = diameter where only 10% of the grain sizes are finer, also known as the **effective grain size** (Chapter 3).

If the C_u ratio is less than 4 it is considered to be well sorted, and if it is greater than 6 it is considered to be poorly sorted (Fetter 1994). Walton (1962) suggests that in heterogeneous aquifer materials having a C_u greater than 6 the screen should retain 50% (allow 50% passing or use the d_{50} as the appropriate slot size) if the overlying materials are unconsolidated and collapsible, or as little as 30% (70% passing or use the d_{70} as the

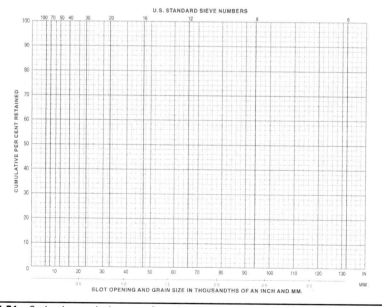

Figure 15.51   Grain-size analysis paper from Johnson Well Screens.

slot size) if the overlying materials are firm (cemented is an example). In homogenous materials that are well sorted, if the overburden is soft, the screen should retain 60% (allow 40% passing or use the d_{40} as the slot size), and if firm the screen should retain 40% (60% passing or use the d_{60} as the slot size). The reason for indicating percent retained or percent passing is because the graph paper is presented both ways. It is just as easy to plot the data using a spreadsheet program. A comparison of grain sizes versus sieve size is shown in Table 2.2.

Example 15.11 In southwestern Montana, an 8-in (20 cm) irrigation well was drilled, and the driller asked the author what slot size would be appropriate for the well screen. The driller had collected representative samples every 5 ft (1.5 m) between 180 and 210 ft (56.1 and 64 m). Approximately 150 to 200 g of sample were placed into tins and cooked at 105°C for 24 h. Each sample was weighed and ran separately through a series of sieves (Figure 15.52) in a ro-tap machine for 15 min. The results are shown in the partial spreadsheet of Table 15.6 and plotted in Figure 15.53.

The grain sizes for the sediments between 185 and 210 ft were more uniform, and the 180 to 185 sample was too fine grained and different to be included in the production zone. The d_{60} ranges

Figure 15.52 Author performing sieve testing on soils samples. An application of this method is described in Example 15.11.

Sieve No.	Size (mm)	Size (in)	180–185	185–190	190–195	195–200	200–205	205–210	185–210
5	4leve	0.157	0	6.22	10.548	10.862	27.272	23.673	10.887
10	1.981	0.078	1.354	9.94	19.605	10.866	17.045	17.618	11.987
18	0.991	0.039	4.426	20.96	31.14	15.104	32.77	29.175	19.004
40	0.425	0.0165	20.06	30.95	34.398	21.628	33.169	23.21	21.833
60	0.25	0.0097	23.38	9.31	14.365	14.02	12.387	7.502	8.831
100	0.149	0.0058	16.84	2.91	8.205	8.485	7.049	4.09	5.031
200	0.074	0.0029	7.568	1.783	5.233	3.184	4.727	2.578	2.851
<200			2.06	1.65	3.472	1.765	3.497	1.767	1.82
		Total	75.688	83.723	126.966	85.914	137.916	109.613	82.244
		Sample weight	76.76	86.13	127.16	85.85	138.04	110.16	82.36
		Screen#	Cum.%	Cum.%	Cum.%	Cum.%	Cum.%	Cum.%	Cum.%
		5	0.00	7.43	8.31	12.64	19.77	21.60	13.24
		10	1.79	19.30	23.75	25.29	32.13	37.67	27.81
		18	7.64	44.34	48.28	42.87	55.89	64.29	50.92
		40	34.14	81.30	75.37	68.04	79.94	85.46	77.47
		60	65.03%	92.42%	86.68%	84.36%	88.93%	92.30%	88.20%
		100	87.28%	95.90%	93.14%	94.24%	94.04%	96.04%	94.32%
		200	97.28%	98.03%	97.27%	97.95%	97.46%	98.39%	97.79%
		<200	100.00%	100.00%	100.00%	100.00%	100.00%	100.00%	100.00%
	Slot Size	Screen #	% passing	% passing	% passing	% passing	% passing	% passing	% passing
	157	5	100.00	92.57	91.69	87.36	80.23	78.40	86.76
	78	10	98.21	80.70	76.25	74.71	67.87	62.33	72.19
	39	18	92.36	55.66	51.72	57.13	44.11	35.71	49.08
	16.5	40	65.86	18.70	24.63	31.96	20.06	14.54	22.53
	9.7	60	34.97	7.58	13.32	15.64	11.07	7.70	11.80
	5.8	100	12.72	4.10	6.86	5.76	5.96	3.96	5.68
	2.9	200	2.72	1.97	2.73	2.05	2.54	1.61	2.21
		<200	0.00	0.00	0.00	0.00	0.00	0.00	0.00

TABLE 15.6 Sieve Analysis of Samples from an 8-in Irrigation Well Production Zone between 180 and 210 ft (from Figure 15.53)

between 1.1 and 1.9 mm (coarse to very coarse sand) and the d_{10} ranges between 0.2 to 0.3 mm (fine to medium sand) with a C_u around 6. By evaluating Figure 15.53 at a 50% passing, a slot size of 40 was chosen.

Grain sizes with a d_{10} smaller than 0.25 mm and are fairly well sorted should probably have a gravel packing (Williams 1981). The well in Example 15.11 was successfully developed without gravel packing because of the coarser fraction present.

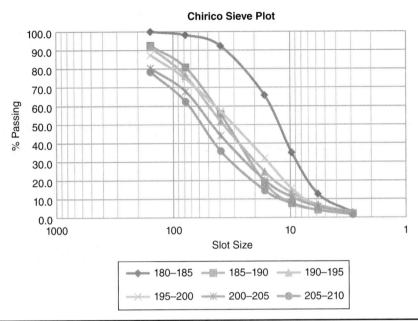

FIGURE 15.53 Graph of percent passing grain sizes versus slot size to evaluate the appropriate screen slot size, for an 8-in irrigation well in southwestern Montana (compare with Figure 15.51). Note how different the 180- to 185-ft sample is from the rest and was thus not considered as part of the production zone.

When artificial gravel packing is introduced, a different philosophy is employed. Normally, the filter packing materials should be uniform in size with a uniformity coefficient less than 2 (Equation 15.2). The aquifer grain-size distribution curves are evaluated to identify the finest grain-size interval in the screened zone. The d_{30} of the aquifer material is multiplied by a factor to obtain the d_{30} of the gravel packing materials (Williams 1981). The slot size is selected so that 90% of the filter pack material is retained (only 10% passes or the d_{10} slot size).

The thickness of gravel-packing materials measured between the borehole wall and the well screen should be at least 0.5-in thick (1.3 cm), but not greater than 8 in (20 cm) (Williams 1981; Driscoll 1986). Greater thicknesses may actually hinder aquifer development because the energy required to develop the aquifer must be able to reach through the gravel packing to the aquifer. For most production wells, gravel pack thickness between 2 and 6 in (5 to 15 cm) around the well screen is a good target (Williams 1981). Since well casing may not be placed "dead center" in the well bore, this range can usually be accomplished.

Well Screen Criteria

The well screen should be designed to facilitate entry of groundwater from the aquifer into the well bore free of sediments after the well has been developed. One of the greatest sources of well problems is from overpumping (Williams 1981). Overpumping causes fines from the aquifer to entrain and creates turbulence near the well bore. Turbulence reduces well efficiency, enhances encrustation, and reduces well life. There is a limitation of how much water can be delivered from an aquifer, so the construction

Criteria
Large open area
Non-clogging slots
Resistant to corrosion
Sufficient strength
Function
Easily developed
Minimizes incrustation
Low head loss
Control of sediment entry

TABLE 15.7 Criteria and Function of Well Screen [Adapted from Driscoll (1986)]

design should complement this maximum capacity. Sufficient funds should be invested for well screens for any production wells that are intended to be used for a long time. Temporary wells or monitoring wells may not require as high a level of design or expense. A list of the screen criteria and functionality is listed in Table 15.7 (modified from Driscoll 1986).

The manufactured opening size is well screen is known as the slot size. It is measured in thousandths of an inch. For example, a 40-slot screen has a slot opening of 0.040 in (1 mm). A table of slot size versus millimeter equivalents is shown in Table 15.8.

Slot Size	Inches	Millimeters
10	0.010	0.246
16	0.016	0.417
20	0.020	0.495
25	0.025	0.635
30	0.030	0.762
35	0.035	0.889
40	0.040	0.990
60	0.060	1.52
80	0.080	2.03
100	0.100	2.54
120	0.120	3.05
150	0.150	3.81
180	0.180	4.57
210	0.210	5.33
250	0.250	6.35

TABLE 15.8 Slot Size in Inches and Millimeters

As water moves toward a well, its velocity increases. If the velocity exceeds a critical limit, finer particles from the aquifer will become entrained and move into the gravel packing, thus reducing the flow properties (Williams 1981). This critical limit is known as the **approach velocity (V_a)**. This is not to be confused with the entrance velocity (discussed next). The entrance velocity is measured where water passes through the well screen into the well, and the V_a is measured at the borehole wall at the damage zone (Chapter 5). The V_a is significant because one should know the approximate pumping rate that would best develop the well once it is completed. Additionally, the V_a is a specific discharge and not a velocity, where the specific discharge is divided by the effective porosity (Chapter 3). It can also be viewed as the discharge divided by circumferential area. For design purposes a conservative V_a (Huisman 1972) can be expressed as

$$V_a = \sqrt{K}/30 \qquad (15.3)$$

where V_a and K are both in units of m/s.

This has also been expressed in terms of grain size (Huisman 1972) from a sieve analysis of the aquifer materials as

$$V_a < 2d_{40} \qquad (15.4)$$

with similar units (m/s).

This assumes that the grain-size distribution is representative of the aquifer materials in the production zone (Williams 1981). Once the V_a is known, it can be multiplied by the circumferential area of the filter's outside surface area to estimate the maximum yield capacity (Q_c). If the distance (r_d) at which fines are removed during development can be estimated, the discharge (Q_d) necessary for proper development can also be estimated from the following relationship (Williams 1981).

$$\frac{Q_d}{Q_c} = \frac{V_a A_d}{V_a A_p} = \frac{r_d}{r_p} \text{ or } \frac{Q_c r_d}{r_p} \qquad (15.5)$$

where Q_c = capacity at the maximum approach velocity
Q_d = discharge during development
r_p = distance to outside edge of filter pack or screen diameter in naturally developed wells
r_d = distance of effective development
A_d = area at the outer effectiveness of development
A_p = area at the outside edge of filter pack or the outside of the screen for naturally developed wells

Williams (1981) points out that although the approach velocity (V_a) cancels out in Equation 15.5, it is necessary for deriving the maximum yield capacity (Q_c).

Example 15.12 Referring to Example 15.11, the d_{40} (40% passing) is nearly equivalent to a slot size of 30. This yields a slot opening for the d_{40} of 0.76 mm. From Equation 15.4 and Table 15.8, the maximum approach velocity is approximately:

$$V_a \leq 2 \times 0.000762 \text{ m} = 1.5 \times 10^{-3} \text{ m/s}$$

From pumping-test data, the hydraulic conductivity (K) is estimated to be about 10 m/day. If this K value is plugged into Equation 15.3 and converted to m/s, another estimate of the maximum approach velocity is calculated to be

$$V_a \le (12 \times 10^{-4} \text{ m/s})^{1/2}/30 = 3.7 \times 10^{-4} \text{ m/s}$$

These are both within a factor of 5 of each other. By using an 8-in telescoping screen with outer diameter of 7.5 in (19 cm) we divide by 2 to obtain the (r_p) distance 0.095 m. The development distance is assumed to be an additional 4 in (10.2 cm) into the formation and the screen length is 25 ft (7.6 m). These can be plugged into Equation 15.5 to obtain a range of discharge values (Q_d). The two areas must first be converted into meters squared:

$$A_d = 2 \times \pi \times r_d \times \text{screen length} = 2\pi(0.1977 \text{ m})(7.62 \text{ m}) = 9.46 \text{ m}^2$$

$$A_p = 2 \times \pi \times r_p \times \text{screen length} = 2\pi(0.095 \text{ m})(7.62 \text{ m}) = 4.55 \text{ m}^2$$

$$Q_c = V_a \times A_p = 1.5 \times 10^{-3} \text{ m/s} \times 4.55 \text{ m}^2 = 6.8 \times 10^{-3} \text{ m}^3/\text{s}$$

$$Q_d = \frac{Q_c r_d}{r_p} = \frac{(6.8 \times 10^{-3} \text{ m}^3/\text{s})(0.197 \text{ m})}{0.095 \text{ m}} = 1.42 \times 10^{-2} \text{ m}^3/\text{s} \text{ (224 gpm)}$$

This means the developmental production rate Q_d should exceed 230 gpm (10 L/s) to mobilize formation fines within 4 in (10.2 cm) of the screen. If the r_p was 6 in (15.2 cm) instead of 4 in (10.2 cm), the Q_d would need to exceed 280 gpm (17.7 L/s).

Using the other approach velocity (Equation 15.3), a value of 3.7×10^{-3} m³/s (58 gpm) was obtained. Actual production rates were approximately 180 gpm (1.14×10^{-2} m³/s), which suggests the grain-size approach of Equation 15.4 may give a better approximation, *if* the production zone samples are representative or the factor in Equation 15.3 should be 10 instead of 30. The beauty of this analysis is that it can be performed if production-zone samples and sieve analyses are performed *prior* to a pumping test or well development.

Screen Entrance Velocity

Although the approach velocity is important for evaluating production rates for well development, the screen entrance velocity is probably one of the most frequently used criteria for sizing well screen. The volume of water pumped from a well is a function of percent open area and the entrance velocity. If the desired well discharge (Q, L^3/T) is known and a minimum entrance velocity is fixed in the design criteria, the only variable is the percent open area. The most commonly used entrance velocity for design purposes is 0.1 ft/s (0.03 m/s) (Williams 1981; Driscoll 1986); however, other screen manufacturers report acceptable entrance velocities as high as 3 ft/s (1 m/s) (Roscoe Moss 1994). A variety of different slot types are available: continuous, slotted, bridge-slot, and louvered (Figure 15.54). The continuous-slot screen has the largest percent open area per length, because a continuous V-wire spacing is welded to vertical supporting rods. The functions of well screen are listed in Table 15.7. As entrance velocities increase, the potential for turbulence and head losses also increase. This results in higher electrical costs, promotes encrustation, enhances sediment mobilization, and thus reduces well efficiencies (Chapter 5).

The open area is increased by lengthening the screen or increasing its diameter. For thin aquifers the only choice may be increasing the diameter; however, only minor percentage gains are made by increasing the diameter. This can be evaluated by comparing some of the continuous-slot values listed in Table 15.9. In unconfined aquifers, lengthening the screen may effectively reduce well yield by reducing the available drawdown.

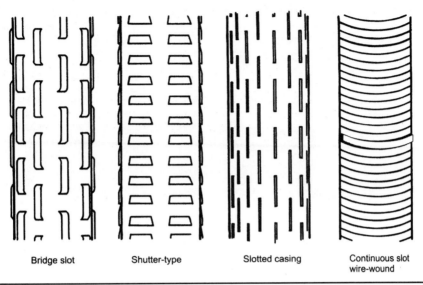

Bridge slot   Shutter-type   Slotted casing   Continuous slot
                                                             wire-wound

FIGURE 15.54 Various types of well screens. [*From NGWA (1989).*]

The opposite may be true for a confined aquifer, as long as the hydraulic head remains above the top of the aquifer (Chapter 4).

Bridge slotting runs one-fifth to one-third less percent open area than continuous-slot screen for similar casing diameters. Louvered slot screen runs less than one-tenth to one-fifth of continuous-slot screen (Driscoll 1986). Vertical slot screens (slotted casing) generally yield a maximum of 3% open area. Other perforations that are handmade may be from torch-cut slots. These are usually used for monitoring deep wells but are known to be used in domestic well completions. This results in sediments coming into the well over the life of the well. Some perforations are made in situ in steel casing using a Mill's knife (Figure 15.55a). This is a device placed downhole that hydraulically punches a

	Slot Size	in²/ft	% Open Area	Screen Diameter	Slot Size	in²/ft	% Open Area
4" ID	20	44	25	12" ID	20	69	14
	40	72	41		40	99	21
	60	90	52		60	135	28
	100	112	64		100	189	39
8" ID	20	45	18	16" ID	20	68	11
	40	77	31		40	124	21
	60	100	40		60	169	28
	100	131	53		100	238	40

TABLE 15.9 Percent Open Area for Stainless-Steel Continuous-Slot Screens (Modified after Driscoll 1986)

Figure 15.55a   Mill's knife tool used to perforate steel casing by hydraulically forcing the cutting teeth outward. Note the cutting teeth are on a wheel to the right of the lowering strap.

Figure 15.55b   Perforations in steel casing made by a Mill's knife.

hole outward in the casing like a can opener (Figure 15.55b). This is commonly done on 6-in domestic wells, but once again may allow sediments to enter the well.

Domestic wells may have their "pump liner" (described in the next section) perforated by a 1/16 in (1.6-mm) or 1/8-in (3.2-mm) slot cut with a circular saw. Generally, this is done by cutting a slot in lengths of approximately 6 in (15.2 cm) to 1 ft (30.5 cm) spaced every 1 ft or so. The casing is turned approximately 120° in successive rows that are staggered to make three rows that have a spiral pattern of perforations. This maintains a certain level of casing strength. The author recommends a maximum depth of 400 ft using this process for schedule 80 PVC screen. He has witnessed the collapse of vertically slotted screens at attempts of greater depths. Deeper wells should be completed with steel.

Example 15.13   This example will refer to the well discussed in Example 15.4, an 8-in production well designed to irrigate a 30-acre cemetery. The production yield was supposed to yield 250 gpm (15.8 L/s). The drilling method used was cable tool. Drilling would proceed 10 ft before welding on another section of casing. During drilling, a clay zone was encountered at approximately 90 ft (27.4 m) with "first water" at 25 ft (7.6 m). Drilling was hampered by heaving sands. The decision was made to continue looking for the next coarse-grained zone. A fine-gravel zone was encountered from 148 and 165 ft (45.1 and 50.3 m) below ground surface (bgs) before encountering another clay zone. A sieve analysis was performed on samples collected from the production zone, assuming a natural gravel pack. The analysis indicated a 46-slot was appropriate for a 50% passing grain size (Figure 15.56). If a 15-ft section of 46-slot screen was used in the production zone, would the entrance velocity be less than 0.1 ft/s? There are two ways to check: (1) either the entrance velocity calculated to be less than 0.1 ft/s using the desired Q or (2) the area and entrance velocity can be used to calculate a Q. As long as the Q exceeds 250 gpm, the design would be adequate. The percent open area from the chosen manufacturer's 46-slot continuous screen was 33%.

The surface area for the 15-ft screen was first determined:

$$\text{Surface area} = \pi \times d \times 12 \text{ in/ft (per ft of screen)}$$
$$= \pi \times 7.5 \text{ in} \times 12 \text{ in/ft}$$
$$= 282.7 \text{ in}^2/\text{ft (for 1 ft of screen)}$$

FIGURE 15.56   A 46-slot stainless-steel v-wire production well screen described in Example 15.13.

Multiplying this by 15-ft yields: 283 in²/ft × 15 ft of screen length = 4,240 in².
The percent open area is 33%, so

$$4,240 \text{ in}^2 \times 0.33 = 1,399 \text{ in}^2, \text{ or}$$

$$1,399 \text{ in}^2/144 \text{ in}^2/\text{ft}^2 = 9.7 \text{ ft}^2 \text{ (ft}^2 \text{ of open area)}$$

$$Q = 250 \text{ gpm, or } 250 \text{ gpm}/(448.8 \text{ gpm per ft}^3/s) = 0.557 \text{ ft}^3/s$$

From the relationship: $Q = V \times A$
We can estimate the velocity moving through the slots, where

$$Q = \text{flow in ft}^3/s$$
$$V = \text{velocity in ft/s}$$
$$A = \text{screen open area in ft}^2$$

The velocity is calculated by $V = Q/A = 0.557 \text{ ft}^3/s/9.7 \text{ ft}^2 = 0.057 \text{ ft/s}$, which is well below 0.1 ft/s. The maximum yield at 0.1 ft/s is given by

$$Q = V \times A = 0.1 \text{ ft/s} \times 9.7 \text{ ft}^2/s = 0.97 \text{ ft}^3/s$$

$$0.97 \text{ ft}^3/s \times 448.8 \text{ gpm}/1 \text{ ft}^3/s = 435 \text{ gpm}$$

The well was over-designed, but the well owner wanted a well that would last more than 50 years, or many years beyond his lifetime. In this case, 10 ft of screen may have been adequate.

Well Completion and Development

Wells are completed either open hole or through a packing procedure similar to the monitoring-well process (Section 15.5). Domestic wells, if drilled in indurated materials, will be drilled open-hole and later have a plastic "pump liner" placed to protect the pump from materials that may slough off from the borehole wall and run the risk of wedging the pump down hole. For example, a 4-in PVC liner might be placed into a 6-in domestic borehole completed "open hole." This type of well completion is not desirable because of the increased potential for downward migration of contaminants from any uncased areas to the pumping level. Most well drillers realize this, but the competition of drilling-by-the-foot results in well completions that are less than desired.

Example 15.14 A local church near Helena, Montana, had a production well completed approximately 330 ft (100 m) deep in a limestone formation. The well was completed in 2001 with about 40 ft of 6-in steel surface casing and the rest of the well was completed with a 4-in PVC liner and a SWL at about 78 ft. Because it was a public water supply, monthly testing for nitrates and other nutrients were required by law. Over a period of a few years a trend was noted that the nitrate levels at the church kitchen spigot were consistently in excess of 10 mg/L (Figure 15.57a; see also "Water-Quality Parameters and Their Significance" online at www.mhprofessional.com/weight3e). Elevated nitrate levels above 10 mg/L (as N) may cause a type of methemoglobinemia ("blue baby syndrome") in infants (Greer and Shannon 2005). An unpublished investigation by the author and students in 2015 helped unravel the source. The church well was pumped, and the discharge was evaluated until groundwater temperature and pH stabilized after about an hour (Chapter 9). At that point in time the nitrate levels were measured via a water-quality multimeter to be about 2.5 mg/L as N, with samples collected and sent off to a certified lab. Neighboring wells were also pumped, evaluated, and sampled in the same manner. A particular parishioner, whose well was topographically and hydraulically up gradient of the church, indicated a nitrate reading of about 50 mg/L as N (confirmed to be 48.3 mg/L as N, by the lab). The neighbor had horses (Figure 15.57b, and the well was drilled in the 1970s) and there was a substantial pile of horse manure near the well. Given that the geology was bedrock, with a similar well completion as the church well (40 ft of 6-in steel surface casing,

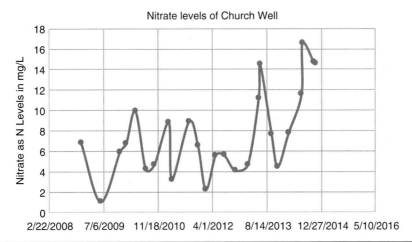

FIGURE 15.57a Nitrate as N levels in mg/L collected from the church kitchen water faucet on a monthly to quarterly basis between December 2008 and October 2014 (from Example 15.14).

FIGURE 15.57b View of neighbor's well with elevated nitrate levels as N likely sourced from a large horse manure pile near the well described in Example 15.14.

a 4-in pump liner, total depth 300 ft, and SWL at 102 ft), it was hypothesized that a plume of elevated nitrates (or ammonia/ammonium that denitrified) was able to spread and travel down gradient to the church's well. The biweekly use of the church's well was hypothesized to pull elevated nitrates down to the intake (perforated zone) but was not enough to remove the influence of the neighbor's ongoing nitrate source, thus putting parishioners who might be using the kitchen faucet at risk.

Larger production wells that require the screen be about the same diameter as the casing should be preconstructed before inserting into the drillhole (Driscoll 1986). The bottom well plate, sump or tailpipe section, and screen should be welded together as a single piece before being attached (welded) to the casing and lowered into the ground.

In this regard a common technique is to package the bottom cap, sump or tailpipe, and well screen plus a short-extension piece so that it "telescopes" inside the casing [OD of well-screen package is ½ in (1.2 cm) less in diameter than the well-casing ID]. There is a "K-packer" constructed of neoprene attached to the short extension piece at the top of the well-screen package that functions to form a seal between the screen package and the casing (Figure 15.58). The whole well-screen package is dropped out of sight and followed in with the drill pipe to "hold" the well-screen package in place while a casing-jack assembly hydraulically pries up the casing, to expose the well screen.

When the well driller needs to pry up the casing to expose the well screen, the first attempt is made by wrapping chains around the casing with an attempt to pull the casing up with the drill rig (using the drawworks). If the casing does not budge then a casing-jack assembly is used. The casing-jack assembly consists of "spiders" and "slips," named after the jacking equipment used to grip the casing (Figure 15.59). The "spider" is a yoke-like feature that allows the casing to slide through the midst of it while resting on top of the hydraulic jack lifts. The "slips" have grooved edges and a wedged shape that *slips* in between the "spider" and the casing to friction-grip on the casing during an 18-in (0.46-m) up-stroke (Figure 15.60). At the top of the stroke the slips are pulled out

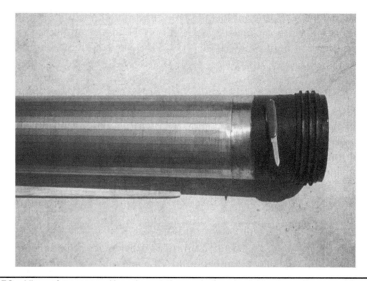

FIGURE 15.58 View of neoprene K-packer at the end of a short blank section used to provide a seal between a telescoped production well screen package that is telescoped inside the driven steel well casing. Pocket knife for scale. Example application is described in Example 15.15.

Figure 15.59 Close-up of two sets of spiders (heavy yoke-like piece) with four slips each (inserted pieces with handles).

Figure 15.60 Casing jacks fully extended, with spider and slips at top gripping the casing on the up-stroke. The driller is on the right-hand side of this reverse-circulation drill rig. The helper is ready to grab the slips during the next up-stroke (see also Figure 15.59).

while the weight of the "spider" helps retract the hydraulics of the jack heads for another thrust. A rhythm is established to jack-up the casing until the desired position is reached. At depth, the well screen is slowly becoming exposed with each stroke, telescoped inside the casing with a seal being made by the K-packer. Once the casing is at the desired position, the jacked-up portion of the casing is cut off with a torch. The hydrogeologist is responsible for making sure that the screen has been exposed properly. If the telescoping screen package is not in the right place, it must be "fished" out and the hole cleaned out, and the process repeated (Example 15.15). A worse-case scenario is if the casing gets jacked past the K-packer resulting in separation! so careful measurements and communication with the driller is critical.

Example 15.15   The 8-in well from Examples 15.4 and 15.13 was drilled to a depth of 170 ft (51.8 m), having penetrated 5 ft (1.5 m) of clay. The telescoping package was designed with a 5-ft tailpipe, 15 ft of screen, and a 5-ft extension equipped with a neoprene K-packer. The tailpipe would be inserted into the lower clay, and the 15-ft screen would fit within the 17 ft of available fine-gravel aquifer. The screen package was dropped out of sight and came to a stop 5 ft short of the goal.

A steel fishing tool was constructed by the driller with two "pointy" metal pieces that could rotate on a bolt that would flange outward to a maximum of ½ in (1.2 cm) smaller than the inside diameter (ID) of the telescoping screen (Figure 15.61). The two "pointy" metal pieces could collapse inward but only flange outward to this maximum distance. This fishing tool was placed inside the well screen to hook onto the lip at the top of the screen. It worked beautifully on the first try.

The hole was cleaned out, and the telescoping package was again dropped out of sight. This time the tailpipe was within 1 ft of the bottom, so the 15-ft screen was situated exactly within the 17-ft gravel zone!

Well development in production wells is performed to correct the damage done to the borehole wall during drilling and enhance the flow characteristics near the screen. With gravel-packed wells, the concepts of approach velocity (V_a) should be used to determine maximum pumping rates (Williams 1981). Proper development pumping rates (Q_d) will mobilize the fines at an appropriate distance from the packing materials. In naturally packed wells, a significant portion of the fines near the screen are washed

Figure 15.61   Fishing tool with flanged wings next to well screen described in Example 15.15.

out during the development process to create a higher transmissivity zone next to the well screen (Chapter 3). There are different methods used to accomplish this.

Overpumping is probably the simplest method of mobilizing fines within the screened zone. In this method pumping occurs at a higher development rate until the sediments cease to entrain. The over-pumped rate is higher than what the production rate is intended for. The problem with this method is that some fines become trapped from just a one-way movement. When the pumping stops, the trapped sediments may become untrapped and then remobilize again when the pump is turned back on. It is better to employ a back-and-forth motion, such as was described earlier in the monitoring-well section (see also Section 10.2).

One effective method of obtaining high volumes of water in a back-and-forth motion is known as **backwashing**. (This term is not to be confused with sharing a soda with a 2-year old.) Using a forward rotary rig, the drill pipe is lowered to a depth a few feet above the well screen. Air is injected into the well and displaces the water up to the surface. This is done for 15 min or so, and then the air is shut off, allowing the water column to rush back out through the screen into the formation. This process is repeated until the water rinses clean (Figure 15.62). This creates a back-and-forth motion in the

Figure 15.62 Clean water using the backwashing development method after 6 h of development. The application is described in Example 15.15.

production zone. It may take several hours to cleanup the well, but it is effective. It is well worth the extra time and cost to make sure wells are properly developed. It should be mentioned that when injecting air from the drill pipe, it should not be within the screened interval because the air flow is too intense. If this occurs, the aquifer materials may become more damaged than from the drilling process.

15.7 Water Witching

If there is to be section on water witching (or dowsing), it seems appropriate to be placed in this chapter since many well drillers and property owners use this service prior to drilling (Example 15.6). There are individuals who claim to have incredible success rates and must therefore have certain divining powers. It seems appropriate therefore that the author makes a few observations that represent his opinion.

The technique involves using welding rods (coat hangers or whatever) approximately 15 to 18 in (38 to 46 cm) long bent approximately 2 in down on one end. Forked hickory or willow sticks are also often used (https://pubs.usgs.gov/gip/water_dowsing/pdf/water_dowsing.pdf). The short lengths of the welding rods are held loosely in the hands while the long ends are allowed to rotate freely. In the case of the sticks, the two forked ends are held in the hands while the single end protrudes outward. The author's understanding is that when "water" is encountered the rods turn inward or outward and the hickory stick points downward (USGS, 1917). This has been used by many people to locate septic tanks or buried pipes on properties. There are mixed reviews on this process compared to technical methods (Mellet 2000; https://pubs.usgs.gov/gip/water_dowsing/pdf/water_dowsing.pdf). Some individuals use this for locating water on properties for a fee. The author is often asked by students and others what he thinks about this. A fairly complete history on this topic is found in U.S. Geological Survey Water Supply Paper Number 416 with a short review at https://pubs.usgs.gov/gip/water_dowsing/pdf/water_dowsing.pdf.

Almost anywhere on the surface of the earth, one will encounter water at depth. In an alluvial valley, almost anywhere in the valley would yield a productive well. The challenge comes when evaluating locations in solid bedrock formations or where the formations may be structurally tilted or disturbed. Even then, water will eventually be encountered; however, the yield may be very low. In hard-layered sedimentary, igneous, and metamorphic rocks, significant well yields may only be derived from secondary fracturing (Chapter 2). It may be that experienced water witches use similar geologic principles as a scientist would. They may recognize structural trends and the orientations of fracture systems and have experience with the information from other drillholes in the area. When this approach is taken, that is one thing; but when water witches predict exact depths and yield rates, no matter how sincere and dedicated these folks may be, it is the author's opinion that this appears to be akin to consulting a Ouija board.

Example 15.16 The following true story took place in July 1993 in southwestern Montana. At the time, local ranchers had hired a driller to perform some wildcat drilling on the property at an hourly rate, prospecting for water. They owned 2,000 acres of land on either side of a small perennial stream. The property extends toward the mountains on either side of the valley. The thought was that they could run more cattle if there was more water available to grow some more grass. Alluvial fans coalesce and slope toward the center of the valley. Drill-site locations were constrained to be within 0.25 mi

(0.16 km) of a power source or a main road. A couple of drilling sites were selected and the lithologies encountered were alternating silty and gravelly materials, with yields only in the 30 to 40 gpm (1.9 to 5 L/s) range instead of the order of magnitude higher that was hoped for. We were just finishing the second, and last, hole when Elmer the neighbor came over.

Elmer was walking along the alluvial fan in a direction parallel with the mountains. As he walked the welding rods would move and cross at times. He was a larger man who said he couldn't always do this. He had worked for the U.S. Forest Service for about 20 years when an accident occurred that affected his neck. He was on disability and didn't charge for his services. He is a nice enough person with an interest in helping. He acknowledged that most people did not believe in "witching."

Elmer asked the rancher if he would like to try. The rods were placed in his hands and Elmer put his hand to the man's neck and wrist. The rancher exclaimed that no amount of squeezing his thumbs could keep the rods from moving. He was amazed and ran to get his daughter. It did not work when the rods were handed to me, even when he put his hand on my neck. The daughter came and it worked for her. Elmer said, "ask how deep it is." The daughter said, "ask who?" (A significant question.) "The rods," Elmer replied. At that point the daughter handed the rods back to Elmer.

Elmer proceeded to demonstrate his skills. He walked in a straight line, and the rods crossed where the water "started." They would cross and move back and forth where the "highest" flow was and then swing around pointing behind him when he walked "past" the water. He turned around and repeated the process. The "water" was approximately 20 ft (6.1 m) wide (the possible width of a gravel channel). He asked the rods, "How deep, how many gallons per minute, and how thick?" The rods crossed as he sequentially called out numbers (5, 10, 15, 20, ... etc.). This was all pretty eerie, but what happened next was especially strange and didn't seem to have much to do with locating water.

Elmer asked, "Where does the water come from?" and the rods pointed toward the mountains, and when asked, "Where is the water going?", they pointed downslope toward the center of the valley. The final question was, "Where is [the rancher's name]?" and the rods swung around and pointed to the client. About that time, I was making tracks for my pickup truck.

The author wishes to apologize to those who may have great faith in this process and may become offended by the above example, but he believes that there is a point where "witches" take this process way too far. After all, they do refer to themselves as witches. Following geologic, hydrologic, or geophysical principles are still the only methods the author chooses to estimate drill sites for water production. People will have to decide for themselves what they have *faith* in.

15.8 Summary

Selection of the appropriate drilling process requires an understanding of the geology and the objectives of the project. Rig safety and getting along with the drilling contractor can be done in a cooperative environment if the hydrogeologist shows common courtesy and common sense. Well construction and development require that accurate samples be collected in the production zone. If it is a monitoring well, then the construction materials need to be resistant to the contaminants involved. For production wells, a sieve analysis, screen design, and development strategy can be chosen that effectively produces a maximum yield.

15.9 Problems

15.1. Your boss has given you a map with a drillhole location to drill a production well for a new school. There will be approximately 300 students, a kitchen or food preparation area (with a dishwasher), six bathrooms, four drinking fountains, four janitorial closets with spigots

for preparing cleaning supplies like mopping, and three outside spigots that hoses could be attached to.
a. Use the reference guide from: http://dnrc.mt.gov/divisions/water/water-rights/docs/forms/615.pdf for U.S. applications (Gallons and acre-feet). What is an appropriate yield (in gpm) needed for this well? What other considerations for water use are appropriate? Justify your answer.
b. Use the reference guide from: http://www.dti.gov.za/industrial_development/docs/fridge/Guideline_Commercial.pdf for South Africa or other SI use areas as examples. What is an appropriate yield (in L/min) needed for this well? What other considerations for water use are appropriate? Justify your answer.

15.2. For the project in Problem 15.1 what are the important safety measures, and supplies one might prepare for prior to beginning drilling?

15.3. Distinguish the basic differences between cable tool, forward rotary, reverse circulation, HSA, and direct push drilling methods.

15.4. For the project described in Problem 15.1 you will be logging the drillhole, identifying where the production zone is and recommending well-completion details.

Review Example 15.8 and describe how you would determine the depths you are at while logging the hole. What should you do when you perceive water is being made from the borehole and you wish to estimate how much water is being produced from the formation?

15.5. Suppose the grain size of the materials from the production zone is according the following Table:

Footage	Clay <0.004 mm	Silt 0.004– 0.063 mm	V. Fine sand 0.063– 0.125 mm	Fine sand 0.125– 0.25 mm	Med. Sand 0.25– 0.5 mm	Coarse Sand 0.5– 1.0 mm	V. Coarse Sand 1–2 mm	V. F. Gravel 2–4 mm	Fine Gravel 4–8 mm	Med. Gravel 8–16 mm
160–165	0	0.5	1.7	3.2	8.5	14.0	21.6	15.2	10.9	8.6
165–170	0	0.3	1.7	1.8	2.9	9.3	30.9	21.0	9.9	6.2

a. Plot the data from the proposed production zone in a spreadsheet or on slot-size versus cumulative-percent-passing paper to observe the shape of the curves.
b. What is the uniformity coefficient? What slot size well screen would you recommend? Why?

15.6. Use Equations 15.4 and 15.5 and refer to Example 15.12 to estimate the approach velocity (V_a) and discharge development rate (Q_d) for this well.

15.7. a. Use Table 15.9 (assuming an 8-in well screen) and the estimated slot size from Problem 15.5 to approximate the open area of a 10-ft length of well screen.
b. What will the entrance velocity be based upon the discharge development rate (Q_d) estimate from Problem 15.6?
c. Assuming an entrance velocity of 0.1 ft/s what is the maximum yield (in gpm) this well could produce?

15.8. Approximately how long would it take for cuttings to reach the surface if the compressor was injecting air at 400 cfm, the drill string is 3.5 in OD, the borehole is 8 in. in diameter, and the depth of drilling is 320 ft?

15.9. Assume you have been given the task of overseeing the drilling and construction of 10 monitoring wells to be drilled in a shallow aquifer (<40 ft each). Assuming you will be using an HSA and each of the wells will be completed with 2-in threaded PVC pipe, with 5-ft, 20-slot screens estimate the total materials you will need and the approximate costs.

15.10. You forgot your e-tape and wish to estimate the depth to water in a well bore. Using a small rock you use your cell phone and estimate that it took 2.7 seconds to reach the water surface. How far it is to water?

15.10 References

Barcelona, M.J., Gibb, J.P., and Miller, R.A., 1983. *A Guide to the Selection of Materials for Monitoring Well Construction and Ground-Water Sampling. Illinois Water State Survey.* SWS Contract 327. https://www.isws.illinois.edu/pubdoc/CR/ISWSCR-327.pdf

Boothman, N., 2000. *How to Make People Like You in 90 Seconds or Less.* Workman Publishing Company, New York, 171 pp.

Crumbling, D.M., 2004. *Summary of the Triad Approach.* U.S. EPA, Office of Superfund Remediation and Technology Innovation, Washington, DC. http://www.triadcentral .org/ref/doc/triadsummary.pdf

Driscoll, F.G., 1986. *Groundwater and Wells.* Johnson Screens, St. Paul, MN, 1108 pp.

Einarson, M., 2006. Multilevel Ground-Water Monitoring. In Neilsen, D.M., ed., *Practical Handbook of Environmental Site Characterization and Ground-Water Monitoring.* Taylor and Francis Group, Boca Raton, FL, pp. 807–848.

Fetter, C.W., 1993. *Contaminant Hydrogeology, 2nd Edition.* Prentice Hall, Upper Saddle River, NJ, 500 pp.

Fetter, C.W., 1994. *Applied Hydrogeology, 3.* Merrill, New York, 691 pp.

Geoprobe Systems, 1999. *The Probing Times*, Vol. 8, No. 1, Summer 1999, Salina, KS, http://www.geoprobesystems.com

Greer, F.R., and Shannon M., 2005. Infant Methemoglobinemia: The Role of Dietary Nitrate in Food and Water. *Pediatrics*, Vol. 116, No. 3, pp. 784–786.

Holt Services Inc. http://www.holtservicesinc.com/clean-water-drilling-projects.htm

HUD.GOV. https://hudgov.prod.parature.com/link/portal/57345/57355/ Article/8231/What-are-the-requirements-for-properties-having-a-shared-well

Huisman, L., 1972. *Groundwater Recovery.* Winchester Press, New York.

Marsh funnel procedure. https://www.wyoben.com/media/magentothem/product-pdf/MarshFunnel.pdf

Mellet, J., 2000. GPR 5. Dowsers Zero. *Fast Times—The EEG Newsletter*, May 2000. The Environmental and Engineering Geophysical Society. 10 pp.

Midwest Drilling, 2004. *The Rotasonic Drill.* A Division of Germac Enterprises Ltd. Winnipeg, MB, Canada.

Neilsen, D.M., Neilsen, G.L., and Preslo, L.M., 2006. Environmental Site Characterization. In Neilsen, D.M., ed., *Practical Handbook of Environmental Site Characterization and Ground-Water Monitoring.* Taylor and Francis Group, Boca Raton, FL, pp. 35–205.

Neilsen, D.M., and Schalla, R., 2006. Design and Installation of Ground-Water Monitoring Wells. In Neilsen, D.M., ed., *Practical Handbook of Environmental Site Characterization and Ground-Water Monitoring.* Taylor and Francis Group, Boca Raton, FL, pp. 639–805.

NGWA, 1989. *Handbook of Suggested Practices for the Design and Installation of Ground-Water Monitoring Wells, EPA 600/4–89/034.* National Water Well Association, Dublin, OH and Environmental Monitoring Systems Laboratory, Las Vegas, NV, 398 pp.

NGWA, 1998. *Manual of Water Well Construction Practices, 2nd Edition.* National Ground Water Association, Westerville, OH, pp. 1–1 to 13–4.

NGWA, 1999. *Water Well Journal*, July. Westerville, OH, pp. 56–58.

NWWAA, 1984. *Drillers Training and Reference Manual.* National Water Well Association of Australia, St. Ives, New South Wales, 267 pp.

Parker, L.V., Hewitt, A.D., and Jenkins, T.F., 1990. Influence of Casing Materials on Trace-Level Chemicals in Ground Water. *Ground Water Monitoring Review*, Vol. 14, No. 2, pp. 130–141.

Precision Sampling, 2006. http://www.precisionsampling.com

Ranney, T.A., and Parker, L.V., 1997. Comparison of Fiberglass and Other Polymeric Well Casings. Part I: Susceptibility to Degradation by Chemicals. *Ground Water Monitoring and Remediation*, Vol. 17, No. 1, pp. 97–103.

Reynolds, G.W., and Gillham, R.W., 1985. *Absorption of Haloginated Organic Compounds by Polymer Materials Commonly Used in Ground-Water Monitors.* Proceedings of the Second Canadian/American Conference on Hydrogeology, Banff, AB, Canada, National Water Well Association, pp. 125–132.

Roscoe Moss, 1994. *Handbook of Ground Water Development.* John Wiley & Sons, New York, 493 pp.

Sara, M., 2006. Ground-Water Monitoring System Design. In Neilsen, D.M., ed., *Practical Handbook of Environmental Site Characterization and Ground-Water Monitoring.* Taylor and Francis Group, Boca Raton, FL, pp. 517–572.

Sonic Drilling. https://www.sonicsampdrill.com/sonic-drilling/how-does-sonic-drilling-work.htm

Spellman, F., and Whiting, N., 1999. *Safety Engineering: Principles and Practices*, Government Institutes and ABS Group Company, Rockville, MD, 459 pp.

USGS, 1917. *The Divining Rod—A History of Water Witching.* U.S. Geological Survey Water Supply Paper No. 416, 59 pp.

USGS. https://pubs.usgs.gov/gip/water_dowsing/pdf/water_dowsing.pdf

Walton, W.C., 1962. *Selected Analytical Methods for Well and Aquifer Evaluation.* Illinois State Water Survey Bulletin 49, 81 pp.

Weight, W.D., 1989. Knowledge-Based Computer Systems for Two Topics in Geology: Micron Gold Deposits and Planning Geological Sampling Procedures, Ph.D. Dissertation, University of Wyoming.

Well Drilling School. http://www.welldrillingschool.com/courses/pdf/DrillingMethods.pdf

Williams, E.B., 1981. Fundamental Concepts of Well Design. *Ground Water*, Vol. 19, No. 5, pp. 527–542.

www.ofite.com, http://www.ofite.com/publications/instructions/92-110-10-instructions/file

APPENDIX **A**

Unit Conversions Tables

Unit	Inch	Foot	Yard	Mile	mm	cm	m	km
1 in	1	0.08333	0.02777	1.5783e-05	25.40	2.54	0.0254	2.54e-C5
1 ft	12	1	3	5,280	304.8	30.48	0.3048	3.048e-04
1 yd	36	3	1	5.682e-04	914.4	91.44	0.9144	9.144e-04
1 mi	63,360	5,280	1,760	1	1.609e-06	1.609e-05	1,609.3	1.609
1 mm	0.03937	3.2808e-03	1.0936e-03	6.2137e-07	1	0.1	0.001	1e-06
1 cm	0.3937	0.03281	1.0936e-02	6.2137e-06	10	1	0.01	1e-05
1 m	39.37	3.2808	1.0936	6.214e-04	1,000	100	1	0.001
1 km	39.370	3,280.8	1,093.6	0.621	1e-06	100,000	1,000	1

TABLE A.1 Unit Conversions for Length

Unit	Feet²	Mile²	Acre	Meter²	km²	Hectare
1 ft²	1	3.587e-08	2.296e-05	0.09291	9.291e-08	9.291e-06
1 mi²	2.788e-06	1	640	2.59e-06	2.59	259
1 acre	43,560	1.5625e-03	1	4,047	4.047e-03	0.4047
1 m²	10.764	3.861e-07	2.471e-04	1	1e-06	0.0001
1 km²	1.0764e-07	0.3861	247.1	1e-06	1	100

TABLE A.2   Unit Conversions for Area

Unit	mL	Liter	Meter³	Feet³	Gallon	Acre · ft	Million gal
1 mL	1	0.001	1e-06	3.531e-05	2.6414e-04	8.107e-10	2.6414e-10
1 L	1,000	1	0.001	0.03531	0.2641	8.107e-07	2.641e-07
1 m³	1e-06	1,000	1	35.31	264.14	8.107e-04	2.641e-04
1 ft³	28,317	28.317	2.8317e-02	1	7.481	2.2957e-05	7.481e-06
1 gal	3,785.9	3.7859	3.786e-03	0.1337	1	3.0691e-06	1e-06
1 acre · ft	1.23335e-09	1.2335e-05	1,233.5	43,560	3.2583e-05	1	0.32583
1e-06 gal	3.7859e-09	3.7859e-06	3,786	1.3367e-05	1e-06	3.0691	1

TABLE A.3 Unit Conversions for Volume

Unit	Seconds	Minutes	Hours	Days	Years
1 s	1	0.01667	2.778e-04	1.1574e-05	3.171e-08
1 min	60	1	0.01667	6.944e-04	1.9026e-06
1 h	3,600	60	1	0.04167	1.1416e-04
1 day	86,400	1,440	24	1	2.74e-06
1 year	3.1536e-07	5.256e-05	8,760	365	1

TABLE A.4 Unit Conversions for Time

Multiply	By	To obtain
1 gal/day/ft^2	0.1337	ft/day
1 cm/s	2,835	ft/day
1 m/day	3.2808	ft/day
1 gal/day/ft^2	4.72×10^{-5}	cm/s
1 ft/day	3.53×10^{-4}	cm/s
1 m/day	1.16×10^{-3}	cm/s
1 gal/day/ft^2	4.07×10^{-2}	m/day
1 cm/s	864	m/day
1 ft/day	0.3048	m/day
1 cm/s	21,203	gal/day/ft^2
1 ft/day	7.48	gal/day/ft^2
1 m/day	24.54	gal/day/ft^2

TABLE A.5 Unit Conversions for Hydraulic Conductivity

Unit	m³/s	m³/day	L/s	ft³/s	Acre · ft/yr	gal/min	Million gal/day
1 m³/s	1	86,400	1,000	35.31	6.255e-07	1,5849	22.822
1 m³/day	1.1574 e-05	1	0.01157	4.087e-04	0.2959	0.1834	2.641e-04
1 L/s	0.001	86.4	1	0.03531	25.57	15.849	0.02282
1 ft³/s	0.02832	2,446.6	28.32	1	724.2	448.8	0.6463
1 acre · ft/year	1.5987e-08	3.3795	0.03911	1.381e-03	1	0.612	8.927e-04
1 gal/min	6.31e-05	5.4526	0.0631	2.228e-03	1.613	1	1.44e-03
1 million gal/day	0.04382	3,786	43.816	1.5473	1,120.2	694.4	1

TABLE A.6 Unit Conversions for Flow

Pressure, mm Hg	Inches Hg	Pascals	Feet	Meters	Calibration
760	29.92	101,325	0	0	100
752	29.61	100,258	278	85	99
745	29.33	99,325	558	170	98
737	29.02	98,259	841	256	97
730	28.74	97,325	1,126	343	96
722	28.43	96,259	1,413	431	95
714	28.11	95,192	1,703	519	94
707	27.83	94,259	1,995	608	93
699	27.52	93,192	2,290	698	92
692	27.24	92,259	2,587	789	91
684	26.93	91,193	2,887	880	90
676	26.61	90,126	3,190	972	89
669	26.34	89,193	3,496	1,066	88
661	26.02	88,126	3,804	1,160	87
654	25.75	87,193	4,115	1,254	86
646	25.43	86,126	4,430	1,350	85
638	25.12	85,060	4,747	1,447	84
631	24.84	84,126	5,067	1,544	83
623	24.53	83,060	5,391	1,643	82
616	24.25	82,127	5,717	1,743	81
608	23.94	81,060	6,047	1,843	80
600	23.62	79,993	6,381	1,945	79
593	23.35	79,060	6,717	2,047	78
585	23.03	77,994	7,058	2,151	77
578	22.76	77,060	7,401	2,256	76
570	22.44	75,994	7,749	2,362	75
562	22.13	74,927	8,100	2,469	74
555	21.85	73,994	8,455	2,577	73
547	21.54	72,927	8,815	2,687	72
540	21.26	71,994	9,178	2,797	71
532	20.94	70,927	9,545	2,909	70
524	20.63	69,861	9,917	3,023	69
517	20.35	68,928	10,293	3,137	68
509	20.04	67,861	10,673	3,253	67
502	19.76	66,928	11,058	3,371	66
495	19.49	65,995	11,455	3,492	65

TABLE A.7 Conversion of Pressure

1 miner's inch = 11.22 gal/min
1 ft^3/sec (cfs) = 40 miner's inches
1 cubic centimeter (cm^3) = 0.061 cubic inch (in^3)
1 cubic meter (m^3) = 6.290 barrels (bbl)
1 bbl = 42 gallons
1 gram (g) = 0.002205 pound (lb)
1 g = 0.0352 ounce (oz)
1 short ton = 2,000 lb
1 long ton = 2,240 lb
1 kg = 2.2046 lb
1 kg/cm^2 = 0.96 atmosphere (atm)
1 kg/cm^2 = 0.98 bar
1 atm = 1.01325 bar
1 g/cm^3 = 62.4 lb/ft^3
T Celsius = [($T°$F − 32] × 5/9
T Fahrenheit = [1.8 × $T°$C] + 32
1 grain of Hardness per gallon = 17.1 mg/L in Total Dissolved Solids

TABLE A.8   Other Combinations

Relationship of Water Density and Viscosity to Temperature

Temp (°C)	Density (g/cm³)	Viscosity (g/s · cm)
0	0.99984	0.01782
2	0.99994	0.01673
4	0.99997	0.01567
6	0.99994	0.01473
8	0.99985	0.01386
10	0.99970	0.01308
12	0.99950	0.01236
14	0.99924	0.01171
16	0.99894	0.01111
18	0.99860	0.01056
20	0.99820	0.01005
22	0.99777	0.00958
24	0.99730	0.00914
25	0.99704	0.00894
26	0.99678	0.00874
28	0.99623	0.00836
30	0.99565	0.00801
35	0.99403	0.00723
40	0.99221	0.00656
45	0.99021	0.00599
50	0.98805	0.00549

Source: Handbook of Chemistry and Physics, CRC Press, Boca Raton, FL, 1990.

APPENDIX **C**

Periodic Table

Periodic Table of the Elements

Atomic masses are based on ^{12}C. Atomic masses in parentheses are for the most stable isotope.

Legend:
- 6 — Atomic number
- C — Symbol
- 12.011 — Atomic mass

Periods	IA	IIA	IIIB	IVB	VB	VIB	VIIB	VIIIB			IB	IIB	IIIA	IVA	VA	VIA	VIIA	VIIIA
1	1 H 1.0079																	2 He 4.00260
2	3 Li 6.941	4 Be 9.01218											5 B 10.81	6 C 12.011	7 N 14.0067	8 O 15.9994	9 F 18.998403	10 Ne 20.179
3	11 Na 22.98977	12 Mg 24.305											13 Al 26.98154	14 Si 28.0855	15 P 30.97376	16 S 32.06	17 Cl 35.453	18 Ar 39.948
4	19 K 39.0983	20 Ca 40.08	21 Sc 44.9559	22 Ti 47.90	23 V 50.9415	24 Cr 51.996	25 Mn 54.9380	26 Fe 55.847	27 Co 58.9332	28 Ni 58.70	29 Cu 63.546	30 Zn 65.38	31 Ga 69.72	32 Ge 72.59	33 As 74.9216	34 Se 78.96	35 Br 79.904	36 Kr 83.80
5	37 Rb 85.4678	38 Sr 87.62	39 Y 88.9059	40 Zr 91.22	41 Nb 92.9064	42 Mo 95.94	43 Tc (98)	44 Ru 101.07	45 Rh 102.9055	46 Pd 106.4	47 Ag 107.868	48 Cd 112.41	49 In 114.82	50 Sn 118.69	51 Sb 121.75	52 Te 127.60	53 I 126.9045	54 Xe 131.30
6	55 Cs 132.9054	56 Ba 137.33	57 La* 138.9055	72 Hf 178.49	73 Ta 180.9479	74 W 183.85	75 Re 186.207	76 Os 190.2	77 Ir 192.22	78 Pt 195.09	79 Au 196.9665	80 Hg 200.59	81 Tl 204.37	82 Pb 207.2	83 Bi 208.9804	84 Po (209)	85 At (210)	86 Rn (222)
7	87 Fr (223)	88 Ra 226.0254	89 Ac† 227.0278	104 Unq (261)	105 Unp (262)	106 Unh (263)												

*Lanthanide series

58 Ce 140.12	59 Pr 140.9077	60 Nd 144.24	61 Pm (145)	62 Sm 150.4	63 Eu 151.96	64 Gd 157.25	65 Tb 158.9254	66 Dy 162.50	67 Ho 164.9304	68 Er 167.26	69 Tm 168.9342	70 Yb 173.04	71 Lu 174.967

†Actinide series

90 Th 232.0381	91 Pa 231.0359	92 U 238.029	93 Np 237.0482	94 Pu (244)	95 Am (243)	96 Cm (247)	97 Bk (247)	98 Cf (251)	99 Es (252)	100 Fm (257)	101 Md (258)	102 No (259)	103 Lr (260)

APPENDIX **D**

Values of *W(u)* and *u* for the Theis Nonequilibrium Equation

N\u	N × E-12	N × E-11	N × E-10	N × E-09	N × E-08	N × E-07	N × E-06	N × E-05	N × E-04	N × E-03	N × E-02	N × E-01	N
1.0	27.054	24.751	22.449	20.146	17.844	15.541	13.238	10.936	8.633	6.332	4.038	1.823	0.219
1.5	26.648	24.346	22.043	19.741	17.438	15.135	12.833	10.530	8.228	5.927	3.637	1.465	0.100
2.0	26.361	24.058	21.756	19.453	17.150	14.848	12.545	10.243	7.940	5.639	3.355	1.223	0.049
2.5	26.138	23.835	21.532	19.230	16.927	14.625	12.322	10.019	7.717	5.417	3.137	1.044	0.024
3.0	25.955	23.653	21.350	19.047	16.745	14.442	12.140	9.837	7.535	5.235	2.959	0.906	0.013
3.5	25.801	23.499	21.196	18.893	16.591	14.288	11.986	9.683	7.381	5.081	2.810	0.794	0.007
4.0	25.668	23.365	21.062	18.760	16.457	14.155	11.852	9.550	7.247	4.948	2.681	0.702	0.0038
4.5	25.550	23.247	20.945	18.642	16.339	14.037	11.734	9.432	7.130	4.831	2.568	0.625	0.0021
5.0	25.444	23.142	20.839	18.537	16.234	13.931	11.629	9.356	7.024	4.726	2.468	0.560	0.0011
5.5	25.349	23.047	20.744	18.441	16.139	13.836	11.534	9.231	6.929	4.631	2.378	0.503	0.00064
6.0	25.262	22.960	20.657	18.354	16.052	13.749	11.447	9.144	6.842	4.545	2.295	0.454	0.00036
6.5	25.182	22.879	20.577	18.274	15.972	13.669	11.367	9.064	6.762	4.465	2.22	0.412	0.0002
7.0	25.108	22.805	20.503	18.200	15.898	13.595	11.292	8.990	6.688	4.392	2.151	0.374	0.00012
7.5	25.039	22.736	20.434	18.131	15.829	13.526	11.223	8.921	6.619	4.323	2.087	0.340	0.000066
8.0	24.974	22.672	20.369	18.067	15.764	13.461	11.159	8.856	6.555	4.259	2.027	0.311	0.000037
8.5	24.914	22.611	20.309	18.006	15.703	13.401	11.098	8.796	6.494	4.199	1.971	0.284	0.000022
9.0	24.857	22.554	20.251	17.949	15.646	13.344	11.041	8.739	6.437	4.142	1.919	0.260	0.000012
9.5	24.803	22.500	20.197	17.895	15.592	13.290	10.987	8.685	6.383	4.089	1.870	0.239	0.000007

APPENDIX **E**

Radial Flow Equation

Refer to Figure E.1.

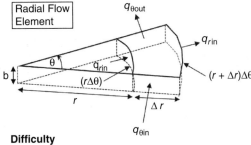

FIGURE E.1 Radial groundwater flow equation diagram based on conservation of mass in polar coordinates in L^3/t, where θ = angle, r = radial distance in the r direction (L), $q_{\theta in}$ = flow into the element in the θ direction (L/t), $q_{\theta out}$ = flow out in the element in the θ direction (L/t), q_{rin} = flow into the element in the r direction (L/t), q_{rout} = flow out in the element in the r direction (L/t), and b = saturated thickness (L).

Conservation of mass in the θ direction:

Mass in = $\rho \times q_\theta \times \Delta r \times b$

where the terms are described in the figure caption and ρ = density of water, M/L^3.

Mass out = $\rho \times q_\theta \times \Delta r \times b + [\Delta(\rho q_\theta b)\Delta r]$

Net change in mass = $-\Delta(\rho q_\theta b)\Delta r = \Delta M / \Delta t$

Conservation of mass in the r direction:

Mass in = $\rho \times q_r \times r\Delta\theta \times b$

Mass out = $\rho \times q_r \times (r + \Delta r)\Delta\theta \times b + [\Delta(\rho q_r b) \times (r + \Delta r)\Delta\theta]$

Net change in mass over time = (including algebraic simplification)

$$\frac{\Delta M}{\Delta t} = -(\rho \times (\Delta r)(\Delta\theta) \times b) - \Delta(\rho q_r b) \times r\Delta\theta - \theta\Delta(\rho q_r b) \times \Delta r\Delta\theta \qquad \text{(E.1)}$$

Assumption 1: 2D flow

$$\frac{\Delta M}{\Delta t} = -\Delta(\rho q_\theta b)\Delta r - (\rho \times (\Delta r)(\Delta\theta) \times b) - \Delta(\rho q_r b) \times r\Delta\theta - \Delta(\rho q_r b) \times \Delta r\Delta\theta \qquad \text{(E.2)}$$

763

Assumption 2: Confined or artesian aquifer conditions. This implies: (a) compressive aquifer, and (b) compressive stresses are vertical:

$$-\Delta(\rho q_\theta b)\Delta r - (\rho \times r\Delta\theta \times b) - \Delta(\rho q_r b) \times r\Delta\theta - \Delta(\rho q_r b) \times \Delta r\Delta\theta = \rho S \frac{\Delta h}{\Delta t} r\Delta r\Delta\theta + \rho W r\Delta r\Delta\theta$$

(E.3)

where S = storage coefficient
   W = source/sink term
   $r\Delta r\Delta\theta$ = area of the element

We now divide by the area ($r\Delta r\Delta\theta$):

$$-\frac{1}{r}\frac{\Delta(\rho q_\theta b)}{\Delta\theta} - \frac{\rho q_r b}{r} - \frac{\Delta(\rho q_r b)}{\Delta r} - \frac{1}{r}\Delta(\rho q_r b) = \frac{\rho S\Delta h}{\Delta t} + \rho W$$

(E.4)

We now take the limits: As $\Delta\theta$, Δr, and $\Delta t \to 0$ [the 4th term on the left-hand side becomes infinitesimally small; physically we have an aquifer filament (of thickness b) passing through an area]

$$\Delta\theta \to 0$$

$$\Delta r \to 0 - \frac{1}{r}\frac{\partial(\rho q_\theta b)}{\partial\theta} - \frac{\rho q_r b}{r} - \frac{\partial(\rho q_r b)}{\partial r} = \rho S\frac{\partial h}{\partial t} + \rho W$$

(E.5)

$$\Delta t \to 0$$

Assumption 3: The fluid is only slightly compressible. All terms with $\partial\rho/\partial r$ are assumed to be insignificant, treating ρ and S as constants.

$$-\frac{1}{r}\frac{\rho\,\partial(q_\theta b)}{\partial\theta} - \frac{\rho q_r b}{r} - \frac{\rho\,\partial(q_r b)}{\partial r} = \rho S\frac{\partial h}{\partial t} + \rho W$$

(E.6)

We divide Equation E.6 by (density) ρ:

$$-\frac{1}{r}\frac{\partial(q_\theta b)}{\partial\theta} - \frac{q_r b}{r} - \frac{\partial(q_r b)}{\partial r} = S\frac{\partial h}{\partial t} + W$$

(E.7)

Recall from the following relationship (product rule from calculus):

$$\frac{\partial(AB)}{\partial r} = B\frac{\partial A}{\partial r} + A\frac{\partial B}{\partial r}$$

$$\frac{\partial(rq_r b)}{\partial r} = r\frac{\partial(q_r b)}{\partial r} + q_r b\frac{\partial r}{\partial r} = r\frac{\partial(q_r b)}{\partial r} + q_r b$$

(E.8)

Multiplying by $-1/r$:

$$-\frac{1}{r}\frac{\partial(rq_r b)}{\partial r} = -\frac{\partial(q_r b)}{\partial r} - \frac{q_r b}{r}$$

(E.9)

We apply the relationship from Equation E.9 and substitute into Equation E.7 to obtain Equation E.10:

$$-\frac{1}{r}\frac{\partial(q_\theta b)}{\partial\theta} - \frac{1}{r}\frac{\partial(rq_r b)}{\partial r} = S\frac{\partial h}{\partial t} + W \tag{E.10}$$

Assumption 4: Darcian flow

$$q_r = -K_r\frac{\partial h}{\partial t} \qquad q_\theta = -K_\theta\frac{\partial h}{\partial t}$$

where K_r = hydraulic conductivity in "r" direction
K_θ = hydraulic conductivity in "θ" direction

$$\frac{1}{r}\frac{\partial\left(K_\theta\frac{\partial h}{\partial t}b\right)}{\partial\theta} + \frac{1}{r}\frac{\partial\left(rK_r\frac{\partial h}{\partial t}b\right)}{\partial r} = S\frac{\partial h}{\partial t} + W \tag{E.11}$$

Assumption 5: Isotropic conditions, or $K_r = K_\theta$
We also recognize that the transmissivity $(T) = K \times b$

$$\frac{1}{r}\frac{\partial\left(T\frac{\partial h}{\partial t}\right)}{\partial\theta} + \frac{1}{r}\frac{\partial\left(rT\frac{\partial h}{\partial t}\right)}{\partial r} = S\frac{\partial h}{\partial t} + W \tag{E.12}$$

Assumption 6: We have radial (1D) flow, so the first term in Equation E.12 drops out:

$$\frac{1}{r}\frac{\partial\left(rT\frac{\partial h}{\partial t}\right)}{\partial r} = \frac{S\partial h}{\partial t} + W \tag{E.13}$$

Assumption 7: No sources or sinks:

$$\frac{1}{r}\frac{\partial\left(rT\frac{\partial h}{\partial t}\right)}{\partial r} = \frac{S\partial h}{\partial t} \tag{E.14}$$

Upon solving Equation E.14 and using dimensionless $u = r^2 S/4Tt$, we arrive at the infinite series known as the Theis equation shown in Chapter 6 (Equation 6.2). Sources for further study of these concepts can be found in Bennett et al. (1990) and Ferris et al. (1962).

References

Bennett, G.D., Reilly, T.E., and Hill, M.C., 1990. *Technical Training Notes in Ground-Water Hydrology: Radial Flow to a Well*. USGS Water Resources Investigations Report 89-4134, 83 pp.

Ferris, J.G., Knowles, D.B., Brown, R.H., and Stallman, R.W., 1962. *Theory of Aquifer Tests*. Geological Survey Water-Supply Paper 1536-E, 174 pp.

Index

Y

Z